FLUIDIZATION

Proceedings of the Second Engineering Foundation Conference
Trinity College, Cambridge, England, 2-6 April 1978

FLUIDIZATION

Proceedings of the Second Engineering Foundation Conference
Trinity College, Cambridge, England, 2-6 April 1978

Edited by

J.F. DAVIDSON
Head of Chemical Engineering Department
University of Cambridge

and

D.L. KEAIRNS
Manager Fluidized Bed Engineering
Westinghouse Electric Corporation

CAMBRIDGE UNIVERSITY PRESS
Cambridge
London New York New Rochelle
Melbourne Sydney

Published by the Press Syndicate of the University of Cambridge
The Pitt Building, Trumpington Street, Cambridge CB2 1RP
32 East 57th Street, New York, NY 10022, USA
296 Beaconsfield Parade, Middle Park, Melbourne 3206, Australia

ISBN 0 521 21943 4

First published 1978
Reprinted 1980

Printed in Great Britain by
Kingprint Ltd, Richmond, Surrey

CONTENTS

CONTENTS

CONTENTS

CONTENTS

PREFACE

Fluidization is a subject that has proved uniquely attractive during the last 30 years. On the one hand academics have been enthralled by the wealth of basic problems suggested by the upward flow of fluid through a bed of particles. On the other hand practical engineers have devised an ever greater variety of applications for fluidized systems. The primary objective of the meeting based on the papers in this volume was to bring together these two groups of workers. The wealth of knowledge generated by basic studies is not fully applied in industry and for this reason there are many industrial processes that have had only the most cursory scientific investigation. The confluence of differing persons at the conference was planned to promote the application of basic science to industrial processes and, conversely, to suggest new basic problems which deserve study.

With these considerations in mind, the Second Engineering Foundation Conference on Fluidization was held in April 1978: it followed from the first of its kind held at Asilomar, California in 1975. Other notable conferences held on the same subject during the last ten years have been

International Symposium on Fluidization, Eindhoven 1967
Symposium on Fluidization (IChemE, Tripartite Chemical Engineering Conference), Montreal 1968
Fluidization and its applications (Sté Chimie Industrielle), Toulouse 1973
Fluidised Combustion (Institute of Fuel) London 1975
International Conferences on Fluidized Bed Combustion USA, 1968, 1970, 1972, 1975, 1977 (primary sponsors include Department of Energy, Energy Research & Development Administration, Environmental Protection Agency).

It was decided not to report the discussion, to allow a frank and uninhibited exchange of views at the meeting. It was hoped that the absence of written discussion would allow industrialists to be a little freer. A further reason for omitting printed discussion was to allow speedy publication. Although we feel sure that the papers herein will be of lasting value, the benefit of having the volume available immediately after the meeting seemed to us to outweigh the doubtful advantage of including printed discussion with the delay caused by that inclusion.

Likewise we decided against the use of rapporteurs. Authors, being human, like to present their own work and the same is true of rapporteurs who inevitably give a personal slant to what they report.

The coverage of subjects reflects current interests. Among the basic topics, bubbles, distributors, particle circulation, elutriation and mixing are still deservedly popular. Centrifugal fluidized beds are a new topic on which several papers present basic information but a practical application has yet to emerge. There are fewer papers than might be expected on reactors: this problem is surely not fully understood in spite of the variety of theories that have been published; it may be that the difficulty of finding satisfactory experiments has deterred university workers. Combustion of coal is rightly a topic of great current interest: the combustion and gasification of coal promises to be one of the most important applications of fluidization. The related topics of sulphur absorption, emission of nitrogen oxides, and heat transfer have also attracted interest. The experimentally difficult question of high pressure operation may have received a stimulus from the need to burn and gasify coal.

We believe that the coal interest alone will stimulate even greater activity in fluidization in the not-too-distant future and we commend this book as a likely indicator of future trends. Another trend in the West is to have more women scientists and engineers. One of the few women to have contributed to our subject remarked 'I fell in love with fluidization as a student and I've been working on it ever since'. Those of us who found the subject attractive as a masculine preserve will welcome this new acquisition of charm and brainpower.

Acknowledgements are the most important part of the preface to a multi-author work. We are very grateful to the authors for their ready acceptance of our complex restraints as to length, format, and time of production. Referees deserve a special mention: their work is usually anonymous but crucial in maintaining standards. We give below a complete list of those who agreed to act; a few were not called upon, but most of those listed refereed one, two or three papers in a short time and we are grateful.

We are also grateful to the Syndics of the Cambridge University Press for agreeing to publish this unusual book.

Workers in the Cambridge University Chemical Engineering Department who helped with the editorial and other work of preparation are Dr.J.R.F.Guedes de Carvalho and Dr.A.M.Xavier.

The following organisations endorsed the conference and their support was valuable:

American Institute of Chemical Engineers
Society of Chemical Engineers (Japan)
European Federation of Chemical Engineering
Canadian Society for Chemical Engineering
Institution of Chemical Engineers (U.K.)
Verein Deutscher Ingenieure.

Financial assistance with payment of registration fees, travel, and the cost of publishing this volume was provided by the following organisations, without whose assistance the conference could hardly have taken place:

Engineering Foundation
Electric Power Research Institute
Department of Energy, Washington D.C.

November 1977.

J.F.D.
D.L.K.

LIST OF REFEREES

Professor H.ANGELINO, Toulouse
Dr.D.H.ARCHER, Westinghouse, Pittsburgh
Dr.S.P.BABU, Institute of Gas Technology, Chicago
Dr.C.G.J.BAKER, University of Western Ontario
Dr.M.A.BERGOUGNOU, University of Western Ontario
Dr.J.S.M.BOTTERILL, University of Birmingham
Dr.J.BRIDGWATER, University of Oxford
Dr.R.CLIFT, University of Cambridge
Dr.R.COLLINS, University College London
Mr.M.COOKE, Coal Research Establishment
Dr.J.P.COUDERC, Toulouse
Dr.R.C.DARTON, Shell, Amsterdam
Dr.N.EPSTEIN, University of British Columbia
Dr.D.GELDART, University of Bradford
Mr.R.B.FIELDES, University of Cambridge
Dr.B.M.GIBBS, University of Sheffield
Dr.W.R.A.GOOSSENS, S.K.E., Mol, Belgium
Dr.J.R.GRACE, McGill University
Dr.J.R.F.GUEDES DE CARVALHO, Cambridge University
Dr.J.S.HALOW, Exxon, Linden, New Jersey
Dr.J.K.HAMMOND, ICI, Runcorn, Cheshire
Dr.D.HARRISON, University of Cambridge
Mr.J.HIGHLEY, Coal Research Establishment
Dr.J.H.HILLS, University of Nottingham
Mr.H.R.HOY, BCURA, Leatherhead
Professor R.JACKSON, University of Houston
Dr.R.D.LANAUZE, Coal Research Establishment
Dr.B.S.LEE, Institute of Gas Technology, Chicago
Dr.J.C.LEE, University College Swansea
Dr.L.S.LEUNG, University of Queensland
Professor H.LITTMAN, Rensselaer Polytechnic Institute
Professor L.MASSIMILLA, University of Naples
Dr.H.MASSON, University Libre de Bruxelles
Dr.K.B.MATHUR, University of British Columbia
Dr.J.M.MATSEN, Exxon, Linden, New Jersey
Professor O.MOLERUS, Universitaat Erlangen
Dr.G.MOSS, Esso Research Centre, Abingdon
Dr.R.A.NEWBY, Westinghouse, Pittsburgh
Dr.K.ØSTERGAARD, Technical University of Denmark
Professor W.RESNICK, Technion, Haifa
Professor J.F.RICHARDSON, University College Swansea
Mr.K.ROBINSON, University of Cambridge
Mr.I.B.ROSS, University of Cambridge
Professor P.N.ROWE, University College London
Dr.D.R.SPINK, University of Waterloo
Mr.A.I.THOMPSON, ICI, Runcorn, Cheshire
Professor W.P.M.VAN SWAAIJ, Twente University of Technology
Dr.D.F.WELLS, du Pont, Wilmington
Professor C.Y.WEN, West Virginia University
Dr.J.WERTHER, Universitaat Erlangen
Dr.D.F.WILLIAMS, Coal Research Establishment
Dr.R.M.WOOD, University of New South Wales
Dr.T.WOOD, University of Sydney
Dr.A.M.XAVIER, University of Cambridge
Dr.W.C.YANG, Westinghouse, Pittsburgh
Professor J.YERUSHALMI, City University of New York.

INDEX OF AUTHORS

Fluidization, Cambridge University Press, 1978

MEASUREMENT OF LOCAL BUBBLES PROPERTIES IN A FLUIDIZED BED

By H. Masson and R. Jottrand

Service de Génie Chimique

Free University of Brussels

50, av. F.D. Roosevelt, 1050-Bruxelles

Belgique

SYNOPSIS

A double light probe permits to measure the three characteristic times, necessary to analyse the properties of the bubbles, at a given point, in a fluidized bed.

The bubbles properties being broadely distributed, there is also a strong dispersion in the measured times.

The weight, to be given to the different measurements is discussed. Uncorrectly weighted averages, may lead to important errors in the estimation of the apparent bubble flow.

The experimental results indicate among others :

- that the relationship between the velocity and the size of the bubbles at a given point is linear;
- that the bubble flow at a given point increases linearly with the fluidization velocity;
- that the bubble distribution changes from one point of the bed to another.

NOTATION

D	column diameter (cm)
d	horizontal bubble diameter (cm)
d_p	particle diameter (cm)
e	exponent of the velocity-size relationship
g	acceleration due to gravity (cm/s^2)
h	vertical position in the bed (cm)
N	total number of bubbles per volume unit (cm^{-3})
N_i	number of bubbles per volume unit, having a velocity v_i (cm^{-3})
n	number of measurements
r	radial position (cm)
t_1	duration of passage of a bubble (s)
t_2	time for the bubble to go δ cm up (s)
$\bar{t}_2^H$	harmonic mean of t_2
t_3	time between successive bubbles (s)
u	fluidization velocity (cm/s)
u_{mf}	minimum fluidization velocity (cm/s)
V''	total bubble flow per unit area (cm/s)
V''_i	bubble flow per unit area due to bubbles having a velocity v_i (cm/s)
v	average bubble velocity (cm/s)
v_i	bubble velocity (cm/s)
X_i	fraction of bubbles detected per time unit having a velocity v_i (s^{-1})
x	total number of bubbles detected per time unit (s^{-1})
x_i	number of bubbles detected per time unit, having a velocity v_i (s^{-1})
α	coefficient of the velocity-size relationship
γ	excentricity factor : ratio of intercepted chord to vertical diameter
δ	vertical distance between the two probe levels (cm)
ε	volume fraction of bubbles
ε_i	volume fraction occupied by bubbles having a velocity v_i
ϕ	shape factor : ratio of vertical to horizontal diameter for a given bubble
θ	angle between the direction of the bubble motion and the vertical (degr.)

INTRODUCTION

The heterogeneous character of the gas flow in a fluidized layer is well known.

The models used to describe their performances take generally into account the flow and the volume repartition between the emulsion and the bubbles; but, quantitative data on this subject are few.

The present work was done to determine those parameters experimentally. The analysis of the results shows the variation of the bubbles characteristics from one point to the other in the fluidized layer and their very broad distribution at a given point.

EXPERIMENTAL TECHNIQUE

A comprehensive review of the different probe techniques developped to measure bubble properties is given in the literature (Werther, 1973). We choose a light probe to have a precisely defined volume of detection.

The experiments were made in a 15 cm diameter column reaching 45 cm in height. The solid used is sieved sand (100 μm < d < 200 μm) with a minimum fluidization velocity of 3.1 cm/s.

The fluidization air is humidified to avoid electrostatic effects. A filtroplast porous plate is used as gas distributor. The bubbles are detected by a double light probe. The probe (fig. 1) consists of two light emitter-detector couples put one above the other two centimeters apart. Each couple includes a light emitting diode and a photodiode separated by a gap of 5 mm. In principle, each couple delivers an "on-off" signal, as the volume included between emitter and receiver ($15.7\ mm^3$) is occupied by a bubble or the solid suspension. The real analogical signal presents a transition between the two extreme levels. It was transformed in a binary signal by fixing the cutting level according to amplitude analysis procedure (Werther & Molerus, 1972; Werther, 1973).

An on line digital computer determines for each bubble passage the 3 following times (fig. 2) :

t_1 : the presence time of the bubble at the lower detection level;

t_2 : the travelling time for the bubble to go from the lower to the upper level;

t_3 : the time interval between two successive bubbles at the lower level.

The measurement were done for different fluidization velocities and different vertical and radial probe positions in the bed. 500 experiments were made. For each experiment, 2000 measurements of each time were taken.

INTERPRETATION

The measured times, for a given experimental condition, present a broad dispersion. Figure 3 gives three typical time histograms. For a vertically moving bubble, t_2 equals δ/v. If the bubble has an oblique motion, $t_2 = \delta.\cos\theta/v$.

The t_2 dispersion results thus from the velocity dispersion and from the bubble direction dispersion.

The velocity dispersion is related to the bubble size dispersion and also to the velocity fluctuations of a bubble of a given size, specially when this bubble approaches another or during coalescence.

It may occur that an obliqually moving bubble is only detected at one measurement level. In this case, t_2 has no simple physical meaning.

In the treatment of the data, we have neglected values of t_2 higher than 200 msec, admitting that those measurements correspond to oblique bubbles cutting only one detection level.

The time t_1 is related to the bubble size and velocity by the relation $t_1 = \phi\gamma d/v$.

The t_1 variations are thus related to the variations of the d/v ratio, to the shape variations and also to γ, the excentrity of the bubbles relatively to the detection point.

The t_3 histogram (fig. 2) reflects the complete randomness of the bubble arrivals at a given point of a fluidized layer.

The measurements of the three times permit to analyse the volume and flow repartition between bubbles of different velocities.

We have the following relationships :

$$x_i = N_i \frac{\pi}{4} d_i^2 v_i$$

$$\varepsilon_i = N_i \frac{\pi}{6} \phi\, d_i^3 = \frac{2}{3} \phi \frac{d_i}{v_i} . x_i$$

$$V_i'' = N_i \frac{\pi}{6} \phi\, v_i\, d_i^3 = \varepsilon_i\, v_i = \frac{2}{3} \phi\, x_i\, d_i$$

and the average bubble velocity is defined by :

$$\bar{v} = \frac{\Sigma V_i''}{\Sigma \varepsilon_i}$$

If we assume a vertical bubble motion, the t_2 distribution gives us a complete image of the velocity distribution. For each t_{2_i} value there is just one v_i value ($t_{2_i} = \delta/v_i$).

On the other hand, the analysis of the t_1 distribution, which results at the same time from the d/v variations, the variations of the intersection coefficient γ and eventually from the shape modifications, is much more complex than the t_2 distribution. It is not possible to attribute a given value of v or d, even not a value of d/v, to a given t_1 value.

If one assumes that there does not exist any correlation between the d/v variations on the one hand and the γ or ϕ fluctuations on the other, the t_1 average value permits to estimate a weighted mean value of the d/v ratio :

$$\bar{t}_1 = \frac{\Sigma t_1}{n} = \Sigma X_i\, \phi\, \gamma \frac{d}{v} = \bar{\phi} . \bar{\gamma}\, \Sigma X_i \frac{d}{v}$$

The probability for the detector to be in a bubble is equal to the total volume fraction occupied by bubbles :

$$\varepsilon = \Sigma\, \varepsilon_i = \frac{\Sigma\, t_1}{\Sigma\, t_3} = \frac{\overline{t_1}}{\overline{t_3}}$$

The t_3 mean value is related to the total number of detected bubbles par time unit by the relation :

$$\overline{t_3} = \frac{1}{\Sigma\, x_i} = \frac{1}{x}$$

Relationship between Bubble Size and Velocity

To estimate the repartition of the bubble volume and flow between the different velocity classes, it is necessary to introduce a relation between the bubble velocity and the bubble diameter.

The data given in the literature are rather contradictory . Godard and Richardson (Godard & Richardson, 1969) interpreting cinematographical measurements at the wall, conclude that the bubble size is only a minor factor in determining the velocity. The values are strongly dispersed and significantly affected by the concentration, spacing and size of the neighbours. Rowe and Matsuno (Rowe & Matsuno, 1970) studied isolated artificial bubbles in a prefluidized bed. They observe a large dispersion of the data and conclude that the two relations $v = k\sqrt{gd}$ and $v = A + Bd$ may both be accepted. Rowe (Rowe, 1971) summarizes our knowledge saying "The bigger they are, the faster they rise." But Werther (Werther, 1974) suggests, without any new experimental data, that local bubble size and corresponding instantaneous rise velocity are stochastically independant. We shall admit the general form $v = \alpha\, d^e$.

To determine the coefficient and the exponent of this law, we have measured for given fluidization conditions, at a given point of the layer, 720 couples of (t_1, t_2) each corresponding to the passage of the same bubble.

These couples have been classified according to increasing values of t_2 and divided into 30 successive classes of 24 measurements. In each class, the arithmetic means of t_1 and t_2 were determined. These results are shown in fig. 4.

One sees that t_2 varies between $6\ 10^{-3}$ and $233\ 10^{-3}$ s whereas t_1 varies between $10\ 10^{-3}$ and $39\ 10^{-3}$ s. It appears that there is no correlation between the variations of t_1 and t_2.

So the ratio $d/v = t_1/(\phi\gamma)$ is not dependent on $v = \delta/t_2$.

In other words, our results are best described as a relation of proportionality between the velocity and the size of the bubbles : $v = \alpha\, d$.

The generally admitted relationship $v = 0.7\sqrt{gd}$ would lead to an inverse proportionality between t_1 and t_2 (represented on fig. 4).

This law, applicable to the motion of gas bubbles in a liquid or for an isolated bubble in a prefluidized layer does not describe satisfactorily the situation in a freely fluidized bed containing a large amount of interacting bubbles.

The indepency suggested by Werther would lead to a direct proportionality between t_1 and t_2 (also represented on fig. 4). This seems to be in contradiction with our experimental results. The measurements of Kang et al. (Osberg, 1975) confirm this conclusion.

Bubble Volume and Flow Repartition according to Bubble Velocity

Once the v(d) law is known, the t_2 distribution and the arithmetic mean values of t_1 and t_3 permit to calculate the volume and flow repartition as a function of the bubble velocity. If one admits the law $v_i = \alpha\, d_i$, and if one admits that the shape coefficient ϕ is independent of v, one may write :

$$v_i = \frac{\delta}{t_{2i}} \qquad d_i = \frac{v_i}{\alpha} = \frac{\delta}{\alpha\, t_{2i}}$$

$$\varepsilon_i = \frac{2}{3}\,\overline{\phi}\,\frac{d_i}{v_i}\,X_i;\ x = \frac{2}{3}\,\frac{x}{\alpha}\,\overline{\phi}\,X_i = \frac{\overline{t_1}}{\overline{t_3}}\cdot X_i$$

$$\varepsilon = \Sigma\,\varepsilon_i = \frac{2}{3}\,\frac{x}{\alpha}\,\overline{\phi} = \frac{\overline{t_1}}{\overline{t_3}}$$

$$\overline{t_1} = \frac{2}{3}\,\frac{\overline{\phi}}{\alpha}$$

$$V''_i = \frac{2}{3}\,\overline{\phi}\,d_i\,X_i\ x = \frac{2}{3}\,\overline{\phi}\,\frac{\delta}{\alpha}\,x\,\frac{X_i}{t_{2i}} = \frac{\overline{t_1}}{\overline{t_3}}\cdot\delta\,\frac{X_i}{t_{2i}}$$

$$V'' = \Sigma\, V''_i = \frac{\overline{t_1}}{\overline{t_3}}\cdot\delta\,\Sigma\left(\frac{X_i}{t_{2i}}\right)$$

If the shape factor is velocity dependent, one must write :

$$\varepsilon_i = \frac{2}{3}\,\phi_i\,\frac{d_i}{v_i}\,X_i\ x = \frac{2}{3}\,\frac{x}{\alpha}\,\phi_i\,X_i = \frac{\overline{t_1}}{\overline{t_3}}\,\frac{\phi_i\,X_i}{\overline{\phi}}$$

$$\varepsilon = \frac{2}{3}\,\frac{x}{\alpha}\,\overline{\phi}$$

with :

$$\overline{\phi} = \Sigma\, X_i\,\phi_i$$

$$\overline{t_1} = \frac{2}{3}\,\frac{\overline{\phi}}{\alpha}$$

$$V''_i = \frac{2}{3}\,\phi_i\,d_i\,X_i\ x = \frac{2}{3}\,\frac{\delta\, x}{\alpha\, t_{2i}}\,X_i\,\phi_i =$$

$$\frac{\overline{t_1}}{\overline{t_2}}\,\delta\,\frac{\phi_i\,X_i}{\overline{\phi}\,t_{2i}}$$

$$V'' = \Sigma\, V''_i = \frac{\overline{t_i}}{\overline{t_3}}\cdot\delta\,\frac{1}{\overline{\phi}}\,\Sigma\,\frac{\phi_i\,X_i}{t_{2i}}$$

If one admits the relation $v_i = \alpha\, d_i^{1/2}$, supposing ϕ independent of v, one must then use the following relationships :

$$v_i = \frac{\delta}{t_{2i}}$$

$$d_i = \frac{v_i^2}{\alpha^2} = \frac{\delta^2}{\alpha^2 t_{2i}^2}$$

$$\varepsilon_i = \frac{2}{3}\bar{\phi}\,\frac{x\,\delta}{\alpha^2}\,\frac{X_i}{t_{2i}} = \frac{\bar{t}_i}{\bar{t}_3}\cdot\frac{X_i/t_{2i}}{\Sigma(X_i/t_{2i})}$$

$$\varepsilon = \Sigma\,\varepsilon_i = \frac{\bar{t}_1}{\bar{t}_3} = \frac{2}{3}\,\frac{\bar{\phi}\,x}{\alpha^2}\,\Sigma\,\frac{X_i}{t_{2i}}$$

$$\bar{t}_i = \frac{2}{3}\,\bar{\phi}\,\frac{\delta}{\alpha^2}\,\Sigma\,\frac{X_i}{t_{2i}}$$

$$V''_i = \frac{2}{3}\,\bar{\phi}\,x\,\frac{\delta^2}{\alpha^2}\,\frac{X_i}{t_{2i}^2} = \frac{\bar{t}_1}{\bar{t}_3}\,\delta\,\frac{X_i/t_{2i}^2}{\Sigma(X_i/t_{2i})}$$

$$V'' = \Sigma\,V''_i = \frac{\bar{t}_1}{\bar{t}_3}\cdot\delta\cdot\frac{\Sigma\,(X_i/t_{2i}^2)}{\Sigma\,(X_i/t_{2i})}$$

EXPERIMENTAL RESULTS

Time Distributions

The average times $\bar{t}_1$ and $\bar{t}_3$ and the t_2 distribution depend on the fluidization velocity and the position of the measurement point.

Fig. 5 represents the ratio $\bar{t}_1/\bar{t}_3 = \varepsilon$ as a function of the gas velocity in two measurement positions (h = 10 cm and h = 25 cm; r = 0).

One may see that the bubble volume fraction is approximatively proportional to the difference $u - u_{mf}$.

Fig. 6 represents $\bar{t}_1$ as function of the gas velocity in the same positions $\bar{t}_1 = \bar{\phi}\,\bar{\gamma}\,(\overline{d/v})$ increases linearly with the fluidization velocity.

This variation may be interpreted as a variation of the shape factor ϕ or a variation of the ratio (d/v).

If one admits a linear law $v = \alpha\, d$, one must admit that α changes with the fluidization conditions: if one admits a law of the form $v = \alpha\, d^{1/2}$, one must admit that the bubble diameter increases with the fluidization velocity, which is quite conform to direct observation.

Bubble Velocity

Fig. 7 represents the average bubble velocity

$$v = \delta\,\Sigma\,(\frac{X_i}{t_{2i}})$$

as function of the fluidization velocity at two measurement points (h = 25 cm; r = 0 and 7 cm). The bubble velocity increase slightly with the fluidization velocity; the average value is higher (+ 23 %) in the column axis than near the walls.

Bubble Flow and Volume Fraction

Fig. 8 represents the volumic bubble flow

$$V'' = \frac{\bar{t}_1}{\bar{t}_3}\cdot\delta\,\Sigma\,(\frac{X_i}{t_{2i}}) = \frac{\bar{t}_1}{\bar{t}_3}\cdot\frac{\delta}{\bar{t}_2^{H}}$$

as a function of the fluidization velocity, at two different points (h = 25 cm; r = 0 and 7 cm). The bubble flow increases approximatively linearly with the fluidization velocity. Generally, the flow per unit area is a little lower than the difference $u - u_{mf}$ (curve 1).

However, above a certain level in the axis of the column the bubble flow per unit area becomes clearly higher than $u - u_{mf}$ (curve 2). The comparison of figures 5, 7 an 8 shows that the increase in bubble flow mainly results from an increase in the volume fraction and secondly from an increase in the bubble velocity.

Fig. 9 and 10 shows the variation of the bubble flow as function of the position of the measurement point, for a given fluidization velocity.

The general trend is qualitatively the same as in previous authors work (Werther, 1973; Park et al., 1969; Grace & Harrisson, 1969).

In the column axis, the bubble flow increases clearly with height. Locally the bubble flow may be greater than three times $u - u_{mf}$.

Bubble Distribution following their Velocity

Fig. 11 shows the bubble distribution at a given point (h = 10 cm; r = 0) for a given fluidization velocity (7,47 cm/sec).

One may notice the broad velocity distribution and the great relative importance of the rapid bubbles in the bubble flow.

CONCLUSIONS

- The analysis of the signals delivered by a double light probe gives information about the motion and the distribution of the bubbles in a fluidized bed.

- From the numerous measurements made with a given solid in a 15 cm diameter column, one may draw the following conclusions :

1) At each point, the bubble velocity is broadly dispersed around the mean value. One may accordingly make an important error by using a mean value in a two-phase fluidized reactor model.
2) The generally admitted relationship $v = 0.7\,\sqrt{gd}$, does not describe satisfactorily the relationship between bubble size and velocity at a given point. Our results are better described by a linear relationship.
3) The bubble fraction at a given point varies proportionally to the difference $u - u_{mf}$. The bubble flow changes linearly with the fluidization velocity.

4) For given fluidization conditions, the volume fraction and the bubble flow change from one point to another in the layer.

REFERENCES

Godard, K., Richardson, J.F. (1969) Chem. Eng. Sci. 24, 663.
Grace, J.R., Harrisson, D. (1969) Chem. Eng. Sci. 24, 497.
Osberg, G.L. (1975), personnal communication.
Park, W.H. et al. (1969) Chem. Eng. Sci. 24, 851.
Rowe, P.N., Matsuno, R. (1970) Chem. Eng. Sci. 25, 1587.
Rowe, P.N. (1971) Fluidization, Chapt 4 (Davidson, J.F. & Harrisson, D., Ed.) Academic Press, New York.
Werther, J., Molerus, O. (1972) Communication presented at Chisa Congress (Praha, 1972).
Werther, J. (1973) Int. J. Multiphase flow 1, 103.
Werther, J. (1974) Trans. Inst. Chem. Engrs. 52, 149.

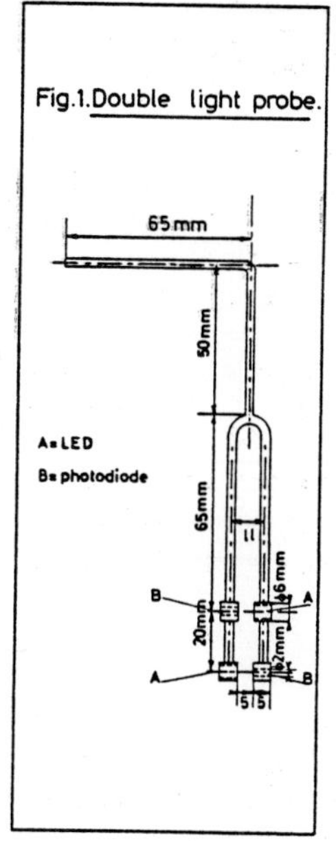

Fig.1. Double light probe.

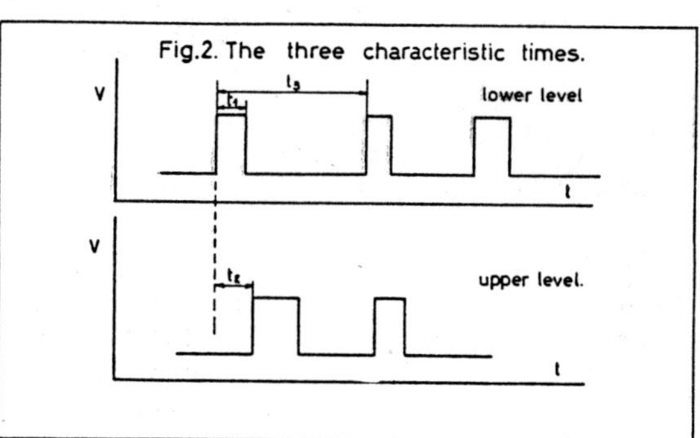

Fig.2. The three characteristic times.

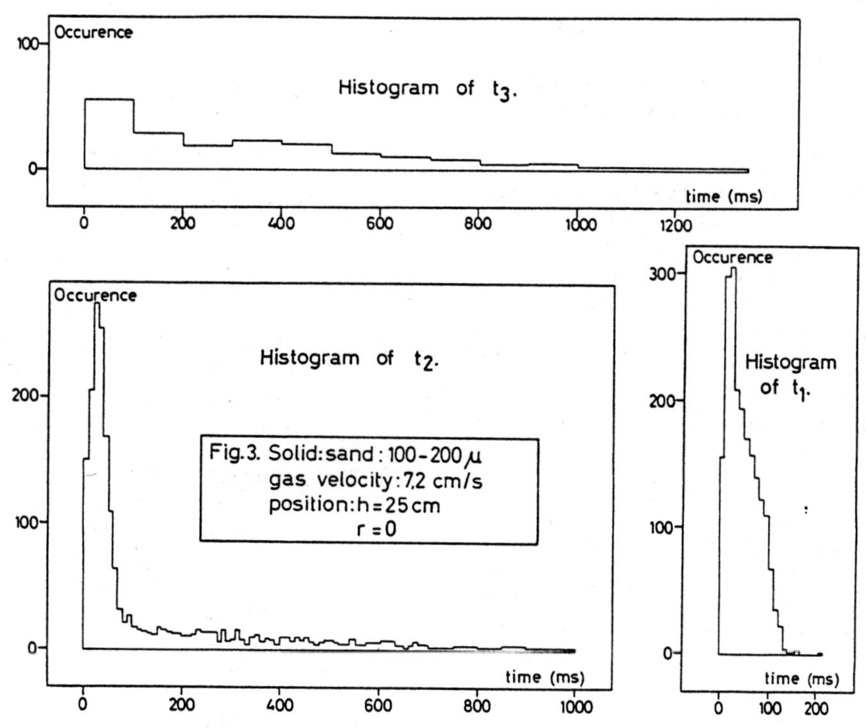

Fig.3. Solid: sand: 100-200 μ gas velocity: 7.2 cm/s position: h=25 cm r=0

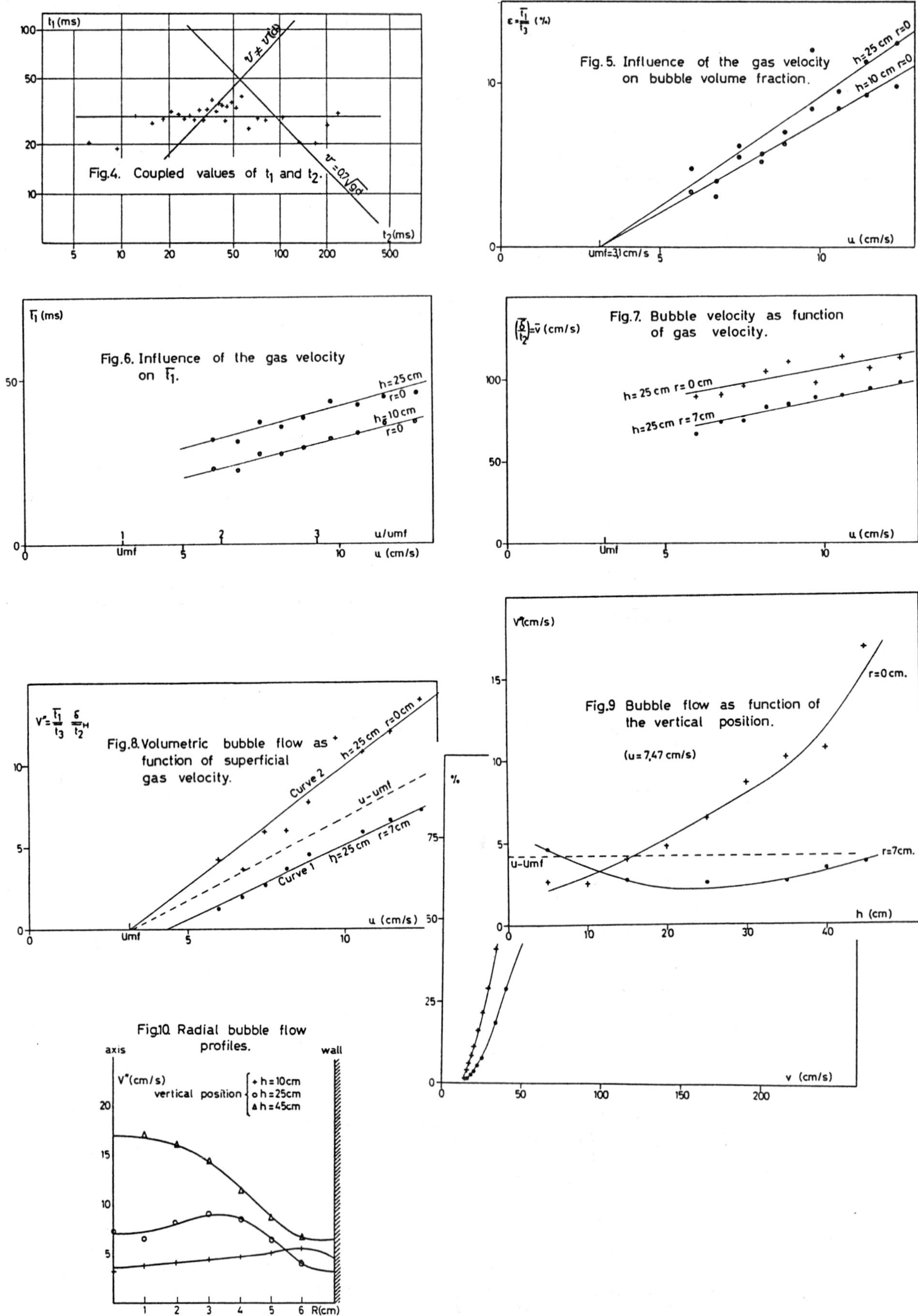

Fig.4. Coupled values of t_1 and t_2.
Fig. 5. Influence of the gas velocity on bubble volume fraction.
Fig. 6. Influence of the gas velocity on $\bar{t}_1$.
Fig. 7. Bubble velocity as function of gas velocity.
Fig. 8. Volumetric bubble flow as function of superficial gas velocity.
Fig. 9 Bubble flow as function of the vertical position.
(u= 7,47 cm/s)
Fig. 10 Radial bubble flow profiles.

INFLUENCE OF THE DISTRIBUTOR DESIGN ON BUBBLE CHARACTERISTICS IN LARGE DIAMETER GAS FLUIDIZED BEDS

By Joachim Werther

University of Erlangen-Nürnberg, W-Germany

present address: BASF AG, D 67 Ludwigshafen, W-Germany

The influence of the distributor design on the local bubble characteristics has been directly measured in a fluidized bed of 1 m diameter using statistically based measuring methods for the determination of local bubble size distributions and local average bubble shapes. 8 different distributor plates, including multihole, nozzle and tuyere-type distributors, have been investigated. Based on a statistical model of bubble coalescence and on other authors' findings about the mechanism of bubble formation at the distributor a method is developed which permits the prediction of bubble growth as a function of height above the distributor, excess gas velocity and distributor design. Predicted local average bubble sizes are in good agreement with measurements for all distributors used in the present work and also for the fluidized bed systems investigated by Whitehead & Young (1967).

NOTATION

d_e bubble eruption diameter, cm
d_h bubble frontal diameter, cm
d_v diameter of a sphere with a volume equal to the local average bubble volume, cm
$d_{v,o}$ initial bubble size, cm
d_o orifice diameter, cm
f_w wake fraction, -
h height above bottom plate, cm
h' equivalent height from (1) for $d_v=d_{v,o}$, cm
h_o height of bubble formation, cm
L jet penetration depth, cm
u superficial gas velocity, cm/s
u_{mf} minimum fluidizing velocity, cm/s
u_o jet velocity, cm/s
$\dot{V}_o$ gas flow through orifice, cm^3/s

INTRODUCTION

Heat and mass transfer properties of gas fluidized beds as well as the performance of fluid bed catalytic reactors are largely determined by bubble characteristics, i.e. in the case of a catalytic reaction the bypass of gas in the bubble phase is limiting the yield of the reaction and in the case of bed-to-wall heat transfer the local bubble size and frequency is governing the surface renewal process at the heat transfer surface. The bubble characteristics in turn are influenced by the design of the gas distributor.

Despite its obvious practical importance details of the influence of the distributor design on the bubble characteristics are largely unknown. This may at least partly be due to a lack of suitable methods for measuring bubble characteristics in fluidized beds of large dimensions. The application of the well-known X-ray technique (Rowe & Partridge, 1965, Rowe, 1971) is restricted to small scale laboratory fluidized beds where only little latitude exists for grid variations as has already been pointed out by Zenz (1968). The bubble eruption measuring technique used by Whitehead & Young (1967) in their investigation of a large scale fluidized bed equipped with a tuyere-type distributor has to be considered rather inaccurate. Furthermore this latter method is restricted to relatively low values of the excess gas velocity $(u-u_{mf})$ where bubble eruptions are still clearly discernible.

In previous publications of the author (Werther & Molerus, 1973, Werther, 1974) measuring techniques have been developed which permit an accurate determination of local bubble properties in beds of large dimensions. These methods are used in the present work to investigate the influence of the distributor design on local bubble properties in a fluidized bed of 1 m diameter.

THEORY

The mechanism of bubble coalescence in gas fluidized beds has been thoroughly investigated by Clift & Grace (1970, 1971). Based on their findings the author (Werther, 1976) has developed a statistical model of bubble coalescence in freely bubbling beds which has been used to interpret bubble measurements in a fluidized bed of 1 m diameter with a porous plate as a distributor thus leading to an empirical correlation for bubble growth

$$d_v = 0.853\left[1+0.272(u-u_{mf})\right]^{1/3}(1+0.0684\ h)^{1.21} \quad (1)$$

where the quantities d_v, h and $(u-u_{mf})$ are in cm and cm/s, respectively. Equ. (1) permits the prediction of the local average bubble diameter d_v as a function of height h above the distributor and excess gas velocity $(u-u_{mf})$, respectively. A large number of measurements of the author and of X-ray photographic bubble measurements by Rowe & Everett (1972) have confirmed the relationship (1) in the following range of variables,

minimum fluidizing velocity	$1 \leq u_{mf} \leq 8$ cm/s
mean particle diameter	$100 \leq d_p \leq 350$ µm
excess gas velocity	$5 \leq u-u_{mf} \leq 30$ cm/s
bed diameter	$D_B \geq 20$ cm
gas distributor	porous plate.

Industrial gas distributors like multihole, nozzle or tuyere-type distributors are introducing the fluidizing gas through a number of discrete openings. According to Davidson & Harrison (1963) the volume equivalent diameter of a three-dimensional bubble formed at an orifice in an incipiently fluidized bed is given by

$$d_{v,o} = 1.3\left[\frac{\dot{V}_o^{\,2}}{g}\right]^{0.2} \quad (2)$$

where $\dot{V}_o$ denotes the volume flow rate of fluid out of the orifice and g the acceleration due to gravity, respectively. For higher gas throughputs the momentum of the gas flow is so high that a gas jet is issuing from the orifice. An empirical relationship for the average penetration depth L of vertical jets has been given by Zenz (1968),

$$\frac{L}{d_o} = 30.2\ \ln\left(u_o\sqrt{\rho_f}\right) - 90.3 \quad (3)$$

where the fluid density ρ_f is in lb/ft^3 and the jet velocity u_o in the orifice of diameter d_o in ft/s, respectively. More recently, Merry (1975) has correlated the results of jet penetration measurements of various authors for vertical jets by

$$\frac{L}{d_o} = 5.2\left[\frac{\rho_f d_o}{\rho_p d_p}\right]^{0.3}\left[1.3\left[\frac{u_o^{\,2}}{g d_o}\right]^{0.2} - 1\right] \quad (4)$$

thus taking into account the influence of particle mean diameter d_p and density ρ_p.

Equ. (2) has been derived for bubbles forming at an orifice. However, Merry (1975) analysed experimental data of Basov et al. (1969) concerning bubble formation at the end of gas jets and concluded that the results provide some justification for using (2) to predict the size of bubbles formed at the end of a jet as well.

On the basis of (1) - (4) a simple concept for the prediction of bubble growth as a function of the distributor design may be formulated:

First of all one has to determine the height h_o above the bottom plate where the bubbles are forming (Fig. 1) for a multihole distributor h_o is identical to the average jet penetration depth given by (3) or (4). In the case of the tuyere-type distributor used by Whitehead & Young (1967) h_o is given by the height of the lower edge of the tuyere cap above the bottom plate whereas for nozzles with horizontal ports h_o may be taken as the height of the openings above the bottom plate. In any case the size of the bubbles initially forming will be given by (2).

Both the exact place of bubble formation and the size of the newly formed bubbles are subject to statistical influences, as has been demonstrated by Markhevka et al. (1971) for bubbles detaching from a vertical jet. The subsequent growth of the bubbles is due to coalescence processes which are again governed by statistical influences (Werther, 1976). It seems therefore reasonable to assume that the growth of the bubbles beginning with the initial size $d_{v,o}$ follows the same laws which are governing the growth of bubbles of the same size in a fluidized bed operated at the same value of the excess gas velocity but equipped with a porous plate distributor (cf. Fig. 2). According to (1) the bubble size $d_{v,o}$ is reached at a height h' above the distributor. For an arbitrary distributor design the bubble growth may then be computed on the basis of the growth correlation for the porous plate (1) where only the height h is replaced by an equivalent height $(h + h' - h_o)$,

$$d_v = 0.853\left[1 + 0.272\ (u-u_{mf})\right]^{1/3}\left[1 + 0.0684\ (h + h' - h_o)\right]^{1.21} \quad (5)$$

where again the quantities h, h', h_o, d_v and $(u-u_{mf})$ are in cm and cm/s, respectively.

EXPERIMENTAL

The experimental unit shown in Fig. 3 consists of a cylindrical fluidized bed of 100 cm dia. with an enlarged freeboard section. External cyclones and subsequent bag filters are used for fines collection. The fluidizing air is delivered by a roots blower. The quartz sand different fractions of which with values of the minimum fluidizing velocity u_{mf} between 1.0 and 1.6 cm/s have been used is a type B solid according to the classification by Geldart (1973).

8 different distributor plates a detailed description of which may be found in Table 1 were used in the course of the present investigation. The tuyeres of the grid no. 4 were of the same construction as those used by Whitehead & Young (1967) (cf. Fig. 1). A sufficiently high distributor pressure drop $\geq$ 30 % of the bed pressure drop was ascertained in the case of the nozzles and tuyeres by means of suitable orifice plates for each nozzle and tuyere, respectively (Whitehead & Young, 1967).

The measuring techniques used which are based on miniaturized capacitance probes and a statistical data processing have been described elsewhere (Werther & Molerus, 1973, Werther 1974) in detail.

RESULTS AND DISCUSSION

Nozzle and tuyere-type distributors

Measurements and predictions of the bubble growth with height h above the distributor are shown in Fig. 4 for the nozzle grid no. 2 together with the bubble growth predicted by (1) for a porous plate distributor.

Fig. 5 gives a comparison between theory and experiment for all nozzle and tuyere-type grids investigated. In computing the initial bubble diameters $d_{v,o}$ from (2) it has always been assumed that each of the 6 horizontal ports of the nozzles and each of the 4 horizontal openings of the tuyeres (cf. Fig. 1), respectively, is acting as a separate bubble generating device. The relative deviation of the theory from the experiments is always less than 25 % which will be sufficiently accurate for most purposes. The deviation may be reduced further by taking interaction effects into account which may occur between gas jets issuing from neighboring nozzles as well as between jets issuing from neighboring ports of the same nozzle (Werther, 1977).

Comparison with measurements of Whitehead & Young

Whitehead & Young (1967) measured bubble eruptions in beds of 2' x 2' and 4' x 4' square cross-section the bottom plates of which were fitted with tuyeres of the type shown in Fig. 1 at spacings of 6" and 12", respectively. Eruption diameters were measured for different bed heights and gas velocities using a sand with u_{mf} = 2.5 cm/s.

With regard to a comparison of these experiments with the present theory bubble eruption diameters d_e have to be converted to bubble volume equivalent diameters d_v. X-ray measurements of Botterill et al. (1966) have yielded the now commonly accepted (Geldart, 1970/71, 1972, Zenz, 1968) relationship between the eruption diameter and the bubble frontal diameter d_h,

$$d_e = 1.5\, d_h \tag{6}$$

The bubble volume is given by

$$V_b = (1-f_w)\pi d_h^3/6 \tag{7}$$

where the wake fraction f_w is defined as the ratio of wake volume to the volume of the sphere forming the spherical cap of the bubble. Values of f_w have been measured by Rowe & Partridge (1965). For the following calculations an average wake fraction f_w = 0.25 is assumed. This same value has been used by Geldart (1970/71) and Merry (1975). Thus it follows for the volume equivalent bubble diameter

$$d_v = \sqrt[3]{1-f_w}\;\; 2\, d_e/3 \simeq 0.606\, d_e \tag{8}$$

In applying equ. (5) to the fluidized beds investigated by Whitehead & Young it has like in the preceding paragraph been assumed that each of the four horizontal openings of a single tuyere (cf. Fig. 1) is acting as a separate bubble generating device.

Fig. 6 gives an example of predicted and measured bubble growth with height above the distributor. Since (1) and therefore (5), too, should only be applied to freely bubbling beds with sufficiently high bubble concentration i.e. with $u-u_{mf} \gtrsim$ 5 cm/s, eruption measurements for lower excess gas velocities have not been included in the comparison between theory and experiment shown in Fig. 7. Fig. 7 covers a total of 81 measurements for 80 per cent of which the deviation of the theory from the experiment is less than 30 %. Two general tendencies are obvious in this plot, namely

(i) for diameters $d_v \geq$ 30 cm i.e. for eruption diameters $\geq$ 50 cm wall effects may already influence bubble growth (this will certainly be the case in the 2'x 2' unit). In this case (5) which is based on unhindered bubble growth will predict bubble sizes which are larger than the experimental values.

(ii) for average bubble diameters below about 15 cm there is an increasing possibility that eruptions of smaller bubbles cannot be observed with sufficient accuracy among the eruptions of larger ones thus leading to an average bubble diameter for the whole bubble population which is too low. This is especially so for higher excess gas velocities.

Taking into account these two effects as well as the general inaccuracy inherent to bubble eruption measurements the agreement between theory and experiment in Fig. 7 may be considered to be satisfactory.

Multihole distributors

For multihole distributors the height of bubble formation h_0 is given by the average jet penetration depth. Fig. 8 shows a comparison between theoretical and experimental bubble growth from which it may be seen that the Zenz correlation (3) which predicts a larger jet penetration than the Merry correlation (4) gives a better fit to the experimental data. This same tendency is obvious from Fig. 9 and 10 where theoretical and experimental bubble sizes are compared for all multihole distributors investigated, Fig. 9 being based on the Merry correlation (4) and Fig. 10 on the Zenz correlation (3), respectively. A total of 37 measurements yields the result that values of d_v predicted on the basis of (4) are consistently higher than the measured bubble sizes whereas the predictions on the basis of the Zenz correlation are giving a good description of the measurements with a maximum relative deviation of 15 %.

The presumption that under the present experimental conditions the Zenz formula gives a better estimate of the average jet penetration depth than the Merry correlation has been confirmed by visual observation of the erosion pattern at the wall of the fluidizing column (Table 2).

CONCLUSIONS

The present investigation of 8 different gas distributors in a fluidizing column of 1 m diameter has shown that the influence of the distributor design on local bubble characteristics is well described by equ. (5) without any fitting parameters being necessary.

In the case of multihole distributors the Zenz correlation (3) for jet penetration is found to yield a better agreement between measured and calculated bubble sizes than the Merry formula (4).

Further evidence of the applicability of the bubble growth correlation (5) to practical situations is given by a comparison with bubble eruption measurements of Whitehead & Young (1967).

REFERENCES

Basov, V.A. Markhevka, V.I., Melik-Akhnazarov, T.Kh. & Orochko, D.I. (1969) Int. Chem. Engng. 9, 263.

Botterill, J.S.M., George, J.S. & Besford, H. (1966) Chem. Engng. Progr. Symp. Ser. 62, 7.

Clift, R. & Grace, J.R. (1970) Chem. Engng. Progr. Symp. Ser. 66 105, 14.

Clift, R. & Grace, J.R. (1971) A.I.Ch.E. Symp. Ser. 67 116, 23.

Davidson, J.F. & Harrison, D. (1963) Fluidized Particles, Cambridge University Press.

Geldart, D. (1970/71) Powder Technol. 4, 41

Geldart, D. (1972) Powder Technol. 6, 201.

Geldart, D. (1973) Powder Technol. 7, 285.

Markhevka, V.I., Basov, V.A., Melik-Akhnazarov, K.Th. & Orochko, D.I. (1971) Theor. Found. Chem. Engng., 80.

Merry, J.M.D. (1975) A.I.Ch.E. J. 21, 507.

Rowe, P.N. & Partridge, B.A. (1965) Trans. Instn. Chem. Engrs. 43, 157.

Rowe, P.N. (1971) Fluidization, Chapt. 4 (Davidson, J.F. & Harrison, D., Ed.) Academic Press, New York.

Rowe, P.N. & Everett, D.E. (1972) Trans. Instn. Chem. Engrs. 50, 55

Werther, J. & Molerus, O. (1973) Int. J. Multiphase Flow 1, 103.

Werther, J. (1974) Trans. Instn. Chem. Engrs. 52, 149.

Werther, J. (1976) Fluidization Technology I (D.L. Keairns et al. Eds.) p. 215, Hemisphere Publ., Washington.

Werther, J. (1977) Particle Technology Nuremberg, preprints I, E 32.

Whitehead, A.B. & Young, A.D. (1967) Proc. Int. Symp. Fluidization Einhoven (Drinkenburg, A.A.H., Ed.) Netherlands University Press, 284.

Zenz, F.A. (1968) I. Chem. E. Symp. Ser. No. 30, 136.

TABLE 1. Gas distributor characteristics

grid	type	construction
1	nozzle	19 nozzles, 23.5 cm triangular pitch, each nozzle with 6 horizontal ports 0.41 cm dia. 0.6 cm above bottom plate
2	nozzle	19 nozzles 3.5 cm dia., 23.5 cm triangular pitch, each nozzle with 6 horizontal ports 0.87 cm dia. 1 cm above bottom plate
3	nozzle	19 nozzles 7.5 cm dia., 20 cm triangular pitch, each nozzle with 6 horizontal ports 0.85 cm dia. 1 cm above bottom plate
4	tuyere	19 tuyeres, 20 cm triangular pitch, tuyere design by Whitehead & Young (1967)
5	multihole	361 openings 0.21 cm dia., 5.2 cm triangular pitch
5 A	multihole	69 openings 0.21 cm dia., for 45 cm dia. bed, 5.2 cm triangular pitch
6	multihole	19 openings 0.95 cm dia., 23.5 cm triangular pitch
7	multihole	73 openings 0.65 cm dia. 10 cm triangular pitch
8	multihole	19 openings 1.3 cm dia. 20 cm triangular pitch

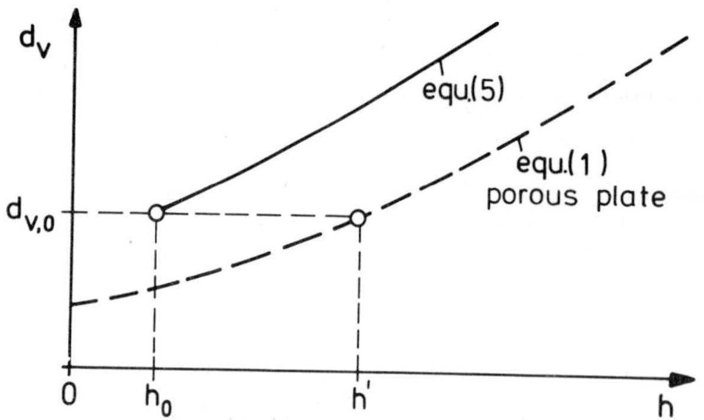

Fig. 2. Influence of distributor design on bubble growth

Fig. 3. Experimental set-up

TABLE 2. Penetration of vertical gas jets

grid no.	5	6
u, cm/s	20	20
jet penetration depth from (3) (Zenz), cm	11.1	47.9
jet penetration depth from (4) (Merry), cm	5.1	25.3
jet penetration from erosion pattern, cm	10-15	50-60

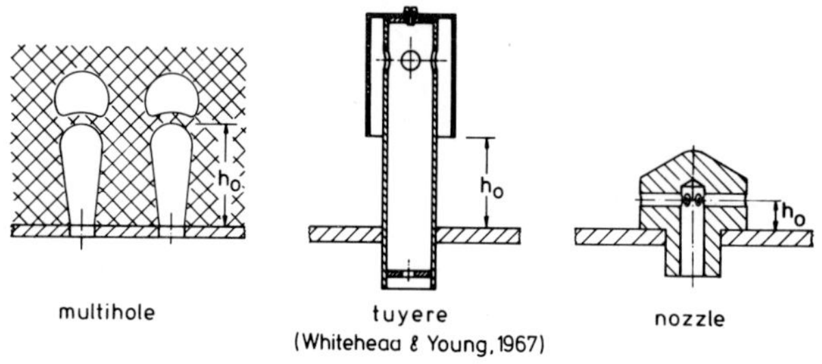

Fig. 1. Types of gas distributors

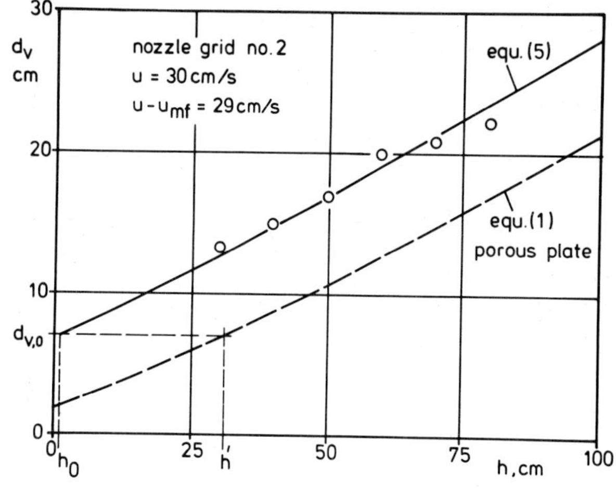

Fig. 4.

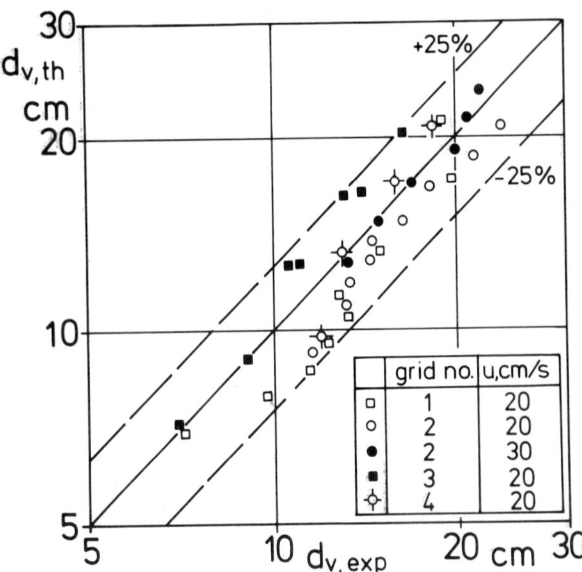

Fig. 5. Theory vs. experiment/distributors with nozzles or tuyeres.

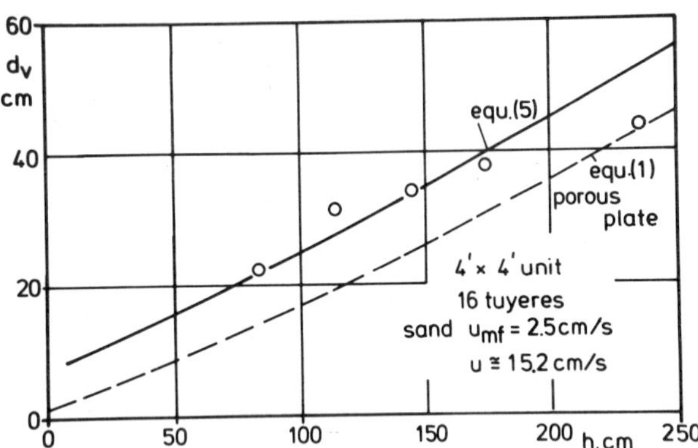

Fig. 6. Bubble growth/experiments of Whitehead & Young (1967).

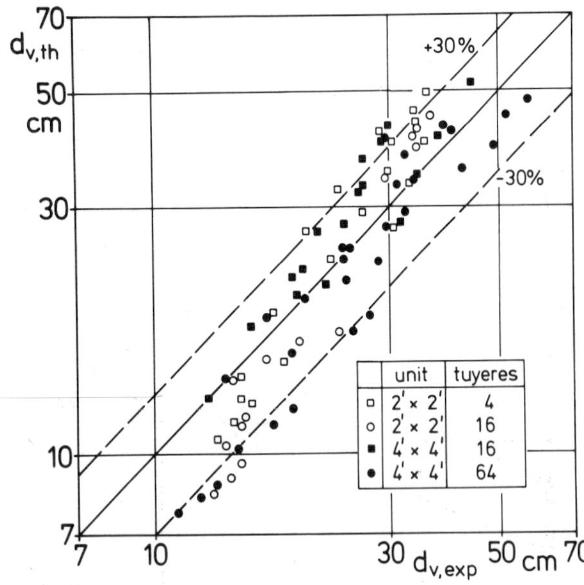

Fig. 7. Theory vs. experiment/measurements of Whitehead & Young.

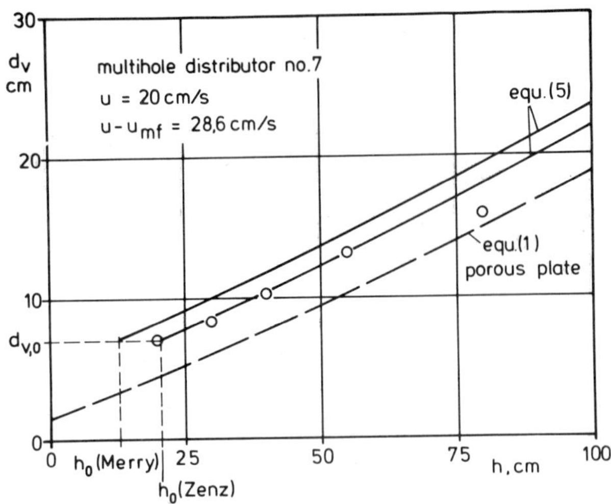

Fig. 8.

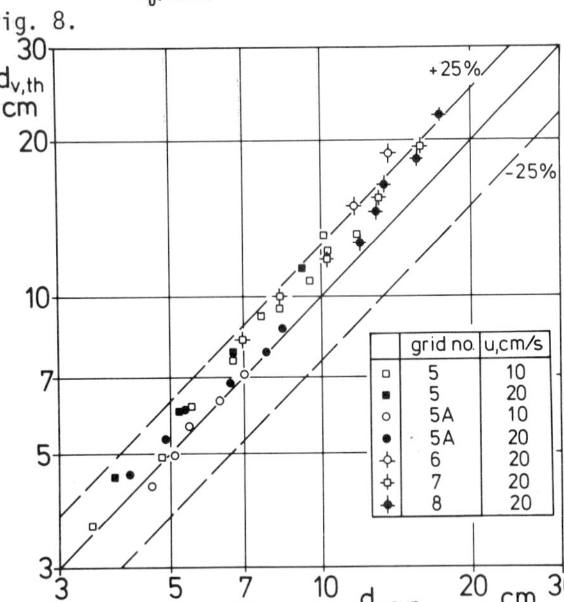

Fig. 9. Multihole distributors/jet length from (4).

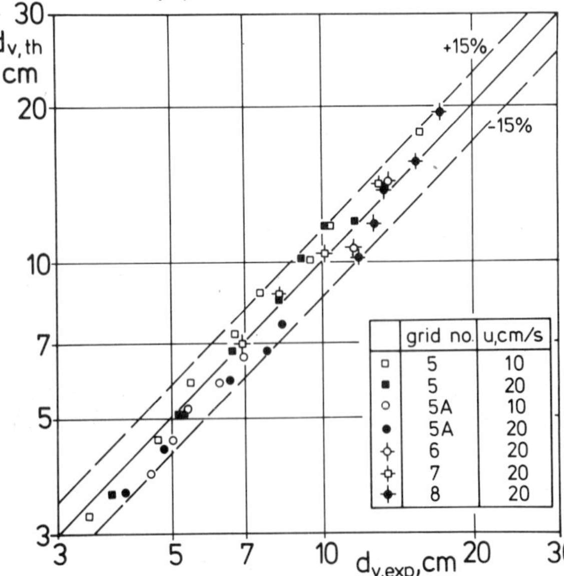

Fig. 10. Theory vs. experiment/multihole distributors, jet length according to Zenz (1968)

SIZE DISTRIBUTION OF BUBBLES IN GAS FLUIDIZED BEDS

By K. Yoshida, K. Nakajima, N. Hamatani and F. Shimizu
Engineering Research Institute, University of Tokyo
2-11, Yayoi, Bunkyo-Ku, Tokyo 113, Japan

SYNOPSIS

Size distributions of bubbles were measured in an air fluidized bed, 40cm I. D., by means of laser beams. Experimental results showed that the distribution has a bi-modal form differing from the distributions previously reported.

A general model for simulation of bubble growth was then proposed. The mathematical equations were formulated based on a proposed mechanism of bubble growth and gave the size distribution of bubbles at a given height. This method elucidates the reason for discrepancy among several data on size distributions.

NOTATION

B	constant of the model, $cm^{-3/2}$
D_b	bubble diameter, cm
D_t	bed diameter, cm
f	number density of bubbles, cm^{-4}
g	acceleration of gravity, cm/s^2
h	height from distributor, cm
N	bubble frequency, s^{-1}
r_b	radius of bubble, cm
r_f	rate of bubble formation, $cm^{-4} \cdot s^{-1}$
r_l	rate of disappearance of bubble $cm^{-4} \cdot s^{-1}$
t	time, s
u_b	rise velocity of bubble, cm/s
u_{mf}	incipient fluidizing velocity, cm/s
u_o	superficial air velocity, cm/s
u_s	rise velocity of slug, cm/s
V_b	bubble volume, cm^3
δ	fraction of cross sectional area consisting of bubbles, -

INTRODUCTION

The major part of gas in fluidized beds passes through the bed as bubbles. These bubbles grow as they rise up in the bed, mainly by coalescence with other bubbles. A number of models of fluidized bed proposed during the last decade depend on how to describe this behavior of bubbles. Therefore, many experimental studies have been carried out so far to determine bubble sizes in fluidized beds. Excluding several studies in two-dimensional beds, Yasui and Johanson (1958) were the first investigators to measure the average thickness of bubbles by use of a probe submerged into the bed. Since then, there have been many works (Baumgarten & Pigford, 1960; Kobayashi et al.,1965 ; Kunii et al., 1967; Whitehead & Young, 1967; Park et al., 1969; Geldart , 1972). In particular, Rowe & Everett (1972) carried out extensive measurements of bubble growth with X-ray technique and presented a semi-empirical correlation (Rowe, 1976). Also, Werther (1974a) studied the bubble behavior in fluidized beds with various diameters.

In most of these works, the bubble size measured at some levels has been quoted as an average value. But the size distribution was seen to be no uniform. Several functions, therefore, have been proposed to describe this distribution; the normal distribution (Park et al., 1969), the Gamma distribution (Argyriou et al.,1971; Rowe & Yacono, 1975) and the log normal distribution (Werther, 1974b). It is interesting to note that Yamazaki et al. (1977) recently showed a radial distribution of bubble size skewed towards bi-modal form under certain conditions.

Theoretical approaches,on the other

hand, to determine bubble size and its distribution have been little published. Miwa et al. (1971) proposed a calculation method based on a hydrodynamic model of bubble coalescence. Their method demands too much calculation effort. Chiba et al. (1973) proposed another calculation method by assuming a simple configulation of bubbles within the bed. This model is much apart from the substance of actual coalescence phenomena.

The purpose of this paper is to propose a general method to simulate the bubble growth by coalescence and to rationalize the many disparate data mentioned above. This method was tested experimentally.

COALESCENCE MODEL

In treating the distribution of some entities accompanying coalescence and breakage the population-balance model (Himmelblau & Bischoff, 1968) is well accepted to be the best tool. This concept was therefore introduced by Argyriou et al. (1971) to analyse bubble growth. However, they simply assumed that collisions between bubbles are propotional to mean projected area of the two bubbles and only obtained the Gamma distribution of bubble size.

Bubbles coalesce not as the result of random walk in the bed, but as the result of the vertical velocity difference between bubbles with different sizes, as described, e.g., in the book of Davidson & Harrison (1963). Therefore, the coalescence process is almost dynamical with random initial conditions.

Furthermore, in the region just above the distributor a two-body collision model is not always adequate, because of the quite large density of bubbles. Therefore, many bubbles coalesce simultaneously before the resultant bubble acquires its steady-state shape. From such a consideration we describe the phenomena by two successive steps.

Let us first consider the step where a number of bubbles of radius r_b and volume V_b (= $4/3\pi r_b^3$) are generated from holes at random position just above the distributor.

Some bubbles are assumed to coalesce into one bubble if the radial distance of two bubbles is shorter than $2r_b$, as shown in Fig. 1. The bubble number produced from the distributor is obtained when the fraction of cross sectional area consisting of bubbles is obtained. Thus, the coalescence phenomena are calculated through bubbles whose generation is simulated by use of a random number table.

After the vigorous coalescence is over, bubbles rise up with a velocity given by

$$u_b = u_o - u_{mf} + 0.711(gD_b)^{1/2} \quad ..(1)$$

Finally, they grow up to a slug which velocity is given by Hovmand & Davidson (1971) as

$$u_s = 0.35(gD_b)^{1/2} \quad ....(2)$$

Therefore, the velocity ranging from a small bubble to a slug may be approximated by a hyperbolic function which smoothly connects the two equations (1) and (2).

While rising up, collisions between two bubbles of volume V_b and V_b' are taking place proportionally to the product of the number densities of the two species, the velocity difference and the mean projected area of two bubbles. The mathematical development in this step almost may follow that of Argyriou et al. That is, the rate of formation of bubbles by coalescence is

$$r_f = \frac{B}{2}\cdot\int f(h,\ V_b',\ t)\cdot f(h,\ V_b-V_b',\ t)\cdot\left|u_b(V_b') - u_b(V_b-V_b')\right|\cdot(V_b'^{\,1/3} + (V_b-V_b')^{1/3})^2\cdot dV_b' \quad ....(3)$$

and the rate of loss of bubbles due to the collision is

$$r_l = B\cdot f(h,\ V_b,\ t)\cdot\int f(h,\ V_b',t)\cdot\left|u_b(V_b) - u_b(V_b')\right|\cdot(V_b^{1/3} + V_b'^{\,1/3})^2\cdot dV_b' \quad ....(4)$$

where $f(h,\ V_b,\ t)$ and B represent the number density of bubbles with volume V_b and the collision parameter, respectively. The factor B is assumed to be a constant.

At steady state, f is governed by the following equation.

$$\frac{u_b\cdot df(h,\ V_b)}{dh} = B\left\{\frac{1}{2}\cdot\int f\ (h,\ V_b')\cdot f(h,\ V_b-V_b')\cdot\left|u_b(V_b') - u_b(V_b-V_b')\right|\cdot(V_b'^{\,1/3} + (V_b-V_b')^{1/3})^2\, dV_b' - f(h,\ V_b)\cdot\int f(h,\ V_b')\cdot\left|u_b(V_b) - u_b(V_b')\right|\cdot(V_b^{1/3} +\right.$$

$$V_b'\ 1/3)2 \cdot dV_b'\Big\} \qquad \ldots\ldots\ldots (5)$$

The solution of eq.(5) will be given by use of the finite differnce method and it will represent the size distributions at a given height of the fluidized bed. Then, the number of bubbles with diameter D_b, frequency N, which pass at a given level of the bed in the unit time is given by

$$N = \pi/2 \cdot D_b{}^2 \cdot f(h,\ \pi/6 \cdot D_b{}^3) \cdot u_b(\pi/6 \cdot D_b{}^3) \qquad \ldots\ldots\ldots (6)$$

EXPERIMENTAL

Fig.2 is a schematic illustration of the apparatus and Table 1 summarizes the experimental conditions used in this study.

Table 1 Experimental conditions

Column	40cm I.D. tube
Distributor	filter cloth supported by a perforated plate
Static bed height	$\sim$ 90cm
Fluidizing velocity	$\sim 6 \times u_{mf}$
Particles used	microspherical catalyst, d_p=166μ, u_{mf} = 1.20cm/s

Entrained solids were collected by a cyclone situated above the bed and returned to the bed through a dip-leg.

Fig.3 shows the probe for detecting bubbles. Two stainless pipes of 6mm I. D. were positioned against each other. Laser beams were introduced through one pipe and received by an optical fiber which was led through the other pipe. Electric currents were raised by a phototransistor, subsequently amplified and recorded as a signal on an oscillograph. The probe system was small so as to allow placement at various level of fluidized bed and distance between two pipes was adjustable.

First, the characteristics of the probe were tested by the arrangement shown in Fig.4. Particles were sent down uniformly in a 8cm I.D. acrylic resin pipe and the permeability of laser beams was measured at the position where particles reached the terminal velocity. The relation of permeability to the volumetric fraction of particles is shown in Fig.5. This figure shows that the laser beam can be transmitted completely through dispersed particles within a bubble of which values (a) and (b) were measured by Toei et al. (1965) and Hiraki et al. (1967).

Then, the permeability of laser was measured in fluidized beds at incipient fluidization. The ratio of current output through various thickness of solid layer to the empty tube was found to follow the Beer's law and to become zero above 4mm in thickness.

From these characteristics bubble sizes were obtained as follows:

At a given position in the bed frequencies of bubbles were measured by changing the distance btween the two pipes. Differentiating the obtained curve the size distribution at this position gives. All size distribution curves obtained from the same measurement at 2.5cm intervals in the radial direction are superposed and normalized to yield the size distribution of the total cross section.

RESULTS AND DISCUSSION

Figs.6 and 7 show two representative cases of the change in distributions of bubble size as bubbles rise up in fluidized beds; the former is the case where the size distribution just above the distributor is given by a smooth curve and transformed gradually into a normal distribution with a wide stretch as the level from the distributor goes up, and the latter is the case where some amounts of bubbles having a quite large size are produced above the distributor and the size distribution is deformed into the bi-modal type of distribution.

When a number of bubbles are generated from the distributor, the size is smaller, the probability of collision is smaller. In the case where the same number of bubbles with relatively large size are produced, one or two bubbles of quite large size are produced because of their big collision probability. Such phenomena are shown in Fig.8, where 1000 bubbles are generated randomly with different sizes which are given by the volume fraction δ occupied by bubbles.

Once a large bubble is produced, it absorbs many small bubbles while rising up with high velocity.

Experimental data on size distribution at several levels are shown in Fig.9. In this experiment, always bi-modal forms were observed.

Such distributions of bubble size similar to our experimental results were also presented by Yamazaki et al. (1977). They developed an electronic system for the measurement of bubbles in three-dimensional beds. The probe showed the passage of a bubble by a lamp matrix connected with a hot-wire assembly. A result obtained in 8.1cm I.D. fluidized bed of glass beads is shown in Fig.10.

Comparison of distributions observed by both experiments with theoretical curves are fairly well, as shown in Figs.9 and 10. In both cases an isolated bubble with large size is produced above the distributor in the same way as shown in Fig.7. In our case bubbles were found to tend to coalesce in the central position of the bed, and in Yamazaki's case the aspect ratio was so high that slugging was observed. It may be considered that these factors would accelerate the production of a bubble with large size and lead to skewed distributions towards bi-modal form.

On the contrary, in the case where a number of fine bubbles are generated uniformly from the porous distributor, the size distribution of bubbles may be expressed by either Gamma or normal function according to the measured level, as in the results of Park et al. (1969) and Rowe & Everett (1972).

CONCLUSIONS

A model named coalescence model was proposed to simulate the bubble growth in fluidized beds. Bubbles were assumed to coalesce simultaneously just above the distributor and the resultant bubbles with large size grow up further by coalescence in axial direction while rising.

Mathematical equations were formulated and solved. This method could simulate even the bi-modal distribution of bubble size and therefore elucidate the reason for disparate data previously obtained.

Bubble size and its distribution were then measured in an air fluidized bed of 40cm I.D.. A new probe consisting of laser beams and an optical fiber was developed and the obtained results are in good agreement with the proposed model.

ACKNOWLEDGEMENT

One of us, K. Yoshida, is indebted to the Kawakami Memorial Foundation for its financial support.

REFERENCES

Argyriou, D.T., List, H.L. & Shinnar, R. (1971) A. I. Ch. E. J. 17, 122.

Baumgarten, P.K. & Pigford, R.L. (1960) A. I. Ch. E. J. 6, 115.

Chiba, T., Terashima, K. & Kobayashi, H. (1973) J. Chem. Eng. Japan 6, 78.

Davidson, J.F. & Harrison, D. (1963) Fluidised Particles, Chapt. 2, Cambridge Univ. Press.

Geldart, D. (1972) Powder Technol. 6, 201.

Himmelblau, D.M. & Bischoff, K.B. (1968) Process Analysis and Simulation, pp. 59-88, Wiley & Sons, New York.

Hiraki, I., Yoshida, K. & Kunii, D. (1965) Kagaku-Kogaku 29, 846.

Hovmand, S. & Davidson, J.F. (1971) Fluidization, Chapt. 5 (Davidson, J. F. & Harrison, D., Ed.) Academic Press, New York.

Kobayashi, H., Arai, F. & Chiba, T. (1965) Kagaku-Kogaku 29, 858.

Kunii, D., Yoshida, K. & Hiraki, I. (1967) Proceedings of the Int. Symposium on Fluidization, pp. 243-253, Netherlands Univ. Press, Amsterdam.

Miwa, K., Mori, S., Kato, T. & Muchi, I. (1971) Kagaku-Kogaku 35, 770.

Park, W.H., Kang, W.H., Capes, C.E. & Osberg, G.L. (1969) Chem. Eng. Sci. 24, 851.

Rowe, P.N. & Everett, D.J. (1972) Trans. Instn. Chem. Engrs. 50, 42, 49, 55.

Rowe, P.N. & Yacono, C. (1975) Trans. Instn. Chem. Engrs. 53, 59.

Rowe, P.N. (1976) Chem. Eng. Sci. 31, 285.

Toei, R., Matsuno, R., Ishii, H. & Kojima, H. (1965) Kagaku-Kogaku 29, 851.

Werther, J. (1974a) A. I. Ch. E. Sym. Ser. 70(141), 53.

Werther, J. (1974b) Trans. Instn. Chem. Engrs. 52, 149.

Whitehead, A.B. & Young, A.D. (1967) Proceedings of the International Symposium on Fluidization, pp. 284-293, Netherlands Univ. Press, Amsterdam.

Yamazaki, M., Ito, N. & Jimbo, G. (1977) Kagaku Kogaku Ronbunshu 3, 272.

Yasui, G. & Johanson, L.N. (1958) A. I. Ch. E. J. 4, 446.

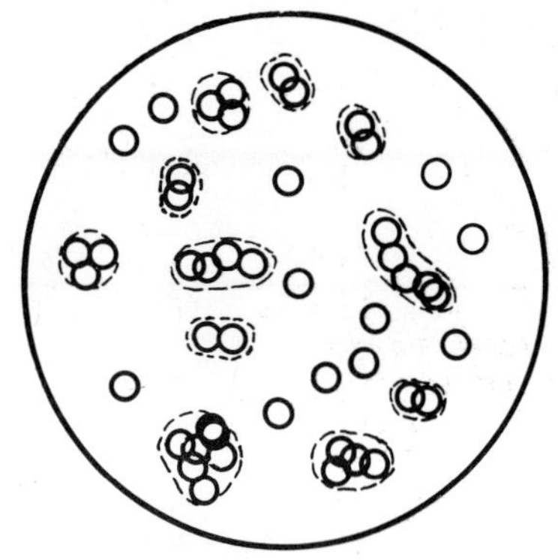

Fig.1. Vigorous coalescence just above the distributor

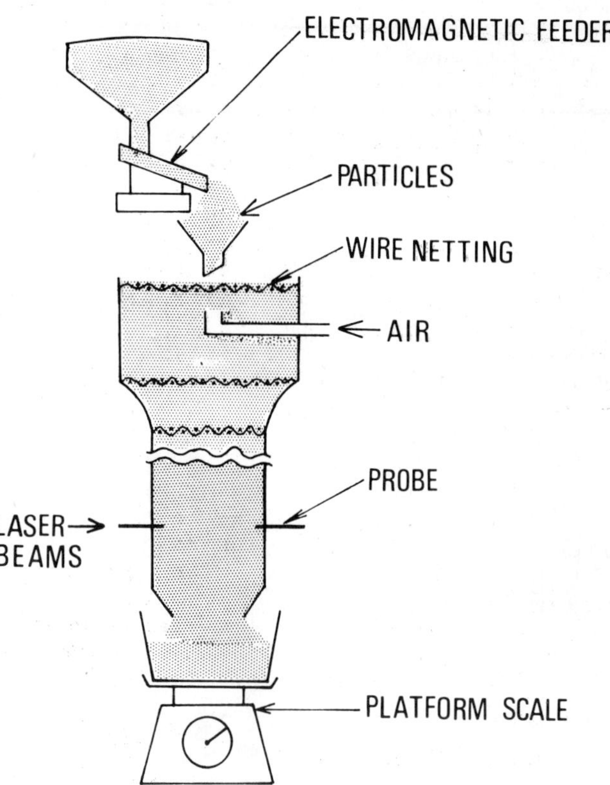

Fig.4. Arrangement for measurement of permeability of laser beams through solid layer

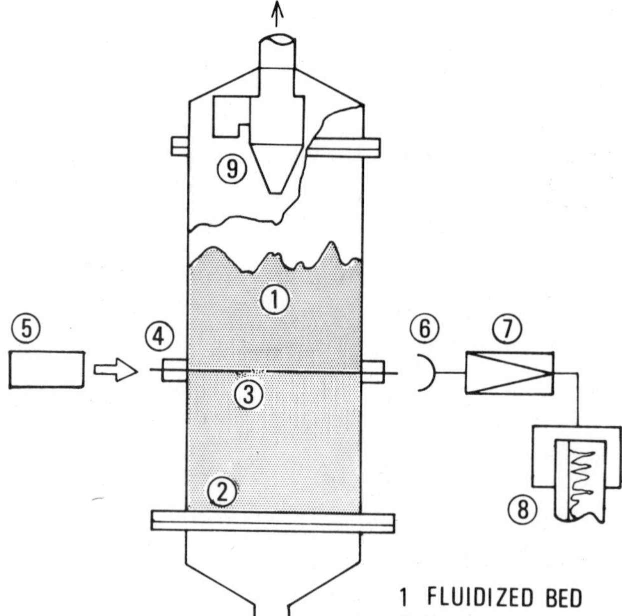

Fig.2. Experimental apparatus

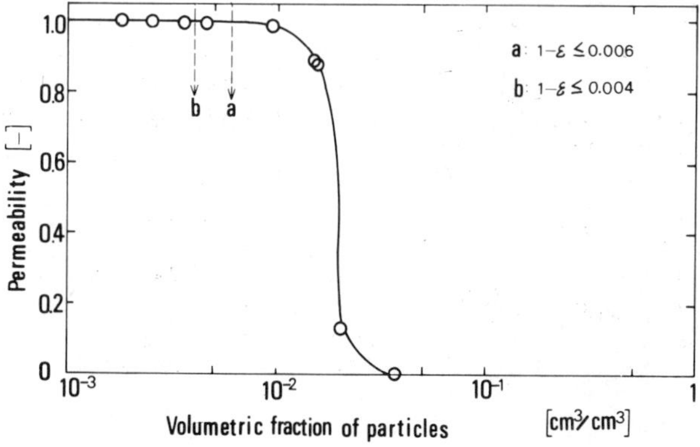

Fig.5. Characteristics of laser beams

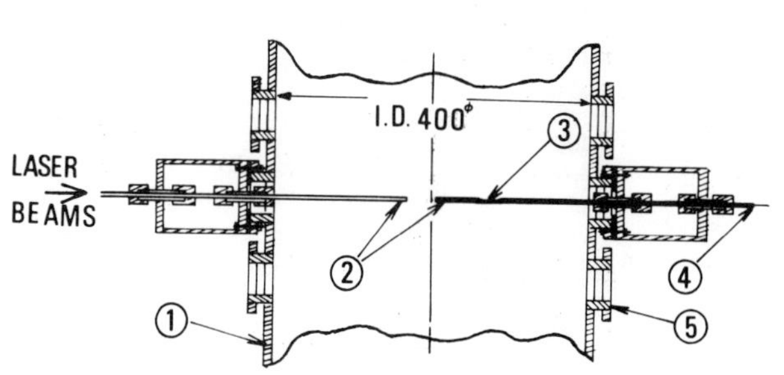

Fig.3. Detecting probe

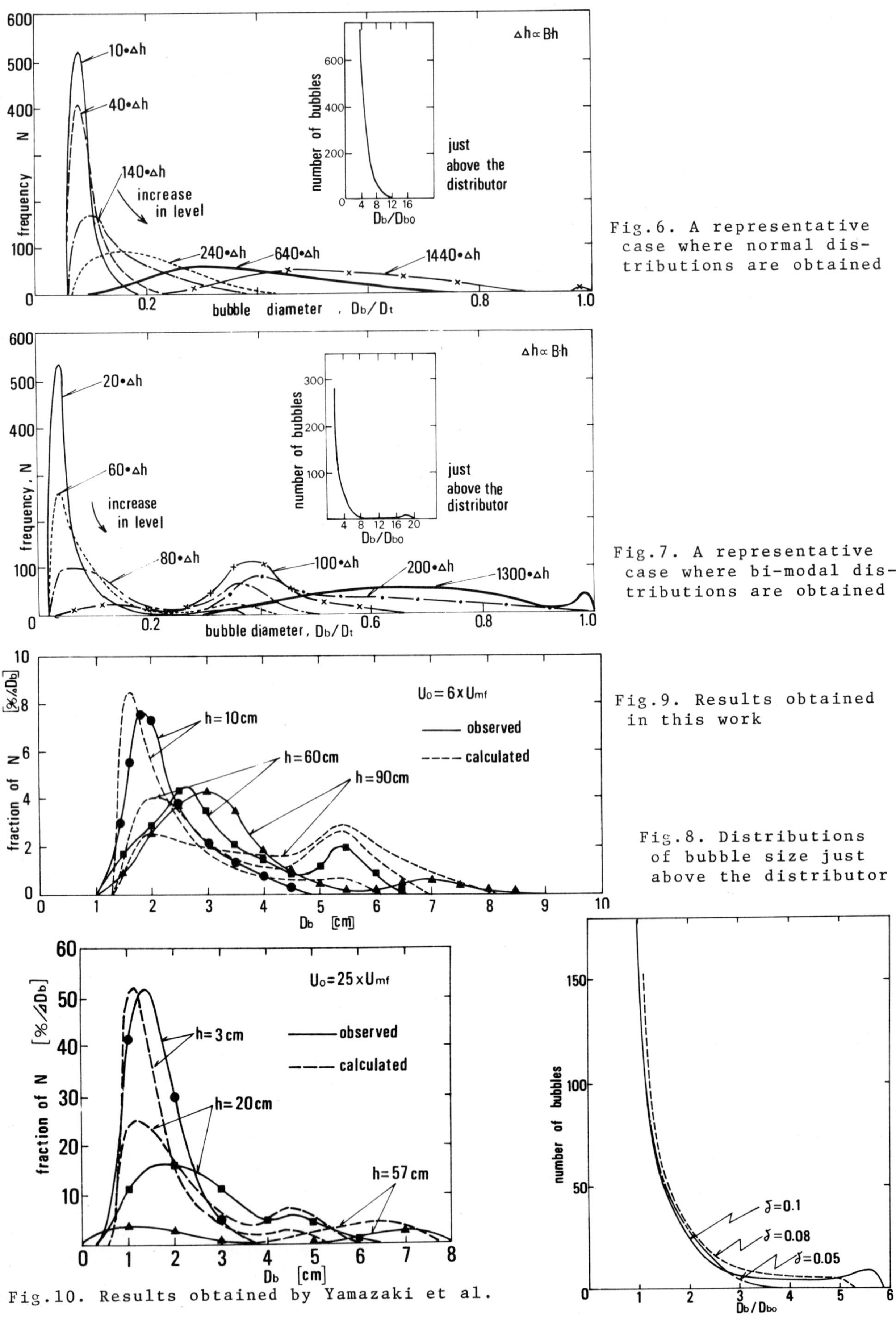

Fig.6. A representative case where normal distributions are obtained

Fig.7. A representative case where bi-modal distributions are obtained

Fig.9. Results obtained in this work

Fig.8. Distributions of bubble size just above the distributor

Fig.10. Results obtained by Yamazaki et al.

FORMATION OF BUBBLES AT AN ORIFICE IN FLUIDIZED BEDS

By T.P. Hsiung and J.R. Grace

Department of Chemical Engineering
McGill University
Montreal, Canada H3A 2A7

SYNOPSIS

It is shown that bubble formation in gas fluidized beds can take place under constant pressure or intermediate conditions, although constant flow formation has been assumed in previous work. It is shown experimentally that the mean frequency of bubble formation depends on the volume of the chamber feeding the orifice. At high flow rates, the mean frequency becomes approximately constant at about 20 Hz. Bubble formation is quite regular at low flow rates, but there is evidence of pairing and other complex phenomena at higher flow rates.

NOMENCLATURE

d_{Bo}	diameter of bubble formed at orifice
d_{or}	orifice diameter
d_p	particle diameter
f	frequency of bubble formation
g	acceleration due to gravity
H_{mf}	bed height at minimum fluidization
Q	flow rate through orifice
U_{mf}	superficial velocity at minimum fluidization
u_{or}	mean gas velocity through orifice
V_{Bo}	volume of bubble formed at orifice
V_c	chamber volume
ΔP_{Bed}	pressure drop across the bed
ΔP_D	mean pressure drop across distributor
ρ_{mf}	bed density at minimum fluidization

INTRODUCTION

Gas is generally introduced into the base of fluidized beds through a series of perforations, nozzles or tuyeres forming a distributor. Many properties of fluidized beds are strongly influenced by the size-distribution and spatial-distribution of gas bubbles in the bed. These distributions are in turn affected by the manner in which bubbles are formed at the gas distributor. In addition, there is evidence (Hovmand & Davidson, 1971; Behie & Kehoe, 1973; Chavarie & Grace, 1975) that interphase exchange and reaction are relatively fast in the region immediately above the grid. Hence it is of considerable practical importance to understand the mechanism of bubble generation at orifices in fluidized beds.

At high gas velocities through the grid, jets or spouts form at many or all of the orifices, with bubbles detaching from the ends of these jets or spouts (Basov et al., 1969; Behie et al., 1971; Chiba et al., 1972; Merry, 1975). At lower velocities bubbles form directly at the grid holes without jet formation. In either case it has been common practice (e.g. Kato & Wen, 1969; Geldart, 1972; Merry, 1975) to base the prediction of bubble size at the bottom of fluidized beds on the theory given by Davidson and Schuler (1960) (often with a modified constant). The theory is based on constant-flow bubble formation of gas bubbles in liquids of low viscosity. Recent studies (Haribabu et al., 1976; Clift et al., 1978) have shown that viscous effects should be included and become important when the bubbles being formed are small. In the present paper we present experimental data relating to the assumption of formation under constant flow conditions. It is shown that this assumption is not always justified. In addition, we consider the frequency distribution as well as the mean frequency of bubble formation in order to examine the boundary condition of regular bubble formation used in bubble coalescence models (e.g. Clift & Grace, 1972; Leung, 1972).

CONDITIONS FOR BUBBLE FORMATION

In practice there are many types of grid plates used in fluidized bed applications. The simplest and most common type is a flat or dished plate with cylindrical holes drilled or punched in a regular pattern. The pressure drop across the grid holes is recognized as being a design variable of considerable importance. In some processes, the pressure drop across the distributor is as little as 3-4% of the pressure drop across the bed, whereas in other processes a ratio close to unity is used. A $\Delta P_D/\Delta P_{Bed}$ ratio of 0.1 to 0.3 is most commonly employed.

As a bubble is forming at an upward-facing orifice, its centre of gravity rises. Hence the pressure on the downstream side of the orifice decreases with time. Since the pressure drop across the orifice therefore tends to change, the flow through the orifice may also change with time. For bubbles forming in liquids, a further cause of pressure variation inside forming bubbles is the decrease in the surface tension pressure increment. In fluidized beds, the pressure variation during formation is approximately $\rho_{mf} g d_{Bo}/2$ where ρ_{mf} is the bed density at minimum fluidization. Hence the ratio of the pressure variation during formation to the pressure drop across the distributor is approximately $(d_{Bo}/2H_{mf})/(\Delta P_D/\Delta P_{Bed})$. For very deep beds ($H_{mf} >> d_{Bo}$) or large values of $\Delta P_D/\Delta P_{Bed}$, the pressure variation during formation is small and constant flow conditions may be assured. However, for shallow beds and small $\Delta P_D/\Delta P_{Bed}$ ratios, substantial variations in flow can occur.

When variable flow conditions prevail, a further complication arises if the volume, V_c, of the plenum chamber feeding the orifice is of comparable size with the volume of bubbles being formed. In this case, the pressure in the chamber fluctuates during each bubble formation cycle as the exit flow rate through the orifice varies; these are said to be "intermediate conditions". If $V_c >> V_{Bo}$ on the other hand, the pressure in the plenum chamber is essentially steady and "constant pressure conditions" are said to prevail. An aim of the present study was to determine whether any effect of chamber volume could be detected for bubble formation in fluidized beds. While such an effect appears unlikely for the size of plenum chamber employed in industrial-scale fluidized beds, the chamber volume could be important for laboratory units with low pressure drop distributors.

EXPERIMENTAL APPARATUS AND PROCEDURE

The experimental set-up is shown schematically in Fig. 1. The fluidized bed (F) was 8 x 10 cm cross-section by 45 cm tall. It was constructed of 2.6 mm thick aluminum to facilitate X-ray observations, although an X-ray unit was not available for the work described herein. Bubbles were formed at different single orifices of inside diameter 0.63, 0.48, 0.32 and 0.16 cm mounted flush with a porous plate gas distributor (D) and covered with a 400 mesh screen to prevent backflow of solids. A brass tank (A) of volume 4800 c.c. served as a plenum chamber. This was connected to the orifice by a smooth tube of inside diameter 0.95 cm and total length 56 cm containing two 90° elbows. The volume of air in the chamber was varied from 0 to 4800 cm^3 by adding or removing water. The background superficial gas velocity to the bed through the distributor was maintained at 1.15 U_{mf} for all types of particles. The gas flow rate to the orifice was varied between 15 and 115 cm^3/s giving a u_{or} range of approximately 0.6 to 60 m/s. The particles used were glass ballotini with size ranges of 88 to 125 μm, 149 to 250 μm, and 250 to 420 μm (U_{mf} = 1.6, 3.6 and 8.1 cm/s respectively). The total bed height was about 3 cm. The maximum jet penetration predicted (Merry, 1975) was 2.6 cm and no jets were observed to break the bed surface for the conditions studied.

The frequency of bubble formation was determined by means of a sensitive pressure transducer (I) of capacitance type (DISA 51D02-Pu2a) connected (H) to a point just upstream of the orifice. Output signals from the transducer were fed to a visicorder (P) and to a Quantech 304 Wave Analyser (N). The latter allowed the frequency spectrum of the signal to be determined for frequencies up to 100 Hz. The sweep time used in the present experiments was 100 seconds. The pressure transducer was operated in a differential mode by blocking off one of the breather holes and connecting the other to a chamber (L) fed by a pressurized air cylinder through regulators (Q).

The recorded pressure traces showed definite peaks of relatively large amplitude in addition to low amplitude noise. In order to confirm that the large peaks corresponded to formation of bubbles, films were taken at 64 frames per second through a small window (E) in the side of the column with an aqueous sugar solution (density 1.35 g/c.c., viscosity 10.8 p) and with a shallow fluidized bed and a low flow rate such that bubble coalescence was unlikely. Sample pressure traces, data for the sugar solution experiments, and further experimental details are presented elsewhere (Hsiung, 1974).

EXPERIMENTAL RESULTS

As shown by Nguyen and Leung (1972), the volume of bubbles formed in a fluidized bed generally differs from the orifice flow rate divided by the frequency of formation. This difference arises from leakage of gas into the dense phase during the formation process. In view of this, we present mean frequencies (number of large amplitude peaks/unit time) and frequency spectra, rather than inferred bubble volumes.

The effects of orifice diameter and flow rate on the mean frequency of formation are shown in Fig. 2 for 149-200 μm particles and a chamber volume of 2000 c.c. For each of the orifices, the frequency increases with increasing flow rate and then levels off at a value of 19 to 25 Hz, in the same range as observed by previous workers (Harrison & Leung, 1961; Botterill & Bloore, 1963). Since the chamber volume is much greater than the volume of gas injected per bubble formed for these results, conditions may be assumed to correspond to constant pressure formation. The relationship

$$f = g^{0.6}/1.138Q^{0.2} \qquad \text{....(1)}$$

derived from the constant flow, inviscid fluid theory (Davidson & Harrison, 1963) is shown to predict considerably higher frequencies at low flow rates.

The effect of chamber volume on mean frequency of bubble formation is shown in Fig. 3 for four different sets of conditions. At low flow rates and relatively large orifices the frequency decreases and then levels out as V_c increases, i.e. as formation passes from constant flow to constant pressure conditions. This implies an increase followed by levelling off of the bubble volume formed, and this is consistent with bubble formation in liquids (Hughes et al., 1955; Davidson & Amick, 1956; Kupferberg & Jameson, 1969). For a larger flow rate (48 c.c./s) and smaller orifice (0.32 cm dia.), the pressure drop is larger and there is no detectable effect of chamber volume.

Frequency spectra are given in Fig. 4 for constant pressure conditions with the 0.48 cm orifice and intermediate size particles. At low flow rates, the frequency spectrum is unimodal, but the spectrum becomes much more complex with increasing flow rate. As shown in Fig. 5, the frequency spectrum also becomes more complex as the particle size is reduced for a constant flow rate and orifice size. Multimodal spectra probably signify departure from single bubble and/or uniform bubble formation. The minimum flow rates (±5 c.c./s) for which bimodal distributions were observed are given in Table 1 and these may be interpreted as the minimum flow rates for non-uniform or "double bubbling" (McCann & Prince, 1971). These are of the same order of magnitude as for liquid systems (Davidson & Amick, 1956; Kupferberg & Jameson, 1969). The spectra with more peaks were favoured by large chamber volume, high gas flow rate and small orifice size, all factors which are expected to promote multiple bubble formation (McCann & Prince, 1971).

Table 1 Onset of Multiple Bubble Formation

d_p (μm)	d_{or} (cm)	Q (cm^3/s)
149-250	0.63	60.5
149-250	0.48	47.2
149-250	0.32	23
149-250	0.16	18
88-125	0.48	50

DISCUSSION

The results presented above and more extensive results available elsewhere (Hsiung, 1974) show that bubble formation depends on the size of the orifice used and on the condition of formation. Formation under constant pressure conditions tends to give a lower frequency of formation than formation under constant flow conditions, other factors remaining unchanged.

Another factor of importance is that bubble formation is by no means perfectly regular, especially for increasing gas flow rates and small orifices. In modelling the coalescence process for vertical chains of bubbles, Clift and Grace (1972) assumed perfectly regular formation, but commented that irregularity of formation would result in a smoother decrease in bubble frequency with height. We have demonstrated (Hsiung, 1974) that this is the case, but the deviations from predictions made assuming regular formation do not appear to be very significant for the conditions studied.

CONCLUSIONS

1. For low pressure drop orifices, the frequency of bubble formation depends on the volume of the chamber feeding the orifice as for bubble formation in liquids. As the flow rate through the orifice increases, the bubble frequency levels off at a constant value between about 19 and 25 Hz.
2. At low flow rates, bubble formation tends to be very regular. With increasing flow rate, bimodal, trimodal and multimodal frequency spectra were observed indicating coalescence before detachment and irregular bubble formation.

ACKNOWLEDGEMENT

We are grateful to the National Research Council of Canada for financial assistance.

REFERENCES

Basov, V., Markhevka, T.K., Melik-Akhnazarov, T.K. & Orochko, D.I. (1969) Intern. Chem. Eng. 9, 263.
Behie, L.A., Bergougnou, M.A., Baker, C.G.J. & Bulani, W. (1970) Can. J. Chem. Eng. 48, 158.
Behie, L.A. & Kehoe, P. (1973) A.I.Ch.E. J. 19, 1070.
Botterill, J.S.M. & Bloore, P.D. (1963) Can. J. Chem. Eng. 41, 111.
Chavarie, C. & Grace, J.R. (1975) Ind. Eng. Chem. Fund. 14, 79.
Chiba, T., Terashima, K. & Kobayashi, H. (1972) Chem. Eng. Sci. 27, 965.
Clift, R. & Grace, J.R. (1972) Trans. Instn. Chem. Engrs. 50, 364.
Clift, R., Grace, J.R. & Weber, M.E. (1978) Bubbles, Drops and Particles, Academic Press, New York.
Davidson, J.F. and Harrison, D. (1963) Fluidised Particles, Cambridge Univ. Press, Cambridge, England.
Davidson, J.F. & Schüler, B.O.G. (1960) Trans. Instn. Chem. Engrs. 38, 335.
Davidson, L. & Amick, E.H. (1956) A.I.Ch.E. J. 2, 337.
Geldart, D. (1972) Powd. Tech. 6, 201.
Haribabu, P., Sarkar, M.K. & Rao, D.S. (1976) Can. J. Chem. Eng. 54, 451.
Harrison, D. & Leung, L.S. (1961) Trans. Instn. Chem. Engrs. 39, 409.
Hovmand, S. & Davidson, J.F. (1971) Fluidization, Chapt. 5 (Davidson, J.F. & Harrison, D., Ed.) Academic Press, London.
Hsiung, T.P. (1974) M.Eng. Thesis, McGill Univ.
Hughes, R.R., Handlos, A.E., Evans, H.D. & Maycock, R.L. (1955) Chem. Eng. Progr. 51, 557.
Kato, K. & Wen, C.Y. (1969) Chem. Eng. Sci. 24, 1351.
Kupferberg, A. & Jameson, G.J. (1969) Trans. Instn. Chem. Engrs. 47, 241.
Leung, L.S. (1972) Powd. Tech. 6, 189.
McCann, D.J. & Prince, R.G.H. (1971) Chem. Eng. Sci. 26, 1505.
Merry, J. (1975) A.I.Ch.E. J. 21, 507.
Nguyen, X.T. & Leung, L.S. (1972) Chem. Eng. Sci. 27, 1748.

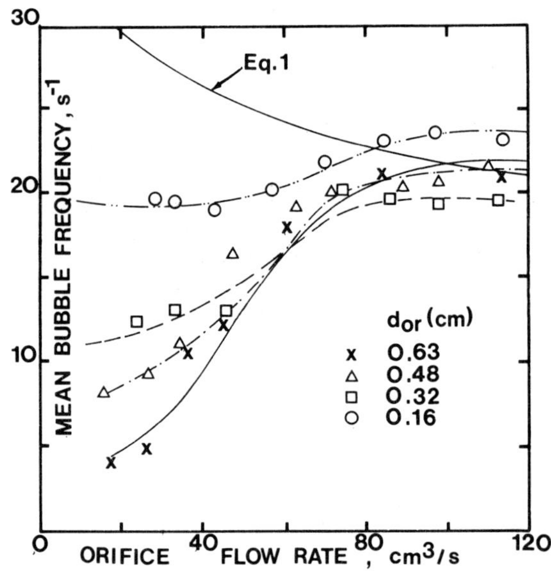

Fig. 2. Effect of orifice diameter and flow rate on mean bubble frequency.

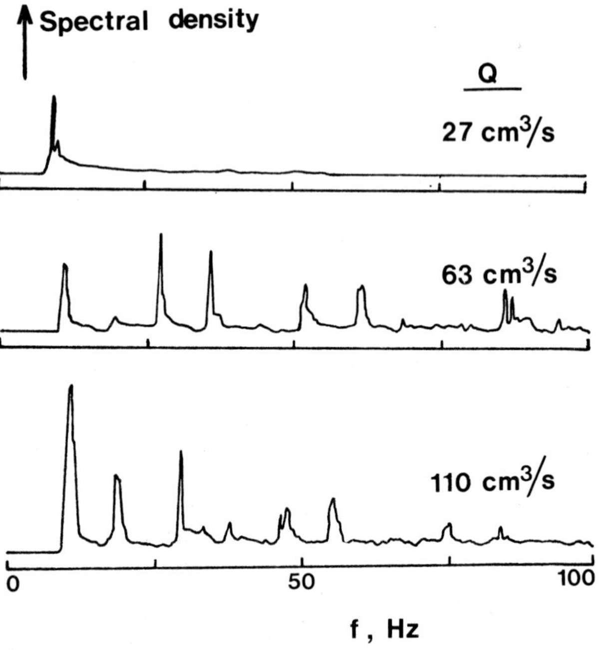

Fig. 4. Frequency spectra for different orifice flow rates. Orifice diameter: 0.48 cm; Particles: 149-250 μm; V_c = 2000 c.c.

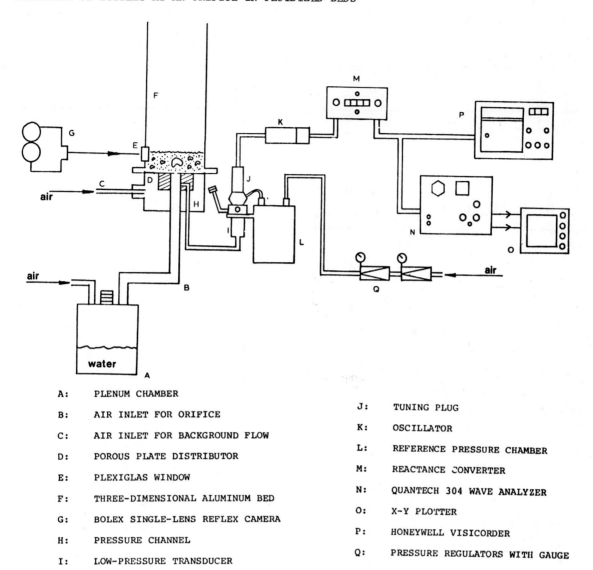

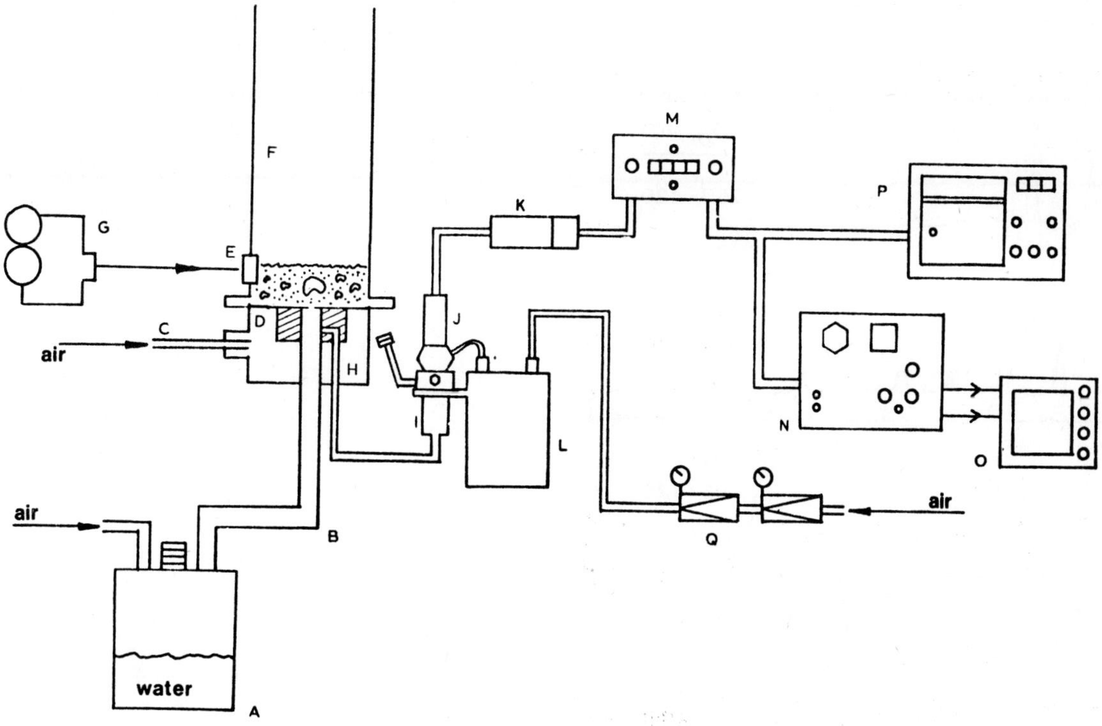

A: PLENUM CHAMBER

B: AIR INLET FOR ORIFICE

C: AIR INLET FOR BACKGROUND FLOW

D: POROUS PLATE DISTRIBUTOR

E: PLEXIGLAS WINDOW

F: THREE-DIMENSIONAL ALUMINUM BED

G: BOLEX SINGLE-LENS REFLEX CAMERA

H: PRESSURE CHANNEL

I: LOW-PRESSURE TRANSDUCER

J: TUNING PLUG

K: OSCILLATOR

L: REFERENCE PRESSURE CHAMBER

M: REACTANCE CONVERTER

N: QUANTECH 304 WAVE ANALYZER

O: X-Y PLOTTER

P: HONEYWELL VISICORDER

Q: PRESSURE REGULATORS WITH GAUGE

Fig. 1. Schematic diagram of apparatus.

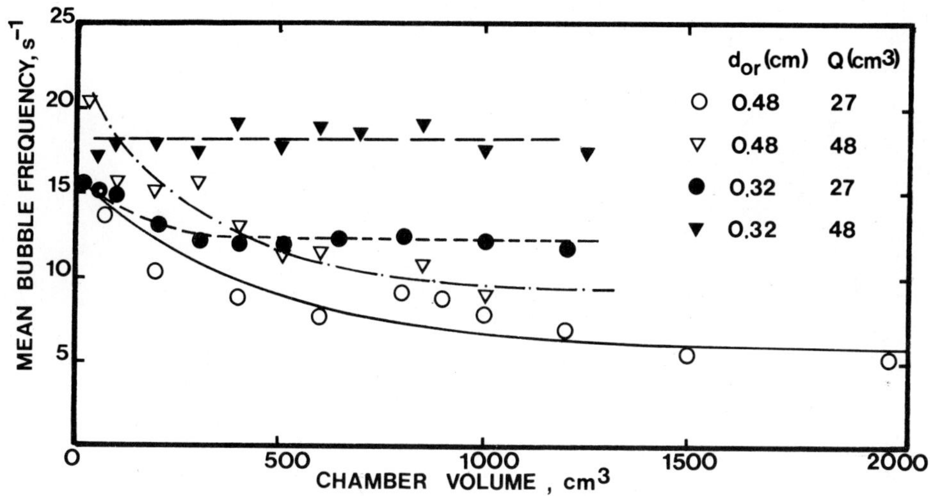

Fig. 3. Effect of chamber volume on mean bubble frequency.

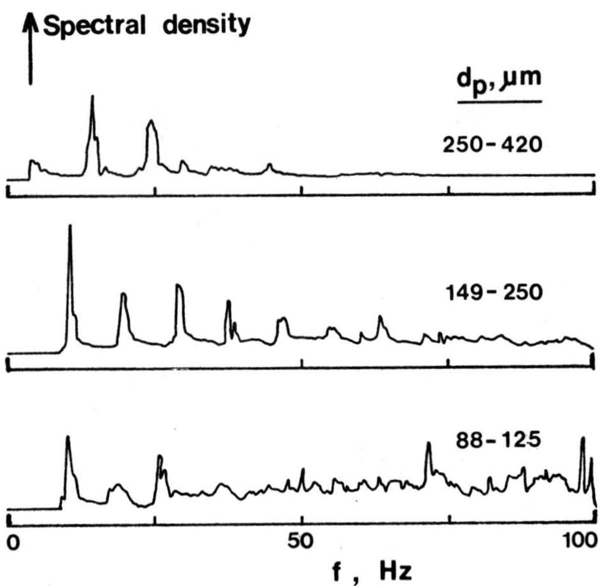

Fig. 5. Effect of particle size on frequency spectra. Orifice diameter: 0.48 cm; V_c = 2000 c.c.; Flow rate: 86–89 c.c./s

THE INFLUENCE OF GAS DISTRIBUTOR ON BUBBLE BEHAVIOUR
COMPARISON BETWEEN "BALL DISTRIBUTOR" AND "POROUS DISTRIBUTOR"

by C. YACONO* and H. ANGELINO

Institut du Génie Chimique
Laboratoire Associé C.N.R.S. n° 192
Chemin de la Loge
31078 TOULOUSE CEDEX (France)

* Present address : Centre de Recherches
Cie Française de Raffinage
Sté TOTAL
76700 HARFLEUR (France)

SYNOPSIS

Bubble behaviour has been examined by X-rays in three-dimensional air fluidised beds fitted with "ball distributors". Results have been compared with those of a previous study carried out with a standard porous distributor. Bubble behaviour was found to be qualitatively similar in both cases although quantitative differences were observed. It was thus found that the bubble size was larger and the bubble concentration smaller with a "ball distributor" than with a porous plate and that, for a "ball distributor", the bubble size became larger and the bubble concentration smaller with an increase in ball size. Various relationships are given.

1. INTRODUCTION : THE GAS DISTRIBUTOR PROBLEM

The choice and design of the gas distributor are essential to the successful industrial application of fluidised bed techniques. The requirements which must be met concern mainly the design of the apertures. Their number, size and shape must be such that the leakage of solids from the distributor is prevented, the particle-attrition is minimized and the fluidisation thus promoted is uniform and stable. (ZUIDERWEG, 1967 ; GREGORY, 1967 ; KUNII and LEVENSPIEL, 1969 ; BOTTERILL, 1975). These basic properties must often be completed by others, specific to the process considered. For example if the gas contains dust the distributor must be permeable so as to prevent blocking ; if the fluidised bed works at high temperature or with corrosive gases the material constituting the distributor must be chosen with care.

With respect to the promotion and the maintenance of uniform and stable fluidisation the pressure drop across the distributor ΔP_D seems the essential factor (BOTTERILL, 1975). It is often expressed in relation to the pressure drop across the bed ΔP_B with the ratio : $X = \Delta P_D/\Delta P_B$. There is no universal minimum value of X, X_{min}, which will ensure homogeneous fluidisation in all working conditions (GREGORY, 1967). The value of X_{min} depends in fact, not only on the distributor type, but also on the particles fluidised, the bed depth , the gas velocity and even the percentage of uneven distribution which can be tolerated (HIBY, 1967 ; WHITEHEAD et al., 1970 ; GVOZDEV et al., 1963 ; DAVIDSON, 1968 ; KELSEY, 1968 ; WORMALD et al., 1971 ; SIEGEL, 1976).

Many designs of gas distributor have been proposed (KUNII and LEVENSPIEL, 1969). The most commonly used for laboratory scale experiments are the densely consolidated porous media. These devices give rise to a very large number of infinitely small bubbles in the immediate distributor neighbourhood (WERTHER, 1974), thereby promoting excellent contact between gas and solid with minimum "entrance effect" (BAKKER and HEERTJES, 1960). They are very seldom used in industry as they have little mechanical strength, are very easily clogged by dust and subject the gas to very high pressure drops.

This paper is mainly devoted to the behaviour of bubbles formed from a new type of distributor : the "ball distributor". The experimental results are compared with those of a previous work carried out with a porous plate (ROWE and YACONO, 1976 ; YACONO, 1975, 1977).

2. THE "BALL DISTRIBUTOR"

A "ball distributor" is a very simple device which consists of two parts (HENGL et al., 1975) : a high porosity support having negligible pressure drop(e.g. a grid or a perforated plate with a great number of holes) and a variable amount of high density balls the upper level of which is approximately horizontal. The particles to fluidise rest on top of the balls.

The principal advantage of this system is to submit the gas to a very small pressure drop while promoting, under certain conditions, a very homogeneous fluidisation characterised by a uniform spatial repartition of small sized bubbles close to the upper surface of the balls. Other interesting features include the high permeability to dust transported by the gas stream, the possibility to use at high temperatures or with corrosive gases if an appropriate material is chosen for the balls, the simplicity of manufacture and operation.

The main drawback is the danger of destroying the distributor by accidentally fluidising the balls. Another disadvantage, common to all low-pressure drop system, is the possibility of spontaneous oscillations linked to a periodic formation of bubbles, (BAIRD and KLEIN, 1973).

A preliminary study has revealed(HENGL, 1975) the different parameters defining the system "column-fluidised bed-ball distributor" and has shown that fluidisation is uniform when the following inequality is respected :

$$\Delta P_D > \text{Max.}\ (0.10\ \Delta P_B \text{ and } 3\text{-}4 \text{ mm } H_2O) \quad \ldots. \ (1)$$

Further investigation (HIQUILY, 1977) shows that this condition,which may be too conservative, should also be related to the value of the gas velocity.

With respect to possible uses of "ball distributors" note that they are very appropriate to multistage fluidisation (HIQUILY, 1977).

3. EXPERIMENTAL EQUIPMENT AND TECHNIQUE

This section describes very briefly the apparatus and the experimental procedure. Full details can be found elsewhere (YACONO, 1975, 1977).

The essential piece of equipment was an X-ray medical diagnostic unit coupled to a cine camera the framing rate of which was adjusted to approximately 25 f/s.

The beds under investigation were contained in two rectangular aluminium vessels 5.10 cm thick x 30.40 cm wide x 90 cm high and 14.40 cm thick X 30 cm wide x 90 cm high. In both cases the vessels were positioned so as to present half their width to the X-ray beam. The region of bed from which bubble measurements were taken had a height of 4 cm for the thinner vessel and 8 cm for the thicker one (2 or 4 cm above and 2 or 4 cm below the level considered) with a cross sectional area given by the product(vessel width/2)x vessel thickness.

The support of the ball distributors was made of three thicknesses of grid : the bottom two were relatively coarse ("12 mesh" grid, size of holes 1400 μm) and were used only to bear the weight of the balls and the particles, the top one was sufficiently fine ("120 mesh" grid, size of holes 125μm) to prevent the leakage of the balls through the support.

Three sizes of nearly spherical steel balls were used. Some of their properties are listed in Table 1.

The balls placed in bulk above the grids formed a layer of approximately constant depth. This depth, ranging from 5cm to 26cm, was different for each series of experiments and was chosen so that the corresponding pressure drop was at least sufficient to ensure an homogeneous fluidisation throughout the bed. No attempt was made to minimize the ratio $X = \Delta P_D / \Delta P_B$ (Cf section 1). The extreme values observed were 5 % and 33 %.

The particles composing the beds were two sizes of silicon carbide. Their main properties are grouped in Table 2.

The minimum fluidisation velocities of these particles were found to be independent of the size of the balls and equal to those previously obtained with a porous distributor (ROWE and YACONO, 1976). The bed depths, ranging from 12 cm to 40 cm, were selected in relation to the height of the level considered.

Three heights above the top of the balls (h = 6 cm, 18 cm and 32 cm) and three gas velocities ($U-U_{mf}$ = 1.57 cm/s, 2.75 cm/s, 3.93 cm/s)were examined. Two of these heights (h = 18 cm and 32 cm) had already been studied with the same beds at the same gas velocities but with a porous distributor (YACONO, 1975). The third height (h = 6 cm) was chosen in order to analyse the bubble behaviour in the vicinity of the balls with the thinner vessel.

There were many parameters involved in this work and only a few of their possible combinations were considered. For example, the material SC 150 was used only in the thicker vessel with the balls of 282μm. A total of 33 experiments were carried out, 4 of which were discarded as they corresponded to situations where the gas distribution was uneven (see next section). For each experiment on the one hand the diameter d_B, the external surface area S_B, the volume V_B and the velocity U_B of approximately 200 bubbles were recorded while on the other hand the average bubble concentration N_{BD} was measured. The technique of analysis of the X-Ray films is described elsewhere (YACONO, 1975,1977).

4. RESULTS AND DISCUSSION

Uniformity of fluidisation

The present work gives some information about the possible effects of gas maldistribution on bubble behaviour. It is a well established fact that, at any level in an homogeneously fluidised bed, the average bubble size increases with gas velocity. As this "rule" was not followed for a particular system "ball distributor + bed of particles" another series of experiments was carried out with the same ball distributor but with approximately half the amount of particles. The average bubble size was then found to vary as expected with gas velocity. Furthermore the average bubble sizes at the same level in both beds became equal when the gas velocity reached the highest value considered ($U-U_{mf}$=3.93 cm/s). It was concluded firstly that in the first series of experiments the bed was not homogeneous at the two lowest gas velocities. Secondly that this homogeneity depended on the ratio (bed height)/(balls height) and on the gas velocity and thirdly that when gas distribution is not uniform bubbles may benefit from the surplus of gas unused in the dead zones and may therefore be larger than in a well fluidised bed.

Bubble shape

The arithmetic average bubble volume $\bar{V}_B$ is plotted against the volumetric average bubble diameter $\bar{d}_{Bv}$ in fig. 1. $\bar{d}_{Bv}$ is calculated, for a given set of experimental conditions, with the formula :

$$\bar{d}_{Bv} = \left(\frac{1}{N_S} \sum_i d_{Bi}^{3}\right)^{1/3} \quad ...... (2)$$

where N_S is the number of bubbles in the sample analysed and d_{Bi} is the diameter of the ith bubble in that sample. The straight line corresponds to the following relationship fitted to the results obtained with the porous distributor:

$$\bar{V}_B = 0.527\ \bar{d}_{Bv}^{\ 2.94} \quad .... (3)$$

Figure 1 shows that, for bubble sizes similar to those recorded with the porous distributor ($\bar{d}_{Bv} > 2$ cm), there is little scatter of the points about (3). This confirms that the average bubble shape is independent of particle size (YACONO, 1975, 1977) and indicates that it is also independent of the gas distributing system. Conclusions previously drawn about the variations of individual bubble shape with height, gas velocity and bubble size remain valid (ROWE and YACONO, 1976). The slight disagreement between (3) and some of the points corresponding to the smallest bubble sizes ($\bar{d}_{Bv} < 1.7$ cm) can indeed be interpreted as an increase in the sphericity of the bubbles as they become smaller.

Bubble velocity

The variations with height of the arithmetic average of the horizontal component of the bubble velocities indicate (YACONO, 1977) that, after travelling vertically in the regions close to the distributors, bubbles move towards the bed centre line.

A detailed analysis of the possible relationships between the vertical component of the bubble centroid velocity U_{GY} and the bubble diameter d_B confirm the result obtained with the porous distributor, namely that there is little correlation between the velocity of a particular bubble and its size. The arithmetic average of U_{GY}, $\bar{U}_{GY}$, is given as a function of the arithmetic average bubble diameter $\bar{d}_{Ba}$ in fig. 2. The relationships between these two variables can be represented by equations of the type :

$$\bar{U}_{GY} = k_a\ (g\ \bar{d}_{Ba})^{\frac{1}{2}} \quad .... (4)$$

where k_a is a dimensionless coefficient and g is the acceleration of gravity.

The ball size has little effect on the value of k_a which decreases with an increase in particle diameter and is smaller with ball distributors than with a porous plate (cf Table 3).

Average bubble size

The average bubble diameter $\bar{d}_{Bv}$, determined from (2), is shown as a function of the ball size d_b in fig. 3. The values calculated from the correlation established with the porous distributor and valid for all sizes of particles (YACONO, 1975) are plotted on the vertical axis for which d_b = 0 μm. Figure 3 indicates that the average bubble diameter is larger with a ball distributor than with a porous plate and that it increases with an increase in ball size.

The variations of $\bar{d}_{Bv}$ with the height h and the gas velocity $(U - U_{mf})$ are clearly described (fig. 4) by the following equation:

$$\bar{d}_{Bv} = \alpha\ h^{\beta}\ (U - U_{mf})^{\gamma} \quad ...... (5)$$

The fitted parameters α, β and γ are given in Table 4.

Average bubble concentration

The average bubble concentration at a certain height above a ball distributor, N_{BD}, is compared to the average bubble concentration at the same height but above a porous distributor, N_{PD}, in fig. 5, N_{PD} being calculated from previously obtained correlations (YACONO, 1975). The two concentrations are proportional : $N_{BD} = K\ N_{PD}$ ..... (6)

The coefficient of proportionality K (cf Table 4) decreases exponentialy with an increase in ball size $\bar{d}_b$ and (6) can be replaced by

$$N_{BD} = e^{-0.534\ 10^{-3}\ \bar{d}_b}\ N_{PD} \quad \ldots\ldots\ldots \quad (7)$$

where $\bar{d}_b$ is measured in μ m

The bubble concentration is therefore smaller with a ball distributor than with a porous plate. Furthermore it decreases with an increase in ball size.

5. CONCLUSIONS

Bubble behaviour is qualitatively similar in beds fitted with ball distributors and with porous plates although there are important quantitative differences mainly concerning the bubble size and the bubble concentration.

The aim of any further investigation should be twofold : firstly to increase the range of experimental conditions and secondly to study in detail the bubble formation at the surface of a ball distributor. The mechanism so far proposed is that bubbles form in the interstices of the upper part of the layer of balls. Another possibility suggested by the fact that a porous plate is probably an extreme case of ball distributor (cf (7)) is that bubbles are created in the immediate vicinity of the distributor surface (WERTHER, 1974), maybe on top of small gas jets (FAKHIMI and HARRISON, 1970).

Acknowledgments : The authors would like to thank Prof. P.N.ROWE, University College London, for providing the X-Rays equipment and all experimental facilities.

REFERENCES

BAIRD, M.H.I. & KLEIN, A.J. (1973) Chem. Eng. Sci, 28, 1039.

BAKKER,P.J. & HEERTJES, P.M. (1960) Chem. Eng. Sci.,12, 260.

BOTTERILL, J.S.M. (1975) "Fluid bed heat transfer", Academic Press, London.

DAVIDSON, J.F. (1968) Tripartite Chem. Eng. Conference, Montreal, Inst. Chem. Engrs., London, p. 3.

FAKHIMI S.,& HARRISON, D. (1970) Chemeca'70, Melbourne and Sidney, Inst. of Chem. Eng. Symp. Ser., n° 33, Session 1, p. 29.

GREGORY, S.A. (1967) Proc. Intern. Symp. on Fluidisation, Eindhoven, Netherlands Univ. Press, Amsterdam, p. 751.

GVOZDEV, V.D., SAL'NIKOV, A.A., FOMICHEV, A.G., TIKHONOV, V.A. & VASIL'EV A.S. (1963) Int. Chem. Eng., 3, n° 4, 562.

HENGL, G., (1975) Thèse de Doctorat d'Université Paul-Sabatier, Toulouse.

HIBY, J.W. (1967) Chem. Ing. Tech., 39, n° 19, 1125.

HIQUILY, N., (1977) Thèse de Docteur-Ingénieur, Institut National Polytechnique de Toulouse.

KELSEY, J.R. (1968) Tripartite Chem. Eng. Conference, Montreal, Inst. Chem. Engrs., London, p. 86.

KUNII D., & LEVENSPIEL, O. (1969) "Fluidization Engineering", John Wiley and Sons Inc., New York.

ROWE, P.N., & YACONO, C. (1976) Chem. Eng. Sci., 31, 1179.

SIEGEL R. (1976) Am. Inst. Chem. Engrs.J., 22, n° 3, 590.

WERTHER, J. (1974) Am. Inst. Chem. Engrs. Symp. Ser., 70, n° 141, 53.

WHITEHEAD, A.B., GARSIDE, G. & DENT, D.C. (1970) Chem. Eng.J., 1, 175.

WORMALD, D. & BURNELL, E.M.W. (1971) Brit. Chem. Eng., 16, n° 415, 376.

YACONO, C. (1975) Ph.D. Thesis, University of London.

YACONO, C. (1977) Thèse de Doctorat ès-Sciences, Institut National Polytechnique de Toulouse.

ZUIDERWEG, F.J. (1967) Proc. Intern. Symp. on Fluidisation, Eindhoven, Netherlands, Univ. Press, Amsterdam, p. 739.

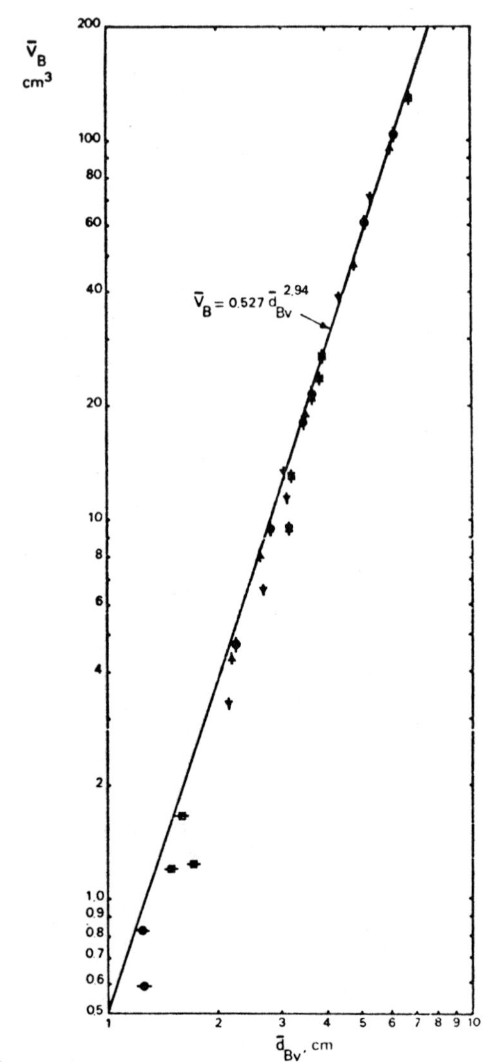

Fig. 1 - Average bubble volume as a function of average bubble diameter

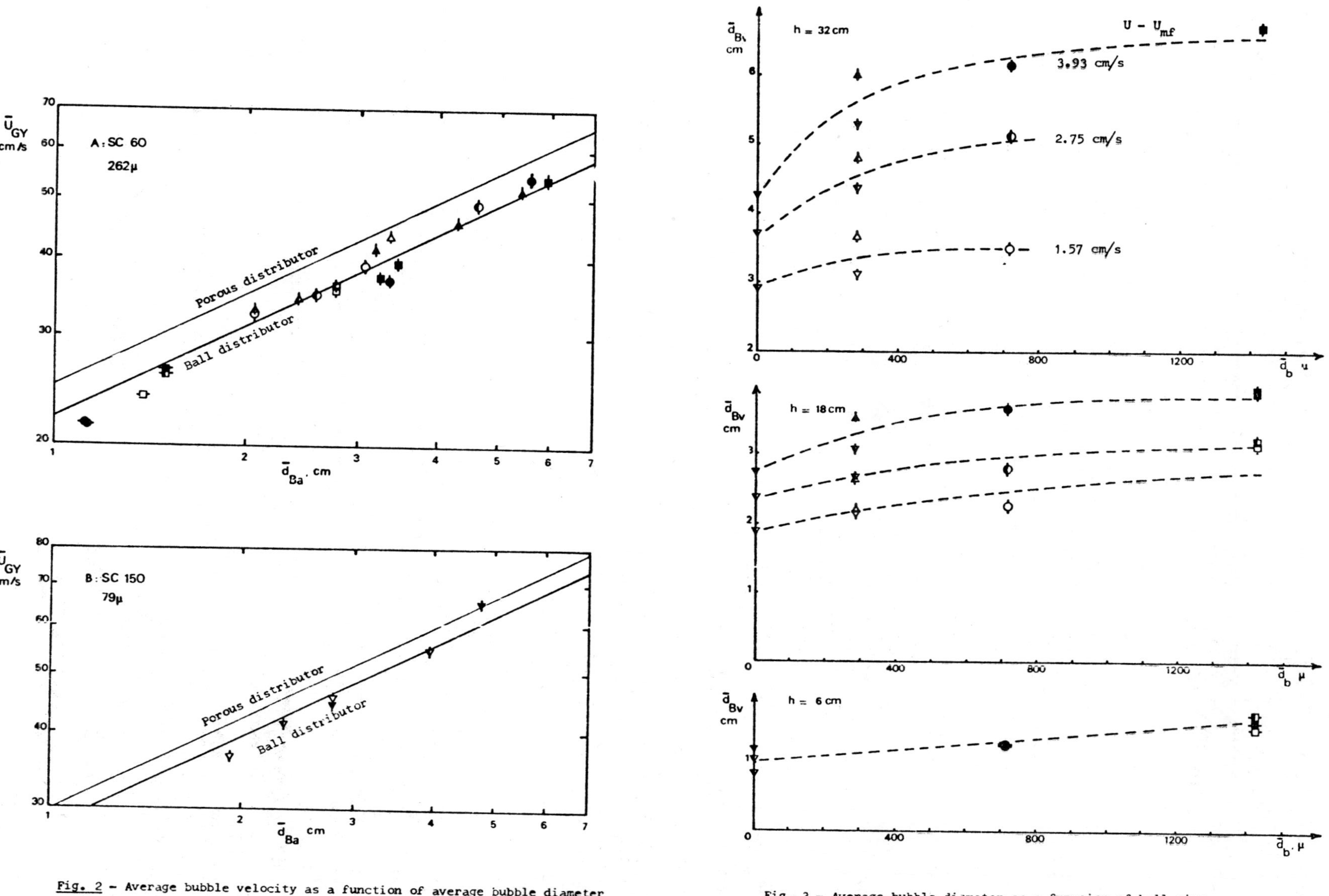

Fig. 2 - Average bubble velocity as a function of average bubble diameter

Fig. 3 - Average bubble diameter as a function of ball size

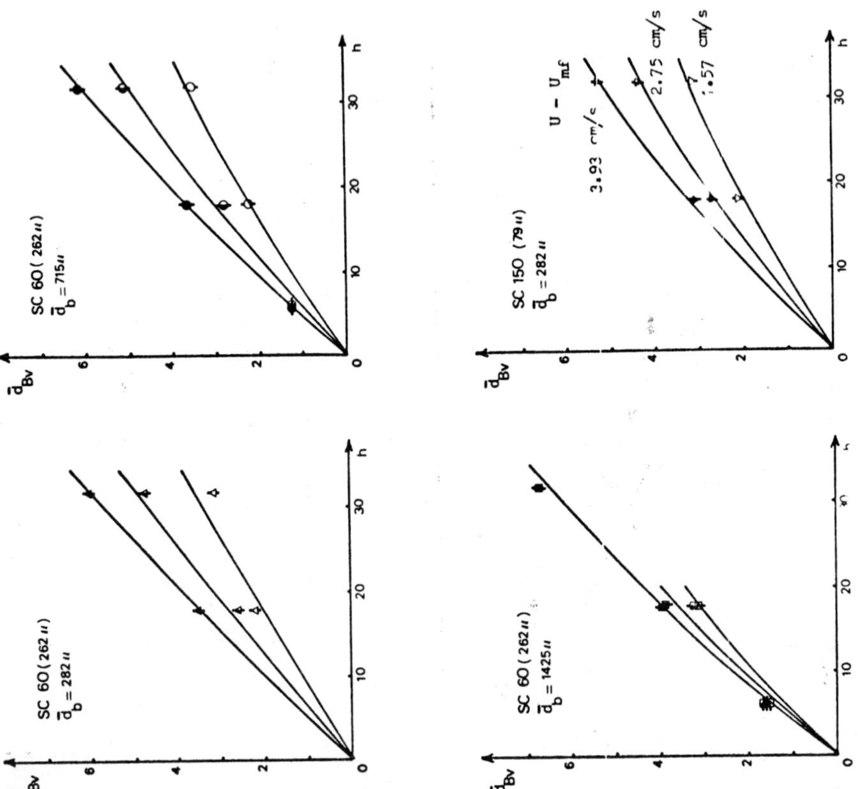

Fig. 4 - Variation of average bubble diameter with height

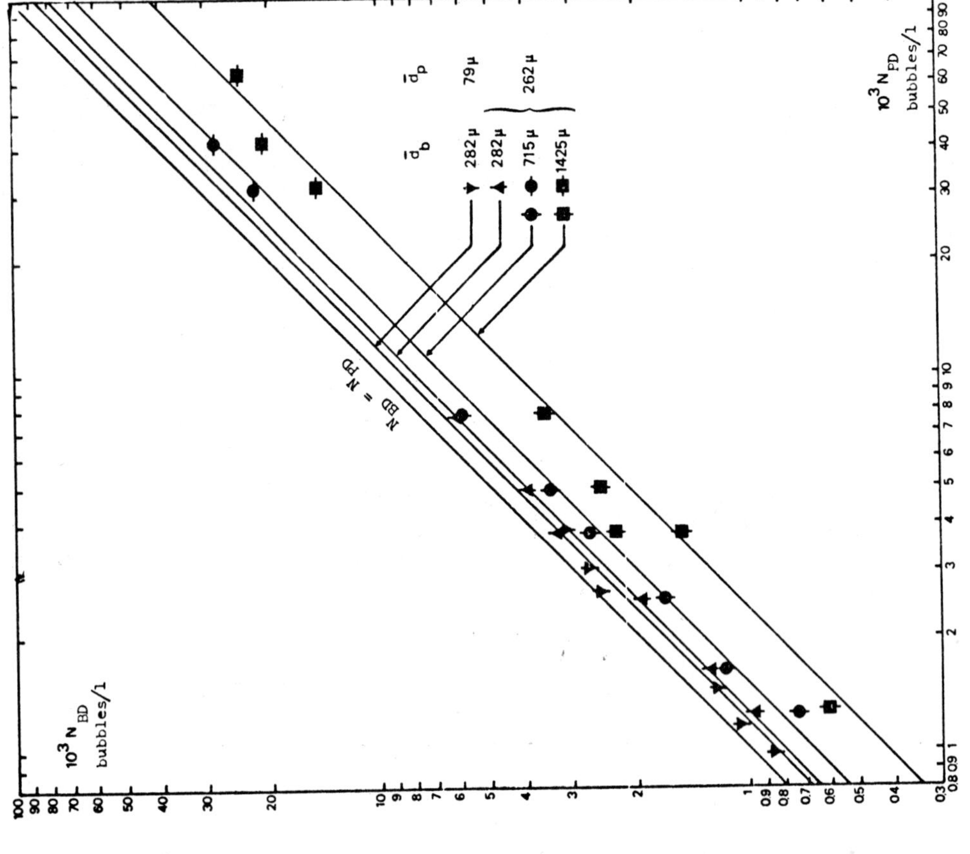

Fig. 5 - Comparison of bubble concentrations with ball distributor and porous plate

Table 1

Average ball size $\bar{d}_b$, μm	Size range, μm	Minimum fluidisation velocity U_{mf} (calculated from Kunii et al.), cm/s	Density ρ_s, g/cm^3
282.5	250–315	19	7.7
715	630–800	82	7.7
1425	1250–1600	159	7.7

Table 2

Name of particles	Surface mean (Kunii et al.) particle diameter $\bar{d}_p$, μm	Relative spread in particle diameter $\pm s/\bar{d}_p$	Minimum fluidisation velocity U_{mf}, cm/s	Minimum fluidisation voidage ε_{mf}	Density ρ_s g/cm^3
SC 150	79.1	0.16	1.07	0.528	3.186
SC 60	261.9	0.13	9.78	0.504	3.186

Table 3

	k_a ; $\bar{U}_{GY} = k_a(g\,\bar{d}_{Ba})^{\frac{1}{2}}$	
	Material SC 60 $\bar{d}_p$ = 262 μm	Material SC 150 $\bar{d}_p$ = 79 μm
Ball distributors	0.704	0.889
Porous distributor	0.795	0.936

Table 4

	$\bar{d}_b$ = 282 μm			$\bar{d}_b$=715 μm	$\bar{d}_b$=1425 μm
	Material SC 60 $\bar{d}_p$ = 262 μm	Material SC 150 $\bar{d}_p$ = 79 μm	Materials SC 60 + SC 150	Material SC 60	Material SC 60
$\bar{d}_{Bv} = \alpha h^{\beta} (U-U_{mB})^{\gamma}$: α	0.10	0.14	0.12	0.12	0.24
β	0.95	0.85	0.90	0.91	0.84
γ	0.54	0.52	0.53	0.59	0.28
$N_{BD} = KN_{PD}$: K	0.801	0.860	0.815	0.674	0.419

DEAD ZONE HEIGHTS NEAR THE GRID OF FLUIDIZED BEDS

C. Y. Wen, R. Krishnan, R. Khosravi and S. Dutta

Department of Chemical Engineering

West Virginia University

Morgantown, West Virginia 26506

SYNOPSIS

Experiments were conducted using two dimensional and three dimensional fluidized beds to investigate dead zone heights near the grid region. The two dimensional results indicate that gas velocity, distributor type, orifice pitch, orifice diameter and particle size have significant effects on dead zone height. The results from the three dimensional bed confirm the importance of orifice pitch and indicate that the behavior of the two dimensional bed cannot be readily extrapolated quantitatively to three dimensional cases. For the two dimensional bed an empirical correlation is proposed relating the minimum gas flow rate needed to eliminate the dead zone in terms of orifice pitch, orifice diameter and particle size.

NOTATION

d_o	diameter of distributor hole, cm
d_p	average of diameter of particle, μm
h_s	height of dead zone, cm
h_T	maximum height of tracer boundary including the quasi dead zone, cm
n	total number of holes on the distributor, -
P	center to center pitch of adjacent holes, cm
Q_j	gas flow rate per orifice, cm^3/sec
$(Q_j-Q_{jmf})_c$	gas flow rate per orifice at zero dead zone, cm^3/sec
Q_{jmf}	gas flow rate per orifice at minimum fluidization, cm^3/sec
u_{mf}	minimum fluidization velocity, cm/sec
u_o	superficial gas velocity, cm/sec

INTRODUCTION

The most critical feature in the design of a fluidized bed reactor is the gas distributor system. A large number of variables determine the structure of the gas distributor and, consequently, the design of the distributor varies widely. Frequent reports in the literature indicate that particle mixing above the distributor could be very poor when the distributor is not designed properly. The regions of stagnant particles, i.e. the dead zones on the distributor plate, is a serious problem in the design and operation of fluidized bed combustors, gasifiers and exothermic catalytic fluidized bed reactors. In spite of the apparent importance of dead zone formation only limited information is available in the literature (Makshima and Shirai (1969a), (1969b), Shinohara et al. (1975), Horio et al. (1977)).

This paper examines the formation of the dead zone and investigates the major variables that affect its magnitude. The experimental work was conducted over a wide range of operating conditions and bed parameters using two dimensional and three dimensional beds having perforated plates and cap distributors with sand and glass beads as bed materials.

EXPERIMENTAL PROCEDURES

Two Dimensional (2-D) Bed Experiments were conducted using sand and glass beads of varying particle size. The bed was constructed of plexiglass and the cross section of the bed was 1.2 cm x 30.5 cm. Perforated plate and cap distributors of varying orifice pitch and diameter were used in the experiments. Glass wool packing was provided at the bottom of the distributor plate to ensure uniform gas distribution and to arrest any weepage of the particles. The bed was initially filled with black particles up to a height of 4 cm. A sufficient gas flow rate was maintained in the bed in order to prevent weepage of particles through the distributor holes. The remaining height of the bed was then filled with white particles. The gas flow rate was monitored with a rotameter. After steady state conditions were established (about 5 minutes after the onset of fluidization), dead zone heights were measured along the distributor and photographs were taken. Except at the walls, the heights of dead zones were found to be relatively uniform along the distributor. The reported heights are the average values of all the dead zones except those adjacent to the walls. For small values of dead zone heights ($h_s \leq 0.3$ cm), a travelling telescope was used to measure the heights within 0.01 mm.

Three Dimensional (3-D) Bed A cylindrical bed of 27.6 cm I.D. constructed of plexiglass was used in these experiments. Steel perforated grid plates with triangular pitches were used to obtain conditions at which the dead zone disappeared.

Elimination of dead zones in the three dimensional bed was detected from the responses of thermistor probes. Several miniature thermistor beads (Fenwall, 10,000 Ω at 25°C) of 1 mm diameter were embedded on the surfaces of the grid plates at various strategic locations. Probe responses were recorded on a chart recorder as a function of gas flow rate through the bed. Fig. 1 shows a typical chart recording of such responses. The initial almost-steady horizontal line represents the presence of a dead zone at the particular location on the grid plate. The oscillations represent the elimination of dead zone. In these experiments, the gas flow rate was increased in small steps and the flow rate at which the recorder pen started to oscillate was noted as the minimum flow rate for elimination of dead zone.

The operating conditions employed in the 2-D and 3-D bed experiments are summarized in Table 1. The pressure drop across the grid for the 2-D bed ranged typically from 93 to 720 cms of H_2O depending on the pitch and gas flow rate.

RESULTS AND DISCUSSION

Earlier it has been indicated (Wen and Dutta, 1976) that the different modes of mixing occurring over the grid region give rise to three distinct zones (Fig. 2). In the intermittently mixed zone, exchange of particles takes place due to bubble drift at low gas flow rates and due to disturbances caused by oscillating gas jets at higher gas velocities. The particles are stagnant only during the time when they do not encounter bubbles or oscillating jets. In the quasi dead zone, there is very little exchange and the particles merely slide over the crown of the completely dead zone that forms directly beneath it. No mixing occurs in the dead zone and the particles are essentially stagnant. In this work, the height of the dead zone is taken as the height of the fixed dead zone as defined in Fig. 2.

The experiments in the 2-D bed show that for a given solid material the orifice pitch and gas velocity have the most pronounced effect on dead zone height as indicated in Figs. 3 to 5. Fig. 5 presents data for different particle sizes for sand in terms of $Q_j - Q_{jmf}$ to take into account the difference in u_{mf} of the particles. A general feature of Figs. 3 to 5 is that the initial decrease in dead zone is rapid and it becomes increasingly difficult to reduce the dead zone as its height decreases. This tendency is more pronounced for smaller particles and larger pitches (Note: Figs. 3 to 5 are plotted on a semi-log basis for convenience; when plotted on rectangular coordinates the decreasing change in h_s with gas flow becomes quite striking).

Figs. 6 to 8 are photographs of the steady-state bed conditions at different gas flow rates. They show how the dead zones (dark coloured humps of tracer material) decrease in height as the gas flow increases. They also show that for the same extent of dead zone elimination a lower gas flow rate is needed when the pitch is smaller. (compare Figs. 6 and 7).

Fig. 3 shows that increasing the orifice diameter results in smaller dead zone heights for the same pitch and gas flow rate (expressed as volumetric flow rate per orifice). Visual and photographic observations indicate that a large orifice diameter produces a wider jet. These jets do not penetrate deep inside the bed and collapse much faster than those emerging from a smaller orifice diameter. The effect is more pronounced at lower gas flow rates.

Experimental data on the effect of particle size for sand are shown in Fig. 5. With the smaller particles, elimination of the dead zone becomes increasingly difficult particularly at larger values of pitch. Fig. 5 shows that for the 275 μm particles, elimination of dead zone becomes increasingly difficult for P=4.0 and 6.0 cm. Visual observations indicate that the small particles have a cohesive tendency and move in clusters rather than as individual particles. Furthermore, with the small pitch (P=2.0 cm) the dead zones are not distinct. In this case, the dead zone is a continuous layer of stagnant solid material about 1.5 to 2.0 cm in height above the grid.

Typical experimental data showing the effect of bed height on dead zone were presented earlier (Horio et al. 1977) and are not reported here. In summary, it has been found that for a pitch of 3.0 cms and in the range of bed heights investigated (7.5 to 21.0 cm), the effect of bed height is not significant. More data over a wider range of bed heights and for different pitches are needed to establish the bed height effect.

From a design standpoint the most important parameter could be the minimum gas flow rate needed to completely eliminate the dead zone. This parameter denoted here as $(Q_j - Q_{jmf})_c$ represents the critical flow rate through each orifice for no dead zone formation for a specific pitch distance. Values of $(Q_j - Q_{jmf})_c$ have been obtained both experimentally and by extrapolation of the dead zone heights to zero in the gas flow rate curves (Figs. 3 to 5). Similar values have been obtained in both cases. These values are plotted as a function of pitch for different orifice diameters (Fig. 9), for large particles of sand and glass beads (Fig. 10), and for two particle sizes of sand (Fig. 11).

In these figures two regions can be identified, viz. dead zone and no dead zone. Operations conducted at gas flow rates above the line for a given pitch produces no solid dead zone on the grid whereas operations below the line could produce stagnant zones on the grid.

Fig. 9 shows that $(Q_j-Q_{jmf})_c$ is smaller for the larger orifice diameter. The elimination of dead zones is easier with the larger orifice diameter since the larger orifice produces wider jets which frequently collide above the grid resulting in the elimination of the dead zones.

The critical gas flow rates for large sand and glass beads are presented in Fig. 10. The data suggest that there is no significant difference between these two materials.

The particle size effect can be seen in Fig. 11. Although the effect is not so pronounced at smaller pitches, it is clear that $(Q_j-Q_{jmf})_c$ is larger for smaller particles, particularly at the higher pitch distances.

The data shown in Figs. 10 and 11 can be represented by an empirical correlation

$$(Q_j-Q_{jmf})_c = 122\ (P-d_o)^{4.22}\ d_p^{-0.18}$$

This equation is compared with the experimental data in Fig. 12. The term $(P-d_o)$ is employed since when $P=d_o$, the dead zone disappears at $Q_{jc}=Q_{jmf}$. The above empirical correlation is based on the data for sand (d_p=273~1840 μm) and for glass beads (d_p = 1015 μm) and pitches ranging from 1.5 to 6.0 cm and orifice diameter of 0.32 and 0.64 cm. It is obvious that other variables such as the density of the particles and gas must also be taken into consideration in order to obtain a more generalized correlation. The accuracy of $(Q_j-Q_{jmf})_c$ calculated from this equation is roughly within ± 20 percent.

Experiments conducted using a cap distributor shows that the performance of the cap distributor is generally superior to the perforated plate. The caps of the distributor are vertical nozzles (0.64 cm diameter) with sealed tops and have two holes (d_o=0.24 cm) pierced on opposite sides of the nozzles. The holes are directed along the length of the distributor plate with the centers placed 0.12 cm above the bottom of the plate. The total cross section area of the two holes in each cap is equal to the cross section area of each of 0.32 cm holes used in the perforated plate. Fig. 13 shows that for the same pitch and bed height, elimination of dead zone occurs at lower gas flow rates. The cap distributor was found to promote a better mixing of particles above the grid region than a perforated plate for the same orifice area and gas velocity. Preliminary experiments conducted with sand in the 200 to 1000 μm range indicate that the particle size effect is less significant than that observed with the perforated plate. Fig. 14 shows that with the small sand particles (d_p=273 μm) and pitch of 4.0 cm, elimination of dead zone is possible with the cap distributor while it is not possible on the perforated plate.

Preliminary experiments have been conducted in the 3-D bed with 650 μm sand particles to compare its performance with the 2-D bed. The results are summarized in Table 2. It appears that with a smaller Pitch (P=1.5 cm) elimination of the dead zone in the 3-D bed requires less gas flow than in the 2-D bed even though the free area in the 3-D bed is somewhat smaller than in the 2-D bed. However, for larger pitches this may not be true.

CONCLUSIONS

1. The height of dead zone is mainly influenced by the pitch, distributor type, orifice diameter, gas velocity and particle size. The effect of bed height has not been clearly established.
2. An empirical equation is obtained which can be used to estimate gas flow rate required to eliminate dead zone above a given perforated plate for a two dimensional bed.
3. Interparticle cohesion due to surface charge becomes increasingly important as particle size decreases. As a result, elimination of dead zones becomes more difficult with small particles, particularly at large pitch.
4. The performance of the cap distributors is superior to the perforated plate distributor under the same operation conditions in regard to elimination of dead zones.
5. Preliminary experiments in a 27.6 cm I.D. cylindrical bed with a perforated plate distributor indicate that the behavior of the 2-D bed cannot be readily extrapolated to the 3-D cases.

ACKNOWLEDGMENT

This work was partly supported by NSF, Grant No. ENG 76-10160. The authors wish to acknowledge F. Saus, D. Kowalczyk and M. K. Choi for their experimental assistance and Dr. A. G. Fane for his suggestions.

REFERENCES

Horio, M., Wen, C. Y., Krishnan, R., Khosravi, R., & Rengarajan, P., (1977) Proc. Second Pacific Chem. Eng. Congress, 2, p. 1182.

Makshima, S., Shirai, T., (1969a) J. Chem. Eng. Japan, 2, p. 124.

Ibid, (1969b), Preprint of 34th Annual Meeting of the Soc. Chem. Eng. Japan, p. 176.

Shinohara, H. et al. (1975) Preprint of 9th Fall Meeting of the Soc. Chem. Eng. Japan, p. 315.

Wen, C. Y., Dutta, S., (1976) AIChE Symp. Series 73 (161), p. 1.

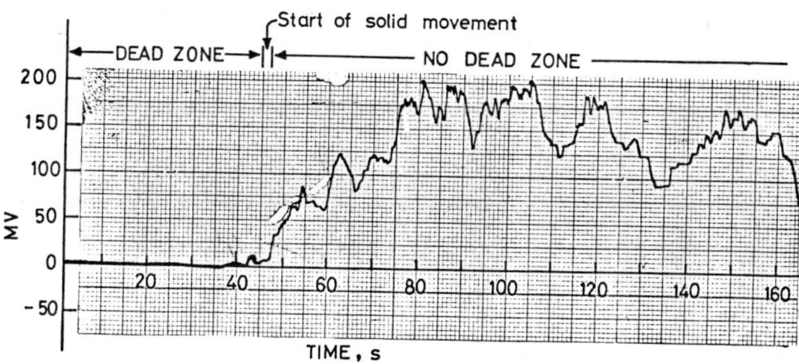

Fig. 1. Response of Thermistor probe

Fig. 4. Effect of pitch for glass beads and sand for large particles.

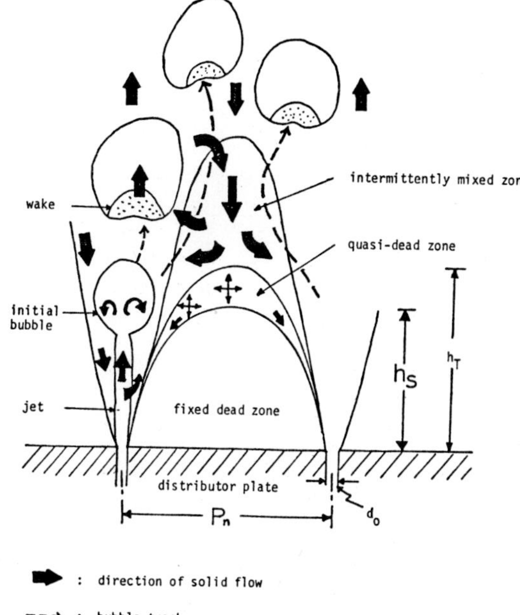

Fig. 2. Illustration of solid mixing zones above grid plate.

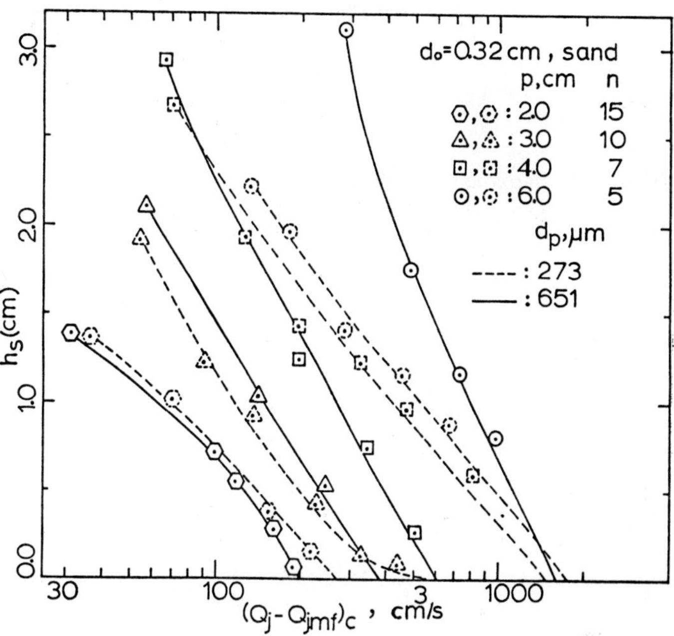

Fig. 5. Effect of pitch and particle size for sand.

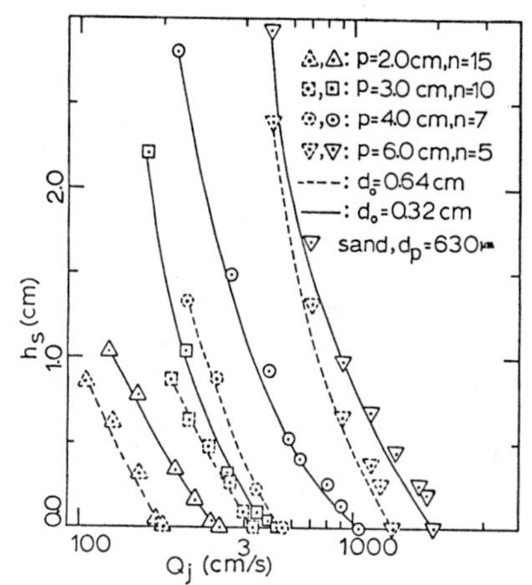

Fig. 3. Effect of pitch and orifice diameter for sand.

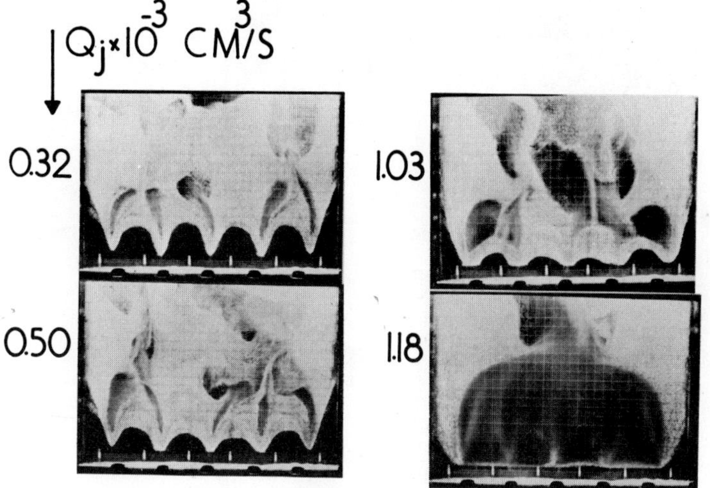

Fig. 6. Decrease of dead zone heights with gas flow rate for large pitch.

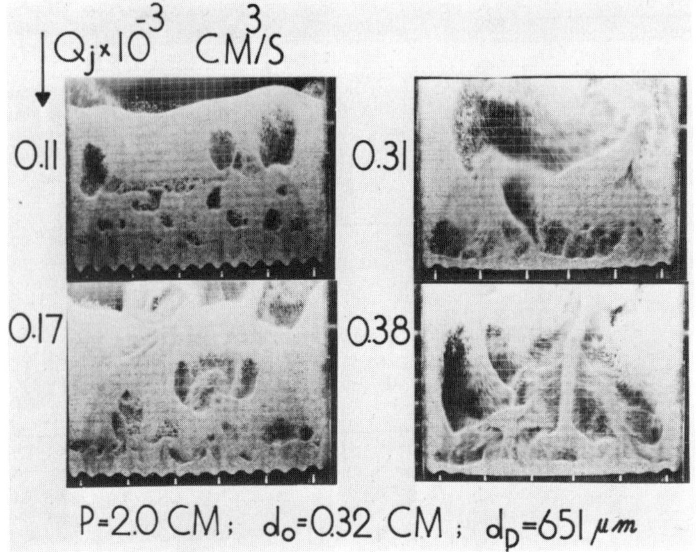

Fig. 7. Decrease of dead zone heights with gas flow rate for small pitch.

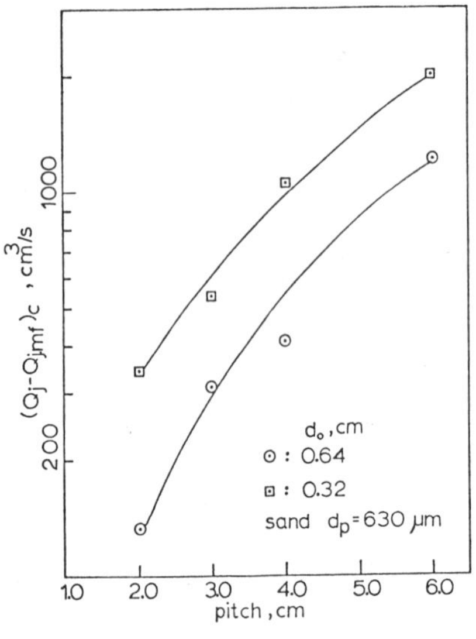

Fig. 9. Minimum gas flow rates for elimination of dead zone for two orifice diameters.

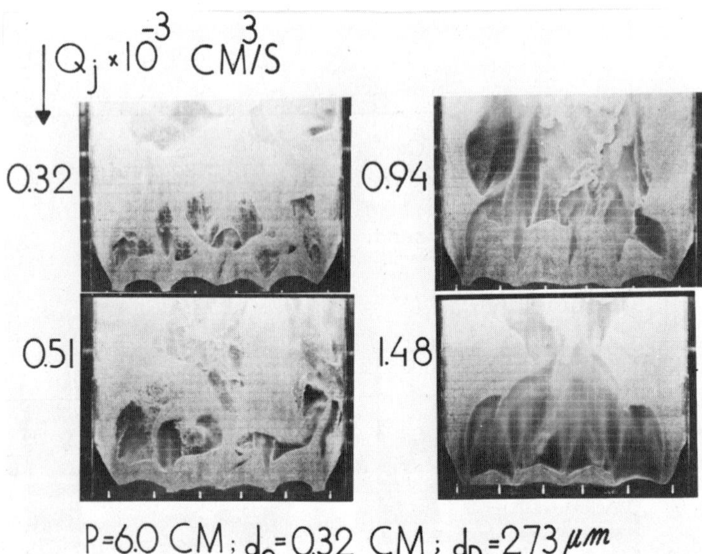

Fig. 8. Decrease of dead zone heights with gas flow rate for small particles.

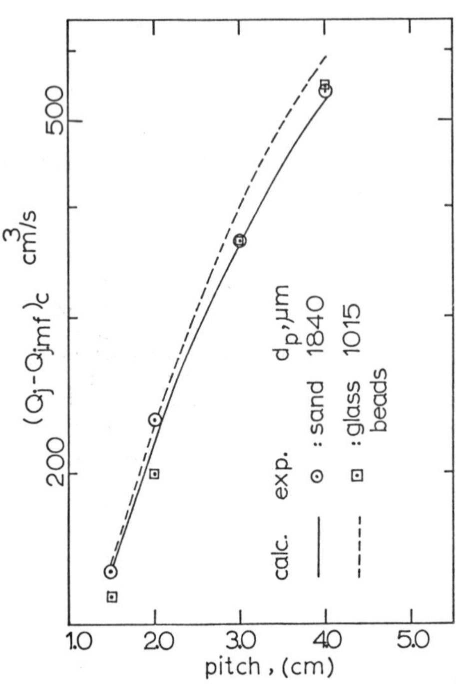

Fig. 10. Minimum gas flow rates for elimination of dead zone for glass beads and sand for large particles.

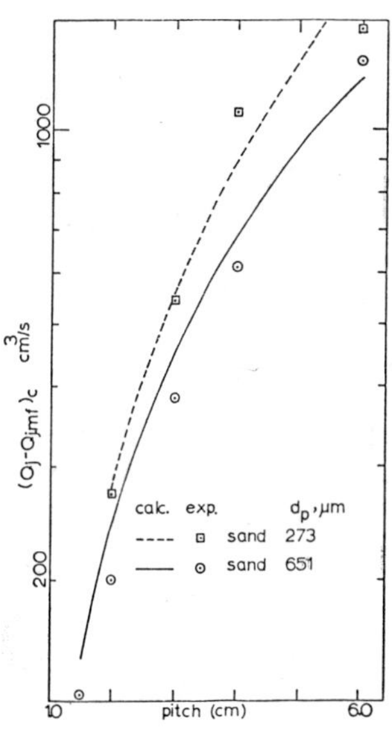

Fig. 11. Minimum gas flow rates for elimination of dead zone for sand particles.

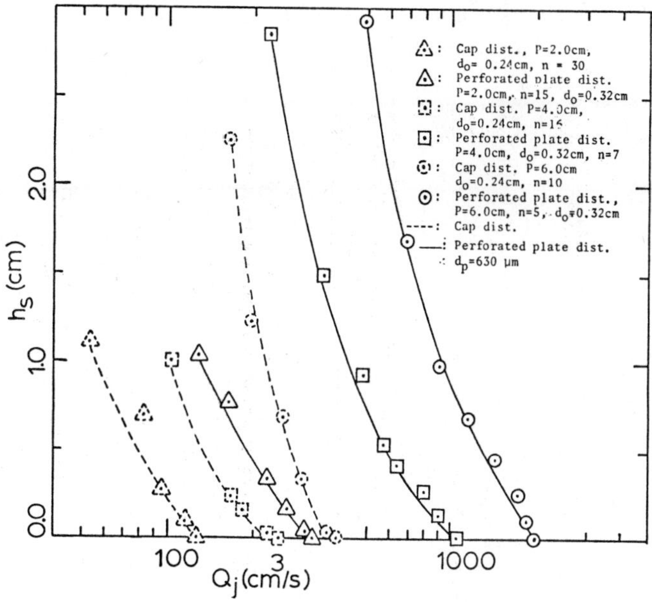

Fig. 13. Comparison of distributor performance for different pitches.

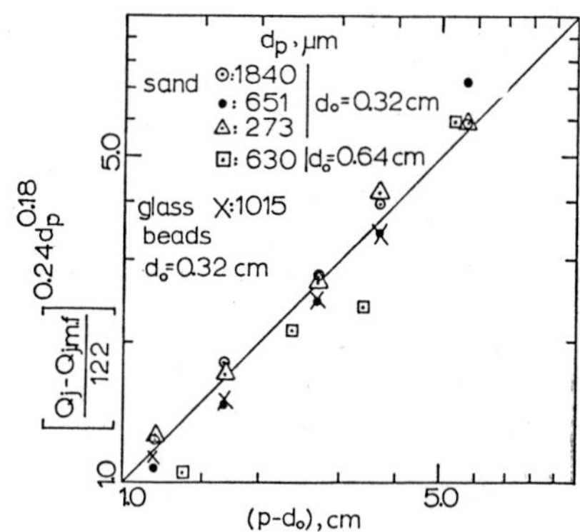

Fig. 12. An empirical correlation for elimination of dead zone.

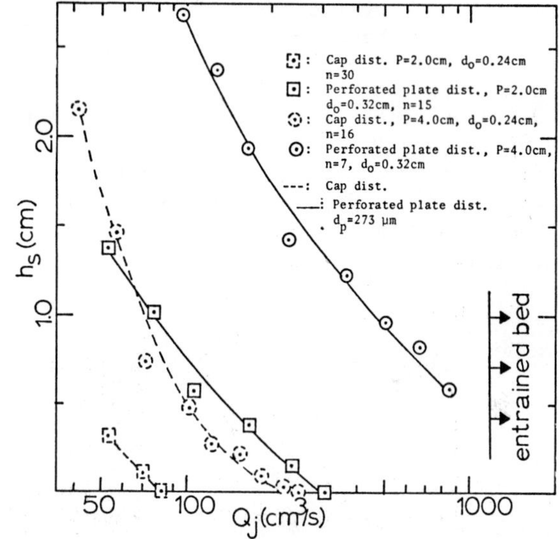

Fig. 14. Comparison of distributor performance for small particles.

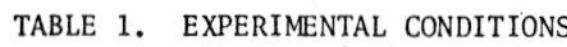

TABLE 1. EXPERIMENTAL CONDITIONS

Bed	2-D (Rectangular)	3-D(Cylindrical)
Bed cross section	30.5 cm x 1.2 cm	27.6 cm I.D.
Thickness of grid plate	0.5 cm	0.15 cm
Orifice diameter, cm	0.32, 0.64, 0.24*	0.32
Pitch, cm	1.5, 2.0, 3.0, 4.0, 5.0, 6.0	1.5, 3.0, 6.0
Particles	Sand Glass beads	Sand
Average particle	Sand 273, 630, 651, 1840 Glass beads 1015	651
Minimum fluidization velocity, cm/sec	Sand 6.7, 31.0, 31.6, 100.8 Glass bedas 55.2	31.6

* Orifice diameter of cap distributor

TABLE 2

COMPARISON OF 2-D and 3-D PERFORMANCE

BED	Q_{jc} (cm³/sec orifice		
	P=1.5 cm	P=3.0 cm	P=6.0 cm
2-D	193	~500	1480
	(4.2)	(2.1)	(1.0)
3-D	96.5**	500~650	*
	(3.0)	(1.5)	(0.8)

** Average value; values near the center, 93~111, midpoint between the center and the wall 108~130; at the wall 197~207.

* Chanelling occurs

( ) percent free area

LEAKAGE OF SOLIDS (WEEPING, DUMPING) AT THE GRID OF A 0.6 m DIAMETER GAS FLUIDIZED BED

By C. L. Briens, M. A. Bergougnou and C. G. J. Baker
Department of Chemical Engineering
The University of Western Ontario
London, Ontario
Canada N6A 5B9

SYNOPSIS

Substantial grid leakage in large industrial fluidized beds is highly undesirable because it can lead, among other things, to grid erosion or plugging. In spite of its importance, virtually no systematic work has been done on this subject.

Experiments have been carried out to study grid leakage using a cracking catalyst (arithmetic mean diameter 65 μm) in a 0.6 m diameter column. The flow pattern under the grid was charted with the help of a hot wire anemometer. Micro-isokinetic probes were developed to study the particle flux returning towards the grid. Particles were sucked in isokinetically, recovered in a filter, weighed and analyzed for particle size in a Coulter counter. Upward particle fluxes at the grid were determined by extrapolation of the flux at various distances under the grid. Grid leakage was studied for grids of various geometries which had holes of ½ inch diameter. Correlations are given.

NOTATION

d_h	hole diameter
$\bar{d}_p$	arithmetic mean particle diameter, μm
d_{pl}	diameter of a particle having a terminal velocity equal to the superficial velocity, μm
$\bar{d}_p^{up}$	$\bar{d}_p$ for the solids re-entrained through the grid holes, μm
$\bar{d}_p^{bot}$	$\bar{d}_p$ for the solids settling at the bottom of the wind-box, μm
$\bar{d}_p^{leak}$	$\bar{d}_p$ for the solids leaking through the grid, μm
e_i	voltage to be applied to the hot wire for the point i
F	total leakage flux (per unit of column cross-section), $kg/s/m^2$.
F^{up}	flux of solids re-entrained through the grid (per unit of column cross-section), $kg/s/m^2$.
F^{bot}	flux of solids settling at the bottom of the wind-box, $kg/s/m^2$.
F_h	leakage flux per unit of hole area, $kg/s/m^2$.
F_{ha}	actual hole leakage flux (per unit of active hole area), $kg/s/m^2$.
k	constant in equation (4)
k_o	constant
l	grid thickness
P	exponent in equation (4)
P_o	exponent
ΔP_{grid}	grid pressure drop, N/m^2
U_i	local gas velocity at point i
V_g	superficial gas velocity, m/s
V_h	hole gas velocity, m/s
V_{ha}	actual hole gas velocity (through the active holes), m/s
x_i	Fraction of solids with a particle diameter between d_{pi} and $d_{pi} + \Delta d_{pi}$
ϕ	grid relative free area
ϕ_a	actual grid relative free area (calculated from the area of the active holes).
ρ_1	gas density in the wind-box, kg/m^3
ρ	gas density at 23°C, 1 atm., kg/m^3.

Good grid design is essential for the satisfactory performance of large industrial fluidized beds. A vexing problem has been the leakage of solids (dumping, weeping) through the grid holes.

Substantial grid leakage can lead to erosion or plugging of the grid, and attrition of the catalyst. In case of an exothermic reaction, it can also expose the grid to temperatures for which it was not designed.

In spite of its importance, little systematic work has been done on this subject. Gregory (1967) stated the problem and gave a few experimental results. Serviant et al (1970a, 1970b, 1972) studied grid leakage in a 29 cm diameter fluidized bed of cracking catalyst. Kassim (1972) reported results obtained with several kinds of solids in an 8cm diameter column. Otero (1974) studied solids leakage through bubble cap distributors. In all these studies, the leakage was evaluated by collecting the solids falling at the bottom of the wind-box. This precludes the study of grid leakage at high superficial gas velocities where a large part of the dumped solids are re-entrained through the grid. Moreover, since the re-entrained solids can erode or plug the grid, it is essential to evaluate their flux.

This study was aimed at providing a reliable measurement method for the flux of solids re-entrained through the grid holes. Because it was carried out in a large diameter column, it is hoped that the results will be applicable to large scale industrial units.

EXPERIMENTAL

Experiments were carried out in a 10m high, 0.61m diameter column, previously described by Behie (1972). Static pressure taps located along the column allowed the determination of bed height, bed density and bed grid pressure drops. The wind-box consisted of a 0.61 m high cylindrical section on top of a 0.47 m high conical section (fig. 1). The fluidizing air was distributed in the wind-box by a conical baffle (chinese hat).

Three grids were successively used. They were 0.0127 m thick aluminum perforated plates with 0.0127 m diameter holes. Their fractional open areas were 0.355% (8 holes), 0.709% (16 holes) and 1.42% (32 holes). The holes were drilled on an approximately square pitch.

The flow of fluidizing air was measured by a rotameter. The oil free air had a temperature of 23°C and a relative humidity of about 15%. All gas velocities mentioned hereafter were computed at 23°C and 1 atmosphere.

The fluidized solids consisted of a 65 μm mean arithmetic diameter cracking catalyst whose apparent particle density was 1130 kg/m^3. The solids loading was maintained at 205 kg per square meter of column cross-section (±15%) throughout this study. The particle size analyses were carried out with a Coulter counter.

The flux of solids re-entrained upward was measured by isokinetic sampling at four horizontal levels under the grid. The probes were designed so that they could be moved to sample the solids flux at any position along a given horizontal diameter. The cylindrical section of the wind-box could be rotated relative to the rest of the column by 45° increments. Measurements could, thus, be carried out along 4 different diameters in a given horizontal plane

The particles entering the isokinetic probe were smoothly accelerated (to prevent choking) through a long conical entry section (mouth diameter 6.4 mm, included cone angle 20°). Such a small probe was chosen to minimize the disturbance to the flow patterns close to the grid. The particles passed through 1.6 mm diameter plastic tubing and were collected in a cellulose filter. The gas flowrate through the probe was regulated by a needle valve and measured with a rotameter.

The total re-entrained flux was computed by integrating the local flux measured at a number of points in the horizontal plane of the probe. The flux at the grid was, then, obtained by extrapolating the integrated fluxes of the four probes to the grid level (fig. 2).

After sampling, the cellulose filter was introduced into a borosilicate glass weighing bottle and left to burn in an oven at 540°C for about 8 hours. After cooling in a dessicator, the solids were weighed on a precision balance.

Solids were also collected in plastic bottles at the bottom of the wind-box and weighed. Samples were taken for moisture determination in order to make it possible to compute the dry weight of the solids collected.

Local gas velocities were measured by means of a hot wire anemometer (Pitot tubes cannot be used at low velocities). Since dust alters greatly hot wire characteristics, (Boothroyd 1967; Mills, 1972), the measurements were made with no solids in the column. The hot wire was used with a constant temperature circuit.

Calibration of the hot wire in a wind tunnel would not have been satisfactory because of the difference in turbulence scales between the tunnel and the wind-box (Mills, 1972). Instead, the calibration was carried out "in-situ" in the wind-box. Experimental data showed that the hot wire voltage was proportional to the velocity, (Mills, 1972). The total gas flowrate through the wind-box was obtained by integration of the local gas fluxes. It was then equated to the flowrate as measured by a rotameter. This gave the proportionality constant between the velocity and the voltage.

RESULTS

1) Choice of sampling space

Fig. 4 shows the various wind-box sectors chosen for the location of the sampling points. Point velocity measurements were made in all sectors (except sectors 2 and 6) for all superficial velocities (0.09, 0.15, 0.24 and 0.30 m/s) and at three probe levels (0.07, 0.17 and 0.27 m from the grid). It was found that the average velocity for combined sectors 4 and 8 was within 12% of the superficial velocity. Similarly, particle flux measurements were carried out at all of the above sampling points at a superficial velocity of 0.15 m/s. The average solids flux in the combined sectors 4 and 8 was within 12% of the average flux over the whole cross-section. Because of these findings and of the great amount of work associated with each experiment, particle flux measurements were usually made only in sectors 4 and 8.

2) Reproducibility and experimental error

Two experiments were carried out, a month apart, on the 16 hole grid(V_g = 0.24 m/s and V_h = 34.4 m/s). The values for the re-entrained particle flux, extrapolated to the grid, were within 10% of each other. Altogether, taking into account the error on the solids collected at the bottom of the wind-box, the error on the total leakage flux through the grid was estimated to be less than 15%. This is rather good considering that an error of 1% on the gas flowrate will give an error of 10% on the leakage flux (fig. 3).

3) Visual observations

At low grid hole velocities, solids dumped rythmically through the whole cross-sectional area of the active grid holes. Dumps were mostly short, random bursts of solids through the grid holes. Dumping switched from one hole to another in a haphazard manner although some holes dumped more than others.

At high grid hole velocities, solids fell only at the periphery of the grid holes (weeping). This phenomenon seemed also to be a random one. It could possibly be described as dumping through an annular area at the periphery of the grid holes.

Experimental observations show that there is no clear-cut transition from one regime to the other. There seens to be weeping throughout the whole range of velocities and dumping even occurred, though rarely, at hole velocities as high as V_h = 43 m/s.

No weeping point (i.e. a hole velocity above which there would be no leakage at all) was observed. This is in agreement with the findings of Serviant et al (1970 a, b; 1972) but in disagreement with Kassim (1972). A glance at Figure 3 shows, however, that this is rather an academic question; the leakage flux decreases so fast with the hole velocity that it soon becomes negligible.

4) Total leakage flux. Variation with hole velocity.

Fig. 3 shows a log-log plot of the solids leakage per unit of grid free area versus the hole velocity for the three grids studied and for grid holes of 0.0127 m diameter. The ratio of grid pressure drop to bed pressure drop varied between 5 and 150%. The ratios usually recommended for commercial operation are between 10 and 40%. It is interesting to see that solids leaked at ratios as high as 150%. The results are similar to those obtained by other researchers (Gregory 1967; Serviant et al, 1970 a, b, 1972; Kassim, 1972); as the hole velocity V_h decreased, the leakage flux increased, reached a maximum, and decreased to zero (for V_h = 0). It can be seen that in the weeping range (i.e. at high velocities) the hole leakage flux was the same whatever the grid free area. Furthermore, in that range, the total leakage flux can be expressed as a power-law function of the hole velocity or $F_h = k_o\ (V_h)^{P^\circ}$

5) Correction for unactive holes

At low velocities, only a portion of the grid holes were active, the other were either plugged by arching solids or were dumping solids. The solids leakage flux F_h per unit hole area had been calculated by assuming that all the N grid holes were active. If only n grid holes were actually active the actual leakage flux F_{ha} per unit of active hole area would be given by:

$$F_{ha} = \frac{N}{n} F_h = \frac{\phi}{\phi_a} F_h \qquad \ldots.(1)$$

Similarly V_{ha}, the actual hole velocity, could be the important parameter:

$$V_{ha} = \frac{V_g}{\phi_a} \qquad \ldots.(2)$$

Van Winkle et al (1958) gave a method to calculate the pressure drop through dry perforated plates. His formula can be used to compute the active relative free area ϕ_a from the experimental grid pressure drop:

$$\phi_a = \left[1 + \frac{2\rho_1\ \Delta P_{grid} C^2\ Y^2}{\rho^2\ V_g^{\ 2}} \right]^{-\frac{1}{2}} \qquad \ldots.(3)$$

where C is the orifice coefficient which depends upon the pitch, the grid thickness and the hole Reynolds number. For the three experiments with the highest hole and superficial velocities, where all holes were active, (i.e. $\phi_a = \phi$), ϕ_a computed from the previous equation was equal to the grid relative area ϕ with an error of less than 3%. This equation therefore applies to the condition of this study where all holes are active.

Figure 5 shows a Log-Log plot of the actual hole leakage flux density F_{ha} against the actual hole velocity V_{ha}.

It can be seen that the experimental points are on a straight line except for points at very low velocities. Therefore, approximately:

$$F_{ha} = k\ (V_{ha})^{p} \quad \ldots.(4)$$

A linear regression on the logarithms gave:

$k = 6.34 \times 10^{9}$ $p = -7.82$ (SI units)

At very low velocities, the actual hole leakage flux F_{ha} appears to be lower than the value given by the above equation. This discrepancy could come from the experimental error in the grid pressure drop but it is more likely that, as the gas velocity decreases, the defluidized zones around the plugged holes become larger and tend to interact more and more with the active holes (by restraining the circulation of solids, for example).

Thus, knowledge of the ratio of active holes $\frac{n}{N}$ either through a correlation from the literature or through experiments, enables the extension of the relationship $F_h = k_o (V_h)^{p^\circ}$ to lower velocities, presumably down to the range of the velocity corresponding to the maximum leakage flux. This could also allow an estimation of the maximum leakage flux and of the corresponding hole velocity.

6) Size distribution of leaked solids

The mean diameter of the solids re-entrained through the grid $\bar{d}_p^{up}$ was estimated by measuring the mean diameter of the solids collected at four horizontal levels and extrapolating it to the grid. The mean diameter of the solids which were collected at the bottom of the wind-box $\bar{d}_p^{bot}$ was also measured. The mean diameter of the solids that leaked was computed from the following material balance:

$$\bar{d}_p^{\,leak} = \frac{1}{F^{up} + F^{bot}} \left[F^{up}\bar{d}_p^{\,up} + F^{bot}\bar{d}_p^{\,bot} \right] \ldots(5)$$

The results of an experiment carried out on the 16 hole grid are shown in the table below:

V_g m/s	V_h m/s	F^{bot} kg/s/m^2	F^{up} kg/s/m^2	$\bar{d}_p^{\,up}$ µm
0.152	21.44	5.18×10^{-4}	1.4×10^{-4}	37

$\bar{d}_p^{\,bot}$ µm	$\bar{d}_p^{\,bed}$ µm	$\bar{d}_p^{\,leak}$ µm	
66	62	60	

It can be concluded from the above data that the solids which leak have the same size distribution as the bed solids. Apparently, the bed falls bodily through the grid holes during dumps.

7) Estimation of total leakage flux from the bottom flux

As seen above, the solids leaking through the grid holes have the same mean diameter as the bed solids. Therefore, by material balance one obtains:

$$x_i^{bed} \left[F^{up} + F^{bot} \right] = x_i^{up} F^{up} + x_i^{bot} F^{bot} \quad \ldots.(6)$$

where x_i is the fraction of the sample having a particle diameter between d_{pi} and $d_{pi} + \Delta d_{pi}$. One can assume that no particle with a terminal velocity higher than the superficial velocity is re-entrained through the grid holes. Then, if d_{p1} is the particle size having a terminal velocity equal to the superficial velocity, one can write:

$$(F^{up} + F^{bot}) \sum_{d_{p_i} > d_{p_1}} x_i^{bed} = F^{bot} \sum_{d_{p_i} > d_{p_1}} x_i^{bot} + 0 \quad \ldots.(7)$$

where the summation is taken on all particles sizes larger than d_{p1}. Thus, by measuring the flux at the bottom of the wind-box and the size distribution of these solids, it is possible to evaluate the flux of solids re-entrained through the grid. The above equation was used for the following experiment with the 16 hole grid:

V_g m/s	V_h m/s	F^{bot} kg/s/m^2	$\bar{d}_p^{\,bot}$ µm	$\bar{d}_p^{\,bed}$ µm
0.152	21.44	5.18×10^{-4}	66	62

	F^{up} kg/s/m^2	F^{total} kg/s/m^2	$\bar{d}_p^{\,up}$ µm
measured	1.4×10^{-4}	6.58×10^{-4}	37
calculated	1.04×10^{-4}	6.22×10^{-4}	41

The error in F^{up}, the re-entrained flux, was 26%, while the error in the total leakage flux was only 5%. If the bottom flux had been taken as representative of the total leakage flux, as previous authors had, the error would have been 21%.

The above computational method could be

very useful in pilot plant work where it is not always practical to use isokinetic sampling to find the re-entrained flux (A quick particle size analysis of the solids collected at the bottom of the wind-box can then give a good estimate of the various leakage fluxes). It also gives an indication that the isokinetic method used to measure the re-entrained flux is reliable.

8) Calculation of the re-entrained flux by means of a choking model

Visual observation and experimental data show that dumps fall down in the wind-box for a distance of about 30 cm and then disintegrate. The particles are re-entrained by the incoming gas flow. As they approach the grid the heaviest particles are stripped off from the flow. What is actually taking place is typical of a choking phenomenon. One can, thus, assume that the flow is at the choking point before entering the grid region.

Using the above assumption the choking load was computed at the superficial velocity of the column by the method of Leung (1976). It was, then, used to calculate the flux of solids re-entrained through the grid from the total leakage flux measured experimentally for four superficial velocities ranging from 0.09 m/s to 0.30 m/s. The errors between the computed re-entrained fluxes and the experimental values were: +15%, +16%, -21%, -4%, respectively. Further details on theory and results for this method will be given in a subsequent article.

CONCLUSION

The following conclusions can be drawn about grid leakage:

- The flux of re-entrained solids can be sampled isokinetically and reliably. The total grid leakage can, thus, be computed at any velocity.
- At high velocities, grid leakage follows a power-law function of the grid hole velocity,
- At a pilot plant level, grid leakage can be estimated from only the solids collected at the bottom of the wind-box.

ACKNOWLEDGEMENTS

The authors would like to thank the Canada Council for a Scholarship to Mr. C. Briens, the National Research Council for a grant in aid of research to Dr. M. A. Bergougnou and the Davison Chemical Company for generously giving the cracking catalyst.

REFERENCES

Behie, L. A. (1972), Ph.D. Thesis, University of Western Ontario, London, Canada

Boothroyd, R. G., (1967), J. of Scientific Instruments, 249-253.

Gregory, S. A. (1967), Proceedings of the International Symposium on Fluidization, Eindhoven, p. 751-758.

Kassim, W., (1972), Ph.D. Thesis, University of Aston in Birmingham, U.K.

Leung, L. S., (1976), IEC Process Des. Dev. 15, 552-557.

Mills, P. E., (1972), M.E.Sc. Thesis, University of Western Ontario, London, Ontario, Canada.

Otero, A. R., Munoz, R. C., (1974) Powder Technology 9, 279-286.

Serviant, G. A., (1970a) M.E.Sc. Thesis, University of Western Ontario, London, Ontario, Canada.

Serviant, G. A., Bergougnou, M. A., Baker, C. G. J., Bulani, W., (1970b), Can. J. Chem. Eng., 48, p. 496-501.

Serviant, G. A., Bergougnou, M. A., Baker, C. G. J. (1972), Can. J. Chem. Eng., 50, p. 690-694.

Van Winkle, M., Smith, P. L., (1958), A.I.Ch.E. J. 4, 266-268.

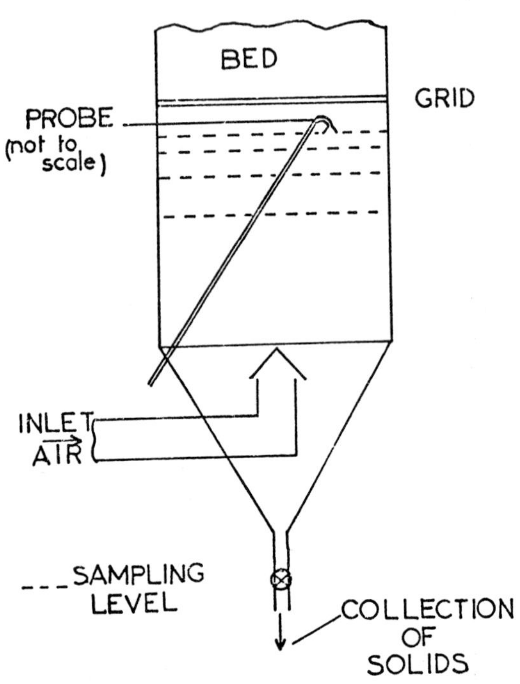

Figure 1: Wind-box

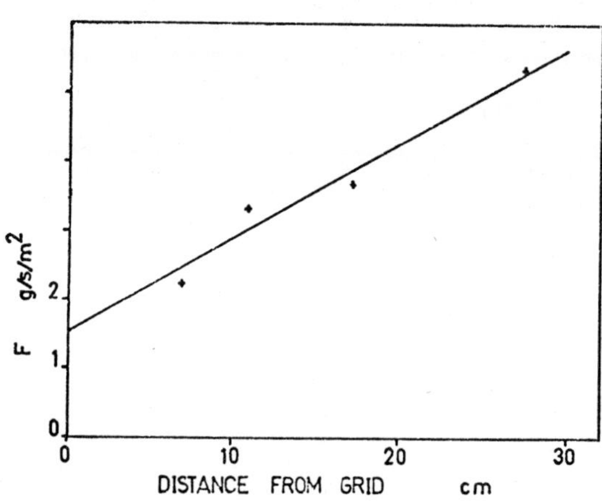

Figure 2: Re-entrained Solids Flux vs. distance from grid.
+ : V_g = 0.15 m/s (32 hole grid)
- : least squares line

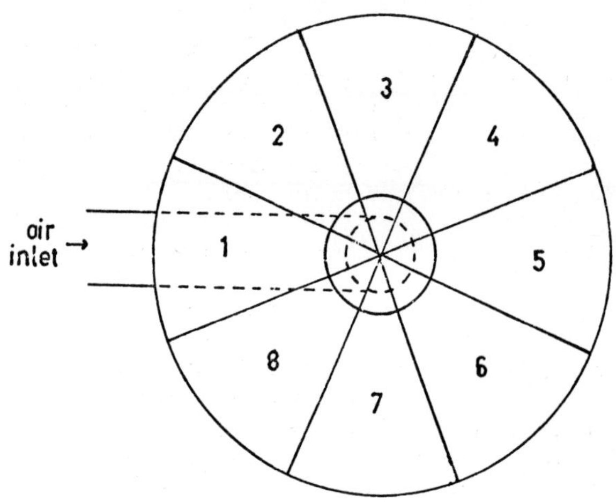

Figure 4: Wind-box; Sampling Sector (View from above)

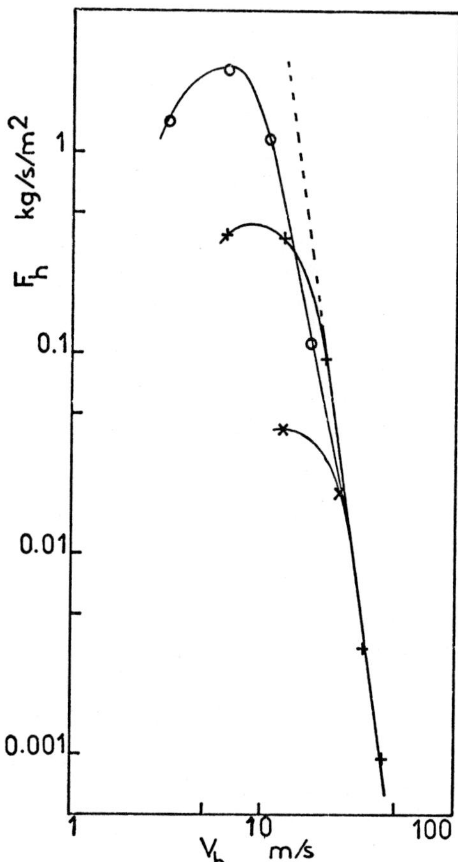

Figure 3: Solids Leakage Through Grid Holes vs. hole velocity
o 32 hole grid ϕ = 1.42%
+ 16 hole grid ϕ = 0.709%
x 8 hole grid ϕ = 0.355%

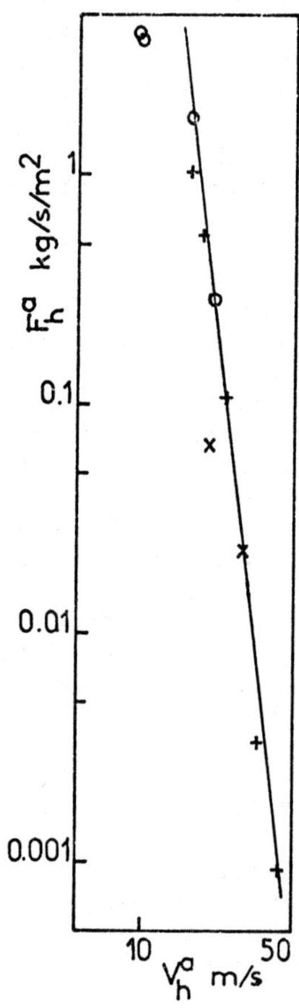

Figure 5: Actual Solids Leakage Through grid holes vs. actual hole velocity.
o 32 hole grid ϕ = 1.42%
+ 16 hole grid ϕ = 0.709%
x 8 hole grid ϕ = 0.355%

SOME EFFECTS OF DISTRIBUTOR SLOPE AND AUXILIARY GAS INJECTION ON THE PERFORMANCE OF GAS-SOLID FLUIDIZED BEDS

By A.B. Whitehead and D.C. Dent

Division of Chemical Engineering

C.S.I.R.O., P.O. Box 312, Clayton, Victoria 3168

Australia

SYNOPSIS

Some characteristics of fluidized beds contained in vessels fitted with horizontal, uniform flow resistance, gas distributors are described. Such beds develop characteristic pressure profiles at their base which are related to the solid and gas movement patterns in the bed. Using a 1.2 m square vessel containing a 1.5 m deep bed of silica sand (U_{mf} 2.5 cm. s-1), the changes in pressure profile resulting from sloping the gas distributor at inclinations of 1:50 and 1:25 have been studied at gas velocities of 4-12 U_{mf}. Changes in pressure profile resulting from the introduction of auxiliary gas at various points along the vertical axis of the same system with the distributor horizontal were also studied.

NOTATION

$\Delta\bar{P}_{B.n}$	time average bed pressure drop above tuyere n. cm W.G.
$\Delta\bar{P}_{c-c}$	mean time average pressure differential between four centre tuyeres and four corner tuyeres. cm W.G.
$\Delta\bar{P}_{o}$	time average pressure drop over a tuyere metering orifice. cm W.G.
$\Delta\bar{P}_{SYS}$	pressure between windbox and freeboard for the whole system. cm W.G.
$\Delta\bar{P}_{T}$	time average pressure drop over a complete tuyere. cm W.G.
H_a	height of auxiliary gas injection point above distributor. cm.
Q_A	auxiliary gas flow injected above the distributor. $\ell.s^{-1}$
U	superficial gas velocity. $cm.s^{-1}$
U_{mf}	incipient fluidizing velocity. $cm.s^{-1}$

INTRODUCTION

A "simple" fluidized bed system may be defined as one possessing the following features: (a) Particle size distribution is such that no segregation occurs when the distributor gas flow is sufficiently in excess of that required for incipient fluidization to allow solids movement. (b) All the fluidizing gas is supplied via a horizontal, uniform flow resistance, distributor. (c) The containing vessel walls are substantially vertical below the expanded bed surface.

Large scale "simple" systems exhibit the following characteristics under certain operating conditions: (d) Bubbles tend to burst at preferred locations at the bed surface, the pattern being related to gas flow rate, bed depth, bed material and distributor type (Whitehead & Young 1967a). (e) Preferred bubble tracks are present within the bed as reported by Werther & Molerus (1973), Calderbank, Toor & Lancaster (1967), and Whitehead & Young (1967b). (f) Solid flow tracks complimentary to the bubble tracks exist allowing the return of solids from the upper to lower regions of the bed. These have been described by Calderbank, Toor & Lancaster (1967), Werther & Molerus (1972b), Schmalfeld (1976) and Whitehead, Gartside & Dent (1970 & 1976). When the rate of solids downflow is sufficiently high, gas trapped on the interstices of the particles is carried downward to the lower regions of the bed. Measurement of this gas backmixing gives further details on solid movement in coherent streams (Nguyen, Whitehead & Potter, 1977 & 1978). (g) Time average pressure non-uniformities (subsequently referred to as pressure profiles) are present at distributor level; the character of the profile being related to the previously described gas and solids flow regimes (Whitehead, Gartside & Dent, 1970; Whitehead & Auff, 1976; Whitehead, Dent & McAdam, 1977).

Pressure profiles are easily determined and form a convenient measure of gas and solids flow regimes in a bed. Comprehensive data are available on profiles in "simple" systems and if modifications to such systems produce significant changes in behaviour, they are reflected in changes of the pressure profile. Thus immersion of an assembly of vertical tubes has been shown to have a major effect on the behaviour of a 150 cm deep bed silica sand (Whitehead, Dent & McAdam, 1977).

Distributors often deviate from the horizontal plane to facilitate complete discharge of the bed, or to accommodate thermal expansion or to give structural rigidity.

Auxiliary gas can be introduced into the bed as a reactant or a heat source, e.g. submerged combustion. Thus both these modifications are of practical significance.

This present paper gives data on the effect on pressure profile of two such modifications to "simple" systems: (a) inclining the distributor at slopes of 1:50 and 1:25; (b) introducing an auxiliary supply of gas at various points along the vertical axis of the bed.

The results presented here are discussed in relation to other work, where deliberate attempts were made to influence bed behaviour.

EXPERIMENTAL

The apparatus, shown in Figs. 1, 2 & 3 has been described in detail elsewhere (Whitehead, Dent & McAdam, 1977) and consisted of a square cross-section vessel, 1.5m^2 area, fitted with a 36 tuyere distributor and containing a 1.5 m deep bed of silica sand (designated S.1. in previous work. U_{mf} 2.5 $cm.s^{-1}$). Time average differential pressure across each tuyere metering orifice, ΔP_o, was measured using transducers coupled to a high speed data processing system. Each tuyere was calibrated for overall pressure drop, ΔP_T, versus metering orifice pressure drop, ΔP_o, at an absolute pressure equal to that prevailing at distributor level with a 1.5 m bed present. Overall pressure drop between the windbox and freeboard, ΔP_{SYS}, was also measured, thus it was possible to obtain the pressure drop between individual tuyere cap outlets and the bed surface, $\Delta P_{B.n}$. Data were collected for the following operating conditions using a 6 min. sample period:

(a) Time average bed drops, $\Delta P_{B.n}$, for tuyeres located on the two diagonals (Fig.2) were determined, first with the distributor horizontal, and then inclined at slopes of 1:50 and 1:25. (Tuyeres 6 & 36 were raised relative to tuyeres 1 & 31). The experiments were conducted with both high and low resistance gas distributors, the mean values of $\Delta P_{B.n}$ then being calculated for pairs of tuyeres situated at the following locations: 6 & 36, 11 & 29, 16 & 22, 15 & 21, 8 & 26, 1 & 31. The procedure was repeated at three different fluidizing velocities.

Inclining the distributor causes a variation in pressure gradient at distributor level in addition to that associated with circulation patterns. The means values of $\Delta P_{B.n}$ for the pairs of tuyeres identified above were corrected for this effect using previously determined information on bed expansion and pressure drop over beds of known weight. The corrected means of the two diagonals and the gas velocities used are shown in Fig.4. Variations in bed pressure drop were due to slight alterations in bed depth and also the inherent changes due to fluidizing velocity (Whitehead, Dent & McAdam, 1977).

(b) The mean time average differential pressure between the centre and corner tuyeres, $\Delta\bar{P}_{c-c}$, where $\Delta\bar{P}_{c-c}$=

$$\frac{(\Delta\bar{P}_{B.21}+\Delta\bar{P}_{B.22}+\Delta\bar{P}_{B.16}+\Delta\bar{P}_{B.15})}{4} - \frac{(\Delta\bar{P}_{B.31}+\Delta\bar{P}_{B.36}+\Delta\bar{P}_{B.6}+\Delta\bar{P}_{B.1})}{4}$$

was measured when an auxiliary gas supply, Q_A, was injected downward through a 5 cm ID tube located at various levels on the vertical axis of the bed as shown in Fig.3. The main fluidizing air was maintained at 255$\ell.s^{-1}$, equivalent to a superficial velocity of 17 $cm.s^{-1}$. Data for two values of Q_A, namely 98 and 225$\ell.s^{-1}$ are shown in Fig.5.

DISCUSSION

Effect of distributor inclination

Fig.6 shows the relationship between $\Delta\bar{P}_{c-c}$ and U previously obtained (Whitehead, Dent & McAdam, 1977) with the same system as that used in the present investigation. The data are restricted to conditions where all the tuyeres were in a continuous operating mode (Whitehead & Dent, 1967c) and the distributor was level; data obtained in the present investigation for similar conditions have been also included. In the present work the effects of two distributor inclinations on pressure profile were investigated at various fluidization velocities known from previous work to give differing modes of bed operation.

$U \sim 15$ cm.s^{-1} - The gas and solids movement pattern at this velocity is very stable for the system under consideration and has been documented in some detail (Whitehead, Gartside & Dent, 1976). The major solids downflow passes down a central path at about 20 cm.s^{-1}, a minor amount descending close to the vessel walls at about 4 cm.s^{-1}. $\Delta\bar{P}_{C-C}$ is about 3 cm. WG., the particular value recorded being dependent on fluidizing velocity (Fig.6).

Inclining the distributor at a slope of 1:50 did not significantly affect the value of $\Delta\bar{P}_{C-C}$ or the shape of the pressure profile (Fig.4, F&G); this applying with both high and low pressure drop distributors.

Further inclination to 1:25 (Fig.4, D&E) produced an asymetrical profile with the low resistance distributor but the high resistance system remained symetrical. The value of $\Delta\bar{P}_{C-C}$ was significantly reduced, especially with the low resistance distributor. With the high flow resistance distributor at inclination of 1:25 the gas flow through the highest tuyeres was approximately 10% greater than through the lowest. With the low flow resistance distributor, the highest tuyeres had 80% greater gas flow than the lowest. The shape of the pressure profile was not affected as much as might be expected by this gas maldistribution.

Conditions were such that all the tuyeres were in an "operating" mode (Whitehead & Dent, 1967c) for the data shown in Fig.4, D, F, G, H, & I. When the low resistance distributor was inclined at 1:25 (Fig.4E), tuyeres in the lowest row, i.e. 1,7,13,19,25 & 31, showed transient "non-operability".

If the relationship between pressure profile and solids downflow is similar in both level and inclined systems, it can be concluded that a distributor inclination of 1:50 has little effect on solids movement. However, an inclination of 1:25 does affect bed behaviour, especially with a low resistance distributor. Some solids downflow appears to be diverted from the centre to the side of the vessel having the lowest tuyeres (i.e. 1,7,13,19,25 & 31). However, the central stream still persists.

It seems probable that the inherent solids downflow pattern of a "simple" system could be restored by progressive variations in distributor flow resistance from the low to the high side of this distributor. This would ensure that the lower tuyeres remained in a continuous operating mode even with a low resistance distributor.

$U \sim 10$ cm.s^{-1} - The low resistance distributor could not be operated at this low velocity without many tuyeres being in the "non-operating" mode. Considering the higher resistance system; a level system operating at this velocity has a similar solids downflow pattern to that noted at $U \sim 15$ cm.s^{-1} but the central core moves somewhat slower and is restricted in area. Inclination of 1:50 and 1:25 both gave reduced values of $\Delta\bar{P}_{C-C}$ and the diagonal profile became slightly skewed (Figs.4, A&B) indicating that the normally central core of downflowing solids was displaced off-centre in the direction of the lower side of the vessel.

All the tuyeres remained in the "operating" mode.

$U \sim 32$ cm.s^{-1} - The solids circulation pattern in a simple system operating at this gas flow rate can be broadly considered as the inverse of that prevailing at 10 - 16 cm.s^{-1}. Massive solids downflow in coherent streams occurs near the walls of the vessel, rather than the centre. The form of the solids downflow is more complex than the central downflow pattern prevailing at lower velocities. This is shown by a study of gas backmixing patterns given elsewhere at this Conference (Nguyen, Whitehead, and Potter, 1978).

Both distributor inclinations gave an asymetric profile but the value of $\Delta\bar{P}_{C-C}$ was not reduced. It is inferred that solids downflow above the lowermost row of tuyeres was increased whilst that at the opposite side of the vessel was reduced.

The changes in pressure profile shape that occurred did not alter, the operability of either of the distributors.

The work reported here is confined to a bed with a fixed ratio of depth to width operated at low distributor inclinations. In view of similarity criteria established elsewhere (Whitehead & Auff 1976) it is likely that the trends in behaviour reported here would also be present in systems with similar relationships between $\Delta\bar{P}_{C-C}$ and U, even though they had slightly different aspect ratios.

Systems with very shallow beds, or those using distributors with much greater inclinations, need to be the subject of separate investigation.

Effect of auxiliary gas injection

It was anticipated that the disturbance of the bed surface caused by the injection of large amounts of gas at the 120 cm level might have a major effect on bed stability for the reasons outlined by Merry & Davidson (1973). However, this did not occur, and the slight reduction in pressure profile indicates that the mode of solids circulation below the gas injection point was not significantly changed from that prevailing in a "simple" system operated at the same distributor gas flow rate.

As the injection point approached the distributor the effect increased, but did not become large enough to affect distributor stability for the range of conditions investigated, i.e. tuyeres were not made "inoperative" and no localised high or low pressure zones were detected.

It is noted that these results were obtained with a low pressure drop distributor. It is known from other experiments not reported here that the introduction of large amounts of gas into a bed immediately adjacent to a low pressure drop distributor will affect stability, and back flow of solids into the windbox may be induced. The need to avoid the possibility of solids entering the windbox restricted the minimum value of H_a that was investigated, particularly at the higher gas flow rate.

The system investigated here is remarkably tolerant to the injection of large amounts of auxiliary gas into the upper 2/3 of the bed at the particular distributor gas flow rate studied.

Using a similar system to that described here, Whitehead, Gartside & Dent (1970) attempted to reduce the pressure profile generated by the central core of downflowing solids by injecting extra gas into the four central tuyeres. It was found that the profile was only significantly affected when flow to the centre tuyeres was greater than 300% of that to the outer tuyeres.

In order to destroy the natural circulation pattern existing in the type of system investigated here it appears that a large imbalance of gas flow must be used. This applies independently of whether extra gas is introduced at the distributor level; or within the bed itself. This is in contrast to the enhancement of circulation that can be easily induced by adding extra gas to the peripheral areas of a distributor (Merry & Davidson, 1973).

CONCLUSIONS

Pressure profiles of the type previously recorded in "simple" fluidized beds fitted with horizontal distributors have been found to persist when the distributor was inclined, suitable corrections having been made for changes in "hydrostatic" head. The profiles may be modified in form and magnitude depending on conditions. However, system operability was only significantly affected when high inclination (1:25) was used with a low pressure drop distributor.

Injection of large amounts of auxiliary gas into the upper regions of a fluidized bed has a smaller effect than might be expected. Distributor stability was not affected by gas injection in amounts equal to 85% of that flowing through the main distributor, providing the injection point was located on the central axis of the bed at least 60 cm from the distributor.

REFERENCES

Calderbank, P.H., Toor, F.D. & Lancaster, F.H. (1967). Proc. International Fluidization Conf. Eindhoven, pp.658-660.

Merry, J.M.D. & Davidson, J.F. (1973). Trans. Instn. Chem. Engrs. 41, p.361.

Nguyen, N.V., Whitehead, A.B. & Potter, O.E. (1977). Trans. Amer. Inst. Chem. Engrs. In press.

Nguyen, N.V., Whitehead, A.B. & Potter, O.E. (1978). This Conference.

Schmalfeld, V.J. (1976). V.D.I.-Z 118, No.2 Jan. pp.65-77.

Werther, J. & Molerus, O. (1973). Int. Jour. Multi Phase Flow 1, pp.103-138.

Whitehead, A.B. & Young, A.D. (1967a). Proc. International Fluidization Conf. Eindhoven, pp.284-292.

Whitehead, A.B. & Young, A.D. (1967b). ibid. pp.294-302.

Whitehead, A.B. & Dent, D.C. (1967c). ibid. pp.802-819.

Whitehead, A.B., Dent, D.C. & Young, A.D. (1967). Powder Technology 1, pp.143-148.

Whitehead, A.B., Gartside, G. & Dent, D.C. (1970). Chem. Engr. Journ. 1, pp.175-185.

Whitehead, A.B., Gartside, G. & Dent, D.C. (1976). Powder Technology 14, pp.61-70.

Whitehead, A.B. & Auff, A.A. (1976). Powder Technology 15, pp.77-87.

Whitehead, A.B., Dent, D.C. & McAdam, J.C.H. (1977). Powder Technology, In press.

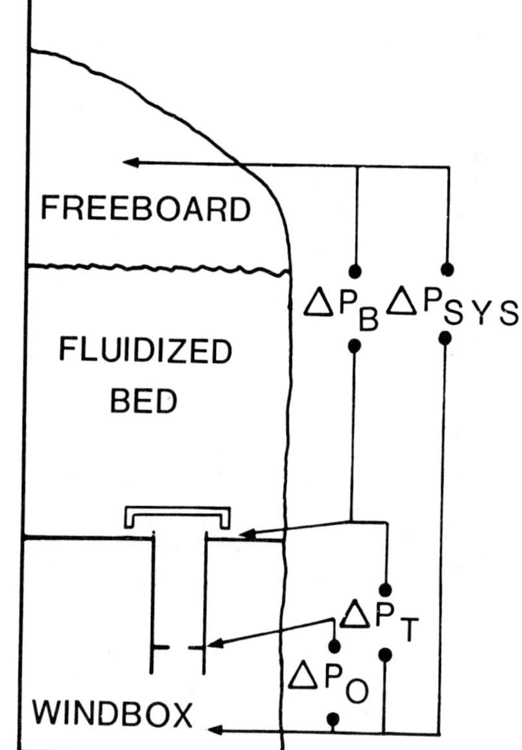

Fig.1 Apparatus showing location of differential pressures measured and derived.

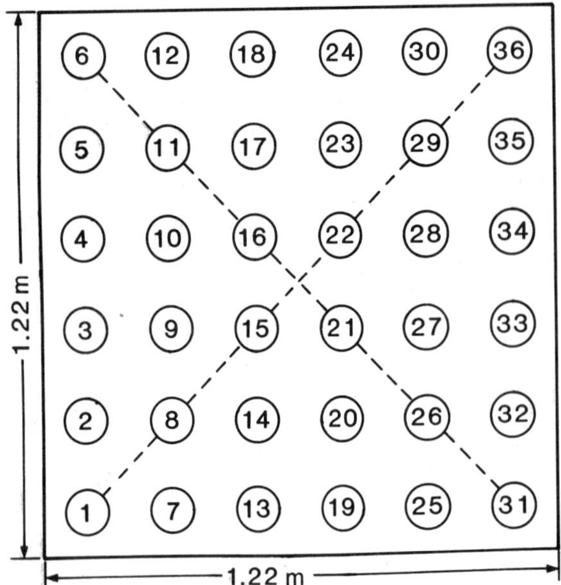

Fig.2 Tuyere layout. When the distributor was inclined at a slope of 1:25 tuyeres in row 6-36 were in the higher zone and 1-7 in the lower zone.

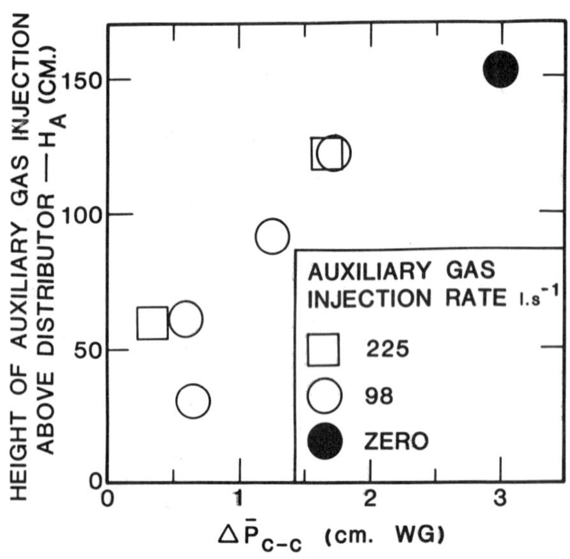

Fig.5 Effect of auxiliary gas injection on the mean time average pressure differential between four centre tuyeres and four corner tuyeres, $\Delta\bar{P}_{C-C}$.

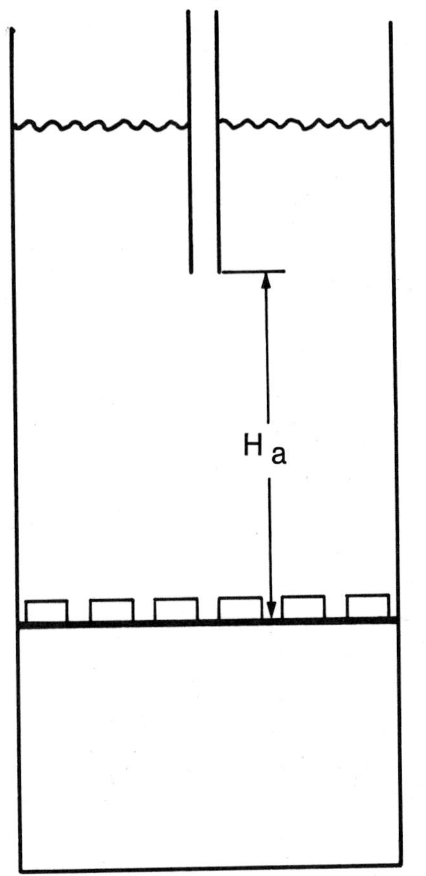

Fig.3 Location of auxiliary gas injector.

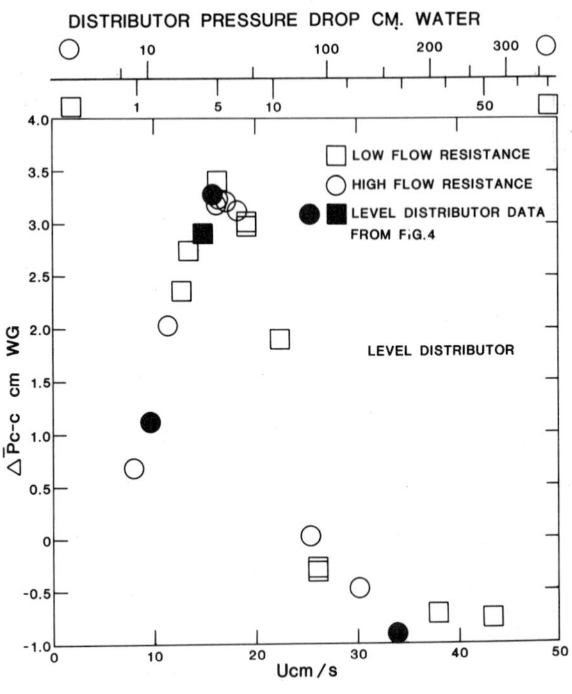

Fig.6 Relationship of fluidizing velocity and mean time averaged pressure differential between four centre tuyeres and four corner tuyeres for low and high resistance level distributors, as used in this investigation.

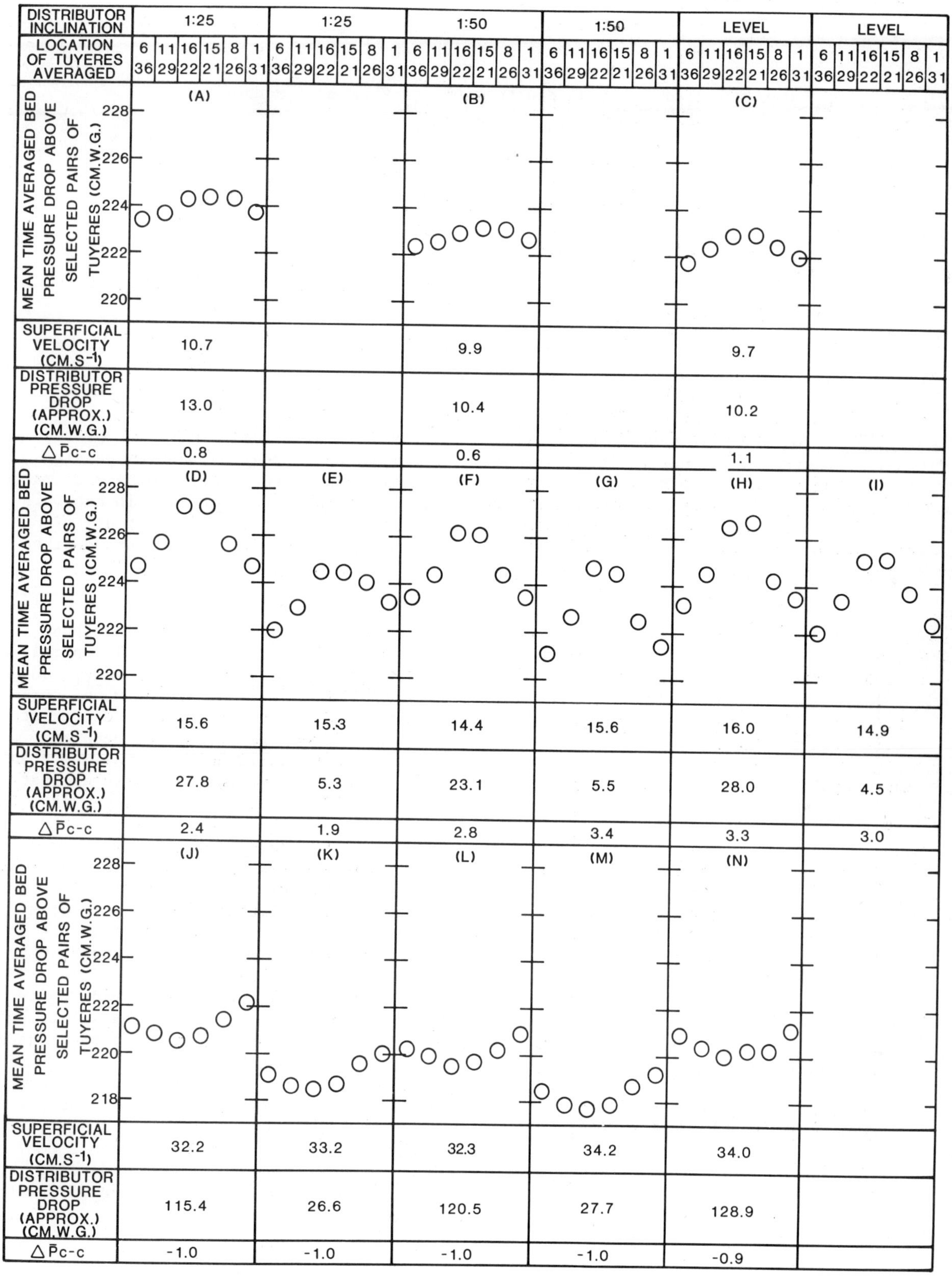

DISTRIBUTOR INCLINATION	1:25	1:25	1:50	1:50	LEVEL	LEVEL
LOCATION OF TUYERES AVERAGED	6/36, 11/29, 16/22, 15/21, 8/26, 1/31	6/36, 11/29, 16/22, 15/21, 8/26, 1/31	6/36, 11/29, 16/22, 15/21, 8/26, 1/31	6/36, 11/29, 16/22, 15/21, 8/26, 1/31	6/36, 11/29, 16/22, 15/21, 8/26, 1/31	6/36, 11/29, 16/22, 15/21, 8/26, 1/31
MEAN TIME AVERAGED BED PRESSURE DROP ABOVE SELECTED PAIRS OF TUYERES (CM.W.G.) (220–228)	(A)		(B)		(C)	
SUPERFICIAL VELOCITY ($CM.S^{-1}$)	10.7		9.9		9.7	
DISTRIBUTOR PRESSURE DROP (APPROX.) (CM.W.G.)	13.0		10.4		10.2	
$\Delta \bar{P}c$-c	0.8		0.6		1.1	
MEAN TIME AVERAGED BED PRESSURE DROP ABOVE SELECTED PAIRS OF TUYERES (CM.W.G.) (220–228)	(D)	(E)	(F)	(G)	(H)	(I)
SUPERFICIAL VELOCITY ($CM.S^{-1}$)	15.6	15.3	14.4	15.6	16.0	14.9
DISTRIBUTOR PRESSURE DROP (APPROX.) (CM.W.G.)	27.8	5.3	23.1	5.5	28.0	4.5
$\Delta \bar{P}c$-c	2.4	1.9	2.8	3.4	3.3	3.0
MEAN TIME AVERAGED BED PRESSURE DROP ABOVE SELECTED PAIRS OF TUYERES (CM.W.G.) (218–228)	(J)	(K)	(L)	(M)	(N)	
SUPERFICIAL VELOCITY ($CM.S^{-1}$)	32.2	33.2	32.3	34.2	34.0	
DISTRIBUTOR PRESSURE DROP (APPROX.) (CM.W.G.)	115.4	26.6	120.5	27.7	128.9	
$\Delta \bar{P}c$-c	-1.0	-1.0	-1.0	-1.0	-0.9	

Fig.4 Effect of distributor slope and pressure drop on pressure profile along diagonal tuyeres shown in Fig.2. Bed depth approximately 1.5m.

BUBBLING BEHAVIOUR OF FLUIDIZED BEDS AT ELEVATED PRESSURES

By M.P. Subzwari, R. Clift*, and D.L. Pyle

Department of Chemical Engineering and Chemical Technology
Imperial College
London, S.W.7.

* Present Address: Department of Chemical Engineering, University of Cambridge

SYNOPSIS

The behaviour of cracking catalyst with mean size 60 μm fluidized by air at pressure from 1 to 7 bars was observed by cinephotography in a two-dimensional column. A series of runs was performed with constant gas superficial velocity. The size and number of bubbles passing through the bed decreased markedly as the pressure was increased, the change being most marked around 4 bars. The decrease in bubble flow rate was accompanied by an increase in the void fraction of the particulate phase. The dominant mechanism of bubble splitting at all pressures was by division from the roof, and the maximum stable bubble size decreased with increasing pressure.

NOTATION

A cross-sectional area of column

G_B volumetric flow rate of bubbles across horizontal section

H mean depth of bed

H_{mf} bed depth at minimum fluidization

n coefficient in modified two-phase theory

U superficial velocity of gas in bed

U_{mf} superficial gas velocity at minimum fluidization

ε void fraction in particulate phase

ε_B volume fraction of bed occupied by bubbles

ε_{mf} void fraction at minimum fluidization

INTRODUCTION

It has frequently been claimed that the "quality" of gas-solid fluidization improves at elevated pressures (e.g. Yasui and Johanson (1958), Hoffert et al (1959), Lee et al (1970), Botterill and Desai (1972), Katosonov et al (1974). "Improved quality" has not been precisely defined, nor is it always clear whether the comparison has referred to fluidization at equal superficial gas velocity or equal mass flow rate. Guedes de Carvalho and Harrison (1975) have attributed changes in fluidization characteristics at elevated pressure to smaller bubbles, associated with a reduction in maximum stable bubble size. Their treatment assumes that bubbles divide by entry of particles from the lower surface of the bubble, as proposed earlier by Harrison et al (1961). However, more recent studies have shown that, at least around ambient pressure, the dominant mechanism of bubble splitting is by entry of a dividing "curtain" of particles from the roof, as a result of a kind of Taylor instability (Clift and Grace, 1972; Upson and Pyle, 1974; Clift et al, 1974). No direct evidence is available to indicate whether the splitting mechanism changes at higher pressures (Guedes de Carvalho, 1976).

The present experimental study was undertaken to examine the behaviour of fine particles fluidised at constant air superficial velocity over a pressure range from 1 to 7 bars. A two-dimensional column was used, so that the bubble flow rate and mode of splitting could be determined.

EXPERIMENTAL DETAILS

Subzwari (1976) describes the experimental equipment in detail. A bed of silica-alumina cracking catalyst was used, with solids density 950 kg/m^3 and mean particle size 59 μm, the maximum size present being 90 μm. The superficial air velocity at minimum fluidization (U_{mf}) was determined in the normal way at atmospheric pressure as 0.5 cm/s. Further experiments (Gatt, 1977) were also carried out using glass ballotini (solid density 2860 kg/m^3) with mean size 100 μm and 210 μm and minimum fluidising velocity 1.5 and 5.5 cm/s respectively. The effect of pressure on U_{mf} is negligible for particles as small as those used here (Creasy, 1971; Botterill and Desai, 1972). A two-dimensional column was used, with internal dimensions 46 cm wide x 67 cm high x 1.5 cm, with a porous plastic distributor. The faces of the bed were made of 1.3 cm thick toughened glass, reinforced by a steel framework to minimise distortion and reduce the unsupported area. A bed charge of 1 kg was used throughout. The bed depth at minimum fluidization was 31 cm, corresponding to a void fraction $\varepsilon_{mf} = 0.51$. The bed was fluidized at pressures from 1 to 7 bars, the superficial air velocity at bed pressure (U) being 3.2 cm/s for most of the experiments with cracking catalyst. The bed behaviour was recorded by cinephotography using a Bolex single-lens reflex camera at a constant measured speed of 68 frames/s. Back-illumination was used to show bubbles, with a little front-illumination to render a length scale visible. The films were analysed frame by frame, to determine:

1. Visible bubble flow rate (G_B) at 0.16 m above the distributor, by measuring the volumes of bubbles crossing the reference level (see Grace and Clift, 1974).

2. Mean fraction of bed volume occupied by bubbles (ε_B) in the region from 0.15 to 0.17 m above the distributor.

3. Bubble frequencies and frontal diameters at several levels in the bed.

Some salient results are summarised in Table 1. The bed depth was also recorded at several values of U at each pressure.

The bubble frequency recorded at 0.13 m above the distributor was invariably lower than that at 0.17 m. This anomaly appeared to result from the presence of bubbles too small to be recorded on the film, although these bubbles would contribute little to the bubble flow rate. Around 0.16 m, the observable bubble frequency was a maximum, and the mean bubble size was almost independent of pressure. Therefore this level was selected for determination of G_B and ε_B, since any underestimation of these parameters due to the presence of bubbles which were not larger than the bed thickness would be the same for all pressures.

DISCUSSION OF RESULTS

The results summarised in Table 1 and given in more detail by Subzwari (1976) indicate that raising the gas pressure to 6 bars has relatively little effect on bed expansion for these particles, but further increase in pressure causes significant expansion. This behaviour is similar to that observed by Guedes de Carvalho and Harrison (1975), who report increased expansion above 6 bars for particles of density 890 kg/m^3 and mean size 48 μm. However, the other measurements show that this expansion cannot be explained by their proposed mechanism of higher ε_B due to smaller, slower rising bubbles. In fact, ε_B decreases significantly above 3 bars (see Table 1 and Figure 1). Therefore the increased expansion observed at higher pressures must result from an increase in the void fraction of the particulate phase. The particulate phase void fraction is given by:

$$\varepsilon = 1 - H_{mf}(1-\varepsilon_{mf})/H(1-\varepsilon_B) \quad \ldots.(1)$$

Table 1 gives resulting values estimated at U = 3.2 cm/s, taking the value of ε_B at the 0.16 m reference level as an estimate of the mean value in the bed. At 1-3 bars the resulting values (0.47-0.48) are lower than the value at minimum fluidisation (0.51); i.e. the particulate phase contracts when bubbles are present, as commonly observed for beds of fine particles (Geldart, 1973). At higher pressures, the void fraction increases modestly.

This trend is clearly not consistent with the two-phase theory of fluidisation (Toomey and Johnstone, 1952; Davidson and Harrison, 1963). Similar conclusions follow from the measured bubble flow rates. Figure 1 shows the measured G_B values expressed as fractions of the prediction of the two-phase theory, $A(U-U_{mf})$. Up to 3 bars the bubble flow rate is slightly less than that predicted by the two-phase theory, as commonly observed (Grace and Clift, 1974). At higher pressures, the deficit is substantial. Following Lockett et al (1967), several workers have ascribed the deficit in G_B to through-flow inside bubbles, leading to the "modified two-phase theory" (Grace and Harrison, 1969):

$$G_B/A = U - U_{mf}(1+n\varepsilon_B) \quad \ldots.(2)$$

From the measured values of G_B, U_{mf} and ε_B, the mean value of n in the range 1-3 bars is 5.5, typical of the values observed by other workers (see Grace and Clift, 1974). However, Eqn (2) does not readily account for the increased deficit at higher pressures. If U_{mf} is assumed constant, then the apparent value of n rises to over 40 at 7 bars. Alternatively, n can be taken as constant at 5.5, and U_{mf} interpreted as the mean superficial velocity in the expanded particulate phase. Values of U_{mf} calculated on this basis rise with pressure to 1.2 cm/s at 7 bars. The increase is very much larger than predicted from the estimated expansion of the particulate phase via the correlation of Richardson and Zaki (1954). Values of $G_B/A(U-U_{mf})$ and ε_B from additional experiments with 100 μm glass ballotini (Gatt, 1977) are also shown in Figure 1, and show the same trend as the 59 μm catalyst. However, 200 μm ballotini showed very little variation of these groups with pressure.

Geldart (1973) has proposed a classification of powders which explains qualitatively the differences in fluidization behaviour resulting from changes in particle size and density. The cracking catalyst is in Geldart's 'Group A' and changes in pressure do not bring it near the boundaries of this group. The 100 μm ballotini is very close to the boundary between Groups A and B; increasing the pressure moves the classification in the direction of Group A. The 200 μm ballotini are well within Group B.

MECHANISM OF BUBBLE SPLITTING

The dominant mechanism of bubble splitting, for all particles and irrespective of pressure, was by division from the roof. Coalescence was often immediately followed by splitting, as typically observed at ambient pressure. A few bubbles were seen to split from the wake but this phenomenon was very rare (12 such events out of approximately 5×10^4 bubbles observed) and occurred at all pressures. Thus the changes in fluidization behaviour at elevated pressure are not associated with a change in the mechanism of bubble splitting.

The maximum observed bubble frontal diameter decreases significantly above about 3 bars for cracking catalyst (see Table 1). At 1 bar the mean bubble diameter increased steadily up the bed, implying that the maximum stable size was not reached. At higher pressures, the mean bubble diameter was almost constant above about 0.2 m from the distributor. There is therefore clear evidence of lower stable bubble sizes at elevated pressure for these particles. Bubbles in the 100 μm ballotini were larger and showed less decrease in size with increasing pressure. Unfortunately, the data are too scant to indicate the magnitude of dense phase expansion, although there was less expansion than with the cracking catalyst. Analyses of stability of the upper surface of a bubble in a fluidized bed show that the growth rate of disturbances increases if the effective kinematic viscosity of the particulate phase is reduced, whereas there is very little effect of gas density and throughflow (Upson and Pyle, 1974; Clift _et al_., 1974). Expansion of the particulate phase will reduce its effective kinematic viscosity (Hetzler and Williams, 1969). Thus the Taylor instability mechanism explains qualitatively why the bubble size is reduced as the particulate phase expands at higher pressures. Quantitative predictions cannot be made in the absence of reliable estimates for the particulate phase viscosity.

CONCLUSIONS

The "quality" of fluidization of a bed of fine particles changes at elevated pressures, the bubble size, frequency, and flow rate all decreasing while the particulate phase expands. These changes cannot readily be explained by any available theory of aggregative fluidization. The mechanism of bubble splitting is unaffected by pressure, the dominant process being division from the roof. The reduction in stable bubble size at higher pressures is qualitatively consistent with the Taylor instability analysis of bubble splitting, which predicts reduced stability if the particulate phase expands.

REFERENCES

Botterill, J.S.M. & Desai, M. (1972) _Pow. Tech._ _6_, 231.

Clift, R. & Grace, J.R. (1972) _Chem. Eng. Sci._ _27_, 2309.

Clift, R., Grace, J.R. & Weber, M.E. (1974) _Ind. Eng. Chem. Fund._ _13_, 45.

Davidson, J.F. & Harrison, D. (1963) 'Fluidised Particles', Cambridge University Press.

Gatt J., (1977) B.Sc. Research Project report, Dept. of Chem. Eng., Imperial College, March 1977.

Geldart, D. (1973) _Pow. Tech._ _7_, 285.

Grace, J.R. & Clift, R. (1974) _Chem. Eng. Sci._ _29_, 327.

Grace, J.R. & Harrison, D. (1969) _Chem. Eng. Sci._ _24_, 297.

Guedes de Carvalho. J.R.F. (1976) Ph.D. Thesis, University of Cambridge.

Guedes de Carvalho, J.R.F. & Harrison D. (1975) _Inst. of Fuel Symp. Ser._ No. 1, B1.

Harrison, D., Davidson, J.F. & de Kock, J.W. (1961) _Trans. Instn. Chem. Engrs._ _39_, 202.

Hetzler, R. & Williams, M.C. (1969) _Ind. Eng. Chem. Fund._ _8_, 668.

Hoffert, F.D., Kelly, E.A. & Squires, A.M. (1959) _Mech. Eng._ _81_, 27.

Katosonov, I.V., Men'shov, V.N., Zuer, A.A. & Anokhin, V.N. (1974) _Z. Prikl. Khim._ _47_ 1861.

Lee, B.S., Pyrcioch, E.J. & Schora, F.C. (1970) _Chem. Eng. Progr. Symp. Ser._ No. 105, _66_, 152.

Lockett, M.J., Davidson, J.F. & Harrison, D. (1967) _Chem. Eng. Sci._ _22_, 1059.

Richardson, J.F. & Zaki, W.N. (1954) Trans. Instn. Chem. Engrs. 39, 212.
Subzwari, M.P. (1976) M.Sc. Thesis, Imperial College, University of London.
Toomey, R.D. & Johnstone, H.F. (1952) Chem. Eng. Prog. 48, 220.
Upson, P.C. & Pyle, D.L. (1974) Proc. Int. Symp. on Fluidn. (Toulouse), 207.
Yasui, G. & Johanson, L.N. (1958) A.I.Ch.E.J. 4, 445.

TABLE 1 Measured Properties of Fluidized Bed of Cracking Catalyst

Pressure (bars)	1	2	3	4	5	6	7
(1) $U = 3.2$ cm/s							
At 0.16 m above distributor:							
Visible bubble flow (cm^3/s)*	168	164.4	168	156.5	144	125	120
Bubble volume fraction (ε_B)	0.109	0.103	0.096	0.078	0.070	0.056	0.044
Maximum observed bubble diameter in bed (cm)	3.8	3.75	3.1	2.6	3.0	2.55	2.4
Bubble frequency (s^{-1}) at levels							
.13m	110	115	123	114	69	80	63
.17m	140	146	140	145	105	110	82
.20m	135	118	110	113	82	93	63
.25m	99	84	117	85	84	52	52
(2) Bed depth (cm) (cf. $H_{mf} = 31.0$ cm)							
$U = 3.2$ cm/s†	32.0	32.3	31.9	32.4	32.5	32.6	33.2
$U = 5.5$ cm/s	33.0	33.5	33.8	34.0	34.5	35.0	36.0
(3) Particulate phase void fraction for $U = 3.2$ cm/s, by Eqn. (1)							
ε	.47	.48	.47	.49	.50	.51	.52

* cf. value predicted by two-phase theory, $A(U-U_{mf}) = 186\ cm^3/s$
† interpolated

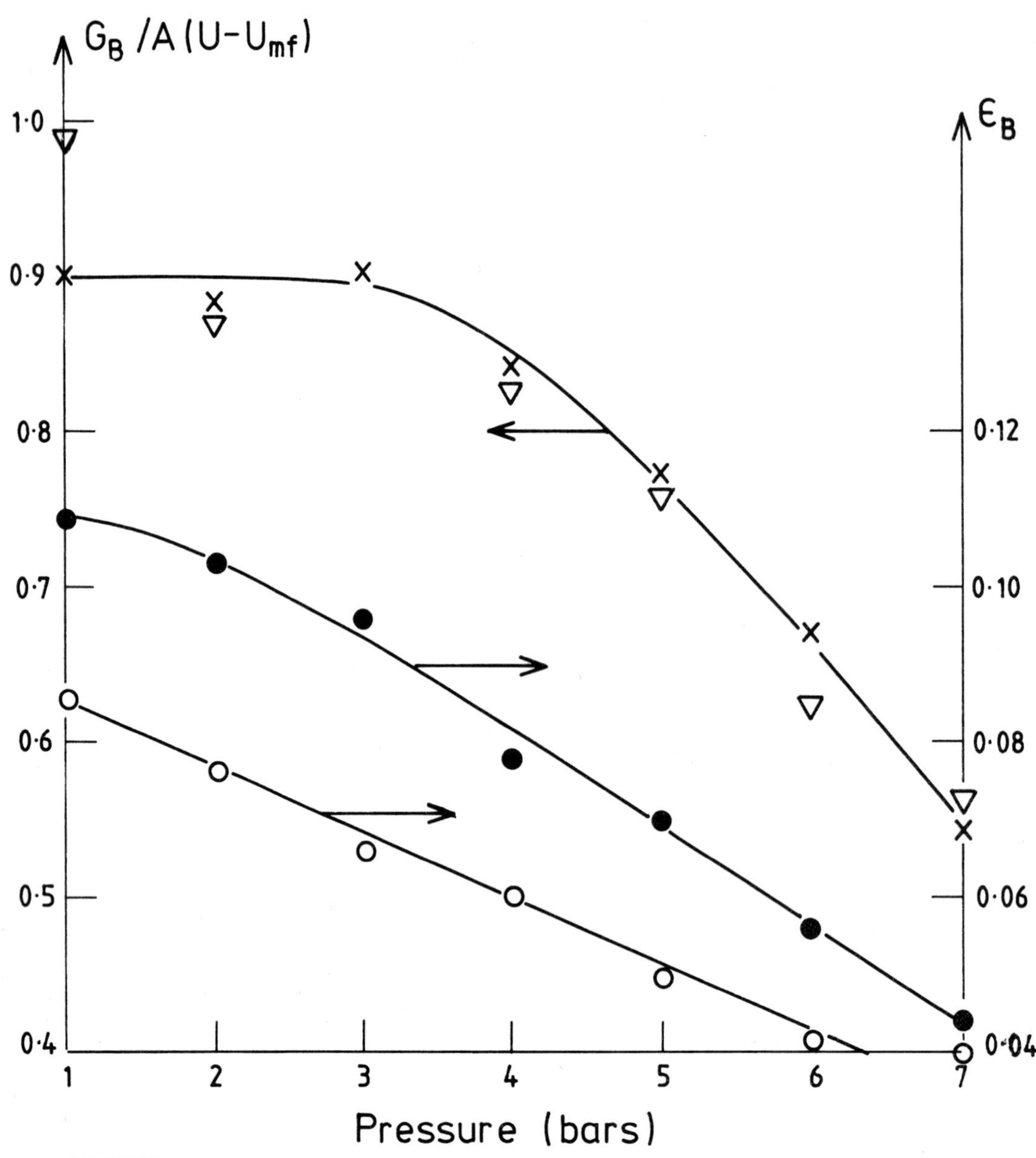

Fig. 1 Bubble volume fraction (ε_B), and bubble flow rate expressed as fraction of value predicted by two-phase theory ($G_B/A(U-U_{mf})$) at 0.16m above distributor, with $U-U_{mf}$ = 2.7 cm/s.

HIGH PRESSURE FLUIDIZATION IN A TWO-DIMENSIONAL BED

By T. Varadi and J.R. Grace

Department of Chemical Engineering

McGill University

3480 University Street

Montreal, Canada H3A 2A7

SYNOPSIS

Previous work on fluidization at high pressures has established that bubbles tend to be smaller at higher bed pressures. An investigation of the underlying causes, in particular the incidence and mechanism of bubble splitting and the visible bubble flow, has been undertaken in a two-dimensional column at pressures up to 2200 kN/m^2. Preliminary results show that the gap between minimum fluidization and minimum bubbling conditions widens considerably with increasing bed pressure. There was no evidence of any significant increase in the incidence of bubble splitting as the bed pressure was increased and what splitting did occur was by division from the roof.

NOMENCLATURE

P_B	absolute pressure at the bed surface
U_{mb}	superficial gas velocity at minimum bubbling point
U_{mf}	superficial gas velocity at minimum fluidization

INTRODUCTION

Recent years have given rise to considerable interest in gas fluidization at elevated pressures, especially in connection with combustion and gasification of coal (e.g. see Horsler et al., 1969; Hoke & Bertrand, 1975; Roberts et al., 1975; Vogel et al., 1976). Many of the workers who have studied the effect of pressure have remarked on "smoother fluidization" and smaller bubbles at higher absolute bed pressures (e.g. Lee et al., 1970; Botterill & Desai, 1972; Katasonov et al., 1974). In addition, there is evidence that higher pressures result in higher heat transfer coefficients (Botterill, 1975), increased entrainment (May, 1953), increased bed expansion (de Carvalho & Harrison, 1975) and decreased particle segregation (Chen & Keairns, 1975).

The reason for the smoother fluidization at higher pressures has not been adequately confirmed. Clearly bubbles are smaller, but there are two fundamentally different mechanisms which might explain why this should be so. In the first place, some workers (e.g. de Carvalho & Harrison, 1975) propose that bubbles become less stable as the pressure increases with the result that smaller bubbles are found in high pressure systems. A theoretical model based on saltation of particles from the floor of rising bubbles was shown to predict a decrease in maximum stable bubble size with increasing pressure. An alternate theory of bubble splitting (Clift & Grace, 1972; Clift et al., 1974), based on splitting from Rayleigh-Taylor instability of the roof, predicts virtually no effect of pressure unless there are dense phase voidage changes which alter the effective dense phase viscosity. Unfortunately, X-ray photographs of slugs splitting in high pressure systems (de Carvalho, 1976) are not conclusive as to whether splitting originates from the roof or the base.

The second plausible explanation for the reduction in bubbling is that a smaller fraction of the gas flow enters the bubble phase at higher pressures. While it is commonly believed that the gas in excess of that required to fluidize an aggregate bed flows as bubbles, there is substantial evidence (Grace and Clift, 1974) that the true flow of gas by translation of void units is less, even at atmospheric pressure. At higher pressures where there is evidence (Godard & Richardson, 1968; Mogan et al., 1971) that the gap between minimum fluidization and minimum bubbling is increased with increasing pressure, it seems plausible that the "two-phase theory of fluidization" might be even further in error.

The aim of the present study was to investigate which, if either, of these explanations was correct. In order to make clear observations of bubbles and their mode of break-up and rise, the study was carried out in a two-dimensional column. The present brief paper describes the experimental set-up and gives some preliminary results.

EXPERIMENTAL APPARATUS

A two-dimensional column of inside cross-section 31 x 1.6 cm and height 51 cm has been constructed to withstand pressures in excess of 2000 kN/m^2. The walls are made of 4.3 cm thick mild steel plate. There are two pairs of 12.7 cm diameter tempered glass windows set in the front and rear faces, flush with the inside surfaces of the steel plates. The lower pair is centred 11.4 cm and the upper 29.2 cm above the grid. These windows allow the bubbles to be viewed with the help of back-lighting as in conventional two-dimensional fluidized beds. The distributor is a perforated plate with 141 holes of diameter 1.2 mm drilled on a square pitch surmounted by a screen to prevent plugging of the holes by the particles.

A schematic of the experimental set-up is shown in Fig. 1. Air is fed to the bed from five compressed air cylinders connected to a manifold. The flow rate and pressure are controlled by throttling valves and a regulator. A metal-tube rotameter upstream and conventional rotameters downstream of the bed are used to measure the air flow rate, while the absolute pressure in the bed is measured by a mercury monometer at low pressures or by a Bourdon gauge at high pressure. The pressure drop across the bed is read from a specially constructed monometer containing a fluid of specific gravity 1.75. Single bubbles can be injected into the bed just above the distributor through a solenoid valve from a sixth pressurized air cylinder.

The particles used in the present experiments were sand screened to a size range 250 to 295 μm. Additional experiments will be carried out with other powders.

MINIMUM FLUIDIZATION AND BUBBLING VELOCITIES

The minimum fluidization velocity, U_{mf}, was measured in the conventional way at bed pressures, P_B, from atmospheric to 2170 kN/m^2. Pressure drops and superficial gas velocities were recorded for both increasing and decreasing gas flows. The minimum bubbling velocity, U_{mb}, was determined visually as the lowest superficial velocity at which bubbles appeared at the windows.

Experimental values of U_{mf} and U_{mb} are given in Fig. 2. Note that at atmospheric pressure there is no observable difference between U_{mf} and U_{mb}, but with increasing absolute pressure U_{mb} increases while U_{mf} decreases slightly. It is shown in Fig. 2 that $(U_{mb} - U_{mf})$ increases approximately linearly with absolute pressure. The measured U_{mf} values have been compared with a number of the more popular U_{mf} correlations from the literature. These generally predict little or no change in U_{mf} with pressure for small particles and a decrease for larger particles (Creasy, 1971; Knowlton, 1977). The closest predictions for the present case were provided by the correlation of Wen and Yu (1966), and the predicted curve is shown in Fig. 2.

OBSERVATIONS ON SINGLE INJECTED BUBBLES

It was necessary to set the background superficial velocity near U_{mf} in order for the volume of the injected bubbles to remain approximately constant. For a higher background fluidizing velocity, injection of a single large bubble was found to trigger spontaneous bubbling throughout the bed. At lower background velocities the bubbles tended to decrease rapidly in size, especially at the higher pressures.

The frontal area of injected bubbles was varied between 4 cm^2 and 46 cm^2 by changing the pressure in the reservoir upstream of the solenoid valve. Some splitting of bubbles did occur, but there was no apparent increase in the incidence of splitting at the higher bed pressures. When splitting did occur, it did so from the roof. A tracing of one such sequence is shown in Fig. 3. No filling up or division from the rear was observed. After splitting had occurred, many of the bubbles recombined by absorption of gas from one bubble into the other as noted by Grace & Venta (1973).

FREELY BUBBLING CONDITIONS

Qualitative observations at pressures up to 2200 kN/m^2 confirm that bubbles do tend to be smaller than when fluidization is carried out at the same superficial gas velocity at lower pressures. Measurements of the visible bubble gas flow will be carried out to determine how it depends on P_B for otherwise identical conditions.

CONCLUSIONS

A high pressure two-dimensional fluidized bed has been built and operated at pressures up to 2200 kN/m^2. The gap between the minimum fluidization velocity and minimum bubbling velocity widens considerably with increasing bed pressure. There is no evidence of a maximum stable size, of significantly more frequent splitting, nor of splitting from the rear for the conditions studied.

ACKNOWLEDGEMENTS

We wish to thank David Hoffman and Linus Ng for assistance with the experiments. Financial support from the National Research Council of Canada is also gratefully acknowledged.

REFERENCES

Botterill, J.S.M. (1975) Fluid Bed Heat Transfer, Academic Press, London.

Botterill, J.S.M. & Desai, M. (1972) Powd. Tech. 6, 231.

Chen, J.L.P. & Keairns, D.L. (1975) Can. J. Chem. Eng. 53, 395.

Clift, R. & Grace, J.R. (1972) Chem. Eng. Sci. 27, 2309.

Clift, R., Grace, J.R. & Weber, M.E. (1974) Ind. Eng. Chem. Fund. 13, 45.

Creasy, D.E. (1971) Brit. Chem. Eng. 16, 605.

de Carvalho, J.R.F.G. (1976) Ph.D. Thesis, Cambridge Univ.

de Carvalho, J.R.F.G. & Harrison, D. (1975) Fluidised Combustion Conference, paper B1, London, Inst. Fuel Symp. Ser. No. 1.

Godard, K. & Richardson, J.F. (1968) I. Chem. Eng. Symp. Ser. No. 30, 126.

Grace, J.R. & Clift, R. (1974) Chem. Eng. Sci. 29, 327.

Grace, J.R. & Venta, J. (1973) Can. J. Chem. Eng. 51, 110.

Hoke, R.C. & Bertrand, R.R. (1975) Fluidised Combustion Conference, paper D5, London, Inst. Fuel Symp. Ser. No. 1.

Horsler, A.G., Lacey, J.A. & Thompson, B.H. (1969) Chem. Eng. Progr. 65, No. 10, 59.

Horsler, A.G. & Thompson, B.H. (1968) I. Chem. Eng. Symp. Ser. No. 30, 58.

Katasonov, I.V., Menshov, V.N., Zuev, A.A. & Anokhin, V.N. (1974) J. Appl. Chem. USSR 47, 1909.

Knowlton, T.M. (1977) AIChE Symp. Ser. 73, No. 161, 22.

May, W.G. (1953) North New Jersey A.C.S. Meeting.

Mogan, J.P., Taylor, R.W. & Booth, F.L. (1971) Powd. Tech. 4, 286.

Roberts, A.G., Stantan, J.E., Wilkins, D.M., Beacham, B. & Hoy, H.R. (1975) Fluidised Combustion Conference, paper D4, London, Inst. Fuel Symp. Ser. No. 1.

Vogel, G.J. et al. (1976) Report ANL/ES-CEN-1016, Argonne National Laboratory, Argonne, Illinois.

Wen, C.Y. & Yu, Y.H. (1966) A.I.Ch.E. J. 12, 610.

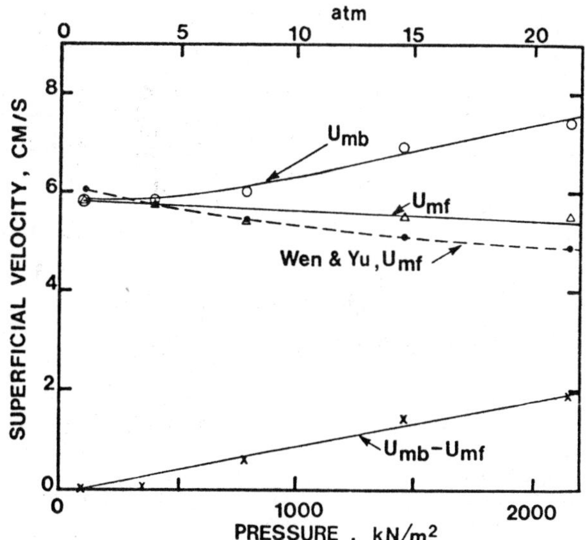

Fig. 2. Minimum fluidization velocity, minimum bubbling velocity and $(U_{mb} - U_{mf})$ plotted against bed pressure.

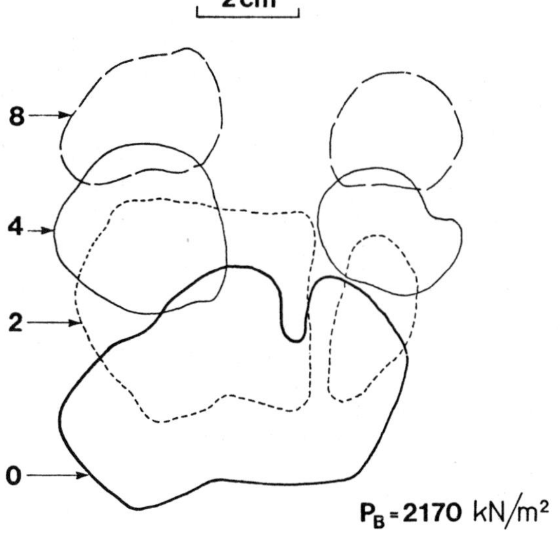

Fig. 3. Tracing of bubble undergoing splitting. Numbers denote number of frames. Film speed was 32 frames per second.

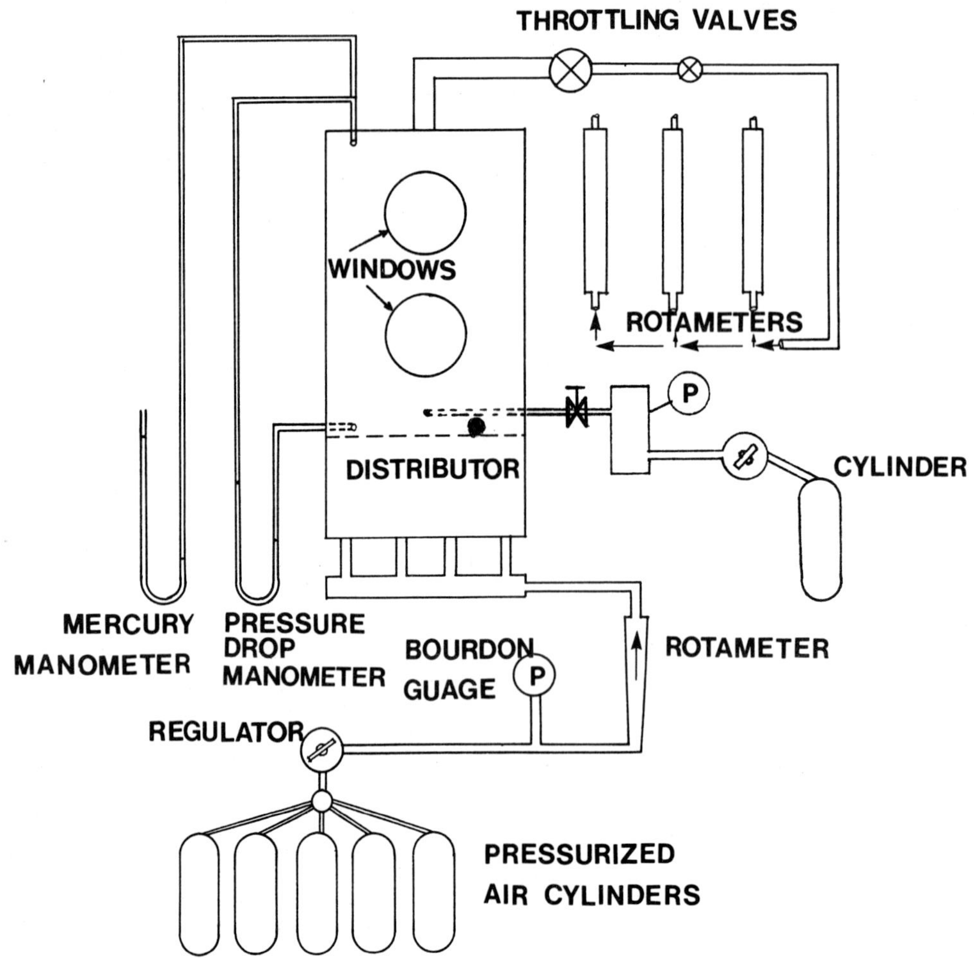

Fig. 1. Schematic diagram of the experimental apparatus.

FLUIDISATION OF FINE PARTICLES UNDER PRESSURE

By J.R.F.Guedes de Carvalho, D.F.King and D.Harrison

Department of Chemical Engineering
Pembroke Street, Cambridge
England CB2 3RA

SYNOPSIS

This work describes the behaviour of three types of particles fluidised by nitrogen and/or carbon dioxide in a 0.1m I.D. perspex column at pressures of up to 28 bar. Superficial gas velocities of up to 0.3 m/s were used to fluidise glass ballotini (mean particle diameters ($\bar{d}_p$): 64μ and 211μ) and sand ($\bar{d}_p$ = 74μ). The variables studied were the bed expansion before bubbling/slugging, the degree of bed level fluctuation, and the overall bed expansion.

The bed of fine ballotini showed a tendency for bubbling to occur with smaller bubbles at high pressures when fluidised with carbon dioxide, but the behaviour of the other fluidised beds did not seem to be influenced by changes in gas pressure. The observed phenomena are discussed in the light of present theories of bubble stability.

NOTATION

$\bar{d}_p$	mean particle diameter
H_{max}	maximum bed height
H_{mf}	bed height at minimum fluidisation
U	superficial gas velocity
U_{mb}	superficial gas velocity at onset of bubbling
U_{mf}	minimum fluidising velocity
ΔH	bed height fluctuation

INTRODUCTION

Leung (1961) and Simpson & Rodger (1961) among others have reported that fluidisation can be made 'smoother' by significantly decreasing the density difference between the solid particles and the fluidising gas. Guedes de Carvalho & Harrison (1975) showed that the fluidisation of cracking catalyst becomes 'smoother' at high pressures even though the density difference between solids and gas was virtually unchanged; and a possible explanation was advanced by Guedes de Carvalho (1976) in terms of a theory of bubble stability based on particle 'pick-up' (or saltation) from the lower surfaces of rising bubbles. There is, however, another explanation in terms of an increased Taylor instability of the roof of the bubbles. On this hypothesis, fluidisation becomes 'smoother' at high pressure because the dense phase then has a higher voidage than at atmospheric pressure, and consequently a lower apparent viscosity which in turn leads to a faster rate of growth of Taylor instabilities (Clift *et al* (1974)).

X-ray photographs of gas slugs in fluidised beds of fine particles (Guedes de Carvalho, 1976) showed that slugs become unstable above a given fluidising pressure; unfortunately though it is not possible to distinguish clearly from the photographs how bubble break-up is initiated.

EXPERIMENTAL

The fluidising gas, initially fed from a pressurised cylinder, was circulated in a closed-loop circuit by means of a rotary blower enclosed in a metal casing in a way similar to that already described by Guedes de Carvalho (1976). The flow-rates corresponding to the points in Figs. 1-6 were determined by three sharp-edged orifice meters. Orifice calibration with air at pressures between 1-5 bar showed no significant deviation from the normal orifice-meter equation, and this equation was used over the full pressure range for both nitrogen and carbon dioxide. The fluidised bed was contained in an 0.1m I.D. column consisting of two tube segments. The upper segment was a

1m length of perspex tube 50mm thick which allowed visual observation of the bed surface; the lower segment was made of aluminium alloy and sufficient to give the necessary bed height. A high pressure drop porous plate distributor was used, and it was tested for uniformity of gas distribution by fluidising a very shallow layer of the fine ballotini. Bed heights were measured by reference to a tape on the outside of the perspex tube. The fine ballotini and the sand particles were obtained by elutriation from batches of wider size distributions. The 211μ ballotini was used as supplied by the manufacturers. Some characteristics of the particles are given in Table 1. The values of the minimum fluidising velocity (U_{mf}) were determined by separate experiment, and pressure is not expected to have a significant influence on U_{mf} of the finer particles used in this work. Superficial velocity at which bubbling first occurs (U_{mb}) was measured at pressure for all particles and was found in each case to be only slightly larger than U_{mf}. For the finer particles U_{mb} was found to be independent of pressure, whereas for the 211μ ballotini it decreases a little with increase in pressure. Accordingly, $U-U_{mb}$ has been made the abscissa of Fig.2.

Table 1 Particle characteristics

Particles	Ballotini (64μ)	Ballotini (211μ)	Sand (74μ)
Solid density (kg/m^3)	2900	2900	2690
U_{mf} at 1 bar (m/s)	0.013	0.044	0.012

BED EXPANSION

For the particles studied, Table 2 shows the very small effect of pressure on the bed voidage at the point at which bubbling is first observed.

Table 2 Voidage at incipient bubbling as a function of gas pressure

Pressure (bar)	Voidage at incipient bubbling		
	64μ ballotini and CO_2	211μ ballotini and N_2	sand and CO_2
1	0.43	0.38	0.53
2	0.43	0.39	0.53
5	0.43	0.39	0.53
10	0.43	0.39	0.53
15	0.43	0.40	0.53
20	0.44	0.40	0.54
25	0.43	0.40	0.54

The variations in maximum bed height, H_{max}, and bed height fluctuation, ΔH, are shown in Figs.1-6 for the different particles studied. Individual values of H_{max} and ΔH may differ by as much as 3-5 cm as a consequence of uncertainties inherent in the observed phenomena. Figs.1 and 2 show that variations in pressure over the range covered have virtually no effect on the slugging characteristics of beds of sand and the 211μ ballotini; except for some increased lack of definition of the bed surface at the highest flow-rates at high pressures with sand. The surface of the fine ballotini also showed this lack of definition at high pressures. It is noteworthy that whereas the expansion of the bed of glass ballotini followed the theoretical line (Hovmand & Davidson, 1971), the bed of sand expanded less than predicted by theory; the latter behaviour has been observed before for beds of fine particles and seems to be associated with the comparatively low viscosity of such beds.

Experiments with the fine ballotini in a bed with a settled height of about 0.6m seemed to show some signs of a change in the slugging behaviour with change in pressure; and in order to make such differences more measurable the bubbling behaviour of this material was studied in a taller bed ($H_{mf} \simeq$ 1.1m). The description of the appearance of the bed at various pressures may be associated with particular points in Figs.3 and 4. For the two lowest pressures bed expansion lies close to the theoretical prediction of Hovmand & Davidson (1971). This is not typical of the behaviour of systems of fine particles and may be due to the comparatively large fluctuations in pressure above the distributor in this work. Above 5 bar, however, these fluctuations are small compared with the absolute pressure, and the bed expansion then falls below the predictions of theory, in line with the normal behaviour of beds of fine particles.

For pressures in the range 5-20 bar, the fluctuations in the bed level seem to be little affected by changes in gas pressure and slugging is clearly suggested by the high values of ΔH. However, when fluidising with carbon dioxide at 20 bar, a change in the behaviour of the bed surface was observed: the frequency of slug eruption at the surface at a given flow-rate was much smaller than at lower pressures, and between the eruption of successive slugs the bed appeared to be bubbling with a fluctuation in bed level which never exceeded 8-9 cm, however high the flow-rate. In Figs.3 and 4 the points corresponding to this bubbling and occasional slugging at 20 bar are surrounded by a circle and a square, respectively. Finally, at 27 bar the fluctuations in bed level were again below 8-9 cm and independent of the flow-rate but at this pressure no large slugs were observed to disturb the bubbling appearance of the bed. At all pressures the gas flow-rates reached 2.5 or more times the minimum required to satisfy Stewart's slugging criterion (see Hovmand & Davidson (1971)), and thus it is difficult to explain why slugs do not seem to be formed so easily at the highest pressures when the beds of fine ballotini were fluidised with carbon dioxide. The slugs may not be stable at such high gas densities as has been suggested by Guedes de Carvalho (1976). If this is the case (and an X-ray study of fluidisation under these conditions is being planned), increased bubble break-up by Taylor instability would not seem to provide a satisfactory explanation because the beds were not of the type which exhibited significant expansion before bubbling. On the other hand, an explanation of bubble instability which turns on particle 'pick-up' from the lower surfaces of slugs has gas density as one of the determinant variables. Figs.5-6 show the bed expansion (H_{max} and ΔH) of fine ballotini fluidised with nitrogen at pressures up to 28 bar as a function of fluidising velocity. No transition from slugging to bubbling was observed. In this sense the bed behaviour was similar to that found with carbon dioxide at pressures below 20 bar where only slugs were seen to erupt at the bed surface (i.e. ΔH greater than 9 cm). This is in line with the theory of particle 'pick-up' if we note that the density of nitrogen at 28 bar (32.0 kg/m^3; 300K) is less than that of carbon dioxide at 20 bar (39.5 kg/m^3).

CONCLUSIONS

The slugging/bubbling behaviour of fluidised beds of fine particles may change with the operating pressure. No significant effect of pressure up to 28 bar has been detected using 74μ sand and 211μ ballotini. However, when 64μ ballotini was fluidised with carbon dioxide signs of slug instability were apparent above 20 bar. Nevertheless, more experimental data are required to establish conclusively whether or not slug break-up occurs under these conditions.

The experimental observations with fine ballotini are consistent with the predictions of a theory of bubble instability which envisages particle pick-up from the lower surfaces of rising slugs.

REFERENCES

1. Clift R., Grace J.R. & Weber M.E. (1974) Ind.Engng Chem.(Fundls) 13, 45.

2. Guedes de Carvalho, J.R.F. (1976) Ph.D. dissertation, University of Cambridge.

3. Hovmand S. & Davidson J.F. (1971) Fluidization (Ed. Davidson J.F. & Harrison D.). Chapter 5 London: Academic Press.

4. Leung L.S. (1961) Ph.D. dissertation, University of Cambridge.

5. Simpson H.C. & Rodger B.W. (1961) Chem.Engng Sci. 16, 153.

ACKNOWLEDGEMENT: JRFGdeC and DFK would like to thank the Science Research Council for financial support during the course of the research work.

Key to all figures

a -	1 bar with carbon dioxide	**A** -	1 bar with nitrogen
b -	2 bar " " "	**B** -	2 bar " "
c -	5 bar " " "	**C** -	5 bar " "
d -	10 bar " " "	**D** -	10 bar " "
e -	15 bar " " "	**E** -	15 bar " "
f -	20 bar " " "	**F** -	20 bar " "
(f) -	bubbling at 20 bar with carbon dioxide	**X** -	25 bar " "
[f] -	occasional slug at 20 bar with carbon dioxide	**G** -	28 bar " "
x -	25 bar with carbon dioxide		
g -	27 bar with carbon dioxide		

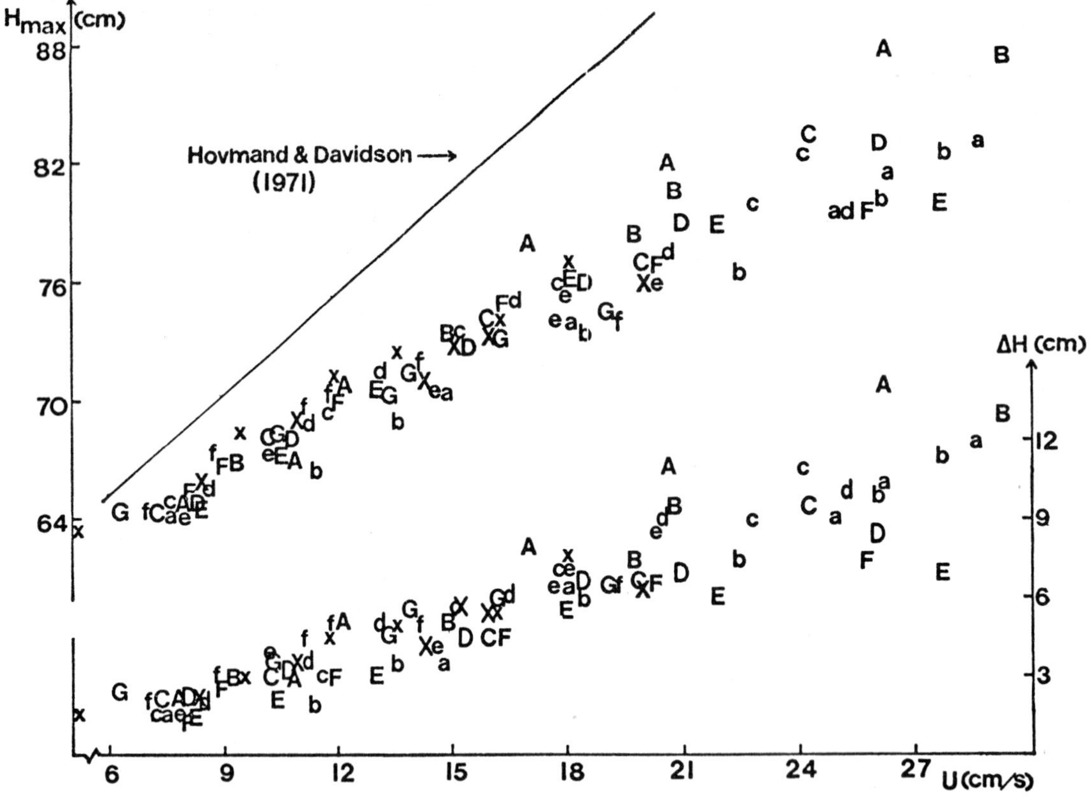

Fig.1 Bed expansion of 74µ sand at various pressures

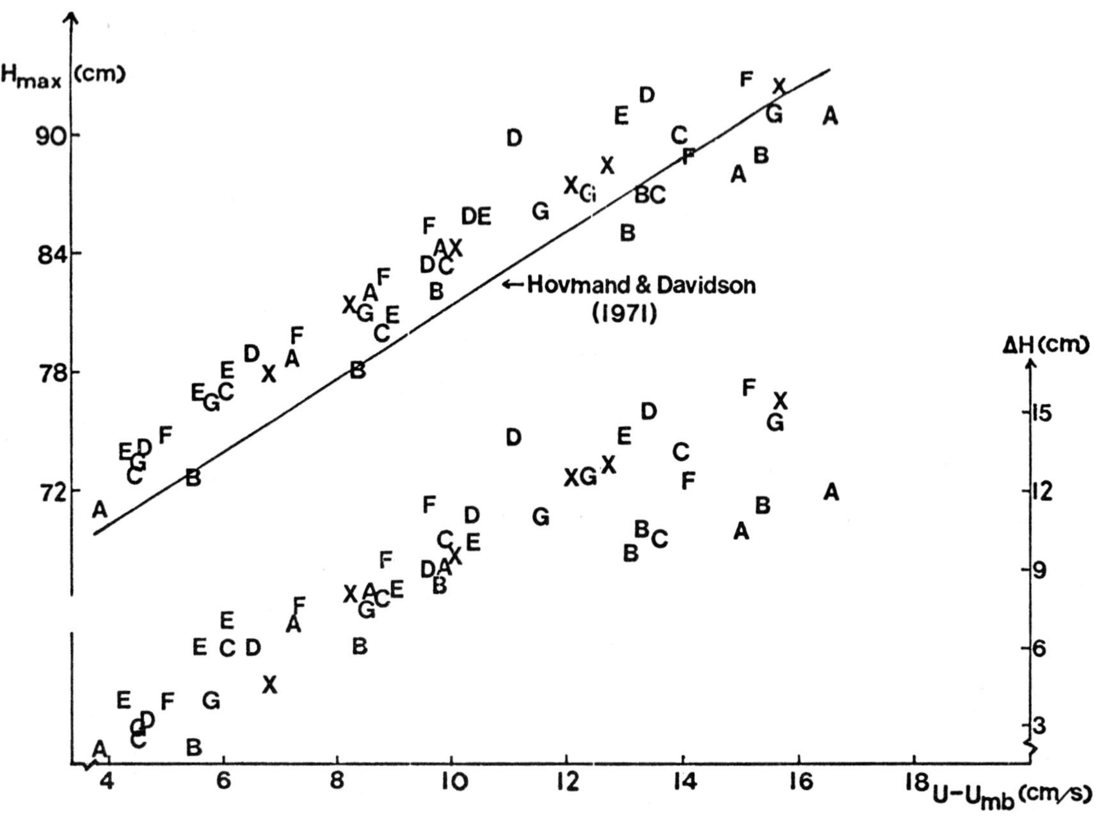

Fig.2 Bed expansion of 211µ glass ballotini at various pressures with nitrogen

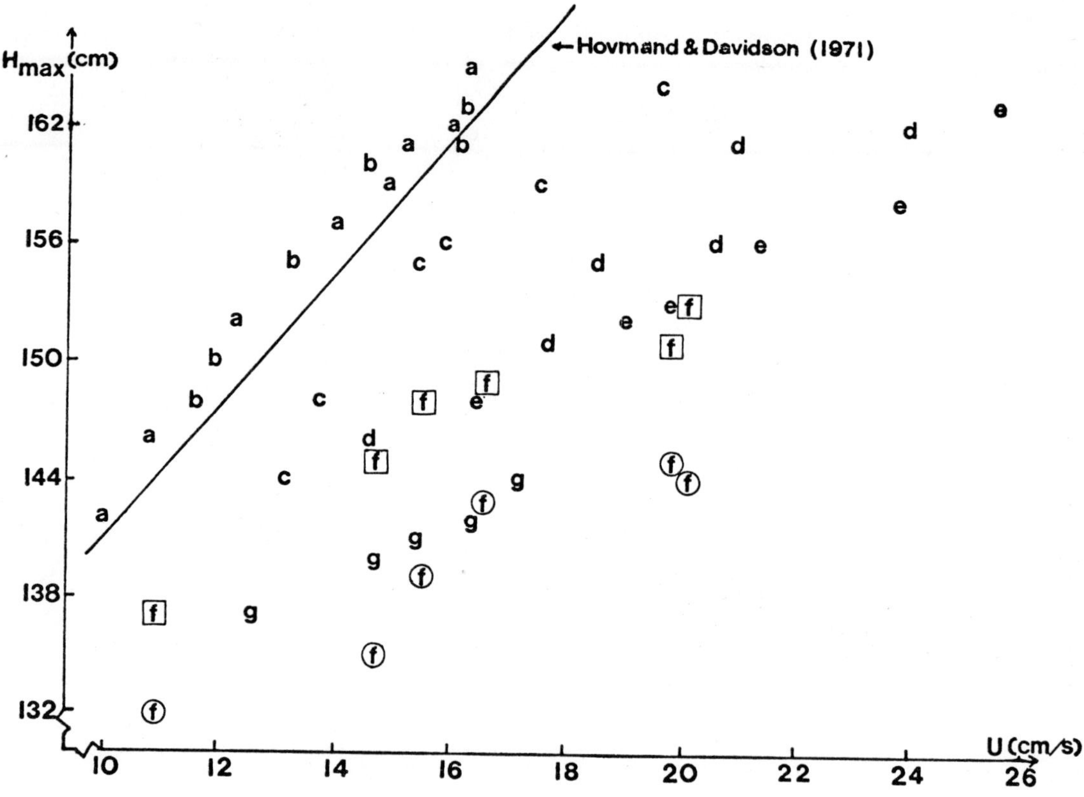

Fig.3 Maximum bed height for 64μ glass ballotini at various pressures with carbon dioxide

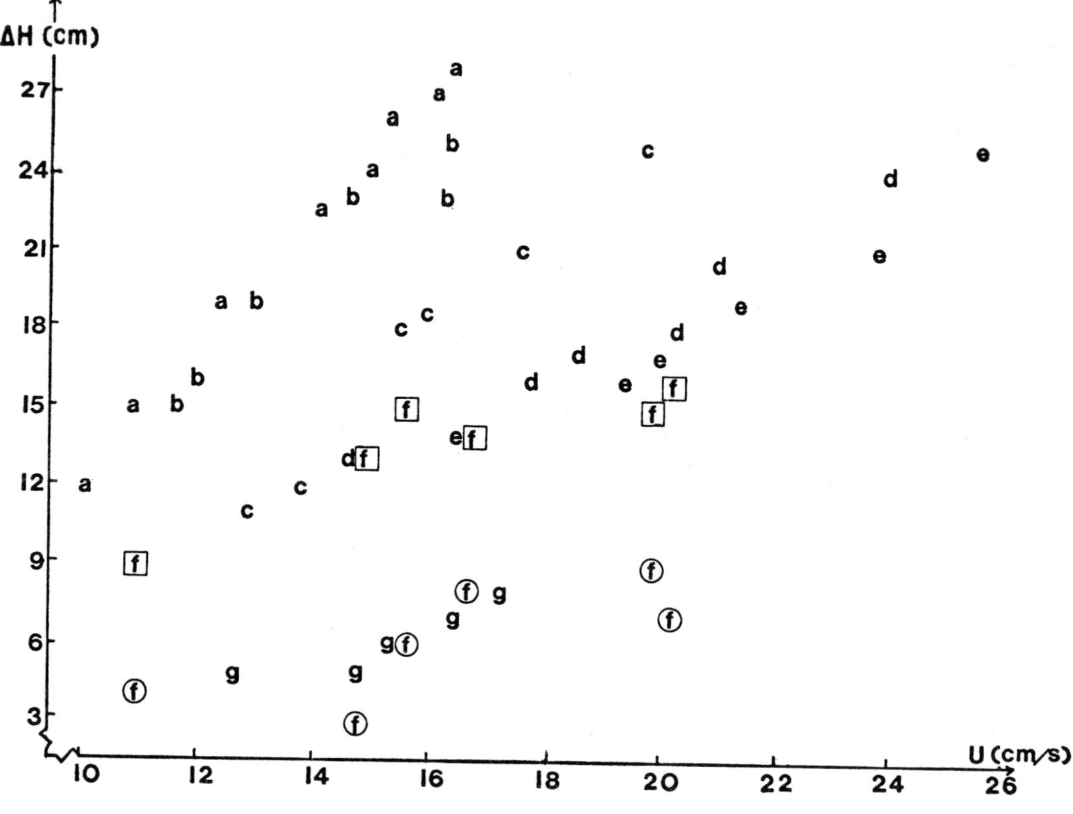

Fig.4 Bed level fluctuation for 64μ glass ballotini at various pressures with carbon dioxide

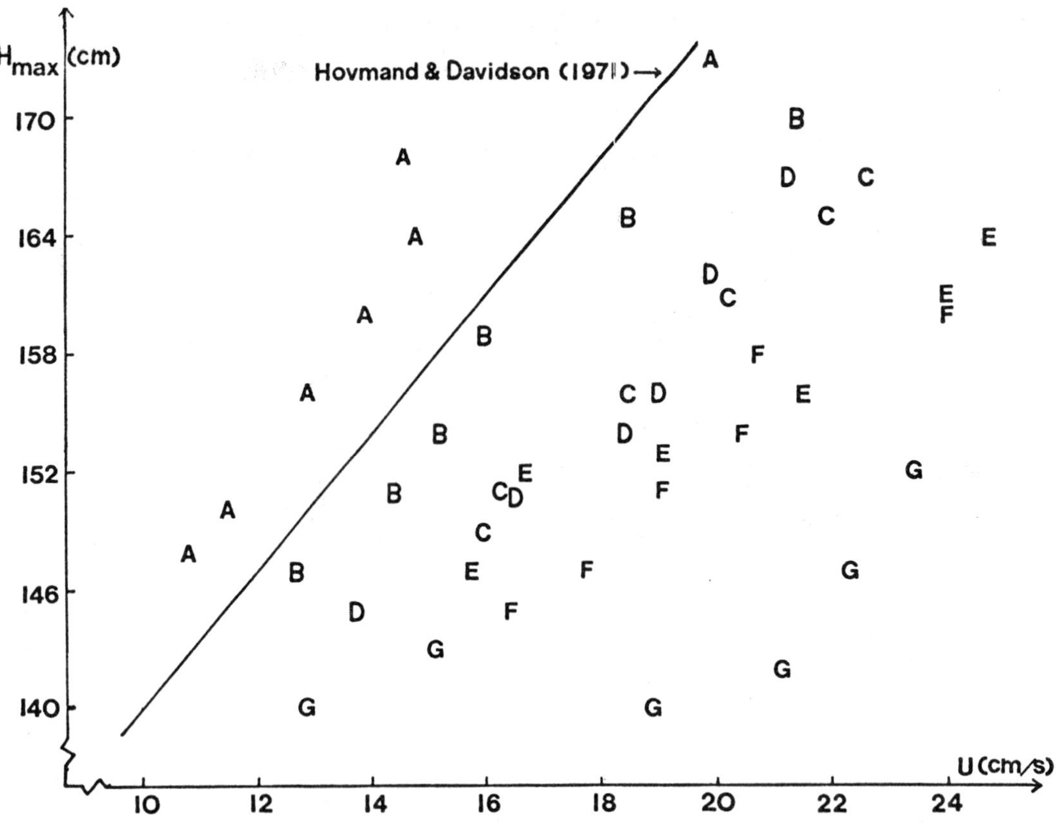

Fig.5 Maximum bed height for 64μ glass ballotini at various pressures with nitrogen

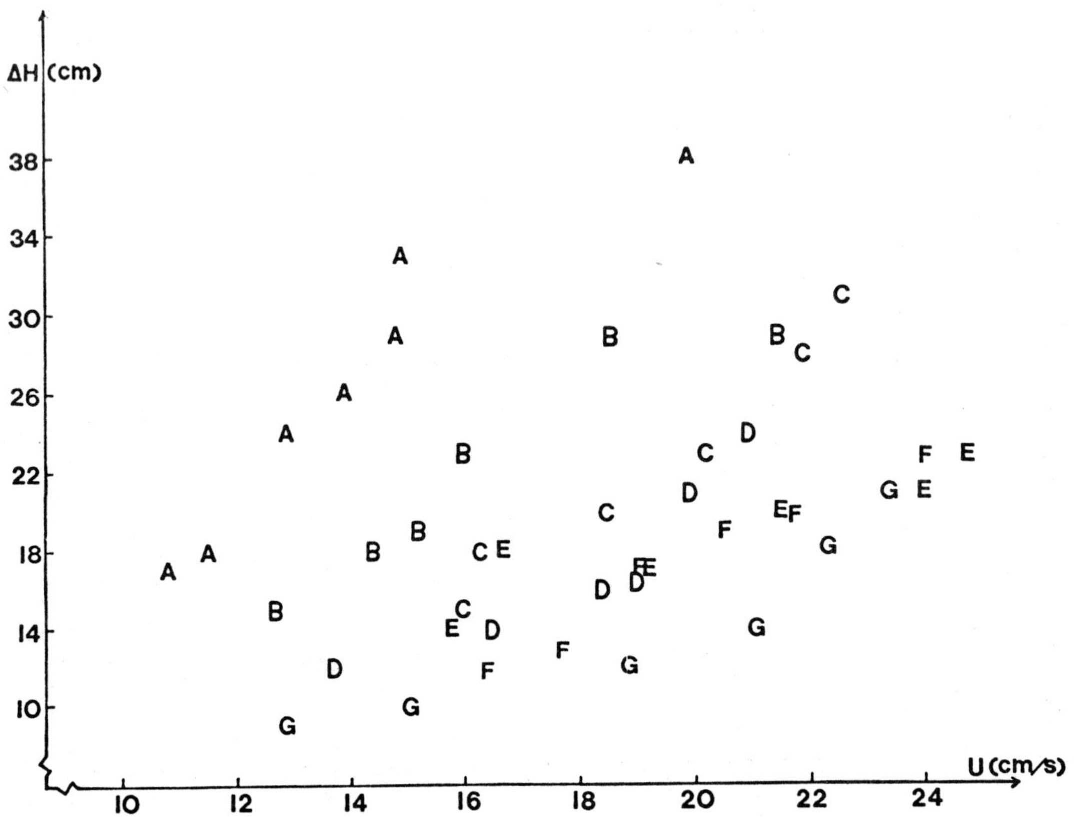

Fig.6 Bed level fluctuation for 64μ glass ballotini at various pressures with nitrogen

FLUIDISATION OF FINE PARTICLES AT ELEVATED PRESSURE

By M.E. Crowther and J.C. Whitehead

Coal Research Establishment, Stoke Orchard, Cheltenham, Gloucestershire GL52 4RZ, UK

Equipment which allows visual observation of a bed of solids fluidised with a high pressure supercritical fluid has been constructed with the objective of predicting the mode of fluidisation and the voidage-flow relationship of such systems. The system investigated comprises fine coal particles fluidised by fluids with physical properties intermediate between those of a liquid and a low pressure gas.

It has been established that at gas densities in excess of 120 kg m^{-3} a bituminous coal crushed to -210 μm fluidises particulately. Data obtained in the particulate region using carbon tetrafluoride at elevated pressure as the fluidising medium have been presented in the form of correlations which relate bed voidage to system properties and superficial fluid velocity.

NOMENCLATURE

- D_p Mean size of solid = $\dfrac{1}{\sum \frac{x}{d_p}}$
- d_p Root mean size of fraction, weight x, between size limits d_1 and d_2, $=\sqrt{d_1 d_2}$
- ϵ Bed voidage
- ϵ_{mb} Minimum bubbling voidage
- ϵ_{mf} Minimum fluidising voidage
- g Gravitational constant
- k Constant = u_t/u_i
- n Exponent in Richardson and Zaki expression
- u Superficial fluid velocity
- u_i Theoretical superficial fluid velocity from Richardson and Zaki plot when ϵ = 1
- u_t Theoretical Stokes Law terminal velocity
- u_{mb} Minimum bubbling superficial fluid velocity
- u_{mf} Minimum fluidising superficial fluid velocity
- x Weight fraction of solid between size limits d_1 and d_2
- ρ_f Fluid density
- ρ_s Particle envelope density
- μ Fluid viscosity.

INTRODUCTION

During the investigation of a process involving the extraction of coal using a compressed supercritical fluid (Whitehead, 1975), it became necessary for plant design purposes to study the fluidisation of coal in fluids having densities in the range 200-600 kgm^{-3} and viscosities in the range 2×10^{-5} - 7×10^{-5} $kg\,m^{-1}s^{-1}$. The coal to be fluidised has a solid density of 1300 kgm^{-3}, and its particle size is limited by other process constraints to a maximum of 210 μm with a mean size of 27 μm. In particular it was necessary to determine whether the system was likely to exhibit particulate or aggregative fluidisation, and to determine the variation of bed voidage with superficial fluid velocity.

A literature survey revealed that little work had been published in this field and therefore it was necessary to carry out an experimental investigation.

LITERATURE SURVEY

Beds of fine particles are known to exhibit particulate fluidisation under certain conditions when fluidised with gas at elevated pressure (Richardson, 1971 and Massimila, 1972). However, at a high fluid velocity defined by u_{mb} (Richardson, 1971) these beds commence bubbling and become aggregative. This phenomenon is known as 'delayed bubbling'. Between the minimum fluidising and the bubbling conditions the bed expands in such a way that log. bed voidage is a linear function of log. superficial gas velocity. Above the minimum bubbling point the rate of bed expansion with fluid velocity falls considerably.

Richardson (1971) has published data illustrating that u_{mb} may considerably exceed the value of u_{mf}, and that the ratio of these values increases with system pressure. Several authors (Leung, 1961; Simpson, 1961) have indicated that "smoother", more particulate, fluidisation occurs at elevated gas pressures.

Harrison (1961) and Carvalho (1976) have explained this phenomenon in terms of a system possessing a maximum stable bubble size which is dependent upon particle size, gas density and viscosity. If this bubble size is equal to or less than the mean particle size the system is considered particulate; if it is greater than ten times the mean particle size the system is considered aggregative. The paper by Harrison is not applicable to particles of less than 100 µm diameter (Carvalho, 1976) and neither explains the phenomenon of delayed bubbling.

Doichev et al (1975) similarly explain the mode of fluidisation adopted by a system in terms of the ratio of maximum stable bubble size to mean particle size. This theory, which equates the energies theoretically associated with particulate and aggregative fluidisation, can predict the presence of a minimum bubbling point but not with sufficient accuracy for this work.

The relationship between bed voidage and superficial fluid velocity has been studied by Richardson and Zaki (1954). They successfully applied the equation $\frac{u}{u_i} = \epsilon^n$ ......(1)

for which n is a function of system properties, to a range of liquid/solid systems. However, it has been found that, for gas/solid systems at elevated pressure, the exponent n may be larger than that predicted, and that u_i can exceed u_t, the Stokes Law terminal velocity (Richardson, 1971).

Values of n greater than those predicted have also been observed by Lewis and Bowerman (1952) for the fluidisation of fine rough particles in liquids, and by Whitmore (1957) in settling experiments on narrow size fractions of methyl methacrylate and on smooth spheres. Whitmore's results are as follows:-

Particle size (µm)	210-178	152-125	104-89	75-53	Smooth spheres
n	6.9	7.5	8.3	9.5	4.8

Mogan et al (1971) attempted to predict voidages of fluidised coal beds at elevated gas pressures but did not find good agreement with experimental observations, at high voidages.

It was considered that none of the correlations or data available in the literature was suitable for use in the accurate prediction of the behaviour of the system to be encountered in the coal extraction process. Therefore a correlation was developed that allowed prediction of system behaviour by direct extrapolation of data obtained on the model system.

THE APPARATUS AND EXPERIMENTAL PROCEDURE

A diagrammatic flowsheet of the apparatus is shown in Fig. 1. The fluidised bed vessel consisted of a cast acrylic tube 27 mm I.D., 35 mm O.D. and 920 mm long, which incorporated a 50 µm mesh stainless steel distributor plate. Flow rate in the system was measured with an integral orifice differential pressure cell with a square root extractor (FM). Maximum superficial velocities of 1-2.5 cms^{-1} were possible depending upon gas density. The humidifier (H), which contained about 0.2 l water, was found necessary to reduce electrostatic effects. Any elutriated solids were removed by a filter candle (F).

The fluidised bed was charged with approximately 50 g of powder and the system sealed. The feed reservoir (A) was pressurised to approximately 120 bar, and then with the outlet valve (V2) closed and the by-pass valve (V1) open the system was pressurised to the desired value (displayed on P) by adjusting the pressure regulating valve (R1). The by-pass was then shut, the outlet valve opened, and the flow rate regulated manually by means of the micrometer flow control valve (V3); the recycle compressor (C), fed from the collection reservoir (B) via pressure regulator (R2) automatically maintained the feed reservoir (A) at about 120 bar.

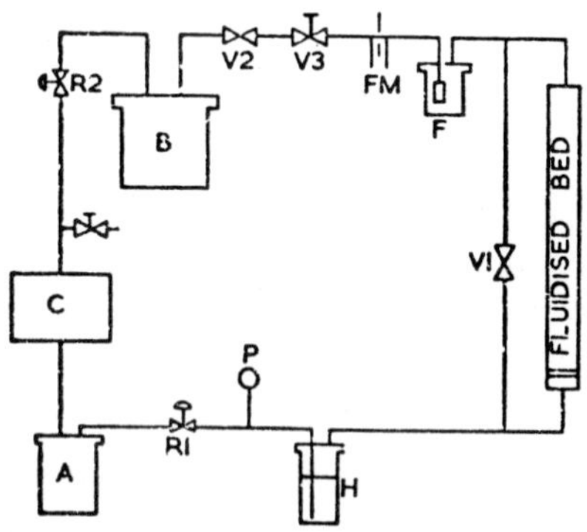

Fig. 1 Flowsheet for high pressure fluidisation rig.

THE PHYSICAL PROPERTIES OF THE GASES AND SOLIDS

Gases

The densities and viscosities of Argon and CF_4 are given in Table 1. The data were calculated from Reid and Sherwood (1966), except for the viscosity of CF_4 below 20 bar for which a relationship supplied by the Du Pont Co. was used.

TABLE 1: Density and viscosity of test gases.

	Ar		CF_4	
Pressure (bar)	ρ (kgm^{-3})	$\mu x10^5$ ($kgm^{-1}s^{-1}$)	ρ (kgm^{-3})	$\mu x10^5$ ($kgm^{-1}s^{-1}$)
14.6	-	2.17	58	1.75
28.2	-	"	120	1.81
42.8	72.3	"	194	1.90
55.4	96.8	"	282	2.10
69.0	121.8	"	388	2.35

Solids

Synclyst (-75 + 53 μm, Dp 63 μm). A catalyst support chosen as a material comprising essentially spherical or obloid particles, in contrast to coal which consists of angular particles. A narrow cut was seived to the above limits and calculated to have a mean particle size (D_p) of 63 μm. ρ_s was considered to be 900 kg m^{-3}.

Coal (-72 + 53 μm, Dp 63 μm). A bituminous coal (Markham Main CRC 802) ground on a hammer mill and then sieved between the above limits.

Coal (nominal -210 μm, Dp 27 um). The wide size fraction from which the previous sample was sieved. Size analysis is given in Table 2.

TABLE 2: Size analysis for coal crushed to -210 μm (D_p 27 μm)

Size (μm)*		Weight %(x)	$\frac{x}{dp} \times 10^2$
	+500	0.6	0.000
-500	+210	3.7	0.011
-210	+125	7.9	0.030
-125	+ 75	15.3	0.157
- 75	+58.8	9.4	0.144
-58.8	+43.3	13.1	0.259
-32.3	+27.2	15.8	0.462
-27.2	+17.3	12.7	0.588
-17.3	+ 9.1	11.2	0.896
- 9.1		10.5	1.153

*Data on sizes less than 75 μm obtained by Coulter count.

Coal (nominal -210 μm, Dp 57 μm). The nominal -210 μm sample was sieved to produce +75 μm and -75 μm fractions. A mixture of 3 parts +75 μm to 1 part -75 μm was prepared to produce this sample.

Coal (-53 μm, Dp 19 μm). This was the -53 μm fraction of the nominal -210 μm coal. ρ_s for all the coal sizes was considered to be 1300 kgm^{-3}.

RESULTS AND DISCUSSION

For most of the powders, voidage/flow characteristics were recorded with the gases carbon tetrafluoride (CF_4) and argon at a range of pressures up to 69 bar. Due to problems associated with channelling, the voidage/flow data reported here were always recorded with reducing flow. Periodic examination of the filter, F, (Fig. 1) showed no significant elutriation occurred.

The transition from particulate to aggregative fluidisation.

The system investigated broadly followed the patterns reported in the literature. At low gas pressures the beds expanded particulately until, at a certain flow rate, bubbles about 1 - 3 mm in diameter (depending upon system conditions) appeared. The rate of bed expansion with increasing gas flow then fell considerably as excess fluid was transported through the bed in the form of bubbles - this is well illustrated by the 21.4 and 35.0 bar data in Fig. 2. On increasing system pressure this transition velocity rose until eventually all the systems reached a pressure above which they no longer exhibited such a transition to aggregative flow, and fluidisation remained particulate. The appearance of bubbles did not cause the overall bed voidage to fall and it would therefore appear that fluidisation did not conform to conventional aggregative fluidisation in that bed particles were fluidised at a voidage approximating to ϵ_{mb} and not to ϵ_{mf}.

A summary of the transitions in terms of minimum bubbling velocities at various pressures is given in Table 3. The bubbling velocity varied between experimental runs, typically over a range of 0.20 cms^{-1} and the values reported are the mean of those obtained. The results marked P show those systems where fluidisation occurred in the particulate mode at all fluid velocities. This behaviour, termed "fully particulate fluidisation" was observed to occur at lower pressures as the mean particle size was reduced. In the case of the -210 μm coal fully particulate fluidisation occurred at pressures down to 28.2 bar with CF_4 and 69 bar with argon. These pressures correspond to fluid densities of about 120 kgm^{-3} and consequently fluidisation of this coal in the extraction process at fluid densities of 200-600 kgm^{-3} is expected to be in the particulate mode.

This data is in qualitative but not quantitative agreement with the theory of Doichev et al, (1975).

The similarity in the results in Table 3 for the two coal samples with similar Dp values - the narrow cut between 75 and 53 μm and the wide cut of Dp 57 μm - shows the presence of very fine particles does not dramatically affect the occurrence of bubbling.

The variation of bed voidage with flow.

Log. voidage vs log. flow plots are presented for the various systems in Figs. 2, 3 and 4. Straight lines have been drawn through the particulate regions of these graphs, thus confirming the applicability of the relationship

$$\frac{u}{u_i} = \epsilon^n \qquad (1)$$

Values of the gradient n at different pressures are given in Table 4 for all the materials examined, and the intercepts, u_i, at $\epsilon = 1$ are given in Table 3. These n values are larger than those predicted for the systems by the Richardson and Zaki (1954) expressions. This phenomenon has been reported previously for fine rough solids fluidised by gases at moderate pressures (Godard, 1968) and by liquids (Lewis et al, 1952). The n values are in reasonable quantitative agreement with those reported by Whitmore (1957) from settling experiments, the values of 7.6-8.9 for coal (Dp 63 μm) compare with 9.5 obtained by Whitmore for rough 63 μm particles and the values of 5.2-7.0 obtained for synclyst (Dp 63 μm) compare with Whitmore's value of 4.8 for smooth spheres. The small discrepancies in the values are probably a result of slightly differing degrees of roughness of the particles.

Values of n may be seen (Table 4) to dramatically increase with decreasing mean particle size; and in a similar way to that observed by Godard, et al (1968) n decreases with increasing system pressure.

A correlation of voidage with gas conditions

The linearity of the particulate regions of expansion when plotted according to equation (1) made it possible to utilise an alternative method of data presentation that would allow direct extrapolation from the density/viscosity region available to the cold model to the extractor conditions. This is achieved as follows:-

The Stokes Law terminal velocity of a spherical particle is given by

$$u_t = \frac{Dp^2 g}{18} \frac{(\rho s - \rho f)}{\mu} \qquad (2)$$

If the assumption is made over the limited range of cold model and extractor conditions that $u_t = ku_i$, where k is constant, substituting (2) in (1) gives

$$u = \frac{1}{k} \frac{Dp^2 g}{18} \frac{(\rho s - \rho f)}{\mu} \epsilon^n \qquad (3)$$

$$\therefore \log. u = \log. \frac{(Dp^2 g)}{18k} + \log. \frac{(\rho s - \rho f)}{\mu} + n \log. \epsilon$$

Thus a plot of $\log. \frac{\mu}{(\rho s - \rho f)}$ against log. ϵ, at constant u, is of the required form as it can be extrapolated to the gas conditions pertaining to the extractor. The results of Figures 5, 6 and 7 confirm that straight line plots are obtained, although the coal data suffers from displaced argon values (these were not used in the construction of the lines).

Table 5 lists the values of the gradient n obtained at different values of u by this approach. Comparison of these values of n with those given in Table 4 shows that the two sets of values are similar, thus confirming the validity of this approach, and the assumption regarding k.

At present it is not possible to establish the reason for the discrepancy between the argon and carbon tetrafluoride data on Figs. 6 and 7. The correlation used for the construction of these plots is based on a very simple model which, for example, takes no account of kinetic energy loses (Ergun, 1952). The latter are predominantly dependant on fluid density and thus under certain conditions it is possible for fluids with the same values of $\mu/\rho s - \rho f$ to produce different bed voidages at the same superficial fluid velocity.

Preliminary results based on pressure drop measurements indicate that the observed voidages in a coal extractor are close to those predicted by extrapolation.

CONCLUSIONS

Fine powders fluidised by compressed gases exhibit particulate fluidisation, especially at low superficial gas velocities. An increase in fluid velocity can cause the system to commence bubbling and the velocity at which this occurs increases with increasing system pressure, until eventually no bubbling transition occurs and fluidisation remains fully particulate at all fluid velocities between u_{mf} and the commencement of bulk transport. Decreasing mean particle size increases the minimum bubbling velocity although it appears that the latter is not greatly influenced by the presence of varying quantities of extremely fine particles (<10 μm).

The occurrence of fully particulate fluidisation appears to be dependent on particle size and system pressure; it is shown that -210 μm coal fluidises fully particulately in systems where gas density is in excess of 120 kgm^{-3}.

The expansion of a particulately fluidised bed of coal or synclyst can be represented by linear plots of log. u vs log. ϵ. The gradient of these, n, is similar to those reported for other fine rough particle systems. Plots of log. ϵ vs log. $\frac{\mu}{(\rho s - \rho f)}$ at constant fluid velocity are also linear and have gradients approximating to the n values associated with the voidage/flow plots. In both cases n increases rapidly with decreasing mean particle size.

ACKNOWLEDGEMENTS

This work was carried out with the support of the European Coal and Steel Community; the paper is published by permission of the National Coal Board but the views expressed are those of the authors, and not necessarily those of the Board.

REFERENCES

Doichev, K., Toderov, S., and Dimitrov, V. (1975). Chem. Eng. Sci. 30, 419.

Ergun, S. (1952). Chem. Engr. Prog. 48, 89.

Godard, K. and Richardson, J.F. (1968) Tripartite Chem. Engr. Conference, Montreal.

Guedes de Carvalho, J.R.F. (1976) Ph.D. Cambridge.

Harrison, D., Davidson, J.F. and de Kock, J.W. (1961). Trans. Instn. Chem. Engrs. 38, 202.

Leva, M., (1959) "Fluidisation", McGraw-Hill.

Lewis, E.W. and Bowerman, E.W. (1952). Chem. Engr. Prog. 48, 603.

Leung, L.S., Ph.D., Cambrdige (1961).

Massimila, L., Doussi, G. and Zucchini, C. (1972). Chem. Eng. Sci. 27, 2005.

Mogan, J.P., Taylor, R.W. and Booth, F.L. (1971), Powder Technology, 4, 286.

Reid, R.C. and Sherwood, T.K. (1966). "The Properties of Gases and Liquids", 2nd Ed. McGraw-Hill.

Richardson, J.F. and Zaki, W.N. (1954). Trans. Instn. Chem. Engrs. 32, 35.

Richardson, J.F. and Meikle, R., (1961), Trans. Instn. Chem. Engrs. 39, 348.

Richardson, J.F. (1971) in "Fluidisation" by Davidson, J.F. and Harrison, D., Academic Press.

Simpson, H.C. and Rodgers, D.W., (1961) Chem. Eng. Sci. 16, 153.

Whitehead, J.C., and Williams, D.F. (1975), J. Inst. Fuel, 48, 182.

Whitmore, R.L., (1957). J. Inst. Fuel, 30, 238.

TABLE 3: Minimum bubbling and u_i velocities (cms^{-1})

Fluidising gas		CF_4					Ar	
Pressure (bar)		21.4	28.2	41.8	55.4	69.0	55.4	69.0
Gas density kgm^{-3}		88	120	194	282	388	97	122
Synclyst, narrow cut, Dp 63 μm,	u_{mb}	0.8	0.8	0.9	1.1	P	0.9	1.0
	u_i	13.8	9.7	6.9	5.1	4.0	7.8	7.8
Coal, narrow cut, Dp 63 μm	u_{mb}	NA	1.1	1.3	P	P	NA	1.3
	u_i	NA	11.0	8.2	5.0	4.4	NA	8.6
Coal, wide cut, Dp 57 μm	u_{mb}	NA	1.7	1.8	P	P	NA	NA
	u_i	NA	8.1	7.1	5.8	5.3	NA	NA
Coal, wide cut, Dp 27 μm	u_{mb}	2.1	P	P	P	P	2.2	P
	u_i	NA	6.3	4.9	3.4	2.6	NA	7.0
Coal, wide cut, Dp 19 μm	u_{mb}	NA	P	P	P	P	NA	P
	u_i	NA	NA	4.9	3.6	2.9	NA	NA

P indicates fully particulate fluidisation
NA indicates data not available.

TABLE 4: Summary of n values from plots of log. ε vs log. u

Fluidising gas	CF_4				Ar
Pressure, bar	28.2	41.8	55.4	69.0	69.0
Synclyst, narrow cut, Dp 63 μm	6.5	6.0	5.4	5.2	6.3
Coal, narrow cut, Dp 63 μm	8.9	8.7	7.6	7.8	8.7
Coal, wide cut, Dp 57 μm	7.1	8.0	8.1	7.6	NA
Coal, wide cut, Dp 27 μm	11.5	11.1	10.0	9.5	13.0
Coal, wide cut, Dp 19 μm	NA	19.7	17.4	12.2	NA

TABLE 5: Summary of n values from plots of log. $\frac{\mu}{(\rho s - \rho f)}$ vs log. ε

Fluid Velocity cms^{-1}	0.63	0.79	1.00	1.26
Synclyst, narrow cut, Dp 63 μm	10.0	8.7	8.1	6.4
Coal, narrow cut, Dp 63 μm	NA	8.0	7.5	6.9
Coal, wide cut, Dp 57 μm	NA	NA	5.8	6.7
Coal, wide cut, Dp 27 μm	12.3	11.6	9.3	8.5
Coal, wide cut, Dp 19 μm	NA	32.0	27.0	23.0

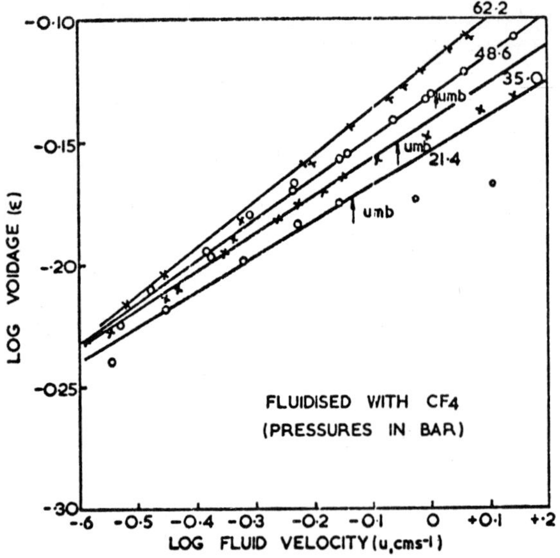

FIG. 2 Synclyst, narrow cut, Dp 63 μm.

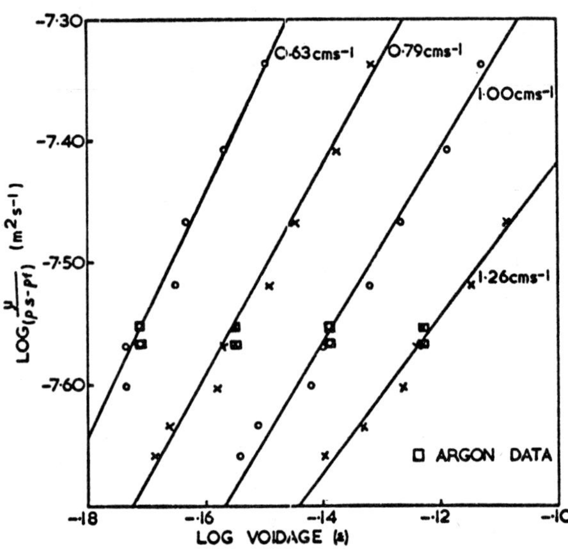

FIG. 5 Synclyst, narrow cut, Dp 63 μm.

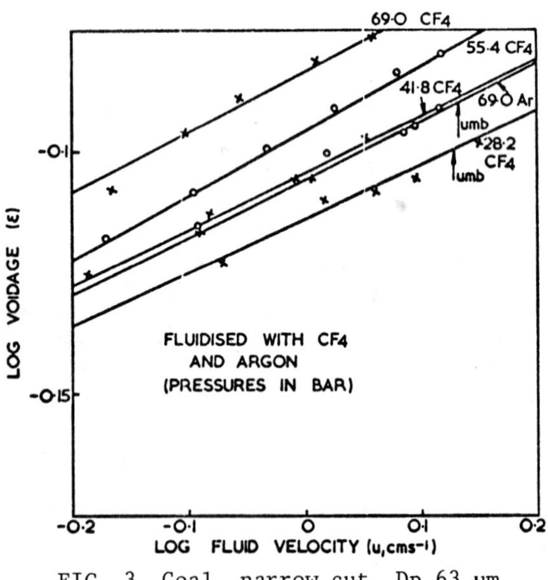

FIG. 3 Coal, narrow cut, Dp 63 μm.

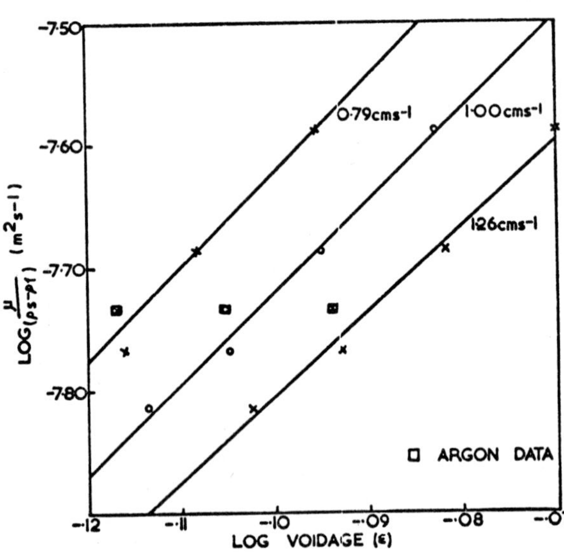

FIG. 6 Coal, narrow cut, Dp 63 μm.

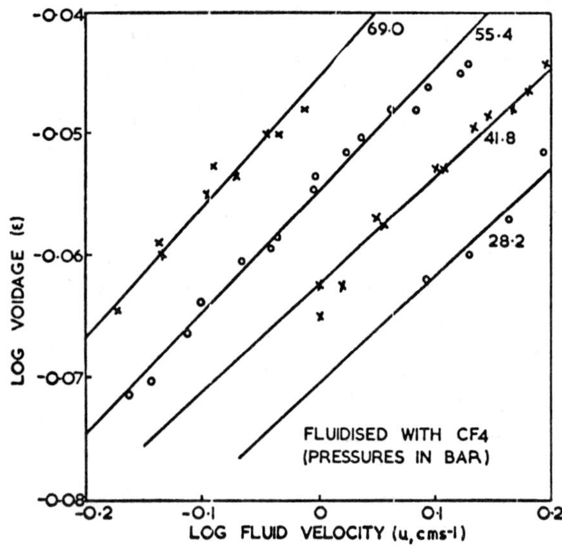

FIG. 4 Coal, wide cut, Dp 27 μm.

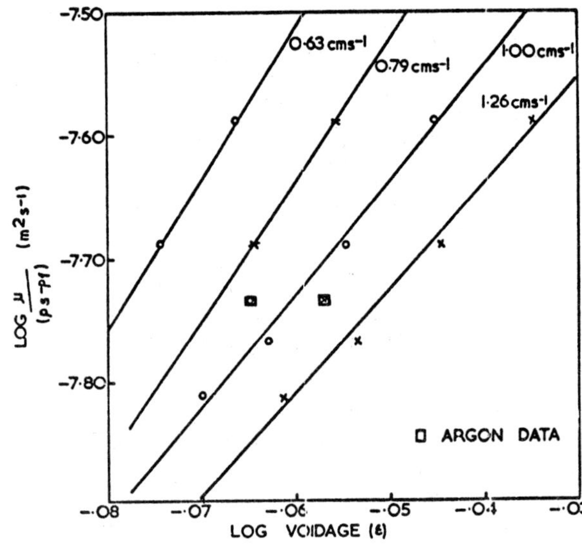

FIG. 7 Coal, wide cut, Dp 27 μm.

MINIMUM FLUIDIZATION AND STARTUP OF A CENTRIFUGAL FLUIDIZED BED

By E. Levy, N. Martin, and J. Chen

Lehigh University; Bethlehem, Pennsylvania 18015, U. S. A.

SYNOPSIS

Experiments were performed with a 0.3 m diameter, 0.15 m high centrifugal fluidized bed operating with atmospheric, room temperature air with glass particles. Experimental and theoretical results are presented for minimum fluidization velocity and bed pressure drop as functions of angular velocity. Experiments on startup of the bed from a stationary slumped position also are described.

NOTATION

d_p particle diameter

$Ga = (\rho_s/\rho_f - 1)\omega_o^2 r_o d_p^3/\nu^2$

P pressure

$Re = u_o d_p/\nu$

r_i inner radius of bed

r_o outer radius of bed

r_o radial velocity at r_o

V_t tangential velocity

γ, β see Fig. 3

ε void fraction

α taper angle, see Fig. 5

ϕ sphericity

ρ_f gas density

ρ_s solid density

μ absolute viscosity

ν kinematic viscosity

ω_o angular velocity of bed

INTRODUCTION

A centrifugal fluidized bed is cylindrical in shape and rotates about its axis of symmetry (Fig. 1). As a consequence of the circular motion, the bed material is forced into the annular region at the circumference of the container; and fluid flows radially inward through the porous surface of the cylindrical distributor, fluidizing the bed material against the centrifugal forces generated by the rotation. Radial accelerations many times one "g" can be generated with modest speeds of rotation, permitting much larger gas flow rates at minimum fluidization than are possible with a conventional fluidized bed operating against the vertical force of gravity. For similar reasons the fluid velocities required to cause entrainment of the bed material also are higher in the case of a centrifugal fluidized bed.

The emphasis in the present investigation is on the use of centrifugal fluidized beds (CFB) for combustion applications, where with bed material of dolomite or limestone to capture SO_2, the centrifugal combustor could be used to burn high sulfur coal or coal char. The rotational nature of the CFB gives it some interesting operational characteristics for the combustion application. The relatively high fluidizing velocities provide for compact combustors, resulting in relatively easy start-up and fewer problems with solids feed and bed mixing. By varying the speed of the rotation of the bed, the bed temperature and the fluidizing velocity, the power output of the device can be varied over an extremely wide range. The concept of centrifugal fluidized coal combustion shows excellent promise for applications where large capacities are needed and/or where relatively small particles must be used (Levy et al., 1976; Shakespeare et al., 1977; Levy & Chen, 1977).

Limited information is available in the literature on centrifugal fluidized beds.

Experiments on minimum fluidization and bed pressure drop were performed at Brookhaven National Laboratory (BNL, 1971 & 1972) and the Air Force Aerospace Research Laboratory (Anderson et al., 1972). Other CFB studies at ARL were made on the heat transfer coefficients between the particles in the bed and the fluidizing gas (Howard, 1975). Work is also underway in Great Britain, where data were reported on bed pressure drop and minimum fluidization and combustion experiments with propane were performed (Metcalfe & Howard, 1977).

The present paper describes experimental and theoretical work at Lehigh University on start-up and minimum fluidization of a centrifugal fluidized bed. This work is part of a program to determine the feasibility of operating a centrifugal fluidized bed in a continuous mode, where analyses and experiments are underway on the effects of parameters such as bed geometry, particle size, angular velocity, and distributor design on the operational characteristics of the bed.

APPARATUS

The apparatus consists of a 0.3 m diameter, 0.15 m high distributor contained in a square plenum. The porous surface of the distributor is fabricated from reinforced fine mesh screen, and the top and bottom end walls are transparent plexiglass for visualization of the flow. A variable speed motor is used to rotate the distributor by means of a shaft connected to the bottom end wall of the test section. Room temperature air is supplied to the plenum from a compressor.

In all experiments, the apparatus was batch operated, where a given charge of bed material was loaded into the test section, and the distributor was accelerated from rest to the desired angular velocity.

BED STARTUP

The first set of experiments, performed with a cylindrical distributor with 500 μm glass beads, indicated serious difficulties in starting up the bed from a slumped position. When stationary, the bed material was at the bottom of the test section. As the distributor was accelerated, the bed material flowed part way up the vertical wall, reaching a limiting height as the angular velocity ω_0 was increased (Fig. 2). As the distributor was decelerated, the bed material stayed on the wall, until a sufficiently low value of angular velocity was reached at which the material slid back down under the influence of gravity.

The experiments described above were run without air flow. When the air was turned on, most of it passed through the bare distributor above the bed, causing localized fluidization and elutriation of the bed material which had climbed the highest along the distributor. The radial air velocities near the bottom of the test section were too low to achieve a fluidized state in that region.

To try to achieve a more uniform distribution of bed material over the grid, experiments were initiated with tapered (conical) distributors oriented with the larger diameter at the top (Fig. 5). Tests with cone half angles from $\alpha = 2°$ to $22°$ indicate no difficulty in achieving flow of solids up the sides of the grid. In experiments with the larger cone angles and with the fluidizing air turned off, the solids moved slowly up the grid as the system was accelerated. When the critical value of angular velocity was reached, the bed material suddenly flowed to the top of the distributor.

The existence of a critical value of angular velocity is analogous to the concept of the angle of slide, θ_s, for bulk solids flow in a normal gravity field. The quantity θ_s is defined (Zenz & Othmer, 1960) as the angle to which the surface supporting the material must be inclined from the horizontal before the solids slide off due to the force of gravity (Fig. 3a).

In Fig. 3a, the angle γ between the sliding surface and the direction of the body force is $\gamma < \pi/2$ and the material tends to slide freely provided $\beta \geq \theta_s$. In a rotating system the orientation of line A-A, which is the normal to the body force vector, depends on angular velocity; and it approaches the vertical as $\omega \to \infty$. With a vertical distributor (Fig. 3b), $\gamma \geq \pi/2$ and the material cannot flow to the top of the test section. In Fig. 3c, where the distributor surface is inclined at angle α to the vertical, $\gamma < \pi/2$ and the material tends to slide freely provided $\beta > \theta_s$. Using tapered distributors with small cone angles ($\beta < \theta_s$), excellent distribution of the bed material can be achieved by use of fluidizing air as shown in Fig. 4. This experiment with $\alpha = 4°$ was started with the system stationary, the air turned off, and the bed slumped. The angular velocity was increased to 24 rad/s and the air flow rate was increased gradually. At point A on the solid curve, the bed began to flow vertically up along the grid, and it was well distributed and fluidized at point B. As the air flow rate was reduced to zero, the bed pressure drop followed the dashed curve, exhibiting a pressure drop — flow rate behavior typical of most fluidized beds.

MINIMUM FLUIDIZATION AND BED PRESSURE DROP

Generally for a given bed geometry and particle density, the important variables are the total weight of the bed material, distributor angle, distributor pressure drop, angular velocity of the bed, and the particle diameter. These variables interact in a complex manner to govern bed start-up, minimum fluidization and particle elutriation. An analytical model was developed which accounts for vertical variations in bed thickness and air velocity (see Fig. 5) by treating the system as a series of n elements, each of height Δz. The total pressure drop across the grid and bed is given for each element as

$$\Delta P_T = \Delta P_{GRID} + \Delta P_{BED} \tag{1}$$

where if the bed is packed locally

$$\Delta P_{BED} = \left\{150(1-\varepsilon)^2/\varepsilon^3(d_p)^2\right\}\mu u_o r_o \ln(r_o/r_i) + \frac{1.75(1-\varepsilon)}{\varepsilon^3\phi d_p}\rho_f(u_o r_o)^2\left[\frac{1}{r_i}-\frac{1}{r_o}\right] \tag{2}$$

If the bed is fluidized locally and the voidage ε is independent of radius, the pressure gradient in the bed is

$$dP/dr = (\rho_s - \rho_f)(1-\varepsilon)\, V_t^2/r \tag{3}$$

Information on the radial variation of tangential velocity is needed in computing the bed pressure drop. Since no measurements of velocity are available, two possible velocity profiles are postulated here. If the tangential velocity in the bed is independent of radius, equation (3) can be integrated to yield

$$\Delta P_{BED} = (\rho_s-\rho_f)(1-\varepsilon)\omega_o^2 r_o^2 \ln(r_o/r_i) \tag{4a}$$

If the bed material is in rigid body motion the expression for bed pressure drop is

$$\Delta P_{BED} = (\rho_s-\rho_f)(1-\varepsilon)\omega_o^2\,[r_o^2-r_i^2]/2 \tag{4b}$$

For thin beds, the radial velocity at minimum fluidization, calculated at the outer radius of the bed, reduces to

$$Ga = [150(1-\varepsilon)/\varepsilon^3\phi^2]Re + [1.75/\phi\varepsilon^3]Re^2 \tag{5}$$

The analytical procedure assumes a value for ΔP_T and uses equation (1) to compute $u_o(r_o)_j$ for each of the n elements. The total mass flow rate through the distributor is

$$\dot{m}(\Delta P_T) = \sum_{j=1}^{n} \rho_f u_{oj} 2\pi r_{oj} \Delta z_j \tag{6}$$

Fig. 6 shows results for the variation of bed pressure drop with air flow rate for a 3.5° distributor with 1.5 kg of 351 μm diameter glass bed material. The dashed curves are computed for three values of ε_{mf} and the circles are experimental data. Points A, B and C correspond to conditions where the bed is totally packed (A), partially fluidized (B) and totally fluidized (C). This is seen more clearly in Fig. 7 where the computed vertical variations of radial velocity are shown for points A, B and C. In this case, the bed is thicker at the top; causing the radial velocities to decrease with height. The bed first becomes fluidized at the bottom and the transition from the packed to the fluidized region progresses upward as the air mass flow increases. The effect of rpm is shown in Fig. 8 where the data are compared to the theoretical values for both the uniform tangential velocity and rigid body cases.

The analysis described above requires information on the shape of the inside surface of the bed. A separate analysis was performed to determine the vertical variation of the inner bed radius $r_i(z)$. Results in Fig. 9 show the effect of charge weight and rpm on bed thickness with a 2° tapered distributor. For the range of rpm considered in the study, very serious nonuniformities in bed thickness can occur, and the effects are particularly severe at the lower charge weights.

In Fig. 10, equation (5) for minimum fluidization is compared to minimum fluidization data from (BNL, 1972) and from Lehigh University experiments. The results show excellent agreement over a wide range of operating conditions. In Fig. 11 the bed pressure drop criterion given by equation (4a) is compared to data from (BNL, 1972). Here, too, the agreement between theory (solid lines) and experiment is good.

SUMMARY

Centrifugal fluidization is a relatively new concept which shows good promise for a variety of applications. The experimental data indicate the system has gross fluidization characteristics which are similar qualitatively to those of conventional beds operating vertically against gravity. The theoretical expressions for bed pressure drop and minimum fluidization velocity are in good agreement with available data. The experimental program is still in progress, with data being gathered on the effects of particle size, bed thickness, and distributor design on bed characteristics.

ACKNOWLEDGEMENT

This work is supported by the National Science Foundation and the U. S. Energy Research and Development Administration under Contract No. E(49-18)-2516. By acceptance of this article, the publisher and/or recipient acknowledges the U. S. Government's right to retain a nonexclusive, royalty-free license in and to any copyright covering this paper.

REFERENCES

Brookhaven National Laboratory (1971) BNL Report 50361.
Brookhaven National Laboratory (1972) BNL Report 50362.
Anderson, L. et al., (1972) J. Spacecraft, 9, No. 5, p. 311.
Howard, J., (1975) U.S. Air Force Aerospace Research Laboratory Report ARL 75-0115.
Levy, E. et al., (1976) ASME paper 76HT68.
Levy, E. and Chen, J., (1977) Proceedings of the International Powder and Bulk Solids Handling and Processing Conference/ Exhibition, Illinois.
Shakespeare, W. et al., (1977) Proceedings of the Intersociety Energy Conversion Engineering Conference, Washington, D.C.
Metcalf, C. I. and Howard, J. R., (1977) Applied Energy, 3, p. 65.
Zenz, F. and Othmer, D., (1960) Fluidization and Fluid Particle Systems, Reinhold Publishing Corp., N.Y.

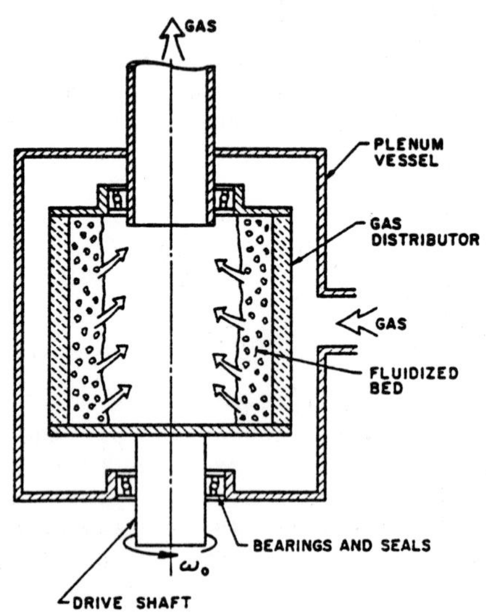

Fig. 1. Sketch of a centrifugal fluidized bed.

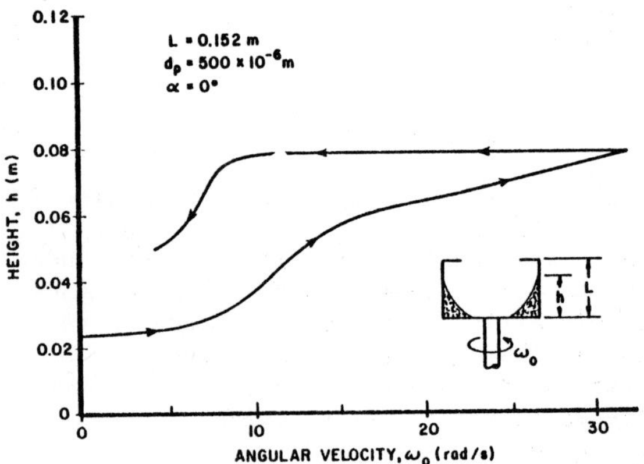

Fig. 2. Height of packed bed in cylindrical distributor.

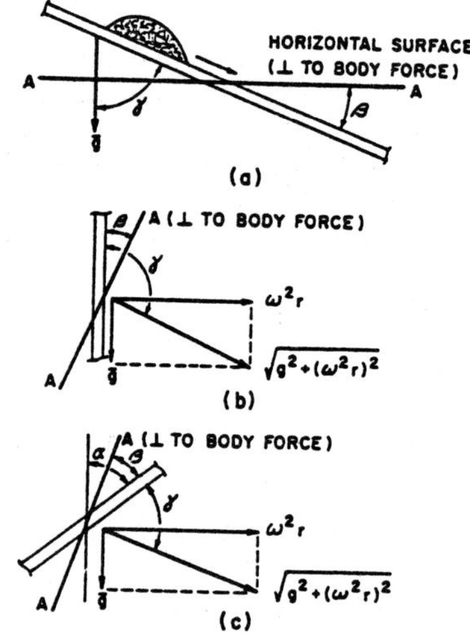

Fig. 3. Effect of grid angle on bed startup: (a) surface inclined to the horizontal - no rotation; (b) vertical distributor in rotating system; (c) tapered distributor in rotating system.

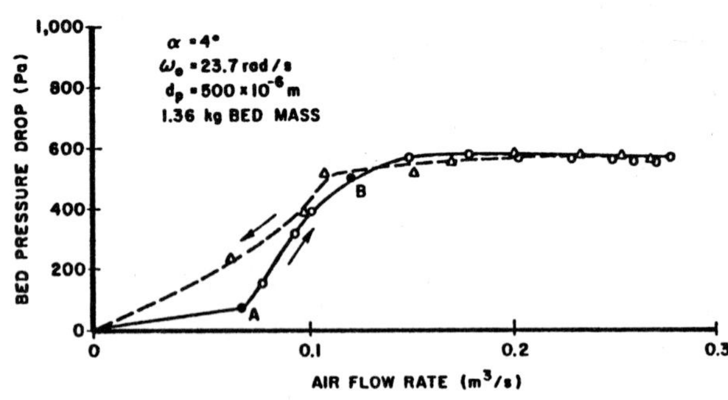

Fig. 4. Bed pressure drop with tapered distributor.

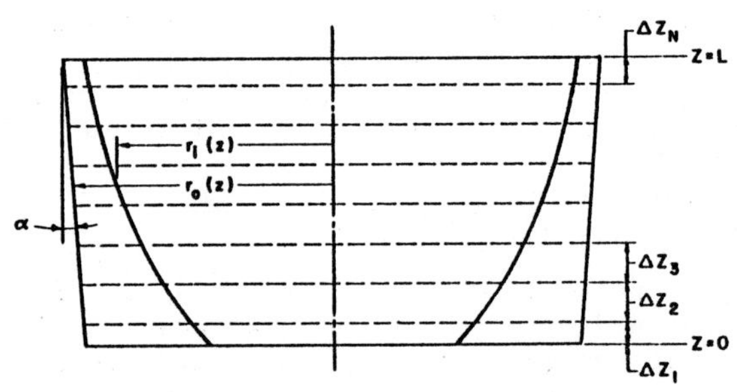

Fig. 5. Model of bed for analysis of minimum fluidization and pressure drop.

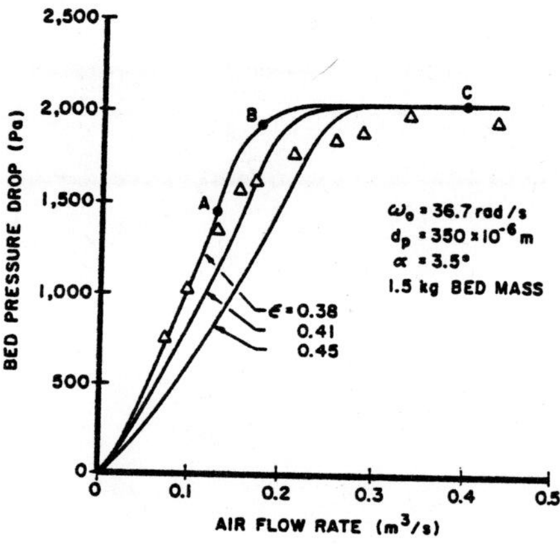

Fig. 6. Variation of bed pressure drop with air flow rate near minimum fluidization.

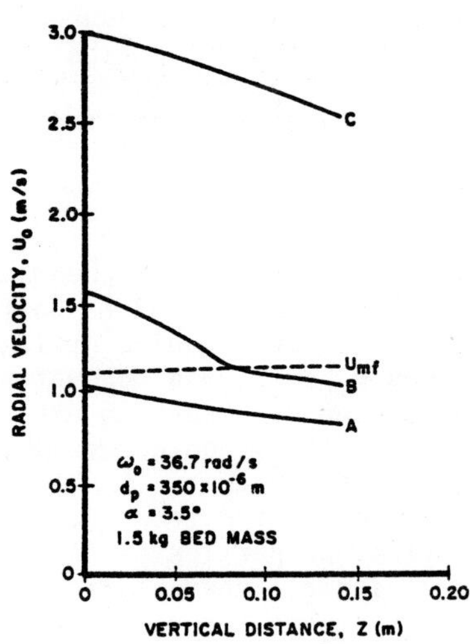

Fig. 7. Axial variation of radial velocity.

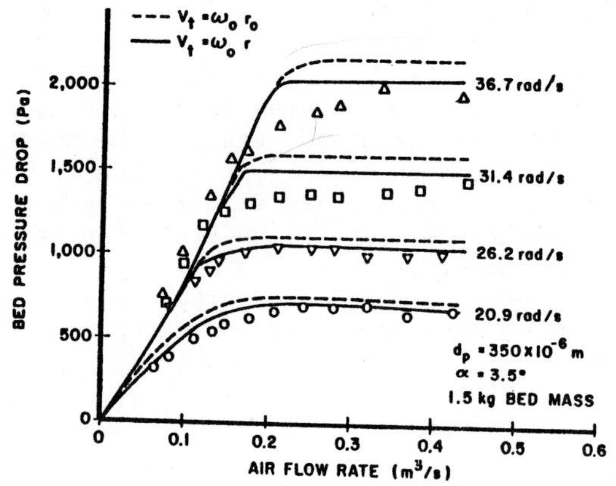

Fig. 8. Effect of angular velocity on minimum fluidization and pressure drop.

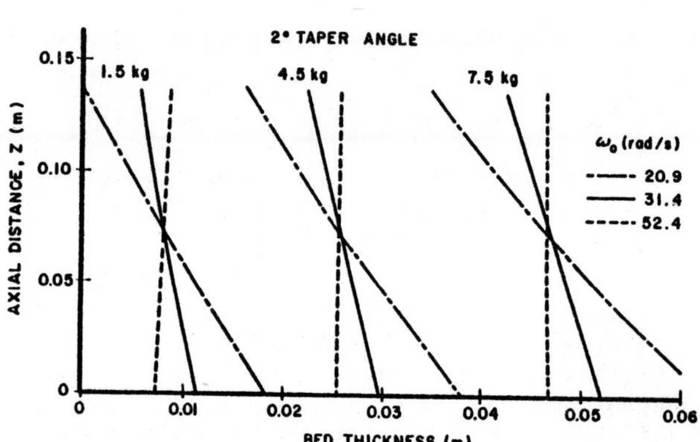

Fig. 9. Computed variations of bed thickness.

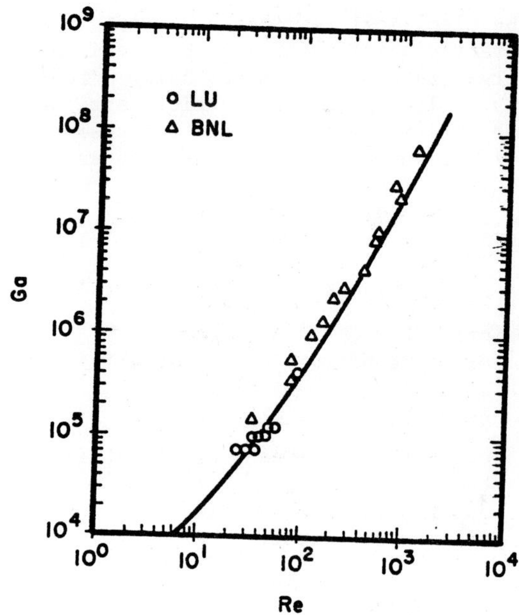

Fig. 10. Minimum fluidization in centrifugal fluidized beds.

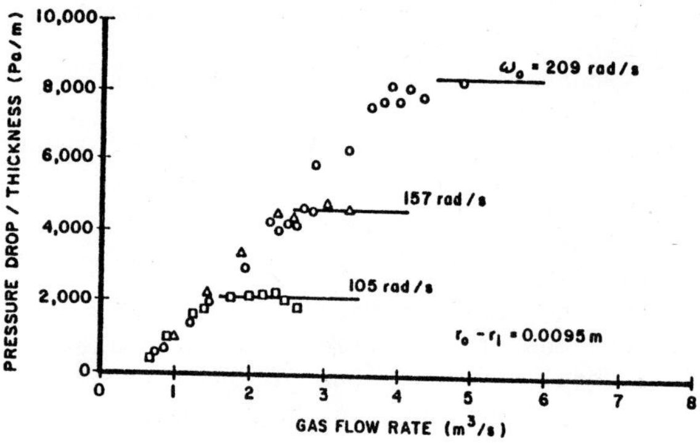

Fig. 11. Bed pressure drop; comparison between data from (BNL, 1972) and theory.

THE ELECTRICAL RESISTANCE OF GAS FLUIDIZED BEDS

By Allen H. Pulsifer and Thomas D. Wheelock

Department of Chemical Engineering and Engineering Research Institute
Iowa State Universtiy
Ames, Iowa 50011, USA

SYNOPSIS

The electrical resistance of fluidized beds of conducting particles is of both theoretical and practical interest. The bed resistance of carbonaceous solids has been measured in several different laboratory systems using a four-point probe technique. It is a function of gas velocity, column diameter, particle size and other properties of the bed material. Also, as temperature is increased, the bed resistance decreases continuously at least in the temperature range between room temperature and 700°C.

NOTATION

D_b	column diameter, cm
I	current flow in bed, amps
L	distance between measuring electrodes, cm
R_b	bed resistance, ohm
S	cross section area for current flow, cm^2
V	superficial gas velocity, cm/sec
V_{mf}	minimum fluidization velocity, cm/sec
V_{fs}	velocity at which bed is fully supported, cm/sec
ΔV	voltage drop between measuring electrodes, volts
ρ	bed resistivity, ohm-cm
ρ_m	minimum fluidized bed resistivity, ohm-cm
ρ_s	bed resistivity with no gas flow, ohm-cm
σ	standard deviation of particle size, microns
Θ	angle of repose of bed material, degree

INTRODUCTION

The electrical properties of fluidized beds of conducting particles are of interest for several reasons. One of these is that conducting fluidized beds may be used as resistance heaters to supply energy for high temperature endothermic reactions which may be carried out conveniently within such beds. Another reason is that a study of the electrical properties of these beds provides some insight into the general phenomenon of fluidization (Goldschmidt and Le Goff, 1967).

Electrofluid reactors which utilize electrodes to supply power to a conducting fluidized bed have been used in both experimental (Beeson, *et al.*, 1974) and commercial reaction (Shine, 1971) systems. Design of these reactors requires an understanding of the basic electrical properties of fluidized beds. The electrical characteristics of such beds can be predicted for simple electrode arrangements where the bed resistivity is constant and known (Knowlton, *et al.*, 1973).

Current flow in fluidized beds appears to be along continuous chains of conducting particles, at least at low voltages (Goldschmidt and Le Goff, 1963). The resistance, and hence the current flow, fluctuate as the particle chains are broken by passing bubbles. However, the time-averaged resistance appears to be constant at steady state. The total interelectrode resistance is the sum of two resistances, the bed resistance and the contact resistance between the fluidized bed and the electrodes.

The bed resistance obeys the well-known resistance equation (Reed and Goldberger, 1966),

$$R_b = \rho \ L/S \tag{1}$$

and therefore it is the behavior of the resistivity which is of most interest. This paper reports data which illustrate the effects of gas velocity, bed material, particle size, column diameter and temperature on bed resistivity. In particular, the results reported extend the work of Jones and Wheelock (1968) on the effect of column diameter to larger size columns and measurements of the effect of temperature on bed resisitivity are presented for the first time.

EXPERIMENTAL METHOD

The bed resistance has been measured by a four-point probe method which isolates the bed resistance from the total interelectrode resistance (Jones and Wheelock, 1968). With this method, the electrodes used to measure the voltage drop across the fluidized bed are separate from the current supplying electrodes. Since there is no current flow in the voltage-measuring electrodes, the voltage drop between these electrodes is unaffected by any contact resistance.

The bed resistance is determined by applying Ohms Law and the bed resistivity by applying equation (1) modified to reflect the geometry. Details of the equipment used to measure resistivity both at room and elevated temperatures are reported elsewhere (Pulsifer and Wheelock, 1975).

During our investigation, the resistivity of beds of graphite, calcined coke and coal char were measured under a variety of conditions. Nitrogen was the fluidizing gas in all cases, with the reactor pressure being atmospheric.

ROOM TEMPERATURE MEASUREMENTS

Shown in Fig. 1 are bed resistivities for two different materials measured at room temperature in a 15.2-cm. diameter column with current flow in a horizontal direction between a 2.5-cm. diameter center electrode and the column wall. As illustrated in this figure, the bed resistivity remains fairly constant as gas flow rate is increased until the minimum fluidization velocity is reached. As the bed becomes fluidized, the bed resistivity increases sharply and reaches a peak value at the point where the first bubbles are formed and the bed becomes fully supported by the gas stream. The resistivity then decreases somewhat with further increases in gas velocity. The sharp increase in bed resistivity is accompanied by an increase in bed height which presumably decreases the number of conducting chains, thus increasing the resistivity. As bubbles form, the void fraction of the dense phase decreases leading to a small decrease in the resistivity.

Fig. 1 also illustrates the effect of bed material on resistivity. Calcined coke gives a higher bed resistivity than graphite, caused at least partly by the higher resistivity of the coke particles. The inherent resistivity of the particles used in the experimental work was unknown, but the resistivity of petroleum coke is reported to be 4-5 times that of graphite. The difference between the resistivties of the graphite and coke beds with no gas flow was about a factor of four. With gas flow in the bed, the resistivity of the coke bed was 4-8 times that of the graphite bed. Since current flow occurs along chains of particles, this indicates that material properties that affect the resistance of particle-to-particle contacts such as hardness and surface roughness contribute somewhat to the observed difference between beds of graphite and calcined coke.

In Fig. 2, bed resistivity-gas velocity data for calcined coke beds in both 10.2-cm. and 15.2-cm. diameter columns are compared to data collected by Jones and Wheelock (1968) in 5.1-cm. and 10.2-cm. diameter columns. All these measurements were made with vertical current flow in the fluidized bed using two needle probes spaced 3.8-cm. apart in the bed for measuring the voltage drop. The curves in Fig. 2 are aimilar in shape to those in Fig. 1, except that the resistivity of the smaller diameter beds increases significantly at higher velocities beyond the trough which follows the peak resistivity. The difference in these curves due to differing column diameter is not surprising since the rise in resistivity beyond the minimum point in the curves seemed to coincide with the onset of slugging in the bed. The fluidized bed resistivity at the minimum point in the curves fits the empirical correlation developed by Jones and Wheelock (1970) for data taken in smaller columns, namely

$$\frac{\rho_m}{\rho_s} = \frac{38.07\ \sigma^{0.21}}{D_b^{\ 0.765}(\tan\ \Theta)^{3.47}} \tag{2}$$

This equation relates the minimum fluidized bed resistivity to the settled bed resistivity and shows that the relation of one to the other depends on particle properties as well as column diameter.

ELEVATED TEMPERATURE MEASUREMENTS

Fluidized bed resistivity is affected by temperature and this effect has been investigated with a 14.0-cm. I.D. Vycor glass fluidization column over the temperature range from room temperature to 700°C. The current flow in the bed was in the vertical direction and the resistivity was determined from the voltage drop over a 2.5-cm. section of the bed. The glass column was externally heated so that the bed temperature could be varied independently of the current flow in the column.

During the study, the effect of temperature on the resistivity of fluidized beds of several carbon containing materials was investigated. The materials included calcined coke, graphite and several coal chars. Except in one case where no gas flow was used, the resistivity was always measured at a relative fluidization velocity of 1.6 and, generally, the effect on the resistivity of heating and cooling the bed several times was determined.

Typical results, in this case for a partially gasified coal char with an ash content of 13.1%, are shown in Fig. 3. The data tend to be somewhat scattered; however, the relationship between the fluidized bed resistivity and temperature is generally linear with the resistivity decreasing as the temperature increases. This same linear trend between bed resistivity and temperature was observed with all of the bed materials, although the resistivity of the graphite and calcined coke beds were significantly less than those of the coal char beds.

When the bed of char was cooled, the bed resistivity increased in an apparent linear manner but along a line of lesser slope than that obtained while heating. Consequently, the resistivity of the bed at room temperature was less than before the heating and cooling cycle was performed. Upon standing for a time at room temperature, the bed material seemed to regain its former level of resistivity so that when another heating and cooling cycle was carried out, the same resistivity versus temperature path was retraced (Fig. 3). Thus the changes in the bed material which took place during the cycle seemed to be reversible. It is not clear whether these changes which affected the resistivity of the fluidized bed were due to adsorption and desorption of permanent gases or to changes in the structure of the char. However, it may be noteworthy that the beds of graphite did not display as much hysteresis as the beds of char over a heating and cooling cycle although the percentage decrease in resistivity for a given increase in temperature was about the same for both types of beds.

The change in resistivity of fluidized beds of carbonaceous materials is probably due in part to changes in the inherent resistivity of the materials. Thus it is well established that the resistivity of coal chars (Waters, 1963) and petroleum coke (Mrozowski, 1952) decreases with rising temperature in the range of our measurements. Also the resistivity of polycrystalline graphite decreases with rising temperature up to about 450°C before the trend reverses and the resistivity increases (Mrozowski, 1952).

Waters (1963) found that the resistivity of coal char carbonized at a temperature near 800°C, about 70°C less than the carbonization temperature of our material, decreased by a factor of 7-8 when it was heated to 700°C. This is somewhat less than the decrease shown in Fig. 3 indicating that additional factors besides changes in material resistivity affect the resistivity of fluidized beds when the temperature is changed. This also is demonstrated by a comparison of the effect of temperature on the resistivity of a fluidized bed of coal char with that of a fixed bed; the resistivity of the fluidized bed declined much more than the resisitvity of the fixed bed between 25°C and 525°C (Fig. 4). It would appear that temperature affects the particle-to-particle contact resistance which must play a greater role in a fluidized bed than in a fixed bed. The reduction in fluidized bed resistivity at higher temperatures may also be due to an increase in arcing between particles which in effect would reduce the contact resistance.

CONCLUSIONS

The electrical resistivity of a bed of conducting particles increases sharply as the bed passes from the fixed state to the fluidized state and it reaches a maximum as the bed becomes fully supported by the gas. At higher velocities the resistivity declines, but if the velocity is increased to the point where slugs develop, the resistivity will rise again. The resistivity is also affected by the properties of the particles, bed dimensions, and temperature. The resistivity of fluidized beds of carbonaceous materials appears to decrease linearly with increasing temperature, at least up to 700°C. Although this effect of temperature is probably due in part to the influence of temperature on the inherent resistivity of the materials, it may also be due in part to the influence of temperature on the particle-to-particle contact resistance and/or arcing between particles.

ACKNOWLEDGEMENT

Work was supported by the Engineering Research Institute, Iowa State University, through funds made available by ERDA, under contract No. E(49-18)-479.

REFERENCES

Beeson, J. L., Pulsifer, A. H., & Wheelock, T. D. (1974) Ind. Eng. Chem., Process Des. Develop. 13, 159.

Goldschmidt, D. & Le Goff, P. (1963) Chem. Eng. Sci. 18, 805.

Ibid (1967) Trans. Instn. Chem. Engrs. 45, 1967.

Jones, A. L. & Wheelock, T. D. (1968) Instn. Chem. Engrs. Symposium Series, No. 30, 174.

Knowlton, T. M., Pulsifer, A. H., & Wheelock, T. D. (1973) AIChE Symposium Series No. 128 69, 94.

Mrozowski, S. (1952) Phys. Rev. 85, 609.

Pulsifer, A. H. & Wheelock, T. D. (1975) Coal Processing by Electrofluidics: Electrical Properties of Electrofluid Reactors. Interim Report No. 3. Published by ERDA, Washington, DC.

Reed, A. K. & Goldberger, W. M. (1966) Chem. Eng. Progress Symposium Series No. 67 62, 71.

Shine, N. B. (1971) Chem. Eng. Progress 67 (no. 2), 52.

Waters, P. L. (1963) Proceedings of the Fifth Conference on Carbon, University Park, PA, Vol. 2, pp. 131-148, Macmillan Co., NY.

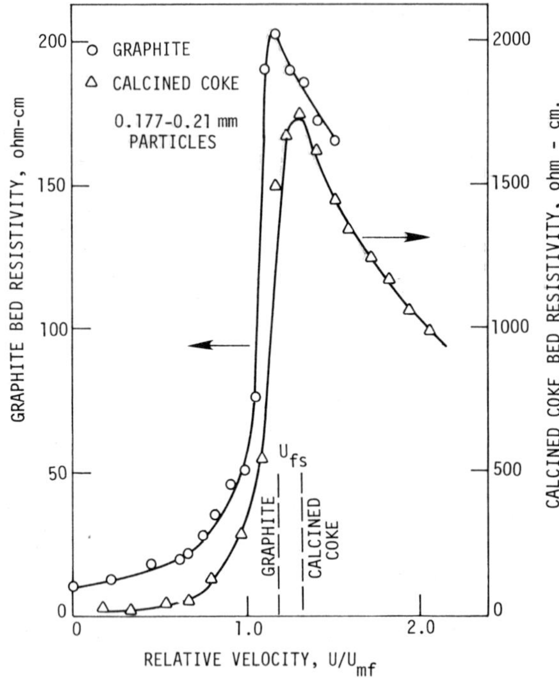

Fig. 1. Effect of relative gas velocity on the resistivity of graphite and calcined coke beds.

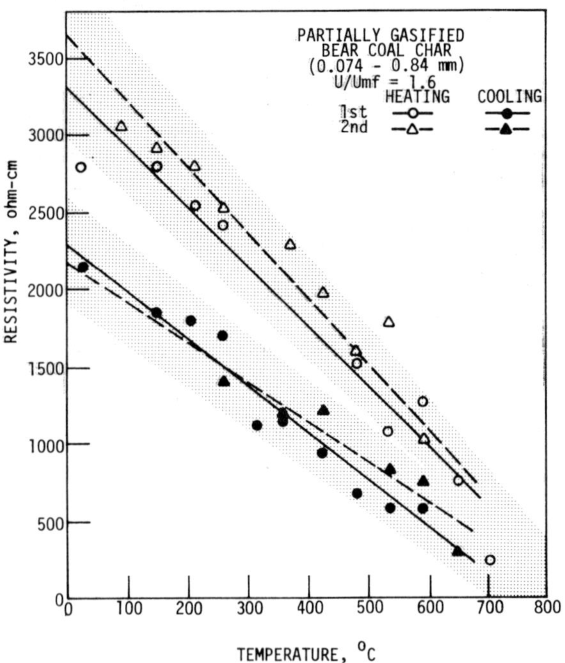

Fig. 3. Effect of temperature on the resistivity of a bed of partially gasified coal char.

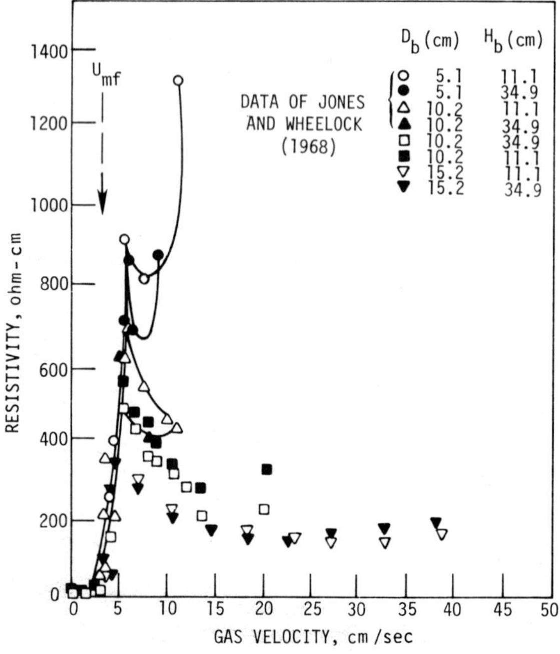

Fig. 2. Resistivity of calcined coke beds fluidized in different diameter columns.

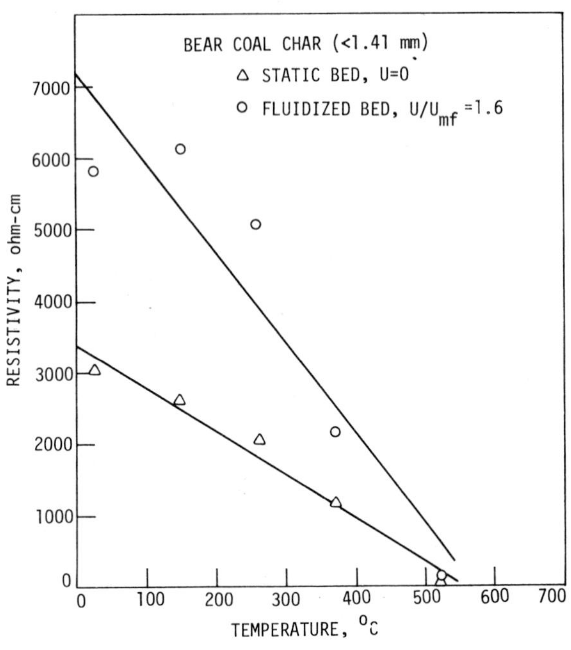

Fig. 4. Comparison of the resistivities of fluidized and static beds of coal char.

THE EFFECT OF GRAVITY UPON THE STABILITY OF A HOMOGENEOUSLY FLUIDIZED BED, INVESTIGATED IN A CENTRIFUGAL FIELD

By K. Rietema and S.M.P. Mutsers

Eindhoven University of Technology

Postbus 513, Eindhoven, Netherlands

SYNOPSIS

Stability of homogeneously fluidized beds is possible if a certain elasticity drives the solid particles back to their rest position upon disturbances introduced from the outside. In the literature it is suggested that such an elasticity might be of hydrodynamic or of mechanic origin. In a so-called human centrifuge to test air pilots fluidization experiments were carried out to determine which theory is correct. In the experiments $g_{apparent}$ was changed a factor 3 and the gas viscosity a factor 2.

The obtained maximum porosity could best be correlated with g^2/μ^2 which points to an elasticity of mechanical origin.

NOTATION

d_p	particle diameter
ρ_d	density of fluidized solids
μ	gas viscosity
E	elasticity of dense phase
g	gravitational acceleration
r	distance from centre
v_o	superficial gas velocity
ε	porosity of fluidized bed
ε_{max}	maximum porosity of homogeneously fluidized bed
ε_d	dense-phase porosity in freely bubbling bed as determined by collapse experiment
ω	circle frequency of centrifuge
Ga_d	Gallilei number for dispersed solids = $\rho_d^2\, d_p^3\, g/\mu^2$
N_F	fluidization number = $\frac{\rho_d^3\, d_p^4\, g^2}{\mu^2 E}$

INTRODUCTION

Since 1963 the stability and the onset of bubbling in gas-solids fluidization has been the issue of many theoretical investigations. The first investigators (Jackson, 1963; Pigford and Baron, 1965; Murray, 1965) all came to the conclusion that in these systems homogeneous fluidization -i.e. expansion of the fluidized bed without bubbles appearing- was not possible. The studies of these investigators were all more or less based on the same principles and on the assumption that solids were free floating so that no momentum transport through the solids phase was possible.

At the first international symposium on fluidization at Eindhoven it was pointed out by Rietema (1967) that this theoretical outcome was not in agreement with the experimental observations which showed that homogeneous fluidization of fine solids was possible. He further suggested that this discrepancy with the theory must be due to the neglection of the interparticle forces which become the more important the finer the solids.

At the same symposium Molerus (1967) came with a different theoretical analysis -but still with the assumption of free floating of solids- according to which homogeneous fluidization would be possible in a certain range of combinative values of the porosity and the Gallilei number (here defined as $Ga_d = \left(\frac{\rho_d^2\, d_p^3\, g}{\mu^2}\right)$. It was shown, however, by Oltrogge (1972) that the analysis of Molerus could not be right since he introduced an exponential perturbation function instead of a harmonic one.

Afterwards Verloop and Heertjes (1970) and Oltrogge (1972) following Wallis (1969) came with a theoretical analysis in which they introduced a certain elasticity of the dense phase which necessitated an additional term in the momentum balance of the solids and which made homogeneous fluidization possible below a certain maximum porosity. The origin of the elasticity -according to these authors-should be found in certain hydrodynamic phenomena in fluidized beds. Because of this reason their analyses again led to a relation between the maximum possible porosity of a homogeneous fluidized bed and the earlier mentioned Gallilei number.

Rietema (1973) pointed out that such a hydrodynamic origin of the elasticity was not well possible at the low Reynolds numbers which generally prevail in homogeneous gas-solids fluidization and he proposed that this origin must be sought in the cohesion forces which are active between the separate particles of the bed. Support for this hypothesis was found in more arguments which indicated the existence of a mechanical structure (however weak it is) of the dense phase.

Rietema and Mutsers (1974) and afterwards Mutsers and Rietema (1977) further elaborated this hypothesis and came to a relation between the maximum possible porosity ε_{max} and a newly defined dimensionless fluidization number

$N_F = \dfrac{\rho_d^3 d_p^4 g^2}{\mu^2 E}$ in which E is the elasticity coefficient. It appears that E depends on the type of solids (especially on their cohesivity) and for the same type of solids is a decreasing function at increasing porosity.

Where Oltrogge's theory predicted a relation between ε_{max} and Ga_d independent on the type of solids the really found strong dependence could now be explained. A direct proof of their theory, however, could not be given.

Such a proof could only be found if both the gravity constant and the gas viscosity could be changed. By changing the gravity constant the most spectacular effect was expected at decreasing gravity constant. Since, however, it is not well possible to decrease the apparent gravity constant at the Earth for a longer time than a few seconds it was decided to seek for a large centrifuge in which the apparent gravity constant could be increased.

Such a centrifuge was found in Soesterberg -Netherlands- at the National Aerospace - Medical Centre where a so-called human centrifuge is available for the testing and training of air-pilots to variation in apparent gravity. It appeared possible to carry out fluidization experiments in this centrifuge.

In these experiments we changed $g_{apparent}$ by a factor 3 and the gas viscosity (using H_2 and N_2 respectively) by a factor 2. Two types of solids were used (spent cracking catalyst and polypropylene). After having measured the fluidization characteristics we tried to correlate ε_{max} with g/μ^2, which would point to a hydrodynamic elasticity, or with g^2/μ^2 which would indicate that the elasticity has a mechanic origin.

EXPERIMENTAL SET UP

Fluidization was carried out in a two-dimensional fluidized bed with glass walls and horizontal dimensions of 25 x 1 cm while the bed height ranged between 40 and 60 cm. The bed was mounted in the cabin of the human centrifuge (see figure 1). A video camera was also installed permitting the fluidization behaviour to be studied quietly from a monitor outside the cabin. All experiments were recorded on tape. A tangential position of the fluid bed with respect to its rotation was chosen so as to reduce the change of undesirable circulation patterns in the bed.

As a result of the rotation the magnitude as well as the direction of the apparent gravity is changed:

$$g_{apparent} = \sqrt{g^2 + \omega^4 r^2}$$

$$\tan\alpha = \omega^2 r/g$$

The change of direction of the apparent gravity inside the cabin was for the greater part compensated by the outward swinging of the cabin. By means of a plummet near the centre of the bed which was observed through the video camera the real direction of $g_{apparent}$ could be measured. Generally it appeared necessary to correct the position of the fluidized bed somewhat so as to let the vertical plane of the bed coincide with the direction of $g_{apparent}$.

The real value of $g_{apparent}$ could be calculated, of course, for each rotation speed. Nevertheless, also a $g_{apparent}$ indicator was installed near the top of the bed. This apparatus consists of a vessel with a volume of about 1 litre filled with air up to an over-pressure Δp, which is connected with a U-tube manometer filled with mercury. The difference in height ΔH between the two mercury-levels is given by

$$\Delta H = \frac{\Delta p}{\rho_{mercury} \cdot g_{apparent}}$$

ΔH again was observed via the video circuit.

The gasflow was regulated and measured by a person sitting close to the shaft of the centrifuge. This person also measured the pressure drop through the bed with a U-tube manometer mounted tangentially to the central shaft. Intercommunication was possible between the man near the shaft and the observers studying the fluidization behaviour from the monitor.

EXPERIMENTAL RESULTS

The experiments were carried out with the following two powders:

- polypropylene ρ_d = 920 kg/m^3
  $\bar{d}_p$ = 40μ

-spent cracking catalyst $\rho_d = 1414\ kg/m^3$

$d_p = 62\ \mu$

These powders have each been fluidized with the following two gases:

-nitrogen $\mu = 1,8 \times 10^{-5}\ N\ sec/m^2$

-hydrogen $\mu = 0,88 \times 10^{-5}\ N\ sec/m^2$

For every gas-solid combination, bed height and pressure drop were measured as functions of the gas flow (both with increasing and decreasing gas flow). At least two collapse experiments were also performed, starting from the freely bubbling bed. All these experiments were repeated for a series of values of the apparent gravity. The main experimental results are presented in tables I and II. In table I the dense-phase porosity in a freely bubbling bed and the bubble-point porosity are to be found for polypropylene (uncorrected columns). The values of the dense phase porosity are always an average of at least two collapse experiments. These experiments being recorded on tape, a very accurate interpretation was possible afterwards. The average values of the packed porosity, which were obtained after the collapse experiments, are also indicated, the porosity values with hydrogen being always lower than those with nitrogen. In table II the corresponding values are given for spent cracking catalyst.

CORRECTION PROCEDURE

For an accurate interpretation of the experimental results the variation of bed porosity with bed height should be taken into account. This variation is caused by the following two effects:

-the gas pressure increases downwards in the bed, and therefore the superficial gas velocity and the porosity decrease (gas viscosity remains constant)

-the apparent gravity increases downwards, so that the porosity decreases downwards even more.

All the porosity values mentioned earlier were averages over the bed height. The correct porosity values, however, refer to the powder surface (if the gas velocity is increased, the bubble-point is reached at first near the powder surface, and only at higher gas velocities also lower in the bed). The porosity values averaged over the bed height have, therefore, to be converted into the values at the powder surface. Because the observed values of the apparent gravity also relate to the powder surface, no additional corrections are necessary for this parameter.

Carrying out the correction procedure, the following simplifying assumptions were made:

-the relative variation in the apparent gravity is small, and this variation is linear with respect to height. Because the bed height is much smaller than the radius of the centrifuge, this assumption is allowable.

-the relative variations of the solids hold-up and gas pressure with height are small. This means that both gas pressure and gas velocity vary approximately linearly with height. The final results of the correction procedure will confirm the validity of these assumptions (total variation of the porosity always smaller than 0.03 and the relative variation in the gas pressure always smaller than 0.11).

If the wall friction is neglected, according to Ergun the following equation holds:

$$\frac{1-\varepsilon}{\varepsilon^3} = \frac{\rho_d\, d_p^2}{150\mu} \times \frac{g_{apparent}}{v_o}$$

Because of the assumptions, mentioned above, the left-hand side of this equation, as well as the porosity itself, vary approximately linearly with height. The total difference in porosity $\Delta\varepsilon$ between the top and the bottom of the bed can easily be calculated (the variations in the apparent gravity and the superficial gas velocity are known). The porosity at the powder surface is then obtained simply by adding $\frac{1}{2}\Delta\varepsilon$ to the average porosities measured.

Although, strictly speaking, this correction procedure is only appropriate for the bubble-point porosities, it was also applied to the dense phase porosities. The corrected porosity values are again presented in tables I and II. In figures 2, 3, 5 and 6 these corrected porosity values are plotted versus $(g_{apparent}/\mu)^2$, while the bubble point porosity ε_{max} is plotted versus $g_{apparent}/\mu^2$ (see figures 4 and 7).

DISCUSSION

From the figures 2 - 4 it is clear that in the case of polypropylene correlation of both ε_{max} and ε_d with g^2/μ^2 leads to very satisfactory results while correlation with g/μ^2 leads to two different lines, one for H_2 and one for N_2.

In figure 2 also a theoretical line is drawn which is calculated from

$$E.N_F = \frac{\rho_d^3\, d_p^4\, g^2}{\mu^2} = E\ \frac{150\ (1-\varepsilon_{max})^2}{\varepsilon_{max}^2\ (3-2\ \varepsilon_{max})}$$

while the relation between E and the porosity is taken from the experimentally determined relation at normal gravitation (see figure 8). The correspondence with this theoretical line is remarkably good. From figure 3 it is clear that also ε_d correlates well with g^2/μ^2 which gives support to the hypothesis that also ε_d is determined by the dynamic equilibrium between constantly introduced hydrodynamic disturbances and the mechanical elasticity of the dense phase structure. ε_d always being smaller than ε_{max} must be caused by the fact that in a freely bubbling bed the disturbances (bubbles) are no longer infinitely small so that the linearization which is applied in the perturbation theory is no longer allowed.

When considering the results obtained with spent cracking catalyst it is striking that again ε_d correlates very well with g^2/μ^2 (figure 6) while the correlation of ε_{max} with g^2/μ^2 is not so good. On the other hand the correlation of ε_{max} with g/μ^2 is even worse (fig.7). Also in this case the theoretical correlation which we would expect on basis of results at normal gravitation (figure 8) is shown and the data of ε_{max} appear to spread along this line which gives us reason to believe that this spread might be caused by the fact that the onset of bubbling in fluidized spent catalyst is more difficult to observe because of two reasons:

a. this powder is black
b. electrostatic phenomena occur between this powder and the bed walls. These phenomena were not observed when fluidizing polypropylene.

We would conclude therefore, that:

1. both ε_{max} and ε_d correlate well with g^2/μ^2
2. the elasticity of the dense phase structure is of mechanical origin
3. the fluidization number N_F is indeed the critical number which predicts both the maximum attainable porosity ε_{max} in homogeneous fluidization and the porosity ε_d of the dense phase in a freely bubbling bed. It might be expected that also for the reological properties of gas fluidized systems this fluidization number will be an important parameter.

If the above mentioned conclusions would be true there is reason to wonder what whould be the effect of temperature and pressure at which fluidization is carried out, the more so since a few experimental data on these effects have recently been reported in the literature:
On the effect of temperature by Yoshida, Fujii and Kunii (1975) and on the effect of pressure by de Carvalho and Harrison (1975).

As to the effect of pressure:
Since the gas density even at pressures up to 20 bar is still neglectable to the solids density it hardly can be of effect on the fluidization behaviour; since, furthermore, the gas viscosity (at temperatures far above the critical temperature) is independent on pressure it is believed that pressure -at constant properties of the solids- should have no effect on fluidization behaviour.

In seeming contradiction to this believe are the observations of de Carvalho and Harrison who found that when fluidizing synclyst catalyst the bed expansion at the same superficial velocity increases remarkably at increasing pressure. On the other hand these autors did find no increase in bed-expansion at increasing pressure when fluidizing glass ballotini.

We therefore believe that this contradiction can be explained very well by the phenomenon of gas adsorption which with the synclyst catalyst must be much stronger than with the glass ballotini and which, furthermore, increases with increasing pressure. As a consequence of this gas adsorption the cohesion between the particles will have increased, hence also the corresponding mechanical elasticity of the dense phase structure and as a consequence of this higher elasticity the dense phase will be more stable at the same porosity, the dense phase expansion will increase while at the same time the reological properties will have improved, the rising bubbles will have decreased in size.

The effect of temperature can be explained two fold: firstly by the increase of gas viscosity and secondly again by the increase of cohesion and hence of the elasticity. Both effects increase the dense phase expansion and improve the reological properties.

In the near future we hope to investigate the effects of temperature and pressure ourselves.

REFERENCES

de Carvalho, J.F.R.Guedes., Harrison, D.(1975) Fluidized Combustion Conference, London 16-17 Sept., Session B, p. B 1.1 - 1.5

Jackson, R. (1963) Trans.Int.Chem.Engrs., 41, 13

Molerus, O. (1967) Proceedings Int.Symp. on Fluidization, Eindhoven, p. 134 - 142

Murray, J.D. (1965) J.Fluid Mechanics, 21, 465

Mutsers, S.M.P., Rietema, K. (1977) Powder Technology, to be published

Oltrogge, R.D. (1972) University of Michigan, Ph.D.thesis

Pigford, R.L., Baron, T. (1965) Ind.Eng. Fundamentals, 41, 81

Rietema, K. (1967) Proceedings Int.Symp. on Fluidization, Eindhoven, p. 154 - 166

Rietema, K. (1973) Chem.Eng.Sci., 28, 1493

Rietema, K., Mutsers, S.M.P. (1973) Proceedings Int.Symp. on Fluidization and its Applications, Toulouse, p. 28 - 40

Verloop, J., Heertjes, P.M. (1970) Chem.Eng. Sci., 25, 825

Wallis, G.B. (1969) One dimensional two-phase flow, McGraw-Hill Book Co.

Yoshida, K., Fujii, S., Kunii, D. (1975) Int. Fluidization Conference, Asilomar, California, Session A, p. 43 - 48

Table 1: Experimental results with polypropylene

		packed bed		porosity at bubble point		dense phase porosity		
gas	g-factor	bed height cm	porosity	uncorrected values	ε_{max} calculated	uncorrected values	ε_d calculated	v_o max cm/sec
N_2	1.00	48.9	0.658	0.717	0.718	0.709	0.710	0.70
	1.39	47.5	0.648	0.703	0.708	0.692	0.697	0.84
	1.60	47.0	0.644	0.698	0.705	0.684	0.691	0.91
	2.03	46.3	0.638	0.689	0.698	0.674	0.683	1.06
	2.50	45.2	0.630	0.672	0.682	0.633	0.673	1.23
	3.07	44.7	0.626	0.668	0.679	0.660	0.669	1.40
H_2	1.00	47.1	0.645	0.699	0.701	0.683	0.684	1.27
	1.39	46.1	0.637	0.677	0.682	0.665	0.670	1.36
	1.60	45.6	0.633	0.669	0.676	0.658	0.665	1.43
	2.03	45.1	0.629	0.667	0.675	-	-	1.67
	2.50	44.7	0.626	0.659	0.669	-	-	1.84
	3.07	44.3	0.623	0.655	0.666	0.633	0.642	2.08

Table II: Experimental results with spent cracking catalyst

		packed bed		porosity at bubble point		dense phase porosity		
gas	g-factor	bed height cm	porosity	uncorrected values	ε_{max} calculated	uncorrected values	ε_d calculated	v_o max cm/sec
N_2	1.00	50.8	0.431	0.544	0.546	0.520	0.522	0.75
	1.18	49.9	0.421	0.523	0.528	0.502	0.507	0.72
	1.47	49.6	0.418	0.493	0.501	0.482	0.490	0.71
	1.74	49.5	0.417	0.493	0.504	0.476	0.486	0.85
	2.10	49.1	0.412	0.475	0.487	0.465	0.477	0.83
	2.78	48.7	0.407	0.451	0.465	0.448	0.463	0.93
H_2	1.00	49.9	0.421	0.512	0.515	0.472	0.475	1.15
	1.18	49.5	0.417	0.503	0.508	0.467	0.472	1.24
	1.47	49.1	0.412	0.477	0.484	0.453	0.461	1.21
	1.74	48.9	0.409	0.473	0.483	0.447	0.457	1.39
	2.10	48.7	0.407	0.457	0.469	-	-	1.39
	2.78	48.4	0.403	0.442	0.456	0.422	0.437	1.63

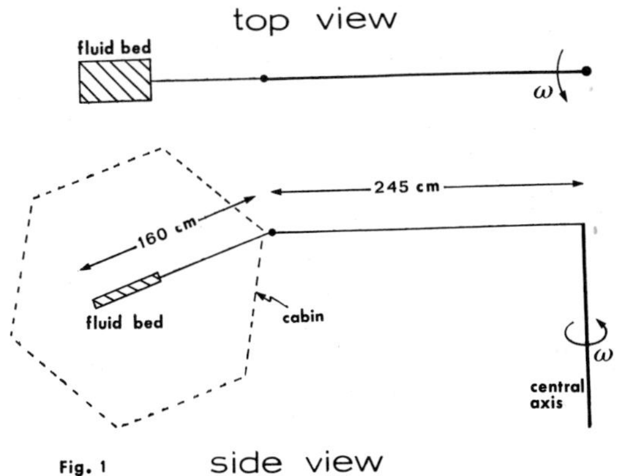

Fig. 1

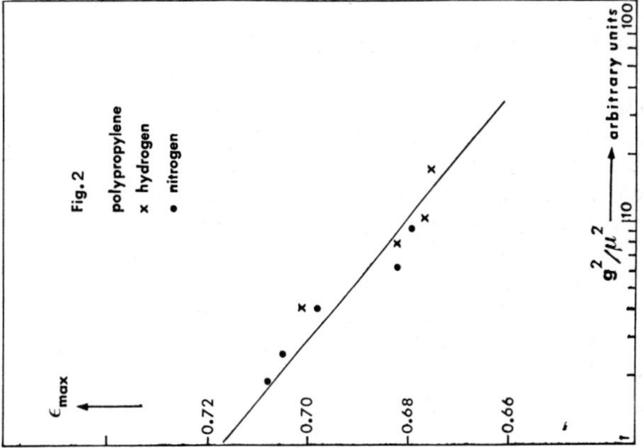

Fig. 2

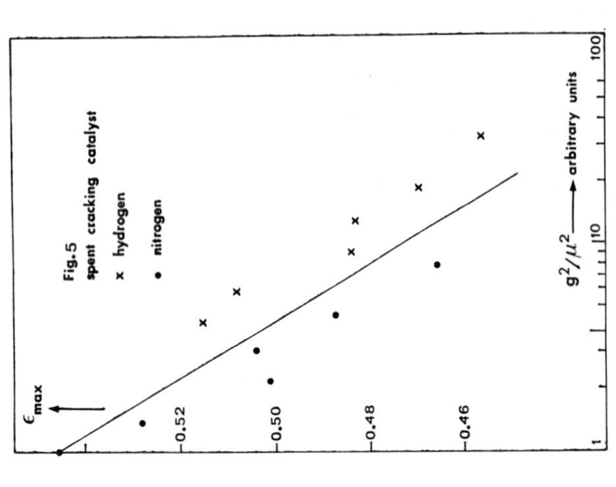

Fig.5

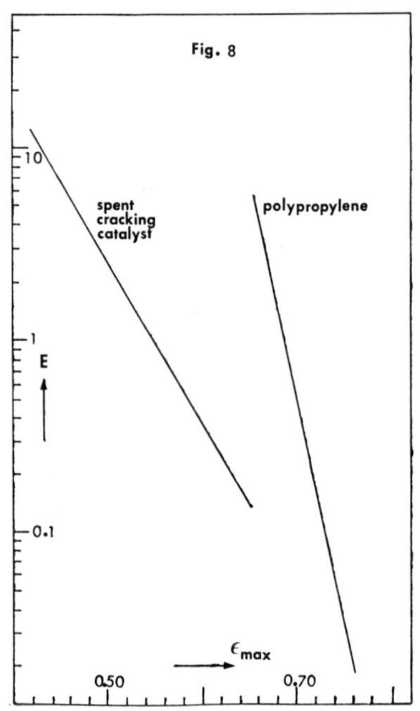

Fig. 8

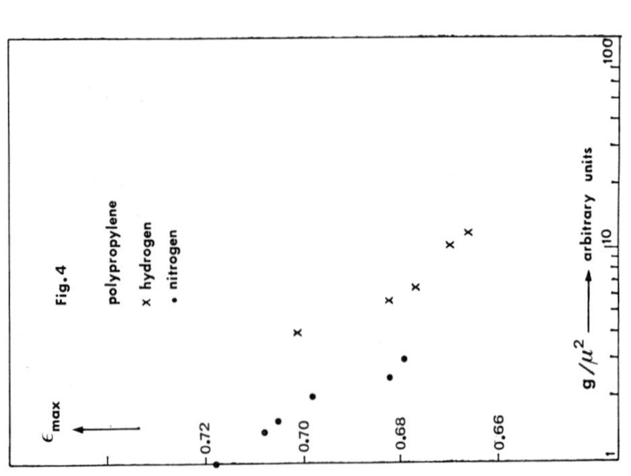

Fig.4

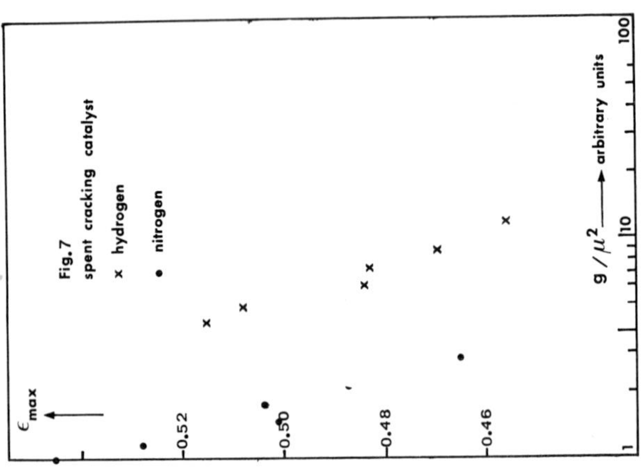

Fig.7

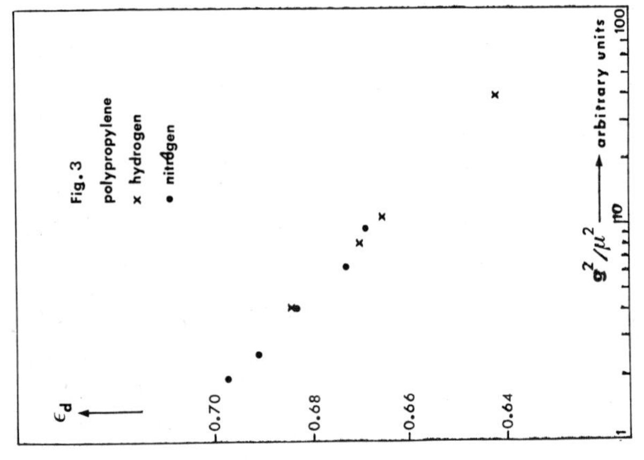

Fig.3

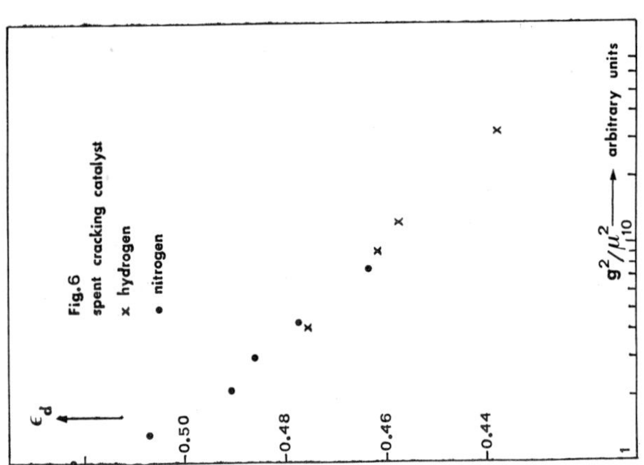

Fig.6

INCIPIENT FLUIDIZATION OF COHESIVE POWDERS BY HORIZONTAL ROTATING NOZZLE

By J.Novosad and E.Kostelková

Institute of Chemical Process Fundamentals

Czechoslovak Academy of Sciences

165 02 Prague 6, Suchdol, Czechoslovakia

SYNOPSIS

A gas distributor for fluidizing cohesive powders in the size range below 0.1 mm and containing larger amounts of fines consisting of a horizontal nozzle rotating around a vertical axis near the vessel s bottom has been developed. Minimum fluidization velocities of cohesive and cohesionless silicon powders were determined in apparatus of 90 mm diameter and effects of speed of rotation, diameter of nozzle opening and cohesion of material were investigated. The minimum fluidization velocity of cohesive materials is larger than calculated velocity of cohesionless material due to the cohesive forces acting in the bed. Experimental results were treated to give a graphical correlation of effects of various parameters on the minimum fluidization velocity.

NOTATION

- C consolidating stress, N/m^2
- d_n diameter of nozzle opening, m
- d_r diameter of rotation of nozzle, m
- f function
- K constant = 1 $N^{0.5}/m$
- n rotations per minute, r.p.m.
- R ratio U_t/U_o
- T tensile stress, N/m^2
- t slope of tensile stress plot, $N^{0.5}/m$
- U dimensionless minimum fluidization velocity = u_{mf}/u_{mfc}
- U_t U for cohesive material
- U_o U for cohesionless material
- u superficial gas velocity, m/sec
- u_{mf} minimum fluidization velocity with rotating nozzle, m/sec
- u_{mfc} calculated minimum fluidization velocity for cohesionless materials, m/sec

INTRODUCTION

Equipment for fluidizing cohesionless particulate solids is widespread in many branches of industry and also the theory treating this process is advanced. If the fluidized material contains larger quantities of fines (particles below about 50 micrometers), then the particulate solid becomes cohesive and many types of classical fluidized bed equipment fail due to the formation of static channels through which the gas flows without coming into contact with the solid. Various equipment has been developed to fluidize cohesive powders as is described e.g. by Gelperin et al. (1967). In many cases this equipment is rather complicated and successful fluidization is not always assured.

Needless to say a simple and reliable equipment for the fluidization of fine (cohesive) powders could furnish considerable advantages to the fluidized bed process as e.g.: possibility of operation at lower gas velocities enabling increased yields in chemical reactions; greater surface area (of importance in catalyst contacting); reduced comminution of particles since with the reduction of particle size there exists a grinding limit when further comminution does not take place; etc. Obviously some disadvantages would be encountered as e.g. increased ellutriation but this disadvantage would probably be counterbalanced by the advantages.

In this paper the minimum fluidization velocity of a cohesive powder has been investigated in a tube of 90 mm diameter using a horizontal rotating nozzle gas distributor which has been developed by Novosad et al. (1977). The minimum fluidization velocity has been selected for study since it could give an indication of the cohesive forces acting in a fluidized bed.

EXPERIMENTAL

The rotating nozzle gas distributor used in the investigation is shown on Fig. 1. The pulley 1 was driven by a DC electromotor allowing to change nozzle velocities in the range 15 - - 560 r.p.m. Dried air was introduced into inlet 2 after throttling by valve and measuring flowrate and pressure. The pressure drop was measured by connecting a U tube to the pressure chambers 5 and 10.

Minimum fluidization velocities were determined from pressure drop vs. superficial gas velocity graphs. A typical grapf is shown on Fig. 2. Below the minimum fluidization velocity as read off on Fig. 2 stagnant zones appear in the bed and after further decreasing gas velocity stationary channels begin to form. The plot on the graph in the vicinity of the minimum fluidization velocity is curved and below the minimum fluidization velocity there is a considerable scatter in experimental points making the determination of the intersection point rather difficult. Also reproducibility was rather bad, being in the region of 10 - 20 %. However, at the given stage of the investigation it was not possible to go into more accurate measurement techniques.

Most of the measurements were performed with a fine silicon powder designated F, with particle size distribution shown on Fig. 3. The diameter of the mean statistical particle was 0.055 mm. In order to evaluate the effect of cohesion three other samples of the silicon powder were prepared by sieving, designated A, B and C with particle size distributions also shown on Fig. 3. The diameters of the mean statistical particles were 0.3 mm, 0.18 mm and 0.065 mm respectively.

The tensile stress of the powders was considered to be a measure of their cohesion. It was measured on a tensile tester described by Novosad et al. (1977) which in principle has the same function as the Warren Spring Laboratory tensile tester described by Ashton et al. (1964). The tensile stress is, however, dependent on the consolidating stress used in sample preparation. Therefore a relationship between tensile stress and consolidating stress was sought which would have a linear relationship. It was found that the dependency between tensile stress and square root of consolidating stress was linear for low consolidating stresses (Fig. 4) and so the slope of the tensile stress plot was considered to be a measure of the cohesion of the powder. The tensile stresses of samples A and B were so small that they were not measurable on the tester used. Thus the tensile slope t of sample F = $= 1.4\ N^{0.5}/m$, of sample C = $0.15\ N^{0.5}/m$ and of sample A and B = 0.

In the present paper the effect on the minimum fluidization velocity u_{mf} of the following parameters was investigated: height of bed, speed of rotation, diameter of nozzle opening d_n and tensile slope t.

Preliminary experiments showed that u_{mf} increased with height of bed up to a height of about 50 cm (measured in non fluidized state) and then it again decreased. This effect was not too pronounced and therefore all measurements in this investigation were performed at a height corresponding to 50 cm at rest.

A dimensionless expression for the minimum fluidization velocity U was obtained by dividing the minimum fluidization velocity with rotating nozzle gas distributor u_{mf} by a calculated minimum fluidization velocity of hypothetically cohesionless material of same particle size, designated u_{mfc}. Thus $U = u_{mf}/u_{mfc}$. For the calculation of u_{mfc} the equation of Ergun (1952) quoted by Richardson (1971) was selected using a voidage value of 0.4. The calculated values of u_{mfc} of the individual samples were A = 0.082 m/sec, B = 0.032 m/sec, C = 0.004 m/sec and F = 0.0025 m/sec.

Values of u_{mf} of samples F, A, B and C were measured for nozzle diameters d_n in the range 1.5 - 3 mm. It was found that with nozzle diameter, u_{mf} was approximately proportional to $(d_n)^{4/3}$.

The speed of rotation had a considerable influence upon U. To eliminate the effect of nozzle diameter the experimental points were plotted in coordinates $U/((d_n)^{4/3})$ and n (Fig. 5). The speed of rotation was varied from 20 to 560 r.p.m. For speeds smaller then 300 r.p.m. the points of sample F can be approximated by the thick line marked F.

In determining the effect of cohesion a function of the slope of tensile stress f(t) was sought by which it would be possible to divide the vertical coordinate in Fig. 5, i.e. $U/((d_n)^{4/3})$, in order to obtain a common plot for cohesive and cohesionless material. It was considered that for cohesionless materials when t = 0, f(t) should be = 1. Therefore an additive argument was selected and from

Fig. 5 it was found that $U/((d_n)^{4/3})$ was linearly proportional to K + 10t at n = 300 r.p.m. where $K = 1\ N^{0.5}/m$.

Thus a final graph can be drawn in coordinates $U/((d_n)^{4/3}.(K + 10t))$ and n (Fig. 6) in which the points of samples A, B and C are included whilst the points of sample F are represented by the line F.

DISCUSSION

Figs 5 and 6 can give some insight into the mechanism of fluidization of cohesive powders by a rotating horizontal nozzle gas distributor. It can be seen that in the fluidized state there is a very strong influence of the cohesion of the powder and that a fluidized cohesive powder will behave quite differently than a cohesionless material.

The minimum fluidization velocity of sample F decreased with speed of rotation up to about 300 r.p.m. and then u_{mf} again started increasing. A possible explanation is that at certain speeds of rotation the jet of air emerging from the nozzle creates an annular cavity along its path and thus the effect of the rotating jet is lost because for its correct function the jet must penetrate into the powder and not into a cavity. This effect will probably not be noticeable in larger diameters.

It is interesting that the rotating nozzle is capable of fluidizing also cohesionless material and minimum fluidization velocities less than calculated can be gained. An explanation could be that the mixing effect of the connecting tube of nozzle together with the jet have an effect like a vibrated bed and also the rotating motion of the nozzle causes local pulsations in the bed.

With decreasing speeds of rotation u_{mf} of the cohesive powder increases, whilst u_{mf} of cohesionless material is fairly constant. This would indicate that the fluidized cohesive powder rapidly gains strength when not disturbed. This effect is very marked at about 20 r.p.m. from which it can be assumed that the gain in strength starts after a period of about 3 - 4 sec without a disturbance by a bubble.

If U for cohesive powders is designated U_t and for cohesionless material U_o then the ratio $R = U_t/U_o$ could give an insight into the magnitude of cohesive forces acting in the material. For sample F, R would be about 25 for 20 r.p.m. and 15 for about 300 r.p.m.

CONCLUSIONS

1. The rotating horizontal nozzle gas distributor can be used to fluidize cohesive powders when the cohesion is caused by the presence of larger quantities of fines. It can also be used to fluidize cohesionless material.
2. The cohesive properties of the material can be characterized by a parameter (K + 10t) where t is the slope of the plot of the tensile strength of the powder versus the square root of the consolidating stress and $K = 1\ N^{0.5}/m$.
3. When fluidizing cohesive powders with the rotating nozzle gas distributor the minimum fluidization velocity is greater than the calcula-

ted minimum fluidization velocity of same particle size. This is due to the cohesive forces acting in the bed and the ratio of the dimensionless minimum fluidization velocity of cohesive material U_t to the dimensionless minimum fluidization velocity of cohesionless material U_o (for same d_r, d_n and n) could be a measure of these forces.

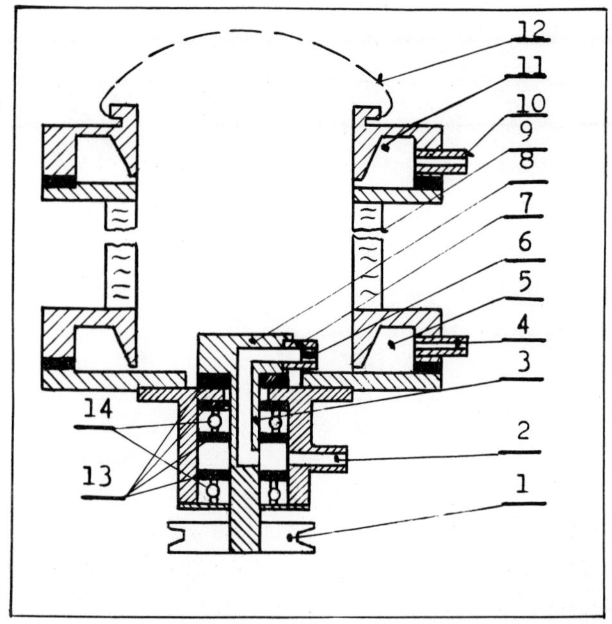

Fig. 1. Experimental apparatus: 1 pulley, 2 gas inlet, 3 gas distributor, 4 pressure withdrawl, 5 pressure chamber, 6 nozzle, 7 connecting tube, 8 head of gas distributor, 9 fluidizing tube, 10 pressure withdrawl, 11 pressure chamber, 12 filter, 13 seals, 14 ball bearings

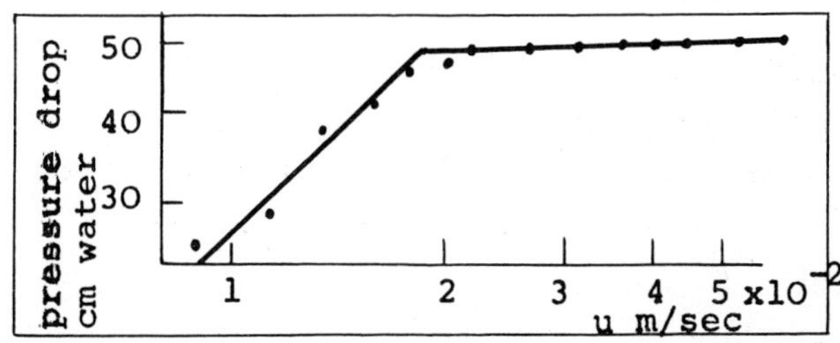

Fig. 2. Determination of minimum fluidization velocity

REFERENCES

Ashton, M.D., Farley, R., Valentin, F.H.H. (1964) J. Scientific Instruments, 41, 763.

Ergun, S. (1952) Chem. Eng. Progr., 48, 89.

Gelperin, N.I., Ainshtein, V.G., Kvasha, V.B. (1967) Fundamentals of Fluidization (Russian), Khimia, Moscow.

Novosad, J., Majzlík, R., Pavlištík, J. (1977) Chemický průmysl, 27, 171.

Novosad, J., Šmíd, V., Šmíd, J. (1977) Preprints of the European Congress "Transfer Processes in Particle Systems" pp. E2 - Ell, Nuremberk.

Richardson, J.F. (1971) Fluidization, Chapt. 2, (Davidson, J.F., Harrison, D. Eds.), Academic Press, New York.

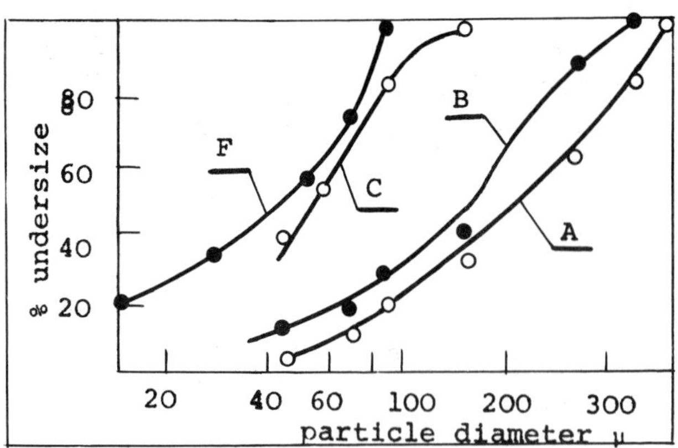

Fig. 3. Particle size distribution of samples

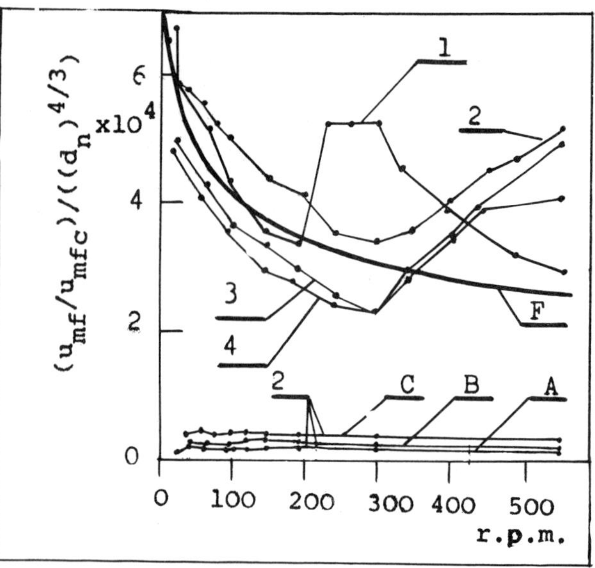

Fig. 5. Experimental results:
d_r = 0.066 m, 1 d_n = 0.0015 m,
2 d_n = 0.002 m, 3 d_n = 0.0025 m,
4 d_n = 0.003 m

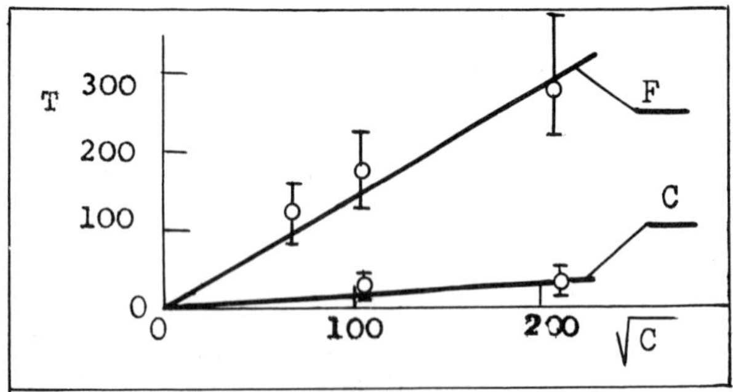

Fig. 4. Tensile strength of samples

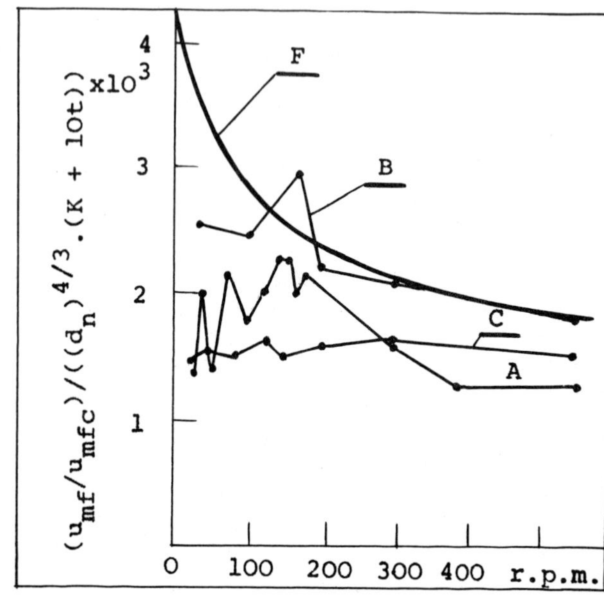

Fig. 6. Final correlation

MEASURING PARTICLE TEMPERATURE AND EMISSIVITY IN A HIGH TEMPERATURE FLUIDIZED BED

By K.E. Makhorin, V.S. Pikashov and G.P.Kuchin

Institute of Gas of the Academy of Sciences
of the Ukrainian SSR, Parkhomenko Street,
Kiev, USSR

SYNOPSIS

Experimental studies of the temperature dependence of particles near the heat exchange surface on various parameters have been carried out to find out the mechanism of the combined heat exchange in a fluidized bed. The influence of conduction-convection and radiation on the degree of particles cooling has been determined. Data on the spectral and integral emissivity of the fluidized bed have been obtained.

NOTATION

ε_{λ} spectral emissivity
ε_{fb} fluidized bed emissivity
t temperature
t_{fb} fluidized bed temperature
λ_{eff} effective wave length
W number of fluidization
d size of particles
τ time

INTRODUCTION

Heat exchange between the fluidized bed and the surface it washes at high temeratures takes place simultaneously through conduction, convection and radiation (Pikashov et al., 1969).

As particles of the fluidized bed are small and pulsate at the heat exchange surface with frequency of several cycles/second it is impossible to measure their temperature with a thermocouple or an optical pyrometer. Under such conditions photopyrometry is the only method for measuring particle temperatures. The method is based on a registration of bodies heat radiation by a photographic film with a subsequent interpretation of the optical density of blackening areas of objects on the film corresponding to their brightness and temperature.

PROCEDURE OF THE EXPERIMENT AND THE EQUIPMENT USED

Special devices - sights (Pikashov et al., 1976) - were arranged to facilitate photographing and filming of the fluidized bed at the heat exchange surface. Two types of sights had been developed, one with a cold window, the other with a hot one, both made of sapphire. The window of the first sight was double with circulating distilled water inside. The external surface temperature of the cold window did not exceed 100°C. The sapphire window of the second sight had been fixed in a frame made of a heat-resistant steel. There was a heat insulating layer between them and the cold wall of the case and that resulted in the window heating up to the temperature of the fluidized particles.

With the help of photopyrometry, making use of the described sight structures, one can measure separately the degree of cooling of the particles both on account of the conduction-covection heat exchange and the heat exchange by radiation. The particles near the hot window will lose heat only by radiating through it into the black cold space insite the sight, and those near the cold window will lose heat through both forms of heat exchange.

To determine the influence of the emissivity of the particles on their cooling by radiation, the following comparative measurement procedure is suggested. Firstly the bed particles of low emissivity at the hot window sight are pyrometred then the same particles are pyrometered after having their surface covered with a thin layer of a blackening coating.

The maximum accuracy of photopyrometry is attained when comparing optical densities of pattern radiation source blackening with that of the investigated object photographed at the same time on the same frame. To this end, there is an ideal black body (IBB) model mounted on the external side of the sight window. The model is a tube with the diameter of 6 mm and 60 mm long made of a heat-resistant steel closed at one of its ends. The open end is fixed closely to the sapphire surface.

The reduction of brightness of the particles at the window in comparison with the brightness of the IBB model may be caused by two reasons - one is the spectral emissivity of the particle ε_λ which is less than 1, the other being a reduction of their temperature which, if measured, is the the brightness temperature. To obtain the real temperature one should introduce a correction for the spectral emissivity. To determine ε_λ a procedure has been developed using direct filming of the fluidized bed particles at the hot sight window (Pikashov et al., 1976).

The determination of the effective integral emissivity of the fluidized bed ε_{fb} was done according to the procedure (Pikashov et al., 1968).

The distinguishing feature of these experiments from those done earlier is that the ceramic frame with a transparent window made of leucosapphire fixed at its ends was put on the probe case zone which was to be submerged into water. Due to the good thermal insulation between the frame and the water cooled probe case, the temperature of the leucosapphire and that of the bed were nearly the same.

CONDITIONS FOR EXPERIMENTS

The tests were carried out on a device with a chamber of 200 mm inner diameter and 600 mm in height. The bed was fluidized and then heated up to working temperature with natural gas combustion products. The height of the packed bed used in the tests was 200 mm. The sights were arranged horizontally at a height of 100 mm from the grill over the burning zone and were introduced inside the chamber at a distance of 30 mm from the window plane to the wall. The hot and the cold sights were fixed at diametrically opposite sides.

In the determination of ε_{fb} , the radiometer probe together with a frame was arranged similarly. The hot junction of a platinum-platinum-rhodium thermocouple in a ceramic sheath was installed in the center at the same height.

The bed temperature t_{fb} during the tests was regulated from 900 up to 1250°C, the fluidization number W was changed from 0 to 4. The materials of the bed were: spherical shaped corundum, blackened fireclay, magnesite and quartz sand.

The bed was filmed through a monochromatic red light filter. The effective wave length λ_{eff} of the light filter together with the film was 0.65 μm. The filming rate was 12 frames per second. A microphotometer was used to measure the optical blackening densities on the film.

RESULTS OF THE EXPERIMENTS

Visual and static analysis of the film showed that the particles had stayed for some time at the window and had formed a low-mobility agglomerate. Then the particles had been shifted aside and had disappeared from the frame. In place of the agglomerate which had been thrown away from the window there was a bubble void or a moving particles vortex caused probably by a bubble running not far from the surface. The bubble was observed to have a trail or a trace of moving particles that had formed a new low-mobility agglomerate. The "packet" phase had three modifications: a low-mobility agglomerate; a flat parallel moving particle agglomerate; an agglomerate of the particles moving in the bubble rear. One should note that hereafter the term "packet" will imply the low-mobility particle agglomerate . Let us call the rest of the modifications of this phase a packet "shift" and bubble "trace". The sequence of phase modifications which is characteristic of the tests is packet, packet shift, bubble void, bubble trace. Sometimes such a sequence was broken and instead of the bubble the vortices appeared or a packet shift was followed by a newly formed agglomerate.

When exposed at the window, the particles of different sizes possess various axial mobility. Observation of small particles (not more than 0.5 mm) showed that from the moment the packet was seen till the moment it had been washed away, a weak rotating particle movement was observed around its axis. The position of the particle rotating axis as regards the window plane was chaotic. The speed of rotation of the particle in the packet was so low that it had been unable to make a full turn around its axis while at the window (up to 0.5 s). The particles of 0.5 to 1.5 mm proved to be even less mobile. In the packet the particles had been arranged in disorder. In some cases they had contacted one another, thus forming vertical chains.

For the most part, clearances were observed between the particles of 0.5 to 3 particle diameters.

The experimental data was processed statistically and the graphical dependence of packet time fraction and bubbles on the fluidization number were plotted, Fig. 1. As the figure shows, during weak fluidization ($W = 1.2$) the packet time fraction was increasing up to 0.4 - 0.6, while that of the bubble was decreasing to 0.2 - 0.08. The increase of the number of fluidization resulted in rearrangement of the relative duration of phase-to-surface contacting. Time fractions of the low-mobility packet and bubble became commensurate for the particles of 0.5 - 1 mm as early as when $W = 1.7$, while the same was observed for the particles with the grain size of 1 - 1.5 mm and 0.25 - 0.5 mm at the number of fluidization 2.3 and 4.2 respectively. With further increase of the W number up to 5-8, a growth of duration of bubble window contact was observed for all the fractions. The relative duration of packet contacting was decreasing simultaneously.

As the bed structure analysis showed, the cycles changed more rapidly as the particle diameters decreased. In the developed fluidized bed ($W = 2 - 3.7$) the averaged filming cycle frequency for corundum particle d=1 - 1.5 mm was equal to 0.93 Hz, 0.5 - 1 mm to 1.25 Hz and 0.25 - 0.5 to 2.3 Hz.

The temperature and ε_λ of the particles in the boundary layer was found from the film for the 1-st row of the phase packet.

ε_λ of the particles, voids among them and bubbles depending on the bed temperature at $W = 3$, is presented in Fig.2. For all the photometered zones, ε_λ increases as the temperature increases. Our data for white corundum stand close to those obtained by Olson and Morris (Goldsmith et al., 1961). At the temperature of 900 - 1200°C, ε_λ, of the blackened corundum is 2.5 - 1.5 times higher than that for the white corundum. The void and the bubble in the fluidized bed may be considered as an approximation to the IBB model. That is why the ε_λ values of the voids and the bubble are always higher than the ε_λ value of the particle. The spectral emissivity of packet voids and bubble voids surrounded by the white particles is lower than of those surrounded by the blackened ones.

The integral emissivity of the fluidized bed cross-section represents a spectrum-, square and time-averaged value. A slight decrease of ε_{fb} was found for the white corundum bed as the temperature increased. The dependence of ε_{fb} on the number of fluidization was investigated for the rest of the materials at constant t_{fb} which was as high as 1100°C, Fig. 3. In these tests, W was altered from 0, the densest packed bed, up to the maximum possible W. As W increases from 0 to 1.5, ε_{fb} for the materials with a lower emissivity increases more rapidly than for the materials with a higher one. Probably in this range of W, growth of ε_{fb} is caused by a decrease of void sizes among the particles due to increase of bed separativeness. With further increase of W, the value of ε_{fb} remains practically the same.

The present data for ε_{fb} are higher by an average of 10 - 20% than those given in the paper (Pikashov et al., 1969). This may be explained by the temperature difference for the probe windows and by the stronger particles cooling under the test conditions in the experiment made by Pikashov et al., 1969.

When investigating the particles cooling in the film, the moment the packet had stopped at the window, was regarded as the zero time value. As Fig. 4 shows, the particles at the hot window, in the course of time, were cooled much less than those at the cold window due to heat losses mainly by radiating inside the sight cavity.

At high t_{fb}, the absolute fall in particle temperature at both sights was more abrupt than that at the low temperature. The cooling rate depends on particle size, too, because fine particles cool quicker than large ones. The less intensive temperature drop of the white particles on account of radiation as compared with the black ones is due to the difference in their emissivity.

A particle cooling rate for sizes of 0.25 - 0.5 mm at the cold window at the first period was τ = 0.08 s; 350 - 860 deg/s. At the second exposition period τ = 0.15 - 0.16 s, the cooling rate dropped to 70 - 100 deg/s. The cooling rate of the coarse particles did not vary so markedly as that of the fine ones. Particle behaviour close to the wall affects the rate of cooling. Analysis of the bed structure showed that the coarse particles maintained contact with the surface during the whole period of the packet exposure, while the fine particles made contact only during the first period. A clearance between them and the wall had then enlarged, resulting in a decrease in their rate of cooling.

Photopyrometering of the bubble and its trace showed that the temperature of these modifications had been different from the measurements made in the nucleus, by the error value of the measurements. Under conditions of the present tests the standard error of temperature determination was equal to ± 3.4°C.

CONCLUSION

As a result of these experiments it becomes possible to explain the mechanism of the radiant and combined heat exchange of a fluidized bed to the surface it washes.
The data obtained on cooling of the boundary layer particles at the surface will allow a new estimation of various theories made earlier for the conductive-convective heat exchange in a fluidized bed to be made.

REFERENCES

Goldsmith A., Waterman, T.E., Hirschhorm, H.I. (Editors) (1961) In handbook of Thermophysical Properties of Solid Materials, N.Y., Pergamon Press, vol. 3.

Pikashov, V.S., Zabrodsky, S.S., Makhorin, K.E., Ilchenko, A.I. (1969) Proceedings of the Academy of Sciences of the BSSR, Physical & Energetics Series, 2, pp. 100 - 109.

Pikashov, V.S., Makhorin, K.E., Kuchin, G.P. (1976) Chem. Technology, 5, pp. 33 - 36.

Pikashov, V.S., Zabrodsky, S.S., Makhorin, K.E., Ilchenko, A.I. (1968) In book of collected works "Heat and Mass Transfer", vol.5, Minsk, "Science & Technology", pp. 303 - 309.

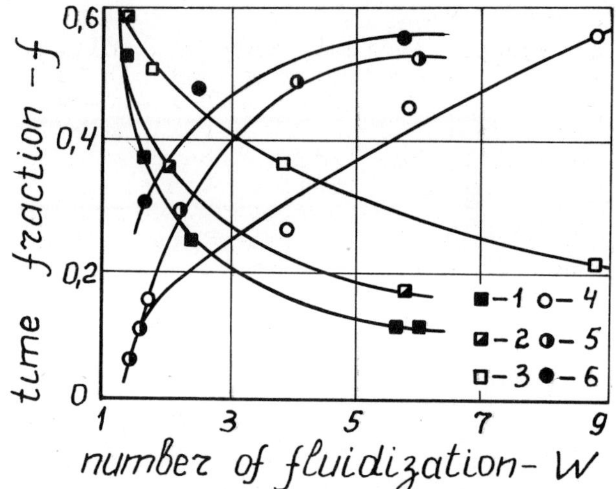

Fig. 1. Dependence of time fraction of "bubble" phase and intermediate state of "packet" phase (low-mobility particles agglomerate) on the number of fluidization. Material of the bed - corundum: 1, 2 and 3 - for particles d = 1 - 1.5; 0.5 - 1; 0.25 - 0.5 mm of packet phase, and 4, 5 and 6 - for the same particles for bubble phase respectively.

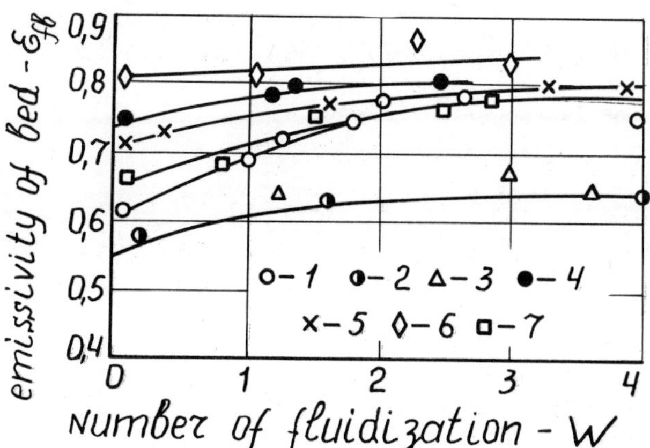

Fig. 3. Emissivity of dense and fluidized bed, $\mathcal{E}_{fb}$, as a function of the number of fluidization W at a temperature of 1100°C: 1, 2, 3, 4 - corundum, respectively, white with d = 0.25 - 0.5; 0.5 - 1 and 1 - 1.5 mm and blackened with d = 1 - 1.5 mm; 5 - quartz sand with size of 0.6 - 1.5 mm; 6 - magnesite, grain, 1 - 2 mm; 7 - fireclay, grain, 1.5 - 2 mm.

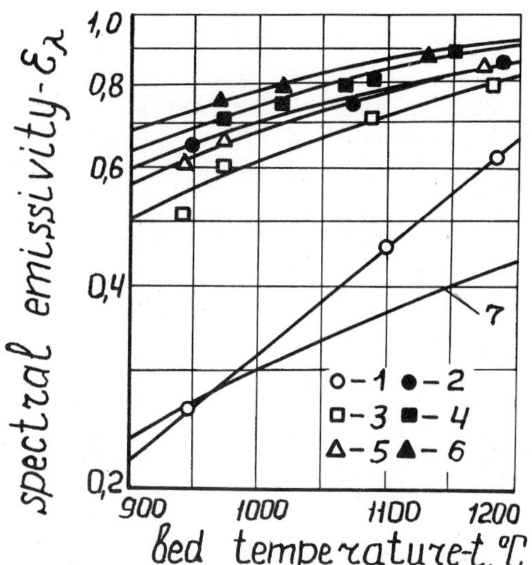

Fig. 2. Influence of the bed temperature t_{fb} on spectral emissivity $\mathcal{E}_{\lambda}$, of corundum particles having diameter of 0.5 - 1 (-1, -2) of voids between them in packet (-3, -4) and of bubble void (-5, -6); 1, 3, 5 - white and 2, 4, 6 - blackened corundum; 7 - data of paper (Goldsmith et al., 1961).

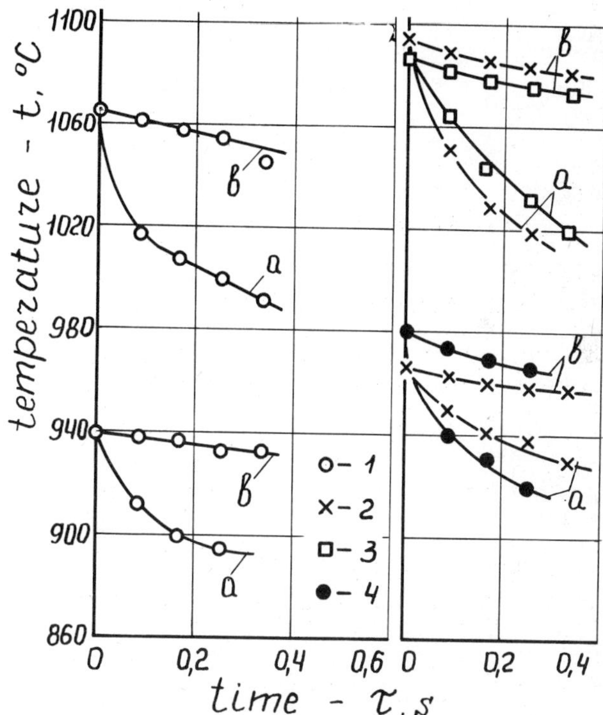

Fig. 4. Temperature of the first row of corundum particles, t, a - at the cold window of the sight, b - at the hot window of the sight, depending on the exposure time of the packet, τ : 1, 2, 3 - white corundum with d = 0.25 - 0.5; 0.5 - 1 and 1 - 1.5 mm respectively; 4 - blackened corundum with d = 0.5 - 1 mm.

NON-DARCY FLOW AND PRESSURE DISTRIBUTION IN A SPOUTED BED

By Norman Epstein and Samuel Levine*

Department of Chemical Engineering
University of British Columbia
Vancouver, B.C., Canada V6T 1W5

*Permanent address: Dept. of Mathematics, University of Manchester, England

SYNOPSIS

The theory of Mamuro & Hattori (1968) for flow and pressure distribution in the annulus of a spouted bed has been reworked using a two-term quadratic equation for relating frictional pressure gradient to velocity, rather than Darcy's law. The results are compared with some experimental data of Lim (1975) [and with the earlier theory]. Agreement is good, provided the experimental data are normalized with respect to real conditions at H_m rather than incipient fluidization conditions. A hybrid analytical method is developed for pressure distribution which gives answers close to those of the more rigorous, but more cumbersome, new theory.

NOTATION

A_a	cross-sectional area of annulus
C_1, C_2, C_3	integration constants
D_c	column diameter
D_i	fluid inlet diameter
D_s	spout diameter
d_{sv}	surface-volume particle diameter
F_1, F_2	constants defined by Eqs.(11a),(11b)
g	acceleration of gravity
H	spouted bed depth
H_m	maximum spoutable bed depth
h	H/H_m
K	Darcy's law constant, $(-dP/dz)/U_a$
K_1, K_2	constants defined by Eq.(6); Ergun values by Eqs.(7a),(7b)
k	Janssen constant, σ_a/σ_r
P	fluid pressure (excluding static head contribution)
P_H	value of P at z=H
$-\Delta P_F$	frictional pressure drop for fluidized bed of height H
$-\Delta P_s$	spouted bed frictional pressure drop
$(-dP/dz)_{H_m}$	frictional pressure gradient at $z=H_m$
$(-dP/dz)_{mf}$	frictional pressure gradient at minimum fluidization
Re_H	$d_{sv}U_{aH}\rho/\mu$
Re_{mf}	$d_{sv}U_{mf}\rho/\mu$
U_a	superficial upward velocity of fluid in annulus at given bed level
U_{aH}	value of U_a at z=H
U_{aH_m}	value of U_a at $z=H_m$
U_{mf}	minimum superficial fluid velocity for fluidization
U_{ms}	minimum superficial fluid velocity for spouting
U_r	superficial radial velocity of spout fluid into annulus
U_s	superficial spouting fluid velocity
x	z/H_m
Y	$-[(\beta+y)/(1-y)]^{1/3}$
y	U_a/U_{mf}
y_H	U_{aH}/U_{mf}
z	vertical distance from fluid inlet
β	flow regime parameter defined by Eq.(21)
ε_a	void fraction in annulus =minimum fluidization void fraction
θ	included angle of conical base
μ	fluid viscosity
ρ	fluid density
ρ_s	solid particle density
σ_a	axial-vertical solids stress
σ_r	radial-horizontal solids stress
X	dz/dU_a

INTRODUCTION

The most coherent non-empirical derivation of fluid flow distribution in a spouted bed, that of Mamuro & Hattori (1968), is based on a force balance over a differential element of annulus and the boundary condition that the top of the annulus is incipiently fluidized at the maximum spoutable bed depth, H_m. The result of this derivation is

$$\frac{U_a}{U_{mf}} = 1 - \left(1 - \frac{z}{H_m}\right)^3 \qquad \ldots.(1)$$

According to Mamuro & Hattori, Eq.(1) is applicable only for a bed with $H=H_m$. They therefore *arbitrarily* modified this equation to

$$\frac{U_a}{U_{aH}} = 1 - \left(1 - \frac{z}{H}\right)^3 \qquad \ldots.(2)$$

for beds having $H<H_m$. This arbitrary modification failed to give good prediction of U_a/U_{aH} as a function of z/H (Mathur & Epstein, 1974). Recently, however, Grbavčić et al.(1976) and Lim (1975) have independently discovered that, for a given column geometry, spouting fluid and solids material, longitudinal pressure gradient and hence superficial velocity, U_a, in the annulus of a spouted bed at any bed level, z, are independent of total bed depth, H. Hence Eq.(1) is applicable to $H \leqq H_m$. For the particular case of z = H,

$$\frac{U_{aH}}{U_{mf}} = 1 - \left(1 - \frac{H}{H_m}\right)^3 \qquad \ldots.(3)$$

Therefore, dividing (1) by (3),

$$\frac{U_a}{U_{aH}} = \frac{1 - [1-(hz/H)]^3}{1 - (1 - h)^3} \qquad \ldots.(4)$$

as proposed by Grbavčić et al. The term $h = H/H_m$ is a constant for a given bed. Eq.(4) gives considerably better prediction of normalized flow distribution along the annulus than does Eq.(2) (Grbavčić et al, 1976), though the match with experimental data of either (1) or (4) (which can be considered different forms of the same relationship) is still quite imperfect (Lim, 1975). Grbavčić et al have used (1) to derive equations for longitudinal pressure profile and total pressure drop across a spouted bed, and these too have given reasonably good, but hardly perfect, predictions of experimental data.

Part of the mismatch with experimental data undoubtedly occurs because of the difficulty of accurately determining the critical parameter H_m without actually measuring it (Mathur & Epstein, 1974). Another difficulty is that of obtaining fully reliable experimental data on both flow and pressure distributions, especially in the anomalous entrance region, where downward gas flow in the annulus has been observed (Van Velzen et al, 1974). The presence of radial pressure gradients in this region also weakens the underlying assumptions in the above derivations that the longitudinal pressure drop across the annulus is the same as that across the bed as a whole. The assumption in the theory that the spout diameter and hence the annulus cross-section is constant with z is also somewhat violated in practice, again especially near the inlet nozzle (Mathur & Epstein, 1974). Finally there is the assumption that Darcy's linear law for flow through porous media prevails in the annulus, which is basic to the derivation of Eq.(1) and its corollaries. This assumption is at odds with most experimental data for flow through spouted bed annuli, in which a typical particle Reynolds number, $d_{sv}U_a\rho/\mu$, half way up the annulus is of the order of 100 (Mathur & Epstein, 1974), i.e. two orders of magnitude greater than that for which Darcy's law is normally applicable in granular beds (Ergun, 1952; Scheidegger, 1974).

In the present paper we rework the Mamuro-Hattori theory and its corollaries by dropping the assumption of direct proportionality between frictional pressure gradient and velocity (Darcy's law) in the spouted bed annulus, and adopt instead a more general relationship between these variables containing both linear and quadratic velocity terms, after Forchheimer (1901) and Ergun (1952). All the remaining assumptions of Mamuro & Hattori and of Grbavčić et al. are maintained.

NON-DARCY FLOW THEORY

We start with the Mamuro-Hattori model of the forces acting in the annulus of a spouted bed, illustrated by Fig. 1. A combination of vertical-axial and horizontal-radial force balances on a differential height dz of annulus leads to (Mathur & Epstein, 1974)

$$k\rho U_r(dU_r/dz) + (\rho_s-\rho)(1-\varepsilon_a)\, g = -dP/dz \qquad \ldots.(5)$$

where the constant, k, which is the reciprocal of the Janssen proportionality ratio (Brown & Richards, 1970) between the horizontal solids pressure, σ_r, and the vertical solids pressure, σ_a, serves to link the two balances.

Instead of using Darcy's law, $-dP/dz=KU_a$, to relate the frictional pressure gradient of the fluid to its velocity in the annulus, we use the more general quadratic equation

$$-dP/dz = K_1U_a + K_2U_a^2 \qquad \ldots.(6)$$

in which, according to Ergun (1952),

$$K_1=\frac{150\mu(1-\varepsilon_a)^2}{d_{sv}^2\varepsilon_a^3} \ (7a) \quad \text{and} \quad K_2=\frac{1.75\rho(1-\varepsilon_a)}{d_{sv}\varepsilon_a^3} \quad (7b)$$

for granular beds. Substituting (6) into (5) and differentiating with respect to z, assuming ε_a is independent of z,

$$k\rho \frac{d}{dz}\left(U_r \frac{dU_r}{dz}\right) = K_1 \frac{dU_a}{dz} + K_2 \frac{dU_a^2}{dz^2} \quad (8)$$

Assuming spout diameter and hence annulus cross-sectional area is independent of z, a fluid balance over the height dz yields

$$\pi D_s U_r = A_a dU_a/dz \quad (9)$$

Substituting U_r from (9) into (8) and rearranging gives

$$\frac{d}{dz}\left(\frac{dU_a}{dz} \cdot \frac{d^2U_a}{dU_a^2}\right) = F_1 \frac{dU_a}{dz} + F_2 \frac{dU_a^2}{dz^2} \quad (10)$$

where $F_1 = \frac{K_1\pi^2 D_s^2}{k\rho A_a^2}$ (11a) and $F_2 = \frac{K_2\pi^2 D_s^2}{k\rho A_a^2}$ (11b)

Eq.(10) is a third order non-linear differential equation which requires three boundary conditions for its solution. After Mamuro & Hattori, these are taken as

B.C. (i): $z = 0,\ U_a = 0$

B.C. (ii): $z = H_m,\ U_a = U_{mf}$

B.C. (iii): $z = H_m,\ -dP/dz = (\rho_s-\rho)(1-\varepsilon_a)g$

The first B.C. states that there is no fluid in the annulus at the bed entrance, while the second and third signify that the top of the annulus is just fluidized for a bed of maximum spoutable depth. From Eq.(5), the third condition yields $U_r dU_r/dz = 0$, and hence from Eq.(8) that $(dU_a/dz)(d^2U_a/dz^2)=0$. This means that either $dU_a/dz = 0$ or $d^2U_a/dz^2=0$. In the Mamuro-Hattori derivation based on Darcy's law, it was not necessary to decide which of these two alternatives apply, and indeed both derivatives at $z=H_m$ work out to be zero in the result, Eq.(1). In the present derivation, however, a decision must be made and this was simply that $dU_a/dz=0$ at $z=H_m$. This condition is supported by all the available experimental data (Mathur & Epstein, 1974; Lim, 1975; Grbavčić et al, 1976). We thus write

B.C. (iii)': $z=H_m,\ dU_a/dz = 0$

which becomes the effective third B.C.

Integrating Eq.(10) once,

$$(dU_a/dz)(d^2U_a/dz^2)=F_1U_a+F_2U_a^2+C_1 \quad (12)$$

Substituting B.C.'s (ii)and (iii)'into Eq.(12),

$$0 = F_1U_{mf} + F_2U_{mf}^2 + C_1 \quad (13)$$

Subtracting (13) from (12),

$$(dU_a/dz)(d^2U_a/dz^2)=F_1(U_a-U_{mf})+F_2(U_a^2-U_{mf}^2) \quad (14)$$

Now put $dU_a/dz=1/X$, whence $dU_a^2/dz^2=-(dX/dU_a)/X^3$ and Eq. (14) becomes

$$-dX/X^4=F_1(U_a-U_{mf})dU_a+F_2(U_a^2-U_{mf}^2)dU_a \quad (15)$$

which integrates to

$$1/(3X^3)=F_1U_a^2/2-F_1U_{mf}U_a+F_2U_a^3/3-F_2U_{mf}^2U_a+C_2 \quad (16)$$

Substituting B.C.'s (ii) and (iii)', the latter of which is equivalent to $1/X=0$ at $z=H_m$,

$$0=F_1U_{mf}^2/2-F_1U_{mf}^2+F_2U_{mf}^3/3-F_2U_{mf}^3+C_2 \quad (17)$$

Subtracting (17) from (16),

$$1/(3X^3)=F_1(U_a-U_{mf})^2/2+F_2(U_a^3-3U_{mf}^2U_a+2U_{mf}^3)/3 \quad (18)$$

Note that if the second term on the right-hand side of (18) is neglected relative to the first (i.e. Darcy's law is assumed),then integration using B.C.'s(i) and(ii) leads to (1).

If we put $y=U_a/U_{mf}$, so that $dy/dz=1/(U_{mf}X)$, we can then express (18) more compactly as

$$(dy/dz)^3=F_2(y-1)^2(y+\beta) \quad (19)$$

or

$$dz = \frac{1}{F_2^{1/3}} \cdot \frac{dy}{(y-1)^{2/3}(y+\beta)^{1/3}} \quad (20)$$

where $\beta = \frac{3}{2}\frac{F_1}{F_2}\frac{1}{U_{mf}}+2 = \frac{3}{2}\frac{K_1}{K_2}\frac{1}{U_{mf}} + 2$ (21)

In evaluating β, the constants K_1 and K_2 from Eq.(6) may be estimated by Eqs.(7). Alternatively, they may be determined empirically by measuring pressure gradient as a function of fluid velocity for flow through a random loose packed bed of the given solids, which is the voidage condition in the annulus. The term U_{mf} in Eq.(21) may be estimated by substituting the minimum fluidization condition,

$$(-dP/dz)_{mf} = (\rho_s-\rho)(1-\varepsilon_a)g \quad (22)$$

into (6):

$$(\rho_s-\rho)(1-\varepsilon_a)g=K_1U_{mf}+K_2U_{mf}^2 \quad (23)$$

and solving (23) for U_{mf}, or it may be determined experimentally. If the Ergun values of K_1 and K_2 are used in (21), then β reduces to $(128.6/Re_{mf})+2$. When Darcy's law (creeping flow) holds, $Re_{mf}\to 0$ and $\beta\to\infty$; when inviscid flow prevails, $Re_{mf}\to\infty$ and $\beta\to 2$.

Now define a new variable, Y, by

$$y = \frac{Y^3+\beta}{Y^3-1} \quad (24a) \quad \text{or} \quad Y= -\left(\frac{\beta+y}{1-y}\right)^{1/3} \quad (24b)$$

Then Eq. (20) is equivalent to

$$dz = \frac{1}{F_2^{1/3}} \cdot \frac{3YdY}{1-Y^3} \quad (25)$$

which can be separated into two parts,

$$dz = \frac{1}{F_2^{1/3}}\left(\frac{1}{1-Y} + \frac{Y-1}{1+Y+Y^2}\right)dY \quad (26)$$

and integrated to

$$z=C_3-\frac{1}{F_2^{1/3}}\left[\ln\frac{1-Y}{(1+Y+Y^2)^{1/2}} + \sqrt{3}\tan^{-1}\left(\frac{1+2Y}{\sqrt{3}}\right)\right] \quad (27)$$

Substituting B.C.(ii), according to which $y=1$, $Y=-\infty$ at $z=H_m$,

$$H_m = C_3 - \frac{1}{F_2^{1/3}}\left[0 + \sqrt{3}\left(-\frac{\pi}{2}\right)\right] \qquad (28)$$

Subtracting (28) from (27) and dividing the result by H_m,

$$\frac{z}{H_m} = 1 - \frac{1}{H_m F_2^{1/3}}\left[\ln\frac{1-Y}{(1+Y+Y^2)^{1/2}} + \sqrt{3}\left\{\tan^{-1}\left(\frac{1+2Y}{\sqrt{3}}\right) + \frac{\pi}{2}\right\}\right] \qquad (29)$$

Substituting B.C.(i), according to which $y=0$, $Y=-\beta^{1/3}$ at $z = 0$,

$$0 = 1 - \frac{1}{H_m F_2^{1/3}}\left[\ln\frac{1+\beta^{1/3}}{(1-\beta^{1/3}+\beta^{2/3})^{1/2}} + \sqrt{3}\left\{\tan^{-1}\left(\frac{1-2\beta^{1/3}}{\sqrt{3}}\right) + \frac{\pi}{2}\right\}\right] \qquad (30)$$

The term $H_m F_2^{1/3}$, which contains the Janssen constant, k, may be evaluated from (30) for a given spouting system (fixed fluid, solids and column geometry). Eq.(29), with this value of $H_m F_2^{1/3}$, then represents a solution for the longitudinal distribution of fluid velocity in the annulus.

Pressure gradient distribution can then be predicted by means of Eq.(6), which is conveniently normalized via (22) and (23) to

$$\frac{-dP/dz}{(-dP/dz)_{mf}} = \frac{K_1U_a + K_2U_a^2}{K_1U_{mf} + K_2U_{mf}^2} \qquad (31)$$

Applying (21) and the definition of y to (31), the result is

$$\frac{-dP/dz}{(-dP/dz)_{mf}} = \frac{2(\beta-2)y+3y^2}{2\beta-1} \qquad (32)$$

where y is related to z/H_m via (29) and (30). Note that, as required, the right-hand side of (32) approaches y as $\beta \to \infty$ (Darcy's law) and approaches y^2 as $\beta \to 2$ (inviscid flow).

The frictional pressure drop profile along the bed can be determined by a graphical or numerical integration of (32) in the form

$$P-P_H = \int_P^{P_H} -dP$$

$$= \frac{(-dP/dz)_{mf}H_m}{2\beta-1}\int_{z/H_m}^{h}\left[2(\beta-2)y+3y^2\right]d\left(\frac{z}{H_m}\right) \qquad (33)$$

and the total frictional pressure drop, $-\Delta P_s$, from

$$\frac{\Delta P_s}{\Delta P_F} = \frac{1}{h(2\beta-1)}\int_0^h\left[2(\beta-2)y+3y^2\right]d\left(\frac{z}{H_m}\right) \qquad (34)$$

since $\quad (-dP/dz)_{mf} = -\Delta P_F/H \qquad (35)$

Again y is related to z/H_m via (29) and (30). Since Eq.(29) is explicit in z/H_m rather than in y, the integral in (34) is, for numerical evaluation purposes, conveniently transformed as follows:

$$\int_0^h\left[2(\beta-2)y+3y^2\right]d(z/H_m) = \left[2(\beta-2)y_H+3y_H^{\,2}\right]h - 2\int_0^{y_H}(z/H_m)(\beta-2+3y)\,dy \qquad (36)$$

and similarly for the integral in (33).

COMPARISON WITH EXPERIMENT AND WITH DARCY FLOW THEORY

Table 1 shows some typical normalized flow data obtained by Lim (1975) compared with predictions by the present theory (i.e. Eqs. 29 and 30), with β obtained via Eqs. (7), (23) and (21), as well as by the earlier Darcy's law based theory, Eq.(4). Agreement of the new theory with the unsmoothed experimental data is good, though not much better than that of the older creeping flow theory. It should be noted, however, that normalization of the velocity data with respect to U_{aH} contributes to the agreement and that the absolute experimental velocities are some 10% lower than those predicted by the theory, all the way up to $z=H_m$, where the measured (extrapolated) velocity, U_{aH_m}, was 1.46 m/s, as opposed to the measured U_{mf} of 1.63 m/s (compared with 1.604 m/s via Eq. 23). Such a negative deviation of U_{aH_m} from U_{mf} was observed consistently for all the eight solid materials studied by Lim (1975) under a wide variety of spouting conditions (Lim & Mathur, 1976).

A similar pattern can be observed for the corresponding pressure gradient data in Table 2. Again there is good agreement between the present theory and the unsmoothed experimental data, but only when the latter are normalized with respect to the pressure gradient measured at the top of either the actual or the maximum spoutable bed (note that in Table 1 the normalizing height is H while here it is H_m), rather than with respect to the pressure gradient for minimum fluidization. The measured absolute pressure gradients were some 20% lower than those predicted by the theory, all the way up to $z=H_m$, where the measured (extrapolated) pressure gradient was 13,100 Pa/m, compared with 16,700 Pa/m at minimum fluidization. Littman et al (1977), in order to fit their analysis of H_m to specific experimental data, have also required negative deviations between $(-dP/dz)_{H_m}$ as computed and $(-dP/dz)_{mf}$ by Eq. (22), but these were not nearly as high as 20%.

For the pressure gradients in Table 2, the Darcy's law based theory, which is equivalent to equating $(dP/dz)/(dP/dz)_{mf}$ to U_a/U_{mf} as obtained from Eq.(1), is not in as good agreement with either the normalized experimental data or the present theory as it is in the case of annular velocity in Table 1.

Normalized total pressure drops are recorded at the bottom of Table 2. Here again there is good agreement between the experimental measurement and the present theory when the former are normalized with respect to a bed pressure drop calculated from the measured pressure gradient at $z=H_m$ rather than from the actual or estimated fluidized bed pressure drop. The Darcy's law based theory of Grbavčić et al (1976) gives an answer which is clearly high.

HYBRID METHOD FOR PRESSURE DROP

Since Eqs. (1) or (4) give close agreement with the non-Darcy flow theory for prediction of U_a/U_{mf} or U_a/U_{aH}, even when β is as close as 2.238 to its limiting value of 2 for inviscid flow, it is reasonable to let Eq. (1) represent the velocity distribution even for non-Darcy flow, but to apply (32), (33) and (34) to get analytical expressions for frictional pressure gradient, pressure distribution and total pressure drop, respectively. The actual value of β which characterizes the given system is then used in these equations, rather than β=∞, which corresponds to Darcy's law. This procedure will be referred to as the "hybrid method".

Substituting Eq. (1) for $y(=U_a/U_{mf})$ into (32), the result is

$$\frac{-dP/dz}{(-dP/dz)_{mf}} = \frac{2(\beta-2)\left[1-(1-x)^3\right]+3\left[1-(1-x)^3\right]^2}{2\beta-1} \quad (37)$$

where $x=z/H_m$. The last column in Table 2 was calculated according to Eq.(37) and shows good agreement with the more rigorous non-Darcy theory.

Substituting Eq.(1) for y into (33) and integrating from $z/H_m=x$ to $z/H_m=H/H_m=h$, the result for the pressure distribution is

$$\frac{P-P_H}{(-dP/dz)_{mf}H} = \frac{H_m}{H(2\beta-1)}\Big|2(\beta-2)(1.5x^2-x^3+0.25x^4) +3(3x^3-4.5x^4+3x^5-x^6+0.143x^7)\Big|_x^h$$

or

$$\frac{P-P_H}{-\Delta P_F} = \frac{1}{h(2\beta-1)}\Big[2(\beta-2)\Big\{1.5(h^2-x^2) -(h^3-x^3)+0.25(h^4-x^4)\Big\}+3\Big\{3(h^3-x^3) -4.5(h^4-x^4)+3(h^5-x^5)-(h^6-x^6)+0.143(h^7-x^7)\Big\}\Big] \quad (38)$$

The total frictional pressure drop, $-\Delta P_s$, is then obtained simply by putting x=0 in Eq. (38):

$$\frac{\Delta P_s}{\Delta P_F} = \frac{1}{2\beta-1}\Big[2(\beta-2)(1.5h-h^2+0.25h^3) +3(3h^2-4.5h^3+3h^4-h^5+0.143h^6)\Big] \quad (39)$$

The total pressure drop according to (39) is shown in Table 2 to agree well with the answer by the non-Darcy theory.

It should be noted that for β→ ∞ (Darcy regime), (39) reduces to

$$\frac{\Delta P_s}{\Delta P_F} = 1.5h - h^2+ 0.25h^3 \quad (40)$$

as already derived by Grbavčić et al (1976), the upper limit of which (at h=1) is the well known value of 0.75; while for β → 2 (inviscid regime), (39) reduces to

$$\frac{\Delta P_s}{\Delta P_F} = 3h^2-4.5h^3+ 3h^4-h^5+ 0.143h^6, \quad (41)$$

the upper limit of which (at h=1) is the previously derived value of 0.643 (Mathur & Epstein, 1974). This value agrees well with the 2/π (=0.637) obtained by Lefroy & Davidson (1969) from an entirely different, empirical starting point. For the particular case of h=1, (39) reduces to

$$\frac{\Delta P_s}{\Delta P_F} = \frac{1.5(\beta-2) + 1.929}{2\beta-1} \quad (42)$$

CONCLUSIONS

1. The newly derived non-Darcy theory is a marginal improvement over the earlier Darcy's law based theory as far as flow prediction is concerned.
2. The new theory is a definite improvement over the earlier theory as far as pressure drop prediction is concerned.
3. The hybrid method for pressure drop is far less cumbersome than the new more rigorous method and gives equally reliable results.
4. The crucial boundary condition that the annulus of a spouted bed is incipiently fluidized at $z=H_m$ requires further investigation, as there is evidence that velocities and pressure gradients at $z=H_m$ are often below those for minimum fluidization. Other spout termination mechanisms may be involved (Mathur & Epstein, 1974).

ACKNOWLEDGEMENTS

We are grateful for the help of Mr. Wai Cheung, Dr. Jim Lim and Dr. Kishan Mathur, and for the financial support of the National Research Council of Canada.

REFERENCES

Brown, R.L. & Richards, J.C. (1970) Principles of Powder Mechanics, p.70, Pergamon Press, Oxford.

Ergun, S. (1952) Chem. Eng. Progr. 48,#2, 89.

Forchheimer, P. (1901) Z.Ver.Deuts.Ing. 45, 1782.

Grbavčić, Z.B., Vuković, D.V., Zdanski, F.K. & Littman, H. (1976) Can.J.Chem.Eng. 54,33.

Lefroy, G.A. & Davidson, J.F. (1969) Trans. Instn. Chem. Engrs. 47, T120.

Lim, C.J. (1975) Ph.D. Thesis and auxiliary unpublished data, University of British Columbia.

Lim, C.J. & Mathur, K.B. (1976) AIChE Journal 22, 674.

Littman, H., Morgan III, M.H., Vuković, D.V., Zdanski, F.K. & Grbavčić, Z.B. (March 1977), paper presented at Nuremberg Congress on Particle Technology.

Mamuro, T. & Hattori, H. (1968) J. Chem. Eng. Jap. 1, 1; correction (1970) J. Chem. Eng. Jap. 3, 119.

Mathur, K.B. & Epstein, N. (1974) Spouted Beds, Academic Press, New York.

Scheidegger, A.E. (1974) The Physics of Flow Through Porous Media, 3rd ed., Chapt. 7, Univ. of Toronto Press.

Van Velzen, D., Flamm, H.J. & Langenkamp, H. (1974) Can. J. Chem. Eng. 52, 145.

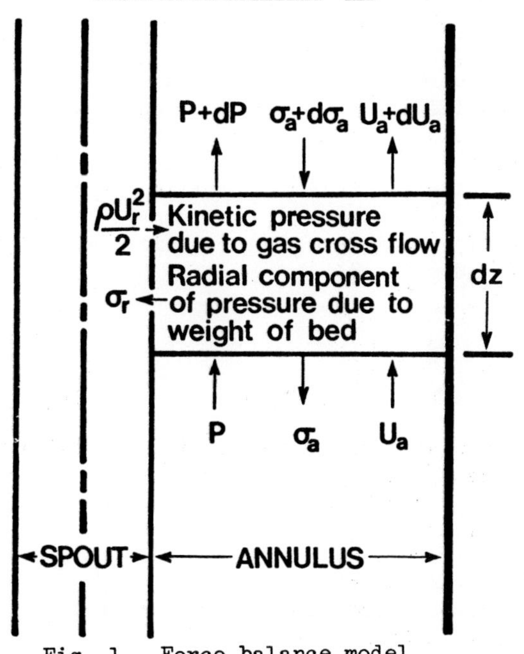

Fig. 1. Force balance model

Table 1

Annulus Flow Data of Lim (1975) Compared with Theory

Solids: glass beads, ρ_s=2.94 mg/mm^3, d_{sv}=2.93 mm.

Fluid: air, ρ=1.20 kg/m^3. Measured U_{mf}=1.63 m/s.

Column geometry: D_c=152.4 mm, D_i=19.05 mm, θ=60°

Bed geometry: ϵ_a=0.42, H=396 mm.

Spouting parameters: U_{ms}=1.87 m/s, H_m=420 mm.

Operating conditions: U_s=1.1 U_{ms}, (U_{aH_m}=1.46 m/s), U_{aH}=1.458 m/s, Re_H=285.

Ergun parameters: K_1=1428 kg/(m^3)(s), K_2=5611 kg/m^4, U_{mf}=1.604 m/s, β=2.238

z/H	U_a/U_{aH} expt.	U_a/U_{aH}* Eq.(29),(30)	U_a/U_{aH} Eq.(4)
0.111	0.333#	0.266	0.282
0.248	0.514#	0.528	0.550
0.354	0.685	0.684	0.705
0.456	0.760	0.801	0.815
0.557	0.838	0.843	0.857
0.658	0.919	0.940	0.946
0.759	0.954	0.975	0.977
0.861	0.981	0.993	0.994
0.962	1.000	0.999	0.999

#Via U_s and Pitot tube measurements in spout; other values of U_a were determined from pressure gradient measurements in annulus.

*U_{aH}(=$y_H U_{mf}$) determined by putting z=H in Eq.(29).

Table 2

Annulus Pressure Data of Lim (1975) Compared with Theory

Same conditions as for Table 1

Same bed levels as for Table 1, except for respective second rows.

z/H_m =x	$\frac{-dP/dz}{(\rho_s-\rho)(1-\epsilon_a)g}$ expt.	$\frac{dP/dz}{(dP/dz)_{H_m}}$ expt.	$\frac{dP/dz}{(dP/dz)_{mf}}$ non-Darcy Eq.(32),(29),(30) β=2.238	$\frac{dP/dz}{(dP/dz)_{mf}}$ Darcy Eq.(32),(1) $\beta=\infty$	$\frac{dP/dz}{(dP/dz)_{mf}}$ hybrid Eq.(37) β=2.238
0.105	-	-	0.097	0.282	0.107
0.239	0.230	0.293	0.322	0.559	0.346
0.334	0.412	0.525	0.500	0.705	0.525
0.430	0.521	0.663	0.664	0.815	0.685
0.525	0.600	0.763	0.796	0.893	0.811
0.621	0.659	0.838	0.893	0.946	0.904
0.716	0.703	0.895	0.954	0.977	0.958
0.812	0.740	0.942	0.987	0.994	0.989
0.908	0.775	0.985	0.998	0.999	0.998

$\frac{-\Delta P_s}{(\rho_s-\rho)(1-\epsilon_a)gH}$ expt.*	$\frac{\Delta P_s}{(dP/dz)_{H_m} H}$ expt.*	$\frac{\Delta P_s}{\Delta P_F}$ non-Darcy Eq.(34),(29),(30) β=2.238	$\frac{\Delta P_s}{\Delta P_F}$ Darcy Eq.(40)	$\frac{\Delta P_s}{\Delta P_F}$ hybrid Eq.(39) β=2.238
0.50	0.63	0.625	0.735	0.637

* $-\Delta P_s$ obtained by graphical integration of (-dP/dz) vs. z in cylindrical portion of column, extrapolating down to z = 0.

ADDENDUM

It should be noted that (29) & (30) reduce to (1) when $\beta \to \infty$ ($Y \to -\infty$). Even for the opposite extreme of $\beta \to 2$, however, it has been shown in a companion paper (Epstein, N., Lim, C.J. & Mathur, K.B. "Data & Models for Flow Distribution & Pressure Drop in Spouted Beds", 27th Canadian Chemical Engineering Conference, CSChE, Calgary, 1977) that (29) & (30) yield values of y averaging only 2% lower than the values of y from (1) over the range x=0.1-0.9. The same paper rationalizes this insensitivity of y to the values of β via series expansions of (29) & (30). Thus (1) is an excellent approximation of (29) & (30) for all values of β and the term "hybrid", used to describe (37)-(42), is better rephrased as "excellent approximation".

The aforementioned companion paper also shows that the Mamuro-Hattori model, unlike that of Yokogawa, A., Ogino, E. & Yoshii, N., Trans. Japan Soc. Mech. Engrs. 38, 148 & 1081 (1972), neglects shear stresses at the spout-annulus interface and at the column wall. An absolute test of the present model could be the value of the Janssen constant, k, which it produces as compared with the corresponding value from the soil or powder mechanics literature. For the glass beads, (30) & (11b) gave k = 3650, but neither soil nor powder mechanics books have yet furnished a number to which this can be compared.

MODELLING OF PARTICLE MOVEMENT IN SPOUTED BEDS

C.J. Lim* and K.B. Mathur

Department of Chemical Engineering
University of British Columbia
Vancouver, B. C., Canada

* Present address: Tree Island Steel Co. Ltd., Richmond B.C.

SYNOPSIS

Mathematical models to describe the movement of solid particles in the annulus and spout regions of spouted beds are formulated. The annulus model enables the calculation of solids flow path and residence time distribution from particle velocity-at-wall data; while for the spout, an improved version of the Thorley force-balance model, together with the Lefroy-Dividson pressure gradient and gas distribution equations, yields longitudinal profiles of particle velocity and voidage.

The models proposed are tested against a wide range of experimental data, their limitations are discussed, and areas requiring further research are identified.

NOTATION

A_a cross-sectional area of annulus
A_s cross-sectional area of spout
C_D drag coefficient $= F/(\frac{1}{4}\pi\, d_p^2)(\frac{1}{2}\rho_f U^2)$
D_c column diameter
D_i fluid inlet diameter
D_s spout diameter
d_p particle diameter
d_v diameter of sphere of same volume as particle
F drag force of fluid acting on a single sphere
G fluid mass flow rate per unit of column cross-section
G_a volumetric flow rate of solids in annulus at any level
G_{aH} G_a at $Z = H$
G_v volumetric flow rate of solids between streamlines J-1 and J in annulus
g acceleration of gravity
H bed depth
H_m maximum spoutable bed depth
I horizontal grid lines in Fig. 1
J particle streamlines in Fig. 1
L length of particle flow path in annulus
m_s mass flow rate of particles in the spout at any level
n exponent in Eq. (17).
n_z cumulative number of rising particles in the spout at level Z
$-dP_T$ total pressure drop along the spout height
$-dP_w$ pressure drop due to solids static head in the spout
$-dP_a$ pressure drop due to acceleration of particles in the spout
$-dP_f$ pressure drop due to frictional loss along the spout-annulus interface
Q_a volumetric gas flow rate in annulus
R radial distance from spout axis, $r_s \le R \le R_c$
R_c column radius
r radial distance from spout axis, $0 \le r \le r_s$
r_s spout radius
S solids cross-flow rate from annulus to spout (cm^2/sec)
$T(R)$ solids retention time in the annulus
t time
U superficial gas velocity
U_a upward superficial gas velocity in annulus
U_{mf} minimum superficial gas velocity for fluidization
U_{ms} minimum superficial gas velocity for

spouting

U_s superficial spouting gas velocity

$\bar{U}_s$ volumetric upward gas flow rate per unit of spout cross-sectional area (cm/sec)

u_s upward interstitial gas velocity in the spout

v_a radial average downward particle velocity in the annulus

v_s radial average upward particle velocity in the spout

v_{sr} upward particle velocity in spout at radial position r/r_s

v_{sc} v_{sr} at $r/r_s = 0$ (at spout axis)

Z vertical distance from fluid inlet

Z' Z/H

ε_a voidage in the spouted bed annulus

ε_{mf} voidage at minimum fluidization

ε_s voidage in the spout

θ_a angle between particle flow path in the annulus and the vertical axis. (Fig. 1)

ρ_f fluid density

ρ_s particle density

INTRODUCTION

Despite the highly systematic pattern of particle movement in a spouted bed, attempts to model it mathematically have met with only limited success. The explanation probably lies in the fact that particle movement is closely interlinked with gas movement as well as bed structure, and therefore any theoretical analysis of solids motion must cover the full hydrodynamic picture.

The work reported here was carried out bearing the above consideration in mind, and constitutes a part of a fairly comprehensive investigation into the hydrodynamics of spouted beds (Lim, 1975). Gas flow models arising out of this work have been published earlier (Lim and Mathur 1974; 1976). In this paper, we extend the modelling approach to particle movement in both spout and annulus, with the objective of formulating mathematical descriptions of particle flow paths, velocities, and voidage distribution in the bed.

PARTICLE FLOW MODELS

Annulus

Let us divide the vertical height of the annulus into M equal intervals and the width of the annulus top into N equal intervals. Let the boundaries between the latter intervals represent flow paths, so that the total number of flow paths equals N, as shown in Fig. 1. Let $G_v(J-1)$ be the flow rate of solids between flow paths J-1 and J. Solids mass balance at any bed level I gives,

$$G_v(J-1) = \pi v_a(I)(1-\varepsilon_a)\left[R(I,J-1)^2 - R(I,J)^2\right] \quad .... (1)$$

where $v_a(I)$ is the downward particle velocity at bed level I, between radial positions R (I,J-1) and R (I,J).

Eq.(1) describes particle flow paths in the annulus from experimental values of $v_a(I)$ with the assumption that the first flow path coincides with the column wall, as suggested by visual observation .

Once the flow paths are known, it is possible to calculate the retention time of solids, T(R), in each flow path by the following equation:

$$T(R) = \int_0^L \frac{\cos(\theta_a)}{v_a} dZ \quad .... (2)$$

where L is the length of the particle flow path, θ_a is the angle between the flow path and the vertical axis, and v_a is the downward particle velocity (Fig.1).

Spout - Model I

Previous workers have used two different theoretical approaches to calculate the longitudinal profile of particle velocity in the spout. The analysis of Thorley et al. (1959) is based on a force balance acting on spout particles, while that of Lefroy and Davidson (1969) considers mass and momentum balances for gas and solids in the spout. Neither model, however, gives acceptable quantitative agreement with observed particle velocities. Certain modifications made by Mathur and Epstein (1974) to the Thorley model improved its agreement with the available experimental results but when tested against more precise data obtained during the present work, the model was found to over-predict velocities by a wide margin. We present here a revised version of the force-balance model which is in better agreement with experimental results.

The balance of forces acting on the spout particles can be expressed as follows (Mathur and Epstein, 1974):

$$\underbrace{\frac{v_s^2}{n_z}\frac{dn_z}{dZ} + \frac{v_s dv_s}{dZ}}_{\text{acceleration}} = \underbrace{\frac{3\rho_f(u_s - v_s)^2}{4d_p^3\,\rho_s} d_v^2 C_D}_{\text{frictional drag}} - \underbrace{\frac{(\rho_s - \rho_f)}{\rho_s} g}_{\text{gravity}} \quad .. (3)$$

where u_s and v_s are the fluid (interstitial) and particle velocities respectively, n_z is the average number of rising particles at a given level, and C_D is the drag coefficient.

Instead of making assumptions about dn_z/dZ for different regions of the spout, as done by the previous workers, we relate the number of particles at any level, n_z, to spout voidage as follows:

$$n_z = D_s^2 (1-\varepsilon_s)/d_p^2 \quad .... (4)$$

where D_s is the spout diameter and ε_s the spout voidage.

Differentiating (4) with respect to Z and substituting the result in (3) with n_z replaced by the r.h.s. of (4), we get

$$-\frac{v_s}{1-\varepsilon_s}\frac{d\varepsilon_s}{dZ}+\frac{v_s^2}{D_s^2}\frac{dD_s^2}{dZ}+\frac{v_s\,dv_s}{dZ}=\frac{3\rho_f(u_s-v_s)^2\,d_v^2\,C_D}{4\,d_p^3\,\rho_s}-\frac{(\rho_s-\rho_f)}{\rho_s}g \quad \ldots (5)$$

For solids mass balance,

$$\frac{d\,[A_s(1-\varepsilon_s)\,v_s]}{dZ}=\frac{dG_a}{dZ} \quad \ldots. (6)$$

where G_a is the volumetric flow rate of solids and A_s the cross-sectional area of the spout. Rewriting (6) in terms of $d\varepsilon_s/dZ$, substituting the result into (5), and using the Richardson-Zaki (1956) equation for C_D for $Re_p>500$, viz $C_D = 0.44/\varepsilon_s^{2.78}$, we get

$$\varepsilon_s = 1.0-\frac{1}{A_s}\left[\frac{v_s}{\dfrac{0.33\,\rho_f(u_s-v_s)^2\,d_v^2}{d_p^3\,\rho_s\,\varepsilon_s^{2.78}}-\dfrac{(\rho_s-\rho_f)}{\rho_s}g}\right]\frac{dG_a}{dZ} \quad \ldots. (7)$$

Equation (7) can be solved in conjunction with the solids mass balance (Eq.6) and the gas mass balance.

$$\frac{d\,(A_s\,\varepsilon_s\,u_s)}{dZ}=-\frac{dQ_a}{dZ} \quad \ldots. (8)$$

to yield vertical profiles of v_s, u_s and ε_s.

The input data required are D_s, G_a and Q_a, each as a function of bed level, in addition to particle properties and fluid density.

According to the above theory, v_s over the entire height of the spout is given by a single equation, viz. (7), rather than by three different equations applicable to the lower, middle and upper zones of the spout, as in the Thorley model. Besides, the revised model also covers spout voidage, which the previous model did not; however, it does require additional data on dG_a/dZ. While generalized expressions for D_s and Q_a are available which can be used for predictive purposes, no reliable correlation, unfortunately, exists for G_a.

Spout - Model II

At attempt has therefore been made to develop the above model further so as to eliminate dG_a/dZ from (7) with the aid of an energy balance first used by Mathur and Gishler (1955). The pressure drop along the spout height (dP_T) is considered to be composed of a solids static head (dP_w), an acceleration pressure drop (dP_a), and a frictional loss (dP_f)

$$dP_T = dP_w + dP_a + dP_f \quad \ldots. \quad (9)$$

where $$dP_w = -\,g\,d[(\rho_s-\rho_f)(1-\varepsilon_s)Z] \ldots.(10)$$

and $$(dP_a + dP_f) = \frac{1}{2\bar{U}_s A_s}\frac{d\,(m_s v_s^2)}{dZ} \ldots.(11)$$

the solids mass flow rate m_s being equal to $v_s\,A_s\,\rho_s\,(1-\varepsilon_s)$. The total pressure drop is well described by the Lefroy-Davidson (1969) equation,

$$\frac{dP_T}{dZ} = -\,(\rho_s-\rho_f)(1-\varepsilon_{mf})\,g\,\sin\frac{\pi Z}{2H_m} \ldots.(12)$$

while the superficial spout gas velocity, $\bar{U}_s$ ($= u_s\,\varepsilon_s$), can be calculated using a gas distribution equation and the gas mass balance given by (8). The distribution equation used in the present work is

$$\frac{Q_a}{A_a} = U_a = U_{mf}\,\sin\left(\frac{\pi Z}{2H_m}\right)^{1/2} \quad \ldots. \quad (13)$$

which follows from (12), but with the assumptions that U_a is independent of H (Grbavic et al., 1976), and that the flow regime in the annulus is inviscid (Epstein et al, 1977).

Substituting (10), (11) and (12) into (9) and combining the result with the force balance equation, (5), we get the final differential equations for spout voidage and particle velocity.

$$\frac{d\varepsilon_s}{dZ'} = \Bigg[(\rho_s-\rho_f)(1-\varepsilon_s)gH\left\{\frac{3}{2}\frac{v_s}{u_s\,\varepsilon_s}-1\right\} + (\rho_s-\rho_f)(1-\varepsilon_{mf})\,gH\sin\frac{\pi HZ'}{2H_m} + \frac{\rho_s v_s^3(1-\varepsilon_s)}{u_s A_s \varepsilon_s}\frac{dA_s}{dZ'} - \frac{0.99}{2}\frac{\rho_f(1-\varepsilon_s)\,v_s\,(u_s-v_s)^2\,H}{u_s\,d_p\,\varepsilon_s^{3.78}}\Bigg] \div \left[\frac{\rho_s v_s^3}{u_s\varepsilon_s} - g\,(\rho_s-\rho_f)HZ'\right] \quad \ldots. \quad (14)$$

$$\frac{dv_s^2}{dZ'} = \frac{2v_s^2}{(1-\varepsilon_s)}\frac{d\varepsilon_s}{dZ'} - \frac{2v_s^2}{A_s}\frac{dA_s}{dZ'}$$

$$+ \frac{0.66\ \rho_f(u_s - v_s)^2 H}{d_p\ \rho_s\ \varepsilon_s^{2.78}} - \frac{2(\rho_s - \rho_f)gH}{\rho_s} \quad \ldots. \ (15)$$

where $Z' = Z/H$

The above equations can be solved for v_s and ε_s using (8) and (13) to get the spout gas velocity, $u_s\ \varepsilon_s$.

EXPERIMENTAL WORK

Apparatus

Three cylindrical plexiglass columns of 15.2 cm (6"), 24.1 cm (9½") and 29.2 (11½") inside diameters, each with a 60° included angle conical base, were used for the experiments. Half-cylindrical versions of each of the above columns were also constructed to observe particle motion and spout shape. Spouting air from the mains flowed through a rotameter and a straightening section before entering the bed through the inlet orifice.

Materials

The solid materials used in this work, were ammonium nitrate, glass beads, polystyrene pellets, millet and wheat (d_p= 1.1 - 3.5 mm, ρ_s= 1.05 - 2.94 gm/cc). Properties of these materials have been published earlier (Lim and Mathur, 1976). The spouting fluid in all experiments was air at ambient temperature.

Techniques

Solids Flow

Annulus particle velocities were determined against the flat face of half-columns by measuring with a stop watch the time taken by a particle to travel down a small vertical distance (4 cm) and were converted to downward volumetric flow of solids, using spout diameter data obtained simultaneously. Spout velocities were measured by cine photography at 2000-3000 frames/sec, using half-columns. Velocities were measured from the film at three locations across the spout radius and radial average values were calculated by integrating the radial profile.

Spout Voidage

Since the downward solids flow in the annulus equals the upward flow in the spout, the bulk density of the gas-solids suspension in the spout could be calculated from the above particle velocity data. The results have been expressed in terms of voidage.

Gas Flow

Annulus gas velocities in the cylindrical part of the bed were determined from measurements of vertical static pressure gradients, and in the lower conical part from spout gas velocities measured with a pitot tube probe. High precision transducers with an accuracy of ± 0.01 mm of water were used for pressure measurements. Details of the apparatus and procedures have been given elsewhere (Lim and Mathur, 1974; Lim, 1975).

EXPERIMENTAL RESULTS AND COMPARISON WITH THEORY

Annulus

Typical data on volumetric solids flow rates, based on radial average velocities, are presented in Fig. 2. In all cases, the average particle velocity and volumetric flow rate of solids were found to decrease linearly from the top of the annulus to $Z=D_c$, a level somewhat above the conical part of the bed (with a 60° cone). All the solids flow data can be described by the following equation:

$$G_a = G_{aH} - S(H-Z) \quad \ldots. \quad (16)$$

where G_a is the volumetric flow rate of solids at any level Z, and S is the cross-flow rate from the annulus into the spout per unit of bed height. Values of G_{aH} (G_a at Z=H) and S varied widely depending upon the solid materials, bed geometry and spouting gas velocity.

Particle flow paths calculated by (1), using experimental results of v_a and G_a, are shown in Fig. 3, and agree closely with the observed flow paths. There was also good agreement between measured and calculated (by Eq.2) residence times for particles starting at different radial position across the annulus surface.

Spout

Radial Profiles

Most of the data obtained in the present work were found to obey the equation

$$\frac{v_{sr}}{v_{sc}} = 1 - \left(\frac{r}{r_s}\right)^n \quad \ldots. \quad (17)$$

with the value of n varying between 1.3 and 2.2. The profiles are thus not alway parabolic, as reported by previous workers (Gorshtein and Mukhlenov 1967; Mikhailik and Antanishin. 1967; Lefroy and Davidson, 1969), except in a qualitative sense.

Longitudinal Profiles

Calculation by Model I (i.e. Eq. 7) was carried out using measured values of G_a, Q_a and A_s. A step-by-step successive approximation procedure was used with the initial conditions v_s = 1.0 at Z = 0, starting with an assumed value of ε_s at each step. The calculated results were generally in good agreement with experimental data as illustrated in Figs. 4 and 5.

Predictions by Model II, where the only experimental parameter used was spout diameter, were less satisfactory. The difficulty arose in the lower-most part of the bed (Z/H<0.2) where the solution of (14) and (15) became very unstable, and unreasonable results were often obtained. Fair agreement between theory and experimental was, however, obtained (Figs 4 and 5) when the initial conditions for solving (14) and (15) were changed from $v_s = 0$ and $\varepsilon_s = 1.0$ at $Z/H = 0$ to the experimental values of v_s and ε_s at $Z/H = 0.2$. This finding confirms previous indications that theory describing particle and fluid motion in the main spouted bed breaks down in the gas entrance region.

DISCUSSION

Calculations by both models have been carried out using the measured vertical profile of spout diameter, rather than an equation such as McNab's (1972) which gives a single mean value of diameter. While the spout diameter is generally considered to reach a relatively constant value above the immediate vicinity of the gas inlet (Mathur and Epstein, 1974), the present results show that this is not always the case. For instance, spout diameters for the two systems of Figs. 4 and 5 increased steadily over the full bed height, by 30% for polystyrene and 47% for wheat between $Z/H = 0.2$ and 1.0. The use of a constant spout diameter for such systems would obviously lead to large errors in particle velocity calculations. Model I additionally relies on experimental data on cross-flow rate of solids from the annulus into the spout (dG_a/dZ), since none of the literature correlations (Lefroy and Davidson (1969), Yokogawa et al. (1970), Van Velzen et al (1974a), were found to correctly describe the wide-ranging data obtained in the present work.

An important advantage of Model I lies in the fact that successive values of v_s calculated by (7) are independent of the results of the previous calculation step, and hence of uncertainties in the difficult gas-entrance region. Since the differential equations involved in the earlier versions of this model, as well as in the Lefroy and Davidson model, do not lend themselves to analytical integration, any errors in calculation accumulate along the spout height.

Model II of this paper does not, in theory, require any experimental input except spout diameter. In practice, however, it runs into difficulty in the entrance region, where the force balance, pressure distribution and gas flow equations (5,6,12 and 13) do not apparently apply. The difficulty with the Lefroy and Davidson model also appears to stem from the entrance region. A spot check showed that predictions by their model too came into fairly good agreement with observed velocity and voidage profiles if the lower-most region ($Z/H < 0.1$) was excluded from the calculations (by taking observed values of v_s and ε_s at $Z/H = 0.1$ as the initial condition). A better understanding of the complex phenomena in the gas entrance region thus seems necessary before the full particle movement can be predicted without experimentation. The height of this region, if judged simply by the distance from the inlet orifice over which the spout diameter undergoes a sharp change, appears to be of the order of 4 D_i.

CONCLUSIONS

1) A simple model to describe the movement of annular solids has been formulated, which enables the calculation of solids flow path and retention time in the annulus from particle velocity-at-wall data.
2) The force balance model of Thorley et al. for spout particle velocity has been revised to include the solids cross-flow rate from the annulus into the spout. The revised equation, when solved using experimental data on cross-flow rate, gas distriution and spout diameter, gave good predictions of longitudinal particle velocity and voidage profiles.
3) The need for knowing the solids cross-flow rates was eliminated by combining the force balance with an energy balance along the spout, and expressing the pressure gradient and gas distribution on the basis of equations proposed by Lefroy and Davidson. The resulting equations for particle velocity and voidage were found to apply in the upper 80% of the spout, but not in the lower gas entrance region.
4) Further work on modelling the cross-flow of particles from the annulus into the spout, on variation of spout diameter along the bed height, and on the hydrodynamics of the gas entrance region, is needed in order to complete the theoretical description of particle movement in spouted beds.

ACKNOWLEDGMENT

We are grateful to Dr. Norman Epstein for his help and to the National Research Council of Canada for financial support.

REFERENCES

Epstein, N., Lim, C.J. and Mathur, K.B.,"Data and Models for Flow Distribution in Spouted Beds", Can. Soc. Chem. Engg., 27th Conf., Calgary, Oct. 1977.

Gorshtein, A.E. and Mukhlenov, I.P., ZH. Prikl. Khim (Leningrad), 40, 2469 (1967).

Grbavcic Z.B., Vukovic, D.V., Zdanski, F.K. and Littman, H., Can. Jr. Chem. Eng., 54, (1976).

Lefroy, G.A. and Davidson, J.F., Trans. Inst. Chem. Eng., 47, T120 (1969).

Lim, C.J., "Gas Residence Time Distribution and Related Flow Patterns in Spouted Beds", Ph.D. thesis, University of British Columbia, 1975.

Lim, C.J. and Mathur, K.B., Can. Jr. Chem. Eng., 52, 150 (1974).

Lim, C.J. and Mathur, K.B., A.I.Ch.E. Jr., 22, 674 (1976).

Mathur, K.B. and Gishler, P.E., A.I.Ch.E. Jr., 1, 157 (1955).

Mathur, K.B. and Epstein, N., "Spouted Beds", Academic Press, (1974).

McNab, G.S., Brit. Chem. Eng. Proc. Tech., 17, 532 (1972).

Mikhailik, V.D. and Antanishin, M.V., Vesti Akad. Nauk. BSSR, Minsk, Ser. Fiz, Takhn. Nauk No. 3, 81 (1976).

Richardson, J.F. and Zaki, W.N., Trans. Instn. Chem. Eng., 32, 35 (1954).

van Velzen, D., Flamm, H.J. and Langenkamp, H., Can. Jr. Chem. Eng., 52, 156 (1974a).

van Velzen, D., Flamm, H.J. and Langenkamp, H., Can. Jr. Chem. Eng., 52, 145 (1974b).

Yokogawa, A., Ogino, E. and Yoshii, N., Trans. Japan Soc. Mech. Eng., 36, 1117 (1970).

Yokogawa, A., Ogino, E. and Yoshii, N., Trans. Japan Soc. Mech. Eng., 37, 1979 (1971).

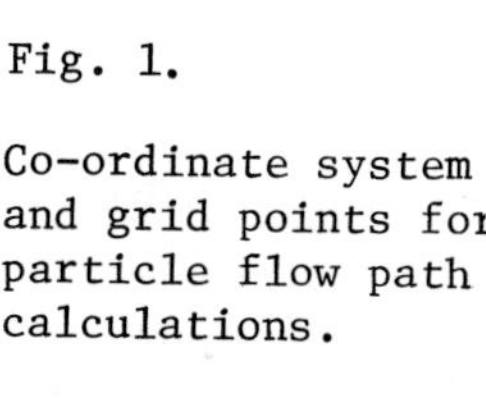

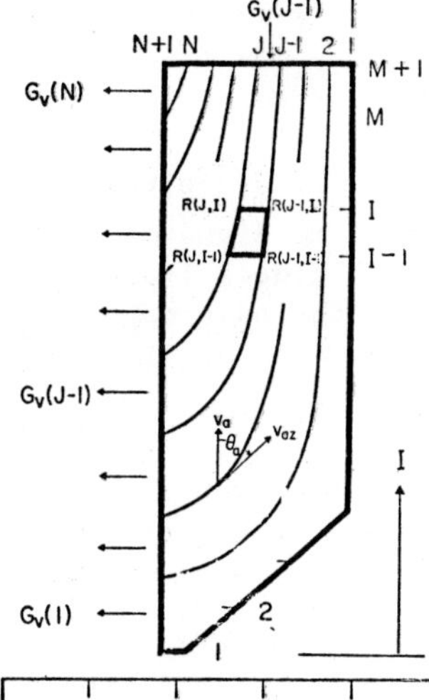

Fig. 1.

Co-ordinate system and grid points for particle flow path calculations.

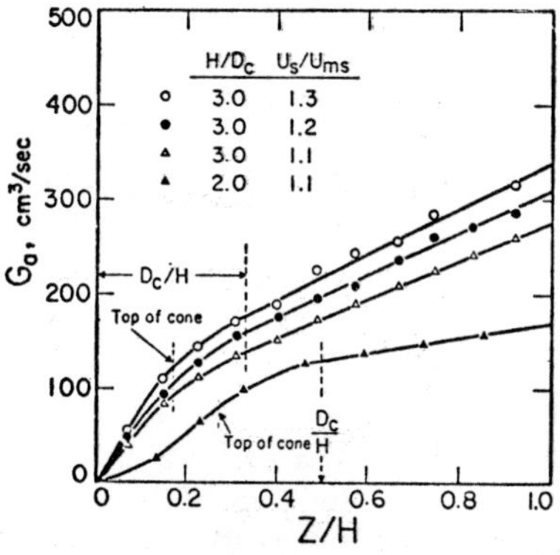

Fig. 2.

Volumetric flow solids in the annulus. Polystyrene pellets D_c = 15.2 cm D_i = 1.9 cm

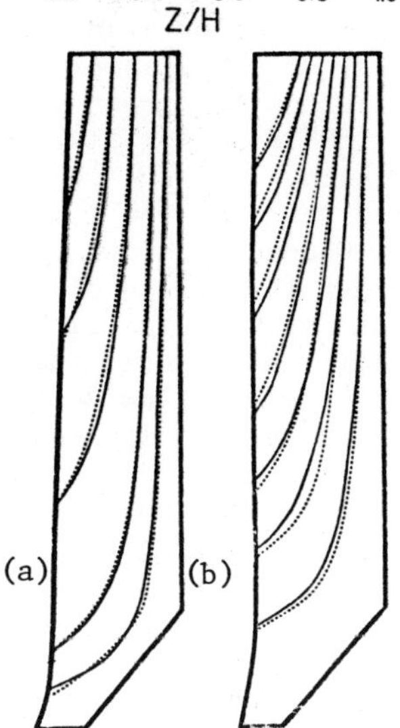

Fig. 3.

Particle flow paths in the annulus of wheat beds. H/D_c= 3, D_i/D_c= 1/8, U_s/U_{ms}= 1.1

——— experimental

- - - - - calculated

(a) D_c = 15.2 cm

(b) D_c = 29.2 cm

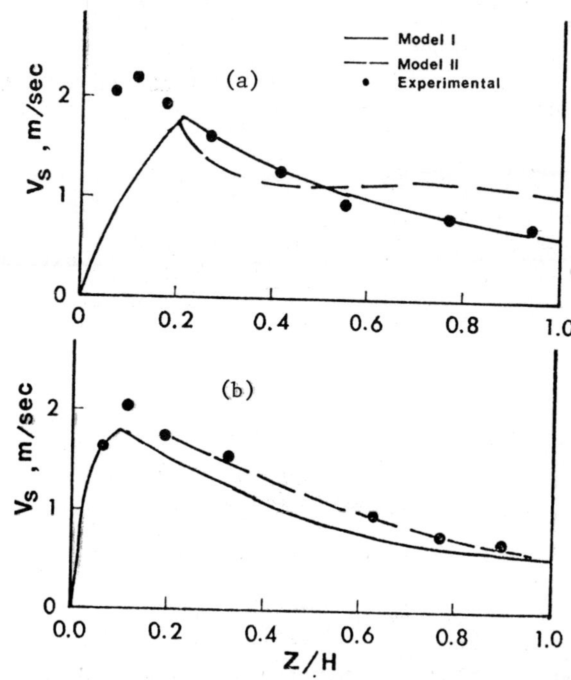

Fig. 4.

Experimental and calculated spout particle velocity profiles. D_c= 15.2 cm, H/D_c= 3, D_i= 1.9 cm, U_s/U_{ms}= 1.1

(a) Polystyrene Pellets

(b) Wheat

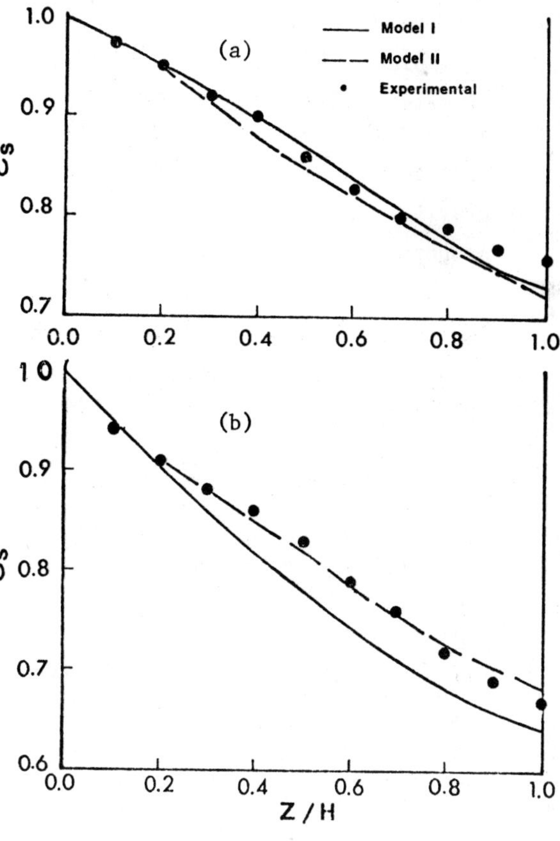

Fig. 5

Experimental and calculated spout voidage profiles for the systems of Fig. 4.

DENSE PHASE FLOW OF A POWDER DOWN A STANDPIPE

By M.R. Judd and P.N. Rowe

University of Natal
Chemical Engineering Department
Durban, South Africa

SYNOPSIS

Experimental data are presented for the flow of fine silica-alumina catalyst (d_p = 103μm) down a vertical hopper-fed standpipe 1,02 m long and 25 mm diameter. The particle/gas mixture behaves just as if it were a true liquid, exhibiting a linear pressure profile and discharge coefficients through orifices of essentially the same magnitude as for liquids.

NOTATION

C_D'	orifice discharge coefficient	
D_o	orifice diameter	(m)
D_T	standpipe diameter	(m)
D_x	diameter of hopper pressure tapping above top orifice	(m)
d_p	surface mean particle diameter	(μm)
g	gravitational acceleration	(m/s²)
H	standpipe height (between orifices)	(m)
ΔP_o	pressure drop across the orifice	(Pa)
p	gas pressure in the flow system	(Pa)
Q_G	volumetric gas rate, positive downwards	(m³/s)
u	superficial gas velocity	(m/s)
u_{mf}	minimum fluidisation velocity	(m/s)
u_s	solids velocity = $W_s/\frac{\pi}{4} D_T^2\rho$	(m/s)
W_s	solids flow rate, positive downwards	(kg/s)
ρ	bulk powder density	(kg/m³)
ρ_s	particle density	(kg/m³)
ε	powder voidage = $(1 - \frac{\rho}{\rho_s})$	
ε_{mf}	voidage at incipient fluidisation	
ε_{FLOW}	voidage in flowing mass of solids	

INTRODUCTION

The crucial function of a standpipe in a circulatory fluidised system is to convey the solids downwards from one vessel to another, generally against a significant adverse gas pressure gradient.

If the flow rate of solids is controlled totally by conditions at the top of the standpipe, such as was done in the work of de Jong (1975), McDougall (1969/70), Pullen (1974), who used small orifices at the top of their standpipes, the solids will not flow in a dense array. This can be readily seen in small diameter glass or Perspex standpipes. It can also be inferred from the very small pressure gradients reported, which to a first approximation reflect a reduced hold up of solids.

Dense phase flow in a standpipe can be realised if the system is effectively bottom restrained by, for example, a bottom orifice. In this paper it is also shown to occur for fine particles when a small constraint is placed at the top of the standpipe, such as an orifice of diameter at least one third of that of the standpipe. This dense state of flow can be either packed bed flow ("slip-stick" flow) or fluidised flow (Leung and Wilson, 1973), depending upon the magnitude and sense of the slip between gas and solids.

Whenever powder falls down a standpipe under the action of gravity it tends to take interstitial gas with it. Any pressure difference across the pipe will help or hinder the gas flow depending on its sense. Relative velocity (slip) between particles and interstitial gas will set up drag forces which, under the conditions discussed in this paper, hinder gravitational forces. An index of the importance of the drag forces is the minimum fluidisation velocity, u_{mf}, and only when the slip velocity is of similar order to u_{mf} is the drag term likely to be important.

Packed bed flow has received some attention and it is the flow regime that is encountered when relatively large particles are studied, that is, when u_{mf} is relatively large. Interparticle and particle to wall frictional forces are predominant and gas drag is only a small part of the overall force balance. Davidson's analysis (1961), when extended to encompass interparticles stresses, can be used to model this situation very well (Altiner, 1975). Packed bed flow is associated with relatively low particle velocities and low pressure gradients. With large particles in packed bed flow the slip velocity needed to produce significant drag force is greater than the downward particle velocity so that gas flow reversal occurs if there is an adverse pressure gradient. That is, the net gas flow is countercurrent to the solids. Dense phase fluidised flow, however, results in low interparticle and particle to wall friction forces, and particle velocities are more than an order of magnitude larger than with packed bed flow. Large pressure gradients are observed with this sort of flow (Kojabasjian, 1958; Judd and Dixon, 1976), and the limited data available suggest that gas flow in the standpipe is downwards. Dense phase fluidised flow is the same phenomenon that results in what has been called "flooding" at hopper outlets (Bruff and Jenike, 1967/68). As might be expected, it is found in systems where u_{mf} is small and the slip velocity is sufficient to cause drag equal to the gravitational force even when the net gas flow is downwards. It should be noted that any change in particle spacing (voidage) will have a profound effect on the drag force for a given slip velocity.

It must also be noted at this point that dense phase flow and packed bed flow have been observed to coexit in the same standpipe (Kojabasjian, 1958; Judd and Dixon, 1976). This has been shown possible by Leung and Wilson (1973) who discuss this phenomenon in some detail. Other effects such as local channelling might also be significant in large industrial scale standpipes and the work which is presently reported (carried out in a small standpipe only 25,4 mm in diameter and 1,02 m long) does not purport to elucidate these and other effects which may well develop in longer and bigger diameter standpipes.

EFFECTS OF INLET CONDITIONS

The effect that inlet conditions have on the mode of dense phase flow has until now received little attention. Kojabasjian (1958) noted that, unless the top of his standpipe was flared, dense phase fluidised flow did not occur. Judd and Dixon (1976) observed that relatively dilute "streamer flow" seems to be the preferred mode of flow when a standpipe is fed from a fluidised bed, although with a long enough bottom restrained standpipe dense phase flow can also occur.

The reason why a fluidised bed at the top of the standpipe tends to result in dilute flow in the pipe is probably related to the observation that the ratio of gas to solids increases significantly as material flows from a hole in a fluidised bed (Massimilla et al, 1961).

Experimentally, however, it is found that if a standpipe is fed from a hopper, then fluidised bed flow occurs readily with small particles.

EXPERIMENTAL APPARATUS

The apparatus used in this study is shown in Figure 1. The solids enter the 25.4 mm diameter Perspex test standpipe from a hopper which in turn is fed from a large fluidised bed storage. Two standpipes are thus involved, the top one is not of interest; it merely serves to convey the solids into the hopper. However, the ratio of gas to solids coming down this top standpipe is large and the excess air escapes through the top layers of material in the hopper as the solids disengage to form the moving fluidised mass of material in the hopper. This ensures a reproduceable inlet condition over a wide range of solids flow rates. It also permits the accurate control of level in the hopper by means of a simple float switch which operates a control valve at the bottom of the fluidised bed storage.

The (lower) test standpipe 1 m long is fitted with pressure tappings at 200 mm intervals and provision is made for inserting orifices not only at the bottom of the standpipe but also at the top.

The solids rate is measured by differentiating the recorder trace from a load cell supporting the bottom hopper. The air leaving the hopper is measured by a miniature turbine meter (which operates satisfactorily even with slightly dusty air) and the air flow rate down the pipe calculated by subtracting the volumetric solids flow rate into the bottom hopper. The solids flow rate was quite large even in the small diameter pipe used, so that a run was of short duration. However, pressure and flow rates came to steady equilibrium values before measurements were made. The particles used in the experiments were a silica alumina catalyst the physical properties of which are given in Table 1. The behaviour of the particles in a 14.5 cm diameter ordinary fluidised bed is shown in Figure 2. The powder exhibits characteristic small particle behaviour (Rowe and Yacono, 1976), over the range tested.

PRESSURE PROFILES AND GENERAL APPEARANCE OF THE FLOW

Typical pressure profiles down the test standpipe are shown in Figures 3, 4 and 5 for various combinations of bottom and top orifices. It is seen that in all cases there is a sharp pressure drop across the top orifice (between points P2 and P3) and, in Figures 3 and 4, a short section of non-linear pressure recovery followed by a linear increase in pressure down most of the pipe to near the bottom orifice. The slope of the linear portion is independent of solids flow rate over the range examined as shown in Figure 6. The observed gradient is a little lower than the value corresponding to minimum fluidisation conditions (3.51 Pa/mm average compared with 3.75 Pa/mm) and the reason for this is probably that the powder has expanded a little to a voidage about 7% greater than ε_{mf}. That is, $\rho_s(1-\varepsilon_{mf})$ = 382 kg/m^3 as measured whilst $\rho_s(1-\varepsilon_{FLOW})$ = 358 kg/m^3 (average) to agree with the observed pressure gradient. This is in accordance with the observation that, when fluidised, the powder expands a little before bubbling as Figure 2 indicates.

When the pressure gradient is linear and near the minimum fluidisation value the material in the standpipe is seen to be homogeneous and flowing smoothly. When viewed by X-rays homogeneity is confirmed and this is a discriminating test because the beam ignores any curtaining of the walls which sometimes misleads visual observation (Davidson and Harrison, 1971). When the top orifice is made small enough (7.5 mm or 8.7% of the tube cross-sectional area in our experiments) then solids flow is dominated by these entry conditions and dense phase flow no longer occurs in the pipe. Discontinuities are plainly visible by eye or on X-ray film but there is no describable pattern such as slugs, plugs or bubbles. Flow of this kind is characterised by near zero pressure gradient as shown in the upper part of Figure 5. As is seen from this figure, the powder recovers its uniform and normal density towards the bottom of the tube and the normal pressure gradient results.

There seems to be a natural law that powders in a gas flow system can only exist at certain concentrations, a fairly narrow range around packed bed values and a wider range corresponding to lean phase transport values and extending to zero. It is analogous to liquid/vapour phase change and is compatible with the phenomenon of bubbling. There is a non-permissible range of voidage (values depending on the type of powder) and it seems that flow in standpipes is no exception to this normal gas/solid behaviour.

FLOW RATE OF SOLIDS

As the fluidised solids flow through either the top or the bottom orifices a characteristic gas pressure drop is observed. The reason for this is still subject to considerable debate (Burbett et al., 1971; Jones and Davidson, 1965), but the data are well modelled by treating the aerated powder just like a liquid and writing Bernoulli's equation over the orifice assuming a constant density.

$$W_s = C_D^{\cdot}\left(\frac{1}{1-\left(\frac{D_o}{D_T}\right)^4}\right)^{\frac{1}{2}} \frac{\pi D_o^2}{4} \sqrt{2\rho\Delta P_o} \qquad (1)$$

This fits the data for the bottom orifice particularly well as seen in Figure 7 (standard deviation of a single observation, s = ±5 g/s (34d/F)) and for the top orifice quite well as seen in Figure 8 (s = ±9 g/s (13d/F)). The straight line in each case has been fitted by the method of least squares and constrained to pass through the origin.

If the powder density, ρ in equation (1) is assigned a value $\rho = \rho_s(1-\varepsilon_{FLOW})$ =358kg/m^3 as concluded above, the discharge coefficient can be calculated from the slope of the fitted line from which C_D' = 0.71 ± .02 for the bottom orifice and C_D' = 0.75 ± .07 for the top orifice. The former value is essentially the same as previously reported for milled iron ore (Judd and Dixon, 1976) and the values are similar to the discharge coefficients to be expected for true Newtonian liquids. This is a rather remarkable fact but a great simplification in engineering fine powder systems. The slightly but significantly higher discharge coefficient for the top orifice is probably a result of better entry conditions from the tapering hopper.

In spite of the close analogy with true liquids, no vena contracta is seen and in fact the solids stream from the orifice appears to diverge. In the case of the top orifice the solids stream diverges to occupy the whole cross-sectional area of the pipe. As is seen in Figure 3 and has already been remarked, the pressure profile deviates from the linear just below the orifice and this extra pressure loss could well be associated with the energy required to cause the solids to diverge. It is not possible to speculate further without some measurements of voidage variation in this region which could be done by careful and detailed X-ray examination.

PARTICLE/GAS SLIP

The mass flow of solids has been measured by direct weighing of the receiving bin as illustrated by Figure 1. This can be converted to a volume flow when a powder density is assumed or deduced. The associated gas flow has also been measured using the turbine meter. Thus, relative velocity or slip between gas and particles can be estimated although not with great precision because it involves a small difference between two imperfectly measured quantities. The measured volumetric gas flow rate is plotted against the measured solids mass flow rate in Figure 9. A "no slip" line is drawn to indicate the gas flow that would occur if there were no relative motion between gas and solids in the flowing mixture. That is to say

$$W_s/\rho_s(1-\varepsilon_{FLOW}) = Q_G/\varepsilon_{FLOW} \qquad (2)$$

To draw this line it is necessary to know the absolute solids density, ρ_s and from this and the powder bulk density, ρ, the absolute voidage, ε_{FLOW}. The alumina particles have internal voidage and pores open to the exterior so that normal liquid displacement cannot readily be used to measure the overall (apparent) particle density. The measurement was made using a very fine starch powder as a pseudo liquid. A measured mass of starch and of alumina was mixed and vibrated so that starch filled all interstitial space but did not penetrate pores. The density of the mixture compared with the density of starch powder gives the required particle density, ρ_s which was found by this method to be 970 kg/m^3 (Whang, 1971). This value compares quite well with that obtained for similar material obtained by other methods.

As seen in Figure 9 the data lie fairly close to the "no slip" line but about 20% less gas appears to be flowing than expected. Pressure change in the standpipe will change the gas volume but in our experiments this could not alter the volume flow by more than 4%. A discrepancy as large as 20% cannot be explained by plausible errors in the estimates of ρ_s and ε_{FLOW} and must surely arise from absolute errors in our measurement of flow rate. The most likely explanation is that fine dust in the exit gas interferes with the calibration of the turbine meter. Less than 0.02% by volume of dust in the gas could cause this sort of discrepancy but we have not been able to check this measurement. Slip of comparable magnitude to u_{mf} would put the observed gas flow below the "no slip" line in Figure 9, as would a lower assumed value for the voidage at the orifice.

Although our measurements are unsatisfactory, it does appear that any slip is slight and the gas flow rate can be estimated from equation (2) to a first approximation. $(Q_G)_{mf}$ = 0.5 cm^3/s so we would not expect to be able to measure such a small rate of slip by this method.

CONCLUSIONS

Fine powders will flow in a dense fluidised state down vertical standpipes provided the bottom outlet orifice controls the flow. This can occur even when the pressure drop across the top or entry orifice is the greater. It appears that the particle velocity should be well in excess of u_{mf}, a condition easily achieved with fine particles, and the powder should be initially in a fluidised condition. If the bottom orifice is too widely open the uniformly aerated powder breaks up into a heterogeneous mixture with virtually zero pressure difference along the pipe.

In the former uniform flow the powder behaves as a pseudo liquid of density close to that of the fluidised powder.

The pressure gradient corresponds to the hydrostatic head and flow through orifices follows the normal hydrodynamic laws with discharge coefficient of about the same value as for a true liquid.

This liquid-like behaviour makes it possible to calculate, for example, the solids flow rate from pressure drop measurements across one or both orifices (but not from the pressure gradient in the standpipe).

There is virtually no slip between gas and solids so that gas flow rates can also be estimated.

It should be noted that the conclusions which we draw are based on small scale experiments and that in industrial standpipes of much greater length and diameter other flow modes may well exist or coexist.

ACKNOWLEDGEMENTS

We are grateful to the CSIR (South Africa) and the South African Coal, Oil and Gas Corporation (SASOL) for financial assistance and to Mr. H.J. MacGillivray and Mr. D.J.Cheesman for enthusiastic help and ideas.

REFERENCES

Altiner, H.K., (1975) Ph.D. Thesis, Cambridge.

Bruff, W. and Jenike, A.W. (1967/68) Powder Technol. 1, 252.

Burbett, R.J., Chalmers-Dixon, P., Morris,P.J. and Pyle, D.L.(1971) Chem. Eng. Sci. 26, 405.

Davidson, J.F. (1961) Trans. Instn. Chem. Engrs. 39, 230.

Davidson, J.F. and Harrison, D. (Eds.) (1971) "Fluidisation", p.130, Academic Press, London.

de Jong, J.A.H. (1975) Powder Technol. 12,197.

Jones, D.R.M. and Davidson, J.F. (1965). Rheologica Acta, 4, 180.

Judd, M.R. and Dixon, P.D. (1976) AIChE 69th AGM Chicago Nov/Dec.

Kojabasjian, C. (1958)M.Sc. Thesis, M.I.T.

Leung, L.S. and Wilson, L.A. (1973) Powder Technol. 7, 343.

Massimilla, L., Betta, V. and Rocca C.D. (1961) AIChE 7, 502.

McDougall, I.R. (1969/70) Proc. Instn. Mech. Engrs. 184, 1.

Pullen, R.J.F. (1974) Ph.D. Thesis, Univ. of Leeds.

Rowe, P.N. and Yacono, C.X.R. (1976) Chem. Eng. Sci. 31, 1179.

Whang, S.T. (1971) Ph.D. Thesis, London Univ.

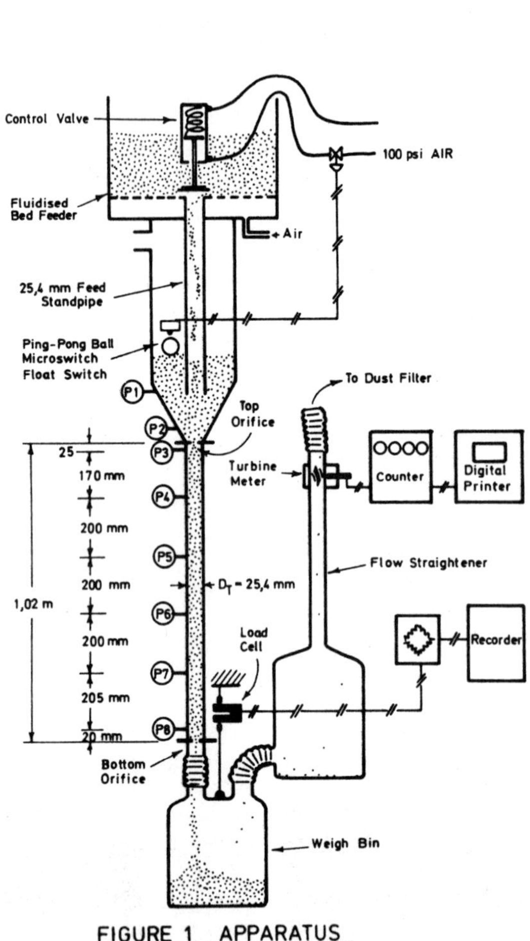

FIGURE 1 APPARATUS

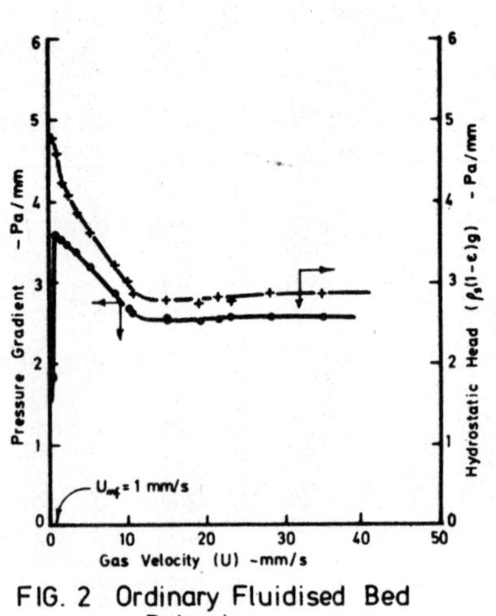

FIG. 2 Ordinary Fluidised Bed Behaviour

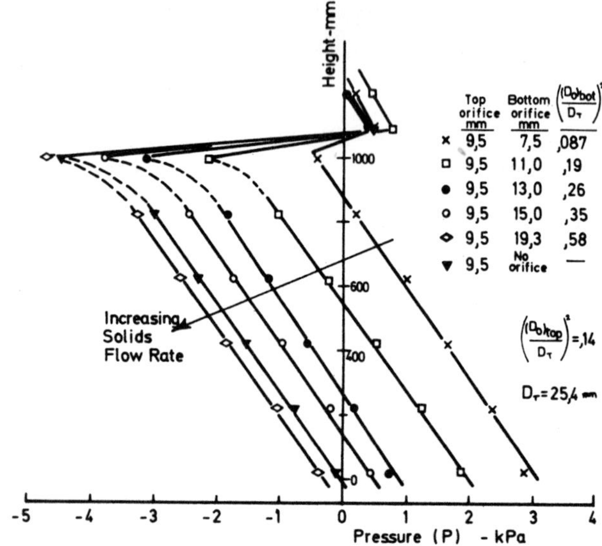

FIG. 3 Top Orifice Constant, Changing Bottom Orifice

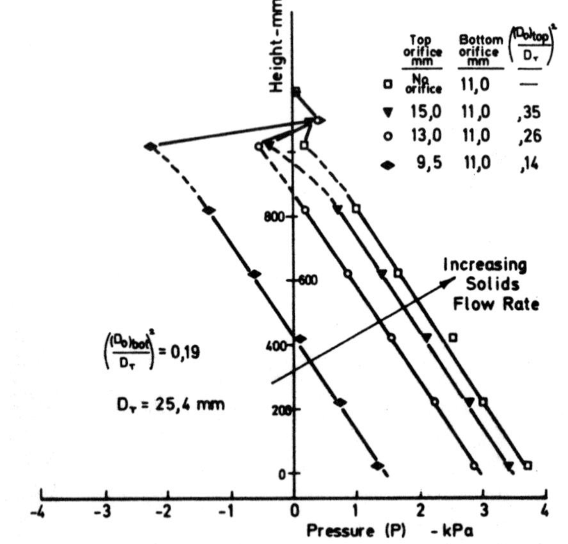

FIG. 4 Bottom Orifice Constant, Changing Top Orif.

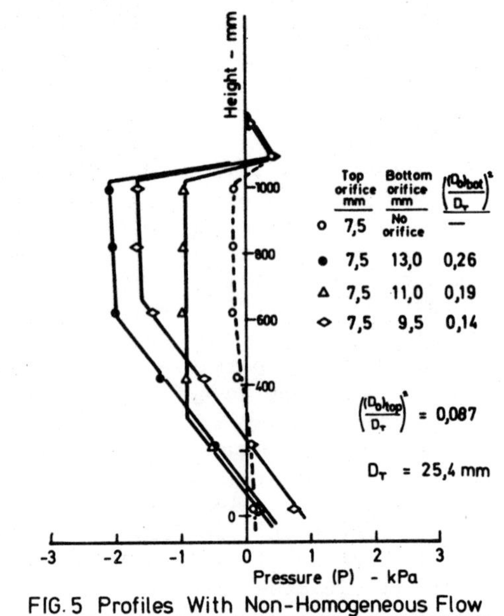

FIG. 5 Profiles With Non-Homogeneous Flow

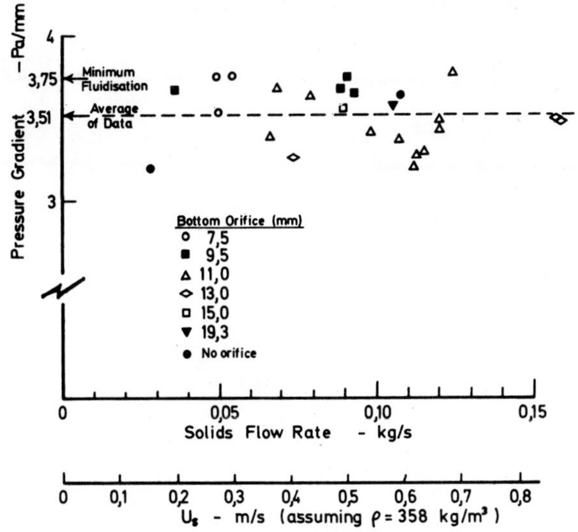

FIG. 6 Pressure Gradient for Dense Phase Flow

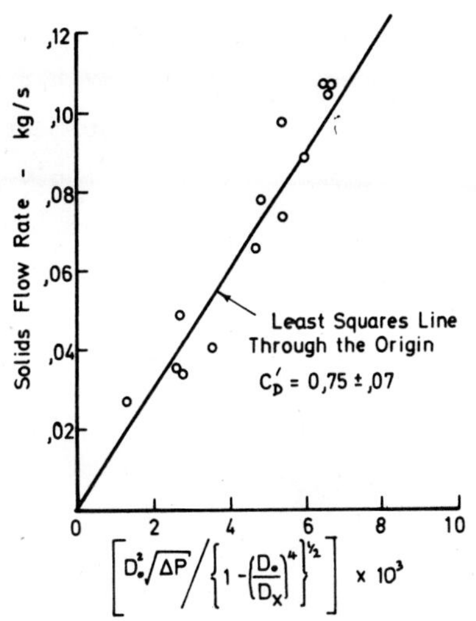

FIG. 8 Orifice Correlation Applied to Top Orif.

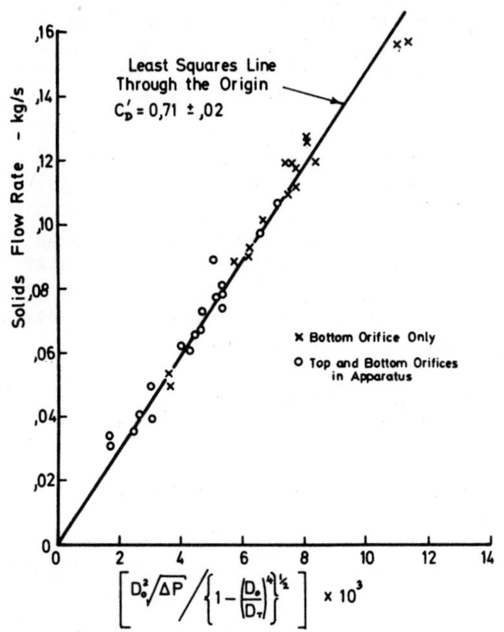

FIG. 7 Orifice Correlation Applied to Bottom Orif.

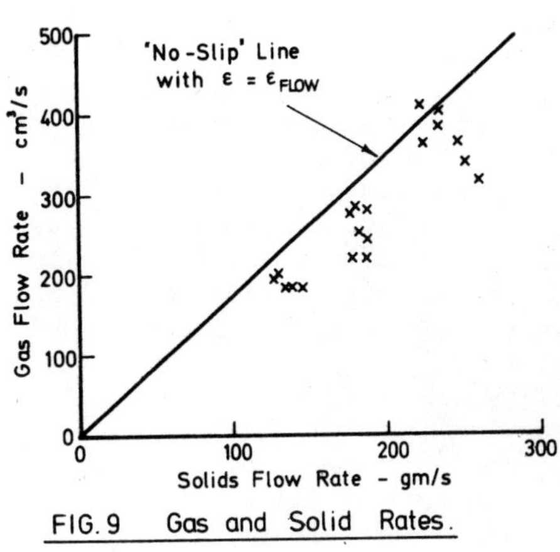

FIG. 9 Gas and Solid Rates.

TABLE 1

B.S. mesh	Aperture (μm)	Weight Fraction
thro' 240	< 63	0.130
- 200 + 240	63 - 75	0.052
- 170 + 200	75 - 90	0.150
- 150 + 170	90 -105	0.128
- 120 + 150	105 -125	0.195
- 100 + 120	125 -150	0.172
- 85 + 100	150 -180	0.097
+ 85	> 180	0.076
		1.000

COEXISTENCE OF FLUIDIZED SOLIDS FLOW AND PACKED BED FLOW IN STANDPIPES

By L.S. Leung and P.J. Jones
Department of Chemical Engineering
University of Queensland
St. Lucia, Queensland, 4067
Australia

SYNOPSIS

For gas-solid downflow in a standpipe, four flow modes can be distinguished: dense phase fluidized solids flow (DENFLO), lean phase fluidized solids flow (LEANFLO), packed bed or moving bed flow (PACFLO) and slip-stick flow (STIKFLO). In this paper a criterion for demarcation between DENFLO and LEANFLO is defined in terms of the different gas compression characteristics of the two flow regimes. For a standpipe with fixed terminal pressures and a slide valve near its lower end, the previously published equations are combined to indicate the conditions at which different flow modes would occur. In particular it is shown that there is a critical aeration rate above which either PACFLO throughout the pipe or DENFLO throughout the pipe is not possible. A hypothesis is proposed that when the operating aeration rate exceeds this critical value, coexistence of DENFLO and PACFLO in the standpipe occurs. For a given set of operating conditions, prediction of the length of the PACFLO section is possible. The analysis further predicts the conditions for transition to LEANFLO which could cause the so-called "pressure reversal" problem in industrial standpipes. While the present analysis appears to give plausible explanation to unusual behaviour in industrial standpipes; further experimental verification is required and is in progress.

NOTATION

A, A_o	cross-sectional area of standpipe, open area of slide valve respectively
C_D	valve coefficient
d	particle diameter
D	diameter of standpipe
D_o	$\sqrt{(4A_o/\pi)}$
f	friction factor ($\simeq 0.003$)
g	gravitational acceleration
$P_{1,2,3}$	pressure at positions 1,2,3 (Fig.3)
ΔP_o	pressure drop across orifice, (P_2-P_3)
U_g	superficial gas velocity - positive in downwards direction
U_{sl}	slip velocity - positive upwards
U_o	minimum fluidization velocity
U_b	velocity of rise of a single bubble
W_s	solid mass velocity - positive downwards
W_{so}	solid mass velocity at $\Delta P_o=0$, positive downwards
W_g	gas mass velocity-positive downwards
z, z_p	length, length of PACFLO section (Fig. 2)
α	angle of internal friction
ε	voidage in standpipe
ε_o	voidage at minimum fluidization
μ	viscosity
ρ_s, ρ_g	solid, gas density respectively
ρ_o	$(1-\varepsilon_o)\rho_s$
ϕ	= sphericity

INTRODUCTION

Although standpipes have been used for many years for transferring solids downwards, their operation is not without some major problems such as flow instability, loss in circulation and pressure reversal (Matsen, 1976; Judd and Dixon, 1976). These problems are caused by the possibility of four different flow patterns and the coexistence of more than one flow pattern in the standpipe. In this paper the four different flow patterns are described and a criterion of demarcation of the two fluidized solids flow modes is defined. The coexistence of dense phase fluidized solids flow (DENFLO) and packed bed flow (PACFLO) as observed by Judd and Dixon (1976) and Leung (1976) is quantitatively discussed. In particular equations are developed for predicting the location of the flow transition point. It is shown that under certain critical conditions change over from coexistence of DENFLO/PACFLO to lean phase fluidized solids flow (LEANFLO) occurs, causing "pressure reversal" and loss in solid circulation.

Analyses of PACFLO throughout a standpipe and PACFLO coexistence with LEANFLO as reported by Kunii and Levenspiel (1969) have been presented in a separate paper (Leung et al 1978).

FOUR FLOW MODES

For downflow of solids, at least three, and possibly four, flow modes can be distinguished, viz.

(i) dense phase fluidized solids flow (DENFLO) in which particles are in suspension often at a voidage close to ε_o. It is also known as "bubbly dense phase flow" to indicate the occasional presence of bubbles. The bubbles usually flow downwards in industrial standpipes.
(ii) lean phase fluidized solid flow (LEANFLO) in which particles flow down at a high voidage sometimes in the form of "streamers" (Judd and Dixon, 1976; de Jong, 1975).
(iii) moving bed flow or packed bed flow (PACFLO) in which solids flow en bloc with little relative motion between gas and solids. The slip velocity here will be less than that at minimum fluidization, and
(iv) slip-stick flow (STIKFLO) in which solids flow is jerky, oscillating between flow and no-flow at a frequency of between about 0.1 and 1 Hz. Although STIKFLO has in the past been classified with moving bed flow, the oscillatory nature of the flow suggests to the authors a system fluctuating between DENFLO and PACFLO. We would therefore classify slip-stick flow separately as the fourth mode. No quantitative analysis of STIKFLO has been published. We suggest, at this stage, that equations for PACFLO may be applied to STIKFLO.

DEMARCATION BETWEEN LEANFLO AND DENFLO

In both LEANFLO and DENFLO, particles are in suspension. Thus the two flow regimes can generally be classified as fluidized solids flow. The demarcation between the two flow modes is diffuse and has not been defined previously. We propose here the following precise definition for the two flow regime:

$$\left(\frac{\partial U_g}{\partial \varepsilon}\right)_{W_s} > \quad \text{for DENFLO} \qquad (1)$$

$$\left(\frac{\partial U_g}{\partial \varepsilon}\right)_{W_s} < \quad \text{for LEANFLO} \qquad (2)$$

In fluidized solids flow, pressure generally increases in the downwards direction and U_g decreases correspondingly due to gas compression. Thus in DENFLO, voidage decreases in the downwards direction and vice versa in LEANFLO. This has important implications when compression and aeration effects are considered.

From the definition given in equations (1) and (2), we can develop a quantitative flow regime diagram. This flow diagram depends on the type of voidage/slip-velocity relationship applicable to a specific gas-solid system. For bubbly gas-solid flow, the following Matsen expansion equation may be applicable (Matsen, 1973).

$$U_g = -[(1-\varepsilon_o)/(1-\varepsilon)][U_b - W_s/\rho_o] + U_b - U_o - W_s/\rho_s \qquad (3)$$

From equations (1), (2) and (3) we have

$$W_s \gtrless U_b\rho_o \text{ for DENFLO and LEANFLO respectively} \qquad (4)$$

In such a system, the flow mode is dependent only on the mass flux of solids.

For a gas-solid system that follows a different slip-velocity/voidage relationship, the equations for predicting flow modes will be different from equation (4). For a hydrocarbon catalytic cracking catalyst for instance, the following voidage/slip-velocity relationship has been obtained experimentally up to a voidage of about 0.75:

$$U_{sl} = 8.4\varepsilon^2 - 6.66\varepsilon + 1.36 \qquad (5)$$

In standpipe flow, the slip velocity, U_{sl} is defined by

$$U_{sl} = -(U_g/\varepsilon) + W_s/[\rho_s(1-\varepsilon)] \qquad (6)$$

From equations (1), (2), (5) and (6), we have

$$W_s/\rho_s \gtrless (1-\varepsilon)^2(25.2\varepsilon^2 - 13.32\varepsilon + 1.36) \qquad (7)$$

for DENFLO and LEANFLO respectively.

A quantitative flow regime diagram for this cracking catalyst is shown in Figure 1. Note that for this system the flow mode may depend on W_s and ε. The curve in Figure 1 for $\varepsilon > 0.75$ is shown as a broken line as it is based on extrapolation of the fluidization expansion data. A diagram such as Figure 1 is useful as a diagnostic tool for determining flow regime inside an industrial standpipe.

STANDPIPE FLOW EQUATIONS

Equations describing DENFLO and PACFLO in standpipe are available in the literature and will be summarised here. The pressure drop equations for fluidized solids flow (i.e. DENFLO and LEANFLO)(de Jong, 1975) and PACFLO

(Yoon and Kunii, 1973) are given by equations (8) and (9) respectively:

$$P_1-P_2=\int_0^z-\rho_s(1-\varepsilon)gdz+[W_s^2/[\rho_s(1-\varepsilon)]]_{at\ z} + \int_0^z [2fW_s^2/[\rho_s(1-\varepsilon)D]]dz \quad (8)$$

$$|(P_1-P_2)/z| = K_1|U_{sl}| + K_2|U_{sl}|^2 \quad (9)$$

where $K_1 = 150\mu(1-\varepsilon_o)^2/[\phi d\varepsilon_o]^2$

$K_2 = 1.75\rho_g(1-\varepsilon_o)/[\phi d\varepsilon_o]$

For PACFLO we have assumed in equation (9) that voidage is given by ε_o, voidage at minimum fluidization. This is an over-simplification and can be refined by recognizing that voidage in a moving bed can vary between ε_o and that for the most closely packed voidage depending on the operating slip-velocity.

For DENFLO, voidage can be obtained from Matsen's equation (i.e. equation (3)), or an experimental equation such as equation (5).

While equation (5) may also be applicable for LEANFLO up to a voidage of 0.75 for the catalyst concerned, reliable equations for predicting voidage in LEANFLO are not generally available. However, since pressure drop in LEANFLO is often sufficiently small, the use of an approximate value of ε in equation (8) may be acceptable. For a more reliable prediction of ε in LEANFLO, we may modify Yang's (1973) equations for vertical pneumatic conveying for this purpose.

SLIDE VALVE EQUATIONS

In most industrial standpipes a slide valve or its equivalent is often present at the bottom of the pipe. The valve can be modelled in terms of an orifice (Leung, 1977). Several equations for flow of fluidized solids through an orifice are available (Massimilla et al., 1961; Jones and Davidson, 1965; de Jong and Hoelen, 1975; Judd and Dixon, 1976). The solid flux can be described by the following equation of Jones and Davidson (1965)

$$W_s = C_D(A_o/A)\sqrt{2\rho_s(1-\varepsilon_o)\Delta P_o} \quad (10)$$

Equation (10) has one minor drawback. It predicts zero solid flow when ΔP_o is zero. In practice, flow persists as a result of gravity (McDougall and Knowles, 1969). It has been suggested (Leung et al., 1978) that the equation may be slightly modified to account for the residual gravity flow as follows:

$$W_s = W_{so} \pm C_D(A_o/A)\sqrt{2\rho_s(1-\varepsilon_o)|\Delta P_o|} \quad (11)$$

The residual solid flow W_{so} can be estimated from an equation for solid flow from an opening with no pressure difference across it such as Zenz's (1976) equation:

$$W_{so}=\rho_s(A_o/A)(1-\varepsilon_o)(gD_o/\tan\alpha)^{\frac{1}{2}} \quad (12)$$

The positive and negative signs in equation (11) refer to positive and negative ΔP_o respectively.

The gas and solid flow through an orifice can be related either by the Carmen-Kozeny equation (Davidson and Jones, 1965) or by the Ergun (1952) equation (de Jong and Hoelen, 1975). Using the latter equation we can derive the following relationship (Leung et al., 1978):

$$|\Delta P_o| = K_3|U_{sl}| + K_4|U_{sl}|^2 \quad (13)$$

where $K_3=K_1D_o(A/A_o)/4$ and $K_4=K_2D_o(A/A_o)^2/24$

Equations (11) and (13) will be used here for the analysis of gas-solid flow through a slide valve for both fluidized solids flow and packed bed flow in the standpipe. This generalization is supported by Pullen (1974) who suggested that equations similar to equation (11) are applicable for both moving bed flow and fluidized solids flow through an orifice. The ratio of particle to orifice diameter is often small for the correction terms involving this ratio in the Zenz and de Jong-Hoelen equations to be neglected

COMPRESSION AND AERATION EFFECTS

For fluidized solids flow along a long standpipe, significant gas compression can occur as a result of the increase in pressure in the downwards direction. The effects of compression have been considered in detail by Kojabashian (1958) and Do et al. (1977). For fluidized solids flow both analyses show that when solid flow rate is greater than a critical value, voidage decreases with z as a result of compression. In terms of our definition of LEANFLO and DENFLO, the critical solid flow corresponds to the transition of the two flow regimes. *Thus ε decreases with z for DENFLO and increases with z for LEANFLO as a result of gas compression.*

In a very tall standpipe, ε may eventually reach the value of ε_o, voidage at minimum fluidization and fluidized solids flow becomes impossible. Thus in DENFLO this transition occurs at the lower end of the standpipe.

In quantitative analysis, compressive effect can be accounted for by expressing U_g in terms of pressure in equations (3) and (8) for fluidized flow and in equation (9) in PACFLO.

Analyses of aeration effects in PACFLO (Leung et al., 1978) and DENFLO (Do et al., 1977) have been reported previously. The main effect of increasing aeration rate at a point is the reduction of the superficial gas velocity down the pipe above the aeration point. In LEANFLO, this leads to an increase in voidage in the section above the aeration point, and vice versa in DENFLO. This subtle difference has not been recognized in industrial practice where aeration is generally believed to increase the voidage. *In DENFLO, the present analysis suggests that increase in aeration leads to a decrease in voidage above the aeration point until fluidized solids flow is not possible* (Leung and White, 1976). In PACFLO, increasing aeration rate increases the pressure gradient, $[(P_2-P_1)/z]$, above the aeration point until transition to fluidized solids flow occurs.

COEXISTENCE OF PACFLO AND DENFLO

The occurrence of one section of a standpipe in fluidized solids flow coexisting with a lower section in PACFLO (Figure 2) is well known. There are two types of coexistence, viz. LEANFLO with PACFLO and DENFLO with PACFLO. The former phenomenon was described by Kunii and Levenspiel (1969) and has been analysed by Leung et al. (1978). Such coexistence occurs in systems where the solid flow rate into a standpipe is restricted from the top. The coexistence of DENFLO with PACFLO was observed in the experiments of Judd and Dixon (1976) and is common in standpipes where the solid flow rate is bottom constrained. Such coexistence occurs in catalytic cracking plant standpipes and can readily be detected by the observation of a discontinuity in the pressure profile such as those shown in Figure 2 (Leung and Wilson, 1973).

This paper is mainly concerned with the DENFLO-PACFLO transition in a standpipe. Judd and Dixon (1976) observed that such a transition occurs at about 0.1 m above the orifice in his 50.8 mm diameter, 4 m long standpipe. No theory, however, is available to predict the conditions under which coexistence of DENFLO and PACFLO would occur, nor is there any equation for predicting the location of the flow transition. These are the two important questions we will now take up.

One Flow Mode Throughout Standpipe

Consider a standpipe system as shown in Figure 3. P_1,P_3, slide valve position and aeration rate are the independent variables, while P_2, W_s, W_g and ε are the dependent variables.

If we assume only one flow mode exists inside the length of the standpipe, the equations presented earlier for DENFLO, LEANFLO and PACFLO can be solved for a given system and the results shown qualitatively in Figure 4. OAM in Figure 4 refers to DENFLO throughout the pipe; OCN for LEANFLO throughout and DBM for PACFLO throughout.

For an operating aeration rate below that corresponding to point O in Figure 4, only one flow mode (PACFLO) is possible. At an aeration rate between that at O and M, Figure 4 shows three possible operating points: point A for DENFLO, point B for PACFLO and point C for LEANFLO. The present theoretical analysis does not predict at which of the three operating points a system would operate. A simple stability analysis (Do, 1976) shows that DENFLO operation (point A) is inherently unstable. However, for hydrocarbon catalytic cracking catalyst for instance, DENFLO operation is generally observed. It is likely that for the Geldart Group A (Geldart, 1973) solids such as cracking catalyst, DENFLO is stable and the system will thus operate at point A. We speculate here that for Geldart Group B solid, the system might prefer to operate in PACFLO (operating point B). Work is in progress in our laboratory to determine if this speculation is correct. LEANFLO operation at point C has been observed in our laboratory and in a commercial standpipe if the operating aeration rate is approached from well above the critical aeration rate. *Thus to summarise, DENFLO operation A normally occurs for cracking catalyst, LEANFLO at C occurs if the aeration rate is approached by reducing from a very high value; there is some uncertainty about when PACFLO would occur and about whether operation at point A depends on the characteristics of the solid.*

Coexistence of two flow modes

When the aeration rate is increased beyond the critical value at M for a given system, Figure 4 suggests that the system will swing over to LEANFLO operation throughout the pipe along DN. This is an oversimplification as it neglects a second more realistic possibility: the existence of DENFLO above a short length in PACFLO, as has been observed in practice (Judd and Dixon, 1976). We shall present below an hypothesis on how the location of the flow transition (i.e. the length of the PACFLO section, z_p) can be predicted.

Figure 5 shows a family of curves calculated for various z_p by assuming fluidized solids flow above z_p and PACFLO in the z_p above the slide valve. Curve MON refers to zero z_p. $M_1O_1N_1$, $M_iO_iN_i$, $M_cO_cN_c$ correspond to increasing values of z_p at z_{p1}, z_{pi} and z_{pc} respectively. Thus for a fixed $z_p=z_{p1}$ say, the present analysis shows that there is again a critical aeration rate at M_1, beyond which DENFLO in the entire section of standpipe above z_{p1} is not possible.

It is our hypothesis that at an aeration rate above the critical aeration rate for zero z_p, the system will operate with DENFLO coexisting with z_{pi} in PACFLO at the critical point M_i. Expressed in a different way, the operating line will follow the locus of the critical points with varying z_p as aeration rate is changed. It should be stressed that the above hypothesis has yet to be tested experimentally although we do have evidence that variation of the aeration rate will result in a change in z_p.

In Figure 5 we show a critical point M_c (corresponding to z_{pc}) on the LEANFLO operating line ON (corresponding to z_p=0). Our second hypothesis is that when the aeration rate exceeds the value at M_c, the system will swing over from DENFLO/PACFLO to LEANFLO operating along the LEANFLO line ON. It must be stressed again that this hypothesis needs to be verified experimentally. We do have some evidence that in an industrial standpipe, under certain operating conditions a very small change in aeration rate can result in change over to LEANFLO. This swing over to LEANFLO at a critical aeration rate has also been observed in our laboratory. Further experimental work is in progress to establish the validity of this hypothesis.

When the aeration rate is reduced from a value greater than the critical value to cause transition to LEANFLO (i.e. point M_c according to our hypothesis), experiments in our laboratory showed that LEANFLO persists at an aeration rate well below the critical value. Such a hysterisis effect has also been observed in a commercial standpipe.

Thus to summarise we hypothesize that the operating line for a standpipe system, such as that described in Figure 3, follows the line DOMM$_i$M$_c$N in Figure 5. Along DO the entire length of the standpipe is in PACFLO; along OM, the entire length is in DENFLO; along MM$_i$M$_c$, DENFLO coexists with an increasing length in PACFLO below the DENFLO section; along M$_c$N, LEANFLO throughout the standpipe occurs. The conditions along the entire operating line can be quantitatively predicted from the equations described in the previous section. Note that the above operating line refers to one fixed set of P_1, P_3 and valve opening in a given system.

DISCUSSION

In the above analysis, we have considered only the effect of variation of aeration rate. Variation of other important parameters, such as (P_1-P_3), valve opening and location and number of aeration points can also be accounted for in the present analysis. One of the possible uses of the analysis is to identify problem areas in a specific standpipe and to arrive at an optimal operating strategy to achieve target recirculation rate with no flow instability. The possible variations of all these parameters need to be considered in arriving at optimal operating conditions. Further, we have restricted this discussion to one type of standpipe system. Three other types of systems are common in industrial plants (Jones and Leung, 1977) with different independent parameters. The present analysis needs to be modified for each different situation.

It should be stressed that the validity of the proposed hypotheses is subject to experimental verification. The tentative operating line, however, appears to give a plausible explanation to some apparently unusual behaviour of industrial standpipes. Our analysis suggests that depending on (P_1-P_3) and valve opening, it is possible to have co-existence of DENFLO/PACFLO at zero aeration rate, as was observed by Judd and Dixon (1976). Examination of their results indicates that the position of inflection in their observed pressure profile (i.e. a measure of z_p) does vary with (P_1-P_3) and valve opening, a point consistent with our hypothesis. Much further experimental supporting evidence, however, is required. Our work continues.

REFERENCES

de Jong, J.A.H. (1975) Powder Technology, 12, 197-200.

de Jong, J.A.H. and Hoelen, Q.E.J.J.M., (1975) Powder Technology, 12, 201-208.

Do, D.D., Jones, P.J., Leung, L.S. and Matsen, J.M. (1977) Proc. Particle Technology Nuremberg, Ed. Brauer, H. and Molerus, O. pp. D23-D46.

Do, D.D. B.E. Thesis, Univ. of Qld. 1976.

Ergun, S. (1952) Chem. Eng. Progr., 48, No. 2 pp.89-94.

Jones, D.R. and Davidson, J.F., (1965), Rheologica Acta, 4, 180

Jones, P.J. and Leung, L.S. (1977). Proc. 5th Australian Chem.Eng. Conference, pp.322-327 Inst. Chem. Engrs. in Australia.

Judd, M.R. and Dixon, P.D. (1976) "The flow of fine dense solids down a vertical standpipe". Paper presented at AIChE Annual Conference, Chicago.

Kojabashian, C., (1958) Ph.D. thesis, Massachusetts Institute of Technology.

Kunii, D. and Levenspiel, O. (1969) "Fluidization Engineering", Chapter 12, J. Wiley, New York.

Leung, L.S., (1977) Powder Technology, 16, 1.

Leung, L.S., Jones, P.J. and Knowlton, T.M. (1978) "An analysis of moving bed flow of solids down standpipes and slide valves" Powder Technology, in press

Leung, L.S. and White, E.T. (1976) Proc. 4th National Chem.Eng. Conference, Inst. Engrs. Aust. pp42-47.

Leung, L.S. and Wilson, L.A., (1973) Powder Technology, 7, 343-349

McDougall, I.R. and Knowles, G.H. (1969) Trans.Inst.Chem.Engrs., 47, T73-T79.

Massimilla, L. et al. (1961) AIChE J. 7, 502

Matsen, J.M. (1973) Powder Technology, 7, 93

Matsen, J.M. (1976) "Fluidization Technology" Vol. II Ed. Keairns D.L. et al. pp135-150, Hemisphere Publishing Co., Washington.

Pullen, R.J.F. (1974) Ph.D. Thesis, University of Leeds.

Yang, W.C. (1973) Ind.Eng.Chem.Fundam. 12,349

Yoon, S.M. and Kunii, D., (1973) Ind. Eng. Chem. Process Des. Dev., 9, 559-566.

REFERENCES

Zenz, F. (1976) "Fluidization Technology" Vol.II Ed. Keairns D.L. et al. pp239-252 Hemisphere Publishing Co., Washington

ACKNOWLEDGMENTS

Financial assistance from the Australian Research Grant Committee is gratefully acknowledged.

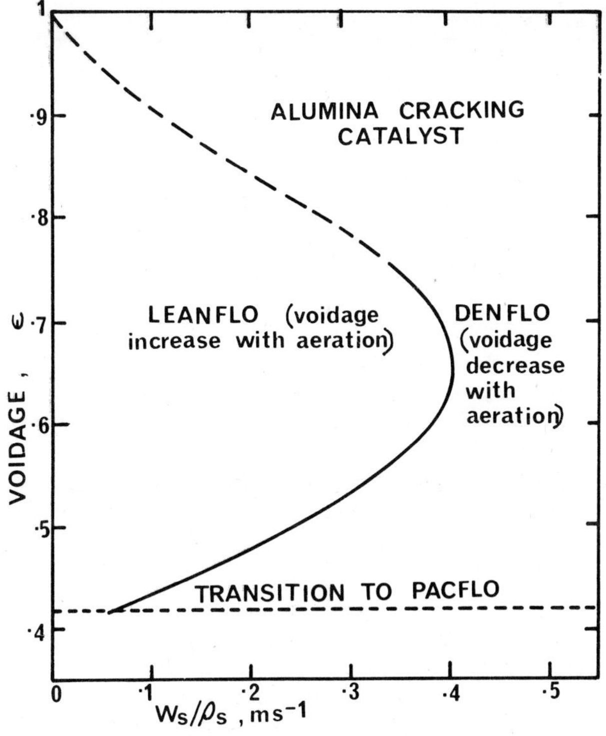

Fig.1. Quantitative flow regime diagram for alumina catalyst

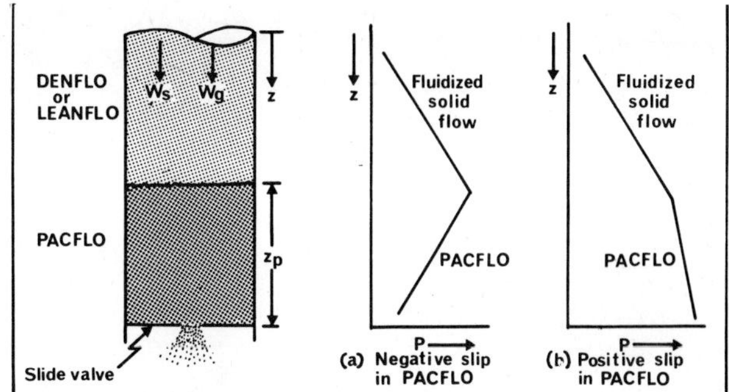

Fig. 2. Coexistence of fluidized solids flow and packed bed flow in a standpipe

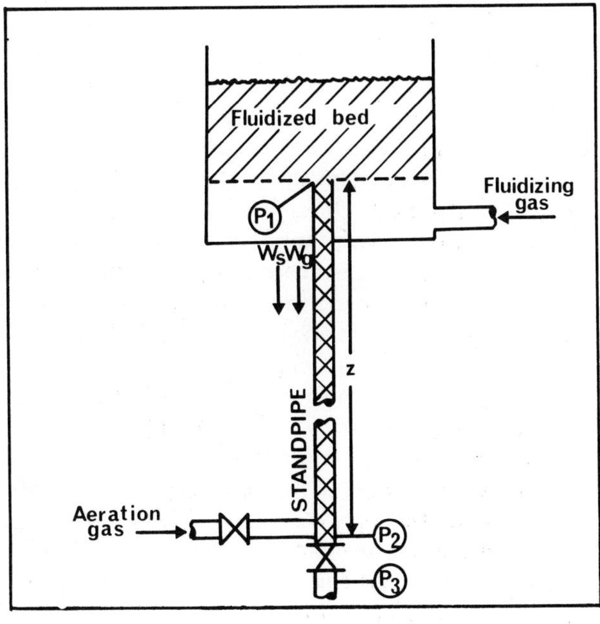

Fig. 3. Standpipe system with P_1,P_3 valve opening and aeration rate as independent parameters

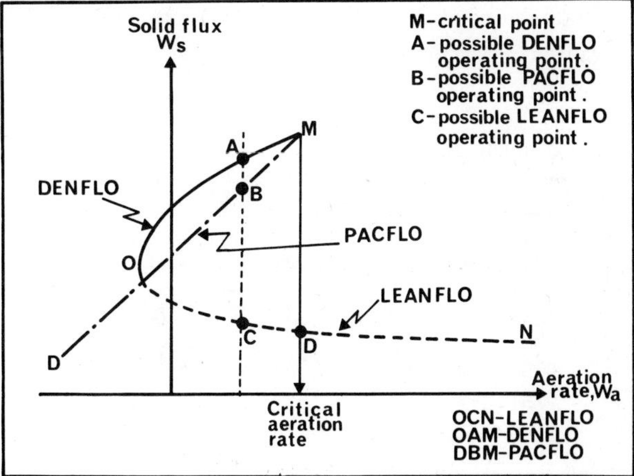

Fig. 4. Effects of aeration on solid flow rate in three flow modes for a fixed P_1,P_3 and valve opening

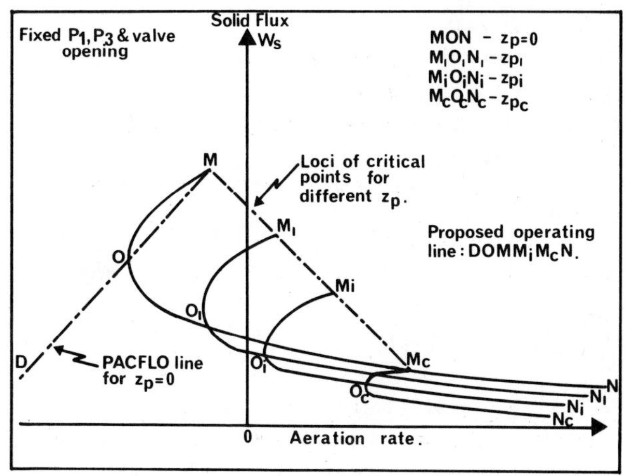

Fig. 5. Fluidized solid flow in standpipe coexisting with various length, z_p in PACFLO Line DOM refers to PACFLO throughout standpipe

THE FLOW OF FLUIDIZED SOLIDS

By J.S.M. Botterill and B.H. Abdul Halim

Department of Chemical Engineering
University of Birmingham
Birmingham
England, B15 2TT

Further experiments on the forced flow of gas-fluidized solids through a horizontal, open channel are reported. The shear stress at the vertical wall was measured directly and that at the distributor estimated. Comparison is made between these measurements for a catalyst of mean diameter 77 μm and the previous experiments with a sand of 200 μm mean diameter. Differences in behaviour reflect the different fluidization behaviour of the two materials. The catalyst expands stably and flows as on an air slide. The flow properties of the sand are strongly influenced by bubble development.

NOTATION

D_e	equivalent diameter (m)
F_d	drag force acting on distributor (N)
F_p	force due to pressure drop ΔP (N)
F_w	drag force acting on vertical channel wall (N)
h	bed height (m)
k	consistency index (N/m^2S^n)
L	length of channel over which pressure drop measured (m)
n	flow behaviour index
U_{mf}	minimum fluidization velocity (m/s)
V	average bed velocity through channel (m/s)
w	channel width (m)
γ	shear rate (I/s)
ΔP	pressure drop along measured length of channel (N/m^2)
τ_d	shear stress at distributor (N/m^2)
τ_w	shear stress at channel wall (N/m^2)

INTRODUCTION

We reported experiments on the open channel flow properties of a fluidized bed of a 200 μm mean diameter sand (U_{mf} = 0.042 m/s) at the Asilomar meeting (Botterill and Bessant, 1975). This type of material belongs to the class which Geldart designated "B" (Geldart, 1973) and with which bubbles develop in the bed as soon as the minimum fluidizing velocity is exceeded. The general flow behaviour showed a pseudo plastic characteristic. At fluidizing velocities of $2U_{mf}$ and for higher bed circulation rates through the open channel, there was a horizontal velocity gradient across the channel but little gradient in the vertical direction so that the flow appeared to be of a semi-plug flow character suggestive of ready slip at the distributor as on an air-slide. At lower bed flow rates and closer to the minimum fluidization velocity, there was also a marked velocity gradient upwards through the bed. There were conditions under which it was apparent that the drag force across the distributor against the direction of bulk solids flow exceeded that at the vertical channel walls. This was probably caused by segregation of larger particles which then became defluidized onto the distributor. In this paper we are reporting further experiments with a catalyst material of 77 μm mean diameter (U_{mf} = 0.003 m/s) which exhibits stable bed expansion with initial increase in gas flow beyond that for minimum fluidization before bubbling begins (Rowe and Yacono, 1973). This material belongs to Geldart's class "A" category.

EXPERIMENTAL

The experimental rig has been described before (Botterill and van der Kolk, 1971 and Bessant, 1973). It consisted of a closed circuit, horizontal, open channel of width 300 mm (Fig. 1). The bed was fluidized over a porous tile distributor. The flow of fluidized solids round the channel was induced by the passage of a series of paddles, P, immersed in the bed along the length of one of the principal straight sides. These paddles were carried on a belt which was driven by a variable speed drive, M, so that it was possible to vary the bulk solids flow rate through the experimental section, E, which was mounted along the other principal side. Channel aspect ratio could be varied by changing the width of the experimental section using spacers of between 100 and 180 mm length. A 100 m length, S, of the outer wall of the experimental section was freely supported and balanced against a strain gauge transducer so that the shear stress at the wall could be measured directly. The maximum force exerted on this section of wall was expected to be $\sim$ 0.05 N. In the experimental situation with the bubbling fluidized bed, reproducibility of measurement was within $\pm$ 10% at the 95% level of significance. The pressure drop driving force due to change in bed depth, ΔP, along the experimental section was measured over a 0.94 m length of the channel, L. The shear stress at the distributor could then be estimated by the difference between the total shear stress represented by the overall pressure drop along the section and that developed at the vertical walls. Thus, in terms of a force balance:

$$F_p = F_w + F_d$$

so that, for the 0.94 m length of channel of width w,

$$\tau_d = \frac{F_p - F_w}{0.94w}$$

Some measurements were again made of the velocity profile across the channel at different heights above the distributor using a small turbine element (Bessant, 1973) but the catalyst particles tended to jam in the rotor's needle bearings and the device was difficult to use. The average bulk bed velocity, V, along the experimental section was measured using a float consisting of three tubes connected together in a triangle and loaded with shot so that the device was submerged to within 10% of the bed's depth from the distributor. Average bed flow velocities up to 0.4 m/s were obtained over a range of fluidizing air flow rates up to 3 U_{mf}.

Cremer and Marchello (1968) have pointed out that many models can be used for the description of non-Newtonian flow characteristics but the simple power law model:

$$\tau = k(\gamma)^n$$

in which n is the flow behaviour index and k is the consistency index, has much advantage because of its simplicity. For a pseudo plastic material, n is less than 1 and can be determined as the slope of the logarithmic plot of $\frac{D_e \Delta P}{4L}$ against $\frac{8V}{D_e}$. The slope is independent of systematic error in the estimate of the values of the characteristic dimension, D_e, and the average velocity, V. Good straight line plots were obtained from the experimental results having defined the equivalent diameter in terms of four times the cross sectional area divided by the wetted perimeter but using a weighting factor in estimating the wetted perimeter according to the relative values of the measured shear stress at the vertical wall and that estimated at the distributor, viz:

$$D_e = \frac{4wh}{\frac{\tau_d}{\tau_w} w + 2h}$$

For a pseudo plastic fluid, the shear rate at the wall is then estimated by:

$$\gamma = \left(\frac{3n+1}{4n}\right)\left(\frac{8V}{D_e}\right)$$

RESULTS

Typical results for the flowing catalyst beds are given in Figures 2 and 3. Comparison is made between the wall and distributor shear stresses at given shear rates for the catalyst and sand beds in Table 1. The shear stresses at the wall reduce (Fig. 2) as the fluidizing gas flow rate is increased. At the distributor they first reduce then tend to rise. At both surfaces they increase with increase in shear rate. As the channel width was increased from 120 to 180 mm there was a marked increase in the shear stresses measured at the wall and estimated at the distributor (Fig. 3). Typical velocity profiles across channels of 100 and 180 mm width are given in Figure 4. It was observed that bubble activity was less influenced by shear rate in narrower channels.

Initially, the pressure drop measurements from the sand experiments had been interpreted on the assumption that the shear stresses at the vertical channel walls and at the distributor were equal (Botterill et al., 1971). It was then realised (Botterill and Bessant, 1975) that, at lower shear rates, the shear stress at the distributor could be larger and, at higher shear rates and fluidizing velocities, smaller than at the walls. Some of the earlier pressure drop measurements have now been approximately reinterpreted in the same way as the catalyst results. Comparative results for given shear rates are given in Table 1. It can be seen that the wall shear stresses are of similar value at 1.75 U_{mf} but those for the sand bed are higher at higher fluidization rates. The shear stress at the distributor is consistently higher for the sand bed, which had the higher mean particle diameter, than for the catalyst bed. Shear stress at the distributor went through a minimum at $2U_{mf}$ with the sand experiments. Whereas the catalyst bed showed measurable stable expansion as the fluidizing gas velocity was increased, there was little change in bed height for the sand bed (e.g. Table 1).

DISCUSSION OF RESULTS

Both the catalyst and sand beds showed pseudo plastic flow properties. The power law indices were 0.31, 0.33, 0.41 and 0.63 for the catalyst bed fluidized at 1.75, 2, 2.5 and $3U_{mf}$ respectively. The indices varied between 0.25 and 1 for the sand experiments. Differences in behaviour between the beds reflect their different fluidization properties and the manner in which the excess gas flows through them. Thus, the catalyst has stable bed expansion characteristics (Rowe and Yacono, 1976) and the bed flows as on an "air slide". The sand bed begins to bubble when the minimum fluidization velocity is exceeded and the greater shear stress at the distributor (Table 1) is an indication of greater particle interaction with the porous tile distributor. Similarly, the increasing wall shear stress as the fluidization velocity is increased beyond $2U_{mf}$ is a consequence of more vigorous bubbling throwing particles into contact with the containing walls and thereby increasing the momentum transfer between the flowing bed and the vertical containing wall. Because bubbles grow as they rise through the bed, the average wall shear stress can be expected to increase with increasing bed depth in bubbling systems and it should be noted that the sand bed depth was less than that of the catalyst bed (Table 1). There was generally noticeable increase in shear stress at the walls and distributor as bed depth was increased for both systems. A bubbling fluidized bed probably possesses different flow characteristics between the vertical and horizontal directions because of the influence of the rising bubbles in generating vertical solids circulation.

The channel width effect with the flowing catalyst bed (Fig. 3) was also observed to a more limited extent with the sand experiments. This is taken to be a consequence of increasing unevenness of glas flow into the bed as the channel width is increased.

The velocity profiles for the flowing sand bed (Botterill and Bessant, 1975) vary between those showing fully developed gradients in both the horizontal direction across the channel and vertically through the bed, to those showing a semi-plug flow condition at higher rates of shear and gas flow with little gradient vertically which would suggest that there was slip at the distributor. The profiles obtained with the catalyst bed (Fig. 4) showed steeper gradients close to the vertical surface consistent with the lower power index generally obtained with this material. These latter profiles are suggestive of a situation in which some slip may also be occurring at the vertical surface. Close to an immersed surface the bed has an increased voidage and it is envisaged that the higher gas flow that this causes locally could give rise to freer bed flow there and hence to apparently steep local bed velocity gradients. With increase in channel width, proportionately less of the fluidizing gas flows up the wall surface and more through the body of the bed with some consequent reduction in the ease of bed flow adjacent to the surface. Tests to examine the effect of surface roughness using sand papers with asperities smaller than the mean particle size in order to reduce the voidage adjacent to the surface as well as with much larger asperities are being undertaken in order to check further on this.

CONCLUSIONS

The degree of non-Newtonian behaviour is related to the fluidization properties of the bed material. Both the stably expanding catalyst and the sand of larger mean diameter display a reducing power law index with decrease in fluidizing gas velocity. The estimated shear stress close to the distributor was markedly greater with the bed of larger mean diameter particles. The shear stress close to the distributor appeared to go through a minimum at about $2U_{mf}$ with the sand experiments.

Velocity profiles suggest that the flow properties of the catalyst bed were sensitive to the gas flow at the bed boundaries. Aspect ratio and surface roughness effects are expected to be related to this through their influence on gas distribution.

Apart from the "air slide" effect with this system, a further complication to be expected is that it will possess different flow characteristics between the vertical and horizontal directions because of the bubbles rising through the bed.

REFERENCES

Bessant, D.J. (1973) Ph.D. Thesis, University of Birmingham.

Botterill, J.S.M. & Bessant, D.J. (1975) International Fluidization Conferences, Asilomar, California, 15-20 June, in "Fluidization Technology", Ed. Keairns, Volume 2, p.7, Hemisphere Publishing Company.

Botterill, J.S.M. & van der Kolk, M. (1971) Chem. Eng. Prog. Symp. Series 67, No. 116, 70.

Cremer, S.D. and Marchello, J.M. (1968) A.I.Ch.E. Journal 14, 980.

Geldart, D (1973) Powder Technology 7, 285.

Rowe, P.N. & Yacono, C.X.R. (1976) Chem. Eng. Sci., 31, 1179.

TABLE 1: Comparable measurements of wall and distributor shear stress, 180 mm channel width

Bed Material	Fluidizing Condition U/U_{mf}	Bed Height mm	Average Bed Velocity mm/s	Shear Rate 1/s	Shear Stress N/m^2 Wall	Shear Stress N/m^2 Distributor
Catalyst 77 μm mean diameter	1.75	95	134	5.6	2.8	0.6
			183	7.7	3.1	0.7
			257	11.2	3.3	0.9
	2	99	127	5.6	2.1	0.7
			169	7.7	2.4	0.9
			253	11.2	2.7	1.2
	2.5	102	159	5.6	1.7	0.3
			204	7.7	1.9	0.5
			283	11.2	2.2	0.75
	3	105	178	5.6	1.35	0.3
			240	7.7	1.6	0.4
			329	11.2	1.35	0.45
Sand 200 μm mean diameter	1.75	77	116	5.6	2.6	2.5
			161	7.7	3.1	3.2
			233	11.2	3.9	3.65
	2	77	111	5.6	2.6	1.15
			150	7.7	3.3	1.5
			235	11.2	4.55	2.0
	2.5	77	120	5.6	2.9	2.0
			160	7.7	4.2	2.4
			233	11.2	6.25	–
	3	77	110	5.6	3.1	2.4
			161	7.7	5	2.4

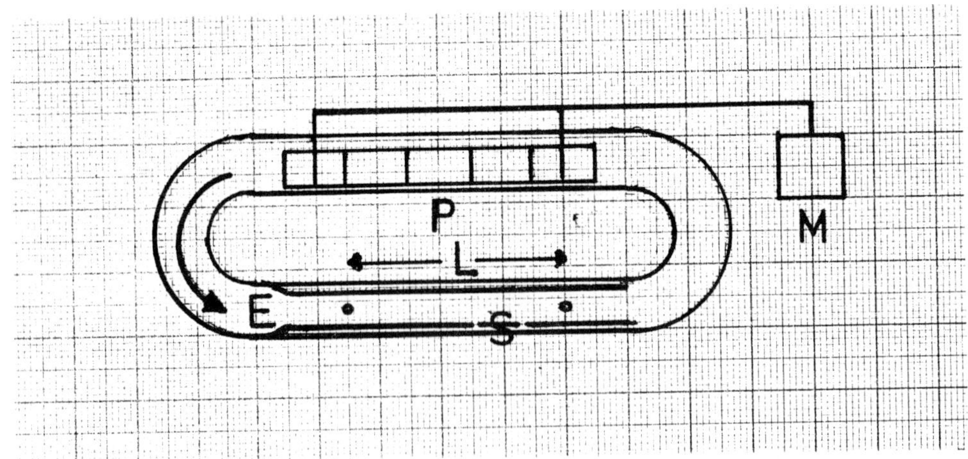

Fig. 1. Outline of rig. Paddles, P, driven by the variable speed motor, M, induce bed flow through the open channel, experimental section, E. Pressure drop is measured over a length of channel, L, and shear stress at a movable section of vertical wall, S.

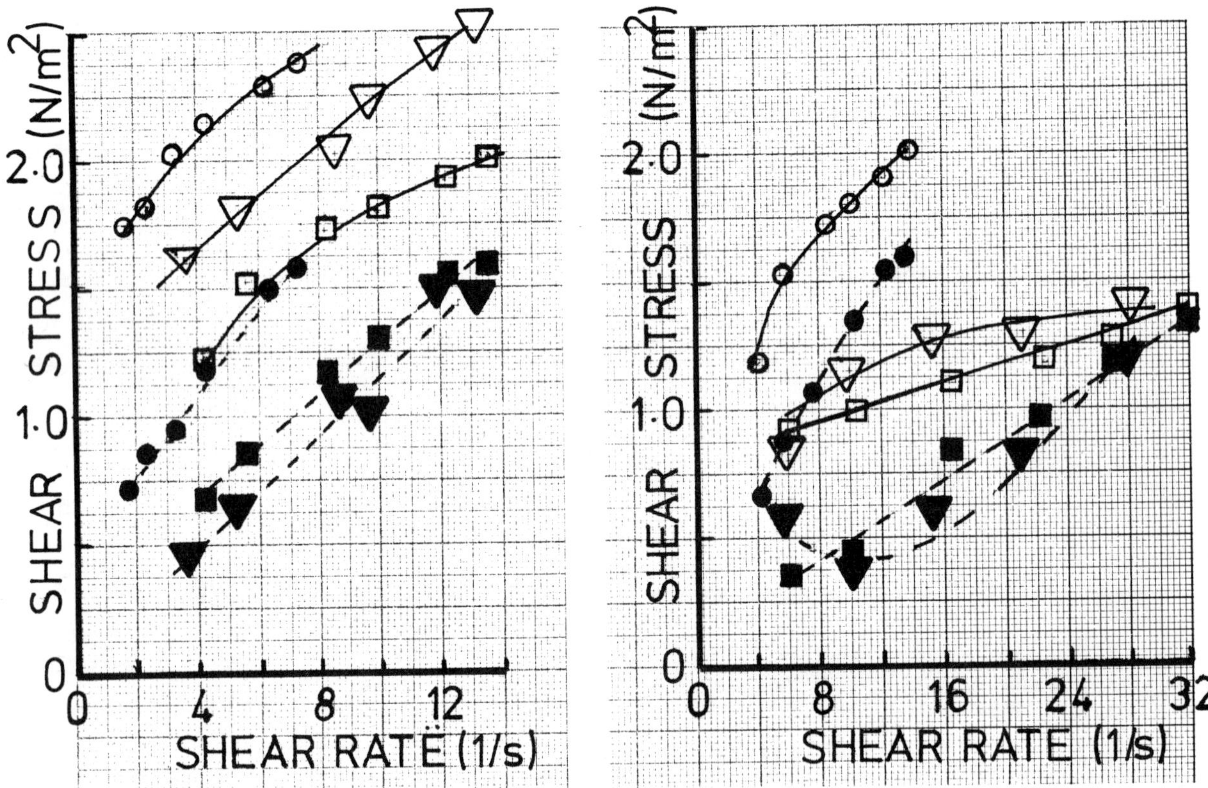

Fig. 2. Effect of fluidizing velocity on wall shear stress (open points) and distributor shear stress (full points); 77 μm catalyst, 180 mm channel width; 0, 1.5 U_{mf}; Δ, 1.75 U_{mf}; $\square$, 2 U_{mf}.

Fig. 3. Effect of channel width on wall shear stress (open points) and distributor shear stress (full points); 77 μm catalyst fluidized at 2 U_{mf}; 0, 180 mm; Δ, 120 mm; $\square$, 100 mm width.

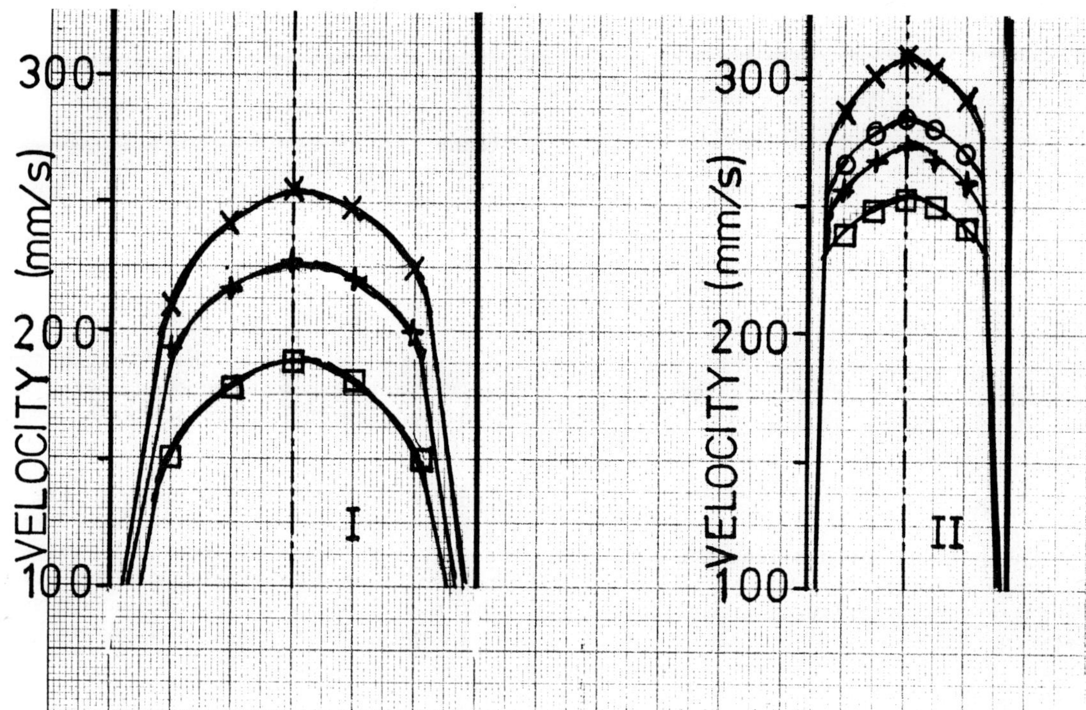

Fig. 4. Velocity profiles across channel; 77 μm catalyst fluidized at 2 U_{mf}. I 180 mm channel width, shear stress 1.8 N/m^2 at wall and 1.3 N/m^2 at distributor, average bed velocity 204 mm/s; height above distributor: x 85 mm, + 45 mm, □ 25 mm. II 100 mm channel width, shear stress 1.3 N/m^2 at wall and 1.2 N/m^2 at distributor, average bed velocity 268 mm/s; height above distributor: x 80 mm, o 60 mm, + 42 mm, □ 20 mm.

THE EFFECT OF AERATION TAP LOCATION ON THE PERFORMANCE OF A J-VALVE

By T. M. Knowlton and I. Hirsan

Institute of Gas Technology
3424 S. State Street
Chicago, Illinois 60616 U. S. A.

and L. S. Leung

University of Queensland
St. Lucia, Queensland
Australia 4067

SYNOPSIS

The effect of varying aeration tap location on J-valve performance was investigated. The maximum solids flow rate obtainable through the J-valve increases as downcomer length increases. The amount of aeration needed to produce a certain solids flow rate decreases with downcomer length.

NOTATION

- A cross sectional area of downcomer, ft^2
- $\bar{d}_P$ average particle size, ft
- d_{P_i} average particle size of a particular screen fraction, ft
- K voidage slope constant, sec/ft
- L downcomer length, ft
- L_{MIN} minimum downcomer length, ft
- ΔP pressure drop, psi
- ΔP_i Pressure drop across i^{th} section of downcomer, psi
- Q_{DC} volumetric gas flow rate in downcomer, actual CF/min
- Q_{EXT} volumetric gas flow rate added externally to J-valve, actual CF/min
- Q_T total aeration gas for J-valve, actual CF/min
- V_G gas velocity, ft/s
- V_R gas-solids relative velocity, ft/s
- V_S solids velocity, ft/s
- W_A mass flow rate of gas in downcomer, lb/s
- W_S mass flow rate of solids in downcomer, lb/s
- X_i wt fraction of a particular screen size, dimensionless
- ρ_G gas density, lb/CF
- ρ_{LB} loosest bulk density of solids, lb/cu ft
- ρ_P particle density, lb/cu ft
- ρ_{VB} vibrated bulk density of solids, lb/cu ft
- μ viscosity of gas, lb/ft-s
- ϕ particle shape factor, dimensionless
- ε voidage in downcomer, dimensionless
- ε_0 minimum packed bed voidage, dimensionless

INTRODUCTION

A nonmechanical solids-flow control valve is a device which uses only relatively little aeration gas to control the gravity flow of particulate solids. These devices possess no moving mechanical parts which are subject to wear or seizure. They are extremely inexpensive and, because they are made of pipe, they can be fabricated "in house". This eliminates the long delivery times sometimes experienced with mechanical valves. Nonmechanical valves can also be used to feed or control solids flow into either a dense- or lean-phase environment, such as a fluidized bed or pneumatic conveying line.

The J-valve is a specific type of nonmechanical valve. Its name derives from its curved shape which is similar to the letter "J". Because of the potential advantages of this valve, a study was undertaken to determine how aeration tap location affected its design and performance. Specifically, the effect of J-valve aeration location on the maximum solids flow rate through the valve and the amount of aeration necessary to cause a particular solids flow rate were investigated.

EQUIPMENT AND MATERIALS

Figure 1 shows a flow sheet of the equipment used in the J-valve investigation. The solids to be tested were charged to the top of the solids receiver. They flowed out of the solids receiver in gravity flow, down through a 3-inch diameter, 29-foot-long Plexiglas downcomer, to the J-valve. The solids flowed around the J-valve and into a dense-phase lift section, approximately 31 inches long. This section was the subject of a separate simultaneous study. At the top of the dense-phase lift section, the solids entered the lean-phase lift section which carried them back into the solids receiver. The air from the lift line passed from the solids receiver into a cyclone assembly and then into an absolute filter. A control valve after the absolute filter kept the system pressure constant. The air passed to the atmosphere via the control valve.

A –40 + 80 mesh size Ottawa sand was used in the investigation. Its physical characteristics are given below.

Ottawa Sand: (—40 + 80 mesh)
Average Particle Size, $\bar{d}_P$:0.000856 ft*
Particle Density, ρ_P: 163 lb/CF
Shape Factor: 0.854
$*\bar{d}_P = \frac{1}{\Sigma x_i / d_{P_i}}$

PROCEDURE

A typical run using the J-valve first involved loading the solids into the solids receiver. After pressure testing, the air flow up the lift line was set at a velocity of 35 ft/s at a system pressure of approximately 8 psig.

Turning on the aeration gas to the J-valve then initiated solids flow around the test loop. Solids were allowed to flow for a few minutes to ensure that the solids in the downcomer were not compacted at the start of the test. Solids flow was then stopped, and the aeration gas rate increased until the solids just started flowing. The aeration gas rate and other loop conditions were recorded. The aeration gas rate was then successively increased until the maximum solids flow rate was obtained. Loop conditions were recorded at every solids flow rate.

Solids flow rates were determined from the lift-line pressure drop. First, the solids wall velocity was correlated with the actual solids flow rate in the Plexiglas downcomer during a solids flow test. In the test, the solids flow rate was determined by collecting the solids in a container over a measured time interval and then weighing the solids. During each solids rate measurement, the solids particle velocity at the wall was measured and correlated with the measured solids flow rate. Then in a separate determination, the solids wall velocity was correlated with the pressure drop across the lift line.

RESULTS AND DISCUSSION

The effect of adding external aeration gas at different locations in the downcomer on the amount of solids flow obtainable through the J-valve was investigated. The locations of the different taps tested are shown in Figure 2. For each tap, the solids flow rate through the J-valve was varied from zero to its maximum value by increasing the aeration gas rate through the tap. Typical results thus obtained are shown in Figure 3, where the rate of external aeration vs. solids flow rate is plotted for taps 3, 5, and 6 (as defined in Figure 2). This plot shows that the external aeration needed to cause a particular solids flow rate increases as the aeration tap location is raised, or equivalently, as the downcomer length is decreased. Also, the maximum solids flow rate dropped sharply when aerating at the upper aeration tap.

The total aeration, Q_T, which causes the solids to flow through the J-valve is not just the external aeration added. It is the sum of the external aeration gas, Q_{EXT}, and the gas carried down the downcomer with the solids, Q_{DC}. Mathematically —

$$Q_T = Q_{EXT} + Q_{DC} \tag{1}$$

The total aeration gas required for a particular solids flow rate is difficult to determine because of the difficulty in determining the amount of gas carried down the downcomer with the solids.

In normal J-valve operation, the solids in the downcomer above the J-valve move downward in packed-bed gravity flow (also called moving-bed and stick-slip flow). For this type of flow to occur, the relative velocity, V_R, between the solids and the gas in the downcomer must be less than the minimum fluidization velocity. When the relative velocity becomes greater than the minimum fluidization velocity, a transition from moving-bed to fluidized-bed occurs. Thus, for moving-bed flow in the downcomer above the J-valve:

$$V_R = V_S - V_G \tag{2}$$

$$= \frac{W_S}{\rho_P A(1-\varepsilon)} - \frac{W_A}{\rho_G A\varepsilon} \tag{3}$$

where W_A is the lb/s of gas carried down by the solids, which are flowing at a rate of W_S lb/s.

The relationship between pressure drop, V_R, and voidage for a packed-bed of solids is given by the Ergun[1] equation. The relative-velocity form of this equation, which applies in the J-valve downcomer, is —

$$\frac{\Delta P}{L} = \frac{150\mu\,(1-\varepsilon)^2 V_R}{(\phi d_P)^2\varepsilon^2} + \frac{1.75\,\rho_G(1-\varepsilon)\,V_R^{\,2}}{(\phi d_P)\varepsilon} \tag{4}$$

In order to determine Q_{DC}, and ultimately Q_T, W_A had to be calculated. However, with only two equations ([3] and [4]) and three unknowns (W_A, V_R, and ε) the amount of gas travelling down the standpipe with the solids could not be determined directly.

The following procedure to calculate W_A in the downcomer above the aeration point was then attempted. A constant voidage value was first assumed in the downcomer section below the J-valve aeration point, for all solids flow rates. Using (2) and (4), V_R and Q_T were calculated for that section. Knowing Q_{EXT}, Q_{DC} (and hence W_A) for the downcomer above the aeration point was then calculated.

In the investigation, pressure drops were measured in four different sections of the downcomer above the aeration point. Using the calculated W_A determined as described above, V_R and ε were calculated using (3) and (4) for all solids flow rates for each downcomer section.

However, the calculated voidages did not fall into the possible ε range for sand at medium to high solids flow rates. The failure of this technique to provide satisfactory downcomer voidage values was noted for every initial voidage value assumed in that downcomer section immediately below the aeration point. The technique of assuming a constant voidage in the downcomer section below the aeration point was clearly not acceptable.

However, an analysis of the above calculations for which acceptable voidage values were obtained, showed that ε in the downcomer increased with increasing V_R. This trend was nearly linear as is shown in Figure 4. Therefore, a linear relationship of the form:

$$\varepsilon = \varepsilon_0 + K V_R \tag{5}$$

was assumed to apply in the downcomer under moving bed conditions.

The range of the packed bed voidage was determined experimentally. The upper packed bed voidage limit was determined by placing the sand in a 4-inch-diameter Plexiglas tube and fluidizing it for several minutes. The gas velocity was then stopped abruptly. The material's voidage at this point was assumed to be at its maximum for a packed bed, or equivalently, at its "loosest" bulk density, ρ_{LB}. From the known weight of the bed material and its height, the "loosest" bulk density could be calculated. The voidage at this condition was calculated using the relationship —

$$\varepsilon = \frac{1-\rho_{LB}}{\rho_P} \tag{6}$$

For the sand material used, the maximum voidage was 0.43.

The lower packed-bed voidage limit was calculated from (6), using the vibrated bulk density, ρ_{VB}, in place of ρ_{LB}. The vibrated bulk density was determined in the laboratory using a vibrator to compact the sand as tightly as possible. Using ρ_{VB}, 0.36 was obtained as the lower voidage limit. This value was used as ε_0 in (5).

The relative velocity just before minimum fluidization was taken to be the velocity at which ρ_{LB}, and thus the highest packed bed voidage, occurred. Experiments in the 4-inch diameter column determined this velocity to be 0.45 ft/s.

Knowing ε_0 and the relative velocity at the maximum packed-bed voidage, the value of K in (5) was determined using —

$$0.43 = 0.36 + K\ (0.45) \tag{7}$$

This gave a value for K of 0.1555, making (5) —

$$\varepsilon = 0.36 + 0.1555\ V_R \tag{8}$$

(3), (4), and (8) could then be solved simultaneously to determine ε, V_R, and W_A in the standpipe. Knowing the pressure drop across the standpipe, the solids mass flow rate, and material physical parameters enable ε, V_R, and W_A to be calculated.

In the actual investigation, pressure drops were measured in 4 different sections of the downcomer. (3), (4), and (8) were applied to each section individually and W_A was calculated for each section. The average value of W_A was then used to calculate the CF/min of gas (Q_{DC} in [1]) carried down by the solids. The volumetric flow rate of gas carried down by the solids was then added to the measured volumetric flow rate of gas added to the J-valve. This gave the total aeration requirement for a particular solids flow rate.

This method was used to calculate Q_T for the three different aeration location test curves shown in Figure 3. The total volumetric flow rate of gas, Q_T, needed to cause a particular solids flow rate was found to be the same for each aeration location. The resulting Q_T vs. W_S curve is shown superimposed on the three Q_{EXT} vs. W_S curves in Figure 3.

The reason that the Q_{EXT} vs. W_S curves are different for different aeration tap locations is basic to understanding nonmechanical valves. It can be explained by looking at the pressure balance around the test loop. The pressure drops measured around the J-valve test loop are depicted in Figure 5 for an aeration tap location comparable to that of tap no. 6 in Figure 2. The high-pressure point in the system shown in Figure 5 is at the aeration tap. Thus, the pressure balance is such that the sum of the lower downcomer, J-valve, dense-phase-lift, lift-line, and bend pressure drops must equal the upper downcomer pressure drop. Mathematically —

$$\Delta P\ (\text{upper downcomer}) = \Delta P\ (\text{lower downcomer}) + \Delta P\ (\text{J-valve}) + \Delta P\ (\text{dense-phase lift}) + \Delta P\ (\text{lift}) + \Delta P\ (\text{bend}) \tag{9}$$

The upper downcomer is the dependent half of the pressure drop loop; therefore, its pressure drop will adjust to exactly balance the pressure drop produced by the sum of the five pressure drop components in the independent side of the loop. However, there is a maximum pressure drop per unit length that the upper downcomer can develop. This value is reached when the solids in the upper downcomer become fluidized. For the sand material used in this investigation, this maximum value was approximately 0.58 psi/ft.

When sufficient aeration is added to the J-valve aeration tap in the downcomer, solids start to flow. As the aeration rate is increased, the solids flow rate increases, which causes the pressure drops across the lower downcomer, J-valve, dense-phase lift, lift line, and bend to increase. At the same time, the upper downcomer develops a balancing pressure drop.

As the solids flow rate through the J-valve is increased, a point is reached where the upper downcomer pressure drop reaches its maximum, and the solids in the downcomer become fluidized. Because of its limited capacity to absorb pressure drop, a short downcomer will reach its maximum pressure drop per unit length at a lower solids flow rate than a longer downcomer. Thus, the maximum solids flow rate through a nonmechanical valve loop system is dependent upon the length of the downcomer above the valve's aeration tap.

The pressure drop in the upper downcomer is generated by the relative velocity between the gas and solids. When V_R reaches that value necessary for minimum fluidization, the transition from packed bed to fluidized bed occurs. Then, any further increase in Q_{EXT} results in gas travelling up the downcomer in the form of bubbles. This was visually observed in the Plexiglas downcomer. The bubbles hinder the flow of solids through the downcomer and decrease the solids flow rate.

The overriding criterion of downcomer operation is that the relative gas-solid velocity in the upper downcomer must cause the pressure drop necessary to balance the pressure drop generated on the other side of the loop. It is not obvious whether gas will be carried down the downcomer or pass up the downcomer for a particular aeration location and solids flow rate. For example, in the run made with the aeration tap in location no. 3 in Figure 2, at low solids flows the gas passed up the downcomer. This had to occur in order to generate the pressure drop needed to balance the pressure drop on the other side of the loop. At these low solids flows, the solids velocity was not fast enough to carry gas down and still develop a V_R large enough to produce the pressure drop needed. At all other solids flow rates, however, gas was carried down the pipe by the solids. In the run using aeration tap no. 6, gas passed up the downcomer at all solids flow rates. Thus, with aeration added at tap no. 3, gas came down the pipe with the solids, and the Q_{EXT} required for a particular solids flow was less than that required in the run using tap no. 6. In the latter run, the external aeration had to supply all of the gas needed for solids flow around the J-valve plus the gas passing up the downcomer needed to produce the pressure drop to balance the other half of the pressure drop loop.

The optimum location of the J-valve aeration tap is as close to the J-valve as possible as (9) shows. Locating the aeration tap at the point where the J-valve bend begins eliminates the pressure drop across the lower downcomer. This effectively increases the length of the downcomer and gives the highest solids flow rate possible around the loop.

How the amount of external aeration added to the J-valve varies as a function of downcomer length is shown in Figure 6. Here Q_{EXT} is plotted against downcomer length for the loop system shown in Figure 1, assuming a constant solids flow rate of 13,000 lb/hr and an aeration location 12 inches above the start of the J-valve bend. This plot was developed by assuming different downcomer lengths and then calculating the change in pressure drop that this would cause. After calculating the pressure drop across the upper downcomer, (3), (4), and (5) were used to determine W_A, and hence Q_{DC}. Knowing Q_T, Q_{EXT} was determined from (1).

The minimum downcomer length, L_{MIN}, which would allow 13,300 lb/hr to pass through the J-valve was determined by this method to be 13.5 ft. At this downcomer length Q_{EXT} is at its maximum value but then drops sharply as downcomer length increases. After a downcomer length of 2 L_{MIN}, the amount of Q_{EXT} is essentially constant with length. The amount of aeration gas added to the J-valve, even at L_{MIN}, is small compared to the lift-line flow required. The values are approximately 2 CFM and 100 CFM, respectively. Therefore, gas requirements for a J-valve are practically negligible in relation to other system gas requirements.

The maximum sand flow rate obtainable through the J-valve test loop was also calculated and is plotted as a function of downcomer length in Figure 7. This is for an aeration location 12 inches above the start of the J-valve bend. This plot shows the important effect of downcomer length on the maximum solids flow rate obtainable through nonmechanical valve systems. Specifically, an increase in downcomer length, from 12 to 13 feet, increases the maximum solids flow rate obtainable from 8,500 to 12,000 lb/hr. It should be noted that this plot assumes that the lift line and downcomer lengths were shortened simultaneously. It does not apply to the case where downcomer length is shortened by raising the aeration tap location.

To design a J-valve, or any nonmechanical valve, to feed a particular solids flow rate, L_{MIN} must be determined. This requires the following steps:

1. Estimate the pressure drops in the independent half of the pressure drop loop, using correlations or experimental data, at the required flow rate.
2. Determine the fluidized $\Delta P/L$ value for the material to be used.
3. Determine L_{MIN} from the relationship

$$(\Delta P/L)\ L_{MIN} = \sum_{i=1}^{n} \Delta P_i \qquad (10)$$

where the ΔP_i are pressure drops in the independent half of the pressure drop loop and $\Delta P/L$ is the value at minimum fluidization.

The gas travelling down the downcomer with the solids, Q_{DC}, can be determined solving (3), (4), and (8) simultaneously. But Q_{EXT} cannot be accurately predicted until more data are taken and a correlation is developed for Q_T. However, Q_{EXT} remains essentially constant with length after 2 L_{MIN} (Figure 5). The ratio of Q_{EXT} to Q_{DC} was constant, at approximately 3:1, at all downcomer lengths greater than 2 L_{MIN}. This result, although certainly not a rigorous design criterion, is a "handle" which can be used to predict the external aeration required for a J-valve.

CONCLUSIONS

1. Downcomer length limits the maximum flow rate obtainable through a nonmechanical valve system.
2. Downcomer length affects the amount of external aeration necessary to produce a particular solids flow rate through the J-valve.

3. The optimum aeration tap location for the J-valve is as close to the J-valve as possible. This aeration location gives the maximum solids flow rate possible through the valve.

FUTURE WORK

The study of J-valves, and nonmechanical valves in general, is being continued. The effect of geometrical and particle parameters, the effect of gas density on valve performance, and a correlation to predict Q_T will be determined and reported on in later articles.

ACKNOWLEDGMENT

The authors would like to thank the American Gas Association, under whose auspices and funding the work was conducted.

REFERENCE

1. Ergun, S. (1952) Chem. Eng. Prog. 48, 89.

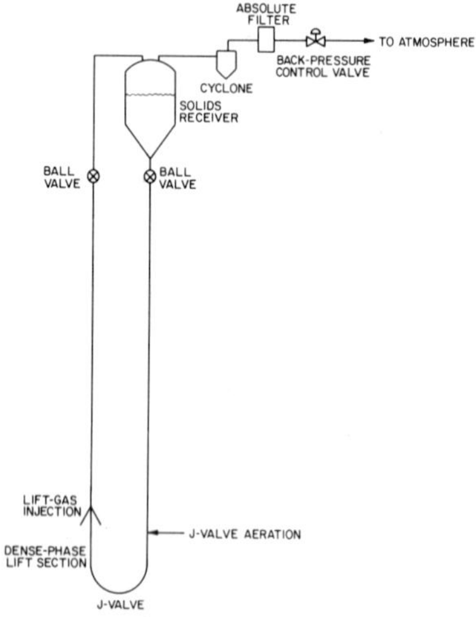

Figure 1. Flow Sheet of J-Valve Solids Recirculation System

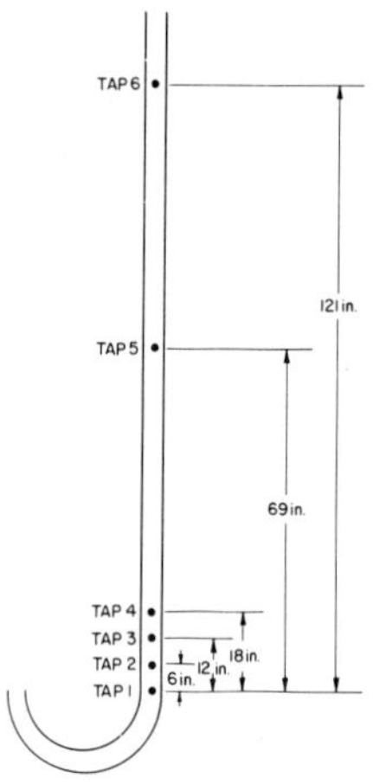

Figure 2. Aeration Tap Locations for J-Valve

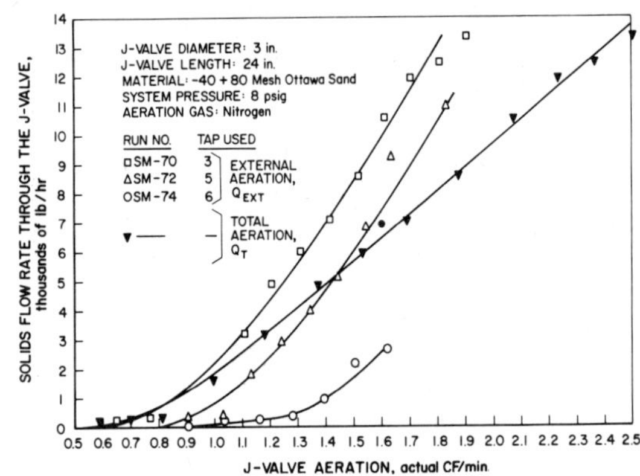

Figure 3. Solids Flow Rate Through J-Valve Vs. Aeration

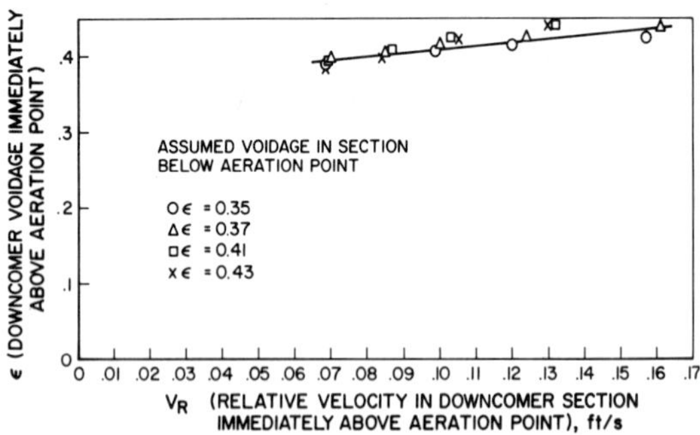

Figure 4. Voidage Vs. Relative Velocity

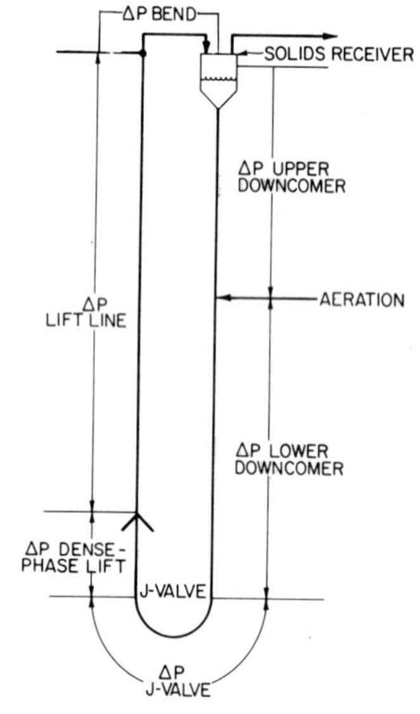

Figure 5. Pressure Drop Around the J-Valve Loop

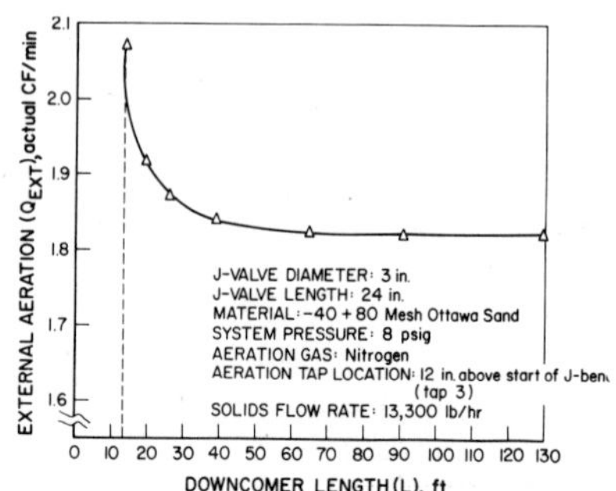

Figure 6. Q_{EXT} Vs. Downcomer Length

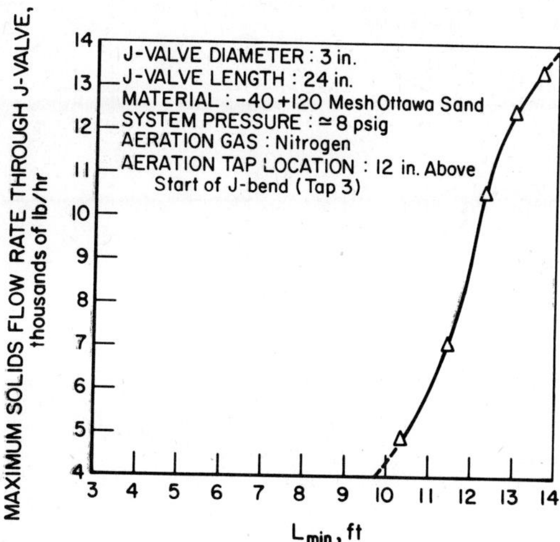

Figure 7. L_{MIN} Vs. Maximum Solids Downcomer Flow Rate

PARTICLE INTERCHANGE THROUGH THIN AND THICK BAFFLE PLATES IN MULTISTAGE GAS FLUIDIZED BEDS

By P. Guigon*, J.F. Large*, M.A. Bergougnou**, and C.G.J. Baker**

*Université de Compiègne
B.P. 233 60206 - Compiègne
France

**Faculty of Engineering
University of Western Ontario
London, Canada
N 6 A - 5 B 9

SYNOPSIS

A comparison of the solid interchange rate between the stages of a baffled fluidized bed has been made with sand and cracking catalyst. Thin plates and plates equipped with nozzles (to simulate thick plates) were used. From experimental results, general trends of variation of interchange rate were deduced. Furthermore, two different types of solid flow, for the same gas velocity, were observed when using long nozzles with a length to diameter ratio greater than four, namely a dilute and a dense phase flow pattern. The latter gives a solids exchange rate several times higher than the former.

NOTATION

d_h hole diameter, cm.

d_p mean particle diameter, μm.

F solids interchange rate based on cross-sectional area of the column, g/cm^2 s.

h_b baffle spacing, cm.

l baffle thickness or nozzle lentgh, cm.

V_g superficial gas velocity, cm/s.

V_h velocity through baffle holes, m/s.

Z height of dilute phase under the baffle, cm.

ε baffle free area

INTRODUCTION

Previous studies (Overcashier et al., 1959; Brauer et al., 1970 ; Nishinaka et al., 1974 ; Guigon et al., 1974) have shown that the use of perforated plates in a fluidization column allows the redistribution of the fluidized bed between a plurality of well-mixed stages. This staging of the bed reduces solids back-mixing, makes it possible to maintain a temperature gradient along the bed, helps keep the bubble size small and improves significantly gas - solid contacting.

The objective of this study was to predict the performance of a baffled fluidized bed through the qualitative and quantitative characterization of the solid exchange between the various stages.

EXPERIMENTAL

The experiments were carried out at constant solids inventory in a column 28 cm in diameter and 260 cm high. The solids exchange rates were measured by means of a heat balance at steady state between a heated lower stage and a cooled upper stage. The experimental procedure has been already described in detail by P. Guigon et al. (1974).

SOLIDS EXCHANGE THROUGH THIN BAFFLE PLATES

The solids used were mainly sand. A few experiments were made with cracking catalyst to compare the results with those obtained by Overcashier et al. (1959) and Nishinaka et al. (1974). The characteristics of the baffle plates are shown in table 1. The solids hold-up in each stage, the height of the dilute phase under each baffle and the solids backmixing were studied.

Results for sand (d_p = 135 and 190 μm)

Fig. 1 shows that gas superficial velocity and baffle open area have a strong influence on the exchange of solids between stages. Similarly the height of the dilute phase depends mainly on the gas superficial velocity (Fig. 2), the interbaffle distance and to a lesser extent on grid geometry and particle size.

Results for fluid cracking catalyst (d_p = 59 μm)

Fig. 3 shows that the graph of the exchange rate versus gas velocity goes through a maximum as was the case with sand. The exchange rate increases with baffle free area but is only slightly affected by the hole diameter. Fig. 4 gives the variation of the dilute phase height Z with superficial velocity. Z varies inversely with baffle free area and is only slightly affected by the hole diameter.

Discussion

The exchange rate is the result of the interplay between a descending flux F_D of solids (plate leakage) and an ascending flux F_E of solids entrained from the bed below. At steady state these fluxes are equal and set the height of the dilute phase. The leakage flux is a function of hole velocity, baffle free area, hole diameter and solids fluidity near the holes. The entrainment flux just under the baffle plate is a function of the gas superficial velocity, the dilute phase height, the solids characteristics and the height of the fluidised bed below the baffle.

At low gas flowrates, the gas velocity and the solids entrainment are too low to counteract the increase in leakage due to the higher fluidity of the bed near the orifice. Thus, at low gas flowrates, an increase in gas velocity leads to a larger downward solids flux. On the other hand, the entrainment flux just under the baffle increases with gas velocity. Therefore the height of the dilute phase below the baffle does not have to vary much to establish itself up in such a way that the ascending and descending particles fluxes remain equal.

At high gas flowrates solids leakage is considerably impeded by upward gas and entrainment flux in such a way that an increase in gas velocity leads to a decrease in leakage rate. The entrainment flux will have a tendency to increase. However, at steady state, this flux at the baffle level has to be equal to the leakage flux and therefore the dilute phase height under the baffle has to increase. These considerations explain the presence of a maximum for the exchange rate versus the gas velocity as seen in Fig. 1, 3 and 5a.

Similarly, the shape of the curve representing the variation of the dilute phase height Z with superficial gas velocity Vg (Fig. 2, 4 and 5b) can be explained. At the level of the baffle the entrainment rate is a function of Vg and Z. Since the exchange rate (and therefore the entrainment flux) goes through a maximum, the curve representing the variation of Z with Vg will have first a small slope (zone AB of Fig. 5b corresponding to the ascending part of Fig. 5a), then a higher slope (zone BC of Fig. 5b corresponding to the decreasing part of Fig. 5a) and finally Z will tend asymptotically to the interstage height.

The observed influence of the gas velocity on the solids exchange rate discussed above seems to infer that leakage is the controlling parameter of the exchange. This is confirmed by the fact that the exchange flux is affected by the same parameters as the leakage flux (baffle free area, hole diameter).

Through the use of two solids having radically different characteristics, namely coarse sand and fine cracking catalyst, it was possible to investigate the overall trends of the leakage curve and of the height of the dilute phase as a function of superficial velocity. Depending on the solid chosen and the operating conditions either a portion or the totality of the curve on Fig. 5a and 5b will be observed.

A comparison of the present result on cracking catalyst with those of Nishinaka et al., (1974) and Overcashier et al., (1959), shows that :

1. the exchange rate found in this study is in between those of the above authors for similar baffle free areas.
2. the variation of the exchange rate with baffle free area is larger. These discrepancies could be due to the effect of column diameter.

SOLIDS EXCHANGE THROUGH THICK BAFFLE PLATES

Such plates are used extensively in industry. It is, thus, interesting to compare their behaviours to those of thin plates. In this study, plate thickness was conveniently simulated by equipping grid holes with cylindrical nozzles of varying lengths (see Table 1).

Results for sand

The variation of the exchange rate with

hole velocity is strongly dependent on the length of the nozzles.

For 2,5 cm long nozzles the results are essentially the same as for thin plates (Fig. 6).

For 7,5 cm long nozzles, a new phenomenon was observed. At low velocities, the exchange rate was four to five times larger than for thin plates. When fluidization velocity was increased, the exchange rate began to increase slightly, then fell suddenly to the same low value as for thin plates (Fig. 7). Similarly, when the velocity was decreased from a high value, the same curve was retraced but the transition did not necessarily take place at the same gas flowrate (P. Guigon, 1976).

This transition point was not strictly reproducible but belonged to a metastable zone in which a slight hydrodynamic perturbation initiated a sudden shift for the exchange rate. It is interesting to observe that for nozzle plate N° 12 it was possible to maintain a large exchange rate up to high fluidization velocities.

Similarly to what happens for thin plates (P. Guigon, 1976), if the ratio $F/\varepsilon^{1,74}$ is given as a function of hole velocity, it can be seen that all the data points fall on the same two curves whatever the value of the baffle free area (Fig. 7). However, on the contrary, the hole diameter has no influence on the exchange rate for both 2,5 cm and 7,6 cm long nozzles.

Results for fluid cracking catalyst

For the sake of comparison, a single experiment was carried out with cracking catalyst for 7,5 cm long nozzles (Fig. 3). The same phenomenon was observed as for sand.

At low gas flowrate the exchange rate was four to five times higher than for thin plates. For a superficial velocity in the neighborhood of 20 cm/s, there was an abrupt discontinuity where the exchange rate fell to the value it had for thin plates.

Overcashier et al. (1959) while studying plates equipped with nozzles did not find the above discontinuity but observed the same behaviour as for thin plates. It could be that the aspect ratio l/d_h of the nozzle is important. Thus, for l/d_h under 2, one would have low interchange rates through the nozzles whereas for l/d_h above 4 high interchange rates would take place at low gas velocities. In Overcashier et al. experiments, l/d_h was equal to 3. One could thus infer that high interchange rate takes place only for ratio l/d_h higher than 3 and that there is some entrance effect for short nozzles.

CONCLUSION

Experiments have clearly shown that there are two regimes for the exchange rate of solids through baffles (Fig. 8a) :

- a dilute phase regime with simultaneous flow of gases and solids through all the holes, gas passing through the central part of the hole and the solids leaking at the periphery : the exchange rate is low. This is the flow regime for thin plates and short nozzles at all velocities and for long nozzles at high velocities (Fig. 8b).
- a dense phase regime where solids leak through a few nozzles and gas flows through the others. This is the flow regime for long nozzles at low velocities (Fig. 8c).

These two types of flow regimes were visually observed in a transparent column. They could be explained either by a maldistribution of gas or by the influence of the thickness of the baffle plate (Medlin et al., 1975).

REFERENCES

Brauer H., Muhle J. & Schmidt M. (1970) Chem. Ing. Techn. 42, p. 494.

Guigon P., Bergougnou M.A. & C.G.J. Baker (1974) A.I.Ch. E. Symp. Series 70 (141), p. 63.

Guigon P. (1976). Thèse de Docteur Ingénieur, Université de Compiègne.

Medlin J. et al. (1975) I.E.C. Fund. 14 (4) p. 315.

Nishinaka M. & Morooka S. (1974) Powder Technol. 9, p. 1.

Overcashier R.H., Todd D.B. & Olney R.B. (1959) A.I.Ch.E.J. 5, p. 54.

TABLE 1 : Baffle characteristics.

Baffle N°	ε	l cm	d_h cm
1	0.116	0.6	1.91
2	0.172	0.6	1.91
3	0.256	0.6	1.91
4	0.116	0.6	1.27
5	0.176	0.6	1.27
6	0.116	0.6	0.63
7	0.116	2.5	1.91
8	0.116	7.6	1.91
9	0.060	7.6	1.91
10	0.118	7.6	1.27
11	0.118	2.5	1.27
12	0.62	7.6	1.27
13	0.62	2.5	1.27

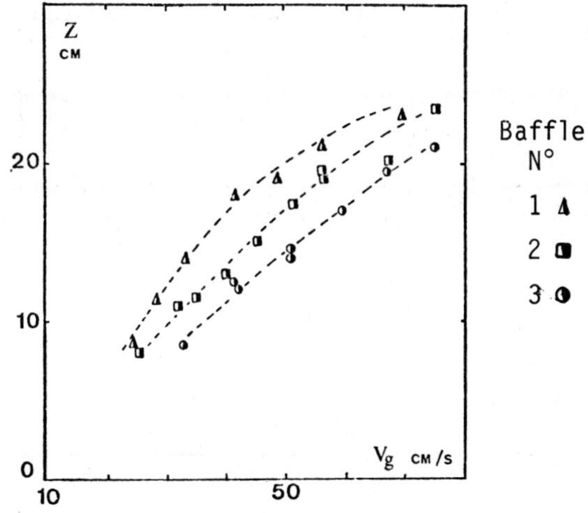

Fig. 2. Height of the dilute phase under the baffle as a function of superficial gas velocity and baffle free area - sand 190 μm

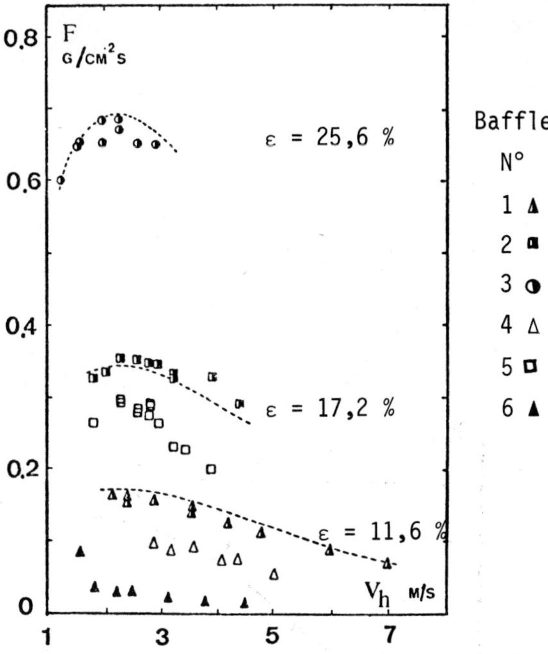

Fig. 1. Effect of hole velocity and baffle characteristics on solids interchange rate - sand 190 μm

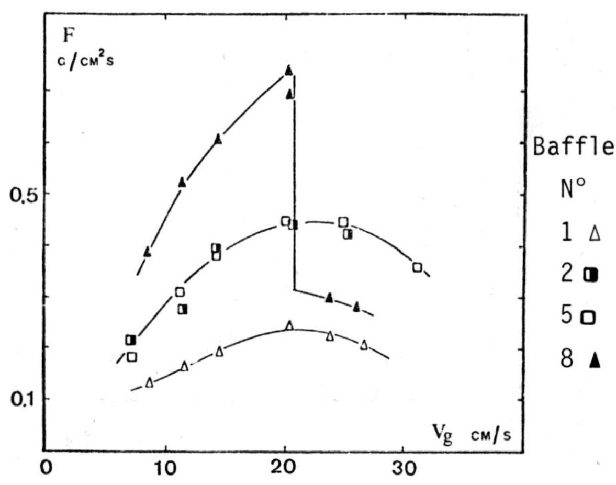

Fig. 3. Effect of gas velocity and baffle characteristics on solids interchange rate - cracking catalyst d_p = 59 μm

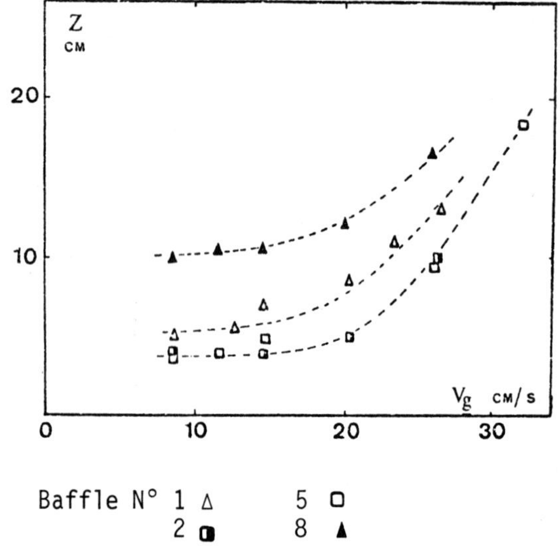

Baffle N° 1 Δ 5 □
2 ◧ 8 ▲

Fig. 4. Height of the dilute phase under the baffle as a function of superficial gas velocity and baffle free area - cracking catalyst 59 μm.

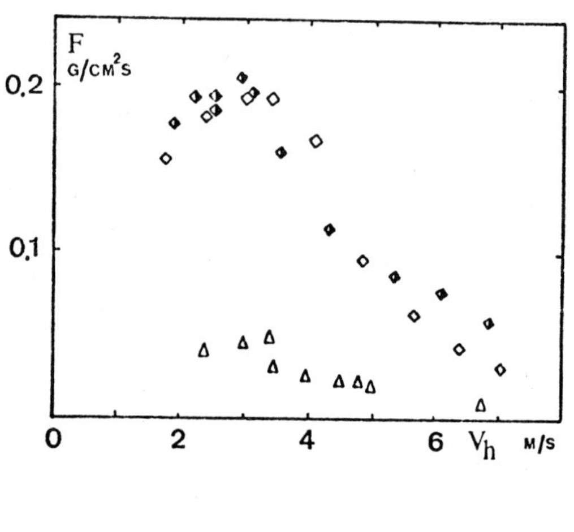

Baffle N° 7 ◆ 11 ◇ 13 Δ

Fig. 6. Effect of gas velocity on solids interchange for 2,5 cm long nozzles plates - sand 190 μm.

a)

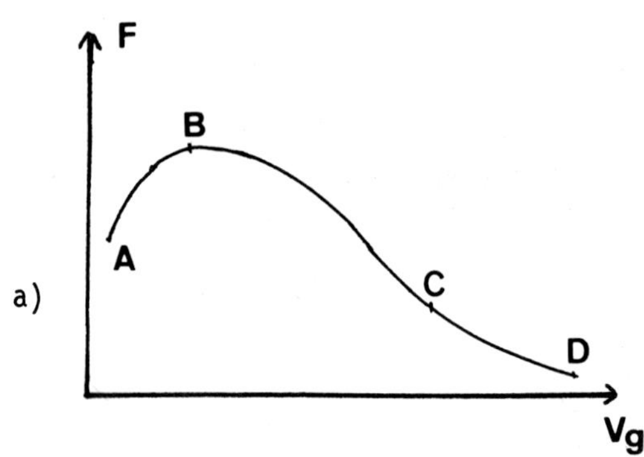

b)

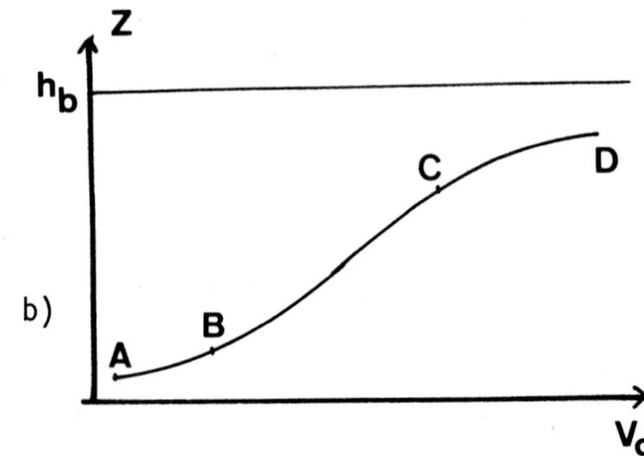

Fig. 5. Shape of the variation of the exchange rate a), and of the height of the dilute b), with the gas velocity.

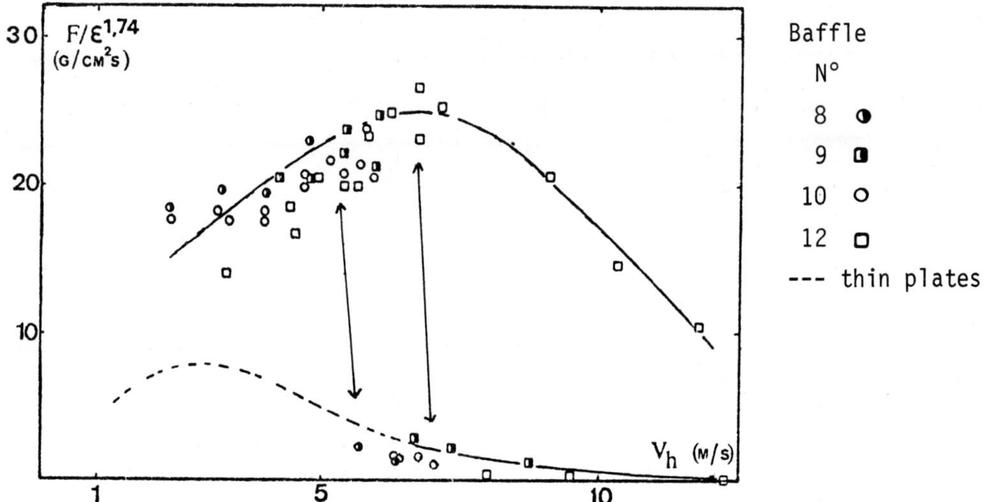

Fig. 7. Effect of gas velocity and baffle free area on solids interchange rate for 7,6 cm long nozzles plates - sand 190 μm

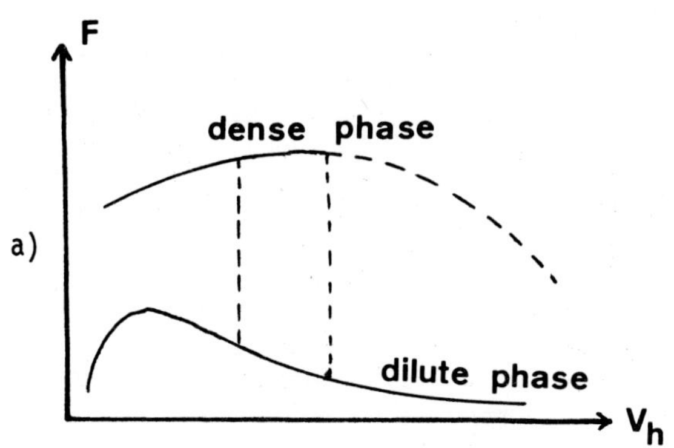

Fig. 8. The two types of flow regime a)
b) dilute phase flow
c) dense phase flow

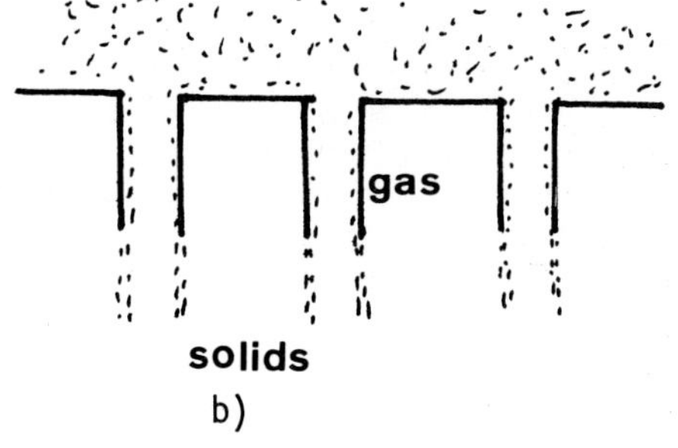

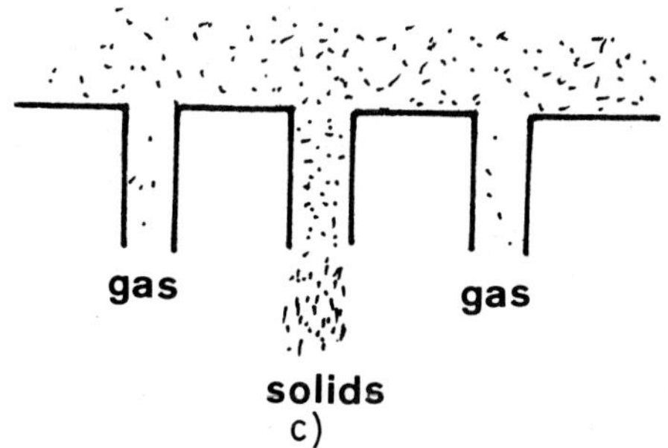

GAS BACKMIXING IN LARGE FLUIDIZED BEDS

By H.V. Nguyen[+], A.B. Whitehead[*] and O.E. Potter[+]

[+] Department of Chemical Engineering
Monash University
Clayton, Victoria 3168, Australia.

[*] Division of Chemical Engineering
C.S.I.R.O.,
P.O. Box 312
Clayton, Victoria 3168, Australia.

SYNOPSIS

Gas backmixing experiments were carried out in a 122 cm square bed under a range of conditions, using tracer gas techniques. As the velocity increased, tracer gas was first backmixed very strongly down the centre of the bed, then went through a transition with the existence of an unstable bubble pattern. Finally, backmixing moved towards the walls. These modes of gas backmixing were completely consistent with other available data on solids and bubble movements within the bed under the same conditions. The prediction of the backmixing model is satisfactory in the upper region of the bed.

NOTATION

C_{av} Average tracer concentration for each bed level

C_H Tracer concentration at each sampling point

C_M mean mixed tracer concentration

D bed diameter

H bed height; H_o static bed height

U fluidizing velocity

U_{CRIT} critical fluidizing velocity for backmixing

U_{gi} Interstitial gas velocity in the dense phase

U_{mf} incipient fluidizing velocity

Greek letters

$\bar{\delta}$ mean differential pressure difference at the bed base ($P_{central} - P_{wall\ region}$).

INTRODUCTION

In an aggregatively fluidized bed, solids are observed to move upwards in the wakes of rising bubbles and descend elsewhere. The solids downflow intensifies as the fluidizing velocity increases and eventually exceeds the gas interstitial velocity in the particulate phase. Therefore, there is a net flow of gas towards the distributor and part of the outgoing gas is backmixed into the bulk of the bed. The velocity at the onset of gas backmixing is defined as the critical velocity U_{CRIT}.

Fundamental studies of gas backmixing have been carried out under various operating conditions, in conjunction with the development of a mathematical model by Stephens et al (1967), Latham et al (1968), Nguyen and Potter (1974, 1975), Fryer and Potter (1976). For systems without chemical reactions, these authors employed a steady-state, tracer-injecting technique. Tracer gas is uniformly injected into the bed near the surface, and if the bed is operated in backmixing conditions, it would be detected at lower bed levels. The results established the increasing trend of backmixing with the fluidizing velocity and a close relationship between bubble activities and the backmixing patterns. Nevertheless, the small size of the experimental equipment employed in these works (beds were from 15 to 30 cm in diameter) restricted the maximum useable fluidizing velocity - as the bubbles readily approach slugs in size at gas rates equivalent to only a few times U_{mf}. It was also believed that in small beds, backmixing was more uniform and this condition may not be satisfied in beds of industrial scale.

The need for improved solids and gas data in larger beds has arisen in the last few years due to the diversified application of fluidization in process and energy industries.

For this purpose, data should be collected either in pilot or real plants. Werther (1974) and Schmalfeld (1976) have reported some findings on solids circulation in beds up to 1 m in diameter. Potter (1971) reviewed information obtained in industrial units. But generally, all these data are still very much incomplete.

It has been established, however, that solids and bubble movements in fluidized beds, large or small, are not results of random processes. In a given system, there exists a solids and gas pattern pertaining specially to a given fluidizing velocity. This inter-relationship has been demonstrated by the work in large beds over the years by Whitehead and co-authors (1967b, 1970, 1976a, 1976b, 1977). Bubble tracks in a 122 cm square bed were located by the use of light sensors (Whitehead et al, 1967b) while the solids downflow was traced by introducing coloured sands and labelled plastic spheres at selected spots and following their movements in the passage of time (Whitehead et al, 1970). With sand S1 (U_{mf} = 2.5 cm/s), starting with a fluidizing velocity of 9.1 cm/s, solids were observed to flow downwards predominantly in a very strong, stable central stream, within which there was no bubble track. In addition, along the walls, solids also downflowed with slow, jerky movements confined to narrow bands. The rest of the bed cross-section was reserved for the flow of bubbles and their wakes of solids. The size and velocity of the central downflow increased steadily with U, and at U = 15.2 cm/s, the maximum velocity of solids in this stream, measured by tracers, reached 21 cm/s. The slow stream near the wall reached only 5.0 cm/s.

As U increased further from 15.2 cm/s, bubble activity was observed intermittently in the centre of the bed and at U = 24.4 cm/s, an unstable, transitional pattern of solids and gas flow was recorded. Bubbles appeared at the centre of the bed for a short period, then alternately switched over to appear in the areas near the bed walls. Consequently, solids downflow was now transient in nature. Any coloured sands added into the bed were found well-mixed almost immediately with the bulk.

This dynamic variation in bed behaviour with respect to the fluidizing velocity was also investigated by measuring the pressure variation at the bed base. Whitehead et al (1970), and Whitehead and Auff (1976a) measured the time average differential pressure difference, $\bar{\delta}$, between the centre and the corners at the base of the 122 cm bed as the fluidizing velocity was being changed. They reported that $\bar{\delta}$ was positive in value in the conditions corresponding to the presence of a stable, persistent central solids downflow and it reached its maximum at the value of U where the central downflow was *strongest*. As U was increased to the value where the transient movements of solids and bubbles had been noted, the differential pressure difference at the bed base fluctuated over a wide range, but the time average reading was close to zero. Beyond this transient regime, at higher U, $\bar{\delta}$ was found persistently negative - i.e. the pressure at the bed base, was now higher at the corners than at the centre.

Results for sand S1 (U_{mf} = 2.5 cm/s) and sand S3 (U_{mf} = 10.4 cm/s), in 122 cm beds are reproduced in figure 1. With sand S1, $\bar{\delta}$ was found to be at its maximum at U = 15.2 cm/s - corresponding to the appearance of the strongest central solids downflow. The transient behaviour occurred at about 24.4 cm/s, where $\bar{\delta}$ approached zero. No solids downflow data were collected beyond U = 24.4 cm/s due to the limitations of the equipment. However, as $\bar{\delta}$ was negative above this velocity, one would expect with a high degree of certainty that the solids downflow now switched over predominantly to the areas near the walls. With sand S3, the strongest solids downflow would be in the range U = 35 → 50 cm/s. Whitehead and Auff (1976a) repeated the measurement of $\bar{\delta}$ with other materials and with varying bed aspect ratios (H_o/D) and found that the relationship between $\bar{\delta}$ and U exhibited a behaviour consistent with beds of different sizes but of the same materials and aspect ratio. Their conclusion was that the measurement of the pressure profile at the bed base could be used as a similarity criterion for the purpose of scaling-up.

A joint project of A.B. Whitehead (CSIRO) and O.E. Potter (Monash University) is being conducted in beds up to 1.22 m in size, with particles of U_{mf} up to 91 cm/s and gas velocities up to 4 m/s with and without bundles of vertical or horizontal tubes. The prime aspect of study is the backmixing of gas under these various conditions, which is related to existing information obtained by Whitehead and co-authors on the performance of large beds. Provision also exists for heat transfer measurements and ozone decomposition will be employed for catalytic fluidized-bed reactor studies.

This paper presents data obtained with the finer particles.

EXPERIMENTAL PROCEDURE

Silica sands of two sizes having incipient fluidizing velocities of 2.5 cm/s (sand S1) and 10.7 cm/s (sand C1) were fluidized with air in a 122 x 122 cm square bed. Static bed heights were 153 cm (sand S1) and 91.1 cm (sand C1). The distributor consists of 36 tuyeres fitted with interchangeable orifice cups and bubble caps. Details on the bed and tuyere design are available elsewhere (Whitehead et al, 1967a).

A mixture of CO_2 and air was used as

tracer gas. The injection was uniform at 132 cm (sand S1) and 71.1 cm (sand C1) from the distributor. A total of 49 time-average samples were taken from each bed level - all lower than the plane of injection. CO_2 concentration was determined by 4 sets of thermal conductivity cells. The design of the tracer injector, sampling probes and analytical equipment has been given elsewhere (Nguyen et al, 1977).

The critical fluidizing velocities were 6.4 cm/s (sand S1) and 30.5 (sand C1). Therefore, 4 operating velocities were employed for sand S1 : 9.1, 15.2, 24.4 and 30.5 cm/s. These values cover the entire range over which the change from one regime of solids and gas movements was most noticeable. For sand C1 - which is essentially sand C3 as used by Whitehead and Auff (1976a), but with less fines - equipment limitations restricted the operating velocities to 42.7 and 48.8 cm/s. These values are within the range of velocities which correspond to the highest reported pressure difference at the bed base between the bed centre and corners.

RESULTS

The ratio of the average tracer concentration at each bed cross section and the mean mixed concentration, C_{av}/C_M for all velocities is plotted in figures 2(a) and 2(b). The prediction of the backmixing model (Nguyen et al, 1977) for sand S1 at U = 15.2 and 24.4 cm/s is included in 2(a). Gas backmixing was found to decline with the distance below the plane of injection, except for a section near the distributor.

Figure 3 presents the concentration contour maps for both sands at some bed height and for all velocities. At each bed height, the 49 C_H/C_M readings were interpolated to produce a three-dimensional plot. From this plot, 10 levels of concentration ratios, from 0 to 9 in increasing order of magnitude, were selected and points representing the same value of tracer concentration ratio C_H/C_M were joined. Figure 3 was then produced by two-dimensional projection. On each contour map, one can recognize the approximate size of the solids downflow and the relative magnitude. Values of each level of concentration ratio within each contour map are included in Table 1.

DISCUSSION

For sand S1, the ratio U/U_{mf} covered in this work was from 3.64 to 12.2. The range is quite significant since it encompasses the lower end of operating velocity in many industrial units, and within it, the bed was entirely in backmixing conditions with the presence of characteristic variations in solids and bubble flows. Figures 3(a) to 3(d) present the three modes of gas backmixing within this range.

(i) From 9.1 cm/s to 15.2 cm/s, backmixing was exceptionally strong in a central core of the bed, reaching the maximum extent at 15.2 cm/s. The solids downflow velocity in this confined area, as reported by Whitehead et al (1970), was between 17 to 21 cm/s, which is several times the interstitial gas velocity U_{gi}. Assuming that the voidage in the dense phase remained the same as at incipient conditions, U_{gi} in this case was around 4.5 cm/s for sand S1. The narrow bands of downflow solids near the walls moved at a similar rate, thus backmixing in these bands was, as expected, almost non-existent.

(ii) At 24.4 cm/s, the CO_2 concentration (figure 3(c)) shows as many as nine peaks, one at the centre, four at the four corners and one at the middle of each side. The peaks, expressed as C_H/C_M are all lower in value than that at U = 15.2 cm/s. This multi-peak configuration is quite consistent with the interchanging bubble pattern between 4 bubble tracks in the medial area, and one track at the centre. The time-average differential pressure difference (figure 1) at 24.4 cm/s was found approximately 0.0 cm H_2O, suggesting that the switching between these two bubble patterns occurred in about the same time interval.

(iii) At 30.5 cm/s, the concentration profile showed maxima along the bed walls (figure 3(d)). This was as expected, since with a negative $\bar{\delta}$ (figure 1), solids downflow should confine to fast moving streams near the walls, with the centre occupied mostly by one bubble track.

For sand C1, the velocities employed (42.7 cm/s and 48.8 cm/s) provided the conditions at which solids downflow was confined to a central stream and at its highest intensity. The contour maps (figures 3(e) and 3(f)) show a smaller central backmixing pattern - compared to corresponding readings with sand S1 at 15.2 cm/s). The fact that the extent of gas backmixing decreases with particle size has been reported before by Nguyen (1975). The ratio U_{CRIT}/U_{mf} was 2.56 for sand S1, and 2.85 for sand C1. Further information on effect of particle size on backmixing will be reported in subsequent publications.

The average concentration profiles (figures 2) show that gas backmixing decays slowly with distance below the tracer injector. This was in line with findings from previous works ((Stephens et al (1967), Latham et al (1970), Nguyen and Potter (1974)) in smaller beds. Near the distributor C_{av} shows a rise in most cases. This could be due to the purging of backmixed gas by incoming air as the solids were picked up in wakes of newly-formed bubbles. The increase in tracer concentration near the distributor was not predicted by the backmixing model.

Below this level, little or no backmixed gas was detected.

In general, the data obtained confirm the intimate relationship between gas backmixing and flow patterns in the bed. As the fluidizing velocity changed, characteristic dynamic variations in solids and bubble movements occurred and these variations were reflected in the change of gas backmixing.

CONCLUSION

Three distinct modes of backmixing were identified in the fluidized bed employed, all three could be directly related to bubble and solids activities within the bed. Backmixing was also found decreasing with particle size.

Gas mixing in fluidized beds thus changes in some stable and predictable manner but by no means uniform and the most favourable operating conditions for a particular application could be chosen with the knowledge of solids and bubble movements, which is now only partially available.

ACKNOWLEDGEMENT

This work has been supported on the University side, by grants from the Australian Research Grants Committee to one of us (O.E.P.).

REFERENCES

Fryer, C. and Potter, O.E. (1976) A.I.Ch.E. Journal 22, 38.

Latham, R.L., Hamilton, C.J. and Potter, O.E. (1968), Brit. Chem. Eng. 23, 666

Nguyen, H.V. (1975) Ph.D. Thesis, Monash University.

Nguyen, H.V. and Potter, O.E. (1974) Chem. React. Eng. II, Hulbert, H.M. (Ed.), Adv. in Chem. Ser. 133, Am. Chem. Soc., 290.

Nguyen, H.V. and Potter, O.E. (1975) Fluidization Technology II, Keairns, D.L. (Ed.), 193, California.

Nguyen, H.V., Whitehead, A.B. and Potter, O.E. (1977). To be published in A.I.Ch.E. Journal.

Potter, O.E. (1971) Chapt. 7 (Davidson, J.F. and Harrison, D., Eds.) Acad. Press, New York.

Schmalfeld, V.J. (1976). V.D.I.-Z, 118, 65.

Stephens, G.K., Sinclair, R.J. and Potter, O.E. (1967), Powder Tech. 1, 157

Werther, J. (1974). Preprint, GVC/A.I.Ch.E. Meeting, Müchen.

Whitehead, A.B. and Dent, D.C. (1967a). Proc. Int. Symp. on Fluidization, Drinkenburg, A.A.H. (Ed.), Netherland Uni. Press, 802, Amsterdam.

Whitehead, A.B., Dent, D.C. and Young A.D. (1967b) Powder Tech. 1, 149.

Whitehead, A.B. and Auff, A. (1976a) Powder Tech. 15, 77.

Whitehead, A.B., Gartside, G. and Dent. G.C. (1970) Chem. Eng. Journal, 1, 175.

Whitehead, A.B., Gartside, G. and Dent, D.C. (1976b). Powder Tech. 14, 61.

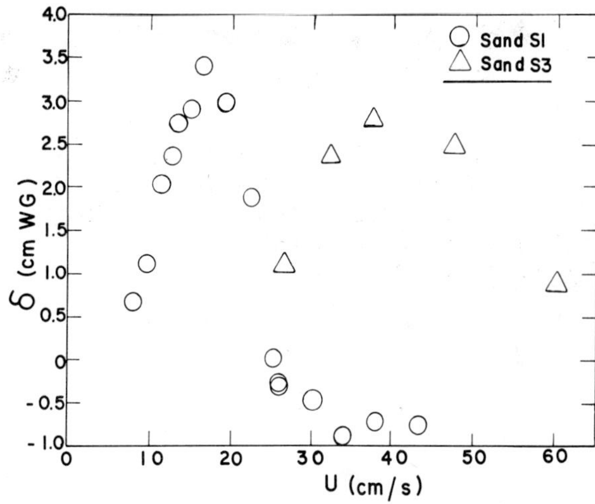

Figure 1: Differential pressure difference at the 122 cm bed base. H_o = 152 cm (S1) and 91 cm (S3).

Table I

Values of $C_H C_M$ for Figure 3

Figure \ Level	0	1	2	3	4	5	6	7	8	9
3(a)	0.22	0.44	0.66	0.89	1.10	1.23	1.55	1.77	1.99	2.21
3(b)	0.65	1.30	1.96	2.61	3.26	3.92	4.57	5.22	5.87	6.53
3(c)	1.36	1.70	2.04	2.38	2.72	3.06	3.40	3.74	4.08	4.42
3(d)	1.38	1.72	2.07	2.42	2.77	3.12	3.47	3.82	4.17	4.52
3(e)	0.35	0.71	1.06	1.41	1.77	2.12	2.48	2.84	3.19	3.55
3(f)	0.44	0.87	1.31	1.75	2.18	2.62	3.05	3.49	3.93	4.36

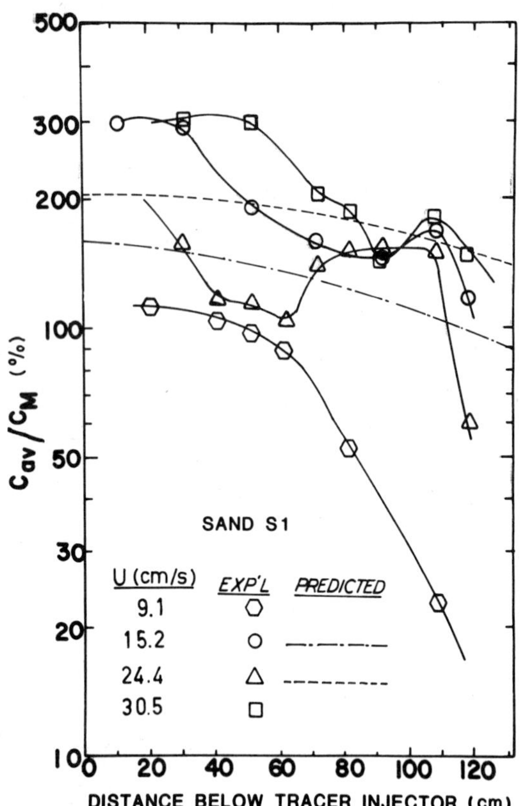

Figure 2(a): Average vertical tracer concentration profiles for sand S1. Tracer injector was at 132 cm from the distributor. Prediction from the backmixing model (Nguyen et al, 1977) is included.

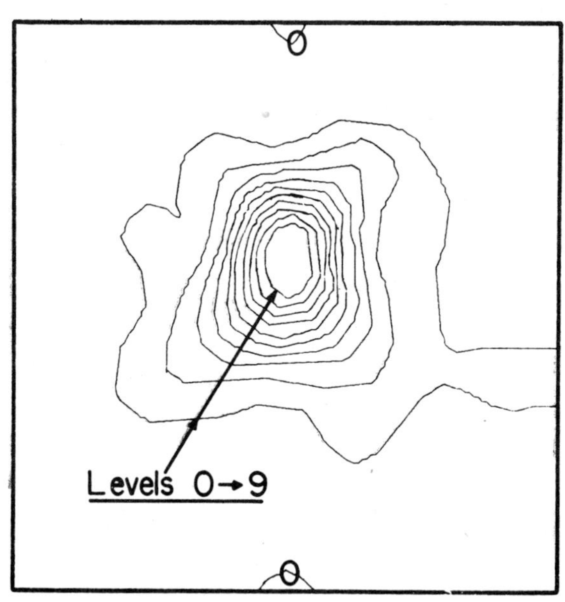

Figure 3(a): Tracer concentration profile at H = 81.3 cm, U = 9.1 cm/s, sand S1

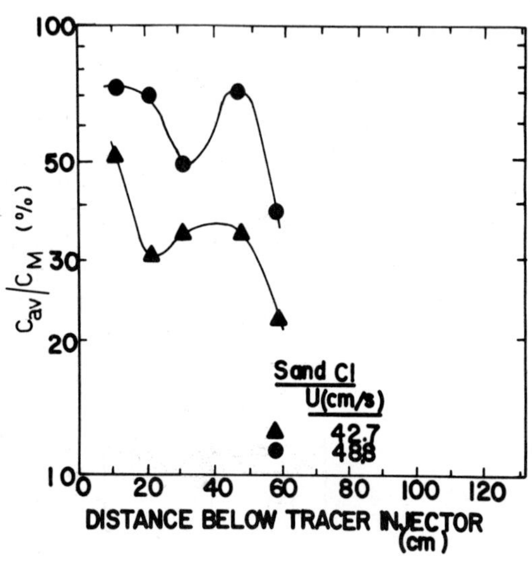

Figure 2(b): Average vertical tracer concentration profiles for sand C1. Tracer injector was at 71.1 cm.

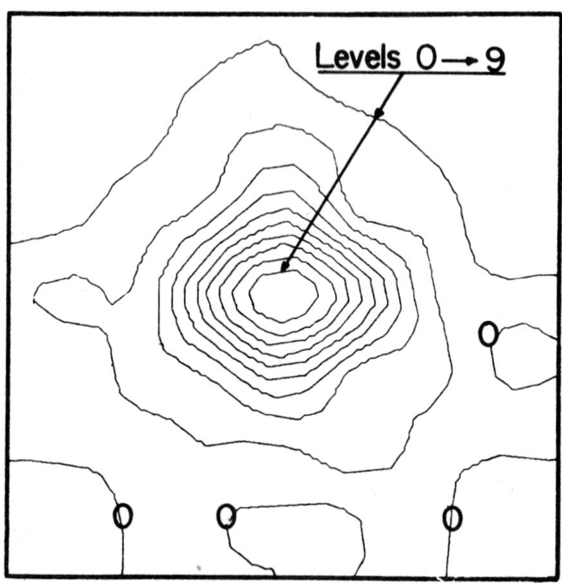

Figure 3(b): Tracer concentration profile at H = 81.3 cm, U = 15.2 cm/s, sand S1.

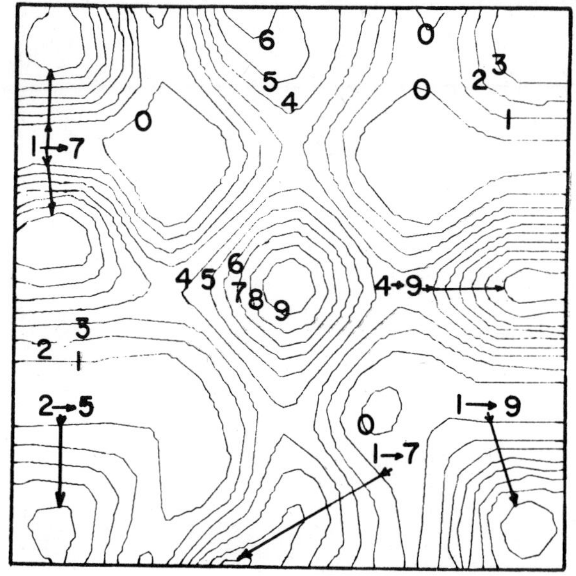

Figure 3(c): Tracer concentration profile at H = 81.3 cm, U = 24.4 cm/s, sand S1.

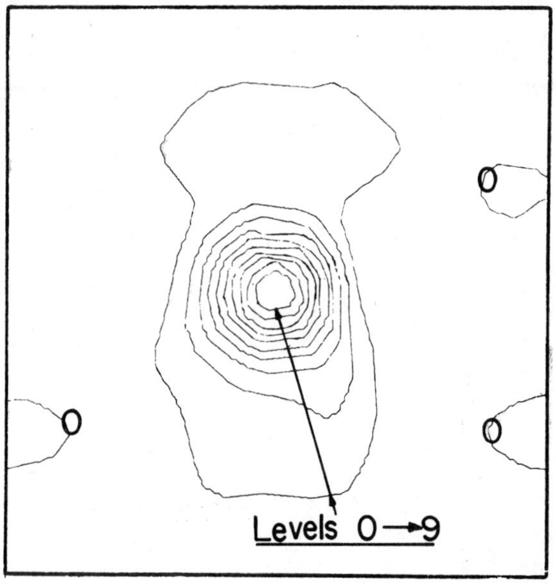

Figure 3(e): Tracer concentration profile at H = 50.8 cm, U = 42.7 cm/s, Sand C1.

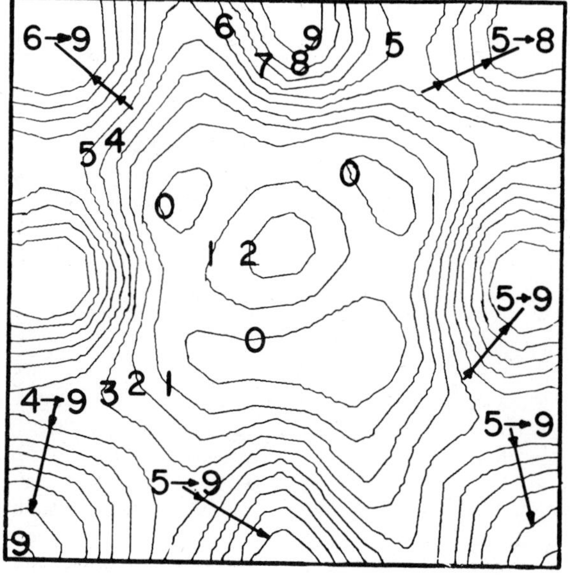

Figure 3(d): Tracer concentration profile at H = 81.3 cm, U = 30.5 cm/s, sand S1

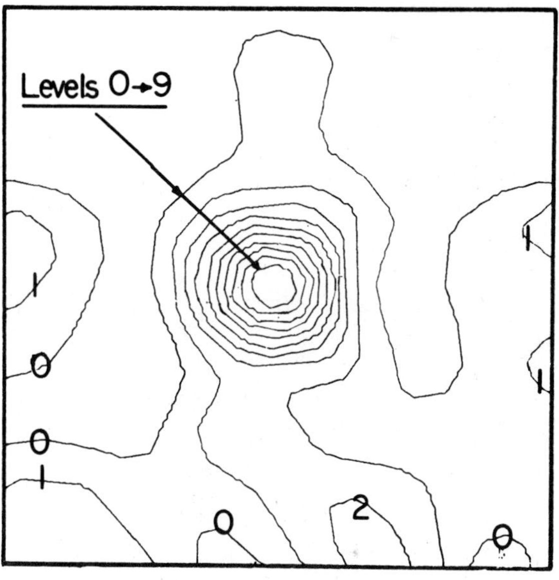

Figure 3(f): Tracer concentration profile at H = 50.8 cm, U = 48.8 cm/s sand C1.

THE MIXING/SEGREGATION BEHAVIOUR OF A DENSE POWDER WITH TWO SIZES OF A LIGHTER ONE IN A GAS FLUIDISED BED

By A.W. Nienow, P.N. Rowe and L. Y.-L. Cheung

Department of Chemical and Biochemical Engineering
University College London
Torrington Place
London WC1E 7JE England

SYNOPSIS

The mixing/segregation behaviour of two close-sieved powders differing in size and/or density is reviewed. Experimental data from a ternary system (two sizes of glass spheres and one of copper) is reported. The two sizes of the same density mix well together as in binary systems. Treating these two components as one, the mixing/segregation behaviour of this ternary system can be predicted from binary system relationships. It is suggested that in this way, binary system relationships for close-sieved powders may be extended to powders of a wide size range.

NOTATION

- D bed diameter-mm
- d_H diameter of the denser (heavier) particle - μm
- d_L diameter of the less dense (lighter) particle - μm
- d_{SM} surface-to-volume mean particle diameter - μm
- d_{ER} shape corrected diameter ratio, $\phi_H d_H / \phi_L d_L$
- H bed height - mm
- H^* reduced aspect ratio, $1-\exp(-H/D)$
- M mixing index, $x/\bar{x}$
- u superficial gas velocity - m/s
- u_C the u_{mf} of a mixture of different-sized, equi-density particles - m/s
- u_F the u_{mf} of the component with the lower u_{mf} (fluid) - m/s
- u_P the u_{mf} of the component with the higher u_{mf} (packed) - m/s
- u_{mf} minimum fluidisation velocity in general - m/s
- u_{TO} the velocity above which mixing takes over (see (2)) - m/s
- x mass fraction of the jetsam in the upper uniform part of the bed
- $\bar{x}$ the overall mass fraction of jetsam in the bed
- Z reduced velocity function, $\left(\frac{u-u_{TO}}{u-u_F}\right) \exp(u/u_{TO})$
- ϕ shape factor, sphericity (=1 for this work)
- ρ_H density of the heavier particle-kg/m^3
- ρ_L density of the lighter particle-kg/m^3
- ρ_R density ratio - ρ_H/ρ_L

INTRODUCTION

When two close-sieved powders differing in size or density are fluidised by a gas they segregate in a characteristic way (Rowe et al, 1972 b). The upper part of the bed attains a fairly uniform composition whilst the component that tends to sink (jetsam) forms a concentrated bottom layer. The concentration of jetsam in the upper constant composition region compared with its overall value is a good measure of the degree of mixing achieved. Fig. 1 illustrates this definition of a mixing index, $M = x/\bar{x}$ that varies from 0 for complete horizontal segregation (Fig. 1 a) to 1 for complete mixing (Fig. 1b). Between these two extremes, the segregation patterns of Fig.1 c and 1 d are obtained at low and high velocities respectively and if M is evaluated over a range of gas velocities, it is found to vary in a logistic fashion as shown in Fig. 2.

If the powders differ in density, the denser becomes the jetsam; if they only differ in size, it is the larger that settles to the bottom (Rowe et al, 1972 a and b). The mechanisms of particle movement that cause denser powders preferentially to become jetsam and that give rise to the observed segregation patterns of Fig. 1 c and 1 d have been described in detail (Rowe et al, 1972 a). One of the main causes of segregation is the out-of-balance forces which arise during the periodic disturbances associated with the passage of bubbles and which are due to differences in density. When there is no density difference, this mechanism is no longer present and powders differing only in size mix fairly readily even when the size difference is large.

These qualitative observations have been quantified as follows. The mixing index M variation with gas velocity is effectively described by a logistic equation (Rowe and Nienow, 1976):

$$M = x/\bar{x} = [1 + e^{-Z}]^{-1} \qquad \ldots.(1)$$

where Z is a reduced gas velocity defined by

$$Z = \frac{u - u_{TO}}{u - u_F} \exp(u/u_{TO}) \qquad \ldots.(2)$$

The use of (2) requires a knowledge of the lower of the two fluidisation velocities (u_F) and the term u_{TO}. u_{TO} has been called the take-over velocity (Rowe and Nienow, 1976), that is, the velocity below which segregation predominates and above which mixing takes over. It is also the velocity at which M = 0.5 (see Fig. 2). Provided u_{TO} is known, then M can be predicted for powders of known u_{mf} for all gas velocities.

For powders of different density, u_{TO} can be predicted from the complex empirical equation (Rowe and Nienow, 1976):

$$u_{TO}/u_F =$$

$$(u_P/u_F)^{1.2} + 0.9(\rho_R - 1)^{1.1}(d_{ER})^{0.7} - 2.2(\bar{x})^{0.5}(H^*)^{1.4} \qquad \ldots.(3)$$

The full details of the derivation of (1),(2) and (3) and their accuracy and ranges of applicability are given elsewhere (Cheung,1976). The terms in the equation for the take-over velocity taken in order account firstly for the ratio of the two minimum fluidisation velocities; secondly for the effect of density and size; and thirdly for the proportion of the jetsam in the bed and the bed aspect ratio. The complexity of the relationship is readily apparent and yet, in obtaining it and (1) and (2), particles were classified into narrow size ranges so that each could effectively be considered to be of constant size.

In practice, of course, the powders will generally contain a distribution of sizes. This paper is a preliminary attempt to extend (1), (2) and (3) to this more general case by using two sizes of one of the powders to simulate a continuum of sizes.

EXPERIMENTAL AND RESULTS

The three component systems used consisted of two sizes of Ballotini glass spheres and one of copper shot. Table 1 gives their basic properties, the dimensions of the fluidised bed and the mass fraction of jetsam, i.e. copper shot, used in each experiment. The proportions of the Ballotini, i.e. the flotsam, were varied as indicated in Table 2.

The experimental method was very simple. Weighed quantities of the three particulate components in the proportions indicated in Table 1 and 2 were loaded into a cylindrical vessel with a porous base and fluidised with air at a chosen velocity for some minutes before shutting off the supply abruptly. The bed at rest was then divided into a large number of horizontal layers each of which was analysed for proportions by weight of the three components. As these differed in particle size by at least one sieve mesh size, the separation was by sieving, for re-use as well as analysis.

Each combination was fluidised at three gas velocities, ranging from just less than the u_{mf} of the larger Ballotini to about twice that value. Earlier work (Rowe and Nienow, 1976) showed that fine material of the size and proportions used here should mix well with coarse of equal density under the present conditions. The layer-by-layer analysis of the bed confirmed that, at all the gas velocities used, a relatively uniform mixture of Ballotini was obtained throughout the bed. Fig. 3 gives examples from each of the three Ballotini mixtures and, in each case, a slightly higher than average proportion of the bigger component is indicated at the bottom of the bed. However, in plotting the mixing/segregation patterns the Ballotini can be justifiably treated as a single component.

With this assumption, the mixing/segregation patterns indicated in Fig. 4 were obtained. They are obviously similar to those for two component systems indicated in Fig. 1 and therefore the same mixing index definition ($M = x/\bar{x}$) can be used. The experimentally determined M values are plotted against gas velocity in Fig. 5 and again the similarily to the results for binary systems as exemplified by Fig. 2 is readily apparent.

DISCUSSION

To test the applicability of (1), (2) and (3) to these three component systems requires suitable values to be estimated for the size (d_L) and minimum fluidisation velocity (u_F or u_P) of the Ballotini pair. The minimum fluidisation velocity of a binary mixture of equal density, u_C, can be obtained from the empirical relationship (Cheung et al, 1974):

$$u_C = u_S \, (u_B/u_S)^{x_B^2} \qquad \text{....(4)}$$

Table 2 gives values of u_C calculated from (4) for each Ballotini mix.

u_C can also be calculated from a theoretical equation provided voidages are known (Rowe and Nienow, 1975). However, for practical purposes, (4) is generally more convenient but the theoretical approach suggests that the surface-to-volume mean diameter, d_{SM} is the correct one to use with multi-sized mixtures. These values are also given in Table 2.

With these values of u_C (for u_F or u_P as appropriate) and d_{SM} (for d_L), the take-over velocity u_{TO} can be calculated from (3) and these values are compared with experimental ones estimated by interpolation from the raw data in Table 2. The agreement is quite good and comparable to that obtained for a wide range of two component systems where the standard deviation between the predicted and experimental u_{TO} values was ±10% (Cheung,1976).

Using these predicted u_{TO} values, the complete mixing index-velocity relationship can be calculated from (1) and (2). Fig. 5 shows these predictions and Fig. 6 is an equivalent plot using the reduced velocity, Z. Again, the comparison between predicted and observed values is about the same as that for a binary system and, in fact, the replotting of only one data point (circled) would greatly improve the agreement between predicted and observed values in both Figs. 5 and 6.

CONCLUSIONS

Empirical equations for predicting the take-over velocity and the mixing index-gas velocity relationship have been developed for close-sieved binary mixture of different density. This work suggests that the use of a composite u_{mf} and a surface-to-volume mean diameter might allow them to be extended to the more general case where each powder extends over a range of particle sizes.

ACKNOWLEDGEMENT

One of us (L.Y.-L. Cheung) would like to thank the Ministry of Technology for support during the course of this research work.

REFERENCES

Cheung, L. Y.-L. (1976) PhD Thesis, University of London.

Cheung, L. Y.-L., Nienow, A.W. and Rowe, P.N. (1974), Chem. Eng. Sci., 29, 1301.

Rowe, P.N., Nienow, A.W. and Agbim, A.J. (1972 a), Trans. Instn. Chem. Engrs., 50, 310.

Rowe, P.N., Nienow, A.W. and Agbim, A.J., (1972 b), Trans. Instn. Chem. Engrs., 50, 324.

Rowe, P.N. and Nienow, A.W. (1975), Chem. Eng. Sci., 30, 1365.

Rowe, P.N.and Nienow, A.W. (1976), Powder Tech. 15, 141.

Table 1. Materials and System Used

	Size, μm	Density, kg/m^3	u_{mf}, m/s x 10^2
Copper shot	273 μm	8860	14.9
Ballotini	195 μm	2940	4.5
Ballotini	461 μm	2940	22.6

Mass fraction copper shot, $\bar{x}$ = 0.1: bed height = 0.1m, bed diameter = 0.145m

Table 2. Ballotini Combinations

	Mass Fraction	u_C cm/s	d_{SM} μm	u_{TO} m/s x 10^2 Predicted	u_{TO} m/s x 10^2 Experimental
A 195 μm 461 μm	0.09 0.81	16.6	405	36.1	29.5
B 195 μm 461 μm	0.45 0.45	6.7	274	29	32
C 195 μm 461 μm	0.13 0.77	14.5	382	34	34

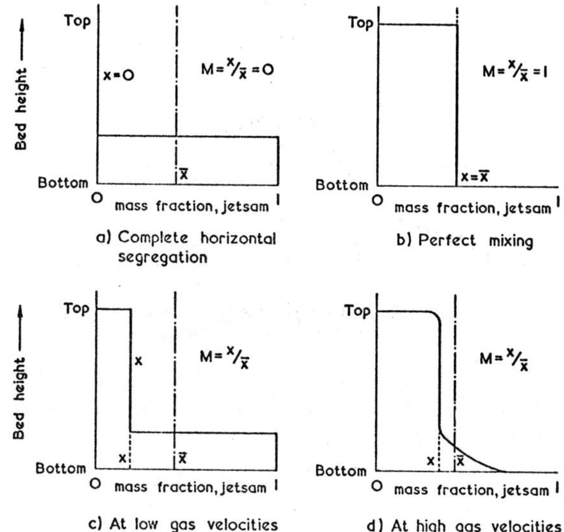

Fig. 1. Segregation patterns and mixing index for close-sieved binary mixtures

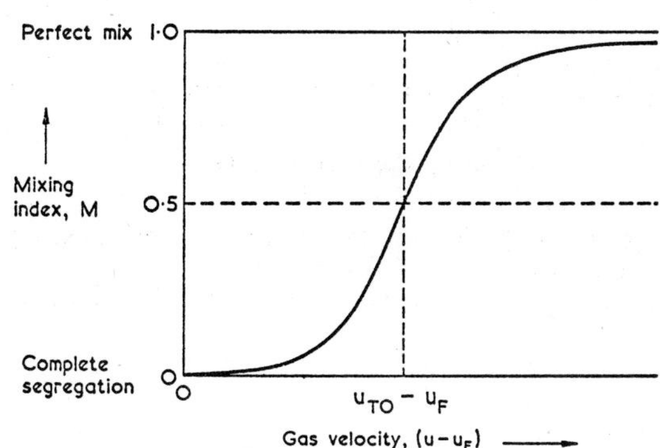

Fig. 2. The variation of mixing index with excess gas velocity

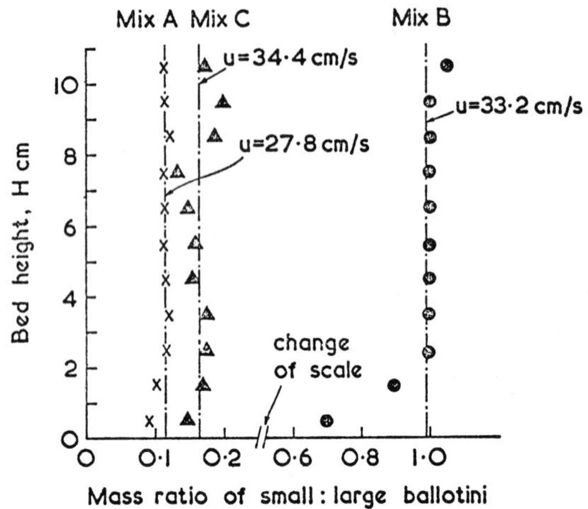

Fig. 3. Mixing of the two sizes of ballotini ($\rho_R = 1$)

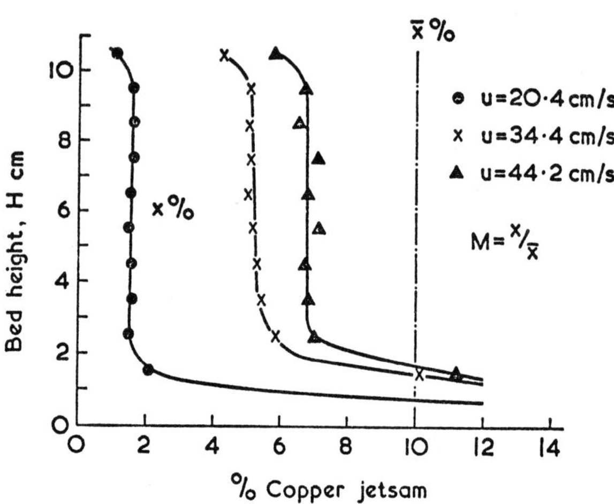

Fig. 4. Segregation patterns, mix C

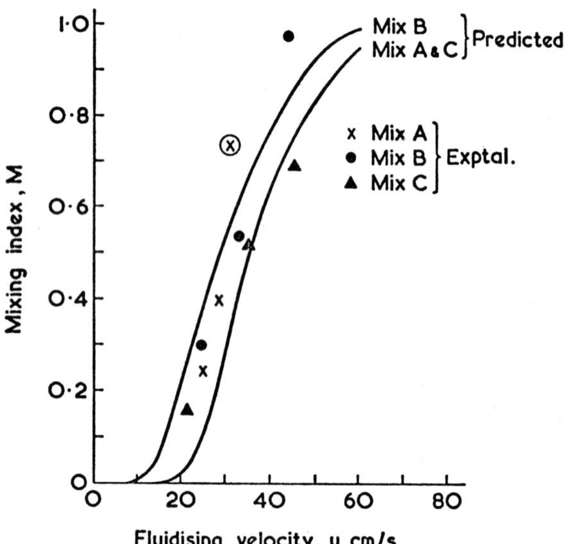

Fig. 5. Experimental and predicted mixing index values

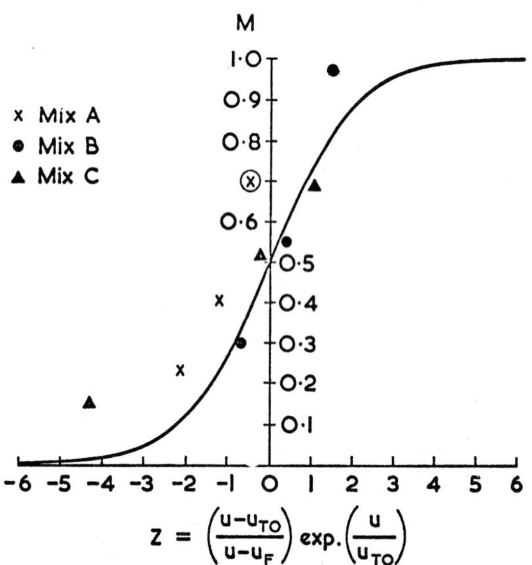

Fig. 6. Comparison of experimental data with the predicted mixing index-reduced velocity relationship

A THEORY OF SOLIDS PROJECTION FROM A FLUIDIZED BED SURFACE AS A FIRST STEP IN THE ANALYSIS OF ENTRAINMENT PROCESSES

T. P. Chen and S. C. Saxena

University of Illinois at Chicago Circle

Department of Energy Engineering

Box 4348, Chicago, Illinois 60680, U.S.A.

SYNOPSIS

A theory is developed for a quantitative treatment of solids projection from a dense fluidized-bed surface. First, it is presupposed that the solids projected into the freeboard are originally contained in the bulges formed when gas bubbles approach the dense bed surface. With the relative movement between dense bed surface and bubbles described by the potential flow theory, an expression for the solids projection rate is delineated. Measurements of the elutriation rate are extrapolated to the dense bed surface for comparison with the theory. Good agreement is observed except in the region of very high gas velocities where slugging prevailed.

NOTATION

- A bed cross-sectional area (cm^2)
- D_c column diameter (cm)
- d_p particle diameter (cm)
- E elutriation rate (g sec^{-1})
- E_o solids projection rate at dense bed surface (g sec^{-1})
- E_s constant elutriation rate above TDH (g sec^{-1})
- g acceleration of gravity (cm sec^{-2})
- h bed height in general (cm)
- H total bed height (cm)
- L freeboard height (cm)
- n_d number of perforated holes on the gas distributor
- r distance between the apex of bulge and the bubble center (cm)
- r_1 r at h=H (cm)
- R_b bubble radius (cm)
- R_{bo} the initial bubble radius at the gas distributor surface (cm)
- R_{b1} the R_b at h=H (cm)
- R_{bm} the maximum attainable bubble radius (cm)
- t time (sec)
- U_o superficial velocity of fluidizing gas (cm sec^{-1})
- U_b the bubble rising velocity (cm sec^{-1})
- U_r the moving velocity of the apex of bulge (cm sec^{-1})
- U_{mf} the minimum fluidization velocity (cm sec^{-1})
- α a constant in eqn (1) (cm^{-1})
- ρ_s particle density (g cm^{-3})
- ε_{mf} the bed voidage at minimum fluidization.

INTRODUCTION

The study of elutriation becomes more essential with the increased application of fluidization processes for solid-gas reactions. For instance, during the past few years there has been extensive development of adapting the coal combustion in fluidized beds as opposed to the conventional pulverized coal boiler. Several pilot-plant test have revealed that unsatisfactory combustion efficiencies resulted because of the elutriation loss of coal fines, Squires (1970), Jonke et al. (1972) and Hammons and Skopp (1972). The particulate loading in the fluidizing gas also contributes to air pollution.

Numerous experimental investigations have been conducted to determine the dependency of elutriation characteristics on different operating conditions such as the gas velocity, particle size and density, freeboard height, etc. It is concluded from reported observations that three major stages can be recognized for the elutriation process, Andrews (1960) and Kunii and Levenspiel (1969).

In the first stage, solids are projected into freeboard from the dense bed surface as gas bubbles erupt. The projected solids are less compacted than the solids in the dense bed and are therefore easily carried up in the freeboard by the fluidizing gas.

In the second stage, solids continue to move upwards but with the less energetic particles falling back to the bed. By assuming that the energy of particles projected from the dense bed surface follows the Maxwell-Boltzmann distribution, Andrews (1960) was able to show an exponential decay of elutriation rate with respect to the freeboard height as also indicated by various experiments. On a different basis of agglomerate dissipation into the dispersed solids phase, Kunii and Levenspiel (1969) could also explain the same experimental results.

In the final stage, the solids carried up by the fluidizing gas are at a concentration such that they travel the length of the freeboard without collapse into a dense slugging mass. This critical concentration was suggested by Zenz and Weil (1958) to correspond to the saturation capacity of fines carried in pneumatic conveying tubes. The freeboard height above which the entrainment rate becomes constant is called the transport disengaging height, TDH.

The elutriation rate E resulting from the above three states can be expressed as:

$$E = \begin{cases} E_o \exp(-\alpha L) & \text{for } L < \text{TDH} \quad ..(1) \\ E_s & \text{for } L > \text{TDH} \quad ..(2) \end{cases}$$

where E_o is the solids projection rate from the bed, α a constant, L the freeboard height and E_s the constant elutriation rate for L greater than TDH.

Different correlations for the elutriation rate have been proposed. However, no general agreement is reached even on the effect of the most important operating variable, i.e. the fluidizing gas velocity. Leva (1951) suggested that the elutriation rate is proportional to the fourth order of the velocity, but Yagi and Aochi (1955) and Wen and Hashinger (1960) considered it to be related to the difference between such velocity and the terminal velocity of particles.

The disagreement results from considering only empirical correlations. In this work an attempt is made to develop a general theory for the first stage of the elutriation process.

THEORY

It is commonly accepted that bubble eruption at the bed surface is responsible for the initial solids release from gas fluidized beds. Three possible projection mechanisms can be envisioned:

1. Solids are raining from the roof of bubble and then gushed away by the gas at the moment of bubble eruption.
2. Solids in the bubble wake are thrown up following the bubble break-up.
3. Solids contained in the leading bulge burst out as the bubble erupts.

Lewis et al. (1962) pointed out that the first mechanism could account for only a small portion of the elutriation. The second mechanism, however, predicts too high an elutriation rate according to the simple calculation of Leva and Wen (1971). A further evidence against the second mechanism comes from the work of Rowe and Partridge (1962). They injected an artificial bubble into a bed which had a layer of dyed solids above the gas distributor with the idea to demonstrate the mixing and shedding of solids behind the rising bubble. Their results showed that the colored solids in the wake were retained in the bed after the bubble erupted from the bed surface. Therefore, it is concluded and assumed here that the third mechanism is the predominant one.

The bulge formation is the response of the bed surface to the approaching bubbles. However, the resulting bulge thickness remains to be determined. To solve this problem, the particulate phase of a fluidized bed is first assumed to be an inviscid fluid permitting the application of potential flow theory. Based on the same postulate, an analytical expression derived by Davies and Taylor (1950) has been successfully extended to the bubble rising velocities in fluidized beds, Davidson and Harrison (1963).

The next assumption to be made is that the surface tension of the dense bed is sufficiently small so as not to damp the bulge formation. This can be justified by the experimental work of Furukawa and Ohmae (1958) where the measured surface tension of fluidized beds was comparable to that of water. The energy required for lifting up the bed surface for this magnitude of surface tension is much smaller than the bubble kinetic energy. As a result, the bubble will rise toward the bed surface as if in an infinite medium.

Shown in Fig. 1 is a schematic description of a gas bubble with radius R_b approaching the surface of the bed velocity U_b. From the potential flow theory, Kunii and Levenspiel (1969), the apex of the bulge (point 0′ in Fig. 1) should move upward at a velocity U_r given by,

$$U_r = U_b\left(\frac{R_b}{r}\right)^3 \qquad \text{......(3)}$$

where r is the distance between point 0 and the bubble center (point 0 in Fig. 1). It is recognized that the potential flow is shown to be altered from the infinite medium case by a top surface boundary in the work of Gabor (1971). Gabor's study was on a bed with fixed boundaries. However, in practice the upper boundary is free and the surface particles bulge above the original flat surface. It is estimated here that the particle movement in the bulge of the free surface can be described by potential flow in infinite medium.

The distance r, on the other hand will vary with time t according to the following relation

$$\frac{dr}{dt} = U_r - U_b \tag{4}$$

If h is denoted as the height of bubble above the gas distributor, then

$$\frac{dh}{dt} = U_b \tag{5}$$

From (4) and (5), the relation between r and h can be expressed as:

$$\frac{dr}{dh} = \frac{U_r - U_b}{U_b} \tag{6}$$

Equation (6) when combined with (3) becomes:

$$\frac{dr}{dh} = \left(\frac{R_b}{r}\right)^3 - 1 \tag{7}$$

Bubbles coalesce and grow as they rise in the bed. A pertinent correlation for the R_b in (7) as a function of h has been recently proposed by Mori and Wen (1975) for a uniformly bubbling fluidized bed,

$$R_b = R_{bm} - (R_{bm} - R_{bo}) \exp -\left(\frac{0.3\ h}{\sqrt{4A/\pi}}\right) \tag{8}$$

R_{bm} and R_{bo} are expressed as:

$$R_{bm} = 0.3125[A(U_o - U_{mf})]^{2/5}$$

$$R_{bo} = \begin{cases} 0.00188(U_o - U_{mf})^2 & \text{for a porous distributor plate} \\ 0.1735[A(U_o - U_{mf})/n_d]^{2/5} & \text{for a perforated distributor plate} \end{cases}$$

All the quantities in the above relations are in CGS units.

Due to the complexity of (8), (7) cannot be solved analytically. However, it can be readily solved by simple numerical methods such as the Runge-Kutta integration for the following boundary condition:

$$\text{at } h = 0,\ r = H \tag{9}$$

where H is the total bed height. The above boundary conditions implies that the bed surface will start to bulge when the bubble is formed at the gas distributor. This result might not be exactly true since a practical bed is actually composed of multiple bubbles, and the bubble interactions would interfere with distinct surface bulges caused by bubbles forming at the gas distributor. Therefore, the boundary condition (9) is only a first approximation. Its adequacy will be tested by a later comparison of the predicted solids projection rate with experimental data.

It is assumed here that the bubbles stop coalescing and growing when their centers are located at a height of $H-R_b$, Fig. 1(c). This particular height can be found by searching the root of the following equation with Newton's method:

$$H - R_b = h = H - [R_{bm} - (R_{bm} - R_{bo}) \exp\left(-\frac{0.15\ h}{\sqrt{A/\pi}}\right)] \tag{10}$$

Therefore, for the rest of the bubble rise, the R_b in (7) is kept constant at a value of R_{b1} corresponding to the bed height, $H-R_{b1}$. It may be pointed out that Do et al. (1972) have observed that bubbles may coalesce right at the bed surface and in such a case solids are ejected with higher velocities. It is believed that the error due to the extra growth of bubbles in the region R_b below the bed surface is not going to influence the calculations significantly as R_b is usually much smaller than H.

Another unknown is the extent of bubble emergence at which the bulge bursts to throw up solids into the freeboard. An experiment carried out by Davidson (1961) showed that an artificially injected tracer bubble retained the surrounding cloud after leaving the bed. This implies that the opening resulting from the bulge break-up must be at least of the same size as the bubble diameter for the bubble to keep its shape when leaving the bed surface. Since no other experimental evidence is available, it is simply assumed that solids projection occurs when the bubble is half way above the bed and the solids in the bulge directly in front of the erupting bubble are thrown up into the freeboard (Fig. 1 (d).

Equations (7) - (8) when solved give the relation between r and h. The value of r at $h=H$, i.e. when projection occurs, is denoted as r_1. The bulge thickness at that particular moment is then r_1-R_{b1}. From this thickness, the amount of solids projected by a single bubble is determined to be approximately $2\pi[(R_{b1} + r_1)/2]^2\ (r_1 - R_{b1})(1 - \varepsilon_{mf})\rho_s$, where the voidage of the bulge is taken to be the same as at minimum fluidization ε_{mf}, and ρ_s is the density of solids.

Finally multiplying the number of bubbles (based on the two phase model) with the amount of solids contained in the bulge at bursting, the solids projection rate is obtained as follows:

$$E_o = [(U_o - U_{mf})A/(\tfrac{4}{3}\pi\ R_{b1}^3)] \left[2\pi \left\{\frac{R_{b1} + r_1}{2}\right\}^2 (r_1 - R_{b1}) (1 - \varepsilon_{mf})\rho_s\right] \quad \ldots(11)$$

COMPARISON WITH REPORTED EXPERIMENTS

Experimental data from various sources are used to assess the theory. Listed in Table 1 are the operating conditions of the reported experiments. Most of them were carried out with varying fluidization velocities. The measurements of Zenz and Weil (1958) were conducted in a two-dimensional fluidized bed. Their experiment is simulated here as a three-dimensional one with the width of the bed assumed equal to the diameter of a cylindrical column.

TABLE 1. Operating Conditions of Reported Experiments

Authors	Mean d_p μm	ρ_s g cm^{-3}	D_c, H cm	U_o cm sec^{-1}	Symb. in Fig. 2
Jolley and Stantan (1952)	76	1.26	5.08, 76.2	15.2	+
				18.3	□
				21.3	×
				24.4	⬡
Nazemi et al. (1970)	60	0.84	61.0, 76.2	9.1	○
				15.3	△
				21.3	■
				27.4	●
				33.5	▲
Lewis et al. (1962)	75	2.34	1.91, 10.2	45.7	◇
				51.8	▽
				57.9	✱
				64.0	⬢
Tweddle et al. (1970)	165	1.37	16.5, 76.0	76.2	▼
Zenz and Weil (1958)	60	0.93	61.0, 50.0	30.5	✸
				45.7	✪
				61.0	◆
				71.6	☆

Due to the vigorous oscillation of the bed surface, these experiments give only the elutriation rate as a function of freeboard height but not a direct measurement of the solids projection rate. However, according to eqn. (1) if the elutriation rate is plotted on a semi-log scale against freeboard height and extrapolated to the bed surface, E_o can be obtained. Figure 2 is obtained following this scheme with the fluidization velocity as the parameter. In this figure, the data points of Tweedle, et al. (1970) and Zenz and Weil (1958) can be represented by horizontal lines at large freeboard height. This is probably because the freeboard height exceeded TDH in these experiments. The measurement of Jolley and Stantan (1952) was carried out in a bed in which only a small portion of the solids had a terminal velocity less than the fluidizing gas velocity. Therefore, only these solids were elutriated and the larger particles were returned to the bed almost immediately after their projection. Their data presented in Fig. 2 are the measured values divided by the weight fraction of the elutriable fines in the bed so that the E extrapolated to L = 0 represents the true projection rate.

It is noted that U_{mf} is also required for the theoretical calculation of the solids projection rate. This can be easily found from the Ergun's correlation (1952) for the operating conditions given in Table 1.

In Fig. 3, the extrapolated values from Fig. 2 are plotted as a function of the superficial gas velocity. The solid lines represent the corresponding predictions based on theory developed here. As can be seen, the agreement between theory and experiments of Jolley and Stantan (1952), Nazemi et al. (1970) and Zenz and Weil (1958) is very good. There is only one data point by Tweddle et al. (1970) and it was obtained for high gas velocity. The bubble size calculated from eqn. (8) indicates that their bed was in a slugging condition for which the solids projection mechanism is very different than that adopted here for a bubbling fluidized bed. This is regarded as the reason for the departure of the experimental data point from the theoretical line. A large disagreement between the experimental data of Lewis et al. (1962) with the predictions of the theory is again attributed to the slugging of their bed which was a small diameter column, (see Table 1). Therefore, in Fig. 3, the solid line representing the theoretical predictions for the experiments of Lewis et al. (1962) extends only up to U_o = 10 cm/sec where the slugging is found to set in the bed. The dashed curve is drawn thereafter only to represent the experimental data.

The dependence of E_o on H is implicit in eqn. (11). Calculations were therefore performed for a representative set (d_p = 75 μm, ρ_s = 0.84 g/cm^3 and D_c = 61 cm) for H ranging from 10.0 to 100 cm and U_o from 8 to 26 cm/sec. These revealed that E_o increases very slowly with H the maximum increase being about 6%. The differences in the operating conditions of Nazemi et al. (1970) and Zenz and Weil (1958) experiments are therefore not likely to show significant differences in the values of E_o and this is substantiated by the results given in Fig. 3. There are significant system differences in the experimental units of Jolley and Stantan (1952) and Nazemi et al. (1970) and E_o/A values differ substantially over the same U_o range. It is to be noted from Fig. 3 that the proposed theory successfully explains these differences.

In conclusion, the developed theory successfully predicts the solids projection rate from the bed surface as long as it is well fluidized without the presence of slugging. However, the present comparison of theory is limited to the experimental data available. Further accurate experimental work is desirable. In addition, theoretical work should be continued on the mechanisms of the second and third stages of elutriation discussed in the introduction. The above theory developed under the assumption that the particulate phase velocity on the axis of the bubble can be calculated as for an ideal flow of an unbounded medium needs to be evaluated by further work in which the Navier-Stokes equations for potential flow should be solved with appropriate boundary condition at the free surface of the bed.

ACKNOWLEDGEMENTS

This research was supported by the United States Energy Research and Development Administration, Washington, D.C., under Contract No. E(49-18) - 1787. We are thankful to Dr. J. D. Gabor of Argonne National Laboratory for a critical reading of the manuscript. Computing services used in this research were provided by the Computer Center of the University of Illinois at Chicago Circle. Their assistance is gratefully acknowledged.

REFERENCES

Andrews, M. (1960) I & EC, 52(1), 85.
Davidson, J.F. (1961) Trans. Instn. Chem. Eng., 39, 230.
Davidson, J.F. and Harrison, D. (1963) "Fluidized particles," Cambridge University Press, New York.
Davies, R.M. and Taylor, G.I. (1950) Proc. Roy. Soc., A200, 375.
Do, H.T., Grace, J.R. and Clift, R. (1972) Powder Technology, 6, 195.
Ergun, S. (1952) Chem. Eng. Progr., 48, 48.
Furukawa, J. and Ohmae, T. (1958) I & EC, 50, 821.
Gabor, J.D. (1971) Chem. Eng. Sci., 26, 1247.
Hammons, G.A. and Skopp, A. (1972) AIChE Sympo. Ser., 68(126), 252.
Jolley, L.J. and Stantan, J.E. (1952) J. Appl. Chem. (London) 2, Suppl. Issue 1, S62.
Jonke, A.A., Vogel, G.J., Carls, E.L., Ramaswami, R., Anastasia, L., Jarry, R. and Haas, M. (1972) AIChE Sympo. Ser., 68(126), 241.
Kunii, D. and Levenspiel, O. (1969), J. Chem. Eng. (Japan), 2(1), 84.
Kunii, D. and Levenspiel, O. (1969) "Fluidization Engineering," p. 113, John Wiley and Sons, Inc.
Leva, M. and Wen, C.Y. (1971) "Fluidization," eds. Davidson, J.F. and Harrison,D., Chap. 14, p. 639, Academic Press, New York.
Lewis, W.K., Gilliland, E.R., and Lang, P.M. (1962), Chem. Eng. Progr. Sympo. Ser., 38(58), 65.
Mori, S. and Wen, C.Y. (1975) AIChE J., 21(1), 109.
Nazemi, A., Bergougnou, M.A. and Baker, C.G.J. (1970) AIChE Sympo. Ser., 70(141), 98.
Rowe, P.N. and Partridge, B.A. (1962) Proc. Symp. on Interaction Between Fluids and Particles, Inst. Chem. Engrs., 135.
Squires, A. (1970) Science, 169.
Tweddle, T.A., Capes, C.E. and Osberg, G.L. (1970) I & EC Process Des. Develop., 6(1), 85.
Wen, C.Y. and Hashinger, R.F. (1960) AIChE J., 6, 220.
Yagi, S. and Aochi, T. (1955) paper presented at the Society of Chem. Eng. (Japan), Spring Meeting.
Zenz, F.A. and Weil, N.A. (1958) AIChE J., 4, 472.

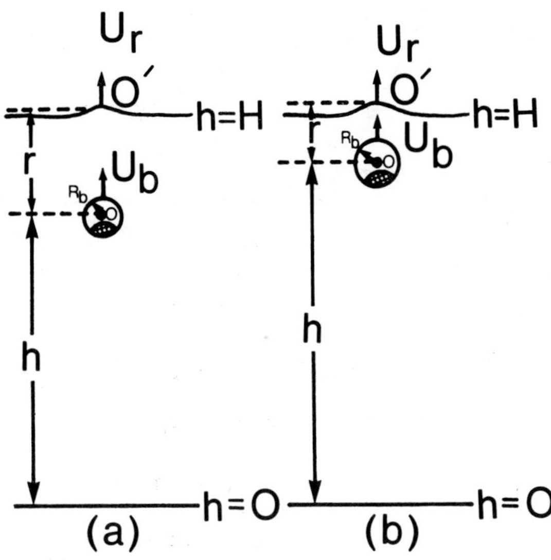

Fig. 1. For caption see over

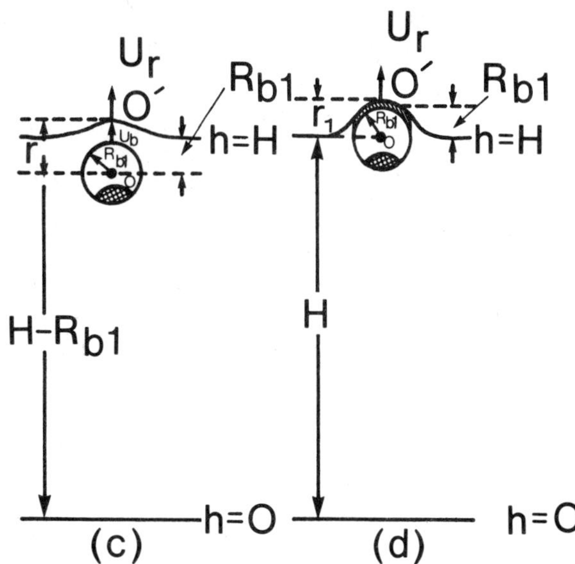

Fig. 1. A schematic representation of the bulge formation.
(a) Bulging starts as a result of bubble approach.
(b) The extent of bulging and bubble size become larger as the bubble rises closer to bed surface.
(c) Bubble approaches the bed surface within a radius of the bubble
(d) Bubble is half above the bed surface and is about to erupt.

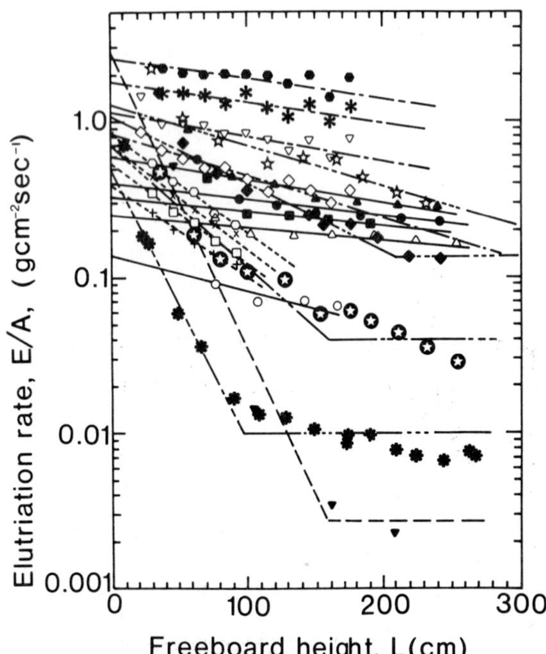

Fig. 2. Reported elutriation as a function of freeboard height. ---: Jolley and Stantan (1952); ——: Nazemi, et al. (1970); — - —: Lewis, et al. (1962); —— —— ——: Tweddle, et al. (1970); —— ---: Zenz and Weil (1958); symbols of data point are given in Table 1.

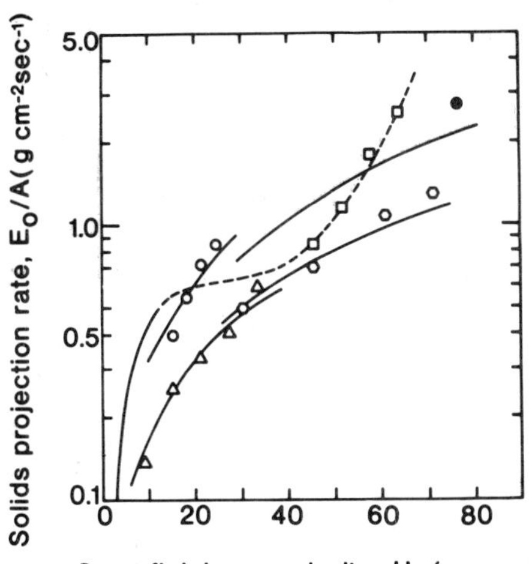

Fig. 3. Solids projection rate as a function of superficial gas velocity; solid lines are theoretical prediction; ○: Jolley and Stantan (1952); △: Nazemi, et al. (1970); □: Lewis, et al. (1962); ●: Tweddle, et al. (1970); ⬡: Zenz and Weil (1958).

MASS TRANSFER FROM A FIXED SPHERE TO A LIQUID IN A FLUIDIZED BED

by J.P. RIBA, R. ROUTIÉ and J.P. COUDERC

Institut du Génie Chimique
Laboratoire Associé C.N.R.S. n° 192
Chemin de la Loge
31078 TOULOUSE CEDEX (France)

SYNOPSIS

Mass transfer in liquid fluidization has been investigated by an electrochemical method using a fixed spherical probe immersed in the bed. Results are in reasonnable agreement with previous data obtained by dissolution of the solid particles. The method allows a local exploration of mass transfer properties and also a measure of the instantaneous fluctuations of the transfer rate. A correlation is proposed and various theoretical developments suggested.

NOTATION

- A area of the electrode
- d_p particle diameter
- F Faraday's constant
- g acceleration of gravity
- I electric current intensity
- k mass transfer coefficient
- r radial position
- R column radius
- s surface renewal frequency
- v fluid velocity
- z number of electrons involved in the electrochemical reaction
- ρ_1 liquid density
- ρ_s solid density
- μ viscosity
- D diffusivity
- σ relative standard deviation
- ε bed voidage
- Re Reynolds number $d_p . v . \rho_1 / \mu$
- Ga Galileo number $d_p^3 . g . \rho_1^2 / \mu^2$
- Mv density number $(\rho_s - \rho_1)/\rho_1$
- Sc Shmidt number $\mu / \rho_1 . D$
- Sh Sherwood number $k d_p / D$
- J_D Colburn factor $Sh/Re\ Sc^{1/3}$

INTRODUCTION

In view of its many actual or future uses, liquid fluidization is a subject of increasing importance in research. The applications which are or could be developed such as dissolution, crystallization, adsorption, ion exchange, leaching... all involve mass transfer in the liquid phase, from or to the solid surface. This work is devoted to a study of mass transfer properties, using an electrochemical technique with a fixed spherical electrode immersed in a fluidized bed of inert particles.

PREVIOUS WORKS

Particle-to-fluid mass transfer in a fluidized bed has been studied only by measuring the dissolution rate of solid particles in a liquid. Moreover nearly all workers selected the binary system benzoïc acid-water, except (Mc Cune and Wilhelm, 1949) who retained the β naphthol-water system (Nanda, Ghosal and Mukerjea, 1975) who used water, cinnamic salycilic and stearic acid system and (Laguerie and Angelino, 1975) who dissolved citric acid particles in its aqueous solutions.

Results are generally presented as correlations between various non dimensional groups

$$Sh = f(Re, Sc, \varepsilon) \qquad (1)$$

or

$$J_D = f(Re, \varepsilon) \qquad (2)$$

Recently (Damronglerd, Couderc and Angelino, 1975), (Vanadurongwan, Laguérie and Couderc, 1976) and (Tournié, 1977) developed a slightly different form of correlation

$$Sh = f(Re, Ga, Mv, Sc) \qquad (3)$$

involving only independant non dimensional groups. It must be noted that void fraction which can be related to the Reynolds number by expansion laws, like Richardson and Zaki's one for example, no longer appears in the equation.

Finally, a model has been proposed by (Nelson and Galloway, 1975) for fixed systems and modified by (Rowe, 1975) for fluidized beds.

Based on a penetration type theory, this model leads to a relation between the Sherwood number, bed voidage and a surface renewal frequency. These developments primarily concern low Reynolds number conditions but can be extended to values as high as 100.

EXPERIMENTAL

Fluidization equipment

For electrochemical reasons it is built entirely of plastic materials.

The electrolytic solution is stored in a 200 l plastic tank and pumped by a centrifugal polyethylene pump into a 94 mm inside diameter fluidization column. The distributor is a porous plate of sintered polyethylene (Vyon). A calming section consisting of a fixed bed of spherical glass beads avoids channeling. The solution temperature is controlled to $1/10^{th}$ of a degree and flow rate is measured using an orifice.

Measuring probes

These consist of brass spheres of diameters 5, 7 and 10 mm, coated by a 5 gold film electrodeposited. The spheres are fixed on thin stainless steel tubes, electrically insulated by a plastic thermoretracatable layer. This stainless steel support is itself stuck in a 6 mm glass tube which gives the probe good mechanical strength.

Mass transfer coefficient measurement

The well-known electrochemical reduction of potassium ferricyanide in sodium hydroxide solutions has been used. It can be recalled that, in a convenient range of potentials, the motion of ions is a diffusion controlled process. The mass transfer coefficient is then related to the limiting current density I_{lim} by the equation

$$k = \frac{I_{lim}}{z \cdot F \cdot A \cdot C} \tag{4}$$

An electrolytic solution 2.10^{-3} M in potassium ferricyanide, 2.10^{-3} M in potassium ferrocyanide (which keeps the equilibrium potential constant)and 0.5 N in sodium hydroxide has been used.

Electrical equipment

The gold spherical cathode is polarised by a Tacussel PRT 10-05 potentiostat. The anode is a large nickel plate situated at the top of the fluidization column. The source of electric potential is given by a saturated calomel electrode.

Auxiliary measurements

* Solution titration :
Potassium ferricyanide concentration is determined by an amperometric method using cobalt sulfate and a rotating disc gold electrode.

* Viscosity :
The viscosity of the fluidization solution has been measured using a falling sphere viscometer.

* Diffusion coefficient :
The mass transfer diffusivity of potassium ferricyanide has been determined by the rotating disc electrode technique. The results obtained are in close agreement with those by(Bazan and Arvia, 1965)(see table 1).

Table 1

t°C	10^{10} x D m^2/s	10^4 x ν m^2/s
25	7.12	1.03
30	7.90	0.92
35	8.77	0.83
40	9.52	0.75
45	10.3	0.68
50	11.2	0.61

* Fluidized particles

Fluidized particles are plastic spheres 5, 7, 9 and 10 mm in diameter ; their density is 1350 kg/m^3.

RESULTS

When the spherical probe is immersed inside a fluidized bed of spheres of equal diameter, for a given value of the voltage, the current measured fluctuates in time about mean values and time smoothed values can be determined. The variations of these time smoothed mean values of the current as a function of voltage show that the polarisation curves present a level of constant current, the limiting current I_{lim} being in the range of voltage 0 - 0.2 Volt. All measurements have then been made at the mean value of the voltage - 0.1 V.

Variations of the mass transfer coefficient in a fluidized bed

For five different radial positions

r/R = 0 - 0.22 - 0.44 - 0.67 and 0.89 the gold probe has been moved from the top of the bed towards the distributor ; the mass transfer coefficient remained constant to about 2 % in the whole volume of the bed except when the probe was in contact with the distributor where the rate of transfer increased significantly (around 50 %). This increase being evidently in connection with a change in the spatial distribution of the solution due to the probe, there remains the general conclusion that the mass transfer coefficient around the probe is constant in the fluidized bed.

Therefore for the following experiments, the probe has been simply positioned at the center of the bed without any particular caution.

Effects of the fluid flow, particle diameter and physical properties on the mass transfer rate.

The influence of various variables has been systematically determined. In particular the temperature of operation has been varied from 25 to 50°C which allows a significant change in physical properties or in the value of the Schmidt number.

Using the form of correlation (3), a multilinear regression method applied to 120 experimental data led to

$$Sh = 0.264\ Re^{-0.11}\ Ga^{0.36}\ Sc^{0.33} \qquad (5)$$

The multiple correlation coefficient is 0.994 and the relative standard deviation 3 %.

As solid or liquid densities have not been varied, the density number $M_v = \frac{\rho_S - \rho_L}{\rho_L}$ does not appear in equation (5); its contribution is included in the numerical constant.

The range of variables covered by equation (5) is

$$265 < Re < 3\ 100$$

$$10^6 < Ga < 3 \times 10^7$$

$$550 < Sc < 1\ 450$$

It is worth while noticing the slight influence of the particle Reynolds number on the mass transfer rate and the value 0.33 of the exponent of the Schmidt number in close agreement with theoretical predictions for the frontal part of a single sphere using the boundary layer theory.

Time fluctuations of the mass transfer coefficient.

As previously said, the instantaneous value of the mass transfer coefficient around the spherical probe fluctuates in time. In our opinion these variations in time can be related to two different phenomena, first the turbulence of the liquid flow and second and probably more important, some random changes in the local void fraction around the probe due to the motion of the surrounding fluidized solid particles.

To try to obtain more insight into this phenomenon of time variation, an autocorrelation function has been calculated using a minicomputer. This autocorrelation, defined by the following equation

$$R(\tau) = \int_{-\infty}^{+\infty} (k(t) - \bar{k})(k(t - \tau) - \bar{k})\, dt \qquad (7)$$

is a decreasing function of time τ. The partucular value τ_o for which the autocorrelation is zero represents an order of magnitude of the time scale of the phenomenon of fluctuations. Some values of τ_o are collected in table 2.

DISCUSSION

In previous work (Mullin and Treleaven, 1962) concluded that the mass transfer rates around a fixed sphere or a mobile sphere are identical in a fluidized bed. So as to confirm this conclusion, our experimental data have been used to calculate eleven different values of the Sherwood number with the eleven different correlations presented in table 3. The deviations between these calculated values and the measured ones are also given in table 3.

Without discussing in detail the significance of the results obtained, it may be noticed that there exist large discrepancies with the oldest results on mass transfer in fluidized beds. Indeed the works by (Evans and Gerald, 1953) or by (Fan, Yang and Wen, 1960) have already been criticized by some authors (Couderc, 1972), (Damronglerd, 1975), (Vanadurongwan, 1976) in particular their use of non spherical particles.

On the other hand, good agreement is found with the more recent results of (Damronglerd, Couderc and Angelino, 1975), (Vanadurongwan, Laguérie and Couderc, 1976), (Tournié, 1977) and (Upadhyay and Tripathy, 1975). It must be added that the range of variables covered is not always the same ; with (Tournié) the best agreement is found with our data for the particles of diameter 5 mm which lie in the range of Tournié's variables.

The best fit is with Upadhyay and Tripathi's work which corresponds to the same experimental conditions.

The autocorrelation analysis of instantaneous fluctuations of the mass transfer coefficient gives us an opportunity to test the theoretical developments proposed by Nelson and Galloway which established the equations

$$Sh = \frac{2\xi + \left\{\frac{2\xi^2(1-\varepsilon)^{1/3}}{1-(1-\varepsilon)^{1/3}} - 2\right\}\tanh\xi}{\frac{\xi}{1-(1-\varepsilon)^{1/3}} - \tanh\xi}$$

and

$$\xi = \frac{2\, d_P}{(1-\varepsilon)^{1/3} - 1}\sqrt{\frac{S}{D}}$$

Our fundamental hypothesis is that $S = 1/\tau_o$, τ_o being the time scale of the fluctuations, represents a good approximation of the surface renewal frequency considered by these authors. The results obtained for the Sherwood number are presented in table 2 and show good agreement with experimental values.

In our opinion this agreement does not allow us to draw definite conclusions. Indeed it must be recalled that Nelson and Galloway's development was primarily devoted to fixed beds at low Reynolds numbers ; here their theoretical proposals have been extended to fluidized systems at much higher Reynolds numbers. Nevertheless the closeness of the mass transfer coefficient predictions is stricking and suggests that further work should be carried out in this direction.

CONCLUSIONS

The good agreement between the data obtained in this work and the various correlations proposed recently by authors studying the dissolution of solid particles fluidized by a liquid show that the mass transfer coefficient around a fixed sphere immersed in a fluidized bed is representative of the mean mass transfer coefficient in a fluidized bed.

This new experimental procedure of investigation thus presents many advantages. First of all it gives local values and a spatial exploration has shown that these local values are constant in the bed. Secondly it allows us to realise instantaneous measurements and gives some insight into the very complex mechanisms that take place around each particle. Finally it offers a good means of studying easily the influence of the solid density. Indeed the fluidized solid particles are simply inert particles and their nature and density may be varied without any particular problems ; this possibility will be examined in the near future.

REFERENCES

Bazan, J.C. and Arvia A.J. (1965) Electrochem. Acta, 10, 1025.

Couderc, J.P., Gibert, H. and Angelino H. (1972) Chem. Eng. Sci., 27, 11.

Damronglerd, S., Couderc, J.P., Angelino, H. (1975), Trans. Inst. Chem. Engrs., 53, 175.

Dwivedi, P.N. and Upadhyay S.N. (1977) Ind. Eng. Chem. Process. Des. Dev., 16, 157.

Evans, G.C. and Gerald, C.F. (1953) Chem. Eng. Progr., 49, 135.

Fan, L.T., Yang, Y.C. and Wen, C.Y. (1960) A.I.Ch.E.Journ., 6, 482.

Laguerie, C. and Angelino, H. (1975) Chem. Eng. Journ., 10, 41–48.

Mc Cune, L.K. and Wilhelm, R.H. (1949) Ind. Eng. Chem., 41, 1124.

Mullin, J.W. and Treleaven, C.R. (1962) Paper Congr. European Fed. Chem. Engrs 3d London B 66.

Nanda, A., Ghosal, S.K., Mukerjea R.N. (1975) Chem. Age of India, 26, 55.

Nelson, P.A. and Galloway T.R. (1975) Chem. Eng. Sci., 30, 1.

Rowe, P.N. (1975) Chem. Eng. Sci., 30, 7.

Rowe, P.N. and Claxton, K.T. (1965) Trans. Inst. Chem. Engrs, 43, 321.

Tournié, P. (1977), Docteur-Ingénieur, University of Toulouse.

Upadhyay, S.N. and Tripathi, G. (1975) J. Chem. Eng. Data, 20, 20.

Vanadurongwan, V., Laguerie, C. and Couderc, J.P. (1976), Chem. Eng. J., 12, 29.

Table 2

d_p mm	ε	τ_o s	$1/\tau_o$ s^{-1}	Sh_{calc}	Sh_{exp}	$(Sh_{exp}-Sh_{calc})/Sh_{exp}$
7	0.7	0.53	1.88	353	337	- 0.05
7	0.8	0.8	1.25	287	308	+ 0.07
7	0.9	1.1	0.91	245	295	+ 0.17
5	0.8	0.95	1.05	192	230	+ 0.17
10	0.8	0.53	1.88	512	462	- 0.11

References	Relations	σ
Evans and Gerald (1953)	$J_D \times \frac{1-\varepsilon}{f'} d_p^{2.6} = 1.66\ 10^{-6}\ Re^{1.33}$ d_p en cm with $J_D \times 1-\varepsilon/f' = 2ShRe/Sc^{1/3} Ga\ Mv$	33 %
Fan, Yang and Wen (1960)	$Sh = 2 + 0.46\ Re^{0.543}\ Sc^{1/3}$	88 %
Fan, Yang and Wen (1960)	$Sh = 2 + 1.51\ [(1-\varepsilon)\ Re]^{\frac{1}{2}}\ Sc^{1/3}$	68 %
Rowe, Claxton (1965)	$Sh = 2/1-(1-\varepsilon)^{1/3} + 0.7/\varepsilon\ Re^n Sc^{1/3}$ with $2-3n/3n-1 = 4.65\ Re^{-0.28}$	17 %
Couderc, Gibert and Angelino (1972)	$Sh = 5.4\ .\ 10^{-2}/\varepsilon^2\ Re\ Sc^{1/3}$	61 %
Damronglerd, Couderc and ANGELINO (1975)	$Sh = 0.301\ Re^{0.038}\ Ga^{0.305}\ Sc^{1/3}$ if $\varepsilon < 0.815$ $Sh = 0.138\ Re^{-0.086}\ Ga^{0.412}\ Sc^{1/3}$ if $\varepsilon > 0.815$	25 %
Vanadurongwan, Laguerie and Couderc (1976)	$Sh = 0.215\ Re^{0.011}\ Ga^{0.309}\ Sc^{0.436}\ Mv^{0.303}$ if $\varepsilon < 0.815$ $Sh = 0.125\ Re^{-0.076}\ Ga^{0.390}\ Sc^{0.436} Mv^{0.33}$ if $\varepsilon > 0.815$	19 %
Tournié (1977)	$Sh = 0.245\ Ga^{0.323}\ Mv^{0.3}\ Sc^{0.4}$	21 %
Nanda, Ghosal, Mukherjea (1975)	$Re' < 1600$ $\varepsilon J_D = 0.281\ (Re')^{-0.257}$ $Re' > 1600$ $\varepsilon J_D = 1.22\ (Re')^{-0.472}$ with $Re' = Re/(1-\varepsilon)$	34 %
Upadhyay, Tripathi (1975)	$J_D = 3.713\ (Re')^{-0.713}$ $Re' < 20$ $J_D = 1.86(Re')^{-0.451}$ $Re' > 20$	13 %
Dwivedi, Upadhyay (1977)	$\varepsilon J_D = 0.765/Re^{0.82} + 0.365/Re^{0.386}$	14 %

Table 3

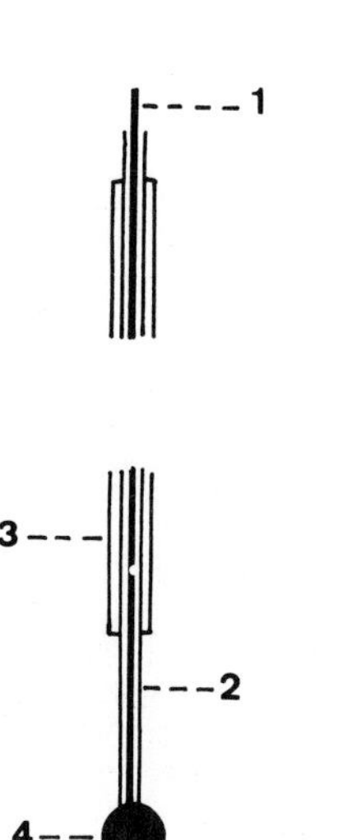

PROBE
1 - Stainless steel tube
2 - Thermoretractable plastic layer
3 - 6mm glass tube
4 - Gold sphere

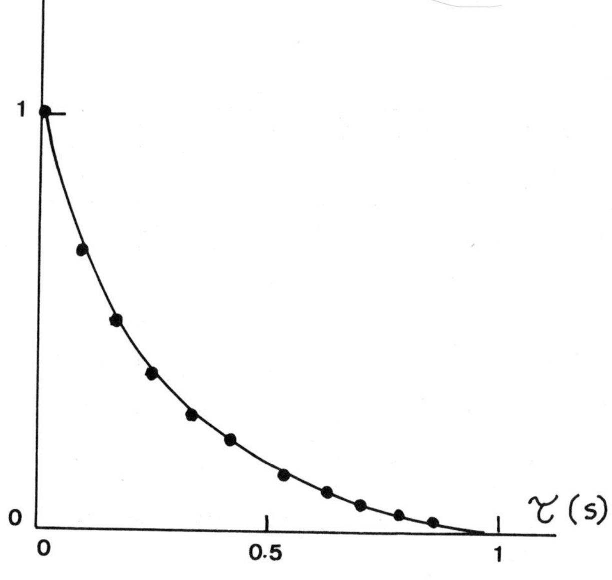

AUTOCORRELATION FUNCTION

$d_p = 5$ mm
$\varepsilon = 0.8$

UNSTEADY MASS EXCHANGE BETWEEN A BUBBLE AND THE DENSE PHASE IN A FLUIDIZED BED

By Yuri P.Gupalo, Yuri S.Ryazantsev and Yuri A.Sergeev

The Institute for Problems in Mechanics
Vernadsky Avenue, 101,
Moscow, 117526, USSR.

SYNOPSIS

The results obtained by the authors for mass transfer from a bubble at large Péclet numbers are compared with experimental data. The agreement is satisfactory in the region near the state of minimum fluidization. The case of high fluidization velocities is discussed.

Approximate solution of the problem of unsteady mass exchange between a bubble and the dense phase in a fluidized-bed reactor for the case of the bulk first order chemical reaction is obtained. The time-dependent reactant concentration inside the circulation region and the mass exchange coefficient are determined by the method of matched asymptotic expansions.

Approximate solution to the problem of bubble growth is obtained for non-linear interphase force and various conditions for the expansion rate and other parameters. The influence of bubble growth on mass exchange between the bubble and the dense phase is discussed.

NOTATION

a	bubble radius
a_c	external cloud radius
c	concentration of matter
D	fluid diffusivity
d_1	fluid density
d_2	solid density
f	interphase interaction function
$\vec{g}$	gravity acceleration
I	mass flux
P	Péclet number
p_1	fluid pressure
p_2	effective solid phase pressure
Sh	Sherwood number
u_b	rising velocity of an isolated bubble
$\vec{v}$	fluid velocity
v_o	fluidization velocity
$\vec{w}$	solid velocity
ε	porosity
ρ	solid concentration
ψ	Thiele modulus

COMPARISON OF PREVIOUS RESULTS WITH EXPERIMENTAL DATA

Recently the authors have considered the steady mass transfer from the isolated bubble to the dense phase. The problem was reduced to the solution of the convective diffusion equation outside the circulation region for the Davidson model. The steady total mass flux to the cloud boundary is found (Gupalo, Ryazantsev and Sergeev, 1976) in the form

$$\mathrm{Sh} = (1 + 2\pi^{-1}P)^{1/2} \qquad (1)$$

Here $P = a_c(u_b - v_o)D^{-1}$ is the Péclet number, a_c is the cloud radius, D is the effective diffusivity of fluid in the dense phase, u_b and v_o are the bubble rising velocity and fluidization velocity, respectively; $\mathrm{Sh} = I/[4\, a_c D(c_o - c_+)]$ is the average Sherwood number and I is the mass flux.

These results can be compared with the experimental data of Drinkenburg and Rietema (1973) only, because the effective diffusivity D was determined only in that work. It was accomplished under the conditions of $P \gg 1$ and for the presence of adsorption of reactants on catalytic particles in the case of equilibrium. For such conditions the adsorption can be taken into account and the results obtained can be given in the modified form as follows

$$Sh = \sqrt{2/\pi}\, P^{-1/2} p^{1/2}; \qquad p = 1 + \rho\varepsilon^{-1}\varepsilon_i \quad (2)$$

Here ε is the porosity of the dense phase, $\rho = 1 - \varepsilon$, ε_i is the adsorption capacity of catalytic particles.

The comparison of results (2) with the experimental data of Drinkenburg and Rietema is given in Fig.1. Here the solid line represents the theoretical results (2). The data bounded by the dashed line are obtained for the upper limits of uniform fluidization; the other data correspond to the minimum fluidization velocity.

Fig.1 shows that the agreement between analytical and experimental results is fair for small fluidization velocities. The discrepancy for high fluidization velocities is due to significant bubble growth. In the latter case the mass transfer from the bubble is governed by the rate of increase of bubble volume.

UNSTEADY MASS TRANSFER FROM THE BUBBLE TO THE DENSE PHASE

The model proposed by Gupalo, Ryazantsev and Sergeev (1976) is developed for the case of bulk first-order chemical reaction and small finite Péclet numbers. A spherical bubble rising at constant velocity $u_b > v_o$ in the semi-infinite reactor is considered.

Due to a chemical reaction, diffusion and convection, a lengthwise concentration distribution is established in the reactor. This distribution must be taken into account when the boundary conditions far from the bubble are considered.

The problem is reduced to the determination of the time-dependent concentration distribution outside the cloud and of the reactant concentration c_+ inside the circulation region (bubble plus cloud). Since solid phase particles are present inside the cloud, it is reasonable to assume a chemical reaction to take place inside that region.

The concentration field outside the closed circulation region is determined by the convective diffusion equation with appropriate initial and boundary conditions; the time-dependent concentration inside the bubble is governed by a mass balance equation.

A Laplace transformation with respect to time is used. The solution of the resulting problem is obtained by the method of matching asymptotic expansion in terms of P-number. Three expansion terms are obtained. The inverse Laplace transformation leads to the expressions for concentration distribution outside the cloud and for time-dependent concentration c_+ inside the circulation region.

The concentration c_+ and Sherwood number Sh defined by the initial concentration are found in the form

$$c_+ = \exp\left[-P^2\eta(1-\delta)^{-1}\tau\right]\xi_+$$

$$\xi_+(\tau) = 1-q+Pq(1-q)\Phi(P^2\tau)+P^2\xi_{+2}+O(P^3)$$

$$Sh = \exp\left[-P^2\eta(1-\delta)^{-1}\tau\right]Sh^o$$

$$Sh^o(\tau) = q+Pq^2\Phi(P^2\tau)+P^2 Sh_2^o + O(P^3)$$

$$\Phi(P^2\tau) = \frac{1}{\sqrt{\pi P^2\tau}}\exp\left[-\frac{(1-2\eta)^2}{4}P^2\tau\right] + \frac{1-2\eta}{2}\,\mathrm{erf}\left(\frac{1-2\eta}{2}P\sqrt{\tau}\right)$$

$$\delta = v_o/u_b; \quad q = \psi_+^2/(\psi_+^2 + 3) \quad (3)$$

Here $\tau = Dt/a_c$ is non-dimensional time, η characterises the lengthwise concentration distribution along a uniform reactor ($c_\infty = c_o \exp(-P\eta x/a_c)$), $\psi_+ = a_c\sqrt{k_+/D}$, k_+ is the effective reaction rate constant inside the circulation region.

The dependence of Sh and c_+ versus time is shown in Fig.2 for various values of parameters.

THE BUBBLE GROWTH IN THE FLUIDIZED BED. NON-LINEAR INTERPHASE INTERACTION

When the state of the dense phase is far from the onset of fluidization the bubble growth during its rising is significant. In that case the problem of unsteady mass transfer from the bubble to the dense phase is reduced to the analysis of the rate of bubble growth.

Now we consider the unsteady motion of a growing spherical bubble. The motion of solid particles (s-phase) and of the fluid (g-phase) is described by the model of two ideal mutually penetrating and interacting fluids. The porosity ε is assumed to be constant.

The momentum and mass conservation equations in a spherical coordinate system connected with the bubble centre are of the form

$$d_1\varepsilon(\partial/\partial t + \vec{v}\nabla)\vec{v} = -\varepsilon\nabla p_1 + d_1\varepsilon\vec{g} + d_1\varepsilon\dot{u}_b\vec{g}/g - \rho\vec{u}F(u)$$

$$d_2\rho(\partial/\partial t + \vec{w}\nabla)\vec{w} = -\rho\nabla p_1 - \nabla p_2 + d_2\rho\vec{g} + d_2\rho\dot{u}_b\vec{g}/g + \rho\vec{u}F(u);$$

$$\mathrm{div}\,\vec{v} = \mathrm{div}\,\vec{w} = 0; \quad \rho + \varepsilon = 1 \quad (4)$$

Here $\vec{w}$ and $\vec{v}$ are s-phase and g-phase velocities, d_1 and d_2 are fluid and particle densities, ρ is particle concentration, p_1 and p_2 are g-phase and s-phase pressures; $u_b(t)$ is the bubble rising velocity;

$\dot{y} \equiv dy/dt$.

The analysis of some specific systems shows the interphase interaction to be non-linear for many real systems. For example, when 1 mm size glass ballotini is fluidized by air the nonlinear term is greater than 20% of the linear one. For a wide range of system parameters the interphase interaction can be represented by the expression

$$F(u) = F_o[1 + \gamma f(u)], \quad \gamma \ll 1$$
$$f(u) = O(1) \tag{5}$$

The case of linear interphase interaction was considered recently by Buyevich (1975).

The condition $d_2 \gg d_1$ usually met in real bubbling systems is used below.

The boundary conditions for the system (4) are as follows

$$r \to \infty, \; p_1 = p_{1\infty} + d_2\rho\vec{g}\vec{r} \; ; \quad p_2 = p_{2\infty}$$
$$w = u_b(t)\vec{g}/g; \quad v = [u_b(t) - v_o]\vec{g}/g$$
$$r = a, \; w_r = \dot{a}, \; \varepsilon v_r = v_{+r} \; ;$$
$$p_1 = p_+(t); \quad p_2 = 0 \tag{6}$$

Here $p_+(t)$ is fluid pressure inside the bubble. $p_{2\infty}$ is s-phase pressure due to particles pulsating.

Assuming the s-phase motion as potential ($\vec{w} = \nabla\psi$) and taking into account the relation $d_2 \gg d_1$ the Cauchy-Lagrange integral for the sum of s-phase and g-phase momentum equations is obtained. All boundary conditions (6) excluding the condition $r = a$, $p_2 = 0$ can be satisfied. The latter together with the Cauchy Lagrange integral leads to the condition

$$r = a, \; d_2\rho - \frac{d\psi}{dt} + \frac{1}{2} d_2\rho(w^2 - u_b^2) + p_+(t)$$
$$- d_2\rho g a \cos\theta - d_2\rho \dot{u}_b a \cos\theta$$
$$= p_{1\infty} + p_{2\infty} \tag{7}$$

which can be satisfied only locally near the front stagnation point by using the Davies-Taylor method. It leads to the system of ordinary differential equations for $a(t)$ and $u_b(t)$.

The latter condition contains the unknown function $p_+(t)$ and the connection between $p_+(t)$ and $\dot{a}(t)$ must be found. Assuming the variables to be of the form

$$p_+(t) = p_+^o(t) + \gamma p_+^1(t); \quad p_1 = p_1^o + \gamma p_1^1(t);$$
$$\vec{u} = \vec{u}^o + \gamma\vec{u}^1 \; ; \quad \vec{u}^o = \vec{v}^o - \vec{w}$$

the problem can be reduced to the boundary value problems for $p_1^o(\vec{r})$ and $p_1^1(\vec{r})$. The Laplace equation for p_1^o and the Poisson equation for p_1^1 are obtained. The solution of these problems and the integral condition for the phases volume concentration balance leads to the relations

$$p_+^o - p_{1\infty} = -\rho\varepsilon^{-2} F_o a\dot{a} \; ;$$
$$p_+^1 = -\frac{5}{9}\rho\varepsilon^{-2} F_o v_o - \frac{df}{dv_o} a\dot{a} \tag{8}$$

The condition (7) together with (8) leads to the bubble growth rate and motion equations

$$\frac{1}{2}\dot{u}_b a + \frac{3}{2} u_b \dot{a} + \frac{9}{4} u_b^2 = ga$$
$$\frac{7}{4} u_b^2 - \frac{p_{2\infty}}{d_2\rho} - a\ddot{a} - \frac{3}{2}\dot{a}^2 =$$
$$= \frac{F_o}{d_2\varepsilon^2}\left(1 + \frac{5}{9}\gamma v_o \frac{df}{dv_o}\right) a\dot{a} \tag{9}$$

That system has the steady solution

$$a_* = 9p_{2\infty}/(7d_2\rho g),$$
$$u_{b*} = \frac{2}{3}(ga_*)^{1/2} \tag{10}$$

similar to one obtained by Buyevich (1975). When $a > a_*$ the bubble's volume increases with time; it decreases when $a < a_*$.

Below we take into account the relation $p_{2\infty}/(d_2\rho v_o^2) = O(1)$. When $a \gg a_*$ and, respectively, $u_b \gg v_o$ the solution of the system (9) is of the form

$$u_b = \frac{2}{3}(ga)^{1/2},$$
$$\dot{a} = \frac{7\varepsilon v_o[1 + \gamma f(v_o)]}{9[1 + 5/9\gamma v_o f'(v_o)]} \tag{11}$$

For systems of large particles we often have $a/a_* = O(1)$ ($u_b/v_o = O(1)$). Assuming $|a - a_*| \ll a_*$ the approximate solution of the system (9) leads to the bubble law in the following form

$$\dot{a} = \frac{1}{6} g^{1/2} a_*^{-1/2}(\xi + 2.1)^{-1}(a - a_*)$$
$$u_b = u_{b*} + \frac{1}{3} g^{1/2} a_*^{-1/2}(\xi + 3.85)$$
$$\times (\xi + 2.1)^{-2}(a - a_*) \tag{12}$$
$$\xi = \frac{3[1 + 5/9\, v_o f'(v_o)]\,(ga_*)^{1/2}}{2\varepsilon v_o[1 + \gamma f(v_o)]}$$

It is worth noting the pronounced influence of the non-linear correction on the bubble growth rate. The above proposed model is applicable for rather high fluidization velocities.

The bubble growth with time is shown in Fig.3. Here the curve 1 represents the bubble growth rate for $a_o = a(t = 0) = 1.1a_*$; the curve 2 shows the bubble shrinking for $a_o = 0.9a$. The bubble velocity versus bubble radius is shown in Fig.4. The values of parameters of fluidization used in Fig.3 and Fig.4 are typical.

It must be noted that when the initial radius of the bubble differs slightly from the critical radius a_*, the growth rate is small and it can be neglected in the investigations of mass transfer processes. The growth rate increases with time and tends to the constant value given by (11) as $t \to \infty$.

REFERENCES

Buyevich Yu.A. (1975) Int.J.Multiphase Flow 2, 337.

Drinkenburg A.A.H. & Rietema K. (1973) Chem.Engng Sci. 28, 259.

Gupalo Yu.P., Ryazantsev Yu.S. & Sergeev Yu.A. (1976) Proc.4th Intern/6th Europ.Symp. on Chem.Reaction Engng 1, 162.

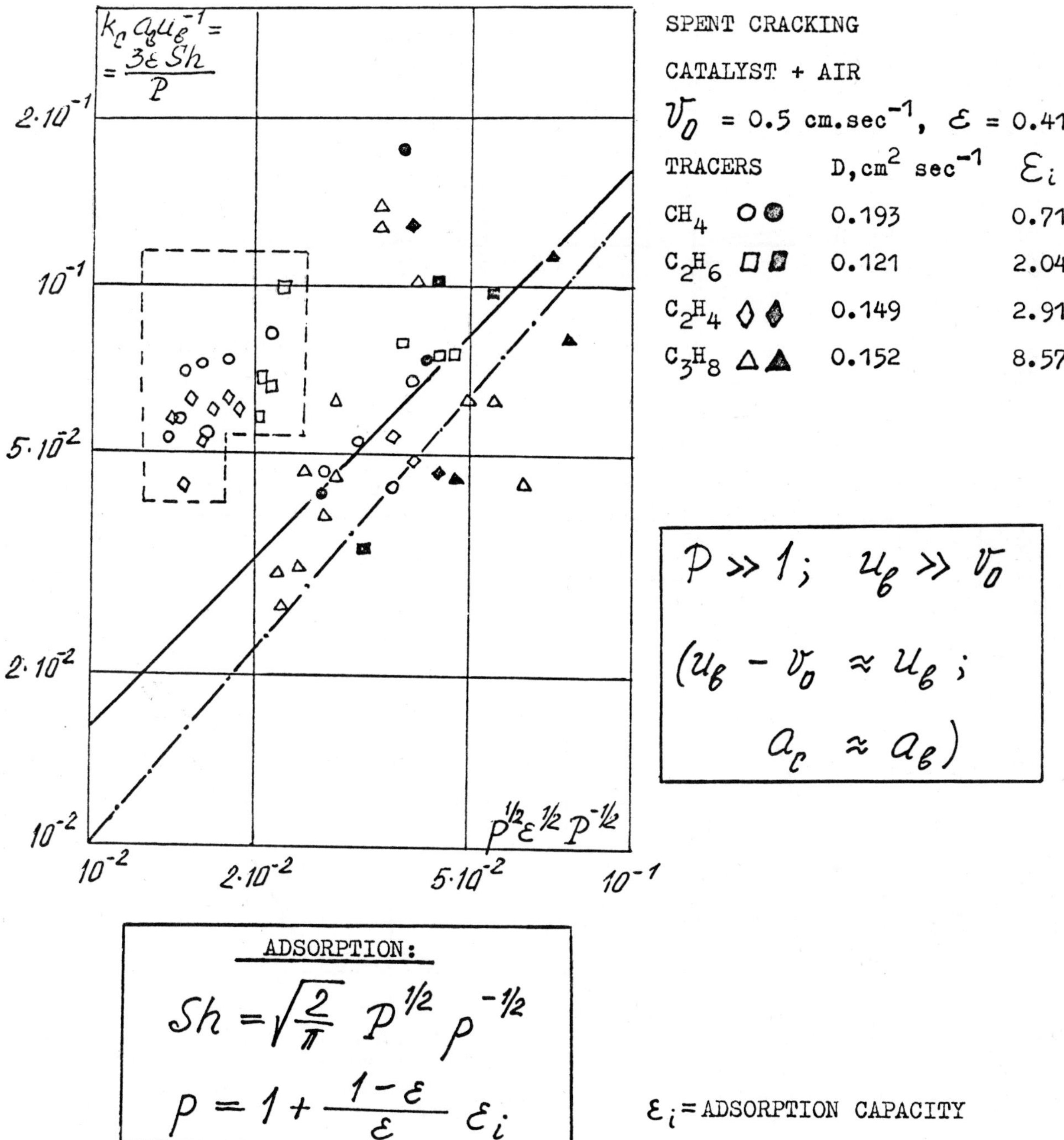

Fig. 1. Comparison with experimental data by Drinkenburg and Rietema. Data bounded by dashed line are obtained for high fluidization velocity. The other data correspond to the minimum fluidization velocity.
.-.-.-.- Numerical results by Drinkenburg & Rietema.

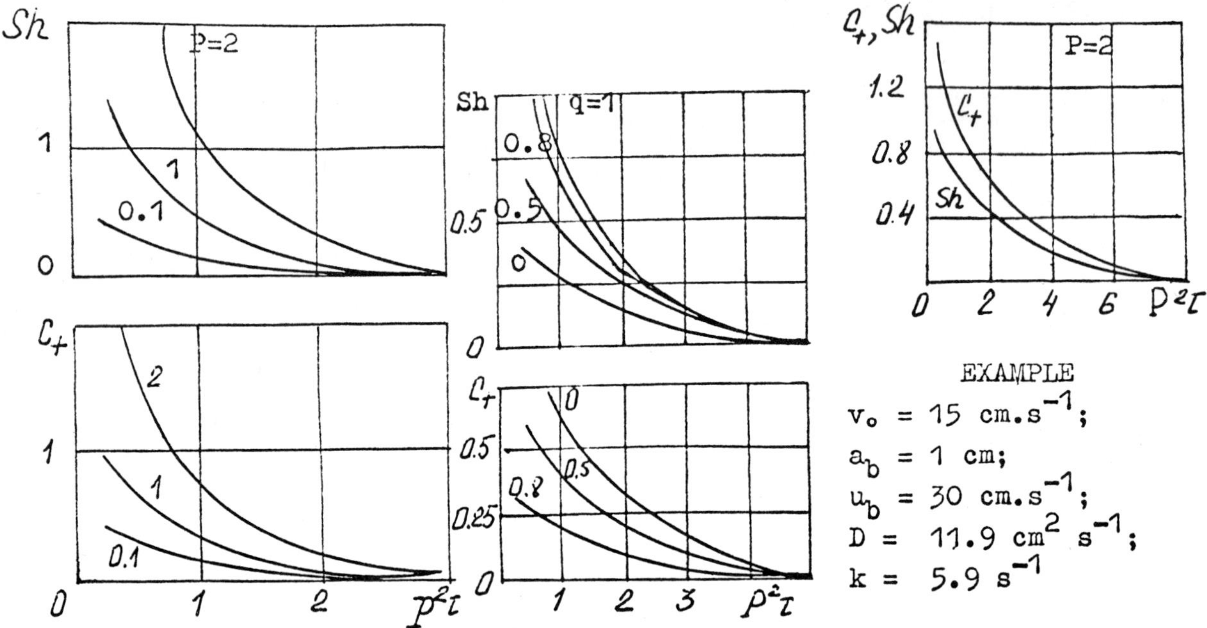

Fig. 2. The dependence of Sh - number and inner concentration on time for various parameters.

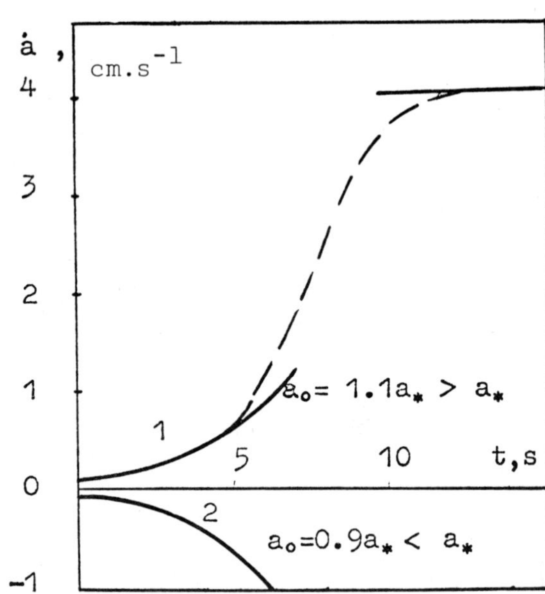

Fig. 3. An example of bubble growth rate versus time for a_*=1cm, v_o = 10 $cm.s^{-1}$, $\mathcal{E}$=0.5, $\mathcal{S}$=9. The dashed line is the interpolation for intermediate bubble radius.

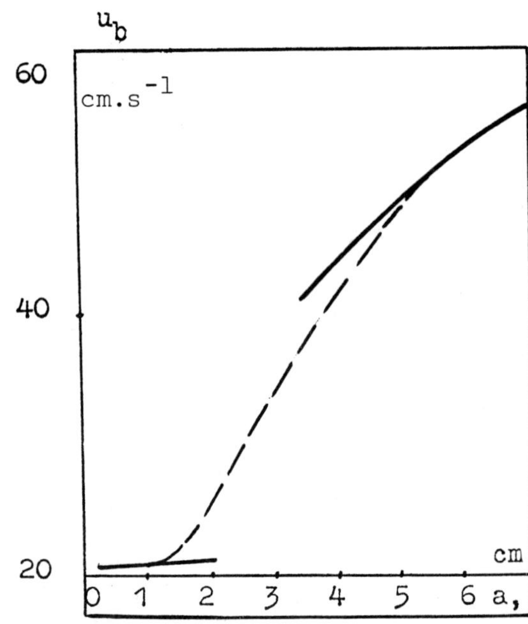

Fig. 4. An example of the dependence of bubble velocity on bubble radius. The parameters are the same as in Fig. 3.

THE INFLUENCE OF GAS ADSORPTION ON MASS TRANSFER AND GAS MIXING IN A FLUIDIZED BED

by W.Bohle and W.P.M.van Swaay
Twente University of Technology
Enschede, The Netherlands

SYNOPSIS

The influence of adsorption on mass transfer and mixing was investigated by steady state and transient experiments with adsorptive gases (argon, methane, ethane, propane and freon–12) and a non-adsorptive gas (helium). Bubble properties were measured in order to compare the results obtained with the two-phase model used with those from bubble models. Gas mixing proves to be proportionally enhanced by adsorption. Mass transfer is increased by adsorption, and does not depend on molecular diffusion. The relation derived by Chiba and Kobayashi between the mass transfer rate and the adsorption equilibrium constant is proved to hold. Mass transfer results agree with the Kunii and Levenspiel bubbling bed model. It is demonstrated that this agreement is probably accidental.

NOTATION

C, c	concentration in the bubble and dense phase (kmol/m^3)
C_0	concentration in the disengaging zone (kmol/m^3)
c_s	surface concentration (kmol/m^3 gas of dense phase)
d_B	equivalent bubble diameter (m)
E_g, E_s	dispersion coefficients of dense phase gas and solids (m^2/sec)
E_l	mean piercing length (m)
F, f	fractional gas volume in the bubble and dense phase (m^3/m^3 bed)
g	acceleration constant, 9.81 (m/s^2)
H_b, H_0	fluidized and fixed bed height (m)
k, k_B	mass transfer coefficient, based on unit volume of bed and bubbles (s^{-1})
k_F, k_f	partial mass transfer coefficient on the bubble and dense phase side of the phase interface (sec^{-1})
m	adsorption parameter, eq.3 (-)
m'	adsorption equilibrium constant (m)
N	equal to $\bar{\theta}$ (-)
N_B	number of backmixing units (-)
N_E	number of dense phase mixing units (-)
N_k	number of mass transfer units (-)
Q_k	mass transfer flow rate (m^3/s)
q	parameter in eq. 6^a (-)
S	mixing slope (-)
S_b	inner and outer solid surface area (m^2/m^3 bed volume)
t	time (s)
u_o	superficial velocity (m/s)
V, v	linear gas velocity in bubble and dense phase (m/s)
V_b	volume of the fluid bed (m^3)
V_B	bubble velocity (m/s)
x	length (m)
γ	fractional flow in bubble phase, FV/u_o (-)
ϵ_b	porosity of fluidized bed, including the particles void fraction (-)
θ	dimensionless time, $tu_o/\epsilon_b H_b$ (-)
$\bar{\theta}$	mean dimensionless residence time (-)
ξ	dimensionless length, x/H_b (-)
σ^2	dimensionless variance (-)
ϕ	fraction of gas in the bubble phase, F/ϵ_b (-)

INTRODUCTION

In the numerous investigations on mass transfer and mixing in gas solid fluidized beds (see reviews of Drinkenburg, 1972; Yates, 1975 and Chavarie & Grace, 1976) interpreted by models ranging from complicated bubble models to simple one and two phase models, relatively little attention was paid to the influence of adsorption of gas phase components on the solid surface (Chiba & Kobayashi, 1970; Wakabayashi & Kunii, 1971; Drinkenburg & Rietema, 1972, 1973; Yoon & Thodos, 1972; Nguyen & Potter, 1974, 1975).
It is, however, obvious that this adsorption can be expected to have an influence on the mass transfer between bubbles and dense phase as well as on the effective axial mixing in the dense phase. This is important in most fluidized bed operations, particularly in fluidized bed reactors. In the present investigation we utilized a combination of steady state and transient measuring techniques to establish the influence of adsorption. The results have been interpreted with the two-phase model as presented by May (1959) and Van Deemter (1961) for non-adsorbing tracers, extended to allow for adsorption. This two-phase model serves as a general framework for comparing the results with those obtained from different bubble models. Although the measuring system can be applied to large scale units

(Bohle and Van Swaay,–) and may be used for analysis of full scale equipment under operating conditions, the present reported results were limited to a small scale unit.

THEORY

The two-phase model we used is an extension of the models described by May (1959) and Van Deemter (1961). The main assumptions are: the solid free bubble phase flows in plugflow and exchanges gas with the gas-solid emulsion phase (dense phase); this mass transfer process is described by one single gas exchange coefficient $k = Q_k/V_b$. The dense phase gas mixing is described by a diffusion-like process with dispersion coefficient E_g. With respect to the adsorption it was assumed that the adsorption isotherm is linear and that adsorption equilibrium is established comparatively rapidly (Drinkenburg, 1972; Chiba, 1970), so that diffusion limitation inside the porous solid particles can be neglected (Drinkenburg, 1972; Moulijn, 1974; Kolk, 1977) and that the dispersion coefficients for the non-adsorbing gas components in the dense phase and of the solids are equal (Bohle,–; De Vries & Van Swaay, 1972; Morooka et al., 1972). Justification of these assumptions is demonstrated in the referenced literature. This two-phase model has proven to be physically acceptable (Nguyen, 1974, 1975; May, 1959; Van Deemter, 1961) leading to reliable, calculated conversions for low-rate reactions (De Vries, 1972; Van Swaay & Zuiderweg, 1973), especially for class A solid material according to the classification of Geldart (1973). Considering the many mechanisms of mass transfer (Drinkenburg, 1972; Chavarie, 1976; Chiba, 1970) the use of one single mass transfer parameter is a simplification. On the other hand, the usual bubble models can be adapted to the two-phase concept to compare the experimental data in relation with their additional assumptions. As in most fluid bed models, axial and radial variation of transport properties is neglected. This condition does not completely hold (El Halwagi & Gomezplata, 1967; Werther & Molerus, 1973; Chavarie & Grace, 1975), which might be considered in the interpretation of the experimental results.
The dimensionless-differential mass balance equations for the extended two-phase model are:

$$\phi\frac{\partial C}{\partial \theta} + \gamma\frac{\partial C}{\partial \xi} + N_k(C-c) = 0 \qquad \ldots (1)$$

$$(m+1)(1-\phi)\frac{\partial c}{\partial \theta} + (1-\gamma)\frac{\partial c}{\partial \xi} - \frac{1}{N_E}\frac{\partial^2 c}{\partial \xi^2} + N_k(c-C) = 0 \qquad \ldots (2)$$

in which m is defined as:

$$m = \frac{c_s}{c} = \frac{m' S_b}{\epsilon_b(1-\phi)} \qquad \ldots (3)$$

and in which $N_k = \dfrac{kH_b}{u_o}$ and $N_E = \dfrac{u_o H_b}{(m+1)fE_g}$ . . . (4)

From (1) and (2) and the appropriate boundary conditions the moments of the RTD ($\gamma = 1$) and the steady-state backmixing slope (slope of semi-logarithmic plotted axial concentration profile during steady-state tracer supply below bed level) can be derived:

First moment, the dimensionless mean residence time: $\bar{\theta} = N = 1+m(1-\phi)$ . . . (5)

Second central moment, the dimensionless variance:

$$\sigma^2 = N^2\left[\frac{2}{N_k}\left\{\frac{(m+1)(1-\phi)}{N}\right\}^2 + \frac{2}{N_E} + \frac{2q}{N_E^{\,2}}\cdot\frac{\left\{e^{\frac{1}{2}N_k(q-1)}-1\right\}\left\{e^{-\frac{1}{2}N_k(q+1)}-1\right\}}{e^{\frac{1}{2}N_k(q-1)}-e^{-\frac{1}{2}N_k(q+1)}}\right] \qquad \ldots (6)$$

with $$q = \sqrt{1+4\frac{N_E}{N_k}} \qquad \ldots (6a)$$

Backmixing slope S, equal to the number of backmixing units N_B:

$$S = N_B = \tfrac{1}{2}N_k(q-1) \qquad \ldots (7)$$

As proved by Van Deemter, S is independant of the amount of tracer that is supplied to each phase, and is equal for both phases. Equations (6) and (7) represent the two basic equations for the calculation of N_k and N_E from steady-state and transient tracer experiments. The adsorption equilibrium can be calculated with (5). Equation (7) can be simplified for strongly adsorbing gases due to the dominating effect of the mixing in the dense-phase on the overall backmixing (small $N_E:N_k$ ratio):

$$N_B \approx \tfrac{1}{2}N_k\left[1+2\frac{N_E}{N_k}-1\right] \approx N_E \qquad \ldots (8)$$

This means that the mixing in the dense phase can in principle also be derived from steady-state backmixing experiments with strongly adsorbing tracers. N_E can also be derived from solids mixing experiments (Bohle, –) to test the validity of the model in an independant way. Once the dense phase mixing and adsorption equilibrium is known, steady-state backmixing experiments should in principle be the only tools needed to measure the gas exchange properties.

EXPERIMENTAL SET-UP

The experimental set-up is shown in fig. 1. The disengaging zone was equiped with a 7x6.5 cm blade stirrer (700 rpm) to ensure reproducible almost ideal mixing in this region. Most of the experiments were performed with a fixed bed height H_0=55 cm at ambient temperature and pressure. A silica/alumina catalyst carrier was used as fluidized solid material. Various properties are given in table 1. Properties of the fluidized bed were derived from these figures and the results of collapse experiments as described by Drinkenburg (1973) and Rietema (1967).
Backmixing concentration profiles were sampled through a small, vertically placed glass filter, while during the RTD measurements the response of the system was picked up in the disengaging zone, where a continuous small flow of gas was withdrawn through a glass filter, connected to the catharometer by a very short tube. The catharometer output was read by a DEC PDP-11/45 computer and stored on tape for calculation on a DEC-10 computer. Mean residence time and variance of the RTD of fluidized bed proper were calculated from the responses on the pulses given below the distribution plate and in the disengaging zone according to the method described by De Groot

TABLE 1

PROPERTIES OF THE FLUIDIZED SOLID PARTICLES

composition		
	:87	weight % SiO_2
	12.9	weight % Al_2O_3
particle diameter distribution (sieve analysis)		
diameter (μm)	weight %	cum. weight %
< 44	7.3	7.3
44 – 75	30.5	37.8
75 – 105	23.9	61.7
105 – 150	36.9	98.0
150 – 210	1.6	99.6
210 – 300	0.3	99.9
> 300	0.1	100.0
mean particle diameter	± 70.10^{-6}	m
skeletal density	2200	kg/m^3
particle density	813	kg/m^3
fixed bed density*	475	kg/m^3
particle void fraction	0.63	–
fixed bed interparticle void fraction*	0.41	–
fixed bed void fraction*	0.78	–
specific surface area	540.10^3	m^2/kg

* settlement of bed after fluidization

(1967). Corrections due to the mixing and adsorption in the plenum chamber and the bottom plate were negligible. It was checked with steady state backmixing tests that no backmixing of tracer below bed level occurred during the pulses in the disengaging zone, except in the case of very strongly adsorbing gases at low gas velocities ($u_0 < 5$cm/s).

EXPERIMENTAL RESULTS

Some typical results from the collapse experiments are presented in table 2. From these results it can be concluded that the dense phase gas velocity is almost independent of u_0 and the fixed bed height; the same holds for the dense phase expansion (9.5 ± 0.5%).
A few examples of the numerous steady state backmixing concentration profiles measured are plotted in fig. 2. The standard measuring position was always halfway between the column wall and the centre. The backmixing slope did not depend appreciably on the radial measuring position, the steady state injection level or the tracer flow rate.

TABLE 2

CHARACTERISTIC FLUIDIZED BED PROPERTIES

property	dim.	u_0 =4.50 cm/s	u_0 =12.50 cm/s
H_b	cm	63.9	70.3
H_b/H_0	–	1.16	1.28
ϵ_b	–	0.81	0.83
F	–	0.05	0.13
f	–	0.76	0.69
ϕ	–	0.06	0.16
V	cm/s	85.2	91.1
v	cm/s	0.24	0.24
γ	–	0.96	0.99
S_b	1/m	2.2 * 10^8	2.0 * 10^8

In general, the C/C_0 ratio was larger close to the wall. The backmixing slope increases within one column diameter from the bottom plate, especially with helium or the weakly adsorbing tracers and close to the column wall. The experimental relation between N_B and u_0 for the different tracers is plotted in fig. 3. The mean residence time in the fluidized bed and the variance of the RTD's for the different tracers calculated from the transient experiments are plotted in fig. 4 and fig. 5. Values of m, N and m′ are tabulated in table 3. The mean residence time for helium from the RTD never deviated more than 5% from the mean residence time calculated from the flow rate. The results of these transient experiments are the average values of five or more measurements under equal conditions.

TABLE 3

ADSORPTION PROPERTIES

TRACER	m′·10^9 (m)	m (-)	N(-) at u_0=5cm/s
Helium	0	0	1.00
Argon	0.51	0.15	1.14
Methane	2.0	0.57	1.53
Ethane	9.9	2.87	3.66
Propane	24.3	7.07	7.56
Freon–12	48.3	14.1	14.1

With these values of N_B, N and σ^2 the values of N_k and N_E were calculated using (6) and (7) by a simple trial and error routine. The resulting quantities k and E_g are plotted in fig. 6 and fig. 7. A few examples of N_k are given in table 4. The accuracy of σ^2/N^2 was not better than ± 10% As the values for all the tracers were within the range 0.90–1.10, a value of 1.0 was used through all calculations for every tracer.

TABLE 4*)

EXAMPLES OF THE N_k, N_E VALUES

TRACER	u_0=5cm/s		u_0=12cm/s	
	N_k	N_E	N_k	N_E
Helium	2.42	1.64	1.71	4.50
Argon	2.53	1.41	1.84	3.38
Methane	2.69	0.85	2.08	2.13
Ethane	3.33	0.53	2.66	1.08
Propane	3.98	0.20	3.36	0.41
Freon–12	4.2	0.18		

*) values from eq. (6) and eq. (7) and σ^2/N^2=1.0

In order to be able to compare our results with the results from other investigations, where the mass transfer rate is related to the bubble volume, some bubble properties were measured with the experimental techniques described by Werther and Molerus (1973, 1974). The results are plotted in fig. 8 and fig. 9. In these plots E_l represents the so-called mean piercing length of the capacitive probe, the property of which can be related to the true bubble diameter (Werther, 1974).

DISCUSSION

The observed independence of the dense phase expansion and dense phase velocity of u_0 was reported earlier (May, 1959; De Vries, 1972; Morooka, 1972). The same applies to the quantitative values of these properties for this kind of solid particles. The bubble measurements suggest a linear growth of the bubble diameter with the height above the distribution plate, but this still depends on the relation between the mean piercing length E_l as measured by the capacitive probe and the equivalent bubble diameter d_B, the relation of which is a function of the unknown bubble size distribution (Werther, 1974). This distribution can change with the distance from the bottom plate. Disregarding the results up to 10 cm above the distribution plate which cannot be quantitatively reliable for a num-

ber of reasons (Werther, 1973), V_B appears to be independent of the bubble diameter. This agrees with the class A characteristics description of Geldart (1973) (V_B is fixed by macrocirculation patterns) but does not agree with the frequently used relation $V_B = u_0 - fv + 0.71\sqrt{gd_B}$. The measured V_B values are lower than the reported linear bubble phase velocities in table 2, probably partly because the V_B measurements are restricted to one radial position and partly due to small deviations in measured bed height leading to large deviations in V from $V = (u_0 - fv)/F$.
The qualitative results of the backmixing experiments are in accordance with those from other investigations: backmixing increases with increasing adsorption of the gas on the solids (Nguyen, 1974; Miyauchi & Kaji & Saito, 1968) and decreases with increasing superficial velocity (El Halwagi, 1967). The larger decrease of the backmixing profile close to the distribution plate was reported by Nguyen (1975) who ascribed this phenomenon to increased mass transfer in this region of small bubbles. Since $N_B < 1$ indicates considerable backmixing, the reported N_B values confirm the good gas mixing due to this class A material, especially with absorbing gases.
The reproducability and reliability of the first RTD moment (θ) was very good (±2%). The spread of ±10% in the variance obtained (σ^2) was reported earlier. (De Vries, 1972), and should be accepted in this experimental situation. Increasing residence time and variance (sec^2) with increasing adsorption and decreasing u_0 was reported for a number of gases by Yates (1973) and is physically expectable. Regarding the backmixing profiles which always existed, even with freon–12, the $\sigma^2/N^2 = 1.0$ value does not indicate an ideal mixing behaviour of the fluidized bed.
Increasing mass transfer due to adsorption of the gas on the solids was theoretically shown by Chiba and Kobayashi (1970). Experimental literature data on this subject were not convincing up till now (Chavarie & Grace, 1976; Drinkenburg & Rietema, 1973). However, the influence of adsorption due to the interchange of particles in and around the bubbles is likely (Miyauchi, 1968), causing k to increase with adsorption and superficial velocity, as is shown in fig. 6. Regarding the weak adsorption behaviour of argon and the large difference between the molecular diffusivities of helium and argon, the mass transfer does not seem to depend on these molecular diffusion coefficients. This conclusion agrees with results obtained by Drinkenburg (1973).
Realizing that:

$$\frac{1}{k} = \frac{1}{k_F} + \frac{1}{k_f} \qquad \ldots\ldots (9)$$

k_F and k_f can be calculated from combinations of tracer tests assuming that neither k_F nor k_f depends on the molecular diffusivities of the gases, that k_F was not influenced by adsorption and that the proportionality $k_f \sim \sqrt{m+1}$ of Chiba and Kobayashi (1970) holds. The resulting values of k_F and $k_f/\sqrt{m+1}$ are shown in fig.10. They confirm that the relation between k_f and adsorption is quite reasonable. Especially for higher u_0 values k_F shows a large spread, partially due to the calculation method. The mass transfer resistance on the bubble side of the phase interface cannot be neglected, as Chiba and Kobayashi (1970) assumed. Assuming $E_1 = d_B$, the mass transfer data for helium can be transformed to $k_B = k/F$ plotted against d_B in fig. 11, where d_B is taken at 0.5 H_b. The results obtained by the equations of Kunii and Levenspiel (1969), Chiba and Kobayashi (1970) and Kato and Wen (1969) are plotted in the same figure.
We used our data on the superficial velocity in and porosity of the dense phase and of the bubble velocity. The agreement with the Kunii and Levenspiel equation seems remarkable but is probably accidental due to the fact that this equation takes the influence of the molecular diffusivities into account. Again it proves that the mass transfer resistance on the bubble side cannot be neglected, when the results from the Chiba and Kobayashi equation are taken into account.
The E_g values in fig. 7, calculated from $E_g = u_0 H_b/(m+1) fN_E$ confirm the theoretical proportionality between gas mixing in the dense phase and the adsorption capability of the gas. The values for the solids mixing dispersion coefficient E_s from the Morooka correlation (1972) (dotted line) for a column diameter of 10.4 cm agree approximately with the E_g values. These E_s values were confirmed by our own measurements (Bohle & Van Swaay,–). Regarding (8) propane could serve as reference gas since the ratio $4N_E/N_k < 0.3$ is found over the full u_0 range used. Calculating E_g from the propane values of N_B and m, N_k can be calculated for the other tracers from:

$$N_k = \frac{N_B{}^2}{N_E - N_B} \qquad \ldots . (10)$$

where N_E is derived from E_g and m. This procedure did not lead to a quantitative satisfactory agreement with the σ^2, N_B method although the results fell within the same range of magnitude. This might be explained from the fact that (10) is very sensitive for deviations in N_B and N_E of the magnitude as shown in fig. 7.
The same conclusion can be drawn from the N_k values obtained with (10) using $E_g = E_s$ from the Morooka correlation, probably due to E_g not being exactly equal to E_s, at least not close enough regarding the sensitivity of (10). In this case a fluidized bed with larger diameter would match $E_g = E_s$ better, due to a larger E_s and about equal slipstream (Bohle & Van Swaay,–).

ACKNOWLEDGEMENT

The authors are much indebted to J.Hoebink for the bubble measurements which were performed at Eindhoven University of Technology.

REFERENCES

Bohle, W., Van Swaay, W.P.M., to be published
Chavarie, C., Grace, J.R. (1975) Ind. Eng. Chem. Fund 14, 75
Chavarie, C., Grace, J.R. (1976) Chem. Eng. Sci. 31, 741
Chiba, T., Kobayashi, K. (1970) Chem. Eng. Sci. 25, 1375
De Groot, J.H. (1967) Proc. Int. Symp. on Fluidization, pp. 348-361, Eindhoven
De Vries, R.J., Van Swaay, W.P.M. (1972) Proc. 2nd Int. Symp. on Chem. Reaction Eng. B9, pp. 59-69, Amsterdam
Drinkenburg, A.A.H., Rietema, K. (1972) Chem. Eng. Sci. 27, 1765
Drinkenburg, A.A.H., Rietema, K. (1973) Chem. Eng. Sci. 28, 259
El Halwagi, M.M., Gomezplata, A. (1967) A.I.Ch.E.J. 13, 503
Geldart, G. (1973) Powder Technology 7, 285
Kato, K., Wen, C.Y. (1969) Chem. Eng. Sci. 24, 1351
Kolk, F. (1977) Thesis, Amsterdam
Kunii, D., Levenspiel, O. (1969) 'Fluidization Engineering', New York
May, W.G. (1959) Chem. Eng. Progress 55, 49
Miyauchi, T., Kaji, H., Saito, K. (1968) J. Chem. Eng. Jap. 1, 72

Morooka, S., Kato, Y., Miyauchi, T. (1972) J. Chem. Eng. Jap. 5, 161
Moulijn, J.A. (1974) Thesis, Amsterdam
Nguyen, H.V., Potter, O.E. (1974) Proc. 3rd Int. Symp. on Chem. Reaction Eng. vol. II, pp. 290-300, Evanston
Nguyen, H.V., Potter, O.E. (1975) Proc. Int. Fluidisation Conference, vol.II, pp. 193-200, Asilomar, California
Rietema, K. (1967) Proc. Int. Symp. on Fluidization, pp. 154-175, Eindhoven
Van Deemter, J.J. (1961) Chem. Eng. Sci. 13, 143
Van Swaay, W.P.M., Zuiderweg, F.J. (1973) Proc. Int. Symp. on Fluidization and its Applications, pp. 454-467, Toulouse
Wakabayashi, T., Kunii, D. (1971) J. Chem. Eng. Jap. 4, 226
Werther, J., Molerus, O. (1973) Int. J. Multiph. Flow 1, 123
Werther, J., (1974) Trans. Inst. Chem. Engrs. 52, 149
Yates, J.G. (1975) The Chem. Engr. 303, 671
Yates, J.G., Constans, J.A.P. (1973) Chem. Eng. Sci. 28, 1341
Yoon, P., Thodos, G. (1972) Chem. Eng. Sci. 27, 1549

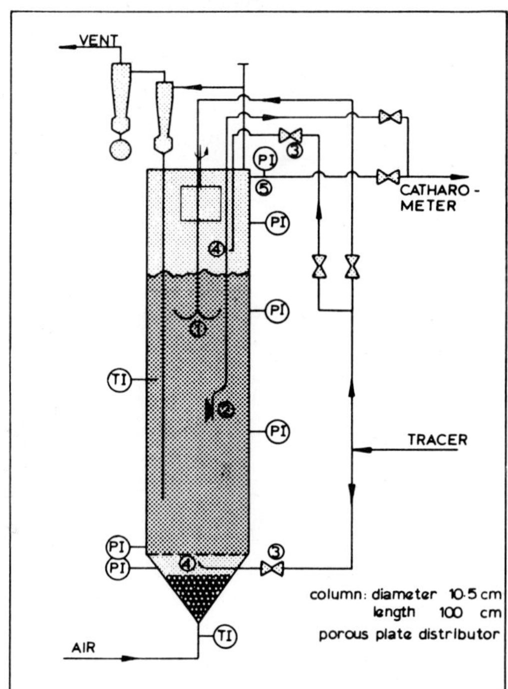

Fig. 1 Flow scheme (central part) 1. Steady-state tracer supply. 2. Intake point steady-state concentration analysis. 3. Magnetic valves. 4. Tracer pulse supply points. 5. Intake point RTD concentration analysis.

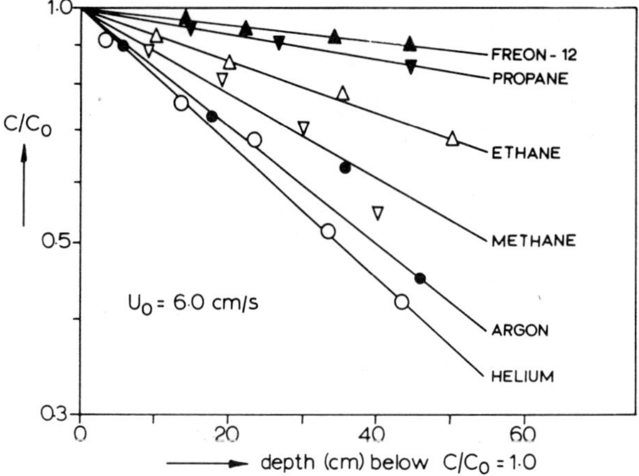

Fig. 2 Steady state concentration profiles at u_0 = 6 cm/s

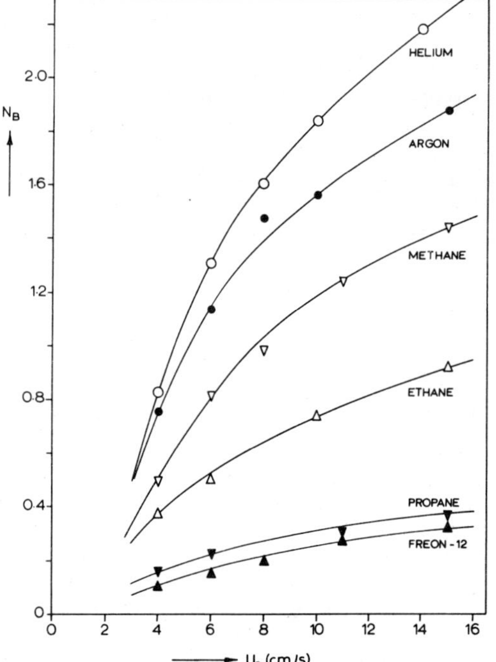

Fig. 3 Number of backmixing units N_B

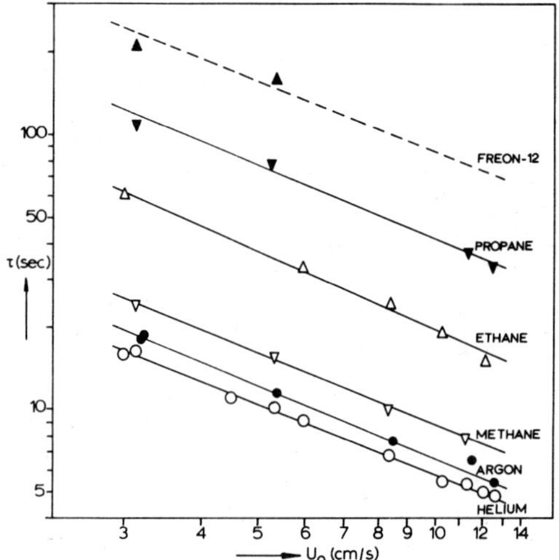

Fig. 4 Mean residence time in the fluidized bed

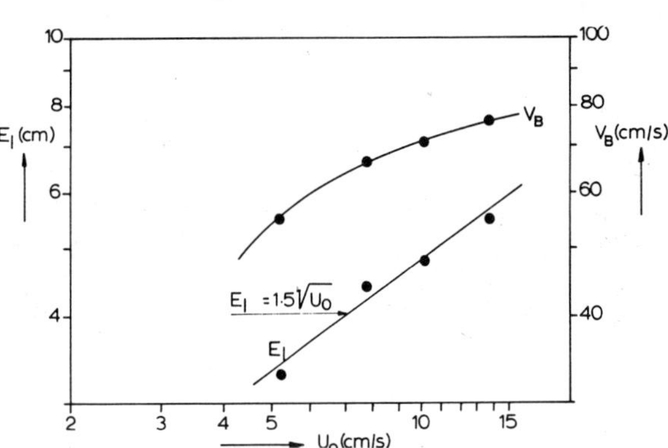

Fig. 8 The mean piercing length E_I and the bubble velocity V_B measured 48 cm above the distribution plate and 3 cm from the wall

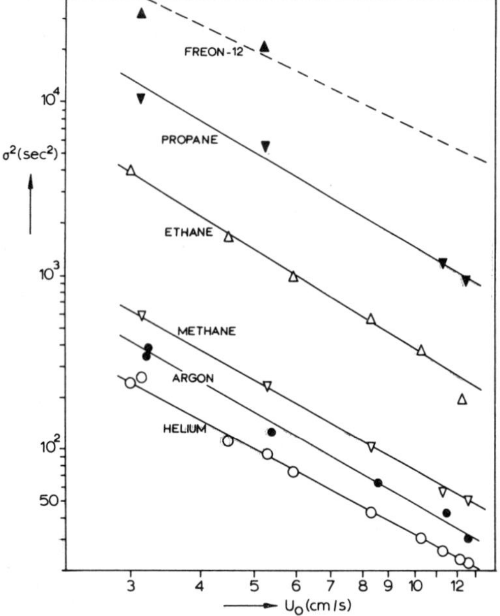

Fig. 5 Variance of the fluidized bed RTD's

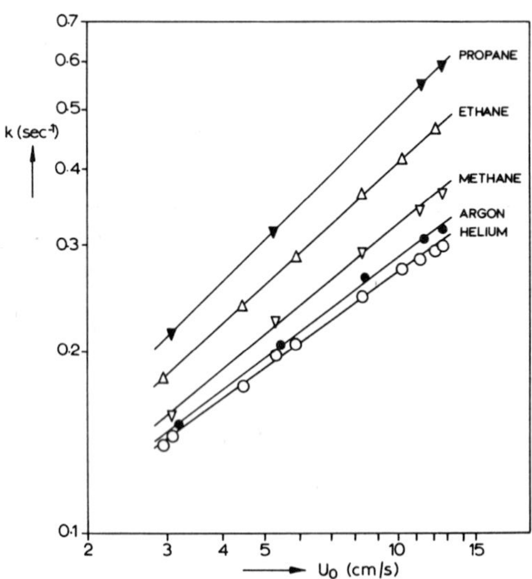

Fig. 6 The mass transfer coefficient k

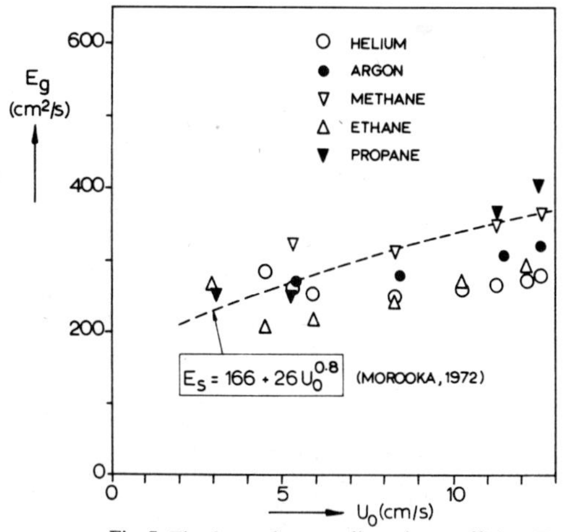

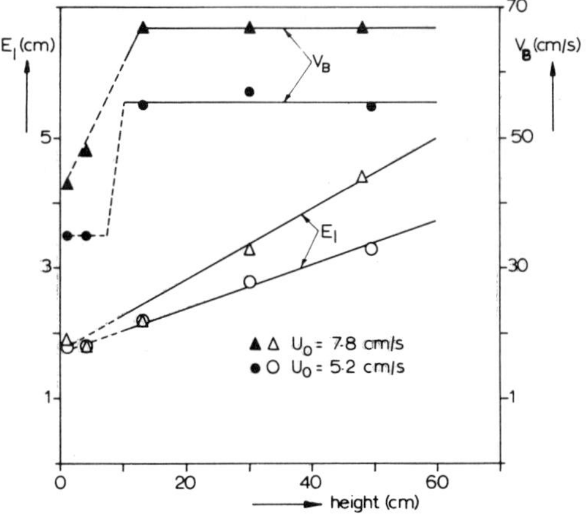

Fig. 9 E_l and V_B as function of the height above the distribution plate

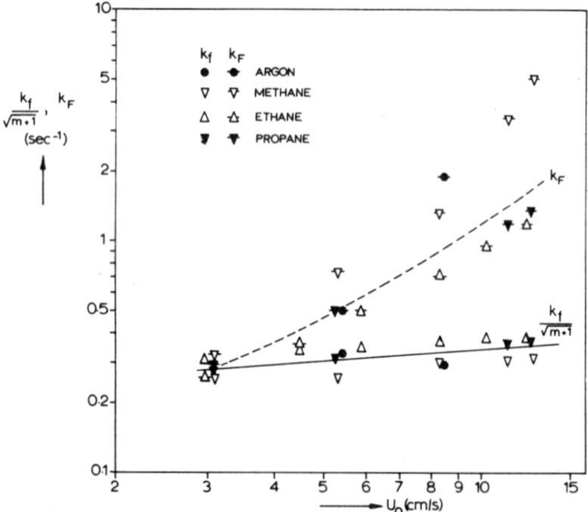

Fig. 10 The mass transfer coefficients k_F and k_f on both sides of the phase interface

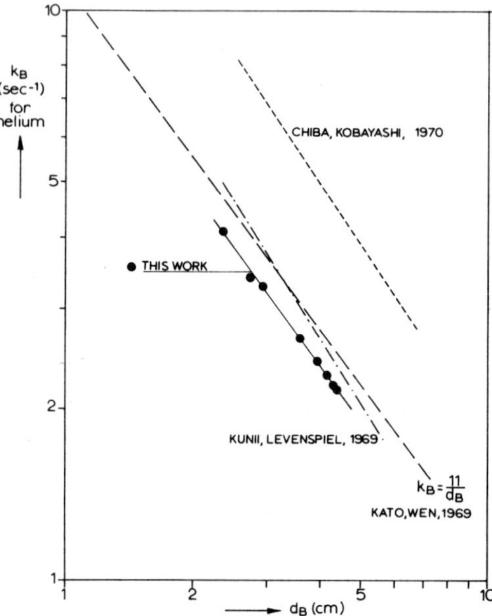

Fig. 11 k_B as a function of d_B for helium from different models.

Fig. 7 The dense phase gas dispersion coefficient E_g

BEHAVIOUR OF THE BUBBLES IN A BENCH SCALE FLUIDIZED BED REACTOR WITH CONVERSION OF SOLIDS IN CONTINUOUS OPERATION AT VARIOUS TEMPERATURES

By K. Wittmann, D. Wippern, H. Helmrich and K. Schügerl

Institut für Technische Chemie der Technischen Universität Hannover
Callinstr. 3, 3000 Hannover
Federal Republic of Germany

SYNOPSIS

Miniature capacitance probe and method for signal processing developed by Werther was applied to characterise the bubble behaviour in $NaHCO_3/Na_2CO_3$ fluidized beds in a bench scale reactor without and with reaction and at various temperatures. The detector signal was processed on-line by computer. An improved method was developed to the estimation of the coalescing bubble fractions. Some experimental results on the radial and longitudinal profiles of the local mean equivalent diameters and rising velocities of bubbles, local coalescing bubble fraction, local bubble gas flow and local bubble volume fraction are reported. The bubble gas flow increases, the size and rising velocity of the bubbles decrease due to the decomposition reaction (CO_2 development). The influence of the temperature on the bubble behaviour is slight in the investigated range between 65 and 200° C.

NOTATION

A	cross section area
$\bar{D}$	diameter of the bubble covering sphere
D_{equ}	diameter of volume equivalent sphere
D	diameter of the bed
F	factor (equ (3))
H	longitudinal distance from the gas distributor
$K, \bar{K}$	local and/or mean coalescing bubble fraction
$\bar{l}$	mean pierced length of bubbles
l_f	distance between the roofs of two bubbles following each other
N	number of bubbles
r	distance from the reactor axis
T_b	temperature of the bed
u	superficial gas velocity
u_{mf}	superficial gas velocity at the minimum fluidization point
$u^+ =$	u/u_{mf}
u_b	bubble raising velocity
$V_b, \bar{V}_b$	local and/or mean bubble gas flow rate
$\bar{V}_{e_m}$	mean gas flow rate in the emulsion phase
ε_b	local bubble volume fraction
ρ_p	particle density

K. WITTMANN, D. WIPPERN, H. HELMRICH and K. SCHÜGERL

INTRODUCTION

Since bubbles play an important role in fluidized bed reactors several investigations were published upon their behaviour (e.g. Rowe (1971)). Werther gave a review on the methods to measure the bubble properties (J. Werther (1972a)(1972b), J. Werther et al. (1973)). The last state of the development of bubble probes is represented by the probes of Calderbank et al. (Calderbank et al. (1975)) and Werther (Werther (1975)).

By means of the first method the properties of the undisturbed bubbles can be measured with high precision. By the second method also bubbles can be investigated which are interacting with others by coalescence. Since at given time often up to 50 % of the bubbles are in coalescing state (Werther (1975)) by the application of the method of Calderbank et al. one would disregard those bubbles which are especially important for the gas exchange between the bubble and emulsion phases.
Therefore the method of Werther was applied in these investigations.

EXPERIMENTALS

Measuring Method

The hard ware of the miniature capacitance probe and its electronics was identical with that of Werther (Werther et al. (1973)). Also the signal processing of Werther was adapted to the investigated systems. However, a new method for the evaluation of the coalescing bubble fraction K was developed. The signal processing was carried out on line by means of a PDP 11-40 process computer.

According to Clift et al. (Clift (1970)) in the precoalescing state two interacting bubbles must come into coaxial position before the second bubble approaches the first one with a constant speed and the coalescence can occur.
To improve the estimation of the coalescing bubble fraction K it was assumed that the two bubbles involved in a coalescence process are equal in size. Therefore, if they are rising coaxially, the momentaneous distance l_f between the roofs of these bubbles is independent of the location of the piercing point of the probe on them. In the precoaxial state the interacting bubbles are far away and the distance l_f depends on the location of the probe piercing point. Under these conditions the probability of finding bubbles in the coalescing state by means of l_f is smaller than in the coaxial state of two equal bubbles.

After the two bubbles came into the coaxial state a deformation of the second (following) bubble occurs due to the suction caused by the first (leading) bubble. With increasing deformation of the second bubble the probability of finding the bubble pair involved in coalescence process decreases. Therefore the frequency distribution $(\Delta N/\Delta l_f) = f(l_f)$ possesses a maximum corresponding to the coaxial state before this deformation occurs, though this frequency should be independent of l_f during the coaxial approach of the bubbles with constant speed.

Thus this maximum appears due to the maximum of the detecting probability for a given constellation of the bubbles. This maximum is the closest estimate for the real number of bubble pairs per interval Δl_f which are in the coalescing state. Therefore in the above frequency distribution $(\Delta N/\Delta L_f)$ is considered to equal $(\Delta N/\Delta l_f)_{max}$ for $l_f \leqslant l_{f\,max}$. To estimate the number of bubbles involved in coalescence process, the start and the end of the coalescence process has to be defined. The start of the coalescence process is defined according to Clift et al. (Clift et al. (1970)) and Werther (Werther (1975)): The coalescence starts, if the stagnation point of the following bubble touches the sphere covering the leading bubble. The coalescence process comes to end, if the stagnation points of the two bubbles overlap.

Hence the number of bubbles involved in coalescence N_K is given by equ (1):

$$N_K = \bar{D}\left(\frac{\Delta N}{\Delta l_f}\right)_{max} \tag{1}$$

where $\bar{D}$ is the diameter of the sphere which covers the bubble.

The fraction of coalescing bubbles K is given by equ (2):

$$K = \frac{\bar{D}(\Delta N/\Delta l_f)_{max}}{N_{o.a.}} \tag{2}$$

where $N_{o.a.}$ is the over all number of detected bubbles.

The mean fraction of coalescing bubbles $\bar{K}$ for one cross section of the bed can be calculated by integration of K over the cross section area A.

The over all gas flow u divides into the bubble and emulsion phases according to equ (3):

$$uA = \bar{V}_b + \bar{V}_{em} = \bar{V}_b + F\, u_{mf}\, A \qquad (3)$$

where $\bar{V}_b$ and $\bar{V}_{em}$ are mean bubble gas flow and mean emulsion gas flow rates.

u_{mf} superficial gas velocity at the minimum fluidization point

F corrections factor for the two phase theory, which were evaluated by numerous research groups (e.g. Pyle et al. (1967), Whitehead et al. (1967), Geldart (1968),(1970), Grace et al. (1969), Mc Grath et al. (1971), Werther et al. (1972b) and mostly values between 1 and 5 were found.

EQUIPMENT AND MATERIAL

The measurements were carried out in a bench scale stainless steel fluidized bed reactor of 19 cm diameter and 250 cm over all height (for details see Goedecke (1974) and Vrbata (1976)). Stainless steel porous plate with mean pore diameter of 200 μm was used as gas distributor.

The decomposition of $NaHCO_3$ to Na_2CO_3 was investigated in this reactor. During the chemical reaction the particle diameter distribution and the mean particle diameter ($d_p \simeq 127$ μm) changed only slightly. The particle density changes with the reaction from $\rho_p = 2.16$ g/cm^3 ($NaHCO_3$) to $\rho_p = 2.52$ g/cm^3 (Na_2CO_3). Because of the low concentration of $NaHCO_3$ in Na_2CO_3 for the bed, approximately $\rho_p = 2.52$ g/cm^3 prevails. u_{mf} is almost independent of the conversion ($u_{mf} = 1.61$ cm/s). The (fixed bed) height to diameter ratio was kept constant (H/D = 3).

The bubble size distributions were measured as a function of the radial coordinate at four different distances x from the gas distributor: at 55 mm (ME6), 165 mm (ME5), 372 mm (ME4) and 565 mm (ME3), at three relative gas velocities $u^+ = u/u_{mf} = 2$, 3 and 4 and at three temperatures T_b = 65, 125 and 194° C without and with chemical reaction with a solid feed rate of 80 g/min and high conversion (e.g. 0.95 at 125° C).

RESULTS

In figure 1 the diameter of the bubble covering sphere is plotted as a function of the distance from the gas distributor at $u^+ = 2$,3 and 4 and compared with the functions calculated according to the semi-empirical correlation of Rowe (Rowe (1976)) with H_o = 4 cm:

$$\bar{D} = 0.16\,(u-u_{mf})^{0.55}\,(H+H_o)^{0.75} \qquad (4)$$

With H_o = 4 cm (instead of $H_o \simeq 0$ expected for porous plate) fairly good agreement was found between the measured and by means of equ (4) calculated diameters $\bar{D}$ for small bubbles. The differences between them for large bubbles can possibly be explained by wall effects.

In figures 2 and 3 the radial local bubble gas flow profiles and the radial bubble volume fraction profiles are plotted for various distances from the gas distributor.

A high local bubble flow causes high coalescence rate as can be seen from figures 2 and 4.

A comparison of the mean coalescing bubble fractions $\bar{K}$ calculated according to Werther (Werther (1975)) and by the improved method (Table 1) indicates, that the latter gives more reasonable values than the former. However, they agree partly fairly well. In general, with increasing gas flow rate and distance from the gas distributor $\bar{K}$ increases up to about 20 %.

Table 1: Comparison of the mean coalescing bubble shares $\bar{K}$ (%) calculated according to Werther and by the improved method.

H(cm)	Method of Werther $\bar{K}$(%)			Improved Method $\bar{K}$(%)		
	$u^+=2$	$u^+=3$	$u^+=4$	$u^+=2$	$u^+=3$	$u^+=4$
56.5	15.3	20.7	21.8	13.0	19.1	24.1
37.2	14.3	19.6	21.2	14.0	17.5	20.1
16.5	13.0	19.1	22.8	8.2	12.4	18.9
5.5	7.1	12.2	12.1	7.0	10.8	12.0

The measured radial profiles of the local bubble gas flows (figure 2) and the bubble volume fractions (figure 3) are comparable to those known from the literature for quartz sand particles (Werther (1972a)(1972b), Werther et al. (1973)):near to the gas distributor the local bubble gas flow is high at the wall in an annular region of the bed. This region gradually shifts to the center of the bed with increasing distance from the gas distributor. The F factors of equ (3), evaluated from the mean bubble flow (Table 2) are comparable with those published in the literature (Geldart (1968), Grace et al. (1969), Mc Grath et al. (1971), Werther et al. (1973)) for sand, Na-aluminate and/or Magnesit.

Table 2: F-factors

H(cm)	$u^+ = 2$	$u^+ = 3$	$u^+ = 4$
56.5	1.48	2.07	2.0
37.2	1.61	1.91	
16.5	1.53	1.71	1.55
5.5	1.65	2.30	2.48

Because during the decomposition of the solid particles CO_2 escapes, the superficial gas velocity increases in the bed (e.g. at the top of the bed at $T_b = 125°$ C and at a conversion of 95 % by 1.2 cm/s). The increase of the bubble gas flow due to the reaction can only be measured in the upper part of the bed.

The rising velocity and the equivalent bubble diameter decreases due to the decomposition reaction. This phenomenon is not understood yet. The temperature has only a slight influence on the rising velocity size and frequency of bubble (Table 3).

Since the rheological behaviour of Na_2CO_3 and $NaHCO_3$ differs significantly from that of common model particles (e.g. sand), it is not allowed to generalize these results. However, it is remarkable that bubbles in fluidized beds of γ-Al_2O_3 particles show similar behaviour: only a slight temperature effect on the bubble behaviour was found.

Table 3: Comparison of the local mean pierced lengths $\bar{l}$ (cm) and of the local raising velocities u_b (cm/s) of the bubbles for different temperatures

H cm		56.5	37.2	16.5	5.5
r cm		2.5	3.1	7.5	7.8
T=65°C	$\bar{l}$	3.12	2.04	1.64	0.58
	u_b	76.3	59.6	66.2	45.6
T=125°C	$\bar{l}$	3.20	2.24	1.41	0.74
	u_b	76.3	68.5	69.5	49.5
T=194°C	$\bar{l}$	3.40	2.49	1.77	0.74
	u_b	78.9	71.7	72.8	49.8
T=125°C with reaction	$\bar{l}$	2.67	1.95	1.28	0.49
	u_b	67.8	55.6	60.8	46.0

CONCLUSIONS

1. The diameters of the bubble covering spheres agree fairly well with those calculated by the empirical relation (1) of Rowe for small bubbles.
2. The local coalescing bubble fraction K depends significantly on the position in the bed. Its distribution is comparable with the distribution of the local bubble gas flow and the local bubble volume fraction. Their maximum gradually shifts from the wall region to the center of the bed with increasing distance from the gas distributor.
3. The bubble gas flow increases, the size and rising velocity of the bubbles decrease due to the decomposition reaction in the bed.
4. The influence of the temperature on the bubble behaviour is slight in the temperature range of 65 to 200° C.

ACKNOWLEDGEMENT

The authors gratefully acknowledge the financial support of DECHEMA, AIF and Stiftung Volkswagenwerk.

REFERENCES

Calderbank, P.H., Pereira, I., Burgess, J.M., (1975) in Fluidization Technology, Ed. D.K.Keairns, Hemisphere Publ.Corp., Vol.1, 115

Clift, R., Grace, I.R. (1970) Chem.Eng. Progr.Symp.Series 66, 14

Geldart, D. (1968) Powder Technology 1 355

Geldart, D. (1970) Powder Technology 4, 41

Goedecke, R. (1974), Dissertation, Technical University Hannover

Grace, J.R., Harrison, D. (1969) Chem. Eng.Sci. 24, 497

Mc Grath, L., Streatfield, R.E. (1971) Trans.Instn.Chem.Engrs. 49, 70

Pyle, D.K., Harrison, D. (1967) Chem. Eng.Sci. 22, 1199

Rowe, P.N., in Fluidization (1971), Ed. J.F. Davidson, D.Harrison, Acad. Press, London 121

Rowe, P.N.(1976), Chem.Eng.Sci. 31,285

Werther, I. (1972a), Dissertation, Universität Erlangen

Werther, I., Molerus, O. (1972b), Chisa Conference, Prague

Werther, I., Molerus, O. (1973), Int. J.Multiphase Flow 1, 103 and 123

Werther, I. (1975), in Fluidization Technology, Ed.D.L.Keairns, Hemisphere Publ.Co. Vol.1, 215

Whitehead, A.B., Young, A.D. (1967), International Symposium on Fluidization. Eindhoven, Ed.A.H.H.Drinkenburg, Netherland University Press 294

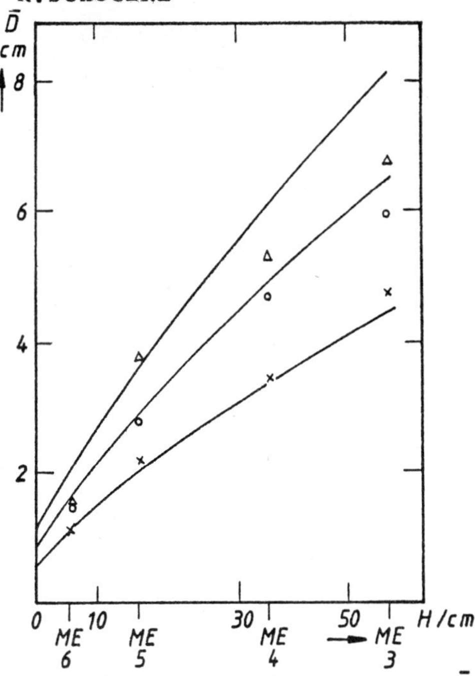

Fig. 1: Comparison of the diameter $\bar{D}$ calculated according to Rowe by equ (4) - and measured for $u^+ = 2$(x), $u^+ = 3$(o), $u^+ = 4$ (Δ)

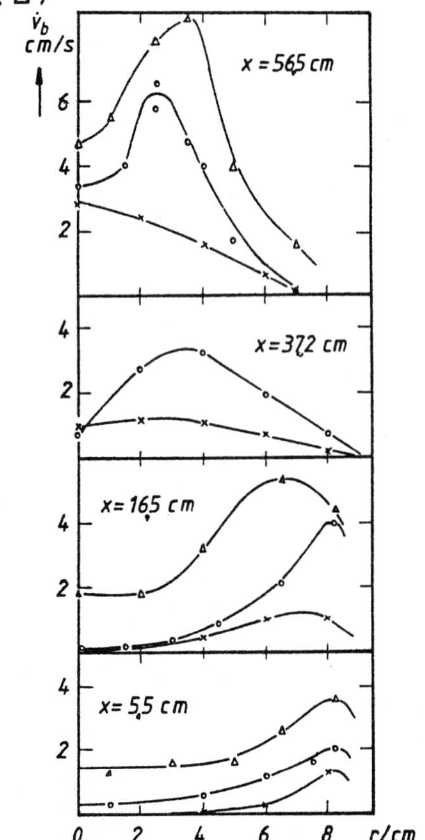

Fig. 2: Transverse profiles of the bubble gas flow v_b
x: $u^+ = 2$, o: $u^+ = 3$, Δ: $u^+ = 4$

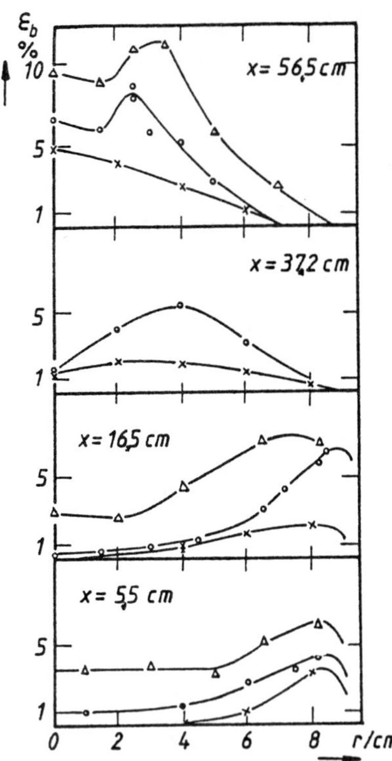

Fig. 3: Transverse profiles of the local bubble volume fraction ε_b (x: $u^+=2$, o: $u^+=3$, Δ: $u^+=4$)

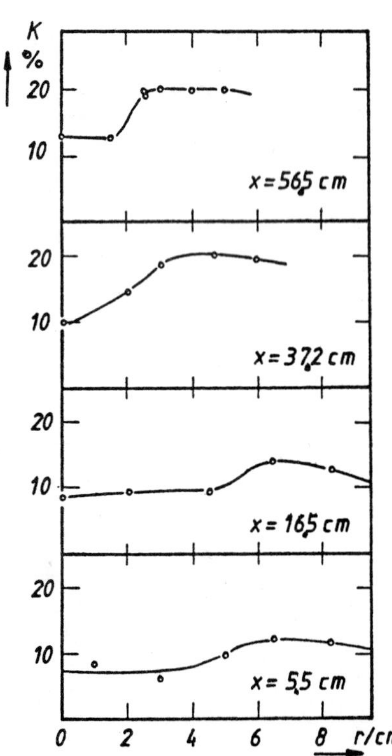

Fig. 4: Transverse profiles of the local coalescence fraction K at $u^+ = 3$

GAS AND LIQUID HOLD-UP AND PRESSURE DROP MEASUREMENTS IN A THREE-PHASE FLUIDISED BED

By V.R. Dhanuka and J.B. Stepanek
Department of Chemical Engineering
University of Salford, Salford
England M5 4WT

SYNOPSIS

The conductivity method for measuring liquid hold-up described by Achwal and Stepanek (1975) for a packed bed, has been tested and successfully used to measure liquid hold-up in a three phase fluidised bed. A modified version of the Darton and Harrison (1975) model has been found to predict the hold-up data satisfactorily. Finally, the theoretical equation for predicting pressure drop has been checked and verified.

NOTATION

- K ratio of liquid wake volume to bubble volume
- L height of the expanded bed
- m exponent in eq. (19)
- n Richardson-Zaki index
- ΔP pressure-drop across the fluidised bed
- U superficial velocity
- V linear velocity
- X ratio of solids content in wake of particulate phase
- ρ density
- ε hold up

Subscripts

- B bubble
- D drift-flux
- F fluidised or particulate
- G gas
- L liquid
- S solid
- Sl slip
- t terminal velocity
- W wake

INTRODUCTION

A three phase fluidised bed is established when a dispersed solid is fluidised by a co-current upward flow of gas and liquid. The industrial importance of three-phase fluidisation has been stressed by many authors and Ostergaard (1968) has published an excellent review of industrially important processes. Studies concerning bed expansion, gas and liquid hold-up, fluid mixing, gas bubble characteristics, models of three phase fluidisation, etc. are available in the literature.

In all these studies, the gas and liquid hold-up was calculated either from the pressure drop or from the average residence time of the phase in question. The latter method tends to give too high values for the liquid holdup as has been experimentally demonstrated by Satterfield (1975) for a trickle bed reactor and by Achwal (1977) for a fixed bed reactor. Hence a different technique has been adopted for measuring hold-up in this investigation and it has been proved that this technique is free from the errors mentioned above.

Measurements of pressure drop in the columns have been undertaken with the aim of testing independently the following pressure drop relation

$$\Delta P = L(\varepsilon_S \rho_S + \varepsilon_L \rho_L + \varepsilon_G \rho_G) \qquad (1)$$

which has been used by a number of authors for the calculation of the gas and liquid hold-up.

EXPERIMENTAL

The experimental system consisted of two sections, each 75 cm long and 5 cm i.d. The top fluidising section was fitted with pressure tappings at every 7.5 cm interval. The solid particles were supported by a wire screen and both gas and liquid entered separately through the lower calming section. The wire screen also acted as one electrode of the conductivity probe. The top section contained a movable second electrode which could be adjusted to coincide with the level of the fluidised bed. The two electrodes were connected to a conductivity bridge. The pressure drop was measured by means of a pressure transducer connected to the pressure tappings (See Fig. 1).

For the purpose of testing the conductivity method, the top electrode was located at the exit and the particles were fluidised to the full length of the top section by varying independently the flow rates of liquid and gas, thereby eliminating the complication caused by the difference in hold-up between the three phase fluidised bed and the two phase region above. The conductivity of the column was noted and the liquid hold-up was measured experimentally by shutting off a specially designed two-port valve which cuts off simultaneously the flow of air and water, followed by the reading of the resultant height of settled liquid.

The procedure was repeated for different concentrations as well as different diameters of solid particles (1.98 mm, 4.08 mm and 5.86 mm glass spheres of density 2.96). A plot of the dimensionless conductivity, defined as the ratio of the conductivity in the presence of gas to that of a bed fluidised by pure liquid to the same height, viz. liquid hold-up on solid free basis showed that all the points lie in a straight line passing through the origin and the point (1,1), which proves that the method can be used with confidence for measuring liquid hold-up in a three phase fluidised bed. The merit of the method is that it disposes of the complication arising from the existence of a two phase region above the fluidised bed.

The method was then used to acquire data on liquid hold-up at various bed expansions and to study the effect of gas and liquid velocities on the gas and liquid hold-up. The measurements were complemented with measurements of pressure drop. For this purpose, a small flow of purge gas (50 - 100 cm^3/min) was introduced into the tapping lines in order to keep them clear of liquid.

THEORY

Apart from purely empirical correlations (Dakshinamurty *et al*, 1972; Kim *et al*, 1972; Ostergaard and Michelsen, 1968; Razumov *et al*, 1973) mathematical models of three phase fluidisation have been proposed by Ostergaard (1965), Bhatia and Epstein (1974), and Darton and Harrison (1975). All three models have been tested in this investigation but none of them gave a good fit with the experimental data.

The model developed here is based on the same principles as those of Bhatia and Epstein (1974) and Darton and Harrison (1975), the similarity between the last two models was recognised by both Darton and Harrison (1975) and Epstein (1976).

It is assumed that the three phase fluidised bed is divided into three parts: the gas phase, the wake phase and the liquid-solid fluidised (or particulate) phase. The last two phases contain no gas, the wake phase may or may not contain the fluidised particles. The volumes are additive in the bed as a whole

$$\varepsilon_G + \varepsilon_L + \varepsilon_S = 1 \qquad (2)$$

and

$$\varepsilon_G + \varepsilon_W + \varepsilon_F = 1 \qquad (3)$$

and in the individual phases, e.g.,

$$\varepsilon_{LF} + \varepsilon_{SF} = 1 \qquad (4)$$

In the (4), the volumetric fractions are related to the volume of the liquid-solid fluidised phase. Combination of the volume balances gives

$$\varepsilon_W = \frac{K\varepsilon_G}{1 - X + X\varepsilon_{LF}} \qquad (5)$$

$$\varepsilon_L = K\varepsilon_G + \varepsilon_{LF}\left(1 - \varepsilon_G - \frac{K\varepsilon_G}{1 - X + X\varepsilon_{LF}}\right) \qquad (6)$$

where $X = \varepsilon_{SW}/\varepsilon_{SF}$ is the ratio of the solid content in the wake to that in the fluidised phase, and $K = \varepsilon_W\varepsilon_{LW}/\varepsilon_G$ relates to the liquid content in the wake phase to the volume of the gas. The former is usually taken as zero and the latter has been correlated with the superficial velocities of gas and liquid by Darton and Harrison (1975).

$$K = 1.4\,(U_L/U_G)^{0.33} - 1.0 \qquad (7)$$

An alternative, however rather complicated correlation for calculating K has been proposed by Bhatia and Epstein (1974). Our calculations have found that Bhatia's (1973) data (and as well as our data) can be fitted satisfactorily with the simple correlation (7) of Darton and Harrison. Equation (6) is one of the equations relating volume fractions in a three phase fluidised bed. A second equation can be obtained from the expression for the relative velocity between the liquid and the solid particles in the fluidised phase.

The material balance for the liquid writes

$$U_L = U_{LF}\varepsilon_F + U_W\,\varepsilon_{LW} \qquad (8)$$

It is assumed that the linear velocity of the wake is equal to that of the gas, thus giving

$$U_W/\varepsilon_W = U_G/\varepsilon_G \qquad (9)$$

The linear velocity of the liquid can be obtained from the combination of (3), (5), (8) and (9) as

$$V_{LF} = \frac{U_{LF}}{\varepsilon_{LF}} = \frac{U_L - KU_G}{\varepsilon_{LF}(1-\varepsilon_G - \frac{K\varepsilon_G}{1-X+X\varepsilon_{LF}})} \quad (10)$$

The continuity equation for the particles in the fluidised and wake phases is

$$V_{SF}\,\varepsilon_{SF}\,\varepsilon_F = -U_W\varepsilon_{SW} \quad (11)$$

By virtue of (3), (5) and (9) it is obtained that

$$V_{SF} = -\frac{U_G KX}{(1-\varepsilon_G)(1-X+X\varepsilon_{LF})-K\varepsilon_G} \quad (12)$$

The relative velocity between the liquid and the solid particles is

$$V_{LS}=V_{LF}-V_{SF} = \frac{U_L(1-X+X\varepsilon_{LF})-KU_G(1-X)}{\varepsilon_{LF}[(1-\varepsilon_G)(1-X+X\,\varepsilon_{LF})-K\varepsilon_G]} \quad (13)$$

The relative velocity can be related to the liquid holdup in the particulate phase by a Richardson-Zaki (1954) relationship

$$V_{LS} = V_t\,\varepsilon_{LF}^{n-1} \quad (14)$$

where V_t can be calculated from the formulae given in Perry (1963). Combining (13) and (14)

$$\varepsilon_{LF} = \left[\frac{U_L\,(1-X+X\,\varepsilon_{LF}) - KU_G\,(1-X)}{V_t\,(1-\varepsilon_G)(1-X+X\varepsilon_{LF}-K\varepsilon_G)}\right]^{1/n} \quad (15)$$

The last equation to close the problem is obtained from the relation for the drift flux

$$V_D = V_{S1}\varepsilon_G(1-\varepsilon_G) \quad (16)$$

Darton and Harrison (1975) suggested that the slip velocity should be the rise velocity of the gas bubbles relative to the fluidising liquid. However, this type of approach failed to fit our experimental data in that the values of the drift flux calculated from experiments exhibited a large scatter with no particular trend, and in some cases a negative value was obtained for the slip velocity. It seems more appropriate to relate the slip velocity to the fluidised bed rather than to the fluidising liquid. It then follows that

$$V_{S1} = U_G/\varepsilon_G - U_L/(1 - \varepsilon_G) \quad (17)$$

The drift velocity is obtained by substituting into (16) as

$$V_D = U_G\,(1 - \varepsilon_G) - U_L\varepsilon_G \quad (18)$$

The drift velocity can be related to the gas holdup by the correlation proposed by Darton and Harrison (1975)

$$V_D = V_B\,\varepsilon_G\,(1 -\varepsilon_G)^m \quad (19)$$

where the value of m lies between zero and two and depends on the type of the mechanism of gas bubble interactions and V_B is the bubble rise velocity and must be known before the model can be solved.

The following procedure is used to solve the model: First gas hold-up is predicted from (18) and (19) using the appropriate value of m (Darton and Harrison, 1975). Then a value of X is assumed and ε_{LF} is calculated by trial and error from (15) using K from (7). For particles with a size larger than 2-3 mm, the value of X is zero and (15) can be solved directly. The value of the liquid hold-up is then calculated from (6) and finally solids hold-up is obtained from (2).

RESULTS AND DISCUSSION

As mentioned earlier, V_B has to be known to solve the model. Since very little information is available at present on bubble rise velocities in a three phase fluidised bed, it was decided to evaluate V_B by (18) and (19) using experimental data: it was found to be 13.35 cm/s for 4.08 mm particles and 8.4 cm/s for 5.86 mm particles and it did not depend on the flow rates of gas and liquid within the investigated range.

The experimental hold-up data is compared with predicted values in Figs. 2 to 6 for 1.98, 4.08 and 5.86 mm particles, respectively. The following trends were revealed: (1) The gas hold-up increases with the gas velocity. (2) The liquid hold-up increases with the liquid velocity and decreases with the gas velocity for all particle sizes. (3) The gas hold-up decreases with the liquid flow rate for the two larger particles. For the 1.98 mm particles, the gas hold-up seems to be independent of the liquid flow rate, however the effect, if any, is to increase the gas hold-up slightly with the flow rate of liquid. The observed trends are in agreement with published reports of other investigators (Bhatia and Epstein, 1974; Michelsen and Ostergaard, 1970). The slight increase of gas hold-up with increasing liquid flow rate in beds with small particles can be detected from the results of Michelsen and Ostergaard (1970) for 1 mm particles and of Bhatia (1973) for 0.25 and 0.5 mm particles, although not reported in their papers.

The full lines in Figs. 3 to 6 representing the predicted values were calculated from the model using X = 0, which was found to give the best agreement with experiments. This indicates that no solid particles are present in the wakes.

Because of the anomalous behaviour of the gas hold-up for the 1.98 mm particles, (18) and (19) were not applicable in this case and the lines in Fig. 2 were calculated by substituting the experimental gas hold-up data directly into (15).

It is seen that the model fits the data

quite well with an average error of around 4-5%, however it still has to rely on experimental determination of gas hold-up for small particles.

The measured values of pressure drop were compared with the predictions calculated from (1) and the maximum deviation was 4%.

CONCLUSIONS

Measurements of liquid and gas hold-up confirmed the known fact that the structure of the bed is different for small and large particles, with the transition taking place at particle size between 2 and 4 mm. For larger particles, the liquid hold-up steadily increases with the liquid and decreases with the gas flow rate and a modification of the Darton and Harrison model can be used to predict the hold-up in this case. For small particles, the gas hold-up tends to increase in the liquid flow rate and no formula has been developed that would account for this phenomenon. However, the derived model can still be used for predicting liquid and solid hold-ups if experimental values of gas hold-up are substituted.

REFERENCES

Achwal, S.K. (1977) Ph.D. Thesis, University of Salford, England.

Achwal, S.K. and Stepanek, J.B. (1975) Chem. Eng. Sci. 30, 1443.

Bhatia, V.K. and Epstein, N. (1974) Proceedings of the International Symposium on Fluidization and its Applications, Paper 4.3, France.

Bhatia, V.K. (1973) Ph.D. Thesis, University of British Columbia, Canada.

Dakshinamurty, P. *et al* (1972) Ind. Eng. Chem. Process Des. Develop. 11, 318.

Darton, R.C. and Harrison, D. (1975) Chem. Eng. Sci. 30, 581.

Epstein, N. (1976) Cand. J. Chem. Eng. 54, 259.

Kim, S.D. *et al* (1972) Cand. J. Chem. Eng. 50, 695.

Michelsen, M.L. and Ostergaard, K. (1970) Chem. Eng. J. 1, 37.

Ostergaard, K. (1968) Adv. in Chem. Eng. Vol. 7, p.71, Academic Press, New York.

Ostergaard, K. (1965) Chem. Eng. Sci. 20, 165.

Ostergaard, K. and Michelsen, M.L. (1968) Preprint 31D, A.I.Ch.E.I.I.Q.P.R. Meeting, Tampa. Florida.

Perry, J.H. (1963) Chem. Engnrs. Handbook p.5-59, McGraw-Hill Book Co., New York.

Razumov, I.M. *et al* (1973) Int. Chem. Eng. 13, 57.

Richardson, J.F. and Zaki, W.N. (1954) Tran. Inst. Chem. Engrs. 32, 35.

Satterfield, C.N. (1975) A.I.Ch.E.J. 21, 209.

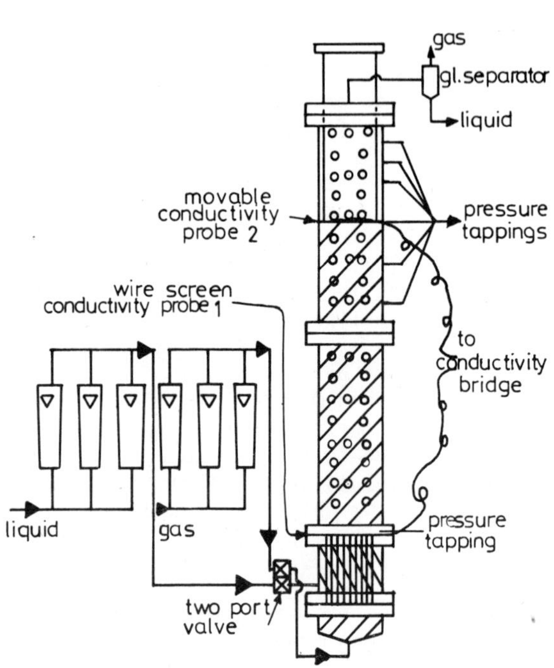

Fig.1. Experimental system.

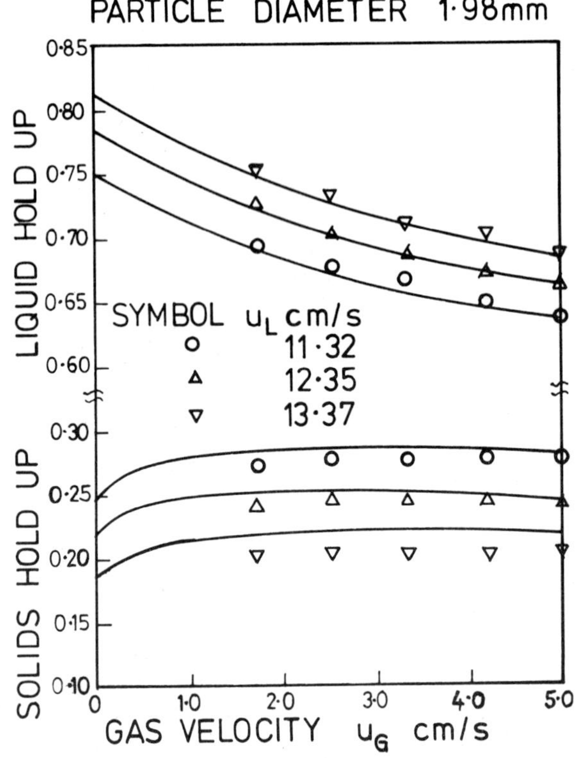

Fig.2. Solids and liquid hold-up as a function of gas and liquid velocities for 1.98 mm glass spheres, ———— predicted hold-up.

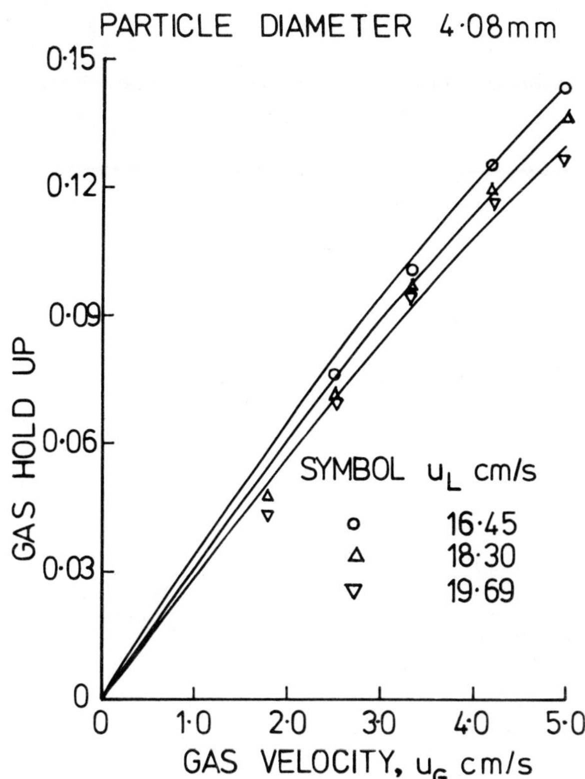

Fig. 3. Gas hold-up as a function of gas and liquid velocities. ——— predicted hold-up.

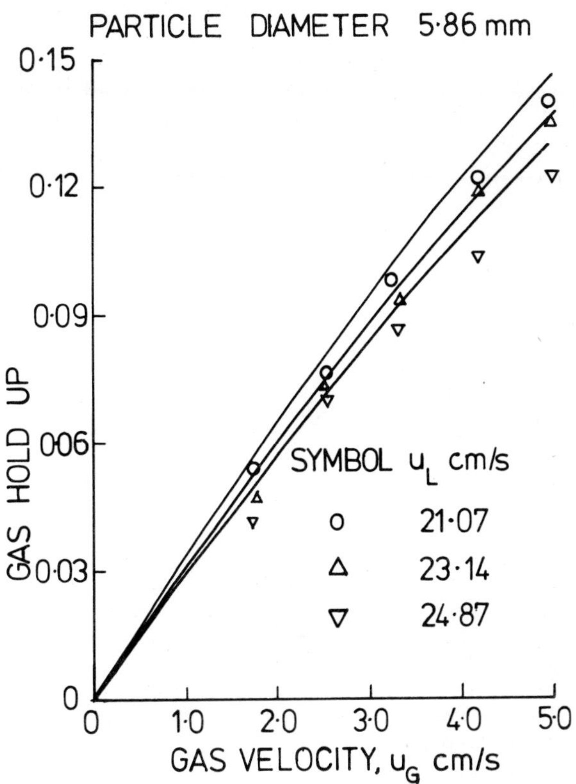

Fig. 5. Gas hold-up as a function of gas and liquid velocities. ——— predicted hold-up.

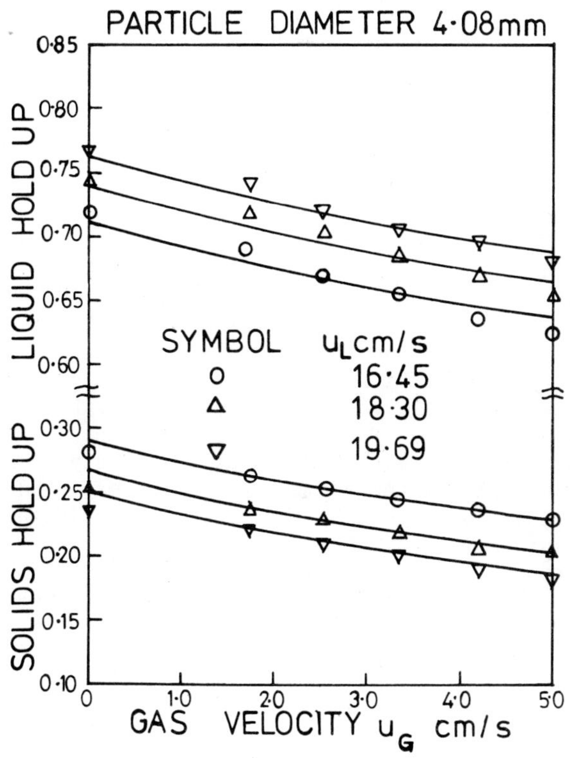

Fig. 4. Solids and liquid hold-up as a function of gas and liquid velocities, ——— predicted hold-up.

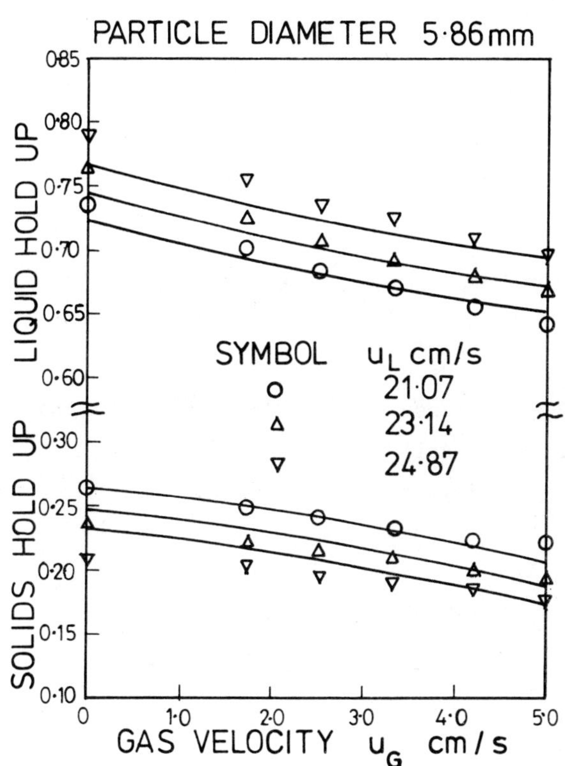

Fig. 6. Solids and liquid hold-up as a function of gas and liquid velocities ——— predicted hold-up.

THREE-PHASE FLUIDIZED BEDS

ONSET OF FLUIDIZATION AT HIGH GAS RATES

By J. C. Lee and N. Al-Dabbagh

Department of Chemical Engineering
University College of Swansea
Swansea SA2 8PP

SYNOPSIS

The increase in pressure drop that occurs when gas is introduced into a liquid flowing upwards through a bed of particles causes fluidization of the bed to take place at a lower liquid rate than with liquid flowing alone. Measurements of pressure drop and of phase volume fraction using γ-ray transmission have been made for beds of 4 mm and 6 mm spherical glass particles with air at mass flow rates up to 0.24 kg/m^2s and water. Pulsations in the gas-liquid flow caused consolidation and channeling in the bed; a very regular packing arrangement of the particles was produced. Nevertheless, values of the liquid flow rate required to fluidize the bed at a given gas flow agreed well with values calculated from the literature.

NOTATION

$f_{\ell g}$	two phase friction factor
G_g	mass flow rate of gas per unit area
G_ℓ	mass flow rate of liquid per unit area
h	height between pressure tappings
H	height of bed
P_1, P_2	pressures at points 1 and 2
ΔP_{12}	pressure drop between points 1 and 2
R	liquid saturation
Re_g	Reynolds number for gas
Re_ℓ	Reynolds number for liquid
W	effective weight of bed
z	two-phase flow parameter
ε_ℓ	volume fraction of liquid
ε_s	volume fraction of solid
ρ_ℓ	density liquid
ρ_s	density solid

INTRODUCTION

A number of investigations have shown that three-phase fluidized beds employing low viscosity liquids may be divided into two main categories, namely beds of large heavy particles which are capable of breaking up the gas flow into a dispersion of relatively small bubbles, and beds of small light particles in which the gas bubbles are considerably larger (Ostergaard, 1971; Lee et al.,1974). The beds of small light particles become fluidized at relatively low liquid flow rates and the particles may even be fully suspended by the disturbances in the liquid caused by the passage of gas alone without any through-flow of liquid (Roy et al., 1964). Beds of larger heavy particles however, require relatively high flow rates of liquid in order to fluidize the bed. It has been observed that where the gas stream is broken up into small bubbles, an increase in the gas flow rate brings about a further expansion of a fluidized bed, suggesting that fluidization is assisted by the gas flow.

The aim of the present paper is to determine to what extent gas flow contributes to the onset of fluidization in a three-phase system where the solid phase consists of moderately large particles, viz. glass spheres of 6 mm and 4 mm diameter. Such beds have applications in catalytic gas-liquid reactors operated in up-flow and in the backwashing of sand filters with aeration in water treatment processes. In a slightly different sense, the system can be regarded as a case of two-phase gas-liquid upwards flow through a fixed but unconstrained bed of particles, a point of interest being to see whether the bed expands when the pressure drop over the bed is equal to its buoyant weight as in the case of fluidization by a single fluid, liquid or gas. This condition implies that the pressure drop associated with friction at the walls of the vessel is negligible. Such an assumption has been previously applied to fully suspended three-phase beds when pressure gradients have been employed as a means of measuring gas hold-ups (Michelsen & Ostergaard 1970; El-Temtamy, 1974). It appears, however, that pressure drops have not previously been measured for unconstrained fixed beds of particles in upwards flow.

Pressure drops for liquid-gas cocurrent downwards flow through fixed beds have been measured and correlated by several authors together with more limited data on upwards flow in constrained beds (Larkins et al., 1961; Turpin & Huntingdon, 1967; Wooding & Morel-Seytoux, 1976). These authors have reported that over a wide range of gas and liquid flow rates there exists a regime of pulsating or slug flow. It has been observed in the present work that the occurrence of pulsations in the flow had an appreciable effect on the character of the fluidization obtained at high gas rates.

EXPERIMENTAL

The column was made from transparent acrylic plastic sheets cemented together to give a square cross-section of 63.5 mm internal dimension. The bed was supported on a porous sintered bronze plate 6.4 mm thick which acted as a distributor for the liquid. Gas was introduced through four nozzles of 2 mm orifice diameter symmetrically arranged and passing through the distributor plate ending flush with the upper surface. There were pressure tappings in the side of the column at 0.15 m, 0.45 m and 0.75 m above the distributor, the region between the 0.45 m and the 0.75 m tappings being regarded as the test section of the column. Constrictions in the manometer lines helped to damp fluctuations in the manometers used to measure the pressure. The fraction of the bed voidage occupied by the liquid i.e. the saturation, was determined by γ-ray transmission measurements using an Americium-241 source. The solids fraction was known from the height occupied by a bed of a given weight and the density of the particles.

Two sets of uniform glass spherical particles were used, one being 4.03 mm diameter, density 2560 kg/m^3, the other 6.08 mm diameter, density 2590 kg/m^3. The weight of particles forming the bed was in both cases 5.10 kg.

DETERMINATION OF PRESSURE DROP AND BED WEIGHT

If P_1 is the absolute pressure at the lower pressure tapping of the test section and P_2 the absolute pressure at the upper pressure tapping situated at a height h above, then the pressure drop over the section allowing for the hydrostatic head of the two-phase gas-liquid mixture in the bed is given by

$$\Delta P_{12} = (P_1 - P_2) - h\left(\frac{\varepsilon_\ell}{1-\varepsilon_s}\right)\rho_\ell$$

neglecting the density of the gas relative to that of the liquid. The pressure drop over the whole bed of height H, which although the bed is termed 'fixed' was somewhat variable as described below was calculated as $\Delta P_{12}H/h$.

The effective weight of the bed was calculated from

$$W = \varepsilon_s\left[\rho_s - \left(\frac{\varepsilon_\ell}{1-\varepsilon_s}\right)\rho_\ell\right]H$$

where again the density of the gas is neglected.

OBSERVATIONS AND RESULTS

The experimental procedure was initially to fluidize the bed of particles with liquid alone, and then reduce the liquid flow to give a randomly packed bed of voidage approximately 0.38. A particular gas mass flow rate was chosen and held constant while the liquid flow was progressively increased, measurements of pressure and γ-ray transmission being made at suitable increments in the liquid rate. Typically it was found that as the liquid flow increased, so the pressure drop and liquid saturation of the bed increased. Visually it appeared that slugs of liquid were passing in rapid succession up the bed but the fluctuations occurred too quickly to be followed easily by the eye. With increasing liquid flow, jumping movements in the top four centimetres of the bed were observed, and as the liquid flow was increased still further, the upper zone in which this jumping was occurring extended further down the bed.

Although the jumping movements were only a few millimetres in amplitude, they were sufficient to consolidate the bed appreciably, the height decreasing from 0.79 m initially to 0.73 m with a corresponding decrease in voidage. The photographs in Fig. 1 show this effect.

At this point it was observed that the gas, which initially had been more or less uniformly distributed, was channeling noticeably through the region of somewhat higher porosity close to the walls. Also quite strikingly the uniformly sized spheres had adopted a surprisingly regular close-packed arrangement as shown in Fig.1. Although the particles had settled into this regular pattern nevertheless the small jumping movements persisted, growing in magnitude with increasing liquid flow rate until eventually the bed could be seen to be fully suspended though in a state of considerable turbulence arising from the gas flow.

On reducing the liquid flow rate from the fluidized condition, the close-packed state of the particles with channeling of the gas near the walls reappeared. The regular pattern of particles then persisted to low liquid rates and remained after gas and liquid flows had been shut off altogether.

Quantitatively, Fig.2 shows the variation of pressure drop through the bed with increasing liquid flow rate for various fixed gas rates. The dips in the curves correspond to the consolidation of the bed into the close-packed arrangement and channeling of the gas near the walls. Fig. 3a shows the pressure drops obtained first with increasing liquid flow up to the point of fluidization, followed by decreasing flow. Also shown are values of the effective bed weight calculated from the measured liquid and gas fractions in the bed (Fig.3b). The curves show that as the pressure drop over the bed approaches the effective bed weight, the jumping movements which consolidate the bed start to occur. Both pressure drop and effective bed weight are diminished by the change in packing and channeling of the gas, but at a higher liquid flow rate pressure drop and bed weight become equal at a point which corresponds approximately to the onset of fluidization judged visually.

COMPARISON WITH LITERATURE CORRELATIONS

The most suitable data in the literature for comparison with the present measurements appeared to be that of Turpin & Huntingdon (1967) who used tabular particles of 8 mm diam. Although data are reported for both upward and downward flow, the values of the liquid saturation obtained by these authors in upflow are much larger than the values measured in the present work, probably because of the difference in particle size. The present values correspond much more closely to the result of Turpin & Huntingdon for downward flow and since pressure drop is likely to be affect-affected by the saturation, the pressure drop correlation recommended by Turpin & Huntingdon for downward flow was therefore adopted as the basis for calculation.

A parameter z is calculated from the relation

$$z = Re_g^{1.167} \Big/ Re_\ell^{0.767}$$

where Re_g and Re_ℓ are Reynolds numbers based on gas and liquid flows respectively. The two-phase friction factor $f_{\ell g}$ is then determined from the expression

$$f_{\ell g} = \exp[\,7.96 - 1.34\,\ell nz + 0.0021(\ell nz)^2 + 0.0078(\ell nz)^3]$$

and the saturation from

$$R = \frac{\varepsilon_\ell}{1-\varepsilon_s} = -\,0.017 + 0.132(G_\ell/G_g)^{0.24}$$

where G_ℓ and G_g are the mass flow rates of liquid and gas respectively.

Results of these calculations in comparison with a further set of experimental data for the 4 mm particles are shown in Fig.4. In one set of calculations, the actual measured bed voidage which varied with the liquid flow rate was used. However, if the aim were to predict the onset of fluidization for design purposes, this variation in bed voidage would not be known, so the pressure drop was also calculated for a constant voidage of 0.39. It can be seen from Fig.4 that the calculated lines for the pressure drop at a constant 0.39 voidage and the effective bed weight intersect almost exactly at the observed point of fluidization although the calculated pressure drops do not coincide with the measured values at lower liquid rates.

Fig.5 shows a similar pattern of results for calculations and experimental data relating to the 6 mm particles. In Fig.6 calculated and observed liquid flow rates at the point of fluidization are compared for both 4 mm and 6 mm particles over the whole range of gas flow rates used in the experiments.

DISCUSSION

The present measurements confirm that three-phase fluidization occurs when the pressure drop over the bed is equal to the effective weight of the particles. Unlike fluidization by a single phase however, fluidization with a two-phase gas-liquid flow is complicated by the pulsating character associated with such flows. It has been shown that these fluctuations can consolidate the bed and lead to channeling of the gas.

In the present experimental arrangement it is possible that changes in the packing of the particles were accentuated firstly by the use of uniform spheres as the solid material and secondly by the square cross-section of the bed. However, it is very likely that in larger beds of other cross-sections with irregular particles, the jumping movements of the bed will cause local variations in porosity and hence channeling of the gas. It appears from the work of Beimesch & Kessler (1971) that distribution of gas and liquid in two-phase flow through a fixed bed is rather sensitive to local bed porosity. They found that the gas channels preferentially through regions of lower porosity as observed in the present work. These findings suggest that if for example a gas-liquid fixed bed catalytic reactor were operated in concurrent upflow at flow rates approaching the fluidization point, gas channeling would be expected which could have a deleterious effect on the reaction products.

Quantitatively it appears that in spite of the consolidation of the bed and channeling which occur some way below the onset of fluidization, the point of fluidization itself can be satisfactorily predicted using correlations available in the literature taking the voidage as 0.39. Presumably as the point of fluidization is approached, the more vigorous particle movements are sufficient to produce a more random packing with less tendency for channeling of the gas.

REFERENCES

Beimesch, W. E. & Kessler, D. P. (1971) A.I.Ch.E.J. 17, 1160.

El-Temtamy, S. A. (1974) Ph.D. Thesis, Cairo University.

Larkins, R. P., White, R. R. & Jeffrey, D. W. (1961) A.I.Ch.E.J. 7, 231.

Lee, J. C., Sherrard, A. J. & Buckley, P. S. (1974) Proceedings of the International Symposium on Fluidization and its Applications, pp.407-416, Toulouse.

Michelson, M. L. & Ostergaard, K. (1970) Chem.Eng.J. 1, 37.

Ostergaard, K. (1971) Fluidization, Chap.18 (Davidson, J. F. & Harrison, D., Eds.), Academic Press, New York.

Roy, N. K., Guha, D. K. & Rao, M. N. (1964) Chem.Eng.Sci., 19, 215.

Turpin, J. L. & Huntingdon, R. L. (1967) A.I.Ch.E.J. 13, 1196.

Wooding, R. A. & Morel-Seytoux, H. J. (1976) Annual Review of Fluid Mechanics, 8, 233.

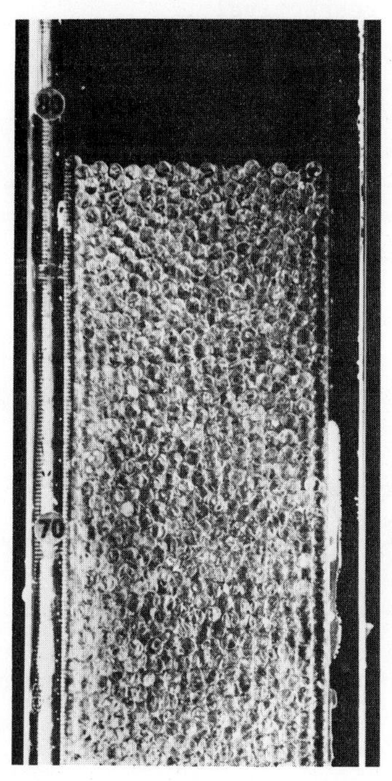

(a)

(b)

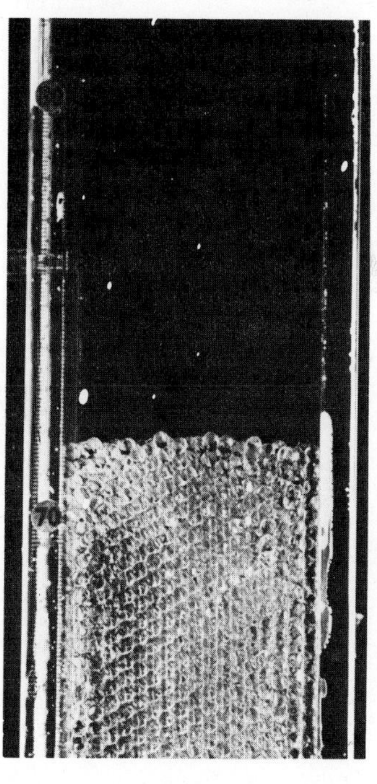

(c)

Fig.1 Change in bed packing accompanying three-phase fluidization. Gas mass flow rate 0.11 kg/m^{2}s (84 mm/s superficial velocity).
(a) Initial packed bed.
(b) Consolidation of bed at flowrates near fluidization point.
(c) Regular pattern of particles remaining after flow shut off.

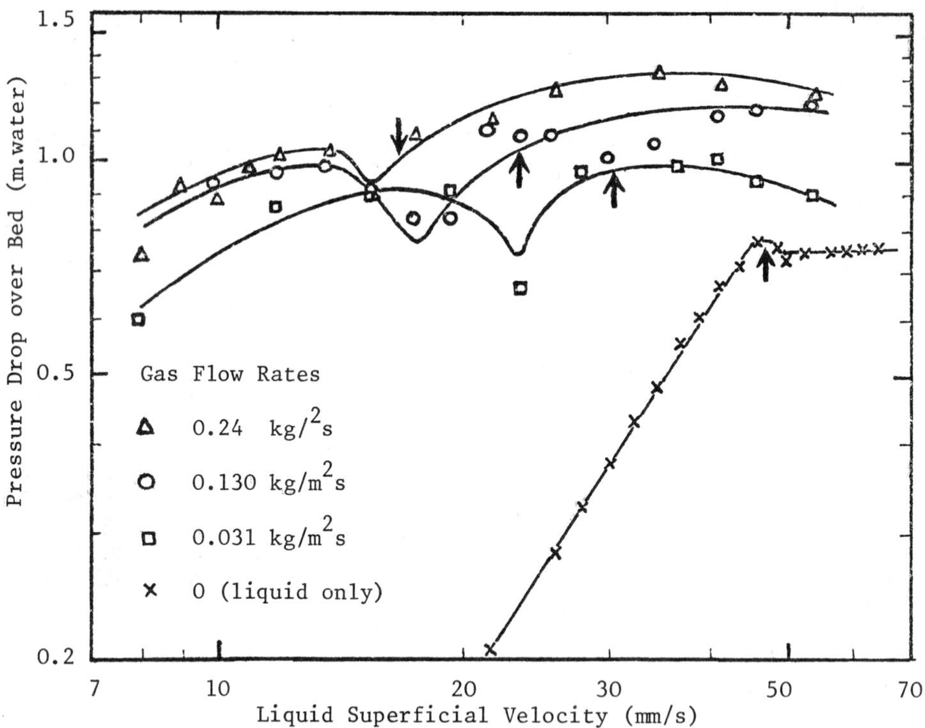

Fig. 2 Variation of pressure drop across bed with increasing liquid flow rate at constant gas mass flow rates corresponding approximately to gas superficial velocities of 178, 98, 24 and 0 mm/s respectively. Arrows show point of fluidization, 4mm particles.

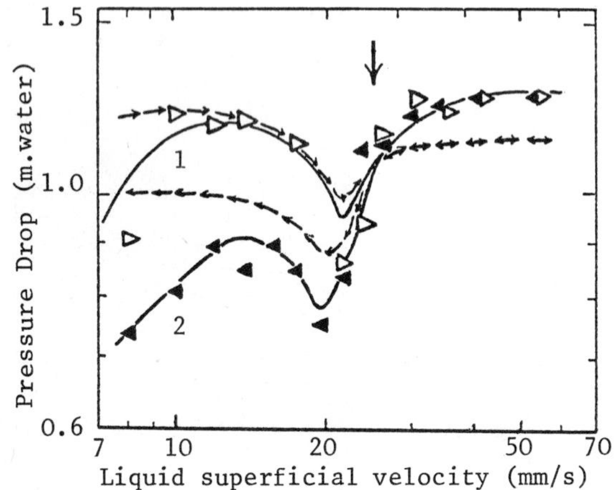

Fig. 3a. Pressure drop over bed and Bed weight for increasing and decreasing liquid flow.
Curve 1 Pressure drop increasing
Curve 2 Pressure drop decreasing
Arrowed lines show Bed weight.
Gas flow rate 0.092 kg/m^{2}s (69mm/s)
4 mm particles.

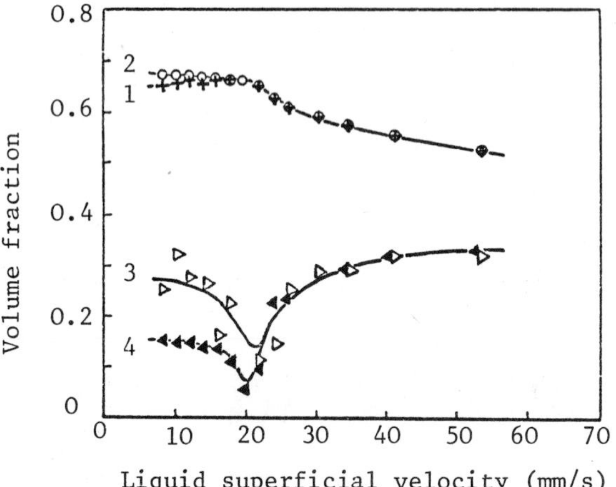

Fig.3b. Results of measurements of gas and solid volume fractions by γ-ray transmission with increasing and decreasing liquid flow.
Curve 1 Solid : liquid increasing
Curve 2 Solid : liquid decreasing
Curve 3 Gas : liquid increasing
Curve 4 Gas : liquid decreasing
Gas flow rate 0.092 kg/m^{2}s (69 mm/s)
4 mm particles.

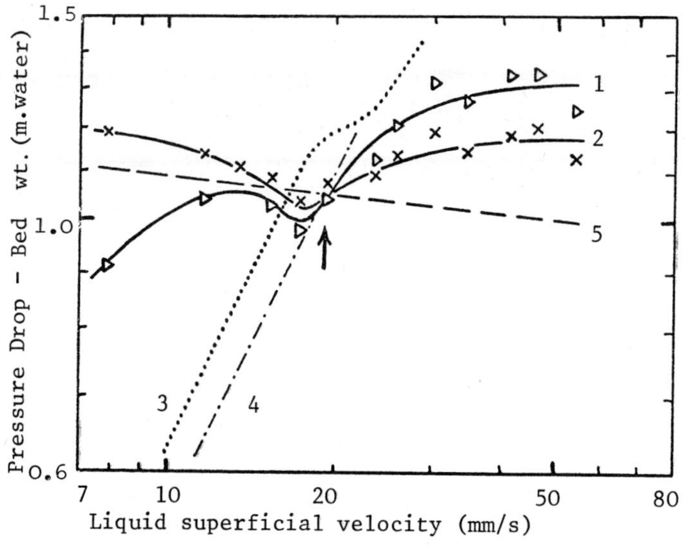

Fig. 4 Comparison between measured and calculated Pressure drop and Bed weight. 4 mm particles.
Gas rate 0.17 kg/m^{2}s (125 mm/s)

Curve 1 Experimental Pressure drop

Curve 2 Experimental Bed weight

Curve 3 Pressure drop calculated for actual bed voidage

Curve 4 Pressure drop calculated for bed voidage 0.39

Curve 5 Bed weight calculated for bed voidage 0.39

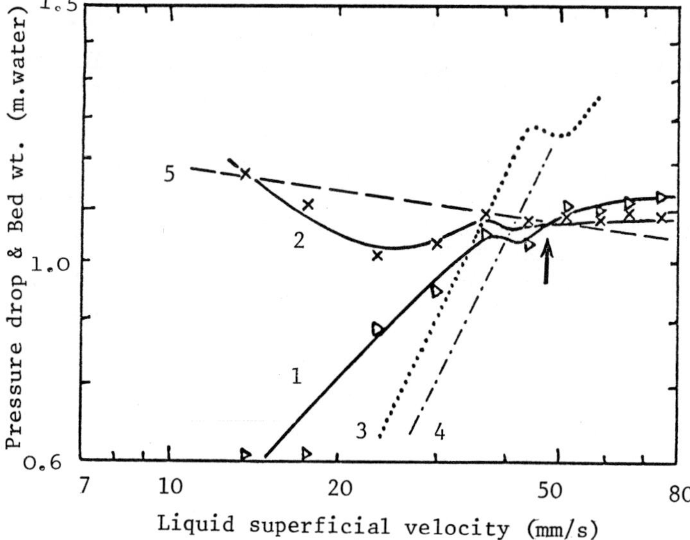

Fig. 5 Comparison between measured and calculated Pressure drop and Bed weight. 6mm particles.
Gas rate 0.11 kg/m^{2}s (84 mm/s)

Curve 1 Experimental Pressure drop

Curve 2 Experimental Bed weight

Curve 3 Pressure drop calculated for actual bed voidage

Curve 4 Pressure drop calculated for bed voidage 0.39

Curve 5 Bed weight calculated for bed voidage 0.39

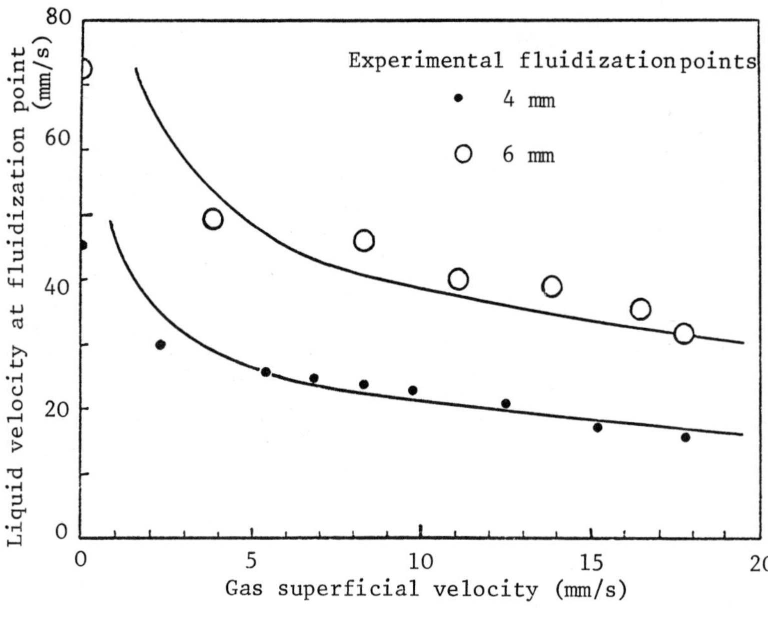

Fig. 6.
Reduction of liquid velocity at fluidization point with increasing gas flow. Curves show calculated Pressure drop - Bed weight intersection points for bed voidage of 0.39

HYDRODYNAMIC CHARACTERISTICS OF THREE-PHASE FLUIDIZED BEDS

By John M. Begovich and J. S. Watson

Chemical Technology Division
Oak Ridge National Laboratory
Oak Ridge, Tennessee 37830

SYNOPSIS

The hydrodynamics of three-phase (gas-liquid-solid) fluidized beds has been studied in two columns of 7.62 and 15.2 cm in diameter. The minimum gas and liquid velocities necessary to fluidize a bed were determined as a function of the particle size and density and the liquid viscosity; no effect of the initial bed height or column diameter was found. An electroconductivity technique was used to measure local concentrations of solid, liquid, and gas in the fluidized bed; these profiles were fitted using the error function. Overall solid holdup data were combined with over 1350 points from the literature to yield a dimensional correlation involving the physical parameters of the systems studied.

NOMENCLATURE

A	cross-sectional bed area	[cm^2]
Ar	Archimedes number = $d_p{}^3\ \rho_L(\rho_S - \rho_L)\ g/\mu_L{}^2$	[-]
d_p	particle diameter	[cm]
erf(x)	error function = $\frac{2}{\sqrt{\pi}} \int_0^x e^{-z^2}\, dz$	[-]
Fr	Froude number = $U_G{}^2/gd_p$	[-]
g	gravitational acceleration	[cm/sec^2]
h	axial column position	[cm]
H	bed height	[cm]
I	inflection point in local holdup vs height curve	[cm]
M	mass	[g]
ΔP	pressure drop across bed	[dyn/cm^2]
P	curve-fitting parameter, defined in Eqs. (7)-(10)	[-]
Re	Reynolds number = $U_L d_p \rho_L/\mu_L$	[-]
U	superficial velocity	[cm/sec]

Greek Letters

ε	local holdup (volume fraction)	[-]
$\bar{\varepsilon}$	average holdup	[-]
μ	viscosity	[g/cm·sec]
ρ	density	[g/cm^3]
σ	standard deviation	[cm]
$\Phi(x)$	probability integral = $\frac{1}{\sqrt{2\pi}} \int_0^x e^{-w^2/2}\, dw$	[-]

Subscripts

G	gas phase
L	liquid phase
mf	minimum fluidization
S	solid phase

Superscripts

''	two-phase
'''	three-phase

INTRODUCTION

Three-phase fluidized beds, containing solid particles fluidized by the upward cocurrent flow of liquid and gas which form continuous and discontinuous phases, respectively, have important present and future applications in hydrocarbon and coal processing and in some biological reactors. However, accurate design of such reactors is complicated by such factors as (1) lack of knowledge of the minimum fluid velocities required to achieve fluidization, and (2) axial variations in reactor properties, particularly distribution of the solid phase. No published data or equations exist for predicting reactor performance under high fluid flow rates where axial variations are important, and only a limited amount of data exists (Vail et al., 1970; Burck et al., 1975; Bloxom et al., 1975) for predicting minimum fluidization (MF) velocities in three-phase fluidized beds.

The minimum fluid flow rates required to achieve fluidization are determined by a plot of the pressure drop across the bed vs the superficial liquid velocity at constant gas flow rate. When fluidized, the pressure drop across the bed will no longer change with increasing liquid flow rate. Thus the flow rates at which a break in the curve occurs correspond to the MF velocities.

The following equations have typically been used to determine the volume fraction (holdup) of each phase in a three-phase fluidized bed:

$$\varepsilon_L + \varepsilon_G + \varepsilon_S = 1 \quad (1)$$

$$\Delta P = gH(\rho_L\varepsilon_L + \rho_G\varepsilon_G + \rho_S\varepsilon_S) \quad (2)$$

$$\varepsilon_S = M_S/\rho_S AH \ , \quad (3)$$

where the bed height in Eqs. (2) and (3) is obtained either visually or from the measured pressure gradient (Kim et al., 1975; Bhatia and Epstein, 1974). At high flow rates, neither method is satisfactory because the indistinct bed height makes visual measurements extremely subjective, while the measured pressure gradient yields a bed height based on an unrealistic homogeneous bed.

Experimental

To provide an alternative equation to Eq. (3), the electroconductivity method of Achwal and Stepanek (1975, 1976) was modified for application to a three-phase fluidized bed. The apparatus used in this study, as well as details of the procedure to obtain the holdup for each phase as a function of axial position within the column, have been described elsewhere (Begovich and Watson). Briefly, however, the experiment involves the use of various solids fluidized by air and water in either a 7.62- or a 15.2-cm-ID column. A series of liquid manometers located at regular intervals along the column walls provided the pressure gradient in the bed. Two small pieces of platinum sheet that were attached to opposite sides of a movable plexiglass ring and connected to a conductivity meter permitted measurement of the electrical conductivity of the bed at any axial position in the column. Using the ratio of conductivities in the bed to those in the liquid alone as the liquid holdup, together with the measured pressure gradient and Eq. (1), all three phase holdups could be determined as a function of height in the column.

The MF velocities required to achieve fluidization were determined from the intersection of the static and fluidized bed pressure drop curves on the plot of bed pressure drop vs superficial liquid velocity at a constant gas flow rate. The range of experimental conditions used in this study are detailed in Table 1.

Table 1. Range of experimental conditions used in three-phase fluidization studies

U_G, cm/sec	:	0 - 17.3
U_L, cm/sec	:	0 - 12.0
Column diameter, cm	:	7.62 and 15.2
Initial bed height, cm	:	22 - 45

RESULTS

Minimum Fluidization

The effects of column diameter and static bed height (or bed mass) on the minimum fluidization velocities for the air-water-4.6-mm-glass beads system are shown in Fig. 1. The minimum liquid velocity required to fluidize the bed with no gas phase present is indicated by the arrow on the ordinate of the plot as calculated from the two-phase correlation of Wen and Yu (1966). Excellent agreement between the calculated and experimental point of this system can be observed, as was the case for each system studied.

Neither the column diameter nor the mass of solids present in the column appeared to have any significant effect upon the MF velocities. Since fluidization of a bed is achieved when the upward inertial and drag forces exerted on the particles by the fluids equals the buoyant weight of the bed, an effect of static bed height on the MF velocities would only be expected if end effects were present in the bed. Likewise, one would not expect the MF velocities to be a function of column diameter unless the size of the gas bubbles approached that of the column diameter or unless channeling occurred.

MF velocities are shown in Fig. 2 for each of the systems studied. As the gas velocity was increased, the minimum liquid velocity required to achieve fluidization in each of the systems decreased. The magnitude of this decrease is considerably different for the plexiglass beads with their small solid/liquid density difference. In their two-phase correlation, Wen and Yu (1966) noted that the MF velocity increases with increasing particle diameter and increasing solid/liquid density difference but decreases with increasing fluid viscosity. The plexiglass beads have the same diameter as the alumina and one of the glass beads and also have a much smaller solid/liquid density difference. Thus they fluidize at lower velocities. The alumina and alumino-silicate beads have approximately the same density, but the smaller diameter of the latter particles causes them to fluidize at lower velocities.

Likewise, the 4.6-mm-diam glass beads fluidize at lower velocities than do the 6.2-mm-diam glass beads. It is of interest to note that although the curves of the alumina and 6.2-mm-diam glass beads start at essentially the same point for zero gas velocity, for increasing gas velocities they rapidly diverge until the gas velocities reach in excess of 8 cm/sec; at this point, the curve for the alumina beads merges with the curve of the 4.6-mm-diam glass beads.

The effect of liquid viscosity on the MF velocities for the 4.6-mm-diam glass beads is shown in Fig. 3. A single smooth curve in Fig. 1 fits all data for the same glass beads in both the 7.62- and 15.2-cm-ID columns. The arrows shown for zero gas velocity in Fig. 3 are calculated values and are shown only to indicate the expected influence of liquid viscosity. For a given gas velocity, the minimum liquid fluidization velocity decreased as the liquid viscosity was increased. However, the influence of the liquid viscosity appeared to decrease for the higher viscosities, particularly at the higher gas velocities. The gas velocity itself did not appreciably affect the minimum liquid fluidization velocity for the more viscous aqueous glycerol solutions studied.

Axial Variation in Holdups

As mentioned previously, bed heights are indistinct at high fluid flow rates, and the holdups calculated using Eqs. (1)-(3) represent those assuming an unrealistic homogeneous bed. Using the electroconductivity of the bed and the measured pressure gradient allows the holdups to be determined. Liquid and solid holdup values are plotted as a function of height, as shown in Fig. 4. The solid holdup, typically, is fairly uniform in the lower section of the bed; however, near the top of the bed, a transition region of slowly decreasing solids concentration connects with a two-phase gas-liquid region above the bed. The liquid holdup remains fairly constant in the bed and then increases to a constant value in the gas-liquid region. Figure 4 also shows that the same relationship between the holdups (solid and liquid) and height in the column was obtained in both the 7.62- and the 15.2-cm-ID columns under identical conditions (i.e., identical gas velocities, liquid velocities, particle type, and static bed height).

The effect of liquid velocity on the axial variation in the 4.6-mm-diam glass bead holdup is shown in Fig. 5 under conditions of constant gas velocity in the 7.62-cm-ID column. As the liquid velocity was increased, the bed expanded, and thus the solid holdup decreased. The calculated bed height, as found from the intersection of the measured pressure gradients in and above the bed, is indicated on the curves for each flow rate. As noted previously by Bhatia and Epstein (1974), this bed height corresponds to the height the same bed would have if the solids concentration in the column were uniformly distributed. The highest position where solids were detected was higher than this calculated bed height, however, since the bed contains a rather wide transition region over which flow changes from a three-phase to a two-phase column. The width of this transition region appeared to remain fairly constant with changing liquid velocity; that is, the solid holdup decreased from the approximately constant value in the bed to zero over about 20 cm of column length.

When the liquid velocity was held constant and the gas velocity was increased, the width of the transition region increased substantially, as illustrated in Fig. 6 which is typical of all of the data. The solid holdup in the lower portion of the bed was decreased slightly by the increase in gas velocity; however, the transition region increased from 20 cm in width to approximately 35 cm. As expected, the calculated bed height for the higher gas velocity indicated a much lower bed height than that observed visually (highest position with solids).

These results demonstrate the shortcomings of assuming a distinct bed height and a uniform bed. The transition region is a significant fraction of the total bed height and must be considered in realistic designs of three-phase systems. In commercial units operating with taller beds, the transition region could be less significant; however, the higher gas rates often employed by such units could cause the transition region to remain a significant fraction of the total bed height.

Overall-Phase Holdups

Figures 5 and 6, in addition to indicating the axial variation of the solid phase holdups, also demonstrate that Eqs. (1)-(3) are approximately correct for calculating the bed height via the pressure gradients (Epstein, 1977). The equivalent homogeneous bed may be sufficient as a model for the actual bed in some applications.

Since all the data necessary to use Eqs. (1)-(3) for predicting equivalent homogeneous bed heights were obtained simultaneously with the measurements for the local holdups, it was a simple matter to calculate overall, or average, phase holdups based on an equivalent homogeneous bed. For the systems studied, increasing the liquid velocity and holding all other conditions constant resulted in: (1) a decrease in the overall solid holdup, and (2) a slight decrease in the overall gas holdup. Since the holdups must sum to unity, the increased liquid velocity thus increased the overall liquid holdup. At constant liquid velocity, increasing the gas velocity caused the overall gas holdup to increase but only slightly decreased the overall solid holdup. Using aqueous glycerol solutions to evaluate the effects of higher liquid viscosities on this same system, Bloxom et al. (1975) showed that the overall solid holdup decreased and the overall liquid holdup increased with increasing viscosity, while the overall gas holdup was unaffected. These results are in good agreement with previous investigators (Burck et al., 1975; Kim et al., 1975; Bhatia and Epstein, 1974; Michelsen and Ostergaard, 1970; Bruce and Revel-Chion, 1974; Dakshinamurty et al., 1971; Efremov and Vakhrushev, 1970; Mukherjee et al., 1974; Ostergaard and Michelsen, 1968; Ostergaard and Thiesen, 1966; Rigby and Capes, 1970; Ostergaard, 1965). All liquid-solid fluidization systems studied in these experiments expanded upon introducing gas into the bed, in agreement with the results predicted by the criterion developed by Bhatia and Epstein (1974) and Epstein (1976).

CORRELATION OF RESULTS

Minimum Fluidization

The minimum liquid fluidization velocities shown in Figs. 2 and 3 were correlated with the system parameters and resulted in the following dimensionless correlation:

$$Re_{mf} = a\,Ar^{b}\,Fr^{c}\ , \tag{4}$$

where the constants and their 99% confidence limits are:

$$a = 5.121 \times 10^{-3} \quad \pm 0.004$$
$$b = 0.662 \quad \pm 0.062$$
$$c = -0.118 \quad \pm 0.048\ .$$

Eq. (4) had a correlation coefficient of 0.94 and an F-value of 440 using a total of 112 points.

Unfortunately, Eq. (4) is not valid for zero gas flow rate. To produce a three-phase correlation that degenerates to an acceptable two-phase correlation as the gas flow rate goes to zero, the MF velocity predicted by the two-phase correlation of Wen and Yu (1966) was used and resulted in the following:

$$U_{L,mf}/U''_{L,mf} = 1 - U_G^{\ a}\ \mu_L^{\ b}\ d_p^{\ c}\ (\rho_S - \rho_L)^{d}\ , \tag{5}$$

where the exponents and their 99% confidence limits are:

$$a = 0.436 \pm 0.088 \qquad c = 0.598 \pm 0.289$$
$$b = 0.227 \pm 0.058 \qquad d = -0.305 \pm 0.146\ .$$

Eq. (5) had a correlation coefficient of 0.93 and an F-value of 179 using a total of 125 data points. The dimensional correlation of Eq. (5) is somewhat less satisfactory statistically, but it behaves correctly as gas velocity approaches zero.

Local Holdups

Figures 4-6 clearly indicate that each of the holdups is fairly constant in two regions: (1) the lower portion of the bed, and (2) the gas-liquid region above the bed. The transition region between these two extremes was seen to depend on the gas velocity and the physical characteristics of the solid particles. An inflection point was observed on each curve with a spread about that point proportional to the width of the transition region. If each curve were differentiated, these two parameters would correspond to the mean and standard deviation of the normalized Gaussian curves. The error function was used to fit the gas and solid holdup curves, and the liquid holdup curve was determined as the residual of Eq. (1). Use of the error function is essentially equivalent to use of the probability integral, since the two are related by the following:

$$\mathrm{erf}(x) = 2\ \Phi\ (\sqrt{2}\ x)\ . \tag{6}$$

Thus the gas holdup curves could be fitted by the following:

$$\varepsilon_G = [(P_G-1)/-2]\varepsilon_G''' + [(P_G+1)/2]\varepsilon_G''\ , \tag{7}$$

where

$$P_G = \mathrm{erf}[(h - I_G)/\sigma_G]\ . \tag{8}$$

The solid holdup was fitted in a similar manner using the error function and the knowledge that the solid holdup in the gas-liquid region of the column is zero:

$$\varepsilon_S = [(P_S + 1)/2]\varepsilon_S'''\ , \tag{9}$$

where

$$P_S = -\mathrm{erf}[(h - I_S)/\sigma_S]\ . \tag{10}$$

The liquid holdup at each point was obtained from the residual of Eq. (1).

Thus knowledge of six parameters — ε_G''', ε_G'', ε_S''', σ_G, σ_S, and I_G — allows one to construct each of the phase holdups vs axial column position curves. Treatment of the experimental data in this way and correlation of the six parameters with fluid and solid properties and experimental conditions is in progress. The fit of the error functions to experimental data is illustrated in Fig. 7. For the system shown, the six parameters are:

$$\varepsilon_G''' = 0.072 \qquad \sigma_G = 2.64\ \text{cm}$$
$$\varepsilon_G'' = 0.129 \qquad \sigma_S = 2.83\ \text{cm}$$
$$\varepsilon_S''' = 0.511 \qquad I_G = 45.7\ \text{cm}\ .$$

Overall Phase Holdups

The overall solid holdups from these studies were combined with 1355 points from the literature (Kim et al., 1975; Bhatia and Epstein, 1974; Michelsen and Ostergaard, 1970; Bruce and Revel-Chion, 1974; Dakshinamurty et al., 1971; Efremov and Vakhrushev, 1970; Ostergaard and Michelsen, 1968; Ostergaard and Thiesen, 1966; Rigby and Capes, 1970; Ostergaard, 1965) to yield the following dimensional correlation:

$$1 - \varepsilon_S = a\, U_L^{\,b}\, U_G^{\,c}\, (\rho_S - \rho_L)^d\, d_p^{\,e}\, \mu_L^{\,f}\, D_c^{\,g}, \quad (11)$$

where

$a = 0.371 \pm 0.017$ $\quad e = -0.268 \pm 0.010$
$b = 0.271 \pm 0.011$ $\quad f = 0.055 \pm 0.008$
$c = 0.041 \pm 0.005$ $\quad g = -0.033 \pm 0.013$
$d = -0.316 \pm 0.011$

Eq. (11) had a correlation coefficient of 0.87, an F-value of 1178, and was based on a total of 2381 points.

Combining the gas holdup with 169 points available from the literature (Kim et al., 1975; Bhatia and Epstein, 1974; Michelsen and Ostergaard, 1970; Efremov and Vakhrushev, 1970; Ostergaard and Michelsen, 1968) resulted in the following correlation:

$$\varepsilon_G = (0.048 \pm 0.010) \cdot U_G^{\,0.720 \pm 0.028} \cdot d_p^{\,0.168 \pm 0.061} \cdot D_c^{\,-0.125 \pm 0.088} \quad . \quad (12)$$

Eq. (12), based on a total of 913 points, had a correlation coefficient of 0.93 and an F-value of 1793. Note that Eq. (11) does not hold for zero gas velocity, for which it would predict a solid holdup of unity. In fact, it is recommended that the equations presented herein not be used for conditions far removed from those tested.

CONCLUSIONS

The minimum gas and liquid velocities required to fluidize various types of solids have been determined as a function of particle size, particle density, and liquid viscosity; no effect of the initial bed height or column diameter was found.

An electroconductivity technique was adapted for use in three-phase fluidized beds and allowed each of the three-phase holdups to be determined as a function of height in the columns. The transition region where the solids concentration drops to zero was found to increase in width with increasing gas velocity, while remaining fairly constant in width with changing liquid velocity.

Using six parameters determined from the local gas and solid holdup profiles, it was possible to fit each of the holdup vs height in the column curves. Correlation of these parameters should give a reactor designer more information concerning vitally important phase distributions than that available using the homogeneous bed model.

ACKNOWLEDGMENT

This research was sponsored by the Department of Energy under contract with the Union Carbide Corporation. The authors are grateful to K. J. Beach, J. D. Cohill, J. D. Hewitt, M. R. McClure, and D. Sellinger for their assistance in the collection of data. Thanks are particularly due C. L. Begovich for much of the computer analysis presented herein.

REFERENCES

Achwal, S. K. and Stepanek, J. B. (1975) Chem. Eng. Sci. 30, 1443.
Achwal, S. K. and Stepanek, J. B. (1976) Chem. Eng. Sci. 31, 69.
Begovich, J. M. and Watson, J. S., to be published in AIChE J.
Bhatia, V. K. and Epstein, N. (1974) Proc. Int. Symp. on Fluidization and Its Applications, Toulouse, p. 380.
Bloxom, S. R., Costa, J. M., Herranz, J., MacWilliam, G. L., and Roth, S. R. (1975) ORNL-MIT-219, Mass. Inst. Technol., School of Chem. Engng. Practice, Oak Ridge, Tenn.
Bruce, P. N. and Revel-Chion, L. (1974) Powder Technology 10, 243.
Burck, G. M., Kodama, K., Markeloff, R. G., and Wilson, S. R. (1975) ORNL-MIT-213.
Dakshinamurty, P., Subrahmanyam, V., and Rao, J. N. (1971) I and EC Proc. Des. Dev. 10(3), 322.
Efremov, G. I. and Vakhrushev, I. A. (1970) Int. Chem. Eng. 10(1), 37.
Epstein, N. (1976) Can. J. Chem. Eng. 54, 259.
Epstein, N. (1977) Univ. of British Columbia, personal communication.
Kim, S. D., Baker, C. G. J., and Bergounou, M. A. (1975) Can. J. Chem. Eng. 53, 134.
Michelsen, M. L. and Ostergaard, K. (1970) Chem. Eng. J. 1, 37.
Mukherjee, R. N., Bhattacharya, P., and Taraphdar, D. K. (1974) Proc. Int. Symp. on Fluidization and Its Applications, Toulouse, p. 372.
Ostergaard, K. (1965) Chem. Eng. Sci. 20, 165.
Ostergaard, K. and Michelsen, M. L. (1968) Preprint 31d, 2nd Joint AIChE-IIQPR Meeting, Tampa, Fla., May 19-22.
Ostergaard, K. and Thiesen, P. I. (1966) Chem. Eng. Sci. 21, 413.
Rigby, G. R. and Capes, C. E. (1970) Can. J. Chem. Eng. 48, 343.
Vail, Y. K., Manakov, N. K., and Manshilin, V. V. (1970) Int. Chem. Eng. 10(2), 244.
Wen, C. Y. and Yu, Y. H. (1966) CEP Symp. Ser. 62, 62, 100.

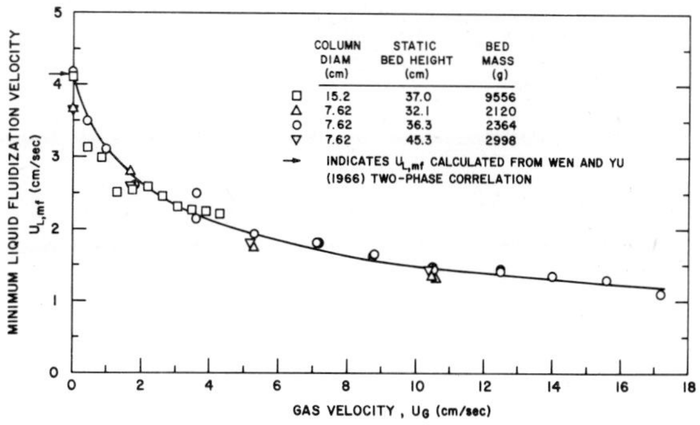

Fig. 1. MF velocities for the air-water-glass beads system.

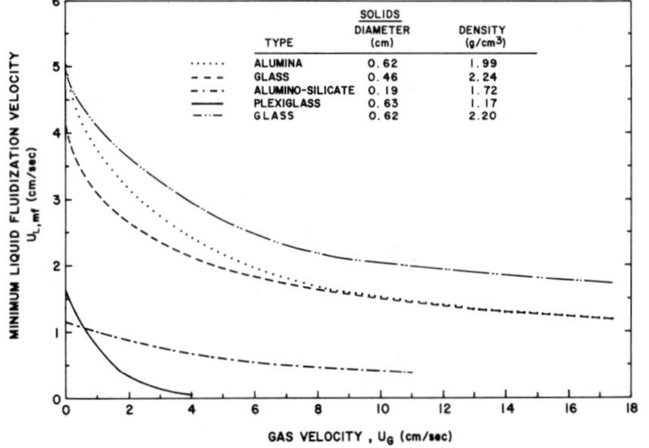

Fig. 2. Effect of particle size and density on the MF velocities.

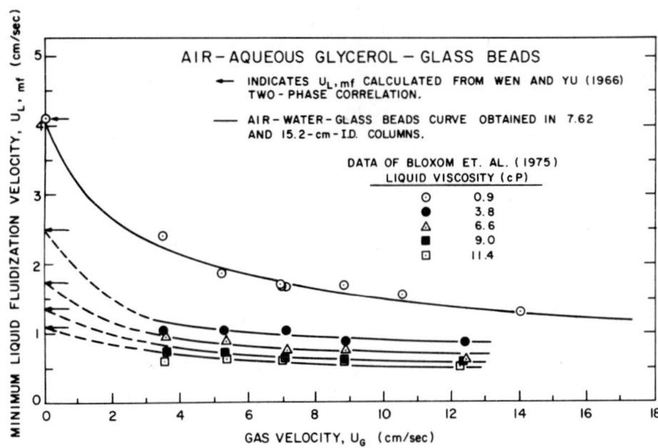

Fig. 3. Effect of liquid viscosity on the MF velocities.

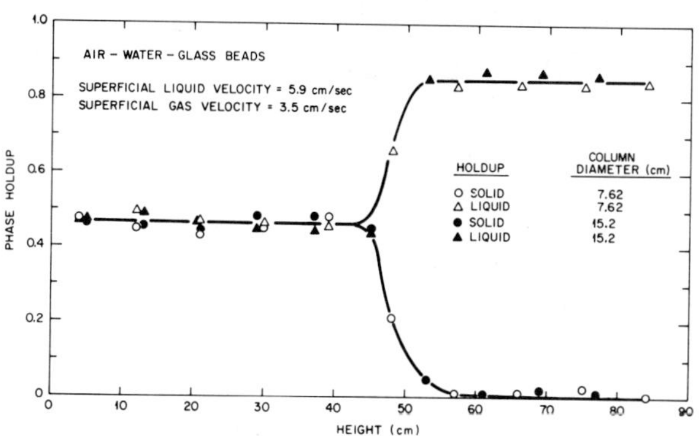

Fig. 4. Effect of column diameter upon the axial variation in holdups.

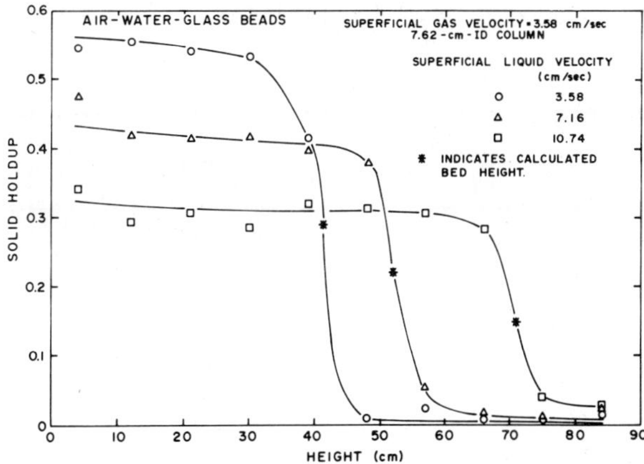

Fig. 5. Effect of liquid velocity upon the axial variation in the solid holdup.

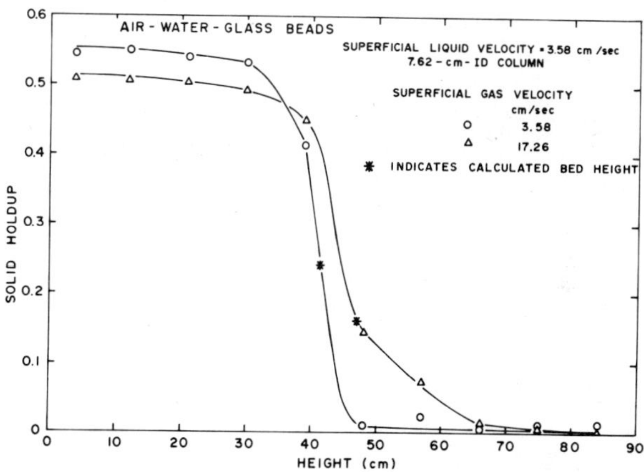

Fig. 6. Effect of gas velocity upon the axial variation in the solid holdup.

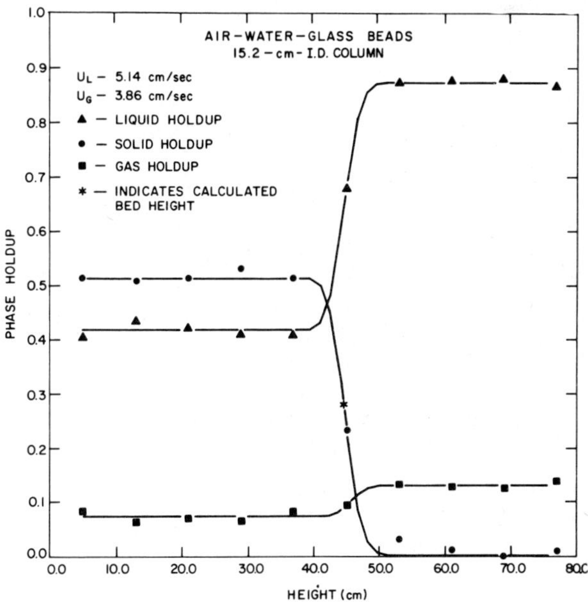

Fig. 7. Fit of local holdup profiles for the air-water-glass beads system.

FLUIDISATION OF MICROBIAL AGGREGATES IN TOWER FERMENTERS

By E.L. Smith, A. James and M. Fidgett

Department of Chemical Engineering,
University of Aston in Birmingham,
Birmingham B4 7ET, U.K.

SYNOPSIS

The behaviour of microbial aggregates in vertical, tubular fermenters has been studied. In such biological reactors the aggregates are fluidised by the action of both the liquid and gaseous phases. Fluidisation-sedimentation experiments have been carried out using flocculent yeasts and filamentous fungi in the absence of a gas phase. These two-phase results are correlated using the Richardson-Zaki type of relationship. The exponent of the voidage term appears to vary between about 7 and 15, depending on the type of microbial aggregate tested. The application of these results to the design of three-phase fermentation systems is discussed.

NOTATION

d_A characteristic diameter of microbial aggregate
f volume fraction of liquid in a microbial cell
g gravitational constant
n exponent in (1)
Re particle Reynolds number ($=\rho_L . U_T . d_A . \mu_L^{-1}$)
U_{SL} superficial liquid velocity
U_T aggregate terminal velocity
x dry weight biomass concentration
x_m maximum settled dry weight biomass concentration
ε voidage between microbial aggregates
ε_A voidage within microbial aggregates
μ_L liquid viscosity
ρ_A microbial aggregate density
ρ_c microbial cell density
ρ_D dried cell density
ρ_L liquid density

1. INTRODUCTION

Microbial aggregates have an important effect on the operation and performance of fermentation systems (Atkinson and Daoud, 1976). The authors have found this to be the case when using flocculent yeasts and filamentous fungi in tower fermenters (Greenshields and Smith, 1971; Smith and Greenshields, 1974; James, 1973).

Tower fermenters are a form of tubular reactor; although they may be divided into compartments and/or fitted with impellers, the authors have used vertical tubes without internals. Agitation and aeration of fermentation broths in such fermenters can be effected by air spargers at the tower base.

Under anaerobic conditions and with a continuous upward flow of liquid, fluidisation of microbial aggregates can occur. Aerobic operation leads to intense mixing of both the liquid and aggregates : although such three-phase systems are difficult to characterise, the fluidisation-sedimentation behaviour of the aggregates is important.

Because of the lack of quantitative information, selected microbial systems have been studied. Preliminary experimental results are presented and discussed in this paper.

2. EXPERIMENTAL WORK AND RESULTS

2.1 Micro-organisms

Because of the authors' interest in the production of biomass and alcohol, the two types of microbial aggregate selected for study were (i) flocs of various yeast strains (mainly *Saccharomyces cerevisiae*), and (ii) pellets of the filamentous mould *Aspergillus niger*. Details of the methods used for preparing samples are given elsewhere (James, 1973).

2.2 Physical Property Measurements

Accurate measurement of the physical properties of living cells is difficult because of cell size and structure and because of the interaction of cells with their environment. However, it is possible to obtain useful information by first determining the densities of (i) dried cells, (ii) intact microbial aggregates, and (iii) filtered or centrifuged aggregates, and then interpreting such results using simple models of aggregate structures.

James (1973) has pointed out that yeast flocs closely resemble beds of packed spheres. Consequently, information about the packing (Gray, 1968) and flocculation of solid particles (Medalia, 1970; Goodarz-Nia and Sutherland, 1975) may be used to build up a model of floc structure. Averaged results for several strains of yeast are summarised in Table 1.

Filamentous mould aggregates can take various forms (Cocker, 1975). In the authors' experiments loose, feathery agglomerates were observed during the early stages of batch fermentations (typically 14-22 hours after inoculation); relatively smooth, spherical pellets developed towards the end of a batch process (about 30-40 hours); and if fermentation was allowed to proceed for about 90 hours, smooth, but hollow, pellets were formed. Different aggregate morphologies have also been observed in continuous fermenters (Pannell, 1976). Such differences lead to variations in physical properties as illustrated by the data in Table 2.

James (1973) has put forward a simple model for characterising mould aggregates which can explain the high values for pellet voidage.

Because mould aggregates are discrete entities and can be readily handled, measurements of pellet terminal velocity can be used to estimate pellet density. Yeast flocs do not form permanent aggregates : nevertheless, it has been possible to carry out some terminal velocity experiments using isolated flocs. Typical values of terminal velocity are given in Tables 2 and 3.

2.3 Fluidisation-Sedimentation Measurements

The behaviour of microbial suspensions was studied in two-dimensional beds, glass columns, and measuring cylinders of various sizes (James, 1973). Fluidisation and sedimentation tests using glass ballotini and perspex spheres were carried out to check the characteristics of the different types of equipment. Results with both non-biological and biological materials were summarised by means of the Richardson-Zaki equation (1954):

$$\frac{U_{SL}}{U_T} = \varepsilon^n \qquad \ldots\ldots\ (1)$$

Regression analysis was used to find both n and U_T.

Fluidisation and sedimentation tests were carried out with four different strains of yeast : the operating temperature was varied from 16 to 40°C and the pH of the liquid phase from 3.8 to 5.4. Because of the problem of determining floc structure, calculations were based on the voidage of settled yeast flocs (about 0.57). Since there was no systematic change in the values of U_T and n with change in pH and temperature, some averaged values are given in Table 3. Reproducibility between batches was not good although within each batch agreement was satisfactory. Fig.1 shows a plot of ln U_{SL} against ln ε for one of the yeasts.

Results of sedimentation and fluidisation experiments using various types of A. niger pellets are summarised in Table 4. Bed voidage was again estimated from settled volumes, assuming random packing of the pellets. Some typical experimental results are plotted in Fig.1.

3. EXPERIMENTAL OBSERVATIONS AND DISCUSSION OF RESULTS

3.1 Yeast Flocs

Floc break-up and coalescence can be observed during fermentation, and such behaviour was also noted during the fluidisation experiments. A decreasing flow-rate technique was used to study fluidisation, but, even so, it took up to 30 minutes before steady-state conditions were achieved at a given flowrate.

In order to relate the fluidisation behaviour of yeast to that of other materials it was assumed that floc voidage for all strains and in all conditions was constant. James (1973) assumed a value of 0.50 in his calculations : data in Table 3 were based on a voidage figure of 0.57, an average value for consolidated beds of yeast. Terminal velocity measurements on isolated flocs of strain NCYC 1119 suggest that floc voidage increases with floc size, and can reach values as high as 0.90 : in the light of the results of Medalia (1970) and Goodarz-Nia and Sutherland (1975) these results are not too surprising. Further work is required to determine how floc voidage varies with yeast strain, floc diameter, and operating/environmental conditions.

3.2 Mould Pellets

Mould aggregates retain their shape and structure for considerable periods, and so assessment of their sedimentation-fluidisation characteristics was not expected to present many problems. However, it will be noted from the results in Table 2 that pellet voidages are high, and so pellet densities are very close to that of water. This low density difference and the tendency of pellets to clump together made it difficult to observe the transition from the packed-bed state to the fluidised state during experimental work : as with yeasts, it frequently took many minutes to achieve steady-state conditions after altering the flowrate in fluidisation tests.

The very high voidage of mould pellets highlights the need to consider the possibility of liquid flow within such structures. This problem has been considered by Sutherland and Tan (1970) who concluded that, even in loose and open flocs, internal flow is likely to be small. In the authors' opinion more research into this problem is required : although it is convenient to neglect interstitial flow, it may be of significance within the outer regions of some types of mould pellet.

3.3 The Fluidisation-Sedimentation Exponent, n

The Richardson-Zaki equation has been widely used to correlate the results of experimental work. When using spheres of uniform size the exponent n varies between 2.4 and 4.7 : in tests with glass and perspex spheres we obtained good agreement between experimental values of n and those calculated from (1).

When particles are non-spherical and the size distribution is not uniform, experimental values of n range from 5 to 10 (Davies and Richardson, 1966; Richardson, 1971). With flocculent materials even larger values of n have been reported (Edeline et alia, 1969; Brown and La Motto, 1971; Davies et alia, 1976). Consequently, the results summarised in Tables 3 and 4 are not so remarkable as would appear at first sight.

Parker (1971), commenting on the work of Brown and La Motto (1971), has pointed out that when analysing results for flocs the value of n depends on assumptions made about floc voidage. This is also true of the authors' results : no direct determinations of voidage were made and calculations were usually based on the settled volumes of the dispersed phase.

When solids of the same density, but of different size are fluidised the bed voidage is usually higher than that calculated on the basis of average particle size. With both yeast flocs and mould pellets the ratio of the largest particle diameter to that of the smallest was usually greater then two, and so, even allowing for changes in aggregate density with size (see Tables 1 and 2), stratification was to be expectea (Wen and Yu, 1966). In fact, stratification was never clearly visible once steady-state conditions had been reached. Nevertheless, this effect may have contributed to the high observed values for n.

3,4 Terminal Velocities of Aggregates

Comparison of terminal velocity (U_T) values obtained by calculation and by experiment provides valuable information both about the experimental method and suspension behaviour.

In the case of isolated yeast flocs U_T values were in the region 10 to 30 mm/s : values obtained by regression analysis are shown in Table 3. An unexpectedly high figure was obtained for strain NCYC 1251 and a somewhat low figure for strain NCYC 1119 : even modifying the assumptions about floc voidage did not markedly affect the results, and at the present time a satisfactory explanation cannot be offered.

The results for the mould pellets appear to be reasonable, although U_T values obtained by regression analysis are somewhat higher than expected.

3.5 Transition to the Fluidised State and Fluidisation Quality

For both flocs and pellets the transition from a consolidated bed of particles to the fully fluidised state appeared to be more gradual than with a suspension of equal-sized spheres. Loose agglomerates of flocs and pellets were observed at the lower liquid flowrates during fluidisation, and so the usual calculations for predicting the minimum fluidisation velocity were not applicable.

In sedimentation tests there was also visual evidence of aggregation at suspension voidages as high as 0.8 to 0.9. With yeasts the formation of very loose flocs made it impossible to analyse sedimentation results using the Richardson-Zaki equation.

Another difficulty arising from the formation of such loose aggregates is the selection of experimental results for regression analysis. Plots of $\ln U_{SL}$ against $\ln \varepsilon$ have been used for this purpose, but frequently it was difficult to distinguish between the fluidised bed region and consolidated bed region.

4. DESIGN APPLICATIONS

4.1 Introduction

Experimental work discussed above concerns two-phase systems : fermentation processes are three-phase systems incorporating a solid (biological) phase, liquid (aqueous) phase, and a gas phase (e.g. O_2,CO_2,CH_4). Although there is published research concerning three-phase fluidisation (Ostergaard, 1971), the effects of the gas-phase are largely unknown and cannot be modelled satisfactorily. Unfortunately this work has been performed only with an "ideal" solids phase, e.g. glass ballotini spheres of constant diameter. It is clear from previous sections that biological solids are far from ideal and difficult to characterise, especially since floc/pellet structure and size may change continually. The fluidisation of biological solids is further complicated by an additional factor : the end-product of many fermentations is a gas and thus gas bubbles are continually forming within the biological flocs/pellets. Also, in fermentations which require air to be bubbled into the reaction vessel (i.e. aerobic conditions) air bubbles tend to attach themselves to the microbial flocs/pellets. These two factors tend to further reduce the small density difference which exists between the solid and liquid phases.

In anaerobic fermentations, e.g. alcohol production by yeast, the effects of the gas phase may be neglected at the preliminary design stage. However, in aerobic fermentations the quantities of gas involved are very much greater and the gas phase appears to have a signigicant effect upon the behaviour of the microbial phase. Aerobic and anaerobic systems are discussed separately below.

4.2 Anaerobic Systems involving Yeast Flocs

For preliminary design purposes a simple, lumped-parameter model can be used to represent a tower fermenter containing fluidised microbial aggregates. Firstly, however, a floc model is required, since terminal velocity depends upon floc density. A diagram of an ideal yeast floc is shown in Fig.2. The floc is assumed to be spherical with characteristic diameter d_A. Individual yeast cells pack closely together to form a floc and so it seems reasonable to assume a floc voidage of 0.5-0.6.

Biological cells consist largely of a dilute aqueous solution. James (1973) found the volume fraction of liquid (f) in a yeast cell to be 0.78 and the density of the remaining dry matter (ρ_D) to be 1320 kg/m^3. Cell and floc densities can thus be calculated as follows:

$$\rho_C = f.\rho_L + (1-f)\rho_D \quad ....(2)$$

$$= 0.78\rho_L + (0.22\times1320)\ \text{kg/m}^3 \quad ....(3)$$

$$\rho_A = \varepsilon_A.\rho_L + (1-\varepsilon_A)\rho_C \quad ....(4)$$

For the particle Reynolds number range 2<Re<500 the terminal velocity of a spherical aggregate can be approximated by the expression (McCabe and Smith, 1976).

$$U_T=1.53\{g(\rho_A-\rho_L)\}^{0\cdot714}.d_A^{1\cdot142}.\mu_L^{-0\cdot428}.\rho^{-0\cdot286} \quad .... (5)$$

$$\text{where}\quad Re = \rho_L.U_T.d_A.\mu_L^{-1} \quad ....(6)$$

Hence

bed voidage (ε) and yeast dry weight concentration (x) are given by:

$$\varepsilon = (U_{SL}/U_T)^{1/n} \quad ....(7)$$

$$x = (1-\varepsilon)\ (1-\varepsilon_A)\ (1-f)\rho_D\ \text{kg/m}^3 \quad ....(8)$$

This modelling approach has formed the basis of a large multi-stage model of a continuous beer tower (Fidgett & Smith, 1975). The model has been successfully used to simulate some of the operational characteristics of continuous beer towers reported in the literature (Klopper et alia, 1965; Royston, 1966; Ault et alia, 1969).

4.3 Aerobic Systems

Visual observation of the behaviour of a filamentous fungus (Aspergillus niger) in an aerobic tower fermenter poses the question : are the microbial pellets truly fluidised or are they merely an extension of the liquid phase? The effect of liquid superficial velocity (U_{SL}) upon solids concentration (x) only partly answers this question. When U_{SL} is increased there is elutriation of solids from the fermenter, indicating some degree of fluidisation. However, when U_{SL} is decreased there is no settling of the solid phase. This is probably due to the considerable agitation which is caused by rising gas bubbles and slugs. The microbial phase appears to mix less freely than the aqueous phase, though, possibly because of hyphal entanglement between neighbouring pellets.

Despite this uncertainty over whether aerobic fermentations exhibit true fluidisation, the Richardson-Zaki form of equation can be used to characterise biomass concentration as a function of liquid superficial velocity. Assuming the pellets to be spherical, the voidage of close-packed spheres to be 0.5, and the maximum dry weight concentration of settled biomass to be x_m, then

$$\varepsilon = 1-(0.5x/xm) \quad ....(9)$$

$$= (U_{SL}/U_T)^{1/n} \quad ....(10)$$

$$\text{or}\quad x = 2x_m\{1-(U_{SL}/U_T)^{1/n}\}\ \text{kg/m}^3 \quad ....(11)$$

The parameters n, U_T, and x_m are specific to the type of micro-organism and are largely dependent upon its morphology, which is governed by environmental factors such as : growth-limiting nutrient; temperature; pH; presence of inhibitors; space for growth; and rate of shear.

The parameters U_T and x_m for a particular morphology and characteristic pellet size may be determined either (i) experimentally, or (ii) theoretically by means of a fungal pellet model (James, 1973; Metz, 1976). Typical values of n for various pellet sizes and morphologies are shown in Table 4.

Fig.3 shows experimental values of steady-state biomass dry weight concentration for the filamentous fungus Aspergillus niger at various superficial liquid velocities (Pannell, 1976). It is worth noting that pellet morphology changes considerably over the range of data included in Fig.3. At low values of U_{SL} ($\sim$0.005 mm/s) the pellets have a fluffy, or feathery, appearance and settle extremely slowly. However, at higher superficial liquid velocities ($\sim$0.05 mm/s) the pellets are smooth spheres which have much higher terminal velocities. Thus the plot can be artificially divided into a number of segments, each corresponding to a particular pellet morphology and characteristic size.

By choosing suitable values for the parameters U_T, n, and x_m, Eq.(11) can be fitted to each segment of the experimental curve (see Fig.3).

Pannell's data can be approximated by just two curves : A and either B or C. This fit, though, is purely empirical since neither n nor U_T are known accurately; further experimental work is required. A reasonable estimate of x_m, however, can be obtained by simple tests : 25 kg/m^3 at low values of U_{SL}, and 12.5-25 kg/m^3 at higher values of U_{SL}. Fig.3 shows that, with 3 variables, curve-fits are possible with a number of choices of parameter values. Even allowing for some variation in n and x_m, values of U_T are approximately two orders of magnitude less than those obtained with two-phase (liquid-solid) systems. This discrepancy is almost certainly due to gas bubbles trapped in the fungal pellets, thus reducing $(\rho_A - \rho_L)$ and U_T. This feature highlights the problems we face.

5. FINAL COMMENTS

5.1 The behaviour of microbial aggregates in tower fermenters in the absence of a gas phase can be described by the Richardson-Zaki form of equation. The exponent of the voidage term appears to vary between 7 and 15.

The presence of a gas phase complicates the situation, but results obtained from a continuous tower fermentation process for mould pellet production can be fitted to a Richardson-Zaki type of equation. The gas phase considerably reduces the apparent terminal velocity of microbial aggregates. Further work on three-phase fluidisation of such aggregates is required.

5.2 Microbial aggregates have a high voidage (about 0.8-0.9) and a similar density to that of water. The fluidisation behaviour of such particles still requires further investigation. Emphasis should be given to (i) the transition from the packed bed to the fluidised state, (ii) bed stability, and (iii) the conditions for stratification.

5.3 A better mathematical model of the structure of microbial aggregates is required. In particular, attention should be paid to the following features: (i) formation and break-up of aggregates; (ii) diffusion of reactants/products into/out of aggregates; and (iii) nucleation and entrapment of gas bubbles within aggregates.

5.4 The design of tower fermentation systems requires some knowledge of the fluidisation-sedimentation behaviour of microbial aggregates : such behaviour can be described by the fluidisation equation in conjunction with an aggregate model. This approach has been used in the simulation of a continuous beer tower fermenter (Fidgett and Smith, 1975). A preliminary study of aerobic systems for biomass production has also been undertaken (Fidgett, 1975).

ACKNOWLEDGEMENTS

We should like to thank both the Science Research Council and the University of Aston in Birmingham for their support in the form of Research Studentships (for A.J. and M.F.) and Research Grants.

REFERENCES

Atkinson, B. and Daoud, I.S. (1976) Adv. Biochem. Engng. 4, 41.

Ault, R.G., Hampton, A.N., Newton, R. and Roberts, R.H. (1969) J. Inst. Brew. 75, 260.

Brown, J.C. and La Motto, E. (1971) J. Sanit. Engng. Div. Am. Soc. Civ. Engrs. 97, SA2, 209.

Cocker, R. (1975) Ph.D. Thesis, University of Aston in Birmingham, U.K.

Davies, L. and Richardson, J.F. (1966) Trans. Instn. Chem. Engrs. 44, T293.

Davies, L., Dollimore, D. and Sharp, J.H. (1976) Powder Technol, 13, 123.

Edeline, F., Tesarik, I. and Vostrcil, J. (1969) 'Advances in Water Pollution Research' (Jenkins, S.H. Ed.) p.523 (Proc. 4th Int.Conf., Prague) Pergamon Press, Oxford.

Fidgett, M. (1975) Ph.D. Thesis, University of Aston in Birmingham, U.K.

Fidgett, M. and Smith, E.L. (1975) paper presented at Inst. Chem. Engrs. Ann. Res. Mtng., Bradford, U.K.

Goodarz-Nia, I. and Sutherland, D.N. (1975) Chem. Eng. Sci. 30, 407.

Gray, W.A. (1968) 'The Packing of Solid Particles', Chapman and Hall Ltd., London.

Greenshields, R.N. and Smith, E.L. (1971) Chem. Engr. (Lond.), No.249, 182.

James, A. (1973) Ph.D. Thesis, University of Aston in Birmingham, U.K.

Klopper, W.J., Roberts, R.H., Royston, M.G. and Ault, R.G. (1965) Proc. Eur. Brew. Conv., 238.

McCabe, W.L. and Smith, J.C. (1976) 'Unit Operations of Chemical Engineering'. 3rd Edn., P.155, McGraw-Hill, New York.

Medalia, A.I. (1970) J. Colloid and Interface Sci. 32, (1), 115.

Metz, B. (1976) Ph.D. Thesis, Technische Hogeschool, Delft, Netherlands.

Ostergaard, K. (1971) 'Fluidization', Ch.18 (Davidson, J.F. and Harrison, D. Eds.) Academic Press, New York.

Pannell, S.D. (1976) Ph.D. Thesis, University of Aston in Birmingham, U.K.

Parker, D.S. (1971) J. Sanit. Engng. Div. Am. Soc. Civ. Engrs. 97, SA5, 793.

Richardson, J.F. and Zaki, W.N. (1954) Trans. Instn. Chem. Engrs. 32, 35.

Richardson, J.F. (1971) 'Fluidization', Ch.2 (Davidson, J.F. and Harrison, D. Eds.) Academic Press, New York.

Royston, M.G. (1966) Brewers' Guard. 95 (Feb), 33.

Smith, E.L. and Greenshields, R.N. (1974) Chem. Engr. (Lond.) No.281, 28.
Sutherland, D.N. and Tan, C.T. (1970) Chem. Eng. Sci. 25, 1948.
Wen, C.Y. and Yu, Y.H. (1966) Chem. Eng. Progr. Symp. Ser. 62, No.62, 100.

Table 1 Yeast Floc Properties

Dry cell density (ρ_D)	= 1320 kg/m^3
Hydrated cell density (ρ_C)	= 1070 kg/m^3
Water in hydrated cell (f)	= 0.78 v/v
	= 0.73 w/w
Centrifuged wet yeast voidage	= 0.316
Floc voidage (ε) (based on settled volume)	= 0.567
Average floc diameter (d_A) (varies with strain)	= 1.5-6.0 mm
Terminal velocity of isolated flocs (U_T)	= 20 mm/s approx

Table 2 Properties of A. Niger Aggregates during a Batch Fermentation

Time (h)	ρ_D (kg/m^3)	ρ_C (kg/m^3)	d_A (mm)	ε_F (from U_T measurements
14	1390	1110	1.3	0.91
19	1380	1102	2.1	0.95
24	1300	1080	3.0	0.96
29	1260	1060	3.4	0.97
38	1240	1055	4.0	0.95
90 (hollow pellets)	1240	1050	4.5	0.92
U_T = 7 mm/s (approx.)				

Table 3 Experimentally Determined Values for n for strains of Flocculent Yeast

Yeast Strain (NCYC code)	n (av.value)	d_A (mm)	U_T (calc.) (mm/s)
NCYC 1119	7.5	3	3.9
NCYC 1260	11	6	39
NCYC 1251	11	2	130
CFCC 34	14	1.5	20

Table 4 Experimentally Determined Values for Terminal Velocity (U_T) and n for A. niger Pellets

Pellet age (h)	d_A (mm)	n	U_T (mm/s) calculated	U_T (mm/s) measured
18	1	9.1	3.7	1.4
40	2	8.8	6.6	7.1
90(hollow)	4	10.3	33	14
40(inhibited growth)	0.7	8.7	2.1	-

FIG. 1 RESULTS OF FLUIDISATION TESTS WITH MICROBIAL AGGREGATES

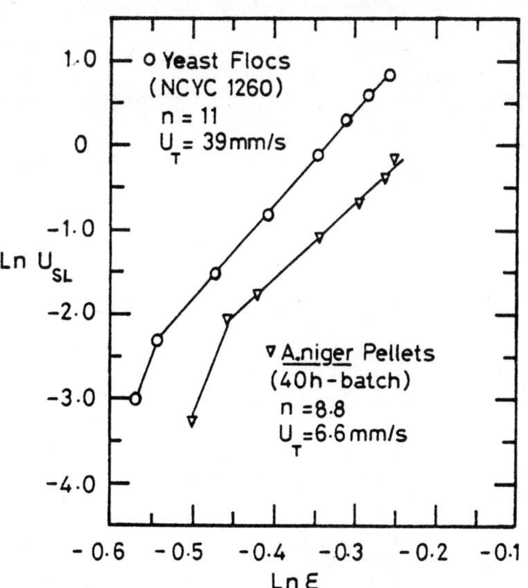

FIG 2 IDEAL YEAST FLOC

d_A

Yeast Cell

5 μm

7 μm

FIG.3 EFFECT OF LIQUID FLOW-RATE ON STEADY-STATE BIOMASS CONCENTRATION

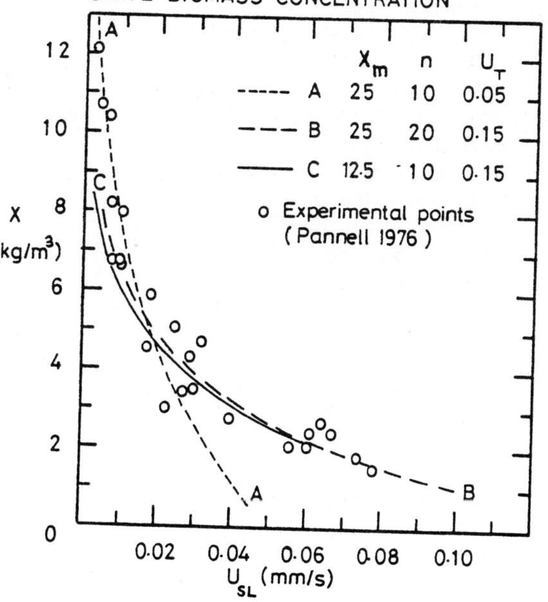

COAL PYROLYSIS, GASIFICATION, AND COMBUSTION IN A MULTI-VESSEL FLUIDIZED-BED SYSTEM

By P.L. Waters and K.McG. Bowling

CSIRO Division of Process Technology, P.O. Box 136
North Ryde, N.S.W., 2113, Australia

SYNOPSIS

Small pilot-plant fluidized-bed reactors were employed to study: (a) low-temperature pyrolysis of coal to obtain tar, fuel gas, and char; (b) gasification of carbonaceous solids in a twin-vessel system to produce hydrogen-rich synthesis gas; and (c) combustion of carbonaceous residues to generate heat. Design data from these studies and mass and heat balances have been used to evaluate a conceptual integrated scheme using circulating bed material to provide the heat for gasification. Such a multi-vessel system for the continuous conversion of coal into liquid and gaseous fuels would be flexible in operation and capable of high overall thermal efficiency.

NOTATION

$\hat{C}$	specific carbon consumption	kg/hm^2
$\bar{d}$	mean particle diameter	mm
dmmf	dry, mineral matter free basis	
F	feed-rate	$kg/h, m^3/h$
$\hat{F}$	specific feed rate	$kg/hm^2, m^3/hm^2$
g	acceleration due to gravity	$9.81\ ms^{-2}$
H	enthalpy (ref. 20°C, H_2O liq.)	MJ
$-\Delta H$	heat of reaction at 1000°C	MJ/kg mole
L	bed height	m
M	mass	kg, t
ΔP	pressure drop	kPa
$\hat{Q}$	specific energy (gross, 20°C, 1 atm)	$MJ/kg, MJ/m^3$
Q	potential energy of combustion	MJ
$\bar{t}$	mean residence time	min, s
T	temperature	°C
T_o	temperature of onset	°C
T_b	bed temperature	°C
U_o	incipient fluidizing velocity	m/s
U	superficial velocity	m/s
$\bar{U}_i$	mean interstitial velocity	m/s
V	volume (20°C, 1 atm; steam 100°C)	m^3
V_b	bed volume	m^3
W_f	power for fluidizing	kW
$\hat{W}_f$	specific power for fluidizing	kW/m^2
y	fractional char yield	-
ε	voidage fraction	-
ρ_f	density of fluidized bed	kg/m^3
ρ_s	density of static bed	kg/m^3
ρ_p	density of particles	kg/m^3
η(T, gas)	gas viscosity relative to air at 20°C, 1 atm.	

INTRODUCTION

It is anticipated that liquid and gaseous fuels derived from coal by hydrogenation, pyrolysis, and gasification will ultimately replace the now rapidly diminishing resources of petroleum and natural gas. Gasification of coal or carbonaceous residues from pyrolysis or hydrogenation, would be a necessary stage in the production of synthesis gas or the hydrogen required for coal hydrogenation and the refining of its products.

The authors have developed on a 20 kg/h pilot-plant scale, a single-vessel system (LOTEC) for the rapid low-temperature pyrolysis of coal (Bowling *et al.* 1961; Bowling & Waters, 1962, 1968a, 1970) and a twin-vessel system (FLUGA) for the gasification of coal, char and coke (Bowling and Waters, 1968b, 1976), using fluidization techniques. A wide range of coking and non-caking coals were tested in the pyrolyser (125 runs) and gasifier (136 runs); both systems were simple in design and proved satisfactory in operation. These studies demonstrated technical feasibility and provided design data on specific feed rates, the effects of temperature and other process variables on the degree of feed conversion, and the yields and composition of products.

A logical development is their combination into a multi-vessel system for the fluidized processing of coal (FLUPROC) as shown in Fig. 1. In this paper, design data are presented and used to calculate the reactor sizes, yields and product compositions for a conceptual 10 t/h gas generator of industrial scale.

Multi-vessel systems using particulate heat-carriers have also been studied by e.g. Standard Oil Development Co. (1948), Rayner (1952), Bloom and Eddinger (1974), Kunii and Kunugi (1975).

DESIGN OF AN INTEGRATED SYSTEM

As indicated in Fig. 1, crushed coal is conveyed pneumatically to the pyrolyser [P] where it is carbonized at 500°C to yield char as the main product, and tar, liquor and gas as byproducts. Char is recycled from the bed to the feed stream (up to 10:1 char/coal ratio) to prevent agglomeration of the coal. Heat for carbonizing is provided by combustion of a small fraction of the hot recycling char by the fluidizing air in preference to that of volatile matter (VM). Fine char elutriated from the bed is efficiently separated from the tar-containing gas stream by cyclones, and returned to the system.

Char from the pyrolyser is introduced into the bed of the gasifier [G] where it is converted by the fluidizing steam at about 975°C into water gas. Heat for gasification is generated by burning a fraction of the char in the air-fluidized combustor and continuously transferring this heat by circulation of the bed material through interconnecting downpipes, which act as gas seals. The bed material, shared between the two reactors, consists of char mixed with sand, ash or firebrick grog. To provide the necessary temperature gradient the combustor operates some 50°C higher than the gasifier.

Combustion is carried out in two stages. The first-stage or main combustor [CI] contains a high concentration of carbon (30-50% w/w) in the bed and operates at sub-stoichiometric conditions with respect to the fluidizing air, as in gas producers; secondary air is required over the bed to complete combustion of the CO and H_2 formed within the bed. Carbon transferred with the ash from the first stage is burnt almost to completion in the second stage [CII], which has a relatively small, shallow bed.

Fine char collected in the cyclones is returned to the base of the gasifier for further conversion during its hindered entrainment through the bed.

To achieve a high overall thermal efficiency, the sensible heat of the product gases is recovered by heat exchangers to preheat the steam and air and to generate power required by the air compressors and other equipment. The relatively small yield of pyrolyser gas may also be burnt in a gas turbine to supplement electrical power from the steam turbine.

REACTIONS

Coal particles entering the system pass through several chemical and physical changes, depending on the temperature of thermal decomposition and the surrounding atmosphere. Most of the coal ultimately forms water gas, which contains most of the thermochemical energy of the coal. The main chemical reactions involved are summarized in Table 1, and their sequence of occurrence is indicated in the multi-reactor model of Fig.2.

TABLE 1 COAL PROCESSING REACTIONS

Reaction	T_o	Equation	$-\Delta H$
Oxidation:			
R1	150	Coking coal → non-coking coal	
Pyrolysis:			
R2	370	Coal → LT char + VM	*
R3	500	LT tar → C + tar + gas	
R4	600	LT char → HT char + gas	
Gasification:			
R5	750	$C + H_2O(g) \rightarrow CO + H_2$	-136
R6	700	$CH_4 + H_2O(g) \rightarrow CO + 3H_2$	-227
	500	$C_2H_6 + 2H_2O(g) \rightarrow 2CO + 5H_2$	
R7	700	$CO + H_2O(g) \rightleftharpoons CO_2 + H_2$	+32.7
R8	850	$C + CO_2 \rightleftharpoons 2CO$	-168
R9	500	$C + 2H_2 \rightleftharpoons CH_4$	+91.7
Combustion:			
R10	500	$C + O_2 \rightarrow CO_2$	+396
R11	570	$CO + \frac{1}{2}O_2 \rightarrow CO_2$	+281
	550	$H_2 + \frac{1}{2}O_2 \rightarrow H_2O(g)$	+249

* Exothermic heat of carbonization approx. 0.6 (±0.1) MJ/kg coal.

In the riser pipe of the pyrolyser, coal particles undergo surface oxidation (R1). Above 370°C, the coal is rapidly carbonized (R2) in the pyrolyser to yield low-temperature (LT) char, primary tar, liquor, and a small quantity of hydrocarbon-rich gas. Above 500°C, primary tar is cracked (R3) into carbon, secondary tar and hydrocarbon gases. Heat is provided by combustion of a small fraction of the char (R10).

Char entering the gasifier is further carbonized (R4) to form high temperature (HT) char, H_2O, H_2, CH_4 and other hydrocarbon gases. The resulting char is subsequently gasified by the highly endothermic reaction, R5. Steam reforming of gaseous hydrocarbons occurs by reaction R6. Carbon monoxide formed by all these reactions, may, in turn, react with excess steam by the reversible, water-gas shift reaction, R7. High reactor temperatures favour CO formation; the reverse applies as the product gas cools. Carbon monoxide may also be formed by the relatively slow Boudouard reaction, R8. Methane is formed by R2 and R3 and some also by R9, which is favoured by high pressure.

In the combustor, HT char is rapidly burnt (R10) just above the air distributors. The CO_2 and moisture formed are partly reduced in the upper portion of the bed by R8 and R5. The CO and H_2 are subsequently burned over the bed by the secondary air (R11).

DISCUSSION OF DESIGN CRITERIA

Conditions shown to be practical in LOTEC, FLUGA and other CSIRO test rigs are summarized in Table 2. Design criteria calculated from these data are shown in the lower part of this Table.

TABLE 2 DESIGN DATA FOR FLUPROC REACTORS

		[P]	[G]	[CI]	[CII]
T_b		500	975	1025	850
Feed, solid		coal	LT char	HT char	
gas		air	steam	air	
Bed material		LT char	sand (ash)		
$\bar{d}$	(mm)	0.7	0.7		
ρ_s	(kg/m^3)	225	1450 (1150)		
$\bar{t}$, char	(min)	14	-	-	-
$U_o(gas, T_b)$	(m/s)	0.06	0.18	0.17	0.18
$\bar{U}$	(m/s)	0.10	0.23	0.50	0.50
$\bar{U}_i$ (T_b)	(m/s)	0.70	2.3	4.0	3.7
$\hat{t}$, gas	(s)	1.5L	0.44L	0.25L	0.3L
$\hat{F}$, solid	(kg/hm^2)	1000L	-	-	
$\hat{F}$, gas	(kg/hm^2)	129L*	490	2170	
	(m^3/hm^2)	106L	817	1800	
$\hat{C}$	(kg/hm^2)	8.0	245†	171‡	
ΔP	(kPa)	1.77L	9.41L	9.41L	
$\hat{W}_f$	(kW/m^2)	0.5L	9.2L	21L	19L

* at 0.03 kg O_2/kg coal reactant ratio
† 75% steam conversion; ‡ 10.5% excess air

Calculations proceeded as follows: Using the correlation of Leva (1959)

$$U_o = 0.0537\, d^{1.82} \rho_s^{0.94} / \eta(T, gas)^{0.88} \quad \ldots(1)$$

where $\rho_s = \rho_p(1-\varepsilon)$. Values of U_o have been calculated as a function of ρ_s and T_b for each of the bed materials used in the reactions for $\bar{d} = 0.7$ mm. Results are given in Fig. 3 together with the normal operating conditions of U and T_b. The fluidizing velocity, U, and hence $\hat{F}$ for the reactant gases in turn determine $\bar{U}_i$ and $\bar{t}(=L/\bar{U}_i)$; these factors, together with T_b determine the degree of conversion and the specific carbon consumption rates, $\hat{C}$ (see Table 2). The density of the fluidized bed (ρ_f) and bed height (L) determine the gas pressure (ΔP) for fluidizing and hence the specific power requirement ($\hat{W}_f$), since

$$\hat{W}_f = U_{(T_b)}\, \Delta P \qquad \text{(Leva, 1959)} \ldots(2)$$

where $\Delta P = \rho_f gL$.

In the pyrolyser, the coal and char are devolatilized to a pseudo-equilibrium value when $\bar{t} \approx 14$ min. A temperature of about 500°C appears to be optimum, since loss of tar yield occurs at higher temperatures due to thermal cracking in the presence of contact material such as char (Hesp and Waters, 1970). This residence time is determined by

$$\bar{t} = 60\, \rho_f V_b / yF = 48\, \rho_s V_b / yF \text{ (min)} \quad \ldots(3)$$

since $\rho_f = 0.8\, \rho_s$ (Bowling *et al.* 1961). Given V_b, a suitable value of L is chosen to provide a practical bed area and aspect ratio. On a unit cross-sectional area basis

$$\hat{F} = 48\, \rho_s\, L / y\bar{t} \quad (kg/hm^2) \quad \ldots(4)$$

In the gasifier, the fluidizing steam is the key reactant rather than the carbon which is retained in the system until it is almost completely converted into gas. Steam conversion is limited by kinetic rather than equilibrium considerations. For high overall thermal efficiency and throughput a steam conversion of > 75% is sought. As Fig. 4 indicates, this may be attained in four ways; operating at relatively high T_b (e.g. ≃975°C); having the fluidized bed sufficiently deep to provide a steam carbon contact time $\bar{t} \gtrsim 1.5$ s (*cf* Jolley *et al.* 1953), e.g. L > 3 m; using reactive chars or activating less active chars with catalysts (Willson *et al.* 1974); and having a reasonably high concentration of carbon in the bed e.g. 40% w/w. (According to Rayner (1952), a variation of 30 to 50% carbon in the bed increases the gas production rate by only 16% and is therefore not critical.)

Product gas compositions in the FLUGA tests indicate that relatively high temperatures favour the formation of CO, mainly by virtue of the Boudouard reaction (R8). Hydrogen content is reasonably constant (54 ± 4% v/v) over a wide temperature range (850 - 1050°C). The composition of the product gas may be adjusted externally by the water-gas shift reaction, depending on the type of synthesis reactions which are to follow.

The char combustion and heat generation rates of the combustors are primarily determined by U (≃ 2.5 (±1.0) U_o) (Waters, 1975). To simplify solids transfer between G and CI the same bed depth should be used in both. However since CI, with excess carbon in the bed, operates under reducing conditions, the gas leaving the bed contains a considerable proportion of CO. Gas compositions from the heat generator of the FLUGA rig indicate an increase of CO with bed temperature from 3% at 950°C to about 20% at 1150°C when burning coke fines. This CO formation reduces the heat available for circulation to the gasifier. The heat must be recovered by burning the CO over the bed with secondary air and transferring it by heat exchangers to the fluidizing air. Preferably, CO formation should be minimized by operating the bed of Stage I of the combustor at the lowest possible temperature consistent with effective steam conversion in the gasifier.

Carbon accompanying the overflow ash from Stage I may be burnt in a shallow (less than 1 m) fluidized combustor at about 850°C with excess air. Heat may be extracted from the bed by submerged tubes to control T_b and to preheat water for the steam generator.

MATERIALS AND ENERGY BALANCES

The above section provided information on feed rates and power requirements per unit cross-sectional areas of the reactors, and on bed height and temperatures for efficient conversion. In order to determine the bed areas and heat requirements for a given coal input, it was necessary to carry out detailed mass and heat balances on each vessel using the reactor model, Fig. 2. Results for a high-volatile bituminous coal (Liddell, NSW) are summarized in Table 3. Tables 4 and 5 give compositions of the coal, char, tar and gas streams that were available or estimated from test-rig and related studies. The main conditions assumed are given at the foot of Table 2, and plant heat losses were based on the estimated surface areas of reactors and equipment for a 10 t/h unit. Steam generation (Stream 30; 39 kWh/100kg coal) was allowed for in the energy calculation to provide the necessary power for fluidizing as indicated in Fig. 1.

For every 100 kg coal feed, 1.2 kg of LT char is burnt in the pyrolyser to yield 39 MJ of heat which supplements the exothermic heat of carbonization (approx. 60 MJ), and so maintains the reactor temperature. The heat required for gasifying 39.3 kg carbon is 445 MJ and this is provided mainly by the transfer of about 540 MJ from the combustor; the combustor burns 22.3 kg of carbon and generates about 736 MJ. Preheating of the steam (to 400°C) and air (to 500°C) largely offsets the sensible heat losses in the gaseous products.

As indicated at the foot of Table 3, the thermal efficiency for the production of cold synthesis gas plus tar is about 84% on the basis of the heating value of the coal feed. Total heat loss is about 9%. Gas from the pyrolyser, which contains 3% of the input energy, has medium specific energy, and could be used with the steam for the combined-cycle generation of about 27 kWh electricity per 100 kg coal feed.

SIZE OF 10 t/h UNIT

From the specific feed-rates (Table 2) and the relative feed ratio (Table 3) of coal, steam, and air to the pyrolyser, gasifier, and combustor respectively, the cross-sectional areas and bed volumes were calculated for a conceptual plant with 10 t/h feed of LT char to the gasifier as indicated in Table 6. Such a unit would be comparable in size to a large industrial water-gas producer.

TABLE 3 OVERALL MASS AND HEAT BALANCES
(Basis: 100 kg Liddell coal)

Stream (see Fig.2)	M (kg)	T (°C)	H (MJ)	Q (MJ)	H+Q (%)
Input:					
1 Coal	100.0	20	-	3430	100.0
2 Ash	8.7	20	-	-	-
3 Air	12.9	20	-	-	-
20 Air	291.9	20	-	-	-
10 Water	95.9	20	-	-	-
4 LT char	77.9	500	48	2633	-
10 Steam	95.9	395	302	-	-
12 HT char	23.6	1000	22	781	-
Heat Carrier	9025	1050	537	-	-
Output:					
6 Tar	14.7	500	13.3	569	17.0
7 Liquor	4.3	500	14.4	-	0.4
8 Pyrol.gas	16.3	500	8.9	93	3.0
16 Prod.gas	126.1	200	45	2306	68.5
17 Moisture	23.7	200	26	-	0.8
29 Waste gas	319.6	200	75	-	2.2
25 Carbon	0.6	850	1	20	0.6
26 Ash	8.7	850	8	-	0.2
Total sensible heat		-	191.6	-	5.6
Total heating value		-	-	2988	87.1
30 Steam	46.9	280	140	-	4.1
Losses	-	-	110	-	3.2

Thermal efficiencies (cold):

16 Product gas (coal input basis)	67.2
16 Product gas (char input basis)	87.6
6,16 Product gas + tar(coal input basis)	83.8

TABLE 4 COMPOSITION OF COAL, CHAR, & TAR STREAMS IN FIG.2
(Liddell coal; % dmmf)

Stream	C	H	O	N	S	VM	$\hat{Q}_c$
1	83.6	5.7	8.2	2.0	0.5	40.4	34.3
4	88.5	3.9	5.0	2.1	0.5	19.5	33.8
6	82.5	8.5	7.6	1.0	0.4	-	38.7
12	96.7	0.3	0.3	1.3	0.3	0.5	33.1

TABLE 5 COMPOSITION OF GAS STREAMS IN FIG. 2 (% v/v)

Stream	H_2	CH_4	C_nH_m	CO	CO_2	O_2	N_2	$\hat{Q}_c$
8	4.5	10.5	4.3	1.3	18.1	-	61.5	7.3
16	54.0	3.9	-	32.1	8.9	-	1.1	11.6
28	4.0	-	-	5.4	18.5	-	72.1	1.1
29	-	-	-	-	17.6	1.9	80.5	-

TABLE 6 SIZE OF CONCEPTUAL 10 t/h UNIT (Liddell coal; pressure 1 atm)

		[P]	[G]	[CI]	[CII]
Solid feed		coal	LT char	HT	char
rate(dmmf)	(t/h)	12.8	10.0	3.03	0.75
Gas feed		air	steam	air	air
rate	(t/h)	1.7	12.3	30.5	7.4
Bed depth	(m)	4.0	4.0	4.0	1.0
Cross-sectional area	(m^2)	3.2	25.1	9.9	3.8
Bed volume	(m^3)	12.8	100.4	39.6	3.8
Pressure drop	(kPa)	7.06	37.7	37.7	9.41
Power requirement	(MW)	0.01	0.92	0.83	0.07
Electricity generated	(MW)	0.7*	-	1.5†	
Gas production(dry)	(t/h)	2.1	16.2	-	-
	(m^3(20°C)/h)	1618	26135	-	-
	(GJ/h)	11.9	296	-	-
LT tar	(t/h)	1.9	-	-	-

* 20% efficiency; † 30%; 2.7 MW combined cycle at 40%.

LT char has been taken as the basic feedstock since its composition and properties (see Table 4) are less divergent than those of the parent coal. The gasifier and combustor should give a reasonably constant performance and gas production rate (i.e. 0.63 Mm^3/d; 7104 GJ/d) using chars (240t/d) from a range of coals, both coking and non-coking. The yield of crude LT tar (incl. light oil) using Liddell coal should be about 46 t/d (4 m^3/d). The yields from other bituminous coals may be predicted from the graphical correlation of tar yield vs. volatile matter (Fig. 5) obtained from LOTEC results. Tannymorel high volatile coal would yield about 26.6% LT tar or 82 t/d from a 307 t/d coal feed (dmmf).

The power requirement of about 2MW, mainly for fluidizing, would be generated within the plant (2.7MW) by combustion of the pyrolyser gas, and by a waste heat boiler. This would also limit the temperature of the fluidizing air to about 500°C, thus simplifying construction of the air preheater.

CONCLUSIONS

1. Fluidization is eminently suitable for thermal coal processing, and could in future play a major role in the production of liquid and gaseous fuels from coal.
2. Integrated fluidized-bed schemes with well designed heat transfer and heat recovery systems are flexible and potentially very efficient, and deserve serious attention as a means of making the best use of available coal resources.
3. Experimental data from carbonization, gasification and combustion studies have enabled the sizes and probable performance of reactors in an integrated 10 t/h plant to be calculated. The results indicate that such a plant would not be unreasonably large and that larger units could be built using proven fluidization technology.
4. Production of liquid fuels from coal by combinations of pyrolysis, gasification plus synthesis, and hydrogenation of coal and crude tar, in quantities equivalent to those now derived from petroleum, would require many large integrated plants, involving high capital costs and coal consumption. Thus although plants of this type would be feasible and probably essential in the near future, they will not obviate the need for major changes in the scale and pattern of energy usage in the longer term.

ACKNOWLEDGEMENTS

Thanks are due to R. Neronowicz, A. Armstrong and R. Hamor for operation of the fluidized-bed test rigs which yielded the data used in this paper.

REFERENCES

Bloom, R.Jr. & Eddinger, R.T. (1974) Sixth AGA Synthetic Pipeline Gas Symposium, pp. 51-70, Chicago.

Bowling, K.McG. & Waters, P.L. (1962) Brit. Chem. Engng. 7, 98.

Bowling, K.McG. & Waters, P.L. (1968a) C.S.I.R.O., M.R.L. Invest. Rpt. 74

Bowling, K.McG. & Waters, P.L. (1968b) Mech. & Chem. Engng. Trans. I.E. Aust., MC4, 234.

Bowling, K.McG. & Waters, P.L. (1970) Fuel 49, 146.

Bowling, K.McG. & Waters, P.L. (1976) Inst. Fuel, Aust. Membership, Biennial Conference Proceedings, Nov. pp.7.1 - 7.14.

Bowling, K.McG., Brown, H.R. & Waters, P.L. (1961) J. Inst. Fuel 36, 99.

Hesp, W.R. & Waters, P.L. (1970) I & EC Product Research and Development 9, 194.

Jolley, L.J., Poll, A. & Stantan, J.E. (1953) 'Gasification and Liquefaction of Coal', pp.60-72, A.I.M.E.

Kunii, D. and Kunugi, T. (1975) U.S. Pat. 3,912,465. (Filed 30 July 1974).

Leva, M. (1959) 'Fluidization'. McGraw-Hill.

Rayner, J.W.R. (1952) J. Inst. Fuel 25, 50.

Standard Oil Development Co. and Arnold, C. (1948) Brit. Pat. 611, 924.(Appln. date 14 Feb. 1946.)

Waters, P.L. (1975) Inst. Fuel Symposium Series No. 1: Fluidized Combustion. C6-1-12.

Willson, W.G., Sealock, L.J., Hoodmaker, F.C., Hoffman, R.W., Stinson, D.L. & Cox, J.L. (1974) 'Coal Gasification'. (L.G. Massey Ed.) Advances in Chemistry Series, 131, pp.203-216. Amer. Chem. Soc., Washington.

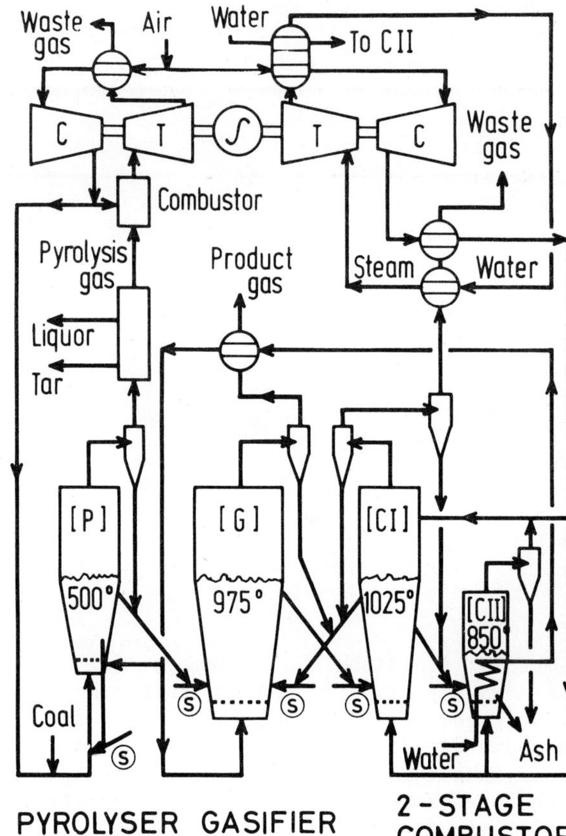

Fig.1 Basic FLUPROC design scheme

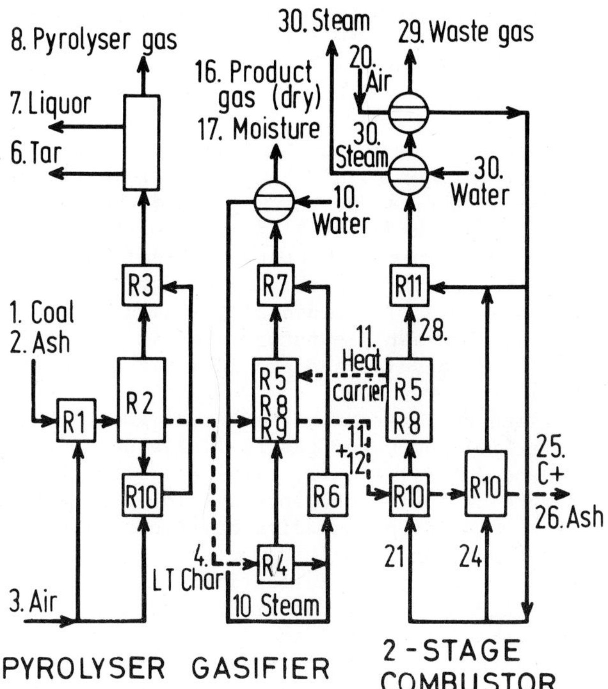

Fig.2 Sequence of reactions (Table 1) and materials flow (Table 3)

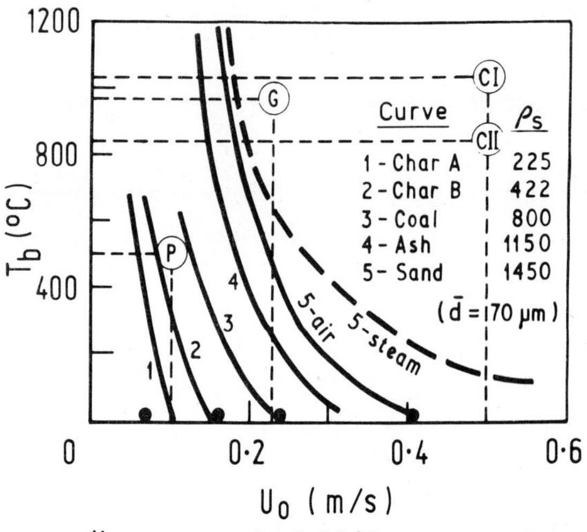

Fig.3 U_0 calculated as function of s and T_b (P,G,CI,CII show operating points for reactors).

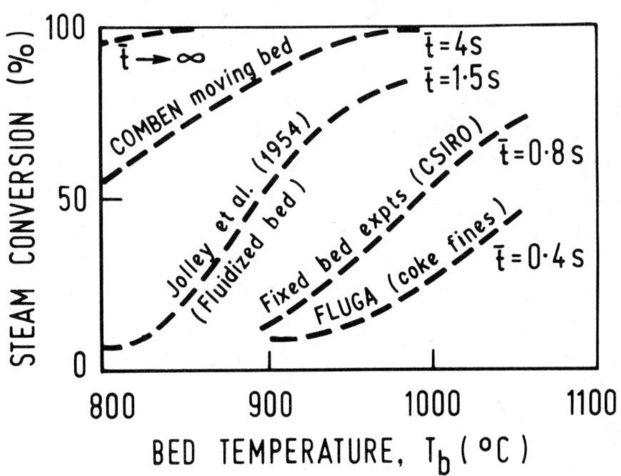

Fig.4 Approximate effect of T_b and $\bar{t}$ on steam conversion

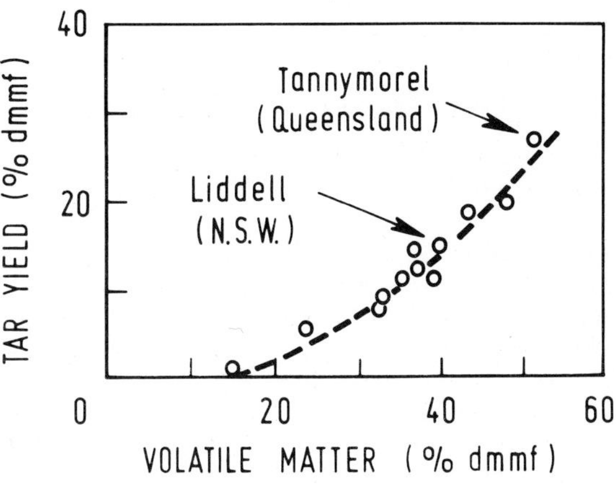

Fig.5 Tar (incl. light oil) yields from 12 bituminous coals (LOTEC results, interpolated for 0.03 oxygen/coal reactant ratio)

DESIGN AND OPERATING PARAMETERS FOR A FLUIDIZED BED AGGLOMERATING COMBUSTOR/GASIFIER

By W.C. Yang and D.L. Keairns

Westinghouse R&D Center

Beulah Road

Pittsburgh, Pennsylvania 15235

SYNOPSIS

Critical design and operating parameters for a commercial fluidized bed agglomerating combustor/gasifier were studied in a 30.5 cm diameter semi-circular transparent cold model, 5.6 m high. The design and operating parameters investigated were size and location of the air tube, air tube velocity, and different operating modes with different flow combinations. The effect of air tube size, air tube location, and air tube velocity on the solid circulation rate between the gasifier and the combustor, and on the performance of the unit are reported. A correlation for jet penetration at different air tube velocities into a fluidized bed is presented.

NOTATION

d_o	inside diameter of the jet nozzle or orifice
d_p	particle diameter
Fr	modified Froude number as defined in (1)
g	gravitational acceleration
L	jet penetration depth
U_o	jet velocity at exit of jet nozzle
ε	voidage
ρ_f	fluid density
ρ_p	particle density

INTRODUCTION

Westinghouse is developing a two-stage fluidized bed coal gasification process to produce low Btu gas for combined cycle power generation (Archer, et al., 1975). The first stage, a recirculating bed coal devolatilizer, has been successfully demonstrated in a pilot scale Process Development Unit (PDU) (Salvador and Keairns, 1976, 1977). Work is being carried out to demonstrate the feasibility of the second stage, a fluidized bed agglomerating combustor/gasifier.

The reactor is divided into three major sections (see Fig. 1). The space immediately above the air tube is the agglomerating combustor. The devolatilized char from the first reactor is combusted and the ash agglomerated in this region. The heat generated during the combustion is carried by recirculating solids into the gasifier. The agglomerated ash particles, being larger and denser, drop into the annular char-ash separator where a fluidized bed operated slightly higher than the minimum fluidizing condition strips out the unreacted char. Air is supplied through the air tube with some steam for temperature control. Additional steam is fed through the conical grid for gasification. The fluidizing medium in the annular char-ash separator is a mixture of steam and recycled gas provided via a sparger ring.

Design considerations for development of a commercial combustor/gasifier have been considered (Merry, et al., 1976) and initial commercial design concepts developed (Lemezis, 1975). Additional studies have been carried out to understand critical design parameters (such as the size and location of the air tube) and operating conditions (such as air tube velocity and the conical grid flow). Experimental results from ambient temperature model simulation studies are reported. The effects of air tube size, air tube location, and air tube velocity on the solid circulation rate between the combustor and gasifier and the effect of air tube velocity on jet penetration are investigated.

EXPERIMENTAL APPARATUS

Semi-Circular Model

The experiments were conducted in a semi-circular column constructed entirely from transparent Plexiglas. The bottom half is 28.6 cm I.D. and 3.05 m high. The column is expanded to an inside diameter of 40.6 cm at the top with a short conical transition section (see Fig. 1). The total column height is 5.6 m. A separate Plexiglas semi-circular piece with a conical grid at the top was inserted into the bottom half of the column to convert the column into the combustor/gasifier configuration. The simulated combustor section is 17.1 cm inside diameter and 1.6 m high. All air tight sealings were accomplished through O-rings and compressible gaskets. A 1.1 cm wide aluminum strip with a dove-tail groove was embedded in the front plate of this section to serve as a guide for the air tube.

Air Tube

Two sizes of air tube were used. They were semi-circular in cross-section and 3.8 cm and 5.4 cm in inside diameter. All were constructed from carbon steel. A 200 mesh screen was soldered onto the top of the air tube to prevent back flow of solid particles into the air tube during shutdown. A dove-tail guide was provided at the top of the air tube so that the air tube could slide along the dove-tail groove on the aluminum strip to guarantee the concentricity of the air tube. For ease of discussion, the air input through the air tube will be called the jet flow.

Conical Grid

The conical grid has 62 vertical holes of 1.6 mm in diameter and an included angle of 60°. The open area in the conical grid is 2.4% of the horizontally projected grid area. The grid surface was also covered with a 200 mesh screen to prevent weeping of bed material into the plenum. The air supplied through the conical grid is designated as conical flow.

Sparger Ring

The sparger ring is a 1.27 cm O.D. tube making a circular arc of $\sim$ 140°. Eighteen holes of 1.6 mm diameter were equally spaced and were alternately directed inwards and outwards at 45° downward from the horizontal center plane of the tube. A 200 mesh screen was also placed over the holes to keep out the particles to prevent plugging. The air supplied via the sparger is called the annular flow.

Bed Material

The bed material used was hollow epoxy spheres ranging from 1.78 to 3.81 mm in diameter with an average particle size of $\sim$ 2800 μm. The particle density is 0.21 gm/cm^3. This gives a particle/gas density ratio of 177 as compared to $\sim$ 200 under the PDU operating conditions of 1090°C and 1520 kPa with char (density $\cong$ 0.8 gm/cm^3) as the bed material.

EXPERIMENTAL PROCEDURES

Two air tube sizes (3.8 cm and 5.4 cm I.D.) and four air tube locations were tried. The top of the air tube was located 0, 15.2, 30.5 and 45.7 cm from the bottom of the conical grid respectively (designated as Y in Fig. 1). Four different flow combinations were tested. They were jet flow alone, jet flow and annular flow, jet flow and conical flow, and combinations of all three flows. For each flow combination, the conical and/or annular flow were held constant while the jet flow was varied. Two levels of conical flow were tried in the last combination.

Both particle velocity and jet penetration depth were observed visually. The tracer particles (similar material with different color) were followed for a distance of 15.2 to 30.5 cm with a stop watch. At least six separate readings were taken for each experiment and the average was taken as the final particle velocity. The particle velocities were always taken midway between the wall and the jet boundary.

Due to the fluctuating nature of the fluidized bed the jet penetration depth was not fixed. The jet penetration usually fluctuated between a minimum and a maximum position. These positions were observed visually and recorded. At higher jet velocities, the jet penetrated through the bed. The jet velocity was varied from 6 m/sec to 69 m/sec during the experiments.

EXPERIMENTAL RESULTS

Typical jet penetration data are presented in Figure 2 in a form suggested by Merry (1975) and Merry, et al. (1976) for the 3.8 cm air tube located at 30.5 cm from the conical grid. The maximum and minimum jet penetration depths are also shown for each data point. The amplitudes of this fluctuation increases with increase in jet velocity and with increase in fluidizing velocity. The earlier model used does not project the jet penetration observed over the full range of operating conditions projected. If the arithmetic mean of the maximum and minimum penetration is taken to be the nominal jet penetration depth, all the data can be presented according to Merry (1975) in Figure 3.

The deviation becomes progressively worse at higher jet velocities. However, the same set of data can be correlated reasonably well with a two-phase Froude number (Wallis, 1969) defined in Equation (1) as shown in Figure 4. The straight line in Figure 4 can be expressed as Equation (2). The d_o in (1) and (2)

$$Fr = \left(\frac{\rho_f}{\rho_p - \rho_f} \cdot \frac{U_o^2}{gd_o} \right)^{1/2} \quad \ldots (1)$$

$$L/d_o = 6.5 \; Fr \quad \ldots (2)$$

is taken to be the diameter of the semi-circular air tube. The correlation is compared with the jet penetration data available in the literature in Figure 5. The conditions under which the data were collected are summarized in Table 1. Three additional groups of data (two with 8 mm orifices and one with 3 mm orifice) by Basov (1969) which were not included in the earlier model development are included in Figure 5. The jet penetration data estimated from jet momentum measurements by Behie, et al (1970) and the data by Wen, et al (1977) for single jets in a two-dimensional bed are also included. Data by Shakhova (1968) with d_o/d_p of only 1.25 can not be correlated with the present correlation, though a straight line also resulted from the similar analysis. An experimental point estimated from thermocouple readings during the actual combustor/gasifier operation by Westinghouse in the PDU at 1038°C and 1520 kPa also compares favorably as shown in Figure 5. The same data points are plotted according to Merry (1975) in Figure 6 for comparison.

The solid circulation rate between the gasifier and the combustor is arbitrarily expressed as the solid particle velocity because the bed voidage was not determined and the cross-sectional area occupied by the emulsion phase in the vicinity of the jet could not be accurately determined. The particle velocities shown in Figure 7 are readings taken midway between the wall and the jet boundary. The discontinuity in the curve occurs at the point where the jet penetrates through the bed.

Higher particle velocities were observed when a larger air tube was used. This is due to the larger solid entrainment rate of the larger jets and smaller cross-sectional area occupied by the emulsion phase in the jet region. Only data for the air tube location at 15.2 cm from the conical grid are presented. Difficulty was encountered in measuring the particle velocity at lower air tube locations for the 5.4 cm air tube due to unstable operation.

DISCUSSION AND CONCLUSIONS

The jet penetration depth into a fluidized bed fluctuates over a wide range. The magnitude of fluctuation increases with increase in jet velocity and with increase in fluidizing velocity in the bed. The average jet penetration depth (an arithmetic average of the minimum and maximum jet penetration depths) can be successfully correlated with a two-phase Froude number, a ratio of jet inertial force to the gravitational force of the bed. Rigorously, the fluid bed density $\rho_p(1-\varepsilon)$ should be used in (1) in place of the particle density ρ_p. However, the accurate bed density is difficult to determine due to pressure fluctuation and experimental observation shows that the bed immediately around the jet region is essentially at minimum fluidization condition and ε can be assumed to be constant. Use of particle density alone, therefore, will not affect the final correlation.

The resulting correlation in this study is also compared with the jet penetration data in the literature. The scatter of the data from different authors may partly be attributed to the different techniques employed. Behie et al. (1970) measured the dissipation of jet momentum to arrive at the jet penetration depth. Basov et al. (1969) used a γ-densimeter to measure the bed density distribution in the zone near the grid. The jet penetration depth was then determined from the bed density measurement for the grids of different designs. The exact interaction between the neighboring jets on the grid was not known. Zenz (1968) performed experiments in a two-dimensional bed and visually observed the depth of the jet penetration. His data agree remarkably well with the present correlation which is also developed with penetration data obtained visually in a semi-circular bed. The effect of particle entrainment into the gas jet on penetration may be another important factor; however, it has not been quantified.

The reasonable agreement between the present correlation and the literature data leads to the conclusion that the jetting phenomenon in a fluidized bed is primarily dominated by the force balance between the inertial force of the jet and the gravitational force of the bed. The correlation is being used to design and to project performance for the air jet in the combustor and alternative gasification design concepts. However, for a case where the bed particle size approximates that of the jet nozzle size (Shakhova, 1968), the correlation is not recommended.

The solid circulation rate between the gasifier and the combustor is expressed with the particle velocity visually measured at a fixed location between the jet boundary and the wall. High speed movies were taken around the jet region for mapping the particle velocity profile in the vicinity of the jet to arrive at an integrated solid circulation rate. Unfortunately, the visibility of the individual particle was poor. The solid circulation rate depends linearly on the jet velocity and also increases with increase in air tube diameter. Different flow combinations also affect the solid circulation rate to some extent. The location of the air tube does not greatly affect the solid circulation rate. If the air tube was located too far down from the conical grid, the operation usually became unstable with slugging between the air tube and the conical grid. The lowest air tube location can be determined by estimating the jet half angle as proposed by Merry (1975) and the jet penetration depth as presented here. These observations and results are used to update and extend the design criteria previously developed.

ACKNOWLEDGEMENT

This work is being performed as part of the Westinghouse Coal Gasification Program and has been funded with federal funds from the Energy Research and Development Administration under contract EF-77-C-1514. The content of this publication does not necessarily reflect the views or policies of the funding agency.

REFERENCES

Archer, D.H., Keairns, D.L. & Vidt, E.J. (1975) Energy Communications 1, 115.

Basov, V.A., Markhevka, V.I., Melik-Akhnazarov, T. Kh. & Orochko, D.I. (1969) Int. Chem. Eng. 9, 263.

Behie, L.A., Bergougnou, M.A., Baker, C.G.J., & Bulani, W. (1970) Can. J. Chem. Eng. 48, 158.

Lemezis, S. "Advanced Coal Gasification System for Electric Power Generation, R&D Report 81-Interim Report 3", August 1975, NTIS No. FE-1514-T-9.

Merry, J.M.D. (1975) AIChE J. 21, 507.

Merry, J.M.D., Chen, J.L.P. & Keairns, D.L. (1976), "Fluidization Technology, Vol. II" (D.L. Keairns Ed.), pp. 423-436, Hemisphere Pub. Corp., Washington, DC.

Salvador, L.A. & Keairns, D.L. "Advanced Coal Gasification System for Electric Power Generation, R&D Report 81-Interim Report 5", October 15, 1976, NTIS No. FE-1514-57.

Salvador, L.A. & Keairns, D.L. "Advanced Coal Gasification System for Electric Power Generation, Quarterly Progress Report", January, 1977, NTIS No. FE-1514-61.

Shakhova, N.A. (1968) Inzh. Fiz. Zh. 14, 61.

Wallis, G.B. (1969) "One-Dimensional Two-Phase Flow", McGraw-Hill Book Company, New York.

Wen, C.Y., Horio, M., Krishnan, R., Khosravi, R., and Rengarajan, P., (1977) "Proceedings of The Second Pacific Chemical Engineering Conference", American Institute of Chemical Engineers, New York.

Yang, W.C. & Keairns, D.L. (1970) AIChE Symp. Ser. 70 (No. 141), 27.

Zenz, F.A. (1968) I. Chem. E. Symp. Ser. No. 30, 136.

Table 1 - Summary of the Literature Data Used for Comparison in Figure 5

Symbol	Bed Material	d_o, mm	d_p, mm	ρ_p, kg/mm	ρ_f, kg/mm	Reference
●	Cracking Catalyst	3	0.14	1000	1.2	Basov, et al. (1969)
▲	Cracking Catalyst	8	0.14	1000	1.2	Basov, et al. (1969)
▼	Cracking Catalyst	14	0.14	1000	1.2	Basov, et al. (1969)
■	Cracking Catalyst	20	0.14	1000	1.2	Basov, et al. (1969)
△	Cracking Catalyst	19.1	0.05	1000	1.2	Zenz (1968)
▽	Wheat	35	3.8	1380	1.2	Zenz (1968)
○	Cracking Catalyst	18.8	0.06	1000	1.2	Behie, et al. (1970)
◑	Sand	28.6	0.5	2640	1.2	Yang & Keairns (1970)
◐	Hollow Epoxy Spheres	28.6	2.8	208	1.2	Yang & Keairns (1970)
□	Lead Shot	19.1	1.0	11750	1000	Merry (1975)
◩	Lead Shot	19.1	2.0	11750	1000	Merry (1975)
★	Agglomerating Coke Breeze	59.0	0.74	1350	5.9	Westinghouse pilot plant data at 1038°C and 1520 kPa (1977)
×	Glass Beads	8	0.28	2410	1.2	Wen, et al (1977)

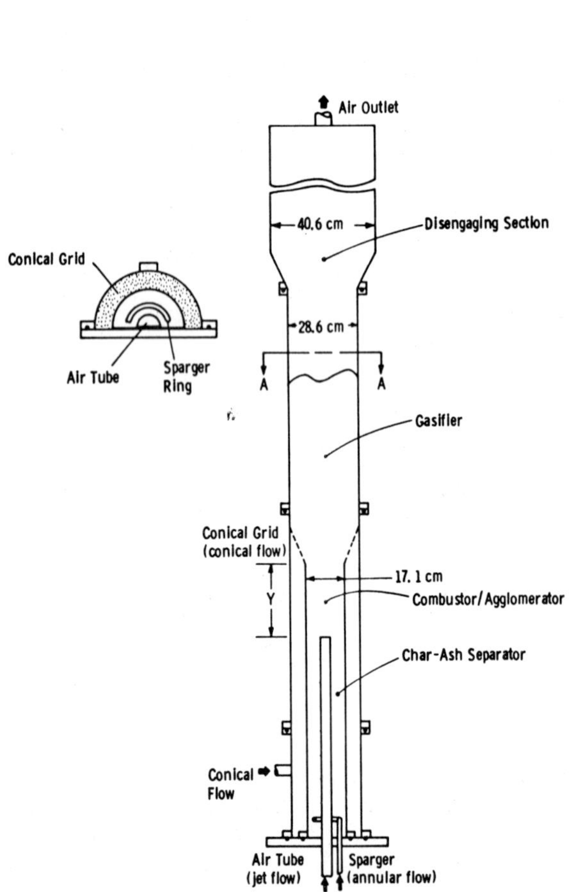

Fig. 1 - Schematic of the Test Apparatus

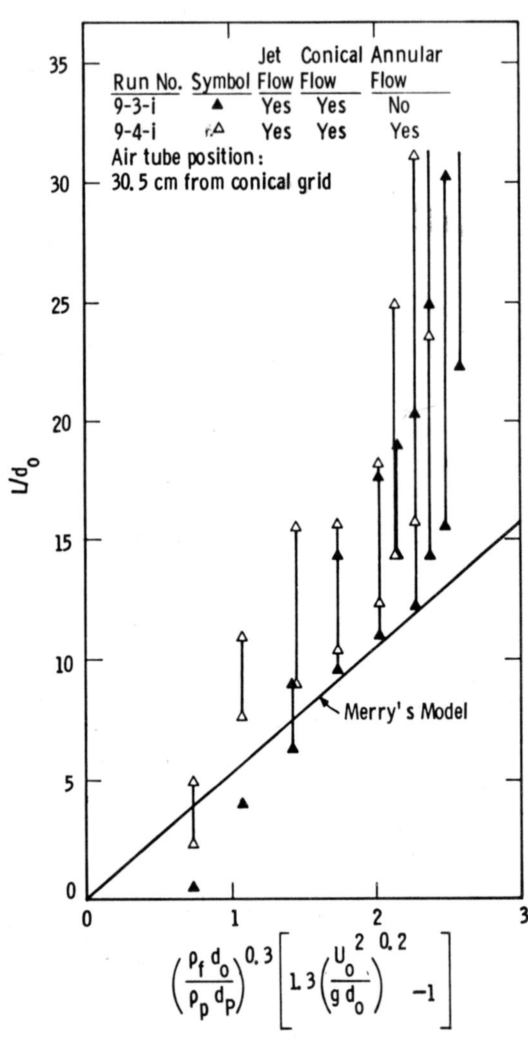

Fig. 2 - Actual Jet Penetration Data Presented as Suggested by Merry

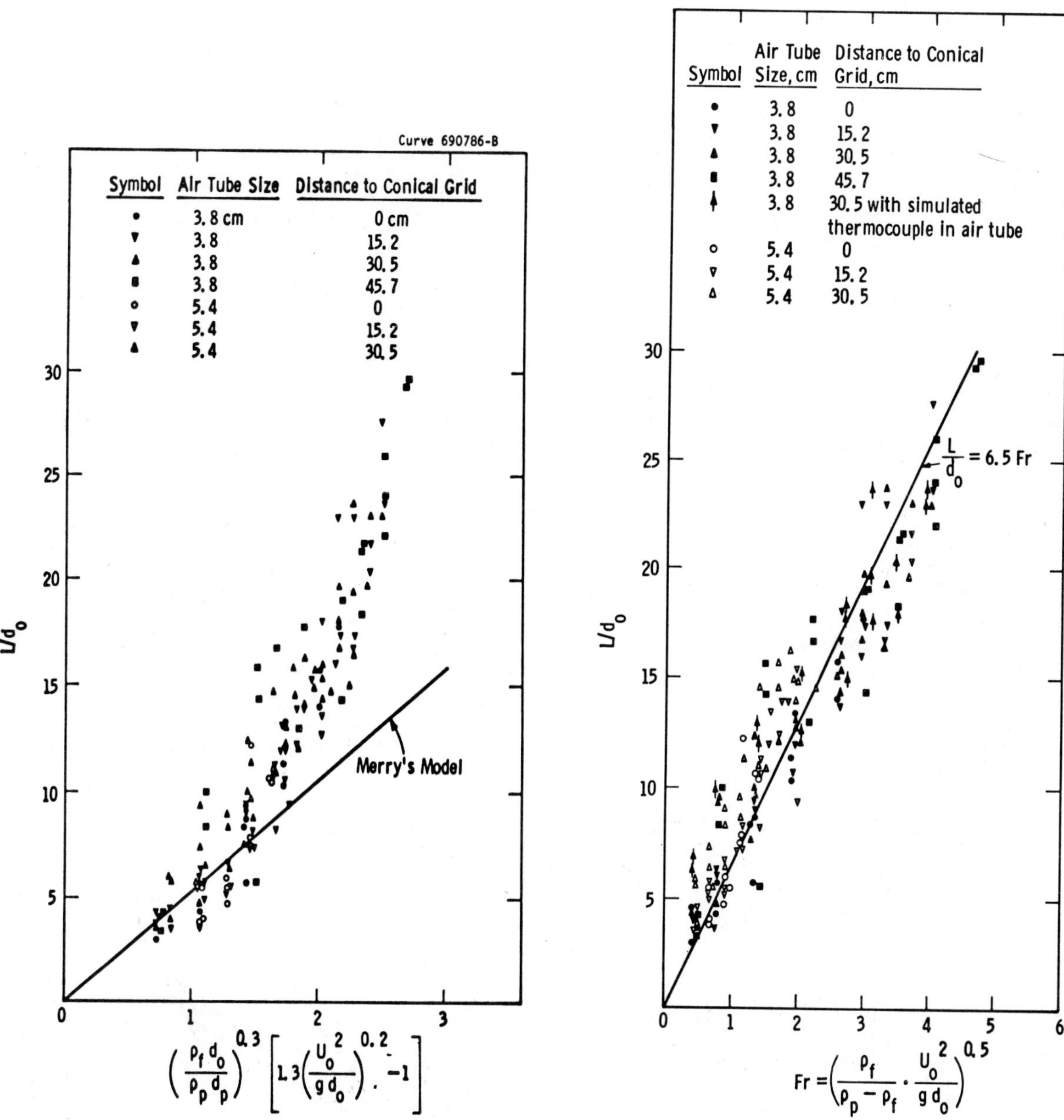

Fig. 3 - Comparison of Average Jet Penetration Depths Obtained in This Work With Merry's Model

Fig. 4 - Comparison of the Experimental Data With the Model Developed in This Work

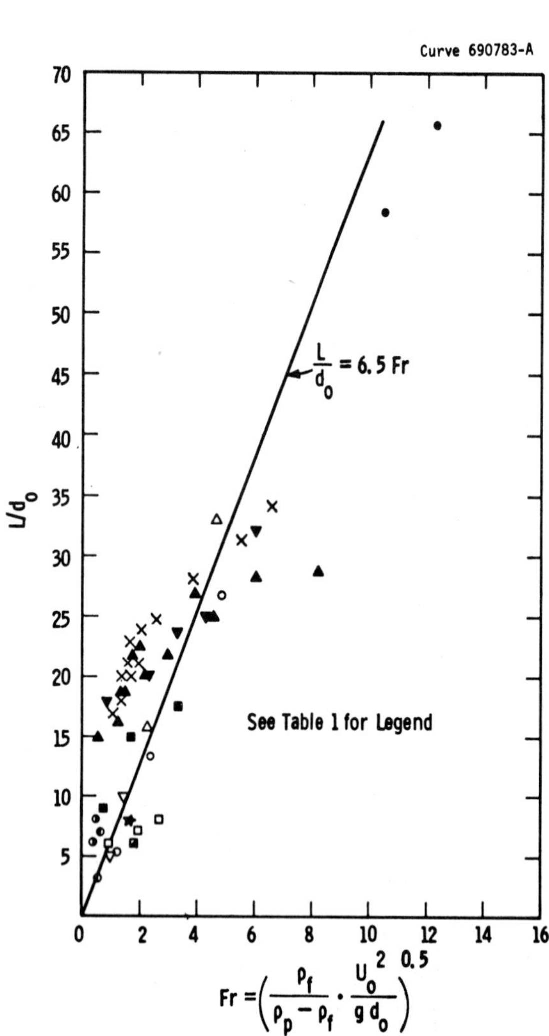

Fig. 5 - Comparison of the Literature Jet Penetration Data With The Model Developed in This Work

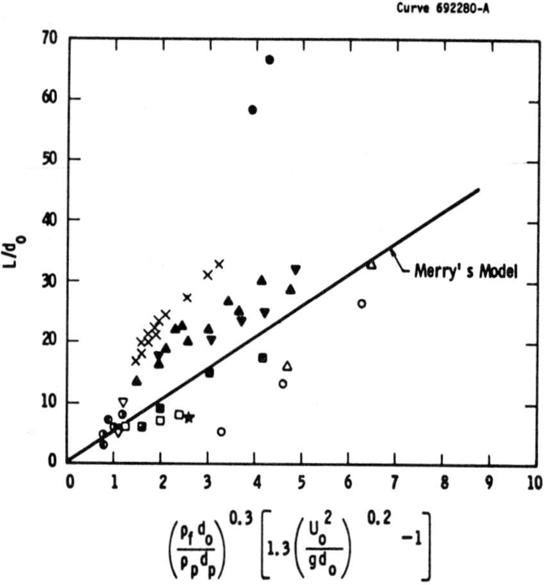

Fig. 6 - Comparison of the Literature Jet Penetration Data With Merry's Model

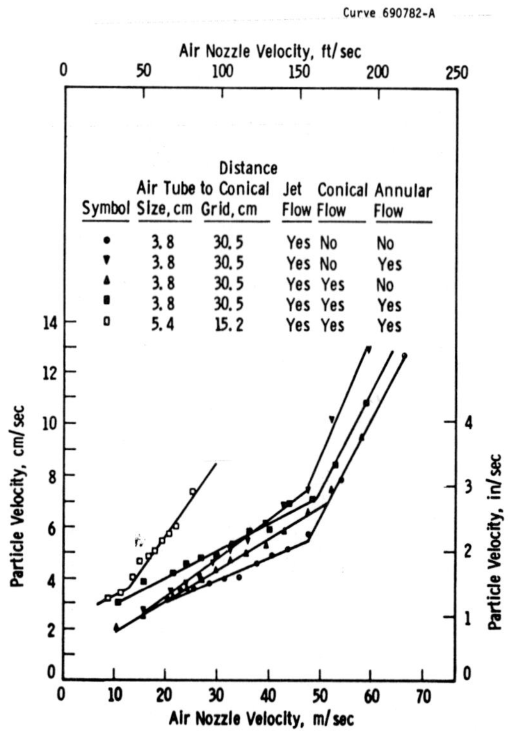

Fig. 7 - Experimental Solid Recirculation Rate Between the Gasifier and the Combustor

THE GASIFICATION OF COAL AND LIGNITE IN A FLUIDISED BED DESULPHURISING GASIFIER

By G. Moss

Esso Petroleum Company Limited

Esso Research Centre

Abingdon Oxfordshire

England OX13 6AE

SYNOPSIS

The Chemically Active Fluidised Bed (C.A.F.B) desulphurising gasifier development programme has reached the demonstration stage and a 20 MWe unit is, at the time of writing, being installed by Foster Wheeler Energy Corporation at the La Palma plant of Central Power and Light at San Benito in Texas. An essential feature of this gasifier is its ability to gasify Texas lignite. The paper describes the experimental work which has been done at Abingdon to establish the feasibility of operating the process with solid fuels. Details are given of the modifications which were made to the pilot plant in order to enable it to run on coal and of the experimental results which were obtained on Illinois No. 6 coal and Texas lignite.

INTRODUCTION

When the concept of the fluidised bed desulphurising gasifier was first developed in 1967, the intention was to provide an economic means for desulphurising fuel oil burned in large boilers. During the latter stages of the development programme the emphasis shifted, first to the utilisation of heavier residues and then to the utilisation of solid fuels, as it became clear that coal must rapidly displace oil in the power generation market.

The first coal gasification tests were made on a batch basis at the Esso Research Centre at Abingdon in 1975, and the results were sufficiently encouraging to justify modifications to the 7MM Btu/hr. continuous unit so that it too could be fuelled with coal. The immediate impetus for this work was provided by a requirement on the part of the Central Power and Light Company of Texas, that the 20 M.W. demonstration unit being installed in their La Palma station at San Benito (Rakes, 1976) should be able to operate on Texan lignite. A supply of this material was therefore shipped to Abingdon to enable a detailed assessment to be made of the problems which the gasification of this fuel might pose. The pilot plant at Abingdon has been described previously (Moss, 1970) (Moss and Tisdall 1973). Briefly it is a two reactor unit, fuel is partially oxidised at about 900°C in the gasifying reactor, within a bed of lime particles which is fluidised with air. Over 80% of the sulphur in the fuel is retained by the lime particles, in the form of calcium sulphide, whilst metals such as vanadium, nickel and sodium are also retained. Bed material is exchanged between the fluidised beds in the two reactors and, within the regeneration bed which is also fluidised with air, the calcium sulphide formed in the gasifier is oxidised to lime and SO_2, at about 1050°C.

Although the unit is conveniently called a gasifier in fact a considerable proportion of the fuel is not gasified but merely undergoes pyrolysis to produce substantial concentrations of condensible hydrocarbons in the fuel gas. When the fuel is oil the coke produced by cracking tends to form on the surfaces of the bed particles and is subsequently gasified. When the fuel is coal, the following factors operate to modify the operating conditions:-

Less hydrogen is available in the fuel for the formation of tars.

The char or coke tends to remain as discrete particles instead of being spread over the surface of the lime.

As they are oxidised the char particles are reduced in size until they are elutriated from the bed and pass through the cyclones to the burner.

The coal contains substantial amounts of

water and ash.

The fine char which is elutriated does not have to be gasified, whilst the water in the coal acts as an oxidant. Consequently, despite the less favourable hydrogen to carbon ratio of coal, the proportion of stoichiometric air which is required to operate the unit on coal is not all that much greater than when oil is used. The presence of ash and water in the fuel also tend to hold down the operating temperature.

The Batch Tests

(a) Equipment

The reactor which was used for the batch tests had an internal diameter of 18 cm and was well lagged with a sleeve of castable insulation between the refractory lining and the casing. A layout of the complete batch rig is shown in Fig. 1. The coal feeder was in fact the apparatus which was normally used to feed bed make-up material into the continuous gasifier. It incorporated a weighed hopper fitted with a vibrating tray feeder, whilst an air blast was used to introduce the coal feed into the gasifier. A lock hopper on top of the casing enabled the weighed hopper to be recharged during the course of a run. Air was fed to the reactor via a gas meter and passed through an electrical heater on its way to the plenum. An automatic steam generator was piped into the air line and drew its water from a weighed tank. A cyclone was provided between the gasifier and the burner but no provision was made for recycling fines.

(b) Test Results

The lack of fines recycle facilities imposed a limitation on coal utilisation. Nevertheless materials balances made using four different coals indicated that the major part of every coal could be gasified at temperatures in the region of 900°C and fluidising velocities in the region of 1.4 m/sec in a bed about 0.4 metres deep. The degree of desulphurisation appeared to approximately match the extent to which the coal was gasified, the best figure, based on the sulphur content of the bed materials was 64%. This figure was obtained using Illinois No. 6 coal under conditions in which 61% of the carbon in the coal appeared to be gasified. A later more detailed review of the data revealed that there had been an appreciable air leak in the air heater, consequently it followed that the extent of gasification had been slightly overestimated whilst the extent of desulphurisation remained unchanged, being based on stone analysis and coal feed rate. The size of the coals used in these experiments varied somewhat but was on the whole rather fine, the average particle size being in the range of 250-600 microns. The bed material on the other hand had an average particle size in the region of 1000 microns so there was a strong tendency for char to be elutriated from the bed and it was common for over 40% of the carbon to be removed with the cyclone fines. This suggested that it would be advantageous to use coarser coal and to install a reliable fines re-injection system. The gas combusted well with a luminous flame, no doubt due to the presence of tars and carbon particles. Some regenerations were accomplished by simply turning off the coal feed. Maximum SO_2 concentrations of up to 8.5% volume were observed during regeneration and also substantial reductions in the sulphur content of the stone, which was subsequently shown to have retained its sulphur absorbing capacity. Overall the results were considered to be sufficiently encouraging to justify a test programme under continuous operating conditions which would enable the long term effects of the coal ash on the reactivity and physical properties of the bed material to be assessed.

Operation of the Pilot Plant on Coal

(a) Equipment

Before the pilot plant could be run on coal two modifications were required. There were naturally no coal handling facilities since the unit had been designed to operate on liquid fuels, also the fines recycle system was intermittent in operation and unable to cope with the additional flow of char and ash which the batch tests had demonstrated.

It was anticipated that the unit would have a coal throughput in the region of 150 Kg/hr. This was rather too much for manhandling so an automatic coal handling system was devised which charged the lock-hopper of the old stone injection system with coal drawn from a 250 Kg dispenser which could be delivered to the unit by a fork-lift truck. One of these dispensers is shown in Fig. 2, it consisted of a 50 gallon drum, the removable lid of the drum being replaced by a conical hopper. This hopper incorporated pipes which enabled the coal to be drawn out of the inverted drum and to be delivered to the lock hopper of the coal feed system (which was the one which had been used on the batch units). The drums were filled with coal and when one was required it was closed with a hopper assembly which was mounted on a drum dolly. The assembled dispenser was then delivered to the unit, where it was inverted, pushed into position and coupled up.

The automatic filling system is shown in Fig. 3. The operating sequence of the system was controlled by two micro-switches fitted to the weight cell recorder. When the low limit switch was actuated it operated the solenoid valve V_3 and shut valves V_1 and V_2. When enough coal had drained into the weighed hopper to actuate the upper limit switch the solenoid was deactivated closing valve V_3 and opening valves V_1 and V_2. On closing, valve V_3 operated a third microswitch which switched on the industrial vacuum cleaner and moved another load of coal from the coal dispenser to the lock hopper. Overfilling of the lock hopper was prevented by the use of a level detector which switched off the vacuum cleaner when the hopper was full. If the level detector was not actuated within a set time after the vacuum cleaner had been switched on then an alarm indicated that the coal dispenser was empty and it was replaced. Since the coal was handled in a closed system a nitrogen purge was used to eliminate the possibility of a dust explosion.

The high flow rate cyclone drainage and fines re-injection system is shown in Fig. 4. This system was designed with the following attributes:-

Provision for the removal from the system of large pieces of coke which break loose from the cyclone walls during burnout.

Provision for fractionating the fines so that a coarse fraction may be re-injected whilst a finer fraction may be discarded.

Provision for re-injecting the coarse fraction of fines at a position remote from the gasifier cyclones and near the bottom of the gasifier bed.

As can be seen in the illustration, fines from the two gasifier cyclones enter a Y-shaped duct which drains into a pot, the contents of which may be fluidised by means of radially disposed gas nozzles. When the bed in the pot is not fluidised the fines accumulate in the Y-shaped duct, but when fluidising gas is provided the contents of the duct drain into the pot and those particles which can be fluidised will overflow a weir into an eductor, being conveyed to two small cyclones in series by a stream of gas which is circulated by a small blower. The fluidisation of the contents of the drain pot is intermittent and the bed is slumped when the height of solids held in the Y-shaped duct falls below a set level. In this way a seal is maintained and it is possible to maintain the pressure at which fines enter the eductor at a higher level than the pressure within the gasifier cyclones. The first cyclone which the collected fines enter has a relatively low efficiency and pressure drop. The coarser fraction of the fines held by this cyclone drains into a second eductor and is re-injected into the gasifier. The remaining fines pass into the second high efficiency cyclone, which is drained externally, and the heated gases are then cooled prior to being recirculated by the blower. Coarse particles entering the drain pot sink to the bottom of the bed and may be removed periodically through a large drainage valve. This system was found to function very reliably, but an abrasion problem arose with the impeller of the blower.

(b) Early Test Results

Two coals, Illinois No. 6 and Texas Lignite, were gasified during the first solid fuel test run of the continuous unit. Analyses of these coals are given in Table 1. In the first instance the unit was operated with heavy fuel oil and slowly switched to coal by decreasing the oil injection rate and increasing the coal injection rate. Subsequently, however, it was found that the unit could be started on coal without difficulty. Visual inspection of the flame through a viewing port at the back of the boiler which was fired by the gasifier revealed that the gasified coal produced a stable, yellow smoke-free flame indistinguishable from that obtained by burning gasified fuel oil. It was also possible to operate the plant with a smokeless stack.

A comparison of the gases produced using fuel oil and coals is shown in Table 2. As can be seen, the nitrogen contents were very similar but the coal gases contained substantially more hydrogen and less gaseous hydrocarbons than the oil based gas. This additional hydrogen no doubt arose from the substantial water contents of the coals.

The coals used in the initial continuous tests were considerably coarser in size than those used in the batch work though the coarsest was still only -1/8". Coal utilisation was reasonably good approaching 90% in one of the lignite tests, and it should be possible to improve on this performance. Desulphurisation performance was as good, if not better, than would have been obtained using fuel oil under similar operating conditions, e.g. bed depth, stoichiometric ratio, bed temperature etc. Table 3 shows a comparison between observed sulphur removal efficiencies and the values which would have been predicted for oil. The results obtained were all well within the E.P.A. New Source Performance Standard of 1.2 lbs SO_2 per million Btu. During these initial tests the carbon content of the bed material was relatively high and it was necessary to slow down the bed transfer rate in order to obtain the right carbon to sulphur balance within the regenerator. Under these conditions and at temperatures in the region of 1100°C, SO_2 concentrations in excess of 7% by volume were obtained, and it was demonstrated that regenerative capacity exceeded the rate at which sulphur was acquired by the gasifier.

The main conclusions drawn from the results of these tests were as follows:-

Both Illinois No. 6 and Texas Lignite can be gasified using the C.A.F.B. process to produce a low sulphur content gas which can be fired in a conventional boiler under smoke free conditions.

A gasifier designed to operate on coal may have to have a bed area about 30% greater than an oil fired gasifier of the equivalent capacity.

Satisfactory bed regeneration can be achieved when the gasifier is run on lignite.

The upper limit on coal feed particle size is in excess of 3000 microns.

At the time of writing further work had been undertaken with the objective of expanding the range of operating conditions under which coal gasifying performance could be assessed, and in particular to examine the long term effects of contact with coal ash on the reactivity of the bed material.

A major portion of the work reported here was funded by the United States Environmental Protection Agency whose support is gratefully acknowledged.

REFERENCES

Moss, G. (1970) Proceedings of the Second International Conference on Fluidised Bed Combustion, Henston Woods, Ohio, 'The Fluidised Bed Desulphurising Gasifier'.

Moss, G. & Tisdall, D.E. (1973) Proceedings of the Third International Conference on Fluidised Bed Combustion, 'The Design, Construction and Operation of the Abingdon Fluidised Bed Gasifier', EPA-650/2-73-053.

Rakes, S.L. (1976) Proceedings of the Joint Power Generation Conference of the ASME, Buffalo, New York, 'Development of the Chemically Active Fluid Bed Process, a Status Report and Discussion' 76-JPGC-Fu-4.

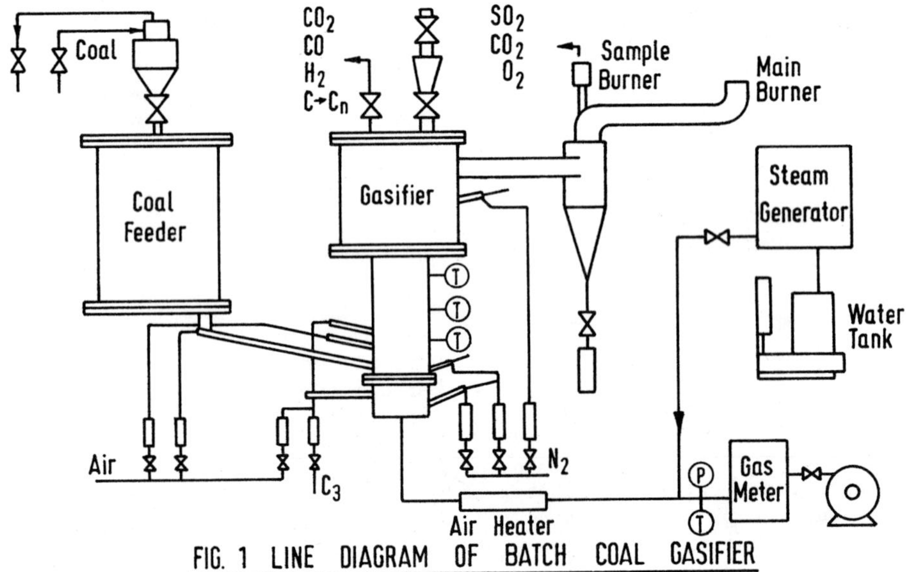

FIG. 1 LINE DIAGRAM OF BATCH COAL GASIFIER

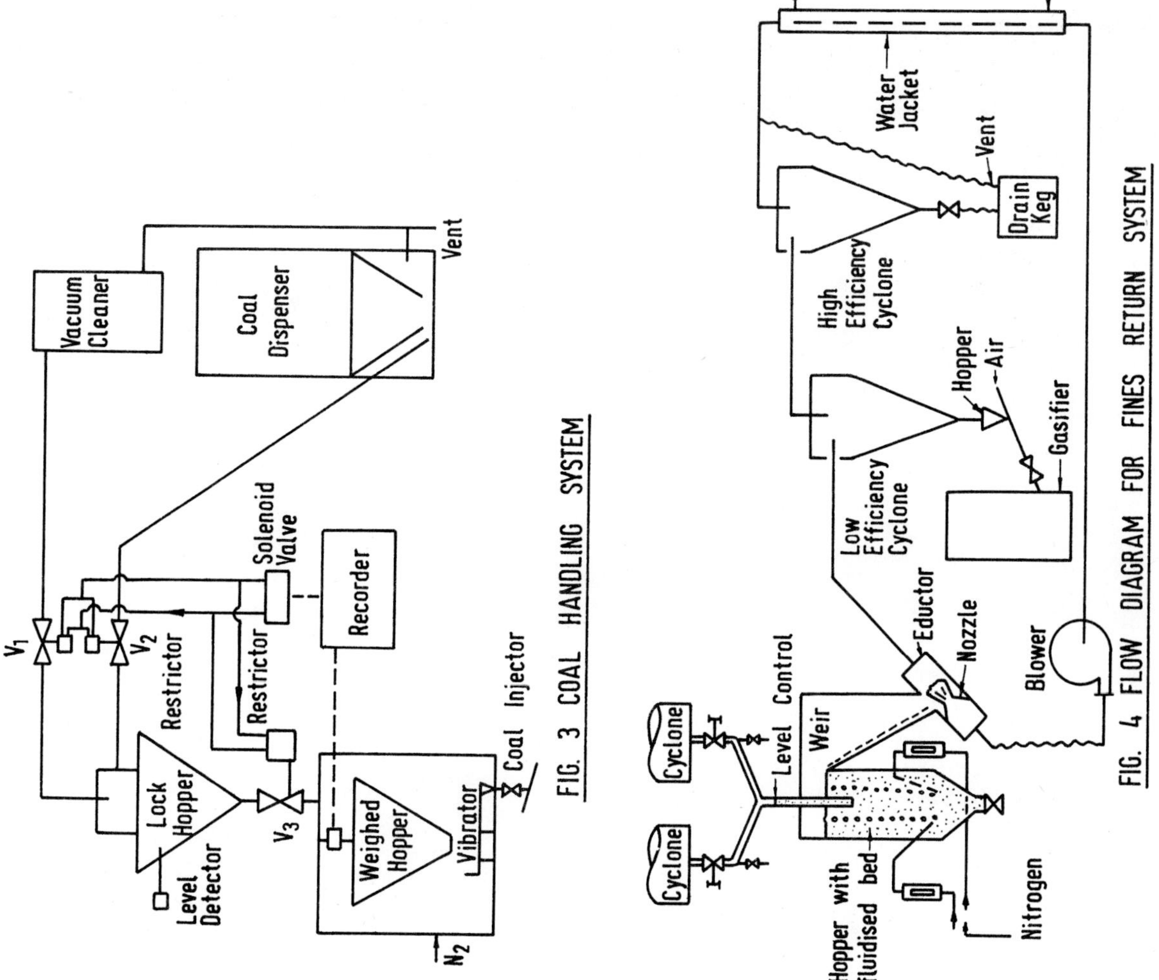

FIG. 3 COAL HANDLING SYSTEM

FIG. 4 FLOW DIAGRAM FOR FINES RETURN SYSTEM

FIG. 2 COAL DISPENSER

Table 1

ANALYSIS OF ILLINOIS NO. 6 AND TEXAS LIGNITE FED TO CONTINUOUS UNIT

	ILLINOIS NO. 6		TEXAS LIGNITE	
	ACTUAL %	DRY BASIS %	ACTUAL %	DRY BASIS %
Moisture	4.63	-	14.02	-
Ash	7.36	7.8	18.5	21.5
Carbon (corrected)	70.3	73.64	51.2	59.5
Hydrogen (corrected)	4.61	4.83	3.6	4.2
Sulphur (total)	1.78	1.9	0.82	0.95
Nitrogen	1.44	1.5	1.0	1.2
Oxygen + errors (by difference)	9.85	10.4	10.9	12.6
Gross calories/GM	6936	7260	5004	5817
Gross BTU/lb	12486	13068	9007	10470
CO_2%	0.16	-	0.51	-

Table 2

COMPARISON OF GASIFIER GAS QUALITY FROM FUEL OIL AND COAL

Gasifier Fuel	Heavy Fuel Oil	Illinois No. 6 Coal	Texas Lignite
Nitrogen + inerts	58.4	59.2	59.0
Carbon Monoxide	10.2	12.2	12.2
Carbon Dioxide	10.2	9.9	12.1
Methane	7.7	4.2	2.2
Ethylene	5.0	0.8	0.7
Ethane	0.1	0.1	0.1
Hydrogen	8.4	13.6	13.7

Table 3

SUMMARY OF DESULPHURISING PERFORMANCE DURING MINI-RUN ON COAL

Coal (type & size)	Mean % S.R.E. on coal	Expected Result on oil (% S.R.E)	SO_2 Emission (lb SO_2 per 10^6 BTU)
Illinois No. 6 (1405μ down)	74.8	73.8	0.73
Texas Lignite (800μ down & 1405μ down)	64.0	68.5	0.76
Texas Lignite (1/8" down)	82.4	67.2	0.42

HYDROGASIFICATION OF RHENISH BROWN COAL IN A FLUIDIZED BED

By L. Schrader and G. Felgener

Rheinische Braunkohlenwerke AG
Stüttgenweg 2, 5000 Cologne
Federal Republic of Germany

SYNOPSIS

In a pilot plant of a throughput of about 100 kg carbon per hour, the hydrogasification of lignite and hard coal in a fluidized bed reactor is investigated regard to the development of a new technology for hydrogasification of coal with high temperature nuclear reactors.

In the course of plant operation in 1976/77 the estimated data were verified. A degree of carbon gasification of over 70 %, corresponding to a degree of coal gasification of nearly 80 % and a methane content in the raw gas up to 40 % vol. were obtained.

NOTATION

a dimensionless	amount of coal gasified in the first reaction step
α dimensionless	degree of coal gasification
d_k cm	particle diameter
d_b cm	bubble diameter
ΔH kJ/mol	heat of reaction
k_d $cm.sec^{-1}$	mass transfer coefficient between fluid and a particle referred to the surface of the particles
$(K_d)_b$ sec^{-1}	overall rate constant for mass transfer referred to the volume of gas in the bubbles
$k_{v,1}$; $k_{v,2}$ sec^{-1}	rate constants for first or second reaction step referred to the volume of gas in the bubbles
$k_{s,1}$, $k_{s,2}$ $cm.sec^{-1}$	rate constants for first or second reaction step referred to the surface of the coal particles
p bar	pressure of gasifying agent (hydrogen)
t min	residence time of residual char
τ_1 ; τ_2 min	residence time needed for total conversion referred to first or second reaction step
Sh dimensionless	Sherwood number
$r_{v,m}$, $r_{v,D}$, $r_{v,c}$ sec^{-1}	rate referred to the volume of gas in the bubbles for mass transfer, pore diffusion or chemical reaction
$r_{s,m}$, $r_{s,D}$, $r_{s,c}$ $cm . sec^{-1}$	rate referred to the surface of the particles for mass transfer, pore diffusion or chemical reaction
D, D_{eff} $cm^2 . sec^{-1}$.	molekular diffusivity and effective diffusity, respecting the porosity of the particles
V_b cm^3	volume of the gas in the bubbles
S_c cm^2	total surface of the particles in the fludized bed
N_g gm-mol	moles of gas
$C_{g,b}$, $C_{g,p}$, $C_{g,s}$ $gm\text{-}mol . cm^{-3}$	Concentration of gas in the bubbles, at the particles or at the interior surface of the particles
ϱ cm	radius of reaction front in the particle

INTRODUCTION

The consumption of natural gas which already comprises a high percentage of the total energy demand is expected to increase much more in the future. On the other hand resources of natural gas cannot meet this demand on a long term basis. Therefore many companies and research institutes all over the world are engaged in improving conventional methods of gasifying coal and also developing new techniques to gasify hard coal and lignite to SNG.

The Rheinische Braunkohlenwerke, Cologne, is engaged in the hydrogasification of lignite and hard coal at high pressure and temperature using heat from high temperature nuclear reactors. This work is carried out in collaboration with four partners as a joint project ("Prototypanlage Nukleare Prozeßwärme (PNP)") which is essentially sponsored by the Ministry of Research and Technology of the Federal Republic of Germany.

A pilot plant with a throughput of about 100 kg carbon/hour for the hydrogasification of hard coal and lignite in a fluidized bed was designed and drafts for commercial scale units have been worked out. The pilot plant was erected in 1974/75 by the Rheinische Braunkohlenwerke AG, Cologne, together with an affiliated company, the Union Rheinische Braunkohlen Kraftstoff AG in Wesseling. The purpose of the pilot plant studies is to test the feasibility of hydrogasification of coal with nuclear heat as shown in Fig. 1 and Fig. 2 (Teggers, 1974; Schrader et al., 1975).

FLUIDIZING CHARACTERISTICS OF COAL

To get basic data for the pilot plant, tests on the fluidizing characteristics of coal were performed in glass tubes of 150 and 200 mm inner diameter using nitrogen als fluidizing agent. Nitrogen is very suitable to simulate the fluidization conditions in the gasification reactor, as this gas at 1 bar and 20 °C has quite similar fluidizing properties like hydrogen at 80 bar and 800 °C. This can easily be demonstrated by calculation of e. g. the minimum fluidization velocity: for both cases one gets nearly the same value.

The tests showed that predried Rhenish lignite (moisture 10-20 %) of a particle size of <1 mm could be fluidized with 2 to 3 cm/s gas velocity to a homogeneous bed almost without bubbles up to a gas velocity of 10 cm/s and a bed height of 400 cm. A number of 3 to 6 nozzles round the lower part of the tube wall proved nearly equivalent to other gas distributing systems. For good homogeneous fluidization an amount of about 30 % fines $<0,1$ mm particle diameter was found to be useful. The entrainment of fines was much lower than the calculated amounts; entrainment was only 2 % related to the whole bed contents.

BASICS OF HYDROGASIFICATION

Hydrogasification is the conversion of coal by hydrogen. This reaction proceeds in two steps at different velocities.

In the first heterogeneous reaction step, the easily separable groups in coal containing hydrogen and/or oxygen are converted or split according to the following reactions as for instance:

$$R\text{-}CH_3 + H_2 = R\text{-}H + CH_4 \quad ...(1)$$
$$R\text{-}COOH = R\text{-}H + CO_2 \quad ...(2) \quad \text{1st stage}$$
$$R\text{-}C{\scriptstyle \lessgtr^{O}_{H}} = R\text{-}H + CO \quad ...(3)$$

Some products of reaction, e.g. CO and CO_2 can react with hydrogen in the homogeneous gas phase to form additional methane.

In the second heterogeneous reaction step, the char formed in the first reaction is converted to methane as follows:

$$C + 2H_2 = CH_4;\ \Delta H = -86\ \text{kJ/mol C} \quad ...(4) \quad \text{2nd stage}$$

For the first heterogeneous reaction step, Wen and Huebler (1965) found at 1.700 °F (927 °C) a rate constant of about 2,0 $(bar.h)^{-1}$. From this for a hydrogen partial pressure of 80 bar a gasification rate of 3 min^{-1} can be calculated, i. e. about 5 % carbon conversion per second according to a first order reaction.

The amount of carbon gasified according to the first reaction step was found to be 0,48 for North Dakota lignite (Wen and Huebler, 1965) and 0,39 for Australian Yallourn brown coal (Birch, 1960); for Rhenish lignite, a value of 0,4 was assumed as its chemical analysis is very similar to Yallourn brown coal.

For the second reaction step, Birch (1960) reported rate constants of the order of 2.10^{-3} to 2.10^{-2} $(bar.h)^{-1}$ at 750 to 950 °C, corresponding to 0,3 to 3 % carbon conversion/minute at 80 bar.

MODELLING OF THE HYDROGASIFICATION REACTION

As the possibility of measuring characteristical data of the fluidized bed under the conditions of gasification was limited, some estimations for these conditions were performed.

Based on the evaluations of Davidson and Harrison (1963), Kunii and Levenspiel (1969) and our own tests the average bubble size in the fluidized bed reactor was estimated. For an average particle diameter of 0,2 mm bubble diameters of about 120 to 200 mm were calculated. Own experients, however, showed that up to a superficial gas velocity of ten times of the minimum fluidization velocity only small bubbles of about 10 mm diameter were observed. The mass transfer coefficients were calculated according to the bubbling bed model of Kunii and Levenspiel (1969).

Table 1 gives a summary of Sherwood numbers and mass transfer coefficients for the expected conditions of hydrogasification of Rhenish lignite.

$Sh_{overall}$ is here the Sherwood number for the overall mass transfer of gasification agent to the surface of all coal particles both in the bubbles and in the emulsion phase. $K_{d,overall}$ is the correspondant mass transfer coefficient based on the surface of the coal particles and $(K_d)_b$ is the mass transfer coefficient based on bubble volume according to the equations:

$$r_{v,m} = -\frac{1}{S_c} \cdot \frac{dNg}{dt} = k_{d,overall} \cdot (C_{g,b} - C_{g,p}) \quad (5)$$

$$r_{s,m} = -\frac{1}{V_b} \cdot \frac{dNg}{dt} = (K_d)_b \cdot (C_{g,b} - C_{g,p}) \quad (6)$$

The following data were used for calculation:

void fraction of fluidized bed: 0,73
void fraction of fluidized bed at minimum fluidization velocity : 0,58
diffusivity D calculated for the system H_2/CH_4 : 0,066 $cm^2 \cdot sec^{-1}$
shape factor of the coal particles : 0,6
average particle diameter dp : 0,2 mm
minimum fluidizing velocity : 1,2 cm/s
superficial gas velocity at gasification conditions: 15 cm/s
density of the coal: 1,3 g/cm^3

The fluidized bed was 300 cm high and had a diameter of 20 cm.
The characteristical data for the gases were calculated for 80 bar and 800 °C.

Under steady-state conditions the rate r_m for mass transfer of the gasification agent from the main gas stream to the exterior surface of the coal particles, the rate r_D of diffusion of the gasification agent in the pores of the coal particle, the rate r_c of the chemical reaction at the interior surface of the coal particle and the corresponding rates of diffusion and mass transfer of the product (here mainly CH_4) must be equal:

$$r_m = r_D = r_c = r \quad (7)$$

These rates can be referred to the volume of the gas in the bubbles, which represents the main part of the total gas stream through the gasifier:

$$r_v = -\frac{1}{V_b} \cdot \frac{dNg}{dt} = (K_d)_b \cdot (C_{g,b} - C_{g,p}) = \mathrm{Deff} \cdot \frac{Sc}{V_b} \cdot \frac{dCg}{d\varrho} = k_{c,v} \cdot C_{g,s} \quad (8)$$

By comparing the coefficients $(K_d)_b$, D_{eff} und $k_{c,v}$ a rough indication about the rate controlling step can be given.

A corresponding comparison can be done for rate equations referred to the total surface of the coal particles.

$$r_s = -\frac{1}{S_c} \cdot \frac{dNg}{dt} = k_{d,overall} \cdot (C_{g,b} - C_{g,p}) = D_{eff} \cdot \frac{dCg}{d\varrho} = k_{c,s} \cdot C_{g,s} \quad (9)$$

This second group of rate equations are applied for the unreacted core model, the first group of rate equations for the continuous-reaction model.

A comparison of the mass transfer coefficients and pore diffusion coefficients with the rate constants for both steps of the chemical reaction shows that, even for large bubble diameters occupying the entire cross-section of the bed of 200 mm diameter, chemical reaction is the rate controlling step, as can be seen from Table 2.

The rate constants for the two chemical reactions were used as the basis to estimate the degree of coal gasification. The calculations were performed for Rhenish brown coal with 4 % moisture content. The oxygen and hydrogen in the coal were assumed to be released with the "volatile matter". Calculations were carried out using the continuous reaction model and the unreacted core model for single particles with the simplifying assumption that first and second heterogeneous reaction steps are running parallel.

The additional calculations for the fluidized bed were performed according to the bubbling bed model of Kunii and Levenspiel (1969) under the simplifying assumptions that coal particles are of uniform size and hydrogen partial pressure is constant in the whole bed.

From these models one gets the following equations for α, the degree of coal gasification:

Continuous reaction model:

$$\alpha = a\frac{k_1pt}{1+k_1pt} + (1-a)\frac{k_2pt}{1+k_2pt} \qquad (10)$$

unreacted core model:

$$\alpha = a\left[3\frac{t}{\tau_1} - 6\left(\frac{t}{\tau_1}\right)^2 + 6\left(\frac{t}{\tau_1}\right)^3\left(1-e^{-\frac{\tau_1}{t}}\right)\right] + (1-a) \cdot \left[3\frac{t}{\tau_2} - 6\left(\frac{t}{\tau_2}\right)^2 + 6\left(\frac{t}{\tau_2}\right)^3\left(1-e^{-\frac{\tau_2}{t}}\right)\right] \qquad (11)$$

"a" is the amount of coal gasified according to the first fast reaction step. Because of the content of volatile matter and moisture for coal the value of "a" is higher than in a corresponding equation for the degree of carbon gasification; instead of 40 % as mentioned above we have now 65 %. For k_1p and k_2p on basis of values from literature as a first approach the values of 1 min^{-1} and 0,01 min^{-1} were assumed. Bench scale investigations yielded that for Rhenish brown coal the first reaction step is nearly finished after seven minutes (τ_1 = 7 min). τ_2 the time needed for total gasification according to the second step was estimated to 700 minutes under the supposition that both the continuous reaction model and the unreacted core model must yield nearly total gasification after the same residence time.

EXPERIMENTAL RESULTS

A series of tests were performed with Rhenish brown coal in the pilot plant with a throughput of about 100 kg C/h from spring 1976 to summer 1977. During this time the pilot plant was on operation for more than 5000 hours of which nearly 1300 hours comprised testruns with coal. The quantity of brown coal intake was more than 200 metric tons.

Fig. 3 gives an example of the operating conditions for one test run. In Table 3 some selected test data are presented.

DISCUSSION

In Fig. 4 a comparison of test results with model calculations is shown. As can be seen from this graph the experimental data can be quite well interpreted by the continuous reaction model. The unreacted core model doesn't fit the experimental data as well, but more test data should be evaluated to confirm this findings. The continuous reaction model also seems suitable because of the porous structure of both dried brown coal and residual char, the high diffusion rate of hydrogen, and good mass transfer in the fluidized bed. On the other hand, the suitability of the model for low residence times, i. e. high coal through-put rate has not yet been investigated.

In the near future a test program for hard coal is envisaged.

CONCLUSIONS

1. The hydrogasification of Rhenish lignite in a fluidized bed can be performed at 80 bar and 800 to 900 °C.
2. The composition of raw gas in respect to content of methane and the degree of coal gasification meet the basis estimations for a commercial scale unit for hydrogasification of brown coal with nuclear heat.
3. The results can be interpreted with a simplified model calculation based on the continuous reaction model and the bubbling bed model.

REFERENCES

Teggers, H. (1974) BNES International Conference, London, November 26-28,

Schrader, L., Strauß, W., Teggers, H. (1975) Nuclear Engineering and Design, Vol. 34, No. 1, pp. 51-57

Wen, C.Y. & Huebler, J. (1965) Ind. Eng. Chem., Process Design and Development, Vol. 4, No. 2, pp. 142-147

Birch, T.J., Hall, K.R., Urie, R.W. (1960) J. Inst. Fuel, Sept. 1960, pp. 422-435

Davidson, J.F., Harrison, D. (1963) Fluidized Particles, Cambridge University Press, New York

Kunii, D., Levenspiel, O., (1969) Fluidization Engineering, J. Wiley & Sons Inc.

Schrader, L. (1975), ICHT, International Seminar, Dubrovnik, Paper 8.1

Table 1. Mass transfer in a fluidized bed for hydrogasification of Rhenish lignite in relation to bubble diameter d_b

d_b	(cm)	0,5	1	5	10	20
$Sh_{overall}$	(-)	$7{,}0 \cdot 10^{-2}$	$3{,}6 \cdot 10^{-2}$	$1{,}2 \times 10^{-2}$	$8{,}2 \times 10^{-3}$	$5{,}7 \times 10^{-3}$
$k_{d,\ overall}$	($cm.sec^{-1}$)	0,26	0,13	0,04	0,03	0,02
$(K_d)_b$	(sec^{-1})	62	45	33	32	32

Table 2. Comparison of rate constants for different possible rate controlling steps

rate controlling step	chemical reaction 1st stage	2nd stage	pore diffusion	mass transfer
rate constant (sec^{-1}) related to volume of gas in the bubbles	$k_{v,1} \sim 10^{-1}$	$k_{v,2} \sim 10^{-4}$	$D_{eff} \cdot \frac{S_c}{V_b} \cdot \frac{2}{d_p} \sim 10^2$	$(K_d)_b \sim 30$
rate constant (cm sec^{-1}) related to surface of the coal particles	$k_{s,1} \sim 10^{-3}$	$k_{s,2} \sim 10^{-5}$	$D_{eff} \cdot \frac{2}{d_p} \sim 1$	$k_{d,overall} \sim 2 \times 10^{-2}$

Table 3. Test run data for hydrogasification of Rhenish lignite

Gasification conditions	Dimension	1	2	3	4	5	6
Gasification temperature	^oC	820	850	865	890	920	920
Gasification pressure	bar	80	80	80	80	80	80
Height of the fluidized bed	m	2,8	2,6	2,7	2,7	2,6	2,7
Input of brown-coal (mf)	kg/h	133	279	300	265	212	252
Grain size	mm	<1	<1	<1	<1	<1	<1
Input of hydrogen	m^3/h	435	376	384	435	382	390
Results							
Coal conversion	%	78,4	61,7	63,2	65,4	71,4	71,1
Carbon conversion	%	73,0	48,0	50,6	54,8	63,2	62,1
Composition of raw gas (dry N_2-free) CH_4	% vol.	24,2	29,0	34,2	31,2	34,3	41,6
C_2^+	% vol.	1,5	4,6	3,4	2,3	2,5	2,1
H_2	% vol.	70,9	59,0	52,8	58,6	55,2	46,5
CO_2	% vol.	0,5	2,3	3,1	1,7	1,9	2,1
CO	% vol.	2,9	5,1	6,5	6,2	6,1	7,7
Yield of hydrocarbons	$\frac{C \text{ in } C_1^+}{C \text{ conv.}}$	0,89	0,84	0,81	0,82	0,83	0,82
Residence time of char	min	53	14	16	19	28	20

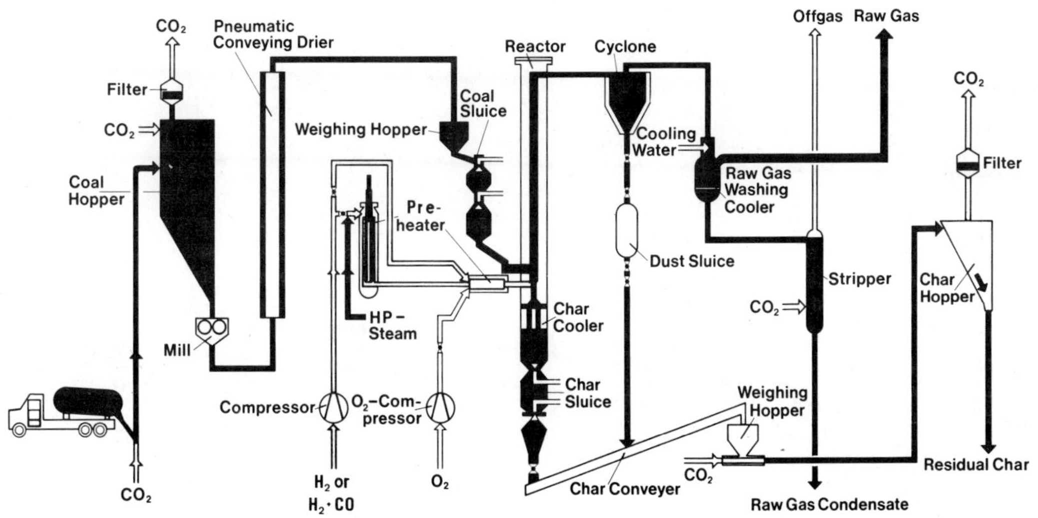

Fig. 1. Pilot plant for hydrogasification of coal

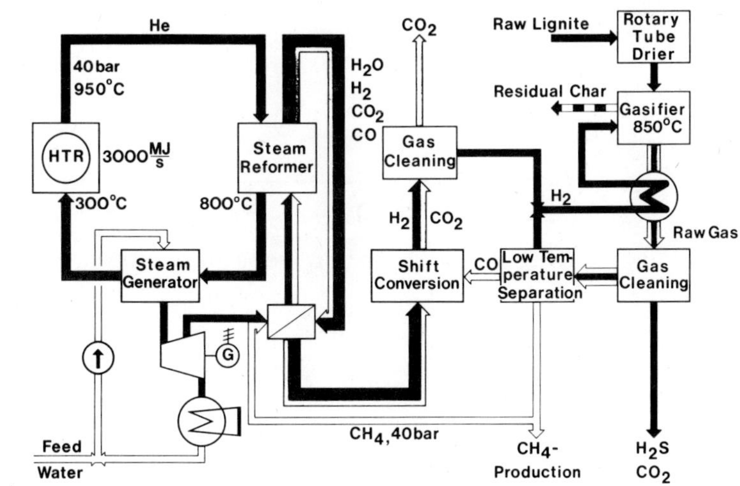

Fig. 2. Hydrogasification of lignite with nuclear heat

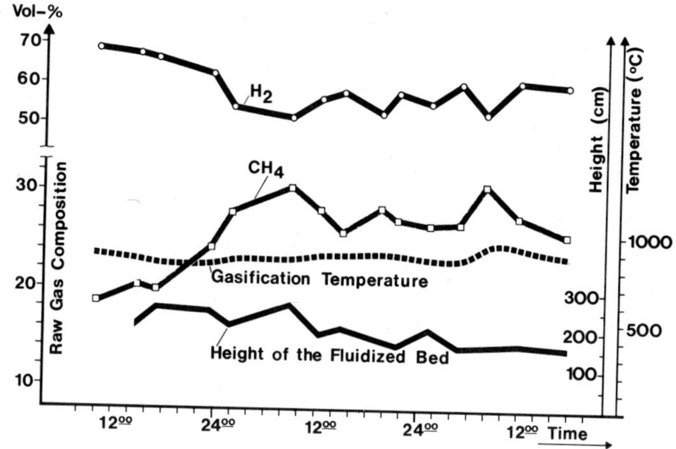

Fig. 3. Example for a test run

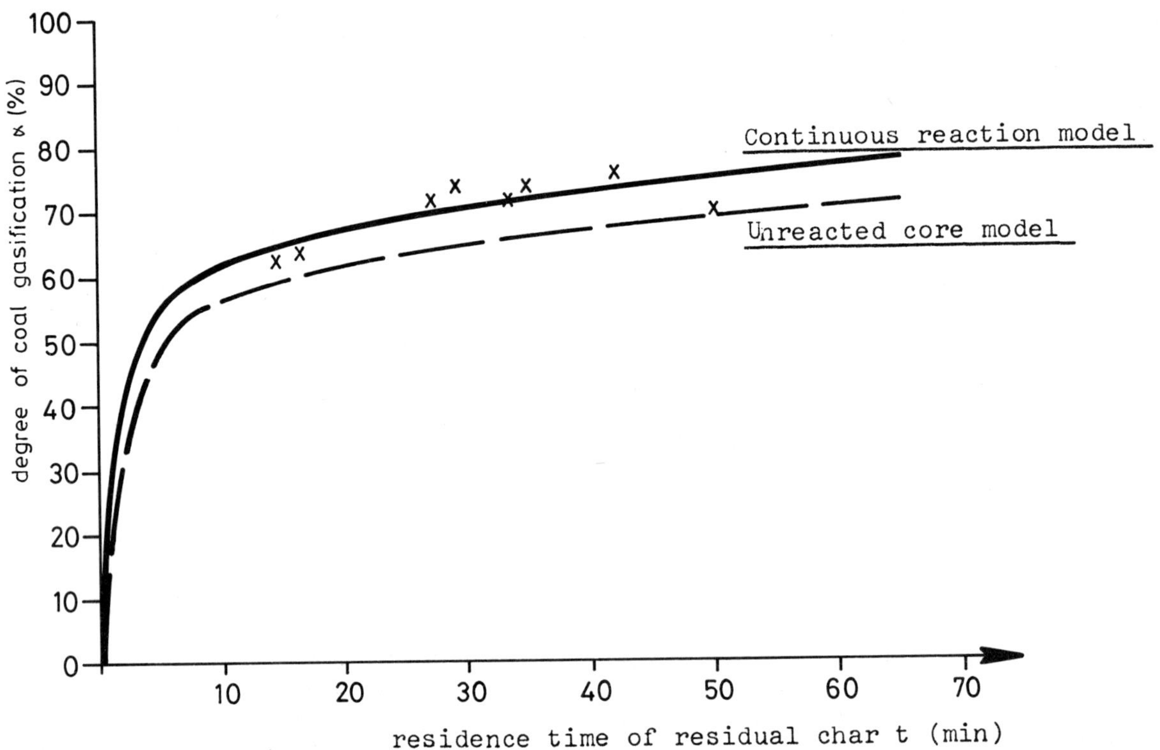

Continuous reaction model:

$$\alpha = a\,\frac{k_1pt}{1+k_1pt} + (1-a)\,\frac{k_2pt}{1-k_2pt}$$

Unreacted core model:

$$\alpha = a\left[3\,\frac{t}{\tau_1} - 6\left(\frac{t}{\tau_1}\right)^2 + 6\left(\frac{t}{\tau_1}\right)^3\left(1-e^{-\frac{\tau_1}{t}}\right)\right] + (1-a)\left[3\,\frac{t}{\tau_2} - 6\left(\frac{t}{\tau_2}\right)^2 + 6\left(\frac{t}{\tau_2}\right)^3\left(1-e^{-\frac{\tau_2}{t}}\right)\right]$$

Gasification conditions: 840 to 860 $^{\circ}$C, 80 bar

Test Results: X

Coefficients for model calculation:

continuous reaction model: k_1p = 1 min^{-1}; k_2p = 0,01 min^{-1}; a = 0,65

unreacted core model: $_1$ = 7 min ; $_2$ = 700 min ; a = 0,65

Fig. 4 : Comparison of test results for hydrogasification of Rhenish brown coal with model calculations

STUDIES OF THE GASIFICATION OF SOLID FUELS IN A FLUIDISED BED AT ATMOSPHERIC PRESSURE

By J.T. Shaw and N.P.McC. Paterson

Coal Research Establishment, Stoke Orchard, Cheltenham, Gloucestershire GL52 4RZ, UK

The paper shows that the fluidised gasification of some chars, coals and coal processing residues provides a means of generating a fuel gas whose calorific value is at least 3.8 MJ/m^3, i.e. is high enough for gas turbine work. The effect of varying the experimental conditions is described. Evidence is adduced that when devolatilised char is gasified in dry air a reaction between carbon and oxygen as well as one between carbon and carbon dioxide can contribute to the output of carbon monoxide.

INTRODUCTION

As a result of the energy crisis, studies of the gasification of solid fuels have been resumed in many countries. At the Coal Research Establishment of the National Coal Board at Stoke Orchard, England, the European Coal and Steel Community has funded work on the production of a low-calorific-value gas suitable for fuelling gas turbines that would drive electrical power generators in combined cycles of high thermal efficiency. The calorific value required is about 3.8 MJ/m^3. There are three main ways of making a fuel gas from the solid. In the first, the reactions take place in a "fixed" (actually, a descending) bed of fuel; in the second, they take place in a fluidised bed; and in the third they take place in an entrained system. The fluidised bed method enables us to use the cheapest fuel readily available in Great Britain without incurring temperatures over about 1000°C, and to obtain the benefits of sulphur retention. For making a gas of low calorific value i.e. about 3.8 MJ/m^3, air rather than oxygen can be used, hence the expense of oxygen making can be saved. We investigated reactions of the fuel with air, and also with carbon dioxide, steam and steam-air mixtures.

The fuels that might be used commercially include coals of the types currently being supplied to electricity generating stations. In continental Europe lignite might be used. Residues from various new processes using coal may become commercially available in the future. We have investigated the use of examples of all these, and some of them were very successful. At the outset, however, in order to be certain of avoiding problems with agglomerating fuel, clinkering in the bed, and tars in the product gas, and to obtain a consistent fuel for kinetics studies, we used Phurnacite breeze which consists mainly of coal char. Though lignite and some coal processing residues are very reactive towards gasification, most coals and the chars to which they give rise are rather unreactive which makes it hard to achieve total gasification without going to an undesirably high temperature. Fortunately, there is evidence that as an economic proposition, combined cycles embodying partial gasification are preferable to those involving total gasification. The ungasified char would be burnt to raise steam or to propel a second gas turbine.

Work up to May 1977 is described on how the gasification of Phurnacite char was affected by the various fluidising gases, temperature, fuel feed rate, fluidising velocity and bed height. The results of our tests using other fuels are summarised. For gas turbine work it is desirable to gasify at raised pressure, but there is not space here to report our preliminary work on the effects of raising the pressure or those of altering major dimensions of the reactor.

EXPERIMENTAL

Fuels

These were Phurnacite breeze, Coalite breeze, lignite, Gedling and Dawmill coals, and the residues from the supercritical extraction of Dawmill coal and of a mixture of Markham Main and Dawmill coals. Analyses of the fuels are presented in Table 1.

Plant

This is shown in Figure 1. The fluidised

bed reactor was a vertical steel cylinder 0.15 m in diameter and 2 m high. The expanded bed height ranged from 0.2 m to 0.9 m and there were no baffles or other fitments in the bed. The fluidising gases were mixed before being led into a windbox beneath the distributor, steam being metered as water which was then boiled. A flat distributor of the fluidising gas was used at first but was changed for a conical one with a central outlet for discarded bed material. This raised by about 100°C the temperature above which clinkering could become a nuisance. The reactor was fitted with electrical heaters and with removable lagging. Fuel was fed from the top through a rotating valve and a dip-leg both of which were cooled. Solids carried up in the gas leaving the bed were caught in a pair of cyclones in series, and the cleaned gas was burnt so that it could be safely vented to the atmosphere. The product gas was sampled from a point between the secondary cyclone and the flare and after further cleaning it was continuously analysed for carbon monoxide, carbon dioxide, hydrogen, methane, ethylene and oxygen. Occasional samples were taken for checking by gas chromatography. There were facilities for determining the tar, hydrogen sulphide and "alkaline aerosol" content of the products, and the height and temperature of the bed.

Experimental procedures

When the bed was fluidised with air its temperature tended to rise, but when mixtures containing little or no oxygen were used the temperature tended to fall despite the electrical heaters and lagging. However, the rates of change of temperature were slow in comparison with the response and lag times of the analytical instruments. The normal procedure was, therefore, to raise the temperature to about 1000°C using air, then to change to some other mixture of gases and to allow the temperature to fall whilst making continuous analyses of the products.

RESULTS AND DISCUSSION

The first five parts of this section deal with Phurnacite breeze and the last part deals with the other fuels.

The carbon-carbon dioxide and carbon-steam reactions

These were investigated in various mixtures of carbon dioxide or steam with nitrogen, whilst the temperature was slowly falling after having been raised to about 1000°C using air. To avoid confusion by pyrolysis reactions, no fuel was supplied after the change of fluidising gas. The results for steam are shown in Figure 2. Those for carbon dioxide showed similar trends but the reactivity was lower. The apparent activation energies were about 110 kJ/mol for steam and about 150 kJ/mol for carbon dioxide. The shapes of the curves in Figure 2 harmonise generally with the reaction mechanism expounded by Ergun (1962).

Gasification in dry air (21% oxygen, 79% nitrogen by volume)

These tests were conducted with fuel that had been devolatilised at 1000°C. No fuel was supplied after the initial charge. In air there was a considerably greater output of carbon monoxide than had been obtained when using a mixture containing 21% carbon dioxide, 79% nitrogen by volume but in otherwise similar conditions. The results are compared in Figure 3. Probing with thermocouples negatived the hypothesis that during gasification in air the bed contained a 'hotter zone' in which the reduction of carbon dioxide could have been accelerated by enough to account for the difference. The possibility that the reduction of carbon dioxide was accelerated in some other way, is not excluded but such an explanation meets difficulties in accounting on its own for the prominent hump in Curve 1a shown in Figure 3. The following explanation accounts for the results quite simply.

An air-blown gasifier contains an oxidation zone and a reduction zone if the bed of fuel is hot enough and deep enough. In the oxidation zone the combustion of carbon gives rise to carbon dioxide but also to carbon monoxide which is not all burnt: some escapes into the reduction zone.

We consider four main reactions. The shape of Curve 1a in Figure 3 shows how their relative rates change with temperature. The reactions are:-

(1) Combustion of carbon to carbon dioxide;
(2) Combustion of carbon to carbon monoxide;
(3) Combustion of carbon monoxide to carbon dioxide;
(4) Reduction of carbon dioxide by carbon to carbon monoxide (Boudouard's reaction).

If combustion does not accelerate Boudouard's reaction in some unknown way, then at a given temperature the difference between Curves 1a and 2 of Figure 3, expressed as a mass flow of carbon monoxide, shows how much monoxide originated through Reaction 2 and survived Reaction 3. In principle corrections are needed because the presence of the combustion zone leaves less room for the reduction zone in a given bed, and because if there is net formation of carbon monoxide in the combustion zone this affects both the driving force (carbon dioxide concentration) and the retarding force (carbon monoxide concentration) of Boudouard's reaction in the reduction zone (Ergun, 1962). These corrections would all show that above about 800°C the real contribution by Reaction 2 is larger than it appears to be on Figure 3. Incidentally, Reaction 2 entails access by oxygen to the carbon surface.

The downslope of the hump in Curve 1a between about 760°C and about 800°C shows that Reaction 2 is being overtaken by Reaction 3; and the subsequent levelling-off and upturn reflects the increasing importance of Reaction 4. (This reaction takes place throughout the bed whereas the other three occur only where oxygen is present).

Arthur (1951) determined the relative rates with which oxygen reacted with carbon to form carbon monoxide and carbon dioxide, at temperatures between 480°C and 900°C, in the presence of an inhibitor of carbon monoxide combustion, as follows:

$$CO/CO_2 = 10^{3.34}\ e^{-51549/RT}$$

where R = 8.31 kJ/k mol °K. Table 2 shows the carbon monoxide/carbon dioxide ratios obtained from this equation at four temperatures of interest, in comparison with the ratios we found. The third column shows the fraction of the amount formed that must have been burnt in our work, if Arthur's equation is assumed to be applicable.

Gasification in air with continuous addition of fresh fuel

The effects on the product gas composition of varying the fuel feed rate, the bed height, and the fluidising velocity whilst keeping the temperature at 1000°C are shown in Figures 4 to 6. The calorific value of the product gas is shown in relation to temperature in Figure 7.

The changes in fuel feed rate hardly altered the output of carbon monoxide; but that of hydrogen which in the absence of steam would come entirely from the pyrolysis of the freshly added fuel, varied more or less in proportion to the fuel feed rate. With a bed height of 0.6 m and a fluidising velocity of 0.46 m/s at 1000°C about 40% of the available hydrogen was burnt. The rest was presumably released in that part of the bed not reached by the air.

The changes of bed height gave curves whose form resembled but was not identical with that expected from Ergun's discussion (1962); the difference may be due to the existence of a combustion zone in our experiments but not in those of Ergun. The results of varying the fluidising velocity showed that as the velocity increased there was a disproportionate decrease in the hydrogen output, showing that more of the hydrogen released by pyrolysis was being burnt in the gasifier.

Gasification in air mixed with steam, with continuous addition of fresh fuel

This was investigated at 1000°C. The calorific values obtained are shown in Figure 8 in relation to the volume percentage of steam in the fluidising gas. To obtain a calorific value of 3.8 MJ/m^3 about 8% of steam would be needed. The percentage of the incoming carbon that would be gasified to carbon monoxide and dioxide would then be 50. Of the other 50% about two-thirds would be discharged from the base and the rest would be elutriated from the bed. The use of steam improved the calorific value of the product but in a combined cycle it would entail a reduction in overall thermal efficiency because the latent heat of both the decomposed and the undecomposed steam could not be recovered.

The water gas shift reaction

Table 3 shows that the water gas shift reaction was not effective in this work.

The sharp rise in the experimentally-determined value of the shift ratio with temperature was caused by the rise in the hydrogen content of the products from nearly zero at 800°C to 7.6% v/v at 1000°C, during gasification in air. But at 980°C during gasification in steam the measured shift ratio was well below the equilibrium value. Therefore the reaction was ineffective in both directions.

The gasification of fuels other than Phurnacite breeze

From Table 4 it is obvious that the various feedstocks covered a wide range of reactivity. The lignite was so reactive that even when gasifying at as low a temperature as 800°C, the calorific value of the products was 5.3 MJ/m^3. Phurnacite breeze was the most unreactive fuel tested. With the coals a useful contribution to the calorific value came from the hydrocarbons which were released by the pyrolysis of the coal volatiles. This is illustrated in Table 5.

CONCLUSIONS

The kinetics of the fluidised-bed gasification of various solid fuels were examined and it was shown that a gas having a calorific value of 3.8 MJ/m^3 or more, i.e. enough to be useful as a gas turbine fuel, could easily be made at atmospheric pressure.

Comparisons between the kinetics of gasification of devolatilised Phurnacite breeze (which consists mainly of coal char) in dry air and in a 21% carbon dioxide, 79% nitrogen mixture showed that the use of air gave a superior product. No hot zone that might account for this result could be found, and accordingly it is proposed that some of the carbon monoxide found in the products was formed in the combustion zone by a reaction of oxygen at the carbon surface.

When other fuels were tried it was found that they covered a wide range of reactivity and treatability. The calorific values of the products obtained from the coals were encouraging as were those obtained from two coal-processing residues. A lignite gave the best results of all.

ACKNOWLEDGEMENTS

The authors thank the National Coal Board for permission to publish this paper. Any views expressed are those of the authors and not necessarily those of the National Coal Board. The European Coal and Steel Community is thanked for financial support.

REFERENCES

Arthur, J.R. (1951). Trans. Faraday Soc., 47, 164.

Ergun, S. (1962). 'Kinetics of the reactions of carbon dioxide and steam with coke'. U.S. Bur. Mines Bull. 598.

Lavrov, N.V., Korobov, V.V. and Filippova, V.I. (1963). 'The thermodynamics of gasification and gas synthesis reactions'. Pergamon Press.

TABLE 1: Analysis of fuels.

Fuel		Phurnacite Breeze	Coalite Breeze	Gedling Coal	German Lignite	Dawmill Coal	Markham Main/ Dawmill Coal Processing Residue	Dawmill Coal Processing Residue
Moisture	% as received	2.1	3.0	7.4	11.1	6.0	2.3	2.2
Ash	"	5.7	9.9	13.0	5.8	10.1	18.1	18.0
Volatile Matter	"	6.4	13.1	34.0	43.5	33.2	17.1	17.2
Fixed Carbon	"	85.6	74.0	45.6*	39.6*	50.7	62.4	62.6

*By difference.

TABLE 2: Formation and combustion of carbon monoxide.

Temperature deg. C	CO/CO_2 according to Arthur (CO combustion inhibited)	CO/CO_2 in the products (present work)	Fraction of CO formed that was then burnt (present work)
600	1.7	0.12	0.82
700	3.5	0.22	0.77
760	5.1	0.44	0.63
800	6.4	0.41	0.66

TABLE 3: Gasification in air with continuous fuel feed: experimental and equilibrium values for the water gas shift ratio $[CO_2][H_2]/[CO][H_2O]$

Temperature deg. C	Experimental value (this work)	Equilibrium value after Lavrov et al (1963)
800	0.06	1.1
850	0.65	
900	1.5	0.79
950	2.2	
1000	2.4	0.60

TABLE 4: Gasification of various feedstocks in air with a fluidising velocity of 0.46 m/s. Calorific value in MJ/m^3, dry basis, calculated as shown in the Note to the Figures.

Feedstock	Feed rate kg/h	Bed height m	Calorific value of the products of gasification at 1000°C MJ/m^3
Lignite	4.4	0.6	Over 7.2
Gedling coal	5.5	0.6	5.2
Dawmill coal	5.7	0.6	6.4
Markham Main/Dawmill coal processing residue	2.5	0.6 to 0.4	4.4
Dawmill coal processing residue	2.3	0.6	4.3
Coalite breeze	3.7	0.6	4.1
Phurnacite breeze	2.8	0.6	2.8

TABLE 5: Gasification of Gedling coal at 800°C in air: contribution of hydrocarbons to the calorific value of the gas. (CH_4, C_2H_4, C_2H_6, C_3H_6).
Bed height 0.6 m; Fluidising velocity 0.6 m/s.

Fuel feed rate kg/h	Calorific value of CO and H_2, MJ/m^3	Contribution of hydrocarbons, to calorific value, MJ/m^3	Total MJ/m^3
0	0.6	0	0.6
3	0.9	0.6	1.5
6	1.4*	1.5**	2.9***
9	1.7	2.0	3.7

*3.8 at 1000°C　**1.8 at 1000°C　***5.6 at 1000°C.

Note. All the diagrams relate to the gasification of Phurnacite breeze, the initial charge of which had been devolatilised at 1000°C. This was not supplemented for Figures 2 and 3 but fresh fuel was added at 2.8 kg/h for Figures 4 to 8. Except where noted otherwise, the fluidising medium was air, the temperature 1000°C, the bed height 0.6 m and the fluidising velocity 0.46 m/s. In Figures 7 and 8 the calorific values are calculated from the CO, H_2, CH_4 and C_2H_4 contents of the gas and their dry gross calorific values at NTP. For Figure 8 the undecomposed steam is regarded as a gaseous diluent.

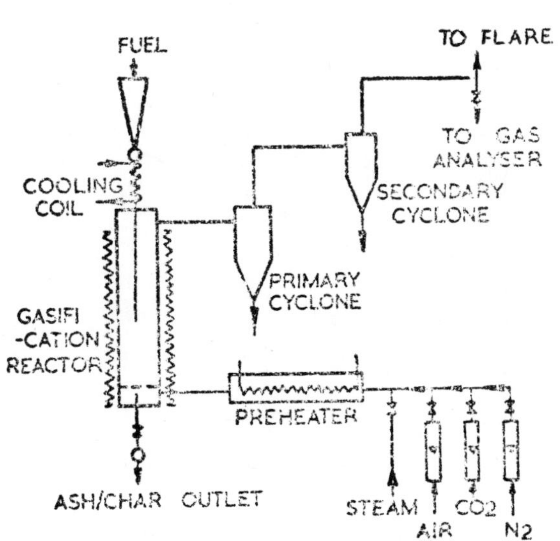

Figure 1. Diagram of the gasifier.

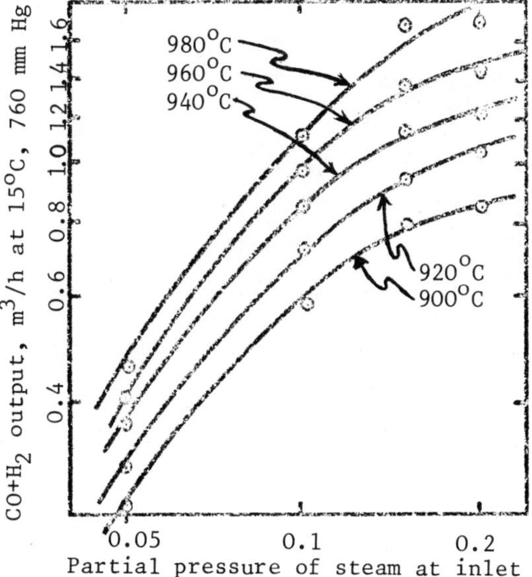

Figure 2. Kinetics of gasification in mixtures of steam with nitrogen.

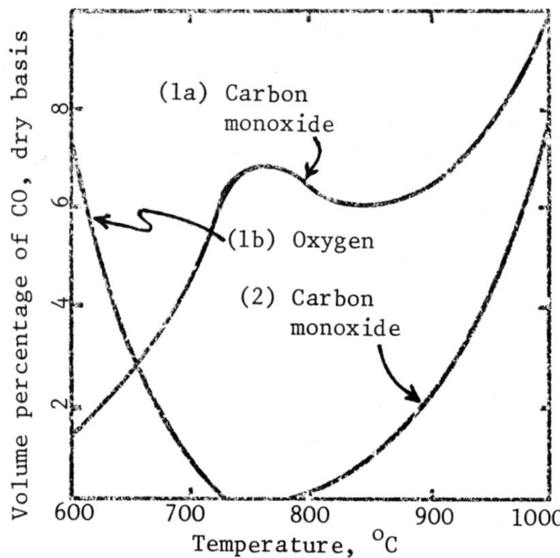

Figure 3. Contrasted product compositions from gasification in dry air (curves 1a and 1b) and in a 21% CO_2, 79% N_2 mixture (curve 2).

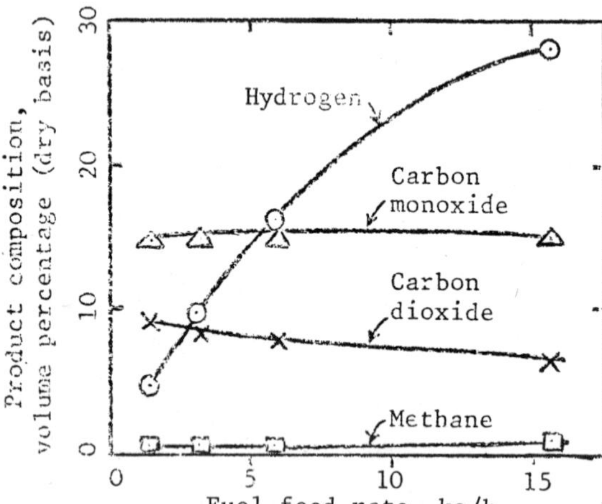

Figure 4. Effect of varying the fuel feed rate.

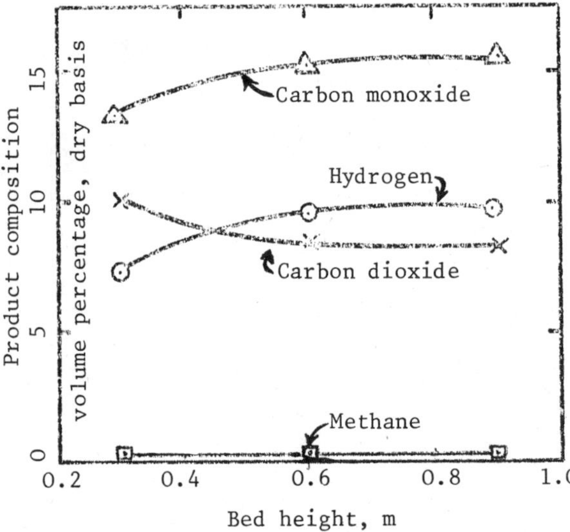

Figure 5. Effect of varying the bed height.

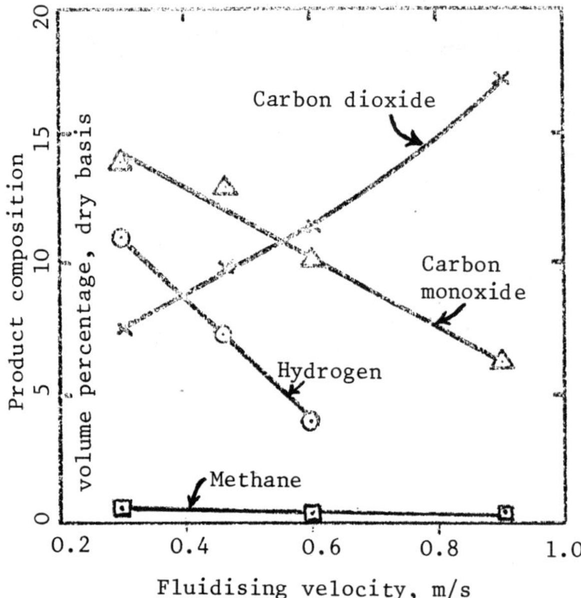

Figure 6. Effect of varying the fluidising velocity.

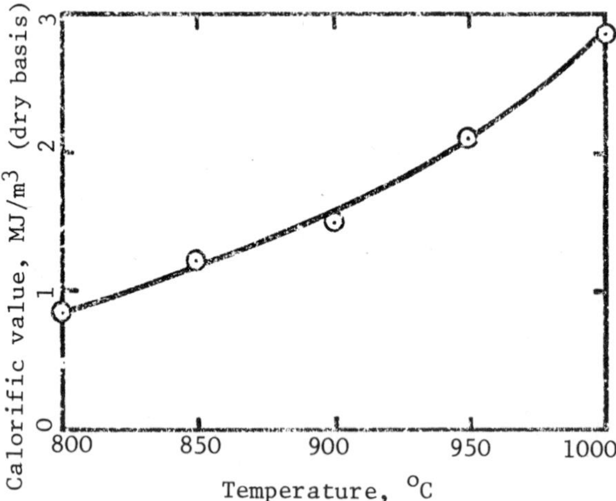

Figure 7. Effect of reaction temperature on the calorific value of the product gas.

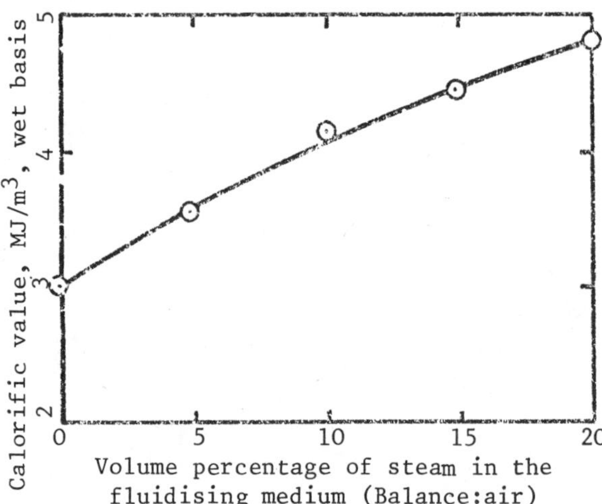

Figure 8. Effect of adding steam to the fluidising air.

COMBUSTION OF LARGE COAL PARTICLES IN A FLUIDISED BED

B.M. Gibbs and A.B. Hedley

Department of Chemical Engineering and Fuel Technology,
University of Sheffield,
Sheffield S1 3JD
England

SYNOPSIS

The potential use of large sized uncrushed coals, typically sized 6-25 mm, in atmospheric fluidised bed combustors has been investigated in a 30 x 30 cm square sectioned pilot scale combustor operating at a bed depth of 60 cm. Experimental measurements of the combustion efficiency, the transient behaviour and the NO emission are reported. Comparison of experimental data are made with corresponding data obtained previously for a < 1.4 mm crushed coal for equivalent operating conditions. The experimental results suggest that both capital and running costs of large scale plant may be reduced, and heat generating capacity for unit bed volume increased by using a large sized uncrushed coal instead of a finely crushed coal as feed to the combustor.

INTRODUCTION

Up to the present time the development of atmospheric fluidised combustors for industrial boiler applications has been mainly concerned with the combustion of crushed coals typically sized < 6 mm. It was found (see for example 1,2) that inorder to achieve the efficient combustion of crushed coals the elutriated combustibles must be recycled back to the bed or as is now preferred burnt in a separate 'carbon' burn-up cell. However, as the heat generating capacity per unit bed volume of the plant and therefore the operating fluidising velocity are increased a limit is reached whereby the elutriation losses even with carbon burn-up become prohibitively high. The use of large sized uncrushed coal feeds as a potential means of reducing freeboard height and increasing the heat generating capacity per unit bed volume, whilst maintaining acceptable combustion efficiences, has been investigated previously by Highley et al (see 3,4) of the N.C.B. (U.K)

The objective of our work was to extend some of the previous work (see 6,7) to the combustion of large coal particles in a fluidised bed and to compare the behaviour and performance of the combustor with that of the same combustor fired by a crushed coal for equivalent operating conditions. Experimental information is reported on the combustion efficiency, bed and freeboard reactions, the transient response and the NO emission.

FLUIDISED COMBUSTOR AND EXPERIMENTS

Experimental Combustor and Ancillaries

The layout of the experimental fluidised combustor and ancillaries is shown in Fig. 1. The combustor was 0.3 m square in section, 1.83 m high and constructed from stainless steel. The walls of the combustor were insulated. Fluidising air was introduced into the bed through a bubble cap distributor plate. Crushed coal could be fed pneumatically to the bed from a sealed hopper via a calibrated rotary valve feeder and injected into the bed through a horizontal tube 12.5mm I.D. Large coal (>3mm) was supplied from a sealed hopper and fed from a screw feeder with variable speed motor to a gravity feed pipe which directed coal onto the bed surface. Start-up of the bed was achieved by pre-heating the bed with an overhead gas burner. The bed temperature was controlled by varying the water flow into a stainless steel coil immersed in the bed. Carryover from the combustor was removed from the combustion products by means of a two-stage cyclone, and could be returned to the bed, if necessary, from the primary cyclone.

Thermocouples and gas sampling probes were located throughout the combustor (see 6,7) to enable bed, freeboard and exit gas compositions or temperatures to be obtained. Gas samples removed from the combustor were filtered and passed to water-cooled probes either to continuous gas analysers (CO_2, CO, NO or O_2) or to standard gas sampling bottles for analysis by gas chromatography. All bed and freeboard temperatures were recorded continuously on a chart recorder, as well as the cooling water or steam exit temperature from the cooling coil immersed in the bed.

Experimental Runs

Experimental runs of up to 8 hrs duration were carried out at steady state conditions using coal sizes of 6-9 mm, and 6-25 mm. The typical proximate and ultimate analysis of the coals used are shown in Table 1; the analysis of the crushed coal (< 1.4 mm size) used in previous comparative runs is also included in Table 1.

In all the runs the bed material was sand sized 500 μm to 1000 μm, the bed depth was maintained constant at 0.6 m and the fluidising velocity at 0.9 m/s. At the end of each run the carryover from the cyclone was weighed and ultimate and proximate analyses carried out. The temperature range studied was 700 to 840°C and the excess air range 0 to 40%.

Transient tests were also carried out by making step changes in either the coal feed rate or in the heat removed from the bed.

EXPERIMENTAL RESULTS AND DISCUSSION

(1) General Combustion Characteristics of large Coal.

No problems were encountered in establishing and maintaining steady state conditions whilst combusting large coal, although it took appreciably longer to reach steady state conditions than when firing on crushed coal (see transient response section). It was thought that accumulation of coal ash in the bed may be a problem but measurement of bed height during a run indicated no significant build up of bed material. Examination of the carryover after a run showed that 99% of the carryover consisted of fines (< 200 μm in size) with a few large ash particles up to 6 mm in size forming the balance. These large ash particles however were extremely friable and disintegrated almost immediately on handling. It is apparent that any large coal ash present in the bed is easily abraded and broken up by the agitation of the bed solids. Measurement of the bed temperature (at 152 mm, 305 mm and 406 mm levels) showed no measureable variation and the bed solids appear to be well-mixed with no apparent segregation due to the presence of larger particles. Large burning coal particles however could be seen periodically on the bed surface and were hotter than the surrounding inert phase there was also some evidence of volatiles burning round these particles. Freeboard temperatures especially 0.6 m above the bed surface were found to be up to 50°C higher than for a crushed coal indicating slightly more freeboard combustion in these regions.

(2) Combustion Efficiency, Excess Air and Bed Temperature.

The combustibles loss for each run was estimated from the proportion of the potential heat of combustion of the coal lost as CO in the flue gases and an unburnt combustibles in the elutriated carryover.

The influence of excess air on the combustibles loss from the combustion of several large sized coals (6-9 mm coal, 6-9 mm char and 6-25 mm coal) and also for comparison a 1.4 mm crushed coal is shown in Fig. 2 for a bed temperature range of 770° to 840°C and no recycling of carryover.

It can be seen from Fig. 2. that the elutriation loss from the combustor when burning large size coal is between a half and two-thirds lower than that obtained under the same conditions for a crushed coal without fines recycling, and is approximately comparable to that obtained for a crushed coal with fines recycling. The elutriation loss is also seen to be almost independent of the size of large coal used. The elutriation loss for the large coal is somewhat higher than might be expected considering the absence of feed fines, it can however be attributed to breakage of the coal in the screw feeder, decrepitation of coal in the bed, abrasion and breakage of coal due to violent agitation in the bed.

The elutriation loss is decreased if the excess air level is increased as shown in Fig. 2.

The elutriation loss from the combustor for large coal combustion (6-25mm) is seen from Fig. 3 to be strongly dependent on the bed temperature, and a marked decrease in elutriation loss takes place as the bed temperature is increased. At bed temperatures much below 775°C the elutriation loss for low excess air firing (5-10% level) is unacceptably high if no recycle is incorporated. At high excess air levels of 30% or more a combustion efficiency of 95% (without recycle) can be achieved at bed temperatures as low as 725°C.

As can be seen from Table 2. a lower CO emission is obtained from the combustion of large coal than from a corresponding crushed coal for comparable operating conditions. It should therefore be possible to operate a combustor burning large coal at a lower excess air level than for a crushed coal resulting in an increased thermal efficiency. At 5-7% excess air and 800°C bed temperature the combustion efficiency for large coal without recycling is about 96% which is adequate,

TABLE 2. CO emission for large coal combustion.

Excess Air, %	CO emission, % dry flue gas	
	< 1.4 mm coal	6-25 mm coal
0	1.0	0.8
2.5	0.5	0.25
10	0.2	0.08
30	0.06	< .04

for many industrial applications. It is expected however that if recycling or carbon burn-up were used a combustion efficiency of at least 99% could be achieved.

(3) Combustion in the Bed.

Axial Gas concentration profiles (CO_2, CO and O_2) were measured in the bed to compare the mixing and combustion of large coal in the bed with that of a crushed coal. Fig. 4 shows a comparision of combustion gas profiles obtained for a range of large sized coal feeds and also a <1.4 mm crushed coal feed, in all cases the bed was at stoichiometric conditions and at 800°C. The CO_2 and O_2 profiles show that as the size of the coal feed is increased less combustion occurs in the lower half of the bed, and more in the upper half of the bed. It is therefore apparent therefore from these profiles that large coal is not as well mixed in the bed as a crushed coal, and the larger the coal feed size the more segregation occurs in the region of the coal feeding plane the bed surface. This apparent segregation of large coal particles in the combustor is consistent with the detailed measurements of large particle 'penetration depths' and 'jetsam' concentrations carried out by Rowe et al. (5).

The CO concentration in the bed for large coal is seen, fig. 4, to be on average much lower than that measured for a crushed coal, and to increase as the bed surface is approached. The latter again indicates that there is a higher concentration of carbon near the bed surface than in the lower regions of the bed.

(4) Freeboard Combustion.

It is apparent from the gas sample compositions obtained above and below the bed surface, shown in Table 3, that the major reactions taking place in the freeboard are the oxidations of CO and traces of CH_4 and H_2. The result of these freeboard reactions was to increase the freeboard gas temperature above the bed surface by some 10-20°C above the bed temperature , almost identical behaviour was observed for crushed coal. However, freeboard temperatures measured some 600 mm above the bed surface were found for large coal to be up to ∿50°C higher than those measured for a crushed coal for equivalent

TABLE 3. Gas sample compositions above and below bed surface.

location	gas composition %			
	$\%CO_2$	CO	CH_4	H_2
150 mm below.	13.0	2.17	.75	.2
450 mm above	15.6	0.62	.1	.1
900 mm above	16.1	0.33	nil	nil

operating conditions. This would indicate that for a large coal 2% more of the total heat available from the coal is released in the freeboard than for a crushed coal. It is thought that surface feeding of the large coal from above the bed surface may contribute to this.

(5) Transient Response Characteristics

In order to investigate the transient response of the combustor to a step change in coal feed rate the combustor was first allowed to reach steady operating conditions with a constant stoichiometric feed rate, fluidising air flow rate and heat extraction from the bed The step change in coal feed was made by suddenly stopping the coal feed supply and observing the flue gas oxygen concentration the bed temperature and the steam temperature. After a given time interval the coal was switched back on again and the responses again observed.

Typical flue oxygen concentration, bed temperature and steam temperature responses measured during the transient period are shown in Fig. 5, for comparison corresponding response measured for a step change in crushed coal feed are shown in Fig. 6. In the case of large coal the transient change in coal burning rate, effectively the heat generation rate in the bed, as measured by the flue gas oxygen has a very large time constant in the order of 720s with a consequential much larger lag in the bed or steam temperature response (see fig. 5). In comparison, for a crushed coal the response in the flue gas oxygen concentration to a step change in coal feed (shown in fig.6) is so rapid, estimated time constant 75s, that a step change in coal feed can be considered to correspond to an equivalent step change in heat generation in the bed (a more detailed description is given by Gibbs (6)). The differences between the observed responses in the flue gas oxygen concentration for large and crushed coal respectively can be attributed to the much larger mass of combustibles in the bed for a large coal feed, it is estimated that mass of combustibles for large coal combustion is approximately 12 times that of a crushed coal, constituting some 3-4% of the total bed mass.

The latter would represent an average large coal burning time in the bed of some 5-7 mins. The relatively large mass of combustibles in the bed for large coal results from the much lower reactivity of the large coal compared to the crushed coal due to the much larger coal particle size.

When the large coal feed is switched back on the response in the flue gas oxygen concentration (hence the coal burning rate in the bed) is greater than that for the corresponding period when the coal was switched off. It is thought that the latter was a result of the greater reactivity of fresh coal feed compared to the residual bed combustibles due to the presence of feed volatiles.

It should also be noted that any theoretical analysis of the transient response of the bed to a step change in the feed rate of large coal will have to include a time dependent burning rate in the transient heat balance, as indicated by Gibbs (6), to enable the bed temperature response to be predicted.

The temperature response of the combustor, operating initially at $\sim 800^{\circ}C$ and stoichiometric conditions, to a step change in heat transfer or extraction from the bed was found for large coal to be identical to that measured for a crushed coal. Fig. 7 shows a typical bed temperature response curve to a step change in heat extraction in the bed. The measured time constant of the bed temperature response was ~ 840 s which compared favourably with the theoretical value of 720 to 1080 s, determined from the thermal capacity of the bed and the volumetric heat exchange of the fluidising gas. (see Gibbs 6)).

(6) No emission and concentration profiles

Fig. 8 shows typical axial and radial NO concentrations obtained in the bed and freeboard for large coal, 6-25 mm size, and for comparison the NO profiles for a crushed 1.4mm coal. In both cases the bed was at $800^{\circ}C$ and stoichiometric conditions. Throughout the bed and freeboard, and also the exit flue the NO concentration is seen, from Fig. 8, to be significantly lower for the large coal than that measured for an equivalent crushed coal. The NO emission from large coal combustion is reduced by approximately 40% when compared to the emission measured from an equivalent crushed coal. These are several factors which could account for the lower NO emission of the large coal; the large particles may be at a lower temperatures in the bed than corresponding crushed coal particles due to their smaller size, the release of volatile nitrogenous components and hence the formation of NO from volatiles is considerably slower for large coal particles than smaller crushed coal which emit volatiles almost spontaneously on injection into the bed, the rate of formation of NO from the char components may be lower for a large particle due to a lower oxidation rate for large particles compared with smaller particles, there may be more reduction of NO by char (see also (7)) due to the much higher combustibles concentration in the bed for large coal combustion.

The axial NO concentration profile in the bed shows that the NO concentration increases from the bottom to the top of the bed in a similar manner to the CO concentration (Fig.4) The differences, apart from the magnitude, of the NO concentration profiles observed for large and crushed coal respectively can probably be attributed to the manner in which the coal is introduced into the bed i.e. injected at the base for a crushed coal, dropped onto the surface of the bed for a large coal. Little difference is observed between NO concentrations measured at the wall and at the centre of the combustor. In the freeboard, the NO concentration profiles shows that there is considerable reduction of NO in the freeboard for both large coal and crushed coals.

CONCLUSIONS

The pilot scale investigation has shown that the combustion and thermal efficiency of atmospheric fluidised bed combustors can be considerably improved if large coal is burnt instead of a crushed coal, and in many industrial applications the need for carbon burn-up or recycling would be removed.

The heat generating capacity per unit bed volume of a fluidised bed combustor can be significantly increased, without excessive elutriation occurring by burning a large sized uncrushed coal in the combustor rather than a crushed coal.

The high combustor efficiencies, the increased throughput, and the elimination of crushing plant show that the operating and capital costs of large scale atmospheric combustors could be considerably reduced by using a large sized coal thus making fluidised combustion even more economically attractive.

The apparent reduction in the NO emission obtained by burning large coal enhance the use of fluidised bed coal burning combustors as a means of minimising the NO pollution from coal combustion.

ACKNOWLEDGEMENT

The author wishes to thank the N.C.B. (U.K) for a grant in aid of this research. The views expressed are those of the author and not necessarily those of the Board.

REFERENCES

(1) Skinner, D.G., 1974, Fluidised Combustion of Coal, Mills and Boan monograph no CE/3.

(2) Ehrlich Shelton, 1975, Fluidised Comb. Conf. Inst. Fuel Symp. Series No.1, paper C4.

(3) Highley, J. et al., 1975, Fluidised Comb. Conf., Inst. Fuel Symp. Series No.1., paper B3.

(4) Highley, J., et al., 1975, Proc. 4th Int. Fluid Comb. Conf., EPA.

(5) Rowe, P.N., A.W. Nienow and A.J. Agbain, 1972, Trans. I. Chem. E., 50, 324.

(6) Gibbs, B.M. and J.M. Beér, 1975, Proceedings of Comb. Inst. Euro. Symp. Paper 1.30, p.169.

(7) Gibbs, B.M., Periera, F.J. and Beér, J.M. 1975, Fluididsed Comb. Conf., Inst.Fuel Symp., Series No.1., paper D6.

Table 1. Analysis of coals used

Component	Ultimate Analysis % dry basis COAL 9mm	CHAR	COAL 6-25mm	COAL <1.4mm
C	77.5	80.2	76.8	62.0
H	4.9	3.9	5.0	3.7
O	8.0	8.7	8.3	7.8
N	1.7	1.5	1.7	1.5
S	1.5	1.7	1.2	1.7

Component	Proximate Analysis % air dried COAL 9mm	CHAR	COAL 6-25mm	COAL <1.4mm
Mixture	6.3	0.7	5.4	6.8
Volatiles	32.9	20.0	33.3	33.4
Fixed Carbon	55.4	75.2	55.5	40.0
Ash	5.4	4.1	5.8	19.8

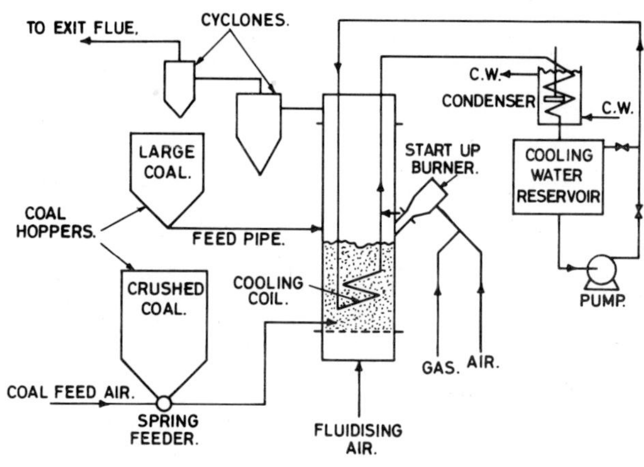

FIG. 1. Fluidised Combustor and Ancillaries

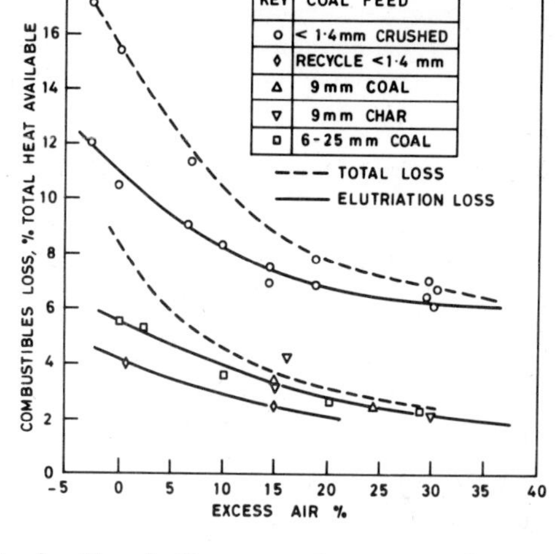

FIG. 2. The influence of excess air on the Combustibles loss

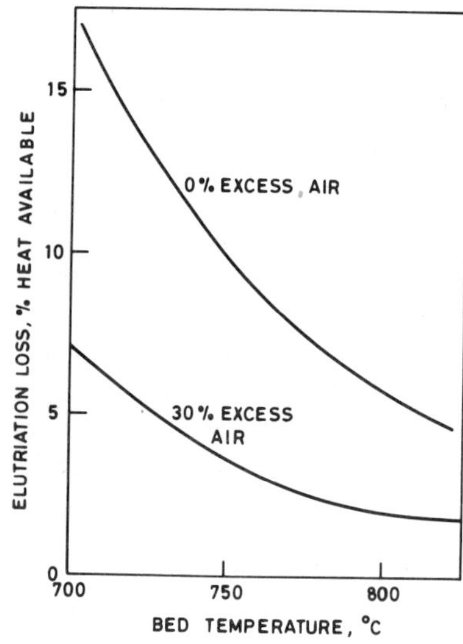

FIG. 3. The effect of bed temperature on the combustibles loss for large coal (6 - 25 mm)

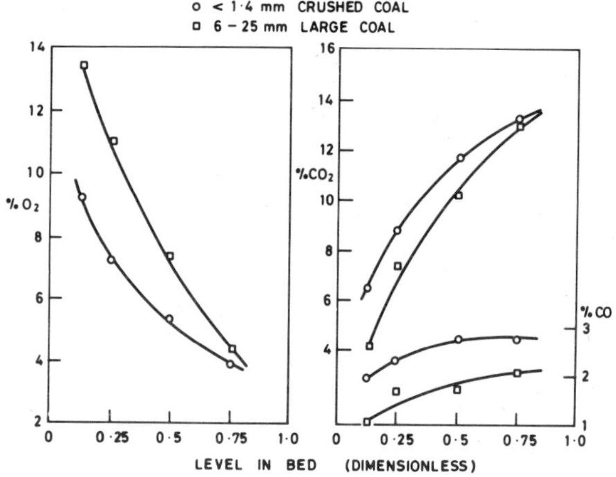

FIG. 4. Axial gas concentrations profiles of CO_2, CO and O_2 measured in the bed for large and crushed coals

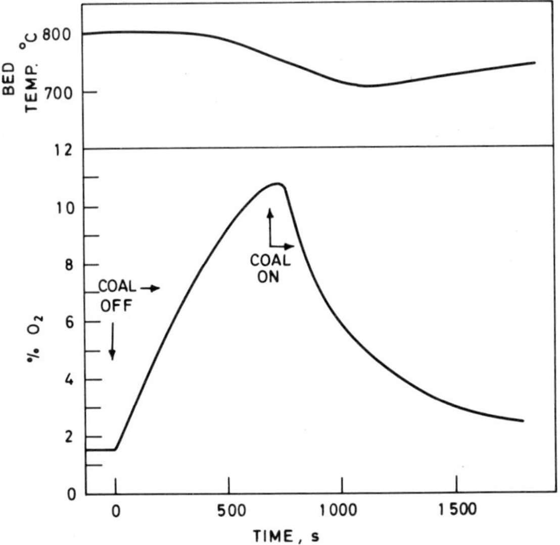

FIG. 5. Transient response of flue gas oxygen concentration and bed temperature to step change in the coal feed of a large coal.

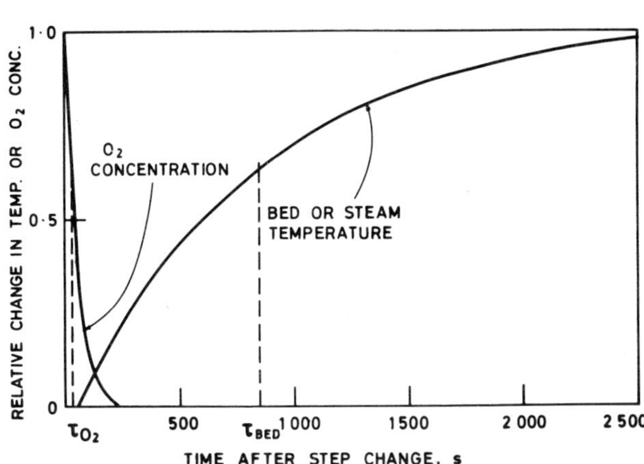

FIG. 6. Transient response to step change in the coal feed rate of a crushed 1.4 mm coal.

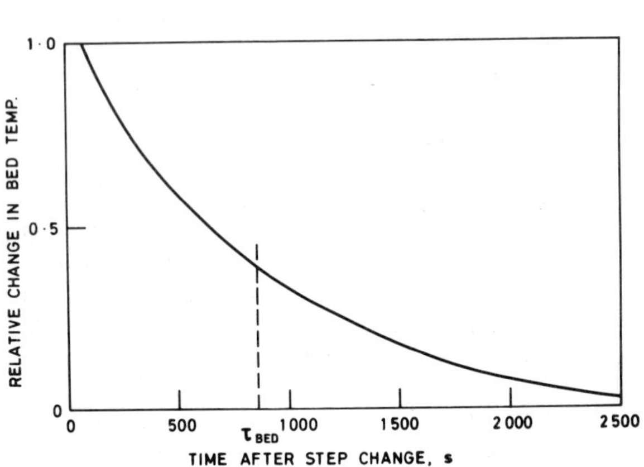

FIG. 7. Transient response to a step change in heat extraction from the bed for large and crushed coals.

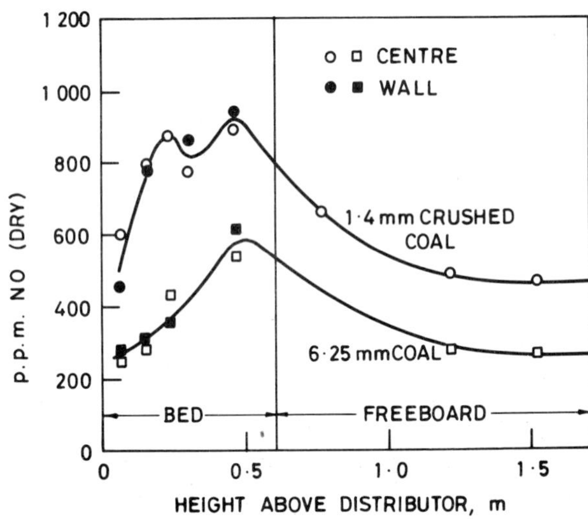

FIG. 8. Axial and radial NO concentration profiles for large and crushed coals.

PARTICLE TEMPERATURES IN A FLUIDISED BED COMBUSTOR

By J.G. Yates and P.R. Walker

Department of Chemical and Biochemical Engineering
University College London
Torrington Place
London WC1E 7JE
England

SYNOPSIS

Experiments are described in which large spherical particles of resin-bonded coal dust were burned in a fluidised bed over a range of temperatures from 1033K to 1233K. The burning rate of the devolatilised particles was measured and shown to vary inversely with bed temperature, an effect thought to have been caused by a temperature dependent increase in the diffusion resistance of the ash layer surrounding the particles. The temperatures achieved by the burning char particles were measured over a limited range by following the history of fusible wire rings embedded inside them.

NOTATION

C_p oxygen concentration in emulsion phase

D_G molecular diffusion coefficient in gas phase

d char particle diameter

E effective diffusion coefficient in emulsion phase

Sh Sherwood number for char particle

INTRODUCTION

Fluidised combustion has become in recent years a subject of widespread interest and it has been increasingly recognised that fluidised beds offer a number of advantages over conventional combustors in terms of scale, emission levels of gaseous pollutants, heat transfer characteristics and ability to deal with low grade fuels. Solid fuel combustion has received particular attention and from small beginnings in the mid-1960's the technology of fluidised coal burning has developed rapidly to the pilot plant scale and in a few instances beyond to the full scale plant. The mechanism of coal combustion in a fluidised bed is however incompletely understood and despite the valuable contributions of a number of authors, recently reviewed by Pyle (1975), much remains to be learned about the details of the process. For example, although it is well known that coal particles admitted to a fluid bed maintained at a temperature above 700°C are rapidly devolatilised, it is uncertain what proportion of the volatiles burn within the bed and what proportion burn in the freeboard region above the bed surface (Pyle, 1975). Furthermore, neither the chemical reactions that occur during the combustion of devolatilised coal or "char" nor the actual temperatures reached by the burning particles are known with certainty. We have attempted to answer some of these questions in the present work. The general burning characteristics of large char particles (diameter 9-10 mm) have been examined and compared with the theoretical predictions of a published model (Avedesian and Davidson, 1973) and the temperatures achieved by the burning particles have been measured over a limited range by following the history of heat-fusible wires embedded inside them.

EXPERIMENTAL

Particle Preparation

Spherical particles of 9.6 mm diameter were prepared from anthracite dust (ash content 21%) containing 16% of a phenolic resin binder. The mixture was compressed in a mould at 20 bar and 413K for 4 hr. A number of particles were made containing centrally symmetrical rings of silver-gold alloy wire melting at 1273K; the wire diameter was 0.15 mm and rings of three sizes were used.

Fluid Bed Combustor

This was made from a silica tube of 100 mm diameter fitted with a porous plate distributor; it was heated electrically by externally wound resistance heaters. The bed material was closely graded sand with particles of mean diameter 225 μm and a minimum fluidisation velocity at STP of 90 mm/s. A static bed depth of 14 cm was used. Coal particles both with and without the fusible metal inclusions were burned in the bed which was fluidised with air over a range of gas velocities at an initial temperature of 1033K-1233K. Incompletely burned particles were removed from the bed from time to time by means of a movable gauze basket, quenched in nitrogen and their weight and diameter measured. The weight and diameter of the unburned core of each particle were measured by removing the ash layer on its surface by subsequently fluidising the particle for 30 minutes in a cold bed at the flowrate corresponding to the combustion experiment. The thickness of the ash layer was generally of the order of 0.25 mm and some 5 minutes agitation was required to reduce it by half.

The resin binder was found to be 46% volatile at 1120K in a bed fluidised with nitrogen and the remaining carbonaceous residue was assumed to be typical of that binding the original coal dust particles together in the char and to burn in a similar manner to the char itself. Following devolatilisation of a particle under typical combustion conditions its density was in the region of 1200 kg/m^3 and it appeared to be little different from a sample of virgin char exposed to the same conditions.

RESULTS

A batch of ten coal particles was dropped into the heated bed from above and was observed visually by means of an angled mirror situated 30 cm above the bed surface. Volatile material was evolved and burned in the freeboard region for a period of up to 45 seconds. During this time the particles appeared to heat up relatively slowly and to remain in most cases at or near to the bed surface. When the devolatilisation process was complete the particles sank below the surface for much of the time and appeared to reach temperatures somewhat in excess of the bed temperature.

The weights of the particles removed from the bed at periodic intervals are plotted in Fig. 1 as a function of particle residence time for bed temperatures of 1083K and 1183K, the superficial gas velocity being the same in each case. It is apparent from this plot that for a short, initial period the particles burn at a rate which increases somewhat with bed temperature but that for the subsequent, considerably longer period the higher the bed temperature the lower is the particle burning rate. This results in the particle "burn-out" time at 1183K being some 250 s or 15% longer than the corresponding time at a bed temperature 100K lower. Burning rates were determined from the slopes of polynomials fitted to the weight-time curves. At any value of bed temperature and flow rate the burning rate was largely independent of whether the initial charge consisted of ten, six or four coal particles. Fig. 2 is a plot of burning rate against particle diameter and illustrates again the effect of bed temperature on the combustion process.

Fluidising gas velocity was found to have an effect on the burning rate, (Figure 3).

The results represent average figures taken from two independent runs for each set of conditions. Reproducibility was always better than 5% on a weight basis.

By burning particles containing metal inclusions of known melting point, it was hoped to determine, albeit somewhat imprecisely, the temperatures attained during char combustion. The results obtained are summarised in Table 1 and Plates 1 and 2.

Table 1

Results of ring fusion experiments.

Bed temperature (K)	Ring diameter (mm)	% Recovery of intact rings
	2.5	100
1030	4.5	100
	7.5	100
	2.5	100
1130	4.5	100
	7.5	100
	2.5	0
1173	4.5	100
	7.5	100
	2.5	0
1213	4.5	0
	7.5	67
	2.5	0
1233	4.5	0
	7.5	0

Metal rings of three different diameters were embedded inside nine separate particles and retrieved at the end of a combustion run by sieving the bed material. Non-retrieval of the rings was taken as evidence that they had attained a temperature of at least 1273K and had fused. It may be seen from Table 1 that the small diameter rings fuse at a bed temperature of between 1130K and 1173K, the medium sized rings between 1173K and 1213K and the large 7.5 mm rings somewhere in the region of 1213K. The results from this last run were rather inconclusive since in addition to the two large rings several fragments of wire of various lengths were retrieved. No wires survived the run at 1233K.

DISCUSSION

Particle Burning Rates

According to a widely accepted mechanism of carbon combustion at temperatures in the region of 1270K (Hougen and Watson, 1947), chemical reactions are fast and burning rates are controlled by the rate of transfer of oxygen through the gas film at the surface. Carbon monoxide formed at the surface diffuses outwards and burns on contact with oxygen in a zone closely surrounding the particle, some of the CO_2 thus formed diffusing back to the surface and being reduced to CO. In their model for char combustion in a fluidised bed based on this mechanism, Avedesian and Davidson (1975) derive the following expression for the molar flow of oxygen, n, to the char surface:

$$n = 2\pi \text{ Sh } D_G \, d \, C_p$$

where

$$\text{Sh} = \frac{2E}{D_G}$$

E being an effective diffusion coefficient for mass transfer through the gas in the emulsion phase. No allowance is made in the model for diffusion through an ash layer on the surface of the burning particle but it is clear from the results obtained here that the rate of carbon loss from a burning particle and hence the rate of transfer of oxygen to the carbon surface begins to decrease at temperatures in excess of about 1070K and it is plausible to attribute the cause of this decrease to ash-limited diffusion. Little is known at present about the composition of the ash in the anthracite used here but it is conceivable that at the higher temperatures a process such as sintering begins which either results in a loss of available reactive surface or causes the ash to adhere more strongly to the char and so to resist the attrition from bed material which at the lower temperatures leads to a more rapid exposure of combustible surface. Visual evidence for the onset of ash fusion at a bed temperature of 1130K is obtained from close examination of the metal rings retrieved from the bed. These reveal the presence on their surface of small, brownish, globular protrusions (Plate 2) features which are absent from (a) the wires recovered from particles burned at lower bed temperatures and (b) a naked wire immersed in the bed at 900°C for 1 hr.

Whatever the cause of the observed reduction in burning rate, it would seem necessary in cases such as this to modify the value of E to take account of a temperature dependent effect limiting the oxygen transfer rate. From a purely practical point of view it would appear to be an advantage to operate a fluid bed combustor at an optimum temperature determined by the physical properties of the ash contained in the burning coal; this temperature may be somewhat lower than that at which the ash normally fuses.

Particle Temperatures

It is difficult to draw many firm conclusions from the metal fusion experiments carried out so far. What can be said with some confidence however is that, under the conditions used with a relatively small charge of coal, for a bed temperature, T_b, of 1173K to 1213K the temperature, T_p, reached by the burning char particles of 7.5 mm diameter is no more than 1273K, while for 2.5 mm diameter particles in a bed at 1130K to 1173K, T_p-T_b is in the range 100K-140K. Furthermore it is clear that the particles become hotter as their diameters decrease. The temperatures indicated above are generally: (i) in excess of those calculated by Zabrodsky and Parnas (1975) for carbon particles of diameter 1.44 mm burning in a fluidised bed with 10% excess air, but (ii) lower than the 200K excess over bed temperature measured for small coal particles fixed to a thermocouple immersed in a fluidised bed of limestone at 900K (Whellock, 1972).

Experiments are continuing with the metal fusion technique and it is intended to investigate the effect on wires of different melting point from the one reported on here. In this way it should be possible to narrow the limits of uncertainty in the temperatures already determined. It is also hoped to repeat the present series of experiments with bigger charges of coal particles.

CONCLUSIONS

Particles of resin bonded coal dust containing 21% ash when burned in a fluidised bed show a maximum rate of combustion at a bed temperature of about 1070K. The reduction in burning rate above this temperature is thought to be due to an increase in the diffusion resistance of the ash layer on the particle surface possibly caused by sintering. Burning particles become hotter as their diameters decrease, the periphery of a particle of 7.5 mm diameter in a bed at 1213K reaching a temperature of 1273K whereas the surface of a 2.5 mm diameter particle is heated to 1273K in a bed at 1130K to 1173K.

REFERENCES

Avedesian, M.M. and Davidson, J.F. (1973) Trans. Instn. Chem. Engrs., 51, 121.

Hougen, O.A. and Watson, K.M. (1947) "Chemical Process Principles Part 3 - Kinetics and Catalysis", (New York, John Wiley and Sons).

Pyle, D.L. (1975) Inst. of Fuel Symp. Ser., No. 1, 2, 6.

Whellock, J.G. (1972) Birmingham Univ. Chem. Engr., 23 (1), 18.

Zabrodsky, S.S. and Parnas, A.L. (1975) Inst. of Fuel Symp. Ser., No. 1, 1, B5-1.

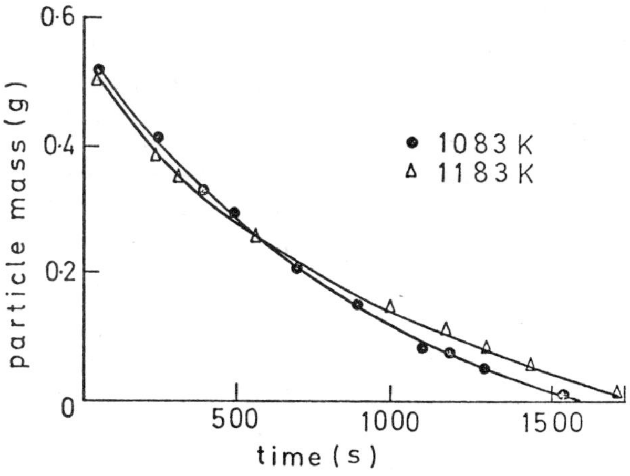

Fig. 1. Change in mass of char particles as a function of their residence time in a fluidised bed at high temperatures.

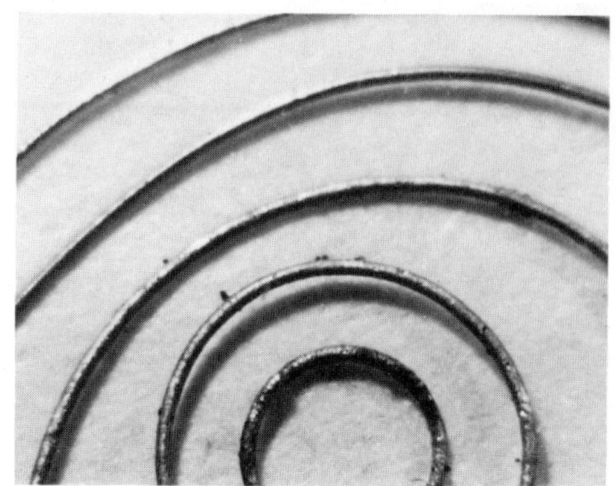

Plate 1. Heat fusible metal rings (in order from the outside):
(a) original wire
(b) naked wire immersed in a bed at 1173K
(c) 7.5 mm diameter ring from coal particle burned at 1033K
(d) 4.5 mm diameter ring from coal particle burned at 1033K
(e) 2.5 mm diameter ring from coal particle burned at 1133K.

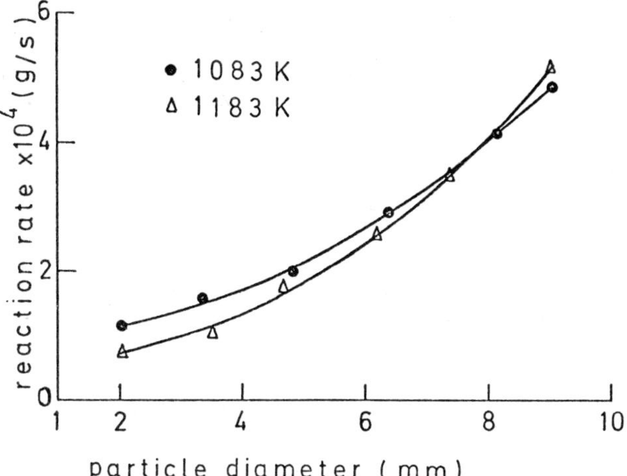

Fig. 2. Variation of the rate of combustion of char particles with diameter.

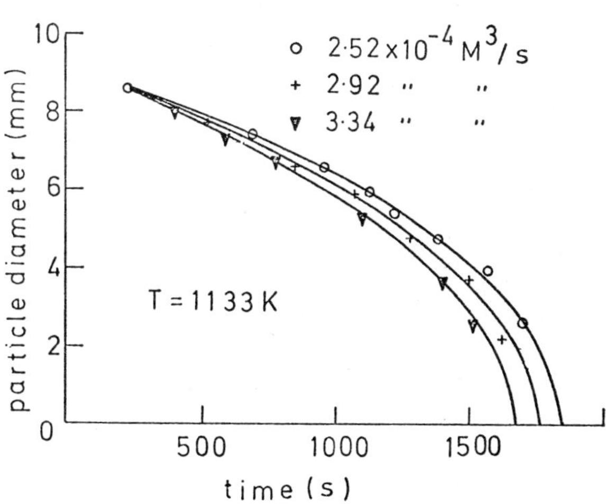

Fig. 3. Effect of air flowrate on char combustion at 1133K.

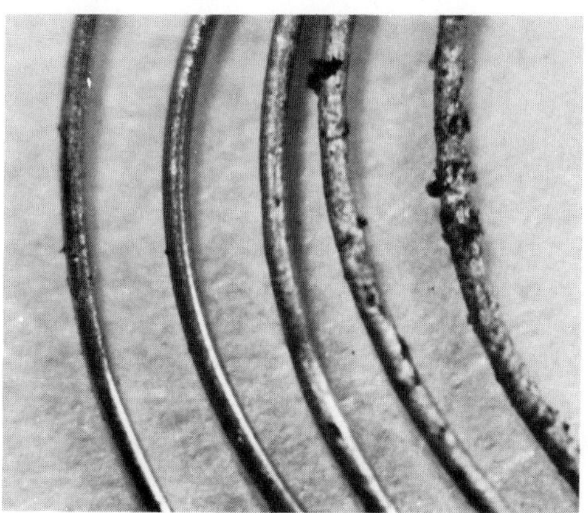

Plate 2. Heat fusible metal rings (7.5 mm diameter) (from left to right):
(a) original wire
(b) naked wire immersed in a bed at 1173K
(c) from coal particle burned at 1033K
(d) from coal particle burned at 1133K
(e) from coal particle burned at 1183K.

COMBUSTION OF COAL IN FLUIDISED BED

T. Kolenko

University of Ljubljana, Ljubljana (Yugoslavia)

R. Collin

The Royal Institute of Technology, Stockholm (Sweden)

SYNOPSIS

Selfsupporting combustion of coal in a fluidised bed of inert particles has been carried out. The diameter of the reactor was 37 cm, it´s static bed height was 30 cm. The fluidising velocity ranged from 50 to 100 cm/s. The experimental results support Gibbs´ mathematical model and indicate that combustion of coal can be carried out with low excess air also in a fluidised bed of limited size.

NOTATION

A_o	bed cross sectional area (cm^2)
c_p, c_o	oxygen concentration in the particulate phase and inlet air respectively (mol/cm^3)
d_b	bubble diameter (cm)
D, D_m	diameter and maximum diameter of a coal particle respectively (cm)
$E(D)$	elutriation constant(s^{-1})
F_o	feed rate of coal to the bed (g/s)
G	molecular diffusion coefficient in the gas phase (cm^2/s)
K_{bp}	gas interchange coefficient between bubble and particulate phase (s^{-1})
M	mass of coal in the bed (g or %)
n	excess air factor
n_{O_2}	molar flow of oxygen (mol/s)
$p_b(D), p_o(D)$	particle size distribution of combustribles in bed, and coal feed respectively (cm^{-1})
$\Delta p_b, \Delta p_d$	pressure drop over the fluidised bed and over the air distributor respectively (Fig. 2, (Pa)
r^2	coefficient of determination
$S(D)$	shrinkage rate due to combustion (cm/s)
Sh	Sherwood Number for burning particles
u_o, u_{mf}, u_b	superficial fluidising velocity (at bed temperature), minimum fluidising velocity, bubble velocity, respectively (cm/s)
v_g	specific waste gas volume (m_n^3/kg)
$\dot{V}$	combustion air flow (m_n^3/s)
X_{mf}, X_f	minimum fluidising and expanded bed height respectively (cm)
$\vartheta_b^1 \ldots \vartheta_b^6$	temperatures in the bed (°C)
ϑ^7, ϑ^8	freeboard and waste gas temperature respectively (°C)
$\vartheta_w^9 \ldots \vartheta_w^{12}$	temperatures in the lining (°C)

INTRODUCTION

A mechanistic model to aid design of fluidised bed combustors has been presented by Gibbs (1975).

The purpose of the present investigation was to compare Gibbs´ model with experiments in a somewhat larger bed, in which the combustion is carried out without support from any external energy source.

The combustion was carried out at different operating conditions with respect to heat load and excess air factor. The intention was to calculate the coal content of the fluidised bed and compare it with the experimental values.

In addition a comprehensive set of data, useful in further development of the model is also presented. A heat balance has been calculated to check the accuracy of the experimental results.

OUTLINE OF THE MATHEMATICAL MODEL

The mathematical model for the combustion of coal in an inert fluidised bed by Gibbs (1975) is based on the two-phase theory of fluidisation by Davidson and Harrison (1963), the material balances for shrinking particles by Levenspiel, Kunii and Fitzgerald (1968/69), and on the diffusion mechanism of the carbon combustion in a fluidised bed by Avedesian and Davidson (1973).

It is assumed in the model that the bed consists of two regions, a bubble and a particulate phase, with gas interchange between phases. Bubbles are in plug flow and of uniform size throughout the bed. Coal particles burn only in the particulate phase of the bed. The burning rate of a coal particle is controlled by the diffusion of oxygen to the particle and is given by Eq. 1 (Avedesian and Davidson, 1973):

$$n_{O_2} = 2 \cdot \Pi \cdot Sh \cdot G \cdot D \cdot c_p \quad \ldots.(1)$$

The volatile component of the coal is assumed to burn at the same rate as the solid carbon. The particle shrinkage due to attrition and particle entrainment due to splashing are neglected because of the lack of experimental data. The corresponding values included in the Gibbs´ model are arbitrarily assumed.

EXPERIMENTS

Equipment

The cooling tubes which occupied 2 % of the bed cross-sectional area could be immersed in the bed at different depths in order to control the bed temperature and to achieve different cooling rates. To feed the coal into the bed the feeding pipe was placed at the top of the equipment. The air distributor design, that was adopted as a result of experiments in a cold flow model (Nordin, 1975), corresponds to the requirements of low pressure drop as well as uniform air flow distribution. At superficial fluidising velocities up to 100 cm/s the pressure drop did not exceed 200 Pa.

Fig. 1 shows the sampling points of temperatures, pressure drop and gas analysis. The waste gas was sampled in the outlet of the combustor and analysed for CO_2, CO and O_2. Following parameters were also measured: cooling water flow and temperature, combustion air temperature and pressure, air humidity, exterior surface temperature of the combustor, feed rate of coal and coal quantity in the bed.

Experimental procedure

The bed of granite sand was first heated with town gas. The addition of coal started at 250° C. When the bed temperature was about 800° C, then the town gas supply was stopped, and the feed rate of coal and the corresponding combustion air flow was adjusted to the desired values. The static bed height was 30 cm. Coal was fed at regular time intervals as a weighed charge at the top of the combustor. The coal charge was contained in a paper bag, to prevent the fines from being elutriated instead of mixed with the bed. The interior of the combustor was observed through an observation window with a quartz pane. The observation was also possible through the feed opening, which was covered by a glass pane. The burning coal particles, passing by the observation window, reflected a bright light against the dark red background.

The coal combustion was carried on at constant feed rate of coal and different excess air rates. At the steady state, established by constant gas analysis and constant bed temperature, bed samples were taken, to determine the coal quantity in the bed. Then another coal feed rate was fixed and the sampling repeated at different excess air rates.

RESULTS

The first column in Table 1 shows the time when the bed reached steady state at specified operating conditions. The temperature ϑ_b is a mean value of the temperatures at points 2, 9, 18 and 27 cm above the distributor (Fig. 1). The excess air factor, n, and specific amount of waste gas, v_g, are calculated from the gas analysis, while the air flow, $\dot{V}$, was measured with an orifice. The superficial fluidising velocity, u_o, is calculated from v_g, F_o, ϑ_b, and the cross sectional area of the bed A_o. In Table 2 the temperatures at the temperature measuring points from Fig. 1 are given. The thermocouples at measuring points 1, 2, 3, 4, 5 and 6 measured temperature of the bed while thermocouples at 7 and 8 of freeboard and waste gas respectively. The thermocouples at measuring points 9, 10, 11 and 12 measured temperatures in the lining.

The minimum fluidising velocity u_{mf} was determined by the conventional pressure drop versus superficial fluidising velocity diagram at room temperature. Its value, 18 cm/s, was multiplied with the ratio of dynamic viscosities at room and bed temperature in order to obtain the value at hot conditions. In this way an average value of 7 cm/s has been obtained.

During the combustion experiments, u_o ranged from 50 to 100 cm/s. The lower value represents the minimum condition to attain a sufficiently developed bubble structure for mixing. The upper value was limited by the fan capacity.

CALCULATIONS

The heat balance in Table 3 was calculated for the steady state conditions at 4,48 o'clock.

Table 3. Heat balance

Heat input	W	%
Combustion heat of coal	83696	97,1
Combustion heat of paper	1912	2,2
Sensible heat of air	515	0,6
Sensible heat of coal	109	0,1
Total	86232	100,0
Heat output	**W**	**%**
Sensible heat of waste gas	33930	39,3
Unburned CO	1594	1,9
Sensible heat in fly ash	363	0,4
Wall losses	7822	9,1
Heat absorbed by cooling tubes	39547	45,9
Radiation through observation window	426	0,5
Unaccounted for	2550	2,9
Total	86232	100,0

The heat balance considers only the lower part of the combustor that contained the fluidised bed.

The heat generated by combustion of the paper bags could not be neglected. The waste gas temperature was assumed to be the same as the temperature, $\vartheta_b 6$, at the top of the bed. The wall losses were calculated at stationary conditions from the bed temperature $\bar{\vartheta}_b$=859°C. The calculated temperature of the external wall surface, 340° C. agrees well with the experimental value, 350° C. The temperature of the fly ash was taken to be the same as the waste gas temperature, $\vartheta_b 6$.

The heat loss through the observation window is calculated as black body radiation from the bed temperature.

The inlet temperature of the cooling water was 9,1° C. The outlet temperatures for the two branches were 26,6 and 50° C, and the corresponding water flow 0,2400 and 0,1282 kg/s respectively.

The item "unaccounted for" contains radiations from the bed into the freeboard.

APPLICATION OF THE MATHEMATICAL MODEL TO THE EXPERIMENTAL RESULTS

The mathematical model was used for interpretation of the data, presented in Table 1.

For the calculation of the coal content of the fluidised bed, M, the expanded bed height X_f, is needed. This is, however, almost impossible to measure experimentally. An iterative procedure was therefore chosen. Value of X_f was varied until the calculated M agreed with the experimental. Table 4 presents basic data used in the calculations, and Table 5 results for some of the steady state conditions.

Table 4.

Reactor cross sectional area	A_o	1053 cm^2
Minimum fluidising bed height	X_{mf}	30 cm
Minimum fluidising velocity	u_{mf}	7 cm/s
Sherwood Number	Sh	1,42

The particle size distribution $p_o(D)$ of the coal feed, according to size analysis, is given in Fig. 2, together with the calculated distribution in the bed $p_b(D)$.

The diagram shows that the bed contains almost no fines, although there is an appreciable amount of fines in the feed. This is a consequence of the high combustion rate. The elutriation constants E(D) have been calculated after experimental correlation given by S. Aigi and T. Aochi (1955 in Kunii and Levenspiel, 1969, p. 315, Fig. 12). It results from the calculation that the elutriation constants are zero for particle diameters larger than 0,025 cm. The elutriation constants of particle size 0,025-0,0015 cm are 0-0,0043 s^{-1}. The integral

$$I(D,D_M) = \exp\left[-\int_D^{D_M} \frac{E(D)}{S(D)} dD\right] \quad \ldots. \ (2)$$

has the values 1 for larger particles than 0,01 cm that cannot possibly be elutriated and 0,999 for particles down to 0,0015 cm. This shows that most particles were burned before they could have been elutriated.

DISKUSSION

The expanded bed height X_f was established for each of the stationary conditions in Table 1, so that the calculated coal content of the bed, M, corresponds to the experimental values. Curve 1 in Fig. 4 is attained through the best fit analysis of X_f, estimated this way. The curve follows the equation:

$$\frac{X_f}{X_{mf}} = 0{,}89824 \cdot \exp\left(0{,}07348 \frac{u_o}{u_{mf}}\right) \ldots \ (3)$$

Extrapolation to $u_o = u_{mf}$ gives

$$\frac{X_f}{X_{mf}} = 0{,}967 \text{ which is}$$

very close to unity. This means that the boundary condition is satisfactorily fulfilled and that the equation indicates a systematic relation between the experiments and the mathematical model.

Curve 2 in Fig. 4 shows the corresponding relationship from experiments in a cold flow model (Nordin, 1975) with a bed diameter of 25 cm, to be compared with 37 cm in the combustor. The cold experiments were carried out in the same range of superficial fluidising velocities as in the hot conditions. In spite of all that, different values of the velocity ratios were obtained. This time they are in the range between 1 and 5,5 as compared to 7 and 15 in the hot experiments.

The courses of the two curves are similar though the curve 2, however, always gives a higher value of X_f/X_{mf}. This is attributed to the smaller vessel diameter used in cold conditions. Last but not least one has to take into account the subjective decision on the location of the violently fluctuating surface. In the case studied the surface of the dense phase fluidised zone was taken for X_f determination.

The value of the integral I (D,D_M) (Eq. 2) is decisive which of both factors, elutriation or burning, prevails. When E(D) is much greater than S(D), the value of the integral approach zero, and the elutriation prevails. When E(D) is much smaller than S(D) the integral value approach 1 and combustion prevails. According to this, the values of the integral I (D,D_M) for all particle sizes show that coal fines were practically completely burnt in the bed. The prevalence of burning over elutriation is confirmed by the heat balance which would show, in the case of elutriation of coal fines, greater deficiency in the item "Unaccounted for". For coal particles smaller than 0,025 cm elutriation is possible. In the extreme case of 100 per cent elutriation a deficiency of 20 % would have taken place.

CONCLUSIONS

Several sets of the interdependent data obtained by the coal combustion in a fluidised bed are reported. Especially important are the accurate data on the waste gas analysis. They show high combustion efficiency at a lower air factor as in conventional practice.

In combination of the experimental and theoretical approach a mathematical relation was found between expanded bed height and superficial fluidising velocity. The equation obtained corresponds to the boundary condition. In this way a feasible outline for predicting fluidisation conditions with the mathematical model by Gibbs is indicated.

The mathematical model shows that the coal fines burn before they could have been elutriated by the gas flow. This is verified by the heat balance. If elutriation would have taken place, a shortage of heat output would result.

ACKNOWLEDGEMENT

This paper is based on a MSc thesis by Mr. T. Kolenko.

The authors are indebted to the Swedish Board for Technical Development and the National Swedish Board for Energy Source Development for economic support, and to the Swedish Institute that enabled Mr. Kolenko's stay in Sweden by means of a scholarship.

Mr. Sven Nordin carried out the cold model experiments, presented in Fig. 3, and assisted in the hot experiments.

REFERENCES

Avedesian, M.M., Davidson, J.F., (1973) Trans. Instn. Chem. Engrs., Vol. 51, No. 2, pp. 121-131.

Davidson, J.F., Harrison, D. (1963) Cambridge University Press, New York.

Gibbs, B.M. (1975) Symposium Series No. 1, Fluidised Combustion Conference, London, pp. A5-1-10.

Kunii, D., Levenspiel, O (1969) John Wiley and Sons, New York.

Levenspiel, O., Kunii, D., Fitzgerald, T. (1968/69) Powder Technology, 2, pp. 87-96.

Table 1. EXPERIMENTAL RESULTS AT STEADY STATE CONDITIONS

Time h.min	Coal feed rate $F_o \cdot 10^3$ kg/s	Coal content in bed, M %	Bed mean temp. ϑ_b °C	Combustion air		Waste gas				
				Flow $\dot{V} \cdot 10^2$ m_n^3/s	Superf. fluid. vel. u_o m/s	CO_2 %	O_2 %	CO %	Excess air factor n	Specific volume v_g m_n^3/kg
17,08	1,817	3,5	834 ± 1,3	1,275	0,509	16,6	2,5	0,40	1,12	7,275
17,25	1,817	3,5	831 ± 0,8	1,261	0,511	16,5	2,6	0,40	1,13	7,322
21,35	2,422	7,2	785 ± 1,3	1,589	0,615	17,7	1,4	0,50	1,06	6,898
21,45	2,422	7,5	797 ± 1,5	1,592	0,627	17,5	1,6	0,35	1,07	6,951
23,18	2,422	4,7	821 ± 1,7	1,856	0,728	15,4	3,9	0,25	1,22	7,901
0,00	2,422	2,2	824 ± 1,4	2,069	0,815	13,9	5,7	0,15	1,37	8,823
0,20	2,422	1,5	805 ± 0,8	2,069	0,814	13,5	6,1	0,15	1,40	8,956
2,10	3,633	4,3	859 ± 0,6	2,236	0,935	18,1	0,5	0,90	1,00	6,538
2,20	3,633	4,1	863 ± 0,5	2,236	0,939	18,1	0,5	0,90	1,00	6,538
4,35	3,633	2,0	868 ± 0,8	2,558	1,074	16,0	3,1	0,50	1,15	7,446
4,48	3,633	1,8	868 ± 0,8	2,525	1,065	15,9	3,0	0,70	1,14	7,388

Table 2. TEMPERATURES AT POSITIONS SHOWN IN FIG. 2

Time	Bed temperatures °C.						Gas temperatures °C.		Wall temperatures °C.			
	ϑ_b^1	ϑ_b^2	ϑ_b^3	ϑ_b^4	ϑ_b^5	ϑ_b^6	ϑ^7	ϑ^8	ϑ_w^9	ϑ_w^{10}	ϑ_w^{11}	ϑ_w^{12}
17,08	808	832	833	834	835	832	562	239	653	530	450	307
17,25	809	831	830	831	832	829	584	243	655	529	449	307
21,35	764	783	785	784	786	772	671	261	647	542	494	375
21,45	755	795	798	796	798	777	603	263	650	540	494	372
23,18	712	819	821	820	823	800	665	307	669	581	535	442
0,00	760	823	824	823	826	802	680	339	674	601	556	462
0,20	766	804	805	805	806	789	676	451	665	593	547	457
2,10	824	859	860	859	860	835	733	460	696	636	595	488
2.20	816	854	855	853	855	831	730	477	699	637	595	491
4,35	853	867	868	868	869	828	723	488	705	642	592	492
4,48	846	868	868	867	869	819	725	508	702	635	584	488

Table 5. RESULTS OF NUMERICAL CALCULATIONS, BASED ON DATA IN TABLES 1 AND 4

Time h.mi	17,250	21,450	23,180	0,200	2,200	4,480
$c_o \cdot 10^6$ mol/cm^3	2,317	2,391	2,338	2,373	2,256	2,242
X_f cm	47,700	53,000	55,000	59,500	72,500	86,000
G (cm^2/s)	1,797	1,708	1,771	1,729	1,883	1,896
f	0,371	0,434	0,455	0,496	0,586	0,651
u_b cm/s	125,600	133,700	151,800	157,100	154,900	160,600
d_b cm	13,400	12,500	14,900	13,800	9,400	7,400
K_{bp} s^{-1}	4,050	4,330	3,600	3,900	6,090	7,940
$c_p \cdot 10^7$ mol/cm^3	1,495	0,946	1,474	5,120	2,470	5,580

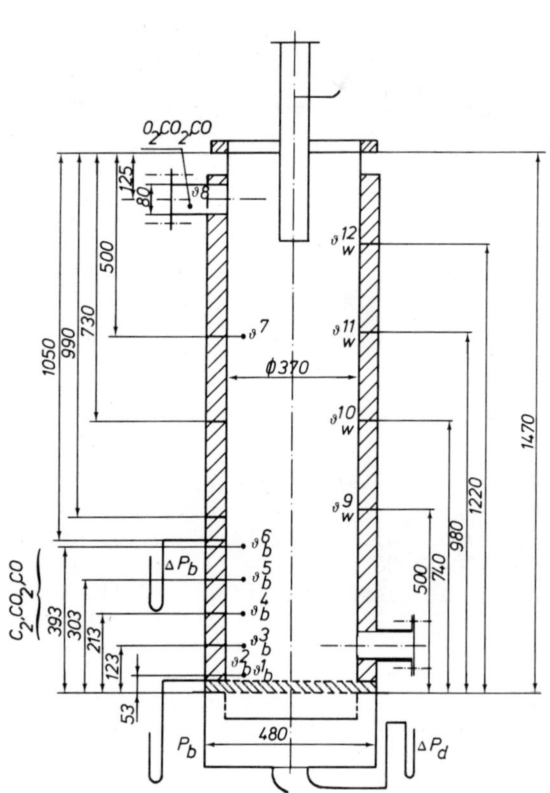

Fig.1. Temperature, ,pressure drop, Δp and gas analysis (O_2, CO, CO_2) sampling positions

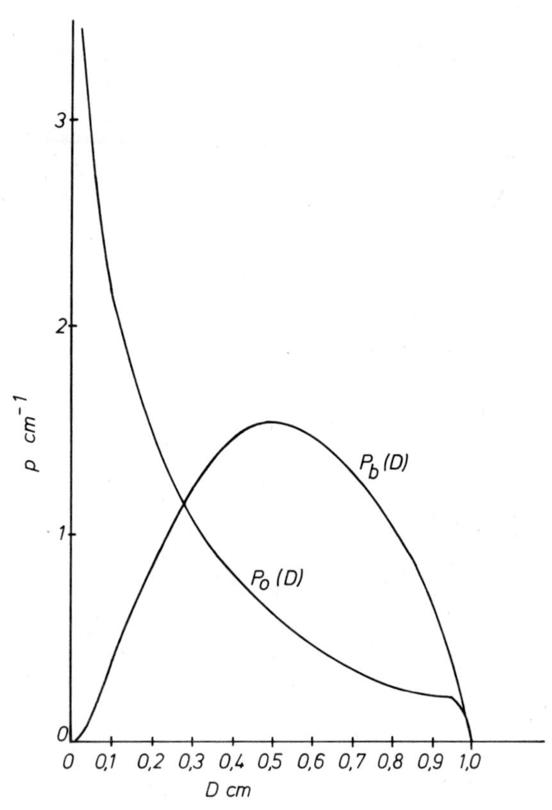

Fig.2. Size distibutions of coal particles in the feed P_o (D) and in the bed P_b(D)

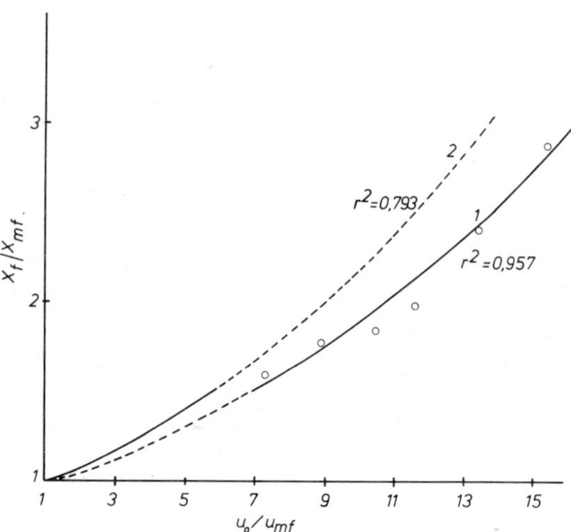

Fig.3. Expanded bed height ratio X_f/X_{mf} versus gas velocity ratio u_o/u_{mf}
1. Hot condition
2. Cold model experiments

PRESSURIZED FLUIDIZED BED COAL COMBUSTION

M. S. Nutkis

Exxon Research and Engineering Company

Linden, NJ 07036

SYNOPSIS

A pressurized fluidized bed coal combustion pilot plant has been in operation at Exxon Research since 1974. The 0.33 m (13 inch) diameter combustor, constructed under EPA contract, has operated at pressures to 1000 kPa (10 atmospheres), superficial bed velocities to 3 m/s (10 ft/sec) and at temperatures of 760°C to 990°C (1400°F to 1800°F). The combustor is capable of burning up to 230 kg/hr (500 lb/hr) of coal at design conditions. Coals containing 2.0% and 4.2% sulfur were combusted in a pressurized fluidized bed of both a limestone and a dolomite sorbent. Combustor bed temperature and pressure, sorbent to coal ratios, superficial bed velocities, expanded bed depth, sorbent particle size and excess air levels were all varied in a test series. Emissions of sulfur dioxide, NO_x and other salient flue gas components, carbon combustion efficiency and heat transfer coefficients were ascertained as a function of combustor operating parameters. Emission levels of sulfur dioxide well within the US EPA standard were readily attained during the test series. NO_x emissions in the PFBCC were very low - in the range of one-quarter to one-half of the current EPA emission standard. Basic pressurized fluidized bed coal combustion process design information has been developed for future boiler design criteria and prediction of sorbent requirements for certain SO_2 and NO_x standards.

INTRODUCTION

Pressurized fluidized bed combustion provides a new boiler technique where coal is combusted in a bed of particles maintained in a state of fluidization by the air required for combustion. The use of limestone, dolomite, or another suitable sorbent as the bed material permits the capture and removal of sulfur dioxide simultaneously with the combustion process. Within the combustor limestone is calcined to calcium oxide (CaO) which reacts with the SO_2 and oxygen in the flue gas to form calcium sulfate.

Part of the heat of combustion is removed by steam coils immersed in the fluidized bed. The immersed coils act to reduce and control the combustor temperature and also provide improved heat transfer rates. NO_x emissions from the combustor are low because of the lower combustion temperature and fortuitous internal NO reactions. The high pressure, high temperature flue gas can be expanded through a gas turbine (after suitable removal of particulate matter) to increase overall power cycle efficiency. Figure 1 is a sketch of the PFBCC cycle.

PROGRAM OBJECTIVES

Under contracts sponsored by the U.S. Environmental Protection Agency, Exxon Research and Engineering Company designed (Nutkis and Skopp, 1972), constructed and operates a pressurized fluidized bed coal combustion facility, the "miniplant," to demonstrate the feasibility of this coal combustion technique for environmentally acceptable energy production (Hoke, et al, 1976).

This FBC program is part of a national development effort sponsored primarily by the United States Energy Research and Development Administration (ERDA), the Environmental Protection Agency (EPA), and the Electric Power Research Institute (EPRI).

The objective of the Exxon coal combustion studies were to characterize the performance and emissions of a PFBCC system and to provide additional engineering and process data for the design of future units and for the estimation of sorbent requirements to achieve certain SO_2 and NO_x emission levels.

DESCRIPTION OF PFBCC MINIPLANT

The Exxon Miniplant combustor is a 10 m (32 ft) tall vessel with a 0.61 m (24 inch) shell refractory lined to a 0.33 m (13 inch) inside diameter. Main fluidizing air for the combustor is supplied at pressures to 10 atmos. and controlled flow rates of 0.28-0.56 m^3s(600 to 1200 SCFM). Coal and sorbent are proportioned, premixed and continuously pneumatically injected into the combustor at a side port 0.28 m (11 inches) above the distributor grid. The air passes up through the distributing grid where it fluidizes the bed material and provides the air required for combustor.

Heat extraction in the combustor is achieved by cooling coils through which water is circulated. The coils are fabricated from type 316 stainless steel 1/2 inch pipe and are arranged in a serpentine, vertical orientation. Each coils consists of 0.55 m^2 (5.9 ft^2) of surface area; the number of coils used can be varied from 1 to 10.

The combustor is capable of operating at pressures to 10 atm. and temperatures to 1060°C (1940°F), superficial velocities to 3 m/s (10 ft/sec) with expanded bed heights up to 6.1 m (20 ft) (Nutkis, 1975).

Flue gas exits the combustor and discharges through 2 stages of cyclones for solids removal. The solids collected by the first stage cyclone are returned to the combustor 0.5 m above the grid. Solids, mainly fly ash, collected in second cyclone is discharged by means of a lock hopper system. The expanded bed height is controlled by continuously rejecting solids through a port at the 2.3 m (90 inch) elevation.

In order to reduce limestone or dolomite feed rates to the FBC and minimize material disposal, a system is under study (Ruth, 1975) in which the sulfated sorbent is regenerated in an adjacent fluidized bed reactor and then returned to the combustor for further absorption of SO_2.

Figure 2 shows the PFBCC Miniplant combustor and regenerator and main support system.

Flue gas component levels of SO_2, CO, O_2, NO_x and CO_2 are continuously monitored with on-line analyzers, and periodic samples are taken for off-line analysis of SO_2, SO_3, CO_2 and O_2 using wet chemical methods. Particulate concentration is evaluated using an isokinetic probe and total filter techniques.

OPERATING PROCEDURES

The Exxon Research PFBCC has been in operation since 1974. It has accumulated over 1500 actual combustion test hours during which over 200 tons of coal were combusted (Nutkis, 1975). The PFBCC has operated at pressures of 5 to 10 atm. absolute temperatures of 680°C to 980°C (1260°F to 1800°F), bed velocities to 3 m/sec (10 ft/sec) and expanded bed heights to 5.5 m (18 ft).

Several series of test runs were made in which the combustor temperature, pressure, superficial bed velocity, sorbent to coal ratio (expressed as Ca/S molar ratio), expanded bed height, excess air level, and sorbent particle size were varied using two types of coals and sorbents. A series of runs was also made with the calcined form of the limestone sorbent. Most runs were made at pressures of 9.0 to 9.5 atm, temperature range of 840°C to 950°C, (1550°F to 1750°F), and superficial velocities of 1.5 to 2.1 m/sec (5 to 7 ft/sec), and expanded bed heights of 2.8 to 3.8 m (9 to 13 ft).

Combustion studies were made with two coal varieties: an Eastern bituminous Pittsburgh seam coal containing nominally 2.0 weight per cent sulfur, and an Illinois No. 6 bituminous coal containing 4.2 weight per cent sulfur. The Eastern coal was screened to a size range of 200 to 2400 µm and the Illinois coal to a 700 to 2400 µm size range. Two sorbents were used in the test series: a Virginia quarry limestone (Grove No. 1359) and a dolomite from Ohio (Pfizer No. 1337). Both were screened to a size range of 840 to 2400 µm. The results presented are based on operation of the combustor only, i.e., without regeneration.

RESULTS AND CONCLUSIONS

Sulfur Dioxide Emissions

Emissions of SO_2 obtained with Eastern and Illinois coals and Pfizer dolomite sorbent are presented in Figure 3. In this figure the percent reduction of SO_2 (i.e., SO_2% captured) is plotted against the sorbent to coal feed ratio expressed as moles of calcium fed in the sorbent to the moles of sulfur present in the coal injected. The stoichiometric Ca/S ratio is unity.

As seen from the single curve drawn Figure 3, there is strong correlation of sulfur retention to Ca/S ratio for all runs, even though the test conditions covered temperature variations from 840° to 950°C (1550° to 1750°F), pressure ranges from 6 to 9 atmos abs., a variation in gas phase residence time from 0.8 to 3 sec, a two-fold variation in dolomite sorbent particle size and a variation in coal source and sulfur content. This infers that the Ca/S ratio is the primary parameter affecting FBC SO_2 emissions; the other variables have a secondary effect. Figure 3 also indicates that very high SO_2 reduction levels are attainable using dolomite at relatively low Ca/S ratios. For example, a Ca/S ratio of 1.5 would yield an 85% SO_2 reduction.

SO_2 emissions measured using Grove limestone as the sorbent are shown in Figure 4. Contrary to the results with dolomite, a noticeable transition temperature effect occurs with increasing temperature, especially above 925°C (1650°F), yielding higher SO_2 removal levels. Also the amount of data scatter is more pronounced and SO_2 retention levels are lower compared to dolomite. These effects are due to the inability of limestone to calcine completely under combustor conditions. Calcination greatly increases the porosity of the limestone, making the interior surface of the stone more accessible to the SO_2 reactant. At the higher temperature conditions, the limestone undergoes extensive calcination and although the sorbent is not as active as dolomite, it is considerably more receptive to SO_2 than at the lower temperatures where the stone is largely in the carbonate form. Figure 5 compares the relative SO_2 activity of dolomite and limestone.

A series of runs was made in which the limestone was calcined outside the combustor and fed to the combustor in the calcined form along with the coal. The activity of the precalcined limestone was found to be significantly higher than that of the uncalcined stone, even when the combustor was operated at the lower temperatures which do not promote in-situ calcination. The precalcined limestone is as active at an equivalent Ca/S molar ratio as dolomite.

Although a fluidized bed utility boiler would normally be expected to operate in the temperature range of about 850 to 950°C (1550 to 1750°F), operation at much lower temperature, i.e. down to about 750°C (1400°F), would be required to turndown the boiler output to match a decrease in electrical power demand. A series of runs was made using dolomite, limestone and precalcined limestone at temperatures near 750°C to determine the behavior of the FBC system at these lower temperatures, and in particular the effect on SO_2 removal. The effect of operating at these low temperatures on SO_2 retention is shown in Figure 5 where the dashed line represents dolomite and precalcined limestone results in the 690-760°C temperature range. As seen, the activity of dolomite and calcined limestone are comparable but slightly lower than the activity measured in the normal combustion temperature range. However, the results with limestone indicate that it is completely inactive at the "turndown" temperatures. The minimum temperature at which combustion was stable was 690°C (1270°F). At temperatures below 690°C, CO concentration in the flue gas increased and temperatures in the fluidized bed become unstable.

The gas phase residence time, which is the ratio of the fluidized bed height to the superficial gas velocity, is a parameter which has some influence on the SO_2 emissions. The magnitude of the residence time effect is such that at a dolomite Ca/S ratio of 1.5, a substantial decrease in residence time from 3 to 0.5 seconds will cause a decrease in the SO_2 retention from 90 to 65%; or in order to maintain 90% SO_2 retention, the Ca/S ratio would have to be increased from 1.5 to about 3.

Other parameters varied in this study had no significant effect on SO_2 emissions. This included total pressure which ranged from 600 to 940 kPa (6 to 9 atm abs), excess air which ranged from 5 to 110%, and sorbent particle size which was varied by a factor of 2 from one batch screened to a size range of 1400 to 2400 μm to a second batch screened to a size range of 790 to 1400 μm.

As a result of the above studies, the sorbent requirements needed to satisfy the current EPA new source performance standards for SO_2 emissions from a pressurized fluidized bed coal fired boiler (1.2 lb SO_2/M BTU coal fired) can be estimated (Table 1). The estimate was based on a gas phase residence time of 2 s and a boiler temperature of 930°C (1700°F). Although dolomite and calcined limestone are more effective than uncalcined limestone on a molar basis, on a weight basis uncalcined limestone is slightly more effective than dolomite with a coal containing 2% sulfur, and equivalent for a 3% sulfur coal. For coals containing more than 3% sulfur, dolomite is more effective than uncalcined limestone even on a weight basis. Calcined limestone, however, is more effective than dolomite for all sulfur levels. Only one-half the weight of precalcined limestone (based on weight of limestone before calcination) is required compared to dolomite for the same sulfur retention capabilities. These observations can be used to establish a design and operating basis for pressurized fluidized bed combustion boilers, where the relationship between sorbent requirements, fluidization velocity and bed depths can be determined.

NO_x Emissions

Measured NO_x emissions varied from 50 to 200 ppm or 0.04 to 0.17 g as NO_2/MJ (0.1 to 0.4 lb/M BTU). Figure 6 presents the NO_x emissions data plotted against percent excess air. Although the combustor operating conditions were varied greatly, the only statistically significant variables influencing NO_x levels were the excess air (or flue gas oxygen concentration) and bed temperature.

The NO_x emissions increased by a factor of 4 from 0.1 to 0.4 lb/M BTU over the 5 to 110% range of excess air. The temperature effect in the 670° to 940°C (1250° to 1750°F) range was secondary and caused only a 25% increase in the emission level. The PFBCC emissions are well below the EPA new source performance standard of 0.7 lb/M BTU at excess air levels of 15 to 100%, the levels most likely to be used by a commercial size boiler.

Other Emissions

CO emissions were very low, generally in

the range of 50 to 200 ppm at combustor temperatures above 825°C (1500°F). At temperatures below 800°C, CO emissions increased sharply to levels of 300 to 800 ppm at 700°C (1300°F).

Flue gas particulate emissions exiting the two-stage cyclone system varied from 0.3 to 2.4 g/m^3 (0.1 to 1.0 gr/SCF), averaging about 1.4 g/m^3 (0.6 gr/SCF). The mass median particle size of the particulates was about 7 μm.

Carbon Combustion Efficiency

Carbon combustion efficiency was found to increase with bed temperature to over 99% at a temperature of 940°C (1720°F). However, at lower temperatures, the combustion efficiency was found to vary considerably between 95 and 98%. No reason for the variation at the lower temperatures was found. Essentially all of the combustible carbon loss was due to carbon particulates removed in the second stage cyclone. This level of carbon combustion is caomparable to that obtained in conventional coal fired boilers.

Heat Transfer Coefficients

Overall heat transfer coefficients in the PFBCC were determined to vary from 250 to 420 w/m^2°K (45 to 75 BTU/hr ft^2 °F). Most values were in the 310 to 340 w/m^2°K (50 to 60 BTU/hr ft^2 °F) range, and increased moderately with increasing temperature and superficial bed velocity, and with finer particle size.

REFERENCES

Hoke, R. C., Bertrand, R. R., Nutkis, M. S., et al, (1976) Report prepared for the U.S. Environmental Protection Agency (EPA-600/7-76-011).

Nutkis, M. S., and Skopp, A. (1972) Proceedings of the Third International Conference on Fluidized-Bed Combustion, Hueston Woods, Ohio Oct. 29-Nov. 1, Session V, pp V-2-1.

Nutkis, M. S. (1975) Proceedings of the Fourth International Conference on Fluidized-Bed Combustion, McLean, Virginia, Dec 9-11, pp 221-238.

Ruth, L. A. (1975) Proceedings of the Fourth International Conference on Fluidized-Bed Combustion, McLean, Virginia, Dec 9-11, pp 425-438.

Table 1. Sorbent Requirements

Coal	SO_2	Ca/S Limestone			Wt. Sorb/100 Wt. Coal Limestone		
S (%)	Retent. (%)	Uncalc.	Calc.	Dolomite	Uncalc.	Calc.*	Dolomite
2	59	1.3	0.8	0.8	8.2	5.0	10
3	73	2.1	1.0	1.0	20	9.4	20
4	79	2.8	1.2	1.2	34	15	29
5	84	3.2	1.3	1.3	51	20	40

Residence Time 2 sec

Temperature >930°F

SO_2 Emissions 1.2 lb/M BTU

* Based on weight of limestone before calcination.

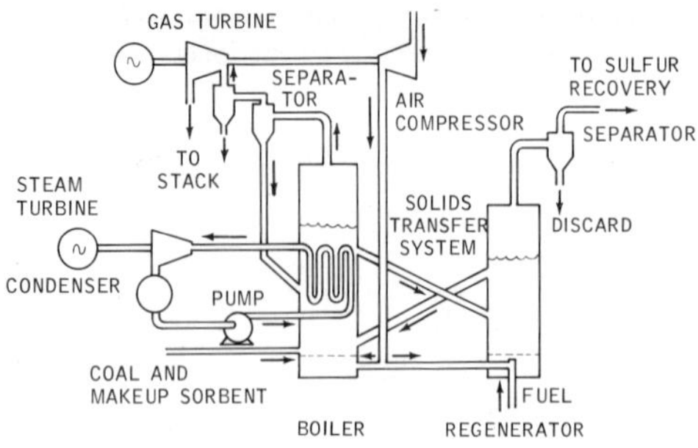

Figure 1. Pressurized Fluidized Bed Coal Combustion System

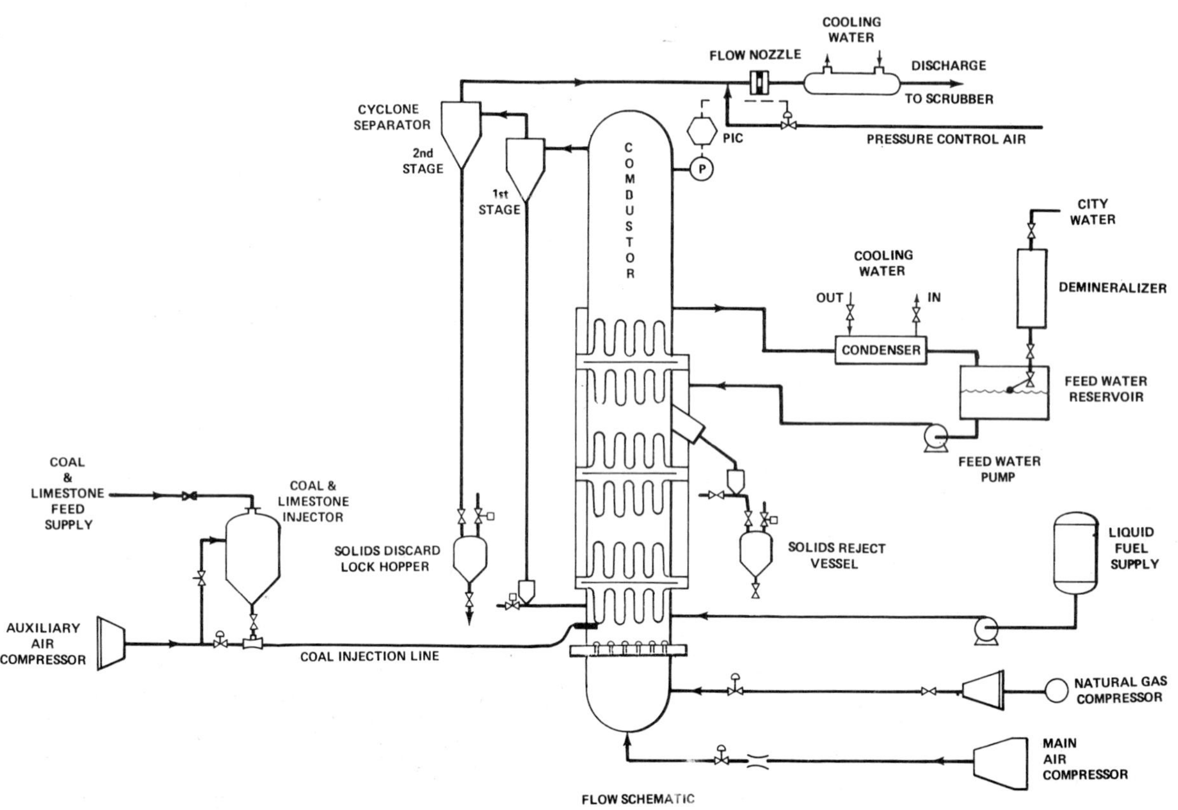

Figure 2. PFBCC Miniplant

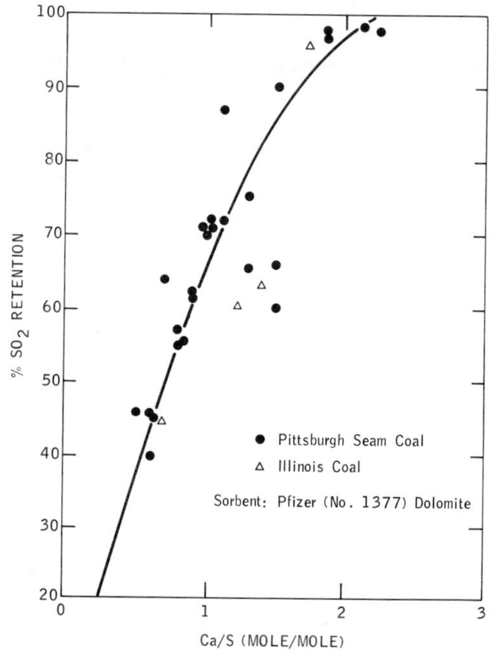

Figure 3. SO_2 Retention with Dolomite Sorbent

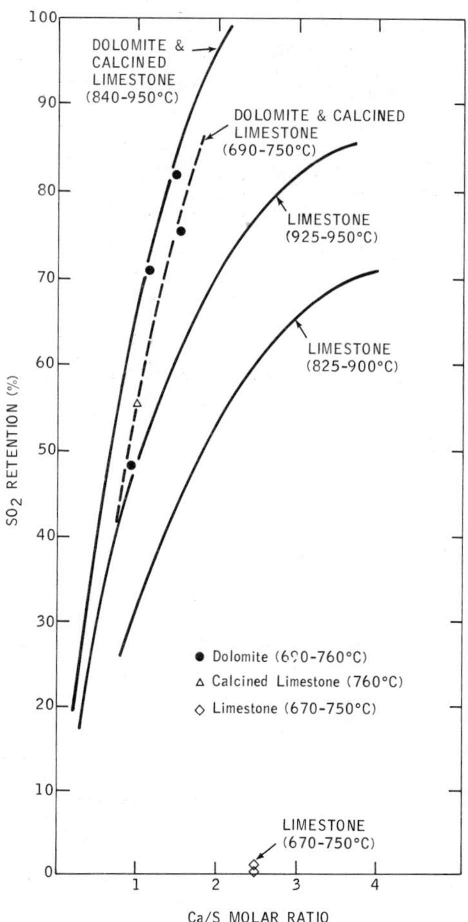

Figure 5. Comparison of Sorbent SO_2 Retention

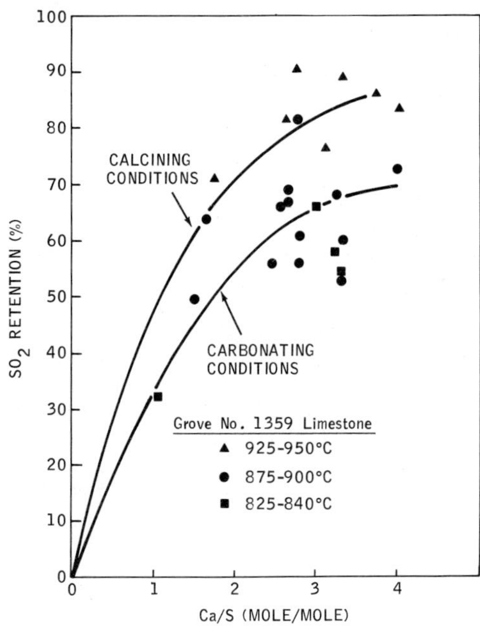

Figure 4. SO_2 Retention with Limestone Sorbent

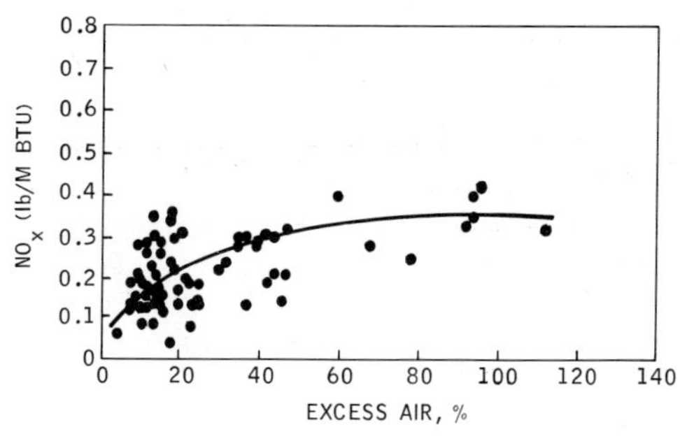

Figure 6. NO_x Emissions

EXPERIMENTAL RESULTS FROM AN 0.46m-DIAMETER FLUID BED PILOT PLANT

By T. E. Taylor

Foster Wheeler Development Corporation

John Blizard Research Center

12 Peach Tree Hill Road

Livingston, New Jersey 07039

SYNOPSIS

A series of combustion tests was carried out in an 0.46m fluid bed combustor to determine the feasibility and operating characteristics of a variety of solid fuels. Once-through combustion efficiencies were found to range from 80% while firing coke breeze, to greater than 99% with oil shale. Carbon carry-over recycle produced significant improvements in combustion efficiency. Tests of overbed fuel feeding produced results consistent with in-bed feed data.

NOTATIONS

η_c	Combustion efficiency based on carbon balance
sm^3/m	Standard meters^3/m
XSA	Excess air
R%	Percent oversize

INTRODUCTION

The utilization of fluid bed combustion for the generation of steam in industrial and utility applications is becoming increasingly attractive as emission regulations become more stringent and clean fuel supplies dwindle. Primary advantages of the process are compact furnace design, reduced slagging and fouling, reduced heat transfer surface requirements and an effective method for the reduction of SO_x emissions. Tempering these advantages is the relatively high conversion rate of fuel nitrogen to its oxides, which although less than in conventional plants does not represent a sigfiicant reduction in emissions.

To gain a better understanding of the fluidized bed combustion and generate design data, a pilot plant program was begun whose major objectives include:

1) Testing of a variety of fuels to determine their compatibility with the fluid bed process.
2) Determination of optimum operating conditions and equipment configuration with respect to combustion efficiency and emissions of NO_x and SO_x.
3) Determination of sulfur capture capacity of limestone bed materials.
4) Measurement of heat-transfer coefficients in submersed and above-bed tubes.

Results from the objectives 1 & 2 will be discussed in this paper.

EXPERIMENTAL SETUP

The fluid bed combustor (Figure 1) consists of a refractory-lined steel cylinder approximately 2.13m in height. The lower half of the furnace has an internal diameter of 0.46m, while the upper half is flared out to an 0.61m I.D. The refractory, a high-alumina castable type, has an average thickness of 0.23m throughout the furnace.

Two types of air distribution plate were used in the tests. The majority of these tests were performed with an angle cover stainless steel distributor plate whose 25.4mm holes are covered by inverted pieces of stainless steel angle which control pressure drop and aid in even air distribution. The other was a perforated plate design.

Air was supplied to the distribution plate and material feeders by a positive displacement-type compressor rated for 10.76sm^3/m at 126 Kpa. Air stream flow rates were measured with orifice plate meters located in the air lines after the surge tank.

Heat removal from the furnace was accomplished with two compound-flow multiple-pass tube bundles, one in the bed zone, the other in the convection pass. The total surface area of each bundle was approximately 1.49 square meters. During the testing, the bed bundle was cooled with steam and convection bundle with water. Flowmeters, installed in each coolant supply line, measured the quantities being fed. Several tubes in each of the compound-flow bundles were outfitted with thermocouples to measure the inlet and outlet temperatures of the coolant stream. After passing through the tube bundles, the coolant was piped through a condenser to eliminate coolant in the vapor phase.

Products of combustion left the furnace through a water-cooled stack and then passed through two stages of cyclone separators. From the outlet of the second cyclone, the gas could either be piped throught a dilution flow mixer and then into the baghouse filter or directly out of the building. The product gases were sampled before the cyclones and after being cooled and dehumidified, passed through continuous, real-time gas analyzers, where the concentrations of CO, SO_x, NO_x, and O_2 were determined. A spare sample point was available for batch testing of the gas for hydrocarbons and CO_2.

Propane gas for start-up was supplied to the unit from two 45.4-Kg cylinders. The gas, after passing through a regulator, rotometer, and shut-off valve, was admitted to the unit through a manifold which fed into the ash removal line.

Bed material and fuel were metered from their respective bulk hoppers to the common feed air lock by variable-speed screw feeders. Load cell systems monitored the weights of each component fed. After mixing in the air lock, the limestone/coal mixture was pneumatically transported into the center of the furnace to a point approximately two inches above the distribution plate. Overbed feeding was accomplished by pneumatically conveying the mixture to a vertical feed pipe which discharged approximately 0.91m above the bed. The 454-Kg capacity weigh hopper was serviced by a coal sizing and handling system.

Spent limestone and coal ash were removed from the bottom of the furnace via a 76mm-diameter standpipe whose output was controlled with a variable-speed screw feeder.

Recycle operation was accomplished via a piping arrangement which routed the cyclone returns back to the feed system air lock. Flow rates were obtained by passing the dust stream into a sample container for a measured time.

Temperatures of the furnace, inlet and outlet gas streams, and some of the peripheral equipment were monitored and recorded on two multipoint recorders.

Pressure-drop data from the bed were monitored on a strip chart recorder powered by a differential-pressure transmitter which was attached to the furnace.

Three water-cooled quartz windows, one at bed level, one in the convection pass, and the third in the top of the furnace, permitted viewing of the furnace durning operation.

EXPERIMENTAL PROCEDURES

The majority of the tests were performed in batch mode to permit changes in feedstock, sorbent, and equipment configuration. Following a light-off, warm-up, and stabilization period of approximately 5 hours, test data taking was begun. Each test was of 1 to 2 hours duration depending upon the gas velocity and prevailing furnace conditions. Furnace stabilization at the test conditions was determined by examination of the bed temperature profile and the stability of the residual O_2 meter.

Overfire feed tests were performed by first starting up the combustor with in-bed feed and then switching over to overbed firing. In both cases the fuel and sorbent were transported dilute phase pneumatically to the feed point.

The effect of dust recycle on combustion efficiency was tested by continuous feeding of the fines collected in the two cyclone separators through the coal feed air lock and into the furnace with the raw coke. Recycle flow rate was measured by diverting the dust stream to a sample container for 5 minutes and then weighing the sample. A collection time of 5 minutes was selected to obtain a representative sample without unduly upsetting the flow rate.

Combustion efficiency was determined by measuring the coal feed rate and collecting all of the carried-over material in the cyclones and baghouse filter. This method afforded the most accurate results as no isokinetic sampling with its inherent problems was necessary. Chemical analysis of the inlet and outlet streams together with the associated flow rates permitted the calculation of a carbon balance.

TEST RESULTS

Results of the pilot plant test series have been tabulated in Table 1 to enable the reader to gain some insight into the operating conditions and the efficiencies obtained with various fuels and equipment configurations. Typical fuel analyses have been included in Table 2 for reference. An understanding of the feed and bed size ranges can be obtained from Figure 2 which is a Rosin Rammler plot of a typical run.

Should comparison of these data to other installations data be tried, several factors unique to this pilot plant should be kept in mind. The first is that the unit is refractory-lined and not surrounded by a waterwall enclosure. This permits one to extrapolate data from this "slice of bed" to larger sizes. It is believed that small units with waterwall enclosures are subject to large wall effects. These wall effects which result in particle quenching and poorer combustion efficiency can be aggravated by the low metal temperatures associated with low pressure saturation temperatures. Another along the same line is the average metal temperature of the submerged heat exchanger surface. The 0.46m pilot plant uses steam to cool the in-bed surface. Typical entrance and exit temperatures were 138 and 538°C respectively which result in an average metal temperature of 371°C (assuming 10°C drop across the tube). This value while modeling a high pressure utility steam generator makes data comparison difficult.

CARBON CARRY-OVER RECYCLE

Tests 1-13 were performed on coke breeze, a high-carbon low-volatile residual fuel. The extremely low volatile content resulted in rather poor once-through combustion efficiencies (81-84%). To improve efficiency without having to resort to a separate carbon burnup cell, reinjection of the fines from the two cyclones was tried. The results of these tests appear both in Table 1 and Figure 3. As is readily apparent from the plot of combustion efficiency vs. XSA, recycling of the carbon fines produced a considerable increase in combustion efficiency. A regression analysis of the data reinforced this by indicating that carbon recycle was the most significant variable affecting combustion efficiency. What is interesting is the fact that it appears that an ideal XSA ratio exists. One can postulate that, as the amount of air flow increased above that necessary for a stoichiometrically correct mixture, more complete combustion will occur primarily due to better mixing. The excess air tends to compensate for the air tied up in bubbles. However, a point can be reached after which increases in XSA will not promote better mixing but will instead tend to quench the combustion process. What remains to be determined is whether a recycle process can be operated on a continuous basis without cyclone overloading and dense phase transport out of the bed.

EFFECTS OF FUEL SIZE

Fuel size was varied primarily by the removal of fines (<0.297mm) in several tests to determine the effect on combustion efficiency. Several different results were noted depending upon the fuel tested and the method of feed introduction.

Run 13, a test of 6.35 x 0.297mm coke breeze, resulted in a combustion efficiency of 91%. This is ≈5% greater than that of Run 1 performed at the same operating conditions with 6.35mm x 0 coke.

The explanation for this might lie in the fact that the residence time necessary for burnout of these extremely low volatile particles is not sufficient before elutriation occurs.

Run 16, on the other hand, was performed with 6.35 x 0.297mm subbituminous. The 96% combustion efficiency obtained ranged from 0-2% less than that of the 6.35mm x 0 runs (18, 19, 20).

Thus, it appears that, at the operating conditions maintained, removal of the fines from a medium volatile fuel results in no increase in combustion efficiency and may in fact result in poorer burnup.

A drop in combustion efficiency was also noted in the overbed fuel feed tests (Run 21) when 9.53 x 6.35mm subbituminous fuel was fired. The 95% efficiency was ≈2% less than values obtained in Runs 15 and 20 where 6.35mm x 0 fuel was fired.

OVERBED FEEDING

Conceptual design studies of utility size fluidized bed steam generators have identified the need for better methods of coal and limestone feeding. If the currently accepted practice of providing one in-bed coal injector for every 2.74-4.88m^2 of bed is scaled to an 800-MW unit, the number of feed points become staggering. Even more unbelievable are the feed and control systems necessary to evenly distribute the flow to the feed points and monitor each line for pluggage. As a first step in dealing with this problem, tests were run to determine whether crushed coal could be successfully introduced above a fluid bed with submerged heat transfer surface. The tests were performed with coal no larger than 9.53mm as it was felt that temperature control problems may occur with larger sizes, due primarily to the larger coal inventory in the bed.

Tests 15, 20, and 21 were run with a subbituminous coal and confirmed that above-bed feeding is a viable approach to the feed problem. The combustion efficiencies listed in Table 1 indicate that within the accuracy of the tests no difference exists.

Two problems remain to be answered before overfire feeding of fluidized beds can be seriously considered for commercial units. The first is the question of temperature control when large fuel sizes are used. Because combustion rate is diffusion-controlled with the larger particle sizes, a considerably larger inventory of fuel must exist in the bed to equal the total surface area of a smaller fuel size distribution and thus maintain constant bed temperature. It has been estimated that a fuel concentration of 15-20% will be necessary if 25.4mm x 0 coal is fired. Further testing will be necessary to determine (1) whether the response lag introduced by fuel inventory can be controlled and, (2) whether the predicted high particle surface temperatures will cause tube slagging and clinker formation. The second problem concerns the design of the overbed feeder. Should a smaller particle size distribution, i.e., 12.7mm x 0 be required, how can the feed be evenly distributed over a 6.1 x 6.1m bed?

NO_x EMISSIONS

Emissions of the oxides of nitrogen are tabulated in Table 1. Fuel nitrogen conversion efficiencies were found to range from 15-50% with higher conversion being favored by higher bed temperature and residual O_2. The NO_x emissions measured were found to be largely NO with a 2-5% contribution by NO_2. An interesting trend in NO_x emissions was noted during the coke breeze tests and is displayed in Figure 4 (Taylor, 1976). It should be noted that the emission rate increases with temperature in the range of 898-906°C. At 960°C, a maximum emission rate is obtained. Operation above this temperature, however, results in lower emission rates.

The two major sources of nitrogen available during the combustion process for oxidation are free nitrogen in the combustion air and chemically bonded nitrogen in the fuel. A review of the literature reveals that the contribution of atmospheric diatomic nitrogen to the concentration of NO_x does not become significant until the reaction temperature is in excess of 982°C. Therefore, NO_x emissions from the fluidized bed combustor operating below 982°C depend primarily on the nitrogen content of the fuel and how much of the released nitrogen is converted to NO_x during combustion.

Many factors influence the conversion of fuel nitrogen to NO_x, namely, the nature of chemical bonding to other constituents of fuel, the time temperature history of the combustion process, etc. Data from Pereira (1974) indicate that the contribution of char to NO_x emissions increases, then starts dropping off at around 815°C. As temperature increases, NO_x emissions from the volatiles increase, but there is a corresponding reduction of NO_x emissions from char. The net effect of char and volatiles is levelling off of NO_x emissions, or very close to it, beyond 871°C or so.

Reduction of NO_x emissions from char at high temperature (beyond 815°C) is due to the heterogeneous reaction of char and already-formed NO_x. A maximum reduction occurs at 1204°C. It is believed that these mechanisms are responsible for the drop in emissions depicted in Figure 4.

CONCLUSIONS

Carbon carry-over recycle is an effective method of increasing combustion efficiency of low-grade fuels.

Overbed feeding into beds containing submerged heat exchange surface appears feasible.

Conversion efficiency of fuel nitrogen to oxides of nitrogen will vary between 20-50% depending upon bed temperature and XSA rate.

Fines content of fuel feeds have varying effects on combustion efficiency, depending upon volatility content and feeding method.

REFERENCES

Anson, D. (1976) Prog. Energy Combust. Sci. Vol. 2, pp. 61-82.

Bartok, W., et al, (1969) Esso Res. & Engrg. Co., Final Report GR-1-NOS-69, Contr. No. PH 22-68-55 (PB 192 789).

First, 2nd & 3rd Intern'l Conf. Fluidized-Bed Combust. sponsored by EPA, Hueston Woods, Oxford, Ohio (1968, 1969, 1972).

Gibbs, B. and Beer, J. M. (1973) paper 104 to Combust. Inst. Eur. Symp., Academic Press.

Jonke, A., et al, (1969) Report ANL/ES-CEN-1001, Argonne National Laboratory.

Pereira, F. J., et al, (Aug. 1974) paper to 15th Int'l Symp. Combust., Tokyo, Japan.

Shaw, J. (1968) paper to 7th Int'l Conf. on Coal Science, Prague.

Shaw, J. (1973) Jnl. Inst. Fuel, p. 170.

Siegmund, C. W. and Turner, D. W. (Oct. 1973) Combustion, pp. 24-30.

Sterling, C. V. and Wendt, J. O. L. (1974) EPA-650/2-74-017.

Taylor, T. E., Biswas, B. K. and Bryers, R. W. (1976) AIChE Annual Meeting.

Wright, S. J., Ketley, H. C. and Hickman, R. G. (June 1969) Jnl. Inst. Fuel, XLII No. 341.

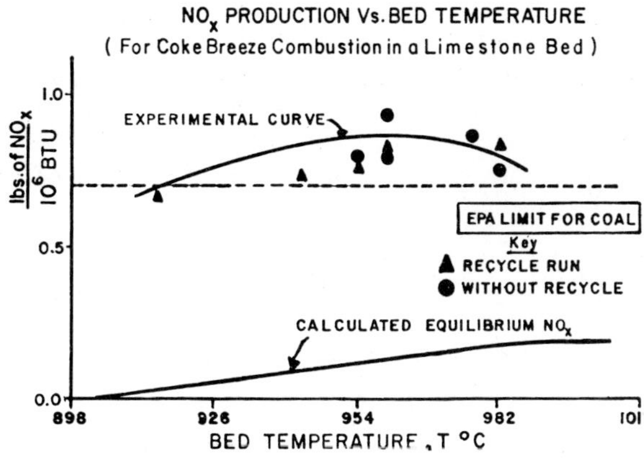

Figure 4

TABLE 1

SUMMARY OF TEST RESULTS

Run No.	Ref.	Fuel	Recycle	Superficial Velocity (m/s)	Average Bed Temp. (°C)	% XSA	η_c (%)	NO_x (ppm/vol)	Fuel Sizes (mm)
1	T1	Coke	No	2.98	960	18	86	680	6.35x0
2	T2	Coke	No	3.03	977	28	80	560	6.35x0
3	T3	Coke	No	3.39	982	31	86	500	6.35x0
4	T4	Coke	Yes	3.04	982	14	89	600	6.35x0
5	T5	Coke	No	2.96	960	16	86	580	6.35x0
6	T6	Coke	No	3.51	954	23	86	590	6.35x0
7	T7	Coke	Yes	3.56	943	15	94	535	6.35x0
8	T8	Coke	No	2.88	1038	12	84	---	6.35x0
9	T9	Coke	Yes	2.69	949	02	89	---	6.35x0
10	T10	Coke	Yes	1.95	960	05	94	660	6.35x0
11	T11	Coke	Yes	3.00	916	18	84	480	6.35x0
12	T12	Coke	Yes	3.05	966	04	91	---	6.35x0
13	T13	Coke	No	2.97	954	08	92	610	6.35x0.297
14	T14	Coke Fly Ash	No	1.39	1079	---	87	1100	-1.7
15	R23	Subbituminous	No	2.43	898	15	97	270	6.35x0
16	R24	Subbituminous	No	2.62	882	15	96	200	6.35x0.297
17	R26	Oil Shale	No	2.62	843	14	99	430	6.35x0
18	C27	Subbituminous	No	2.65	898	15	96	210	6.35x0
19	C30	Subbituminous	No	2.52	896	15	98	230	6.35x0
20	C31	Subbituminous	No	2.82	898	15	97	220	6.35x0
21	C32	Subbituminous	No	2.49	898	15	95	270	9.53x6.35
22	C33	Bituminous 1	No	2.71	871	15	95	280	6.35x0
23	C34	Bituminous 1	No	2.82	896	15	97	---	6.35x0
24	C35	Bituminous 1	No	2.64	898	15	93	---	6.35x0
25	H36	Bituminous 1	No	2.64	871	14	93	---	6.35x0
26	H37	Bituminous 1	No	2.72	890	15	93	---	6.35x0
27	H38	Lignite 1	No	2.71	892	15	96	---	6.35x0
28	H39	Bituminous 1	No	2.63	898	15	98	---	6.35x0
29	H40	Lignite 1	No	2.69	871	15	95	260	6.35x0
30	H42	Lignite 2	No	2.41	896	15	96	---	6.35x0
31	S43	Lignite 2	No	2.60	898	15	98	260	6.35x0
32	S44	Lignite 3	No	2.70	898	15	96	265	6.35x0
33	S45	Lignite 3	No	2.68	898	14	96	300	6.35x0
34	S46	Bituminous 2	No	2.53	849	17	82	420	6.35x0

Static Bed Height 0.61 m

FUEL KEY

Oil Shale	-	Utah
Subbituminous	-	Decker-Montana
Bituminous 1	-	Indiana No. 5
Lignite 1	-	Knife River
Lignite 2	-	Indian Head
Lignite 3	-	Baukol Noonan
Bituminous 2	-	Crow, County Roscomman, Ireland

TABLE 2

REPRESENTATIVE FUEL ANALYSES

Fuel	Decker	Indiana #5	Knife River	Indian Head
Proximate				
Fixed Carbon	43.37	46.50	31.87	30.39
Volatile Matter	39.41	37.73	33.43	31.99
Ash	4.94	7.95	9.52	10.00
Moisture	12.28	7.82	25.18	27.62
Ultimate				
Carbon	61.70	67.48	48.86	41.42
Hydrogen	5.97	4.29	2.86	2.82
Oxygen	13.70	8.87	11.70	16.94
Nitrogen	1.03	0.96	0.60	0.52
Sulfur	0.38	2.63	1.28	0.68
Ash	4.94	7.95	9.52	10.00
Moisture	12.28	7.82	25.18	27.62

Fuel	Baukol Noonan	Cokebreeze	Crow	Oil Shale
Proximate				
Fixed Carbon	33.04	79.86	30.43	---
Volatile Matter	34.32	2.94	14.63	---
Ash	14.46	10.88	52.28	---
Moisture	18.18	6.32	2.66	---
Ultimate				
Carbon	45.18	79.55	33.79	22.45
Hydrogen	3.35	0.59	2.82	2.39
Oxygen	17.50	1.50	7.38	---
Nitrogen	0.62	0.57	0.64	0.56
Sulfur	0.71	0.59	0.43	1.15
Ash	14.46	10.88	52.28	73.45
Moisture	18.18	6.32	2.66	0.00

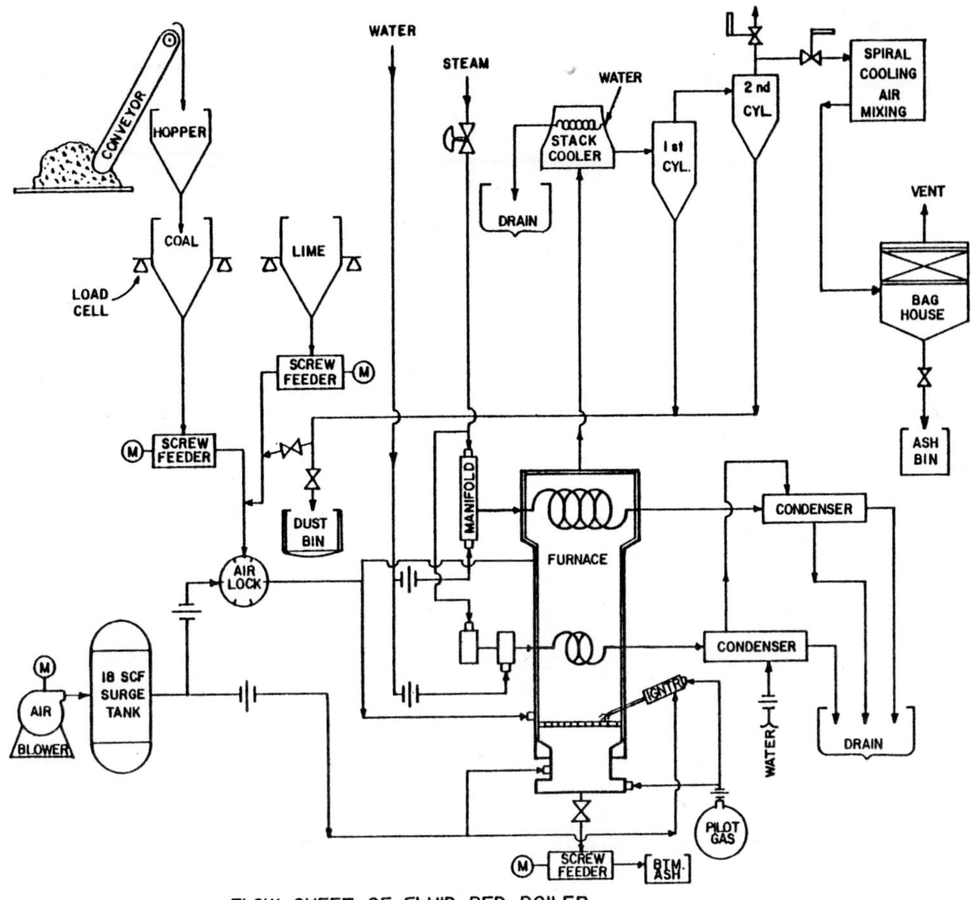

Figure I

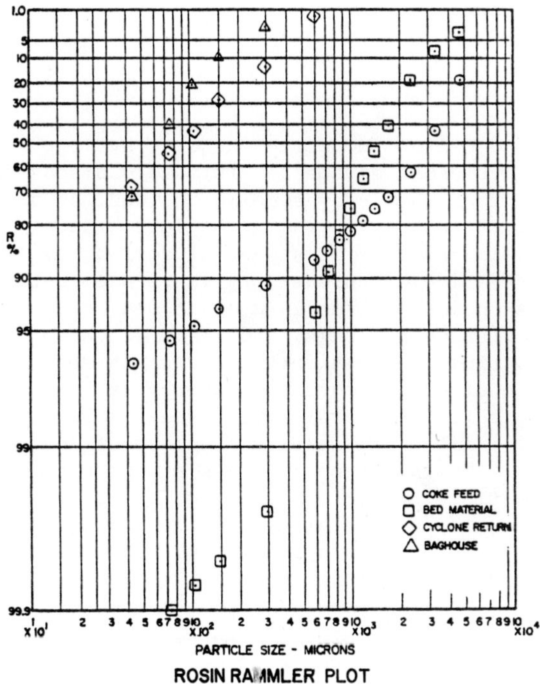

Figure 2

COMBUSTION EFFICIENCY Vs. % EXCESS AIR
(For Coke Combustion in a Fluid Bed, With and Without Recycle)

WITH RECYCLE

WITHOUT RECYCLE

COMBUSTION EFFICIENCY η_c

% EXCESS AIR

KEY
● RUNS WITHOUT RECYCLE
▲ RUNS WITH RECYCLE

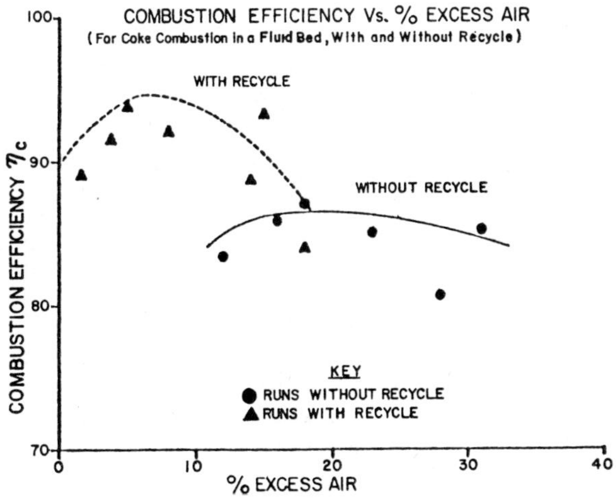

Figure 3

FLUIDIZED BED INCINERATION - A DESIGN APPROACH FOR COMPLETE COMBUSTION OF HYDROCARBONS

By J. Baeyens* and D. Geldart+

*Seghers Engineering NV,
Brussels,
Belgium.

+Postgraduate School of Powder Technology,
University of Bradford,
U.K.

SYNOPSIS

Of the new processes developed to ensure the safe disposal of sludges from municipal and industrial effluent treatment plants, the fluidized bed incinerator is attracting increasing interest. This paper presents a practical strategy for their design and concentrates on the freeboard design necessary to achieve complete burn-out of the combustibles in the sludge.

NOTATION

Ar	Archimedes Number $\frac{\rho_g d_{sv}^3 (\rho_p - \rho_g) g}{\mu^2}$
b	constant determined by stoicheiometry of the reaction
C_A	concentration of the oxygen in the bulk of the gas in freeboard
$\bar{d}$	size of the burning particle at any time
d_i	size of initially elutriated particles
D	diameter of fluidized bed at the distributor
D_{fb}	diameter of the freeboard
H_{fb}	height of the freeboard
k	rate constant for a 1st order reaction
k_G	mass transfer coefficient between gas and a particle
L	burn-out length
Re_o, Re_t	particle Reynolds Number at incipient fluidization and free fall velocities
Sh	Sherwood Number ($k_G\bar{d}/\mathbb{D}_{AB}$)
T	temperature
U, U_o, U_{fb}, U_t	superficial velocity of gas, at incipient fluidization, in the freeboard, and at terminal conditions
α	coefficient in (5)
ρ_g, ρ_p	density of gas and particle
ρ_{MP}	molar density of the sludge particles
μ	gas viscosity
$\mathbb{D}_{AB}$	diffusivity of oxygen in air
τ	burn-out time

INTRODUCTION

Sludge from industrial water treatment and from municipal sewage is ultimately disposed of in four ways - by tipping on land, by disposal at sea, by spreading on agricultural land (in the case of sewage sludge) or by incineration. Increasingly there are restrictions on disposal on land or at sea because of pollution problems, and there is therefore increasing interest in the incineration of sludge alone or together with refuse to yield a hygienic ash of relatively small volume which can be more readily disposed of (Gale 1975). Incineration furnaces may be single or multiple hearth or fluidized beds. The latter are particularly suitable for sludges which have too low a combustible content (high water, high ash content) to be burnt in conventional furnaces. Although a number of large fluid bed incinerators are now in operation, little has been published and this paper is an attempt to provide a basis for design.

GENERAL COMMENTS

Figure 1, published by courtesy of Lurgi Apparate-Technik GmbH, illustrates a typical fluid bed incinerator. A bed of inert material, usually coarse sand, is fluidized at high velocity in a refractory shell which may be cylindrical or slightly tapered. Primary combustion air provides the gas for fluidization and secondary air is admitted above the bed. Depending on the nature and composition of the sludge, the feed may be introduced by pumps or screw feeders into and/or above the fluidized bed. Evaporation of water and partial combustion of the sludge take place in the bed, the combustion being completed in the freeboard. After energy recovery the flue gases are further cooled and cleaned to meet local air pollution control requirements.

The efficiency of the incinerator depends both on geometrical and operational factors such as fuel and sludge distribution, bed material, bed height and diameter, mixing, combustion rates, operating temperature, elutriation etc. These will be considered during the stages of the design procedure outlined in Figure 2.

THE DESIGN PROCEDURE

Each of the stages is numbered and the numbers correspond to the explanatory notes in the text. Wherever possible we have tried to verify the basic design criteria by comparison with large scale data made available to us.

EXPLANATORY NOTES

Note 1 The sludge characteristics (% H_2O, ash, composition of combustibles) and thermodynamic data are generally available or can be estimated with sufficient accuracy. Sludges vary considerably both in concentration (5-20% solids for refinery sludges (Becker & Wall, 1975), 15-40% for dewatered sewage sludges (Gale, 1975)), and the percent combustibles (50-75% of dry solids) and consequently their calorific values vary. Depending on the composition of combustibles, the desired operating temperature and the amount of heat recovery made the minimum solids concentration required for autothermal operation varies. This is illustrated in Figure 3 for typical commercial applications.

Even for complex systems such as municipal sewage sludges, various studies (Tamalet, 1974) have shown that the composition of the organic matter averages to $(CH_{1.65}O_{0.34}N_{0.1})_m$ Depending on the ash content the calorific value lies between 5500 and 800 k cal/kg dry solids containing 100% to 20% combustibles. The calorific value of digested sludges is 10-15% lower.

Note 2 The stoicheiometry enables us to calculate the air requirements for complete combustion allowing for approximately 20% excess air. In general 80% of the total air is introduced through the windbox and 20% into the freeboard. Figure 4 shows the heat balance including heat recovery. Bed temperatures are generally 700-900°C (see Note 4). Air preheating and preconditioning/preheating of the sludges are commonly used, but sludge drying gives rise to the evaporation of malodorous components. These require further degradation by combustion in the fluid bed, and since this constitutes an additional thermal load it is desirable to keep the volume of associated air as small as possible by using indirectly heated sludge dryers. The relative amounts of heat transferred to the air and sludge will depend on the pumping or handling properties of the sludge at various concentrations. The lower limit for readily pumpable sewage sludge is approximately 92% water (Munro & Rolfe, 1975). Sludges of any sort with less than 82% water generally have to be fed by screw conveyor or slinger.

Note 3 The volumetric flow rate of gases, including water vapour, into the windbox, through the bed, and into the freeboard can now be calculated. Special burners are often used at start up to increase the volumetric flow rate of gas because the fan is often not large enough to fluidize the bed with cold air.

Note 4 Generally the inert solids are quartz sand in the size range 0.6-3 mm. Coarse particles are used not only because they have higher thermal and incipient fluidization velocities but also because they are less affected by stickiness due to fewer interparticle contacts and greater particle momentum which breaks incipient bonds.

Special problems occur (Becker & Wall, 1975) when treating chemical sludges or whenever salt concentrations are high. The possible formation of eutective mixtures with a melting point close to the bed operating temperature (e.g. 800°C for $Na_2O.3SiO_2$) can cause defluidization and necessitates a decrease in operating temperature, the use of coarse solids other than silica sand or the addition of clay or special metal oxides in the feed so as to bind the alkali silicates to compounds with higher melting points.

Note 5 The high heat and mass transfer rates required in a fluidized bed combustor are associated with rapid solids mixing and are only achieved within a gas velocity range U_{min} - U_{max}. Our earlier mixing and heat transfer studies (Baeyens, 1973; Baeyens & Geldart 1973a) on powders ranging from 40 to 1850 µm, together with published data enabled us to establish the range of gas velocities and in particular U_{min}. Figure 5 shows U_{min}

and U_{max} together with the operating ranges for some commercial incinerators. Each pair of velocities represents flow with and without the evaporation of water and correspond roughly to the velocities at the distributor and the top of the bed respectively. Most beds operate well above U_{min} in order to reduce the bed diameter, but this is discussed further in Note 6.

The minimum fluidization and particle terminal velocities may be estimated from (Baeyens & Geldart, 1973b):

$$Ar = 1823\ Re_o^{1.07} + 21.7\ Re_o^2 \qquad (1)$$

$$Ar = 18\ Re_t + 0.33\ Re_t^2 \qquad (2)$$

However, it is desirable to measure U_o particularly for large particles.

Note 6 It is common for incinerators to have tapered bed zone to allow for the large extra volume of gases produced by the evaporated water. This taper may be continued into the freeboard (or the freeboard diameter may be increased more sharply) to minimise the elutriation of fines. A high gas velocity in the bed gives a smaller bed diameter but leads to several disadvantages: (a) the coarse inert particles are ejected at high velocity by the bursting bubbles and unless the freeboard is very high some will be carried out. (b) the injected fines will be carried out of the bed zone rapidly and although their residence time in the incinerator can be increased by having a large free board, much of the combustion would then take place above the bed rather than in it. There is a lack of data both on the ejection of coarse particles and on the elutriation of fines injected into a bed of coarse solids. For the present we may note that in practice superficial velocities in the bed are normally limited to < 2 m/s and less than 1% of the bed weight has to be added daily.

In the event of solids being deposited on the bed particles causing an increase in particle size, a solids removal system must be provided at the bottom of the bed, and the design must allow for particle growth.

CALCULATION OF FREEBOARD FOR COMPLETE COMBUSTION

Note 7 The model we shall use to calculate the freeboard height is based on the following assumptions. The sludge, normally in slurry form, is fed uniformly to the bed where its temperature rises rapidly, water is evaporated and combustion starts. The fluidized sand acts as a trap for the sludge particles which are subjected to thermal cracking, gasification and combustion; this together with flaking and abrasion causes the particles to shrink. When they reach a size whose terminal velocity is less than the superficial gas velocity they are liable to be elutriated and continue burning in the freeboard. Particles larger than this critical size will be ejected from the bed but will fall back.

Combustion rates and intensities for some commercial plants have been estimated and most beds are found to operate between 0.6 and 0.85 x 10^6 kcal/h m^3 bed volume.

Combustion of relatively large particles in a fluid bed has been the subject of several recent papers (e.g. Avedesian & Davidson, 1973; Waters, 1975). Commercial fluid bed incinerators operate with bed heights of 0.5-0.6 m which represents a compromise between the need to (a) minimise compressor power and (b) provide an adequate residence time for sludge/oxygen combustion.

Malodorous gaseous products are produced when the sludge liquors evaporate and these require a residence time in the freeboard of approximately 2 s for their destruction at 750°C. Since freeboard velocities are usually 0.7-1 m/s this time is always exceeded. Most municipal and water treatment sludges consist of solids in the range 40-500 μm and it is the burn-out of the elutriated particles which determines the height of the freeboard.

Note 8 Combustible hydrocarbon solids react with oxygen according to the shrinking particle model $A(g) + bB(s) \longrightarrow$ products. For particles larger than 100 μm and temperatures above 1200 K (Field et al., 1965) the diffusion of oxygen to the surface of the burning particle controls the rate of reaction. The mass balance shows that because of the large amounts of water in the feed, the mol fraction of oxygen in the freeboard is approximately 0.025.

The shrinkage rate of a particle is given by:

$$d\bar{d}/dt = bkC_A/\rho_{MP} \qquad (3)$$

Since oxygen diffusion controls, $k = k_G$ and for single particles

$$Sh = k_G\bar{d}/\mathbb{D}_{AB} = 2 \qquad (4)$$

The temperature of the burning particle is 50 to 100 K above that of the gas, (Avedesian & Davidson, 1973) and at 1173 K and 1 bar $\mathbb{D}_{AB}$ is estimated to be 2.1 cm^2/s. Substituting in (4):

$$k_G = \alpha/\bar{d} \qquad (5)$$

where α is 4.2 cm^2/s.
The burn out time τ of a particle of initial size $\bar{d}_i$ is found by integrating (3):

$$\tau = \rho_{MP}\bar{d}_i/2bk_GC_A \qquad (6)$$

To calculate the height of freeboard required for complete burn out of any particle initially of size $\overline{d}_i$ we have to integrate the absolute particle velocity $U_{fb} - U_t$ over the burn out time:

$$L = \int_{t=0}^{t=\tau} (U_{fb} - U_t)\, dt = U_{fb}\,\tau - \int_{t=0}^{t=\tau} U_t\, dt \qquad (7)$$

U_t is a function of $\overline{d}$ and $\overline{d}$ is a function of t. Combining (2), (3), (5) and (7) gives the burn out length L:

$$L = \frac{U_{fb}\,\rho_{MP}}{2b\alpha C_A}\, d_i^2 - \frac{g(\rho_p - \rho)\,\rho_{MP}}{72\mu b C_A \alpha}\, d_i^4 \qquad (8)$$

Municipal sewage sludges can be represented by a molecular weight of 20.5 m. With a particle density ρ_p of 1.7 g/cm^3, ρ_{MP} = 0.083/m gmol/cm^3. From the stoicheiometry it is calculated that b = 0.8/m. $C_A = C_{O_2}$ = (1/22400) x (273/1123) x 0.025 = 2.71 x 10^{-7} gmol/cm^3. Fitting these values into equations (6) and (8) gives:

$$L = 4.56 \times 10^8\, U_{fb}\, \overline{d}_i^{\,2} - 4.8 \times 10^{15}\, \overline{d}_i^{\,4}\ \mathrm{m} \qquad (9)$$

where $\overline{d}_i$ is the initial size of a particle whose terminal velocity is less than U_{fb}. The solution of (9) is given graphically in Figure 7 for particle sizes which would just be elutriated by various values of U_{fb}. The height of the freeboard should obviously be larger than the burn out length L. Freeboard heights for various commercial incinerators are shown on the same figure for comparison.

Note 9 The design approach outlined enables an economic balance to be drawn up based on various bed and freeboard geometries. However, if possible pilot plant tests should be carried out with the actual sludge to finally select the optimum conditions.

DISCUSSION

The model we have used to calculate the burn out length, and hence the height and diameter of the freeboard is capable of improvement though it would become more complicated and require more basic data. For example, we have neglected gasification reactions and these undoubtedly occur when large concentrations of water vapour are present. The inclusion of these would decrease the burn out time. On the other hand, we have assumed that diffusion of oxygen continues to control even when the shrinking particle becomes smaller than 100 μm (see Note 7). The burn out time of such particles is somewhat longer than indicated by equation (6) and this increases the overall burn out length.

Commercial designs need to be conservative and provide more than adequate freeboard heights (Figure 6); consequently the combustibles elutriated with the ash are negligible (Sohr et al., 1965).

CONCLUSIONS

A design procedure is proposed for fluidized bed incinerators; in particular equations are given relating the gas velocity in the freeboard to the freeboard height required for complete burn-out of combustible particles of sludge. Freeboard designs of various commercial incinerators are compared with the model predictions and give reasonable agreement, however more plant data are essential to give greater confidence in design.

REFERENCES

Avedesian, M.M. & Davidson, J.F. (1973), Trans. Inst. Chem. Engrs., 51, 121.

Baeyens, J. (1973), Ph.D. Dissertation, University of Bradford.

Baeyens, J. & Geldart, D. (1973a), Proc. of Conf. on Fluidization and its Applications, Tolouse, 182.

Baeyens, J. & Geldart, D. (1973b), Ibid 263.

Becker, K.P. & Wall, C.J. (1975), Hydrocarbon Proc., October, 88.

Field, M.A., Gill, D.W., Morgan, B.B. & Hawksley, P.G.W., (1967), "Combustion of Pulverised Coal", British Coal Utilisation Res. Assoc., Leatherhead.

Gale, R.S. (1975), Proc. of Conf. on Application of Chemical Engineering to the Treatment of Sewage and Industrial Liquid Effluent, I. Chem. E. Symp. Ser. No. 41, N1.

Munro, C.S.H. & Rolfe, T.J.K. (1975), Ibid O1.

Sohr, W.H., Ott, R. & Albertson, O.E. (1965), Water Wks. & Wastes Engng., September, 90.

Tamalet, M. (1974), Rev. Gen. Therm. Fr. No. 150/151, June/July, 511.

Waters, P.L. (1975), Inst. of Fuel Symp. Ser. No. 1; Fluidized Combustion C6.

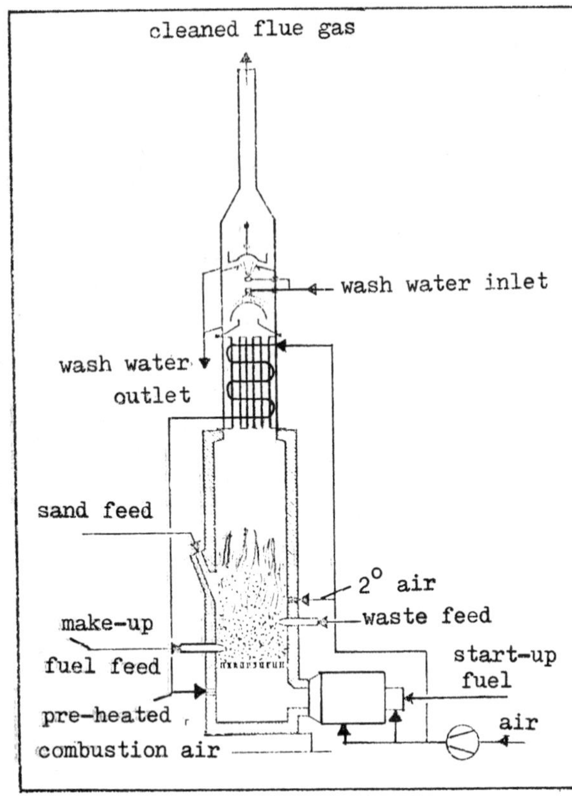

Fig. 1. Sludge Incinceration with Heat Recovery - Lurgi Patent

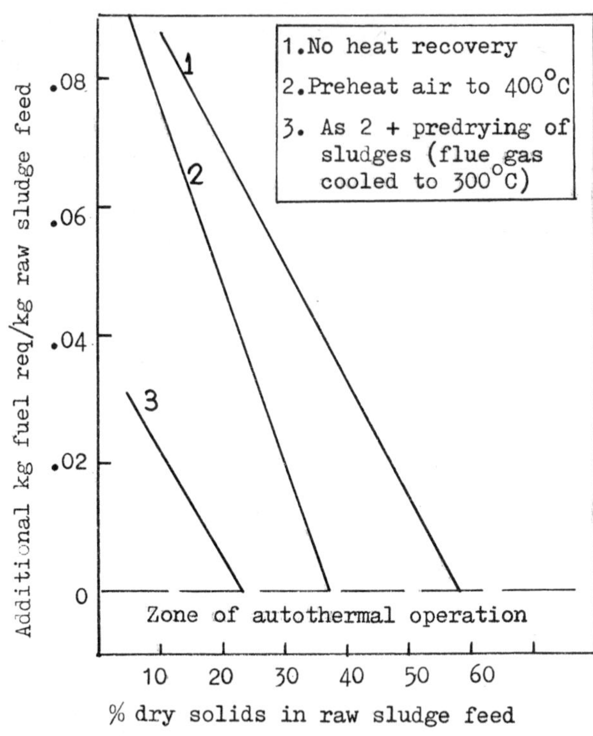

Fig. 3 Fuel required for combustion of sewage sludge with CV of 2000 kcal/kg dry solids, combustible concn 50% of dry solids, temp = 800°C, excess air 20%, fuel 9000kcal /kg.

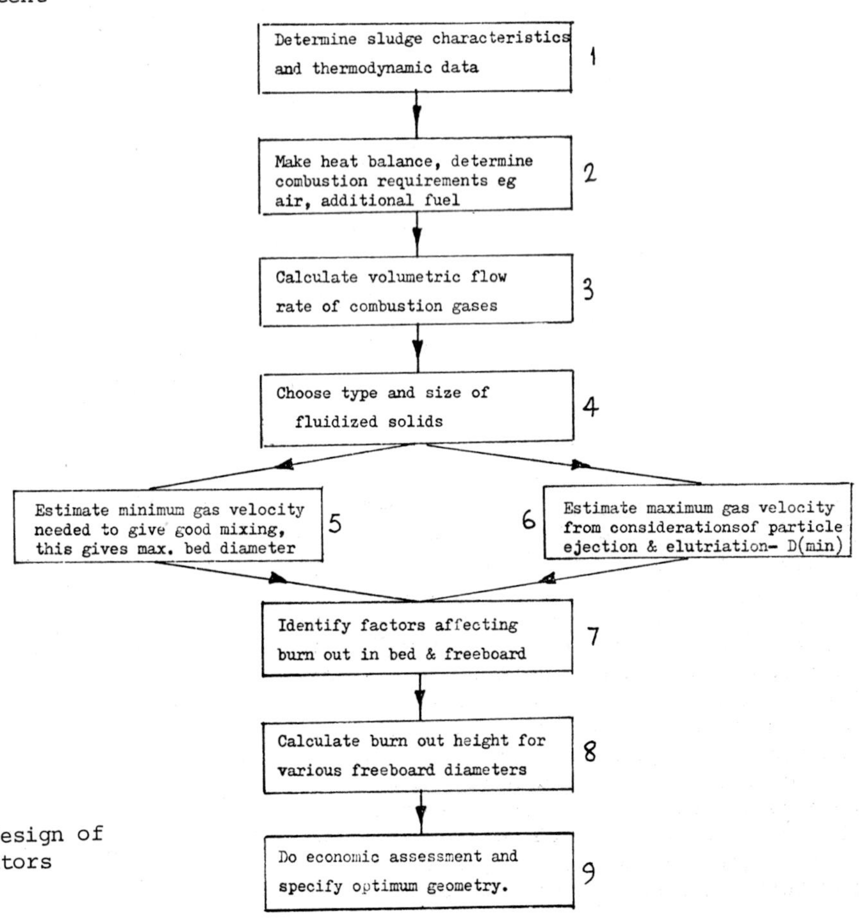

Fig. 2. Flow Diagram for Design of Fluid Bed Incinerators

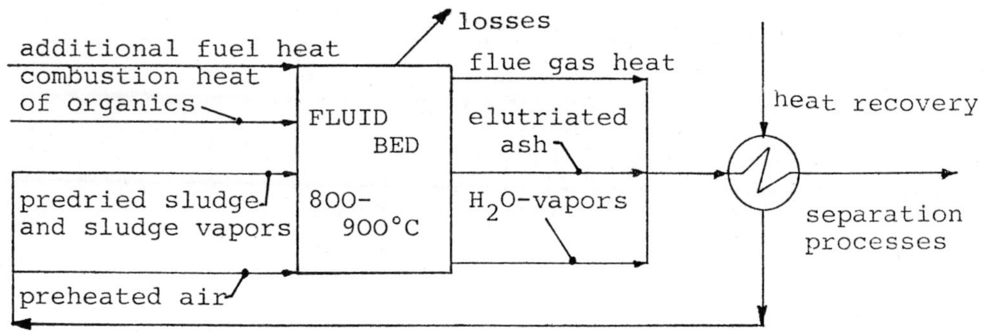

Fig. 4. Heat Balance

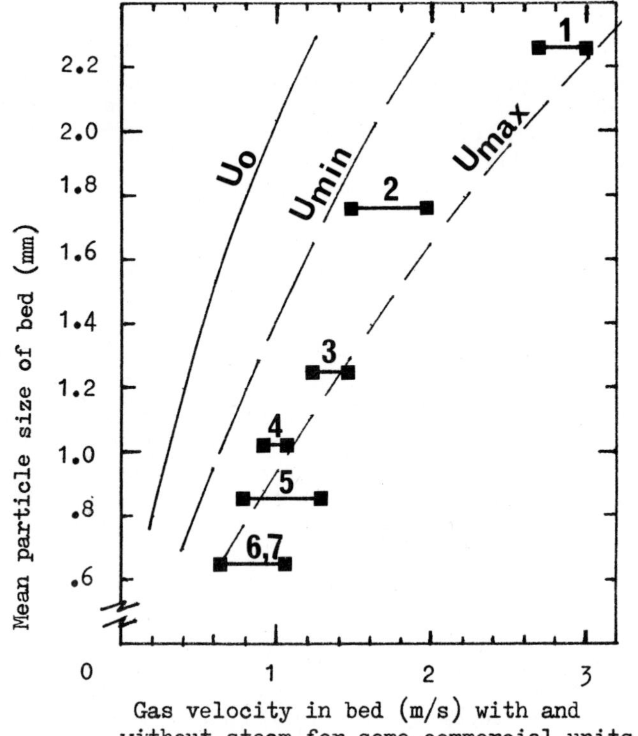

Fig. 5. Optimum Mixing Velocity Limits

Commercial Unit	1	2	3	4	5	6	7
D m	2.5	2.1	1.7	0.6	2.5	3.3	3.3
T ^{o}C	800	750	800	720	800	800	850

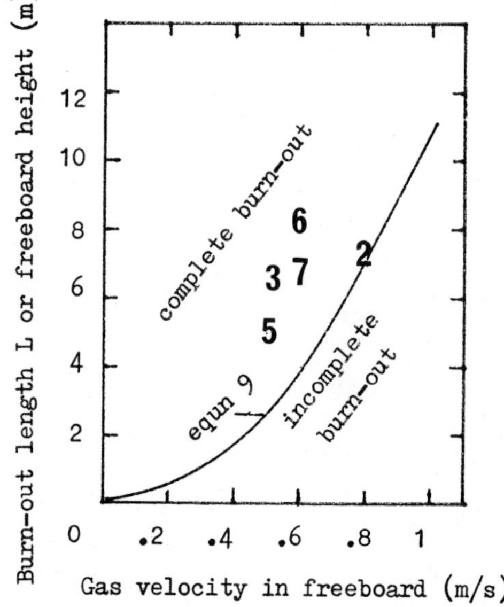

Fig. 6 Theoretical Burn Out Length vs. Freeboard Velocity (Numbers relate to the freeboard heights for units listed on Fig. 5)

ROTATING FLUIDISED BED COMBUSTOR

N. Demircan, B.M. Gibbs, J. Swithenbank, D.S. Taylor

Department of Chemical Engineering and Fuel Technology,
University of Sheffield,
Sheffield S1 3JD
England

A cylindrical rotating fluidised bed combustor was constructed with an inside diameter of 200mm and an axial length of up to 200mm. The fluidising medium was silica sand (0.6-0.7mm) and the bed depth 30-50mm. Twelve water-cooled tubes 25.4mm diameter were immersed in the bed, each extending along its length. The rotational speed was usually between 100-1000RPM (1 -100 g's loading).Propane, gas oil and coal particles (2/4mm diameter) were burned in the bed at temperatures between 700-900°C. Measurements of heat transfer rate, bed temperature and exhaust gas composition were carried out over a range of operating condition. The influence of rotational forces on basic fluidisation characteristics (pressure drop, minimum fluidisation, particle mixing) were also investigated on a perspex model. The results demonstrated that the rotating fluidised combustor can produce considerably higher combustion intensities, has a much better turn-down range, and a more rapid start-up than a conventional fluidised combustor.

NOTATION

D_p	:	particle diameter.
g, g'	:	gravitational acceleration, effective gravitational acceleration.
G	:	flow per unit area.
h	:	height above cylinder base
p	:	pressure.
r	:	cylinder radius.
N	:	number of revolutions, R.P.M.
V	:	radial velocity.
Re_{MF}	:	Reynolds Number = GD/μ_F
R	:	Radius
Ga	:	Galileo Number $= D_p^3\rho(\rho_p-\rho)g'/\mu_F^2$
ρ, ρ_p	:	gas density, particle density respectively.
μ_F	:	gas viscosity.
ω	:	angular velocity, radians/s.

INTRODUCTION

When coal is burned in a fluidized bed, the combustion products can be passed directly through a gas turbine for power production, or additionally in a combined cycle heat transfer tubes can also be immersed in the bed to raise steam. Such plant is considerably more compact than a conventional p.f. fired steam boiler plant, but the power density is still much less than in a conventional gas turbine. The typical combustion intensity of a pressurized fluidized bed is only $\sim$ 10 MW/m^3 compared to a value of > 100 MW/m^3 for a conventional gas turbine combustor (Fig 1). At present there is little data (1) on the mechanism controlling the coal combustion rate in a fluidized combustor, and the maximum throughput is determined by the terminal air velocity at which the bed elutriates. Although it is possible to increase the combustion rate by increasing the pressure, the optimum pressure is typically about 6 atm,dictated by the thermodynamic cycle, and the only additional parameter at our disposal is the gravity field to which the particles are exposed. If the fluidized bed is formed inside a cylindrical distributor plate rotated about its axis, the 'g' loading may be increased to a limit determined by the mechanical strength of the design. The terminal air velocity may therefore be increased by several orders of magnitude at the price of a correspondingly increased pressure drop. An important factor in the design of a rotating fluidized bed is therefore the rotational speed, although the optimum will depend on the specific application.

The use of rotation introduces the 'g' level as an additional variable in the operation of the bed. This is significant since the turndown range of a conventional fluidized bed is often less than 2:1. A wide turn down range is valuable in many applications including transportation power units and electrical generation plant.

The dynamic characteristics of the rotating bed are also superior to a conventional stationary bed, and while this may not be significant in base load plant applications, it will be important for rapid load following and in transportation systems.

Historically, the rotating fluidized bed has been considered as a high surface area heat transfer system for many years, and in the early '60s the concept of a hydrogen cooled rotating fluidized bed nuclear rocket was investigated at the Brookhaven Laboratory (e.g. see Ref. 2). In the early 1970's the Brookhaven team carried out some design studies on the concept of a coal fired rotating bed (Ref. 3).

Other work on rotating fluidized bed combustors known to the authors was carried out at Aston University where the interest has been centred on the combustion of lean mixtures (waste gases) in an uncooled bed (Ref. 4). There has also been some Russian work in this field (Ref. 5), but we are not presently aware of the full extent of their effort. Work at Sheffield University has concentrated on the practical demonstration of the combustion of coal, oil and gas at all mixtures including stoichiometric in a cooled rotating bed. The applications originally in mind were the gas turbine and combined steam/gas turbine systems, however as will be shown, the application of the tumbling mode of operation to incinerators and boilers is also promising.

2. DESIGN AND CONSTRUCTION

To date, two experimental rigs have been built. The first is a hot rig in which combustion tests are carried out, whilst the second is a cold rig, made of transparent plastic, so that the two phase flow can be studied more easily. Since the cold rig is aerodynamically almost identical to the hot rig, the design features discussed below for the hot rig are effectively applicable to both.

The general assembly of the rotating bed is shown in Fig. 2. It is normally mounted with a vertical axis where the air is introduced into the outer cylinder (260mm ID) from below, and the exhaust from the bed is at the top. The air distribution cylinder, which contains the bed, is made from stainless steel 200mm diameter and 5mm thick. Two lengths of 200mm and 100mm respectively have been used. The distributor cylinder is drilled with 1mm diameter holes, uniformly distributed, to give approximately 1% open area. The upper and lower plates confining the bed are water manifolds interconnected by water cooling tubes 25.4mm OD immersed in the bed. Most tests to date have utilized 12 tubes and this permits sufficient heat removal to allow 'stoichiometric' combustion without ash fusion for a 200mm long bed chamber. With the 100mm length, the number of tubes may be decreased to 6 to avoid overcooling. It should be noted that if the bed is scaled up, there will be great difficulty in ensuring adequate cooling at 'stoichiometric' mixtures due to the square /cube law. (NB see Section 5.2) Air and water are supplied to the bed through conventional rotating seals acting on smooth non-corroding surfaces.

In the present arrangement, the whole assembly, including the outer cylinder, is rotated at selected speeds up to 1400 rpm by a 2HP electric motor and belt drive.

Air is supplied by a compressor rated at 0.5 kg/sec at pressures up to two atmospheres gauge. It is metered by a rotameter and the pressure drop across the bed and distributor is measured with a mercury manometer. Gaseous fuel (usually propane) is metered by a rotameter and introduced into the air line about 500mm upstream of the bed.

When required, coal is fed at a controlled rate from a hopper, through a steel pipe mounted on the axis of the bed, and is ejected radially onto the surface of the moving bed, approximately mid way along its length. Alternatively, an oil fuel feeder may be used which again introduces the fuel onto the surface of the bed. This feeder consists of a pump supplying oil through a valve to a calibrated fuel nozzle located at the top of the injection tube. The nozzle sprays a film of oil which runs down the interior surface of the tube. Air is also introduced into the top of the tube, which blows large drops of oil radially out through 4 holes drilled in the lower end of the tube.

Chromel/alumel thermocouples are mounted at various points within the bed and on the distributor plate, and their signals are connected via slip rings onto the display instruments. The flue gas temperature is also measured by chromel/alumel thermocouples. The entry and exit water temperatures are measured by mercury in glass thermometers and the water flowrate is checked by noting the time to fill a known volume.

The cold rig was made of polished perspex with an air distributor cylinder of 200mm diameter; lengths of 200mm, 100mm and 75mm were investigated. Tests were carried out with 0, 6 and 12 simulated water cooling tubes. A scale was mounted on the lower end plate so that the inner diameter of the bed could be checked on the photographs.

For ease of coal and ash handling, it was decided to generally use sand and coal particle diameters of the order of 1mm. This factor led to the choice of 1mm for the air distributor hole size. As is well known, the use of finer particles will increase the heat transfer rate (inversely proportional to particle size) and it was found possible to operate the bed with particles down to 0.65mm.

The bed was operated on sand (and ash)

rather than limestone since our main interest at present is in the combustion rather than sulphur retention characteristics. The quantity of sand used in the bed (which determined the bed depth) was determined by weighing.

3. THEORY

The design velocity through the bed is generally about two-six times its value at minimum fluidization conditions. In the rotating bed, the factors governing the minimum fluidizing velocity are fluid density and viscosity, particle diameter and density, and effective gravitational acceleration. The relationship between these factors is given in Section 4·4 below. The effective gravitational acceleration due to rotation is equal to $\omega^2 R$ where ω is the angular velocity, and the corresponding g-loading is $\omega^2 R/g$. The relationship between g-loading and rotational speed (N) for various bed radii is shown in Fig. 3. For our experiments, the selected bed radius is 0.1m so that at N=100 RPM, loading $\simeq$1g; N=300 RPM, loading $\simeq$10g; N=1000 RPM, loading $\simeq$ 100g.

The pressure drop across the fluidized bed is approximately equal to the weight of the bed (see Section 4.4), so for a bed 50mm deep of bulk density 2600 kg/m^3 (ordinary sand), the pressure drop across the bed at 1g loading is 1.3 kN/m^2 rising to 130 kN/m^2 (18.8 psig) at 100g loading.

For a typical particle size of 650 μm, the operating superficial flow velocity at 1g (100 RPM) was $\sim$0.9 m/s, the corresponding air flow = 0.0565 m^3/s. At 100g loading, the air throughput rises to about 0.5 m^3/sec. At a maximum combustion temperature rise of 700°C, this represents a peak heat release rate of 350 kW.

Most combustion tests were conducted at about 10g loading corresponding to about 100 kW peak heat release rate.

To remove approximately one third of the heat through cooling water, a heat transfer area of about 0.14 m^2 was required. This was provided by the twelve cooling tubes and cooled end walls. The output could be reduced by decreasing the bed depth so that the heat transfer tubes were only partially immersed in the bed.

4. COLD MODEL TESTS - RESULTS AND DISCUSSION

The hot combustion test rig was built first to prove the feasability of building and operating a rotating fluidized bed fired on gas, coal or oil. After about 1000 hours of hot operation with all three fuels during two years, the cold model was constructed to help in the interpretation and to visualise some of the rotating fluidising phenomenon.

4.1. Bed Surface

The variation of the radius of the surface of the bed (r) depends on the height above the base (h) and the rotation speed according to:

$$r^2 = (2g/\omega^2)\,h + \text{const.}$$

Thus the surface follows a paraboloid which is usually truncated by the base of the cylinder. This formula assumes that the particles are rotating at a speed proportional to radius, which is a good approximation, especially when heat transfer tubes are immersed in the bed. The validity of this relationship when the bed was fluidized was verified by means of the scale mounted on the end plate. At high rotation speeds ($\sim$1000 rpm) the bed surface is almost cylindrical.

4.2. Tumbling mode of operation

One of the biggest problems in the operation of conventional fluidized beds is the attainment of adequate transverse mixing. This problem can be completely overcome by the use of a rotating bed in the tumbling mode. This mode is readily produced, for example, when the bed is operated at rather less than 10g with a fairly shallow bed near the upper end plate and sufficient air throughput to lift the bed in this region. The bed then sets up a strong tumbling action with fluidized material rising at the air distributor surface and falling at smaller radii. Typical operating conditions producing this effect were 350 rpm, 0.7-0.8mm particles, 60mm bed depth, 200mm ID x 75mm distributor, hole size, 0.9mm, open area 2%. To some extent the motion is reminiscent of the flow within a cement mixer, and indeed the bed could be operated at an angle to the vertical. In this mode, mixing is so strong that the fuel could be introduced almost anywhere, however for fuels such as refinery sludge and rubbish an optimum point would be by a downward facing nozzle near the bottom end plate. The fuel would then be immediately engulfed by hot bed material and carried directly to the distributor plate where paper and volatiles are quickly consumed leaving the char to burn subsequently. The applicability of this mode of operation to incinerators etc. is clear.

4.3 Bubble formation and elutriation

Short duration flash (1/40,000 sec) photographs were used to study bubble formation and mixing in the bed e.g. Plate 1. The fundamental characteristics of the bed were little changed by the rotation, that is, with increasing air flow the bed first expanded then formed bubbles normally. The major difference was the behaviour of the particles after the bubbles burst. Whereas in a conventional bed they tend to splash high above the bed, in the rotating bed, the particles have a high tangential velocity and tend to proceed in a straight line which transfers them back to the surface of the bed. At the lower rotational speeds, the presence of the vertical gravitational field causes them to return to a lower point on the surface again augmenting the longitudinal mixing within the bed.

At high rotational speeds the radial pressure gradient in the gas above the bed becomes significant ($dp/dr = \rho V^2/r$) and the main body of the gas approximately constitutes a Rankine vortex. The boundary layer on the upper (and lower) walls forms a forced vortex which has a lower radial pressure gradient than in the free vortex region in the body of the gas. The boundary layer therefore has a strong radial inflow in this region and elutriated particles can creep radially inwards in this region. Particles escaping by this mechanism can be clearly seen in Plate 2.

4.4 *Bed pressure drop*

The pressure drop across a rotating fluidized bed has been reported previously (Refs. 2 and 4). The measured pressure drop across the bed and distributor for a typical particle size and rotational speed varied with air flow as shown in the upper curve in Fig.4. Subtracting the distributor pressure drop gave the familiar Δp vs flow curve with a knee at the minimum fluidizing velocity. As the rotational speed is varied the knee moves further up the curve as shown in Fig. 5. This curve clearly demonstrates the ability of the rotating bed to operate over a wide range of throughput. The minimum fluidizing velocity for any operating condition can be calculated from the relationship due to Wen and Yu (6)

$$Re_{MF} = [(33.7)^2 + 0.0408\ Ga]^{\frac{1}{2}} - 33.7$$

The pressure drop across the fluidized bed is approximately equal to the effective weight of the bed/unit distributor area($\rho g'h$).The dashed lines shown on Fig.5 indicate the pressure drop computed from this simple criterion. This criterion therefore provides a good preliminary design tool.

4.5 *Effect of heat transfer tubes*

The effect of the heat transfer tubes was studied photographically. The photographs showed a slight dip in the fluidized surface immediately 'above' the simulated heat transfer tube as shown in Plate 3. This suggests a tendency for the bed to slump along a thin zone in the wake of the tube. However the effect is small, and confirms (see 5.2) that the mixing in the bed was relatively uniform.

4.6 *Fuel mixing*

The tumbling mode of operation discussed previously is an extreme case of mixing. For the typical operational mode, coal was introduced onto the bed surface from a coal feed pipe. Photographs of the feeding process showed that a significant proportion of the coal fell to the bottom of the cavity, Plate 4. Photographs taken a short time later showed the coal distributed throughout the bed, although it was not possible to quantify the uniformity of mixing nor the mixing rate. Nevertheless, the indications are that the longitudinal mixing is not as good as the circumferential mixing unless the bed is operated in the tumbling mode.

5. HOT MODEL TESTS - RESULTS AND DISCUSSION

5.1 *Light-up*

The bed was initially filled with 3-4 kg of sand. The operation was started by turning on the cooling water and starting rotation at the desired speed. Pure propane was first ignited as a diffusion flame at the nozzle exit. Air was gradually increased until, as stoichiometric mixture was approached, the flame flashed back into the fluidized bed chamber. As the mixture throughput is increased to the fluidizing velocity, a thin annular layer of the bed first fluidizes and heats up. The adjacent material then fluidizes until in about 15 seconds from first ignition, the whole bed is fluidized and burning steadily. If it is required to operate on coal or oil, the appropriate fuel feed is started and the gas is gradually phased out, using the thermocouples in the bed (or bed appearance) to guide the operation. Most tests occupied about 1 hour of running after which the bed is simply shut down by terminating the fuel feed and allowing the bed to cool down.

5.2 *Combustion characteristics and heat transfer*

Using the start-up techniques discussed above, the bed could be operated on gas, oil or coal fuel. The several thermocouples immersed in the bed showed that their temperatures were identical, demonstrating that the mixing was surprisingly uniform considering the large size of tubes immersed in the bed. With water flow through the 12 tubes maximum bed temperature with all three fuels was about 950°C. This demonstrates that similar heat release rates were achieved with all three fuels. The range of total heat release investigated extended from $\sim$1kW to > 100kW, with rather less than half the heat to the cooling water and the balance to the exhaust gases. As already pointed out, in some tests, the heat transfer to the tubes was limited by restricting the quantity of sand in the bed, thus the tubes were only partially submerged and the effective area reduced.

With gas fuel, the combustion was clean and the only visible sign of combustion was the yellow glow from the bed. With oil fuel a luminous flame was present in the reactor probably due to some of the finer fuel droplets burning above the bed. With coal fuel, some of the volatiles burned in the freeboard above the bed, and any ash and unburnt carbon was elutriated from the bed.

Flue gas analysis was carried out by a gas chromatograph using samples withdrawn from the freeboard within the rotating chamber. Some typical results at 400 RPM for gas and coal combustion are shown in table 1. From the results it can be seen that combustion efficiency was 100% with propane fuel and over 90% with coal (carbon loss was not measured).

In addition to the combustion of gas and coal, Table 1 includes a corresponding set of operating conditions for oil fuel at 400 RPM. The heat transferred to the water was 16 kW for both gas and coal, and 23 kW with oil fuel. These figures should not be interpreted as an inherently different efficiency since the depth of bed varied between these tests. The 'effective' overall heat transfer coefficients can be calculated (ref.1) and were 90 W/m^2K for gas and coal and 130 W/m^2K for oil. (NB Heat transfer coeff. $h \sim (g')^{0.2}$).

For the conditions given in Table 1, the combustion intensities were similar for all three fuels at 35 MW/m^3 at 400 RPM. The throughput and therefore combustion intensity varies approximately as the square-root of the g loading and is therefore proportional to the RPM. Thus operating at about 100 g gives a combustion intensity of the order of 100 MW/m^3 at atmospheric pressure, almost an order of magnitude higher combustion intensity than a conventional fluidized bed.

6. CONCLUSIONS

The combustion of gas, oil and coal in a rotating fluidized bed with heat removal by cooling tubes has been successfully demonstrated. The rotating bed is mechanically more complicated than the stationary bed but it has the following additional advantages:

High combustion intensity-compact; Wide turndown range-versatile; Easy, quick light up; Rapid response-low thermal inertia; Good mixing-simpler fuel injection; Augmented heat transfer rates; Tumble mode operation.

The potential applications include boilers, open cycle gas turbines, combined gas/steam cycles and tumble mode incinerators.

REFERENCES

1. Basu, P., May 1976, Ph.D.Thesis, Aston Univ.
2. Hendrie, J.M. et al. 1972, BNL Report 50632.
3. Chalchal, S. et. al, Oct. 1974, BNL Report 19308
4. Metcalfe, C.I. & Howard, J.P., 1977, Applied Energy, pp.65-74.
5. Gel'perin, N.J. et al., 1964, Khim.Mash(5), 18.
6. Wen, C.Y., & Yu, Y.H., 1966, Chem.Eng.Symp. Series, Nos.67, 62, 100.

TABLE 1. Fluidized bed combustion conditions at 400 RPM

a) Propane Burning

Air (lt/m)	Propane (lt/m)	Water (lt/m)	Temp °C	Water in temp °C	Water out temp °C	flue gas % by vol. CO_2	CO	O_2	H_2	CH_4
850	27.6	8	760	10	40	14.2	-	3.2	-	-
700	24	8	780	10	40	16	-	1.6	-	-

b) Coal Burning

Air (lt/m)	Coal (gm/m)	Water (lt/m)	Temp °C	Water in temp °C	Water out temp °C	% by vol. in flue gases CO_2	CO	O_2	H_2	CH_4
720	100	8	750	12	42	13	1.2	4	.6	.1
710	100	8	790	12	43	14.5	.5	6.6	.4	.2

c) Gas Oil Burning

Air (lt/m)	Gas oil (lt/h)	Water (lt/m)	Temp °C	Water in temp °C	Water out temp °C
750	4.8	7	750	12	62
800	5	7	770	12	63

PLATE 1

PLATE 2

PLATE 3

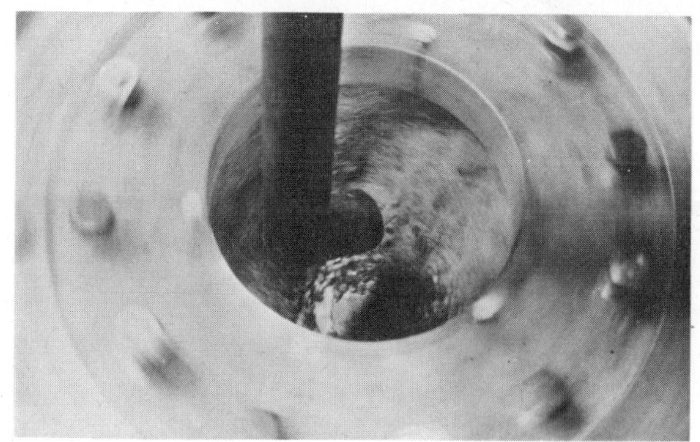

PLATE 4

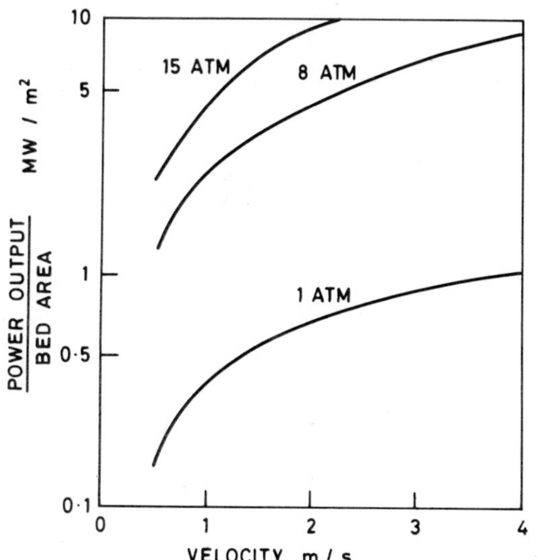

FIG.1. Power Output/unit area as a function of fluidising velocity.

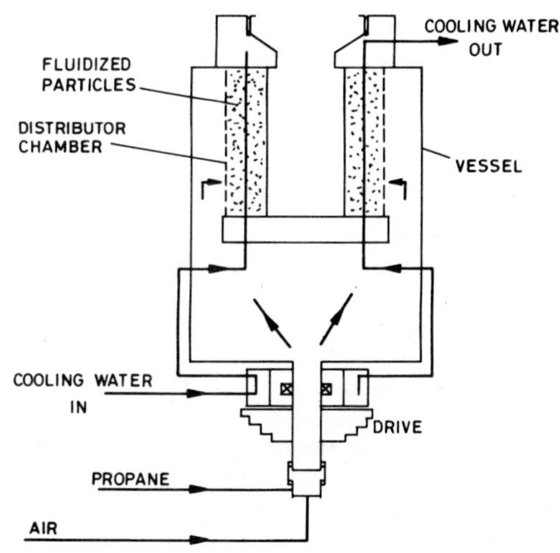

FIG. 2. General Assembly of water-cooled rotating fluidised bed.

FIG.3. Variation in 'g' loading with bed radius and R.P.M.

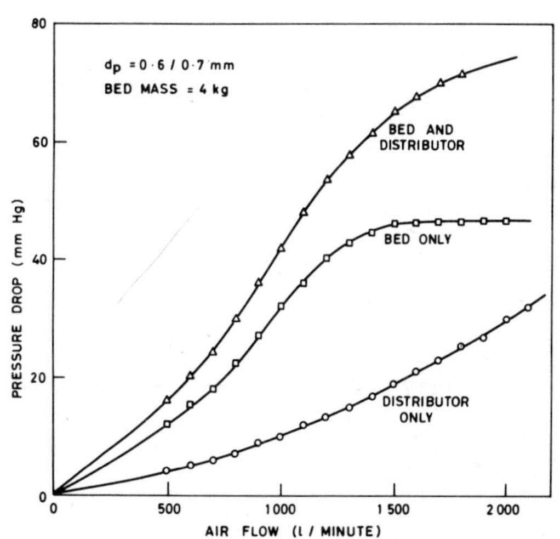

FIG.4. Pressure Drop Measurements at 350 RPM

FIG.5. Bed ΔP at 200 and 400 RPM →

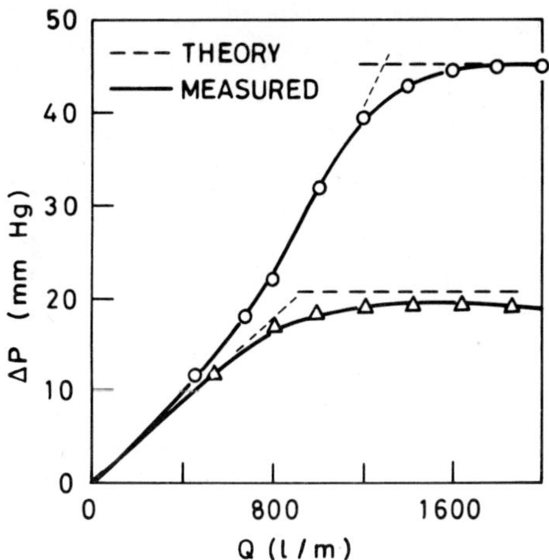

TOWARDS HIGHER INTENSITY COMBUSTION:-

ROTATING FLUIDIZED BEDS

By C.I. Metcalfe and J.R. Howard

Department of Mechanical Engineering
University of Aston in Birmingham
Gosta Green, Birmingham B4 7ET
England

SYNOPSIS

Design, development and experimental operating experience with this type of high-intensity combustor, when burning propane are described.

The experimental work illustrates the operating regions within which stable, complete, high intensity combustion is achievable and defines the practical constraints. Experiments further demonstrated how the range of "turn-down", normally a limitation of conventional gas-fired fluidized bed combustors, can be extended significantly.

INTRODUCTION

The intensity of combustion within a given fluidized bed combustor can be increased by supplying a greater mass flow of air and fuel, subject to strict limitations. A rotating fluidized bed combustor, (Metcalfe and Howard, 1977) overcomes some of these limitations by imposing a high centripetal acceleration on the particles, thus allowing higher fluidization velocities to be employed without causing particle elutriation. Earlier work (Metcalfe and Howard, 1977) showed that satisfactory combustion of propane at an intensity of 85 MW/m^3 of fluidized bed could be achieved when the centripetal acceleration on the particles was 10 x gravity, (10 xg).

A second rotating fluidized bed combustor of improved design, which allowed higher accelerations to be imposed was built. This paper describes experiments performed with this new combustor.

DESIGN OF COMBUSTOR

Fig. 1 shows a schematic arrangement of the re-designed combustor consisting of the water-cooled seal unit, the plenum chamber/distributor and the support frame with the drive shaft.

The propane and air were admitted to the plenum chamber, through four ports, 75 mm diameter in the stationary outer wall, this feature eliminating the restriction to air flow encountered with the earlier design.

Three alternative distributor plates were experimented with; they were made of pierced, thin stainless steel sheet rolled into the shape of a hollow cylinder, 200 mm diameter, 78 mm axial length and seam welded. The distributor was attached to two steel plates top and bottom. The lower plate was bolted to the hollow drive shaft, through which thermocouple leads passed to slip rings. The upper plate was bolted to the exhaust duct. This duct was surrounded by a water jacket, which enabled ordinary oil seals to be used to prevent air and water leakage. The rotating components were driven by a 4 kW variable speed motor through a multiple-wedge belt drive.

Air, propane and water flows were measured by rotameters. Fluidized particles and distributor temperatures were detected by chromel-alumel thermocouples connected to a multi-channel recorder.

Dry exhaust gas analyses were made on samples drawn through a water-cooled probe positioned in the exhaust duct. A paramagnetic oxygen analyser and an infra-red CO/CO_2 analyser were used to measure the concentrations and gas chromatograph calibrated against gas samples of certified composition corroborated these measurements.

FLUIDIZATION EXPERIMENTS

Initially, the distributor plate pressure drop/velocity characteristics were obtained at selected rotational speeds. Pressure drop/velocity characteristics were then obtained with 600 g of sand in the combustor, so that the pressure drop/velocity characteristic of the particulatematter could be found.

The particle size ranges used are shown in Table 1.

Type of sand:- Silicon Sand

Particle density:- 2.63 g/cm^3

Trial No.	SIZE Range µm	Geometric mean particle size µm
1	90 - 180	127.3
2	180 - 250	212.1
3	215 - 300	254.0
4	250 - 355	297.9

Table 1.

Fig.2 shows the pressure drop/fluidization velocity characteristic for Trial No.2, (180-250 µm size range). Since the acceleration imposed on the particles varies with the radius from the axis of rotation, then as the radially inward air velocity increases, the particles at the free surface become fluidized first. As the air velocity is increased further, a greater radial thickness of bed becomes fluidized until eventually the entire bed of particles is supported. Thus the range of air velocity between incipient fluidization and a fully supported bed is dependent on the radial thickness of the bed and is larger than with a stationary fluidized bed, giving a more gradual curvature to the pressure drop/air velocity characteristic. This phenomenon was observed in experiments with the earlier combustor, (Metcalfe and Howard, 1977). Likewise the criterion for fluidization of the entire bed is when the pressure drop across the bed does not change with increase in air velocity, rather than when bubbles appear at the free surface. This criterion was used to establish the minimum operating air velocity at each rotational speed, while the maximum operating air velocity was established by that at which particles were elutriated from the bed.

COMBUSTION EXPERIMENTS

The start-up procedure used for the earlier combustor (Metcalfe and Howard, 1977) and stationary beds (Elliott, 1972) was found to be adequate over the range of rotational speeds explored. In all cases, combustion was established within the bed within 60 s from ignition. Experiments at each rotational speed had two objectives, namely:-

(i) To establish, (a) the weakest mixture at which stable combustion existed and (b) the richest mixture which could be burnt without exceeding a bed temperature of 1150 °C, above which, particle sintering would occur. These mixture strengths were determined by varying the air:propane ratio at various fixed air velocities.

(ii) To determine the maximum and minimum heat output achievable. These data were derived from the results of test (i) above.

Fig.3 shows the limiting air:fuel ratios encountered with flammability limits for propane/air mixtures normally quoted, (Spiers, 1961). This characteristic shows that this type of combustor can burn a significantly weaker propane/air mixture than normal flammability data would suggest. Thus when a rotating fluidized bed combustor is used to incinerate very weak gaseous mixtures such as industrial fumes, less auxiliary fuel is required to effect combustion.

Fig.4 outlines the operating envelope at accelerations up to 30 xg. Drager tubes were used to estimate NO_x pollutants in the combustion products. Within the operating area enclosed by ABCD, the level of NO_x did not exceed 10 ppm.

Fig.5 shows the maximum and minimum heat outputs achieved and illustrates the turndown obtainable at each rotational speed. It should be noted that the range of turndown can be extended by altering the rotational speed in addition to changing the air:fuel ratio and air velocity. Thus on Fig.5 the maximum heat output was 90.6 kW at 30 xg and the minimum heat output 18.3 kW at 10 xg, giving a turndown ratio of 90.6/18.3 = 4.95.

The combustion intensity at the maximum heat output was 239 MW/m^3 of fluidized bed. This compares with a combustion intensity of 10.45 MW/m^3 quoted for a stationary atmospheric pressure fluidized bed gas combustor (Cole and Essenhigh 1972) without heat transfer surfaces in the bed. However three times this value has been quoted (Elliott 1972) when heat transfer surfaces are immersed in the bed.

DISCUSSION

Experiments demonstrated that a bed of particles may be fluidized uniformly when a nominal centripetal acceleration of 30 xg is applied. Choice of distributor is very important in three aspects. Firstly that the hole size must be smaller than the smallest particle in the size range used. Secondly the holes must be uniformly distributed on small pitch to produce fluidization. Thirdly the open area needs to be less than 5%. The high fluidizing velocities encountered in combustion experiments at the higher accelerations may allow a large open area to be used without risk of "burning back", since the velocity of the emergent jets will continue to be greater than the flame velocity. Further exploration of this matter is required because of the advantage of reducing the pressure drop through the distributor.

Other workers, (Ludewig, Manning and Raseman,1974), suggested that with a rotating fluidized bed combustor, distributor plate temperatures could be expected to be low because the high mass flow of air cools the distributor. Measurements of distributor temperature in shallow fluidized bed combustors (Westwood,1975) and Broughton,1975) showed that the distributor temperature could be close to the mean bed temperature. This fact sets the minimum mass flow for a given distributor material in two ways, namely the temperature limitation of the distributor material and the likelihood of combustion taking place within the distributor, (Broughton,1975). Thermocouples in contact with the distributor of the rotating fluidized bed combustor showed the temperature to be between 400 °C and 600 °C when running at bed temperatures between 900°C and 1150°C. The highest distributor temperature was registered when operating with the lowest mass flow of air and richest mixture.

CONCLUSIONS

Further development of the rotating fluidized bed appears to be justified. Clearly this type of fluidized bed has potential for application to fume incineration.

The use of the rotating fluidized bed for burning solid and liquid fuels now requires systematic exploration, because it allows the use of small particles thus increasing the surface area per unit mass of particles which is available for heat transfer or chemical reaction.

ACKNOWLEDGEMENTS

The work reported here has been supported by the National Research Development Corporation.

REFERENCES

Broughton, J., (1975), Applied Energy (1) pp.61-79.

Cole, W.E. and Essenhigh, R.H.,(1972), Third International Conference on Fluidized Bed Combustion, Hueston Woods.

Elliott, D.E., (1972) Third International Conference on Fluidized Bed Combustion, Hueston Woods.

Ludewig, H., Manning, A.J., and Raseman, C.J., (1974) J.Spacecraft, Vol 11, No.2, 65-71.

Metcalfe, C.I., and Howard, J.R., (1977) Applied Energy (3) (1977), 65-74.

Spiers, H.M., (1961), "Technical Data on Fuel", p.260, The British National Committee World Power Conference, London.

Westwood, C.R., (1975), Ph.D. Thesis, University of Aston in Birmingham.

Fig. 1. Schematic layout of rotating fluidized bed combustor.

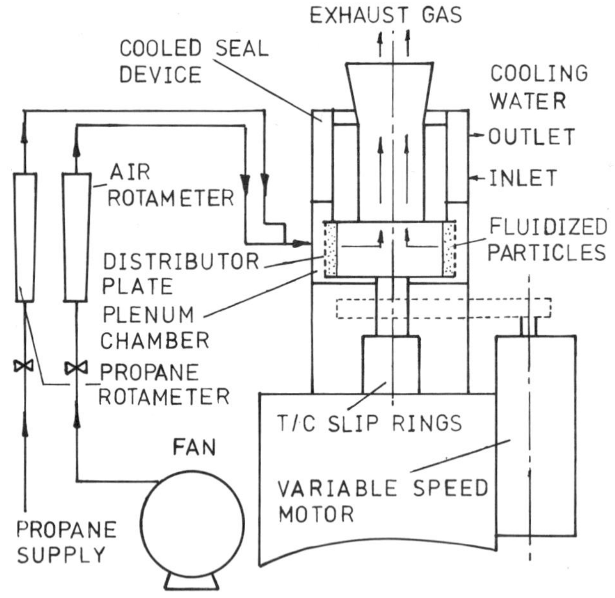

Fig. 2. Fluidization characteristics.

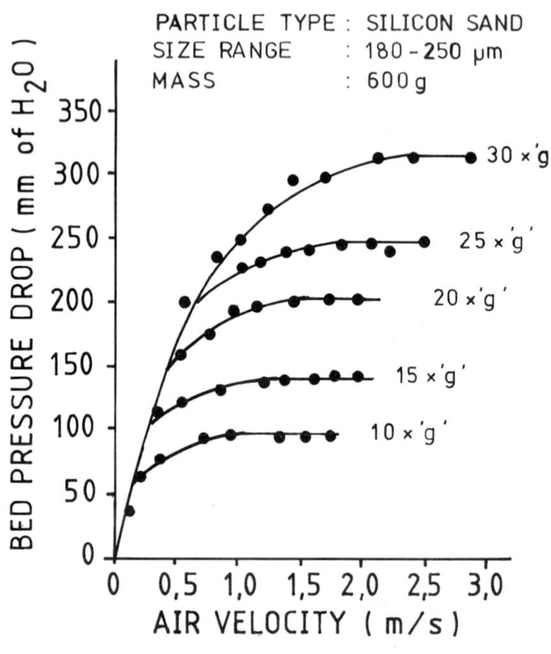

Fig. 3. Overall operating regime.

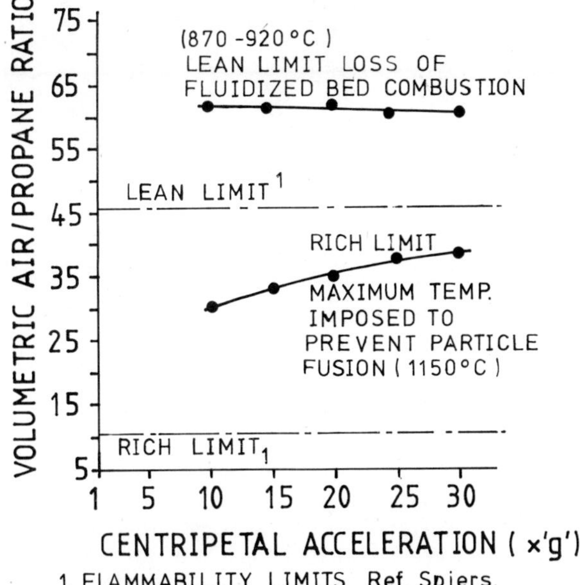

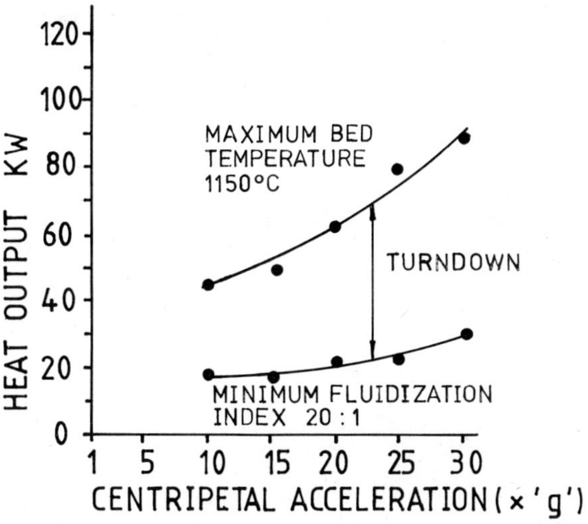

Fig. 4. Operating envelopes at accelerations up to 30 x g.

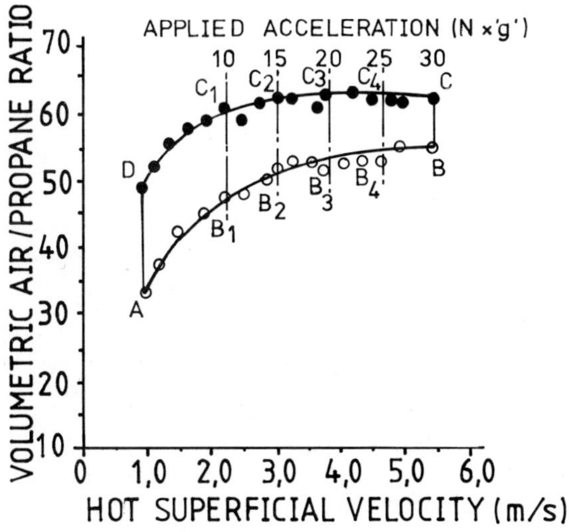

Along curve AB, bed temperature 1150°C; curve CD defines the loss of fluidized bed combustion at 870-920°C.
The vertical lines B_nC_n represents the velocity, for a given g force above which the CO/CO_2 ratio in the flue gases exceeds 2%.

Fig. 5. Heat output characteristics.

FLUIDIZED BED COMBUSTION OF POOR QUALITY COAL

By D.P. Naude and R.K. Dutkiewicz

Department of Mechanical Engineering
University of Cape Town
Rondebosch, 7700
South Africa.

SYNOPSIS

The combustion of poor quality coals in an open top 300mm diameter refractory lined atmospheric fluidized-bed combustor is described. Quantitative relationships are derived for the combustion efficiency in terms of the temperature, superficial gas velocity and static bed height for a coal with a large proportion of fines and considered to be difficult for burning in conventional industrial equipment. A second coal used had an ash content in excess of 35%. The third fuel used was a coke breeze. Low combustion efficiencies have been attributed to to the high fines content of the fuels. Entrainment has been divided into an elutriation rate and a splashing rate component, however only a qualitative assessment of the splashing rate has been possible.

NOTATION

- H static bed height (mm)
- T fluidized-bed temperature (^{o}C)
- u_f superficial gas velocity (m/s)
- u_{mf} minimum fluidizing velocity (m/s)
- V vigorousness of bubbling
- η combustion efficiency (%)

INTRODUCTION

The bulk of the energy requirements of South Africa are and will continue to be drawn from coal. The coal resources though substantial, are smaller than has generally been accepted (Petrick,1975). Most of the coal has a relatively low calorific value and a high ash content, with this ash being more inherent or homogeneously dispersed in the coal body than the ash found in coals in the northern hemisphere. South African coals are therefore difficult to wash to produce a low ash quality material. Proven extractable coal reserves, on the basis of coals having an ash content of less than 35%, have been estimated as 25 000 million tons (Petrick,1975) of which only a small proportion is made up of metallurgical coal, anthracite and low-ash high grade steam coal.

The generation of power accounts for the largest proportion of the coal consumed, with a projected consumption of about 40 million tons in 1980 or 55% of the total coal consumption for that year. The greater portion of coal supplied to power stations is based on the captive colliery policy in which a captive or tied colliery delivers its total output to a power station. Although this policy has distinct economic advantages for both the electricity supply utility and the producer, it leads to a wastage of coal in other commercial collieries which are forced to apply selective mining and preparation in order to comply with the demands of other markets for coal. Smalls coal,i.e. coal sized between 25 and 6,3mm is in greatest demand by the industrial sector and in the production of this size grading, duff or coal less than 6,3mm is produced which is supplied almost exclusively to the brick and cement industries. Due to the captive colliery policy, future power stations will not utilize this grade of coal,the supply of which is presently exceeding the demand. This production of fines or duff "is therefore a problem area where solutions must be found;not only in putting good coal to use,but also by decreasing the pollution of the atmosphere by the burning of the fines on a dump"(Petrick,1975).

Poor quality coal is therefore defined as that which is not easily utilized. Two types of coal are considered, that having an ash content in excess of 30% and the second being duff coal with a high fines content or particles smaller than 0,5mm. Fluidized combustion permits the utilization of both these types of coal and therefore a small combustor was built to study the combustion of these poor quality coals.

EXPERIMENTAL EQUIPMENT

The prime objectives of this work are the assessment of the combustion and entrainment phenomena, and therefore no heat transfer surface has been provided in the experimental rig. The heat generated by the combustion process is removed by the fluidizing gas. Thus about three times the air which would normally be required for stoichiometric combustion has to be supplied to the rig to maintain the bed temperature within acceptable limits. The general arrangement of the major components of the fluidized-bed combustion test rig is illustrated in Fig.1, whilst a flow diagram is given by Fig.2. In view of the disadvantage of high pressure drops through the fluidized bed, only shallow beds were investigated using a low pressure drop distributor.

The combustor vessel is completely insulated having refractory lined walls to limit the amount of heat lost to the surroundings. The vessel has an internal diameter of 300mm, and a height of 800mm. During operation of the rig a cylindrical shield was placed on top of the combustor vessel to limit the loss of bed material caused by splashing as a result of bubbles bursting at the surface of the bed. A perforated plate distributor, having 2,5mm diameter holes and a mean free area of 1,86% is situated at the base of the combustor vessel. A 90mm deep layer of refractory stones graded to be sized between 6 and 9mm was placed on the distributor to limit the back-flow of bed material and to insulate the distributor plate from the fluidized bed. Although it is generally accepted that in order to ensure uniform fluidization and to prevent channelling within the fluidized bed, the pressure drop across the distributor plate should be equivalent to the bed pressure drop(Botterill,) 1975), Wright(1968) has reported satisfactory fluidization with distributor pressure drops of 50 to 75mm Wg and a corresponding bed pressure drop of 400mm Wg.

Coal is screw fed through the side of the combustor to the base of the fluidized bed, i.e. immediately above the refractory stones. The exhaust gases are mixed with the surrounding air immediately above the combustor vessel and are removed via an extraction hood to a cyclone before being exhausted to atmosphere.

EXPERIMENTAL PROCEDURE

The combustion of three types of fuel has been investigated. The first was a typical duff coal having a large proportion of fines, the second was a coal with an ash content of about 39%, whilst the third fuel was a coke breeze obtained from a local gas works and for which there is only a limited market. The gradings of these three fuels are illustrated in Fig.3, whilst the proximate analyses are presented in Table 1. A closely graded silica sand having a mean particle diameter of 0,789mm(the harmonic mean diameter) has been used as the bed material. The size grading of this sand is also included on Fig.3 for easy reference. The minimum fluidization velocity was determined (experimentally by fluidizing with air at 40°C) to be 0,53 m/s. The approximate relationship given by Kunii and Levenspiel(1969) has been used to evaluate minimum fluidizing velocities ranging from 0,25 to 0,21 m/s for bed operating temperatures between 700 and 1000°C.

TABLE 1 : Fuel Properties

	Duff Coal	High Ash Coal	Coke Breeze
Proximate Analysis			
Moisture %	2,96	2,25	5,06
Ash %	15,28	39,40	27,80
Volatiles %	23,86	21,30	11,30
Fixed Carb. %	57,90	37,05	55,84
Gross Cal.Val. MJ/kg	27,37	17,34	20,62

After initial attempts at start-up without resorting to a secondary source of fuel, a gas ignition system was designed and constructed. Gas is mixed with the air in the combustor plenum chamber before passing through the distributor into the combustor vessel, where it is ignited above the bed by means of a pilot burner. This enables the bed temperature to be raised from cold to about 700°C within 15 minutes.

A series of tests were conducted to establish the combustion efficiencies and the entrainment rates under different bed conditions. It was felt that the system variables which would have a major effect on these two parameters and in particular on the combustion efficiency could be limited to bed temperature, superficial gas velocity and bed height. These variables were tested over the following ranges

Bed temperature	:	700 to 1000°C
Gas Velocity	:	0,9 to 1,5 m/s
Bed height	:	150 to 250mm

COMBUSTION EFFICIENCIES

The combustion efficiency of the fluidized bed combustor is defined as the ratio of the rate of heat liberated to the rate of heat input. The difference between these two is the result of incomplete combustion of carbon to carbon monoxide,the combustion of the volatile component of the coal above the bed section, or unburned carbon remaining in the entrained ash or remaining in the bed overflow. As the rig is cooled by the fluidizing air,about three times the stoichiometric air quantity passes through the bed section. This abundance of oxygen ensures complete combustion of the carbon to carbon dioxide and of the volatile component within the bed section. This has been observed during operation of the rig even at temperatures as low as 700^{0}C. The oxygen supply even at the coal feed aperture,is sufficient to ensure that the volatiles are burnt within the bed section rather than in the freeboard. As no bed material was removed as overflow during any of the tests,the loss as a result of unburnt carbon can be deduced from the carbon content of the entrained ash. It has been assumed that the ash particle on leaving the fluidized bed by entrainment in the off gases is quenched such that no further combustion takes place. Therefore,the combustion efficiency has been assessed by determining the quantity of carbon contained in the entrained ash.

A total of forty-seven tests have been conducted under different velocity and temperature conditions for different bed heights. The pressure drop across the distributor plate and refractory stones remained between 50 and 100mm Wg for most of the tests,being dependent mainly on the air flow rate. The pressure drop through the bed varied from 240 to 400mm Wg,as a result of the variation in the static bed height from about 150 to 250mm.

The results have been statistically analysed to establish a correlation between combustion efficiency and the various operating variables. In view of the fact that Ehrlich(1975) used other independent variables in his correlation of combustion efficiency,and further that this efficiency would appear to be dependent upon the quality of fluidization,the effect of five other variables in addition to temperature, gas velocity and bed height was considered. These variables are:coal feed rate,the gas flow rate, the ratio of the bed to the distributor presure drops,the vigorousness of bubbling as defined by equation(1)(Merrick & Highley,1974) and the splashing rate as determined in the section below. The latter three variables were incorporated in an attempt to assess the effect of the quality of fluidization on the combustion efficiency.

$$V = (u_f - u_{mf}) / u_{mf} \qquad (1)$$

Duff Coal

Twenty-four tests were performed with this coal and the results of most of the tests are given in Fig.4 and Fig.5. By applying a regression analysis to the results,it was found that gas flow rate and coal feed could be eliminated without affecting the overall result due to their interrelationship with the remaining six variables. The multiple linear regression analysis on these remaining variables resulted in a multiple correlation coefficient of 0,9679. Eliminating the three variables which describe the quality of fluidization resulted in a multiple correlation coefficient of 0,9422 which can be shown to be not significantly different from the original value. Therefore the combustion efficiency would be best correlated by temperature,velocity and static bed height as given by equation (2).

$$\eta = 36{,}27+0{,}0675\cdot T-17{,}1\cdot u_f+0{,}0650\cdot H \qquad (2)$$

having a standard error of estimate of 2,15%.

Lines of constant velocity have been drawn on Fig.4 and Fig.5 to illustrate equation (2). The combustion efficiency is seen to be strongly dependent on temperature,with velocity and static bed height having secondary effects.

High Ash Coal

Fifteen tests were performed with this coal at velocities of 1,1 and 1,5 m/s and at different temperatures and bed heights. In contrast to the duff coal,which has a higher proportion of fines,no significant correlation of combustion efficiency with velocity could be deduced. By eliminating the values found at 700^{0}C,which may be affected by incomplete combustion,the following multiple regression equation was deduced :

$$\eta = 31{,}57+0{,}0388\cdot T+0{,}114\cdot H \qquad (3)$$

The multiple correlation coefficient is found as 0,9125 whilst the standard error of the estimate is 2,32%. The results have been illustrated in Fig.6 together with two lines of constant static bed height deduced from equation (3). From the analysis it is clear that the combustion efficiency resulting from the use of the high ash coal is less sensitive to temperature whilst it is more strongly affected by bed height than the duff coal.

Coke Breeze

Eight tests have been conducted using coke breeze. Only a single static bed height of 185mm has been used and therefore the effect of bed height was not considered in the regression analysis. From Fig.3 it is seen that the size grading is very similar to that of the duff coal,having a high fines content. Increasing the bed depth is therefore expected to have a similar effect on the combustion efficiency of the coke breeze. The multiple regression analysis results in equation (4):

$$\eta = 3{,}47+0{,}0102\cdot T-12{,}61\cdot u_f \qquad (4)$$

The multiple correlation coefficient is found as 0,9890,whilst the standard error of determination is 1,87%. Two lines of constant velocity determined according to equation (4) are included in Fig.7 to illustrate the effect of temperature on combustion efficiency when using coke breeze. From Fig.7 it is evident that the combustion efficiency is very strongly influenced by the bed temperature,whilst the effect of velocity is less than that experienced when burning the duff coal.

ENTRAINMENT

Particles are entrained from the combustor vessel of the fluidized bed combustion rig and separated from the off gases in a cyclone. The entrainment mechanism can be described by considering it as the sum of two processes, the first due to the elutriation of the fine particles and the other as a result of splashing. The concept of splashing has been introduced to explain the entrainment of those particles too large to be elutriated. This approach is similar to that adopted by Gibbs (1975) in his development of a theoretical model of the fluidized-bed combustion process in which both splashing and elutriation rate constants are defined. Merrick and Highley (1974) have however concluded that the existing empirical and semi-empirical correlations of elutriation are not applicable to their data and have developed a new form of correlation which explains the entrainment of particles larger than those which would normally be elutriated. The approach by Gibbs (1975) however is felt to have more general application .

In the fluidized bed combustion process the inert bed material is reduced in size by attrition whereas the incoming coal feed is continually reduced in size by combined action of combustion and attrition. In the present system the silica sand suffers very little degradation,whereas the ash resulting from the coal or coke is relatively soft and after the initial reduction in size of the coal particle by combustion,the ash particle becomes smaller as a result of attrition until it is finally elutriated. During this process the bed material is continually removed from the bed by the splashing mechanism. In order to be able to assess splashing rate constants,the entrained material must be divided into two components, that lost from the bed by elutriation and that lost as a result of splashing. From Fig.3 it is seen that about 8% of the original bed material has a size less than 0,6mm. Particles above this size can only be removed as a result of splashing and therefore the entrained material has been divided into two size fractions,that below 0,6mm and that above this value. A visual observation of the fraction above 0,6mm has shown that this fraction comprises well over 90% of silica sand. Therefore the fraction retained on the 0,6mm aperture sieve is assumed to have been lost as a result of splashing . The percentage entrainment attributed to splashing is illustrated in Fig.8 for those tests in which the superficial gas velocity was approximately 1,1 or 1,5 m/s. It is evident from Fig.8 that no correlation amongst the parameters is apparent, however some trends can be observed. As the velocity increases the fraction lost from the bed by splashing is increased,whilst increases in this fraction on a more dramatic scale are experienced as the static bed height increases.

DISCUSSION

Combustion Efficiencies

The combustion efficiencies determined in the above series of tests are lower than would normally be expected. In particular, the results for the duff coal are poor. Reference to Fig.3 indicates that about 16% of this coal is less than 0,5mm in size whilst the corresponding value for the high ash coal is only 8%. It is clear that much of the fine coal passes straight through the fluidized bed section and is elutriated before combustion takes place. Although the combustion when burning the high ash coal is only marginally better than when firing the duff coal, the percentage of carbon retained in the ash of the high ash coal after combustion is substantially less than the fraction of carbon remaining after combustion of the duff coal.

The combustion efficiency when burning coke breeze is substantially less than when burning the other two fuels. This is due firstly to the poor reactivity of the coke at low temperatures as a result of the low volatile content,and further due to the high fines content which is in fact higher than that of the duff coal.

The effect of bed height is more pronounced with the high ash coal than with the duff. This is due to the longer residence time of the high ash coal in the fluidized bed as a result of its higher mean size making it less prone to removal by elutriation.

Entrainment

In contrast to the quantitative results obtained from the investigation into the combustion efficiency,no such relationship could be derived in the description of the entrainment rates. It is evident however,that the deeper beds resulted in higher splashing rates as was the result with increasing the velocity. From the slug flow criterion of Stewart and Davidson(1967),the bed should clearly be in the slugging flow regime. However,the shallower beds of 150mm static bed heights would behave as bubbling beds,whilst the initial phases of fully developed slug flow may be manifested in the deeper bed investigated in which the bed height approaches the rig diameter of 300mm. This would explain the large differences in the fraction of the splashed material between bed depths of 150mm and 250mm indicated in Fig.8. In order to arrive at a quantitative result, the use of bubble diameter, bubble velocity, fraction of solids in the bubble wake or other complex phenomena associated with gas fluidized beds may be necessary. As all these parameters are highly complex, it is evident that a considerable amount of work will have to be done in order to quantify the results even for very specific operating conditions.

CONCLUSIONS

1. Quantitative assessments of the combustion of poor quality fuels in shallow fluidized beds with a low pressure drop distributor have been successfully performed.
2. The combustion efficiency is reduced substantially by the increasing of the fines content of the fuel.
3. Fluidized-bed combustion offers a means of utilizing duff coal having a high fines content for which there is at present no demand, whilst coke breeze could be utilized as a fuel provided that the combustion temperature is of the order of 900°C. High ash coals can be burned successfully in a fluidized bed, thereby affording an upward assessment of the coal reserves.
4. The entrainment can be divided into elutriation and splashing rate components. Only a qualitative assessment of the splashing rate has been possible.

REFERENCES

Botterill,J.S.M.(1975),Fluid Bed Heat Transfer Academic Press, London,p.85.

Ehrlich,S.(1975), Inst.Fuel, Fluidized Bed Combustion Conference,London,Vol.1,C4.

Gibbs,B.M.(1975),Inst.Fuel, Fluidized Bed Combustion Conference,London,Vol.1,A5.

Kunii,D. & Levenspiel,O.(1969), Fluidization Engineering,John Wiley & Sons Inc.,p.73.

Merrick,D & Highley,J.(1974),AIChE Symp. Series, No.137,Vol.70.

Petrick,A.J.(1975), Chairman on:Commission of Inquiry into the Coal Resources of South Africa,Dept. of Mines, RP63/1975.

Stewart,P.S.B. & Davidson,J.F.(1967) Powder Technology 1,p.61.

Wright,S.J. (1968) Instn. Chem.Engrs. Symposium Series, No.27.

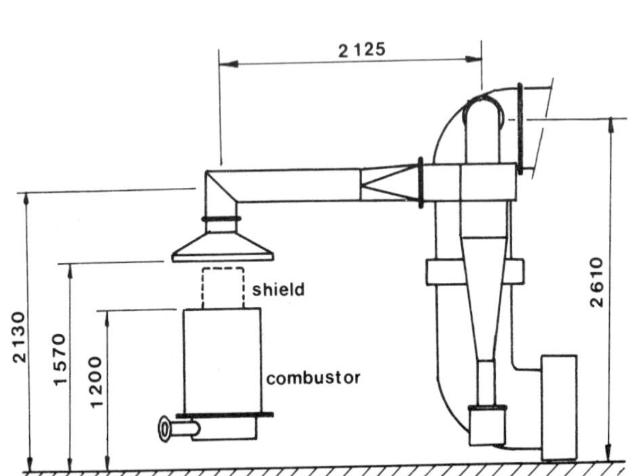

Fig.1 General Arrangement of 300mm I.D. Fluidized-bed Combustion Test Rig.

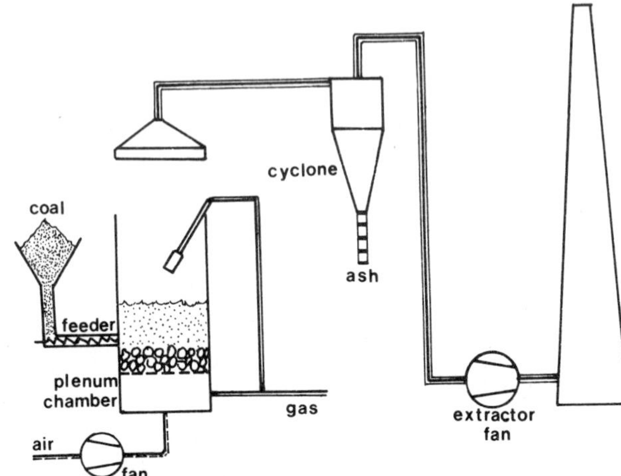

Fig.2 Flow Diagram of 300mm I.D. Fluidized-bed Combustion Test Rig.

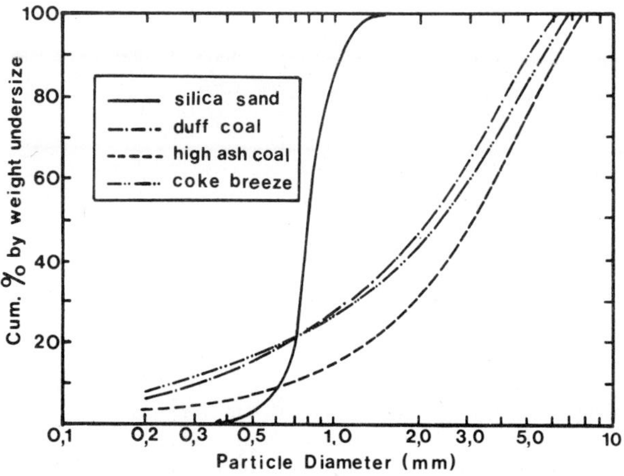

Fig.3 Size Gradings of Fuels and Bed Material.

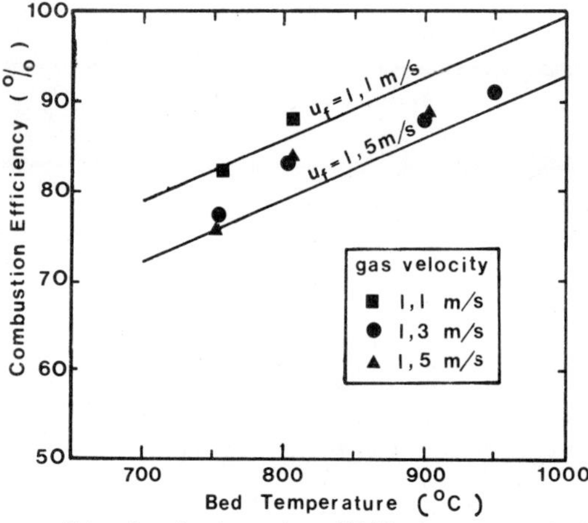

Fig.4 Combustion Efficiencies with Duff Coal in a 220mm Deep Bed.

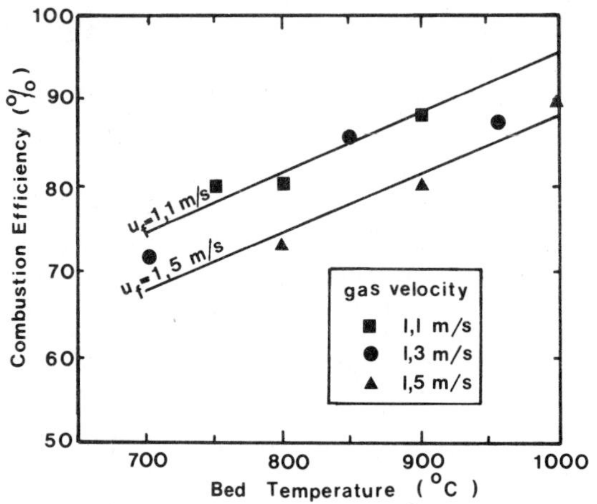

Fig.5 Combustion Efficiencies with Duff Coal in a 155mm Deep Bed.

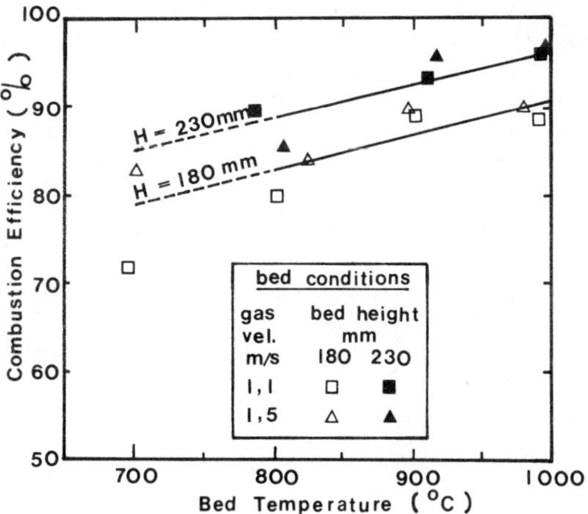

Fig.6 Combustion Efficiencies when Burning the High Ash Coal.

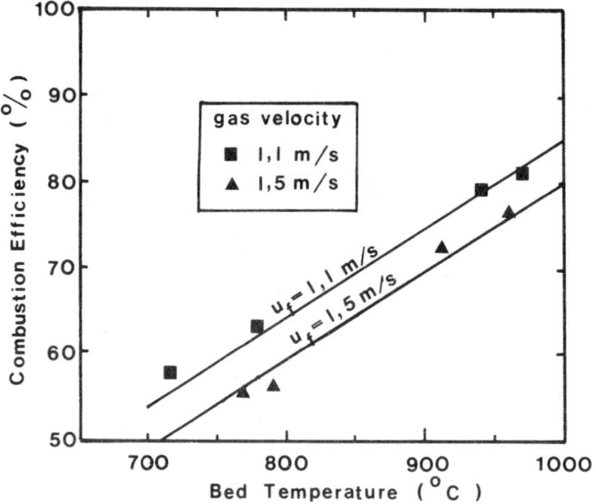

Fig.7 Combustion Efficiencies with Coke Breeze in a 185mm Deep Bed.

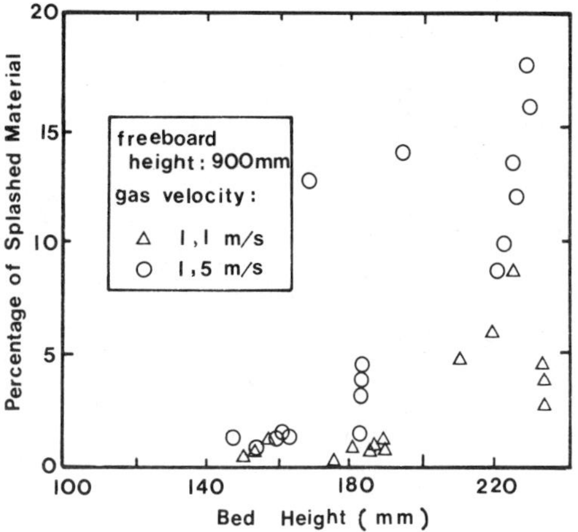

Fig.8 Splashed Material as a Percentage by Weight of the Total Entrainment Flow.

DISPOSAL OF COLLIERY TAILINGS BY FLUIDISED BED COMBUSTION

By A.A. Randell, D.W. Gauld, R.L. Dando and R.D. LaNauze

Coal Research Establishment, Stoke Orchard, Cheltenham, Gloucestershire GL52 4RZ, UK

A process for the disposal of colliery tailings using fluidised bed combustion has been developed at the Coal Research Establishment of the National Coal Board. The paper outlines the process, describes the one tonne per hour pilot plant and discusses plant control. In the process, tailings of typically 65% mineral matter (d.b.), 24% coal (d.b.) and 45% water (a.r.) are sprayed onto the top of a fluidised bed operating at a superficial velocity of 1 m/s and 800°C. The feedstock has a calorific value of about 7500 kJ/kg (a.r.) and generally more than 50% of the solids particles are less than 45 micron. The water content of the tailings is evaporated by heat produced from combustion of the coal component and leaves the combustor as part of the flue gas. Two principal ash products are produced, a granulated coarse bed ash ranging from 0.5 to 1 mm and fine cyclone dusts. Both products are more suitable for disposal than the untreated feedstock. A summary of data obtained during some of the commissioning tests is given. Essential to stable plant operation is control of the mean bed particle size, methods of achieving this are described.

INTRODUCTION

The process of coal preparation requires the run of mine coal to be washed to remove unwanted mineral matter. A by-product of this process is colliery tailings which consists of a fine suspension of the mineral and coal particles in water. The tailings are normally thickened before tipping or lagooning. However, the increasing mechanisation of coal mines together with improvements in coal washing techniques have progressively increased the proportion of very fine material (as much as 35% less than 5 micron, see Table 4) in the tailings. De-watering this material using conventional methods has therefore extended treatment times and is becoming more expensive. Also, the dewatered products can be unstable when tipped and slipping may take place when subjected to heavy rainfall.

In recent years the performance of deep cone plant for thickening tailings has much improved and such units are now capable of producing a thickened discharge with a solids content of up to 70%. The coal component of this discharge, which would otherwise be wasted, is utilised as the principal fuel in the thermal treatment of the tailings by the fluidised bed combustion process. Although the inert products from this process will absorb water they do not regain the same physical form as in the feedstock tailings and can be blended with additional untreated tailings to give a product which is more stable when tipped.

The work described in this paper was carried out at the Coal Research Establishment of the National Coal Board, under a contract from the European Coal and Steel Community. Earlier development work has been reported by Cooke and Hodgkinson, (ref. 1).

DESCRIPTION OF THE PROCESS

In the process developed for the disposal of colliery tailings, Figure 1, tailings are sprayed onto the surface of a fluidised bed operating at a temperature between 800°C and 850°C. The water content of the tailings is evaporated and leaves as part of the flue gas. The mineral component undergoes both chemical and physical changes and is collected as a granulated product from the bed or as dry, fine powders from the gas cleaning cyclones. The bed ash and cyclone products are then cooled and blended with additional untreated thickened tailings to yield a material which is more stable when tipped. Thus, the overall treatment rate is substantially increased.

DESCRIPTION OF PILOT PLANT

Tailings from a washery are delivered by road tanker and stored in a large receiving tank, Figure 1. The tailings are transferred from this tank through a 3 mm screen to a metering tank from where it is pumped to the spray situated in the combustor freeboard.

The combustor vessel, Fig. 2, is cylindrical and has an internal diameter of 1.5 m. Details of pertinent dimensions and nominal throughputs are given in Table 1. The distributor consists of vertical uprisers on a square pitch. Each upriser is sealed at the top and drilled radially with a number of holes. Start up from cold is achieved by two 600 kW gas fired burners directed onto the fluidised bed surface.

A central ash off-take containing a rotary valve is used to remove bed ash and automatically control the fluidised bed level. Similar rotary valves are sited in the cyclone downlegs. All ash products are then blended in the conditioning equipment.

The fines - laden off gases pass through insulated ducting, two cyclones in series and finally to a wet venturi scrubber before emission to atmosphere.

PLANT CONTROL

There are four principal process control parameters: fluidising velocity, bed height, bed temperature and feed rate. Under normal operation the first three are fixed and the feed rate varied to maintain the bed temperature. However, the tailings feedstock from a washery varies in quality and its calorific value can range from 5000 to 15,000 kJ/kg (dry basis), that is both above and below that needed for self-sustaining operation at 800°C; to accommodate such changes, either water or an auxiliary fuel is added to moderate or supplement the combustor feedstock as necessary. A commercial process would not use water addition as a control feature but would utilise the heat available to raise steam or for other uses (ref. 2).

PLANT TRIALS

An initial period of plant trials has been completed. At low tailings inputs, coal was used as a supplementary fuel. All trials were made at 800°C and superficial fluidising velocities close to 1 m/s. This temperature was selected since it ensures both good combustion and a high oxygen availability. Table 2 contains plant data illustrating the range of conditions examined so far. The plant has operated reliably at throughputs up to 1400 kg/h. Blockages in the top of the cyclone discharge legs initially presented a major obstacle to continuous plant operation. This was overcome by removing the disentrainment hoppers fitted to the cyclone outlets and extending the conical sections.

RESULTS

Analysis of a typical tailings feedstock is given in Table 4. A summary of the results from a number of runs is presented in Table 2. An example of a mass and heat balance is given in Table 3.

It is shown that 36% of the non-combustible solids in the feedstock was retained in the bed where it was upgraded from 0.005 mm (average mean particle size) to 0.5 mm. The particles formed in this way are roughly spherical and constitute the bed material. The amount retained in the bed can be increased by adjustment of the spray conditions and as well as ease of handling, this larger material has potential commercial applications.

The efficiency of carbon utilisation was 89% in the reported test, most of the unburnt carbon being present in the primary cyclone fines. This can be considerably improved by recycling the fines product which also has the additional benefit of improving plant stability as described later.

No attempt has yet been made to optimise the use of heat on the plant. In the example given, 87% of the heat released in the bed is contained in the flue gases and a proportion of this could be utilised for example by raising steam and preheating the fluidising air, in this way the throughput of the plant could be increased considerably (ref. 2.).

As coarse ash formed by the granulation process is the most suitable material for handling and disposal, the plant conditions are being varied to maximise the amount of ash retained in the bed consistent with maintaining good fluidisation and uninterrupted operation. This latter point is of considerable significance since these objectives tend to oppose each other; increasing the amount of material retained in the bed by, for example, lowering the spray promotes the formation of large agglomerates in the bed which in turn will lead to de-fluidisation.

This is illustrated in Figure 3 where for constant input conditions the mean bed particle size increased with time. Bed segregation was detected by thermocouples at the base of the fluidised bed as early as point (1) on Figure 3, eventually leading to complete loss of fluidisation at point (2). The mean particle size at this point was 1.65 mm which is close to that predicted from theoretical calculations, 1.7 mm being the largest value that can be fluidised at a velocity of 1.2 m/s.

DISCUSSION

The plant trials have demonstrated the feasibility of using fluidised bed combustion as a means of disposal of colliery tailings. The trials have highlighted one area in particular, the need to be able to predict and control the mean bed particle size. Certain modes of operation can lead to the build up of large particles, Figure 3, to such an extent that defluidisation occurs. Three methods of controlling this problem are under consideration.

(i) The larger particles can be selectively removed from the bed (ref. 2). This does not seem practical in the present system of operation where the increase in particle size of all bed particles occurs slowly rather than a few large agglomerates forming.

(ii) The bed can be continuously seeded with fine particles by recycling part or all of the primary cyclone fines. This simple expediency also increases the combustion efficiency.

(iii) The particle size is controlled by process adjustment.

A method of obtaining a controlled recycle rate is currently under development. A mathematical model is being employed to predict the total bed size distribution and also the fines recycle rate necessary to maintain a specific particle size distribution in the bed. As a typical example of the analysis that is being used, consider the mean particle size growth illustrated in Figure 3. Assuming a coating process in which the bed particles are characterised by the mean bed particle size, D. Then, if the weight of particles W in the bed is kept reasonably constant by controlling the bed height it can be shown that:-

$$\frac{D}{D_o} = \exp \frac{(\dot{m}\, t)}{(3\, W)}$$

where D_o is the initial mean bed particle diameter at time t = 0, and m is the rate of input tailings retained in the bed.

This expression, plotted on Figure 3, shows good agreement with the experimental data.

CONCLUSIONS

The feasibility of treating colliery tailings by combustion within a fluidised bed to form a stable material for disposal has been demonstrated. A 1.5 m diameter plant having a nominal input of 1 tonne per hour has been successfully operated and has simulated the principal features of a commercial scale unit. No scale up penalties for such a plant have been highlighted and the design and construction of such a unit can now be undertaken with confidence.

Commercial exploitation of the ash products is currently under consideration.

ACKNOWLEDGEMENTS

The authors wish to thank their colleagues for their advice and assistance. The work described in this paper is part of a research programme currently being carried out at the Coal Research Establishment of the National Coal Board. The views expressed are those of the authors and not necessarily those of the National Coal Board.

REFERENCES

Cooke, M.J., Hodgkinson, N. (1975), The Fluidised Combustion of Low Grade Materials. Institute of Fuel Symposium Series No.1 September, 1975.

Hodgkinson, N., Thurlow, G.G., (1976), Combustion of Low Grade Material in Fluidised Beds. Paper to 68th Annual Meeting, American Institute of Chemical Engineers.

TABLE 1: Principle plant dimensions and design input data.

Combustor

Height m	4.2
Diameter m	1.5

Shell Construction 10 mm mild steel plate, refractory lined.

Nominal Operating Data

Temperature °C	800-850
Pressure Bar	0.14
Velocity m/s	1.2
Bed height m	0.6

Cyclones

		Primary	Secondary
Gas volume @ 800°C	m^3/s	2.83	2.83
Inlet velocity	m/s	18.5	19.1
Pressure Drop	mb	5	10
Barrell Diameter	mm	864	914
Collection efficiency (no fines recycled)		89	61.2

Venturi Scrubber

Optimum conditions:- 4.25 m^3/s @ 800°C.

		Input	Output*
Dry gas	kg/h	62	62
Water vapour	kg/h	15.6	37.7
Dust	kg/min	1.6	0.014

*2 m^3/s @ 82°C

Nominal throughput rates

Tailing wet feed (to combustor only)	kg/h	1206
Solids	kg/h	662
Total ash products	kg/h	485

TABLE 2: Summary of Plant Data - Commissioning Operations

Run No. 5CT/		7B	8A	8B	9A(R)	9B(R)	9C(R)	13A(R)	13C(R)	13D(R)	14A	14B
Tailings Input	kg/h	201.8	708.5	785.4	337.9	758.4	595.6	1290	1345	1372	1192	1397
Coal or Oil Input	kg/h	72.6	19.1	5.4	49.8	3.6	8.6	-	-	-	27.6	34.7
Bed Temperature (nominal)	^oC	800	800	800	800	800	800	800	800	800	800	800
Spray Probe (height above bed)	m	0.76	0.61	0.61	0.61	0.61	0.61	0.60	0.46	0.39	0.37	0.55
Bed height	m	0.76	0.58	0.70	0.70	0.70	0.73	0.69	0.69	0.68	0.69	0.69
Fluidising Velocity	m/s	0.91	1.0	0.94	0.98	0.98	0.98	1.14	1.10	1.13	1.11	1.11
Flow to 1° cyclone	True m^3/s	1.23	2.50	2.43	2.12	2.50	2.34	3.56	3.54	3.60	3.7	3.92
Pressure drop across base	kPa	1.44	1.64	1.79	1.62	1.67	1.57	2.94	2.88	2.88	3.18	3.20
Pressure drop across 1° cyclone	kPa	0.34	0.20	0.42	0.41	0.44	0.41	0.71	0.67	0.80	0.60	0.62
Pressure drop across 2° cyclone	kPa	0.31	0.32	0.74	0.48	0.59	0.49	1.07	1.11	1.01	1.41	1.57
Tailings calorific value (db)	kJ/kg	14880	14360	14640	14380	14440	14460	17000	16980	16980	15440	15260
Plenum - Metal Base Plate	Temp ^oC	72	72	72	67	73	65	NA	NA	NA	NA	NA
Maximum Shell Temperature	^oC	68	70	68	43	65	65	61	64	65	54	59
1° Cyclone recycle rate	kg/h	-	-	-	714	1224	1236	2399	3609	3829	-	-
1° Cyclone efficiency	%	-	86	88	93	94	95	91	94	97	84	85
2° Cyclone efficiency	%	-	72	75	69	89	90	90	91	91	74	71
Overall Cyclone efficiency	%	-	96	97	98	99	99	99	99	99	95	96
Coarse ash make as % input ash	%	16	36	34	28	28	32	30	34	34	25	17
Excess air	%	102	51	29	94	50	83	32	28	27	17	-1
Solid carbon loss	%	4.6	6.1	5.1	1.0	0.7	1.0	5.8	7.8	7.1	11.3	14.5
Heat loss	% gross input	2.6	1.7	1.6	2.1	1.8	2.2	1.3	1.3	1.3	1.12	0.94

TABLE 3: Mass and heat balance data

Data collected during run 5CT/14A

			MASS BALANCE		
	TOTAL kg/h	ASH kg/h	CARBON kg/h	HYDROGEN kg/h	OXYGEN kg/h
Slurry	1191.58	191.05	147.31	100.83	743.65
Oil/coal	27.62	0.00	23.73	3.73	0.00
Air	2759.4				638.54
Total input	3978.60	191.05	171.04	104.56	1382.19
Accumulation in bed	0.34	0.34	0.00	0.00	0.00
Coarse ash	49.89	49.89	0.01	0.00	0.02
Fine ash Primary	147.87	129.98	17.43	0.37	0.35
Fine ash Secondary	19.28	17.50	1.71	0.04	0.03
Dust	6.70	6.37	0.12	0.01	0.00
Flue gas	3675.56		138.41	104.18	1312.11
Total output	3899.64	204.14	157.67	104.60	1312.52
Loss (in-out)	78.96	-13.09	13.36	-0.04	69.67
Loss (% of in)	1.98	-6.85	7.81	-0.04	5.04

Solid carbon loss is 11.3%
Excess air is 17%
Vol. flow into Primary cyclone 3.7 m^3/s

TABLE 3: (Continued)

5CT/14A

HEAT BALANCE

	W	% of in
Input		
Slurry	1610312	
Oil/coal	356959	
Total	1967271	
Output		
Dry flue gas	650140	33.05
Steam from slurry water	926990	47.12
Steam from combustion	137855	7.01
Ash	51030	2.59
Unburnt carbon in ash	163151	8.29
Unburnt carbon as CO	0	0.00
Heat loss	21980	1.12
Total	1951140	99.18
Unaccounted for	16125	0.82

TABLE 4:

TYPICAL ANALYSIS OF THICKENED COLLIERY TAILINGS

Chemical			Size μm	Size % undersize
Water	% a.r.	45	3175	100
Ash	% d.b.	65	1680	99.7
Carbon	% d.b.	24	500	88.4
Hydrogen	% d.b.	1.8	250	72.0
Nitrogen	% d.b.	0.5	125	61.0
Sulphur	% d.b.	1.6	75	54.4
Chlorine	% d.b.	0.1	63	54.4
Carbon dioxide	% d.b.	1.6	45	54.2
			31	51.8
Ash fusion temperature			22	49.5
Deformation	1450°C		16	46.1
Hemisphere	1480°C		11	42.7
Flow	1500°C		7.8	39.2
			5.5	35.5

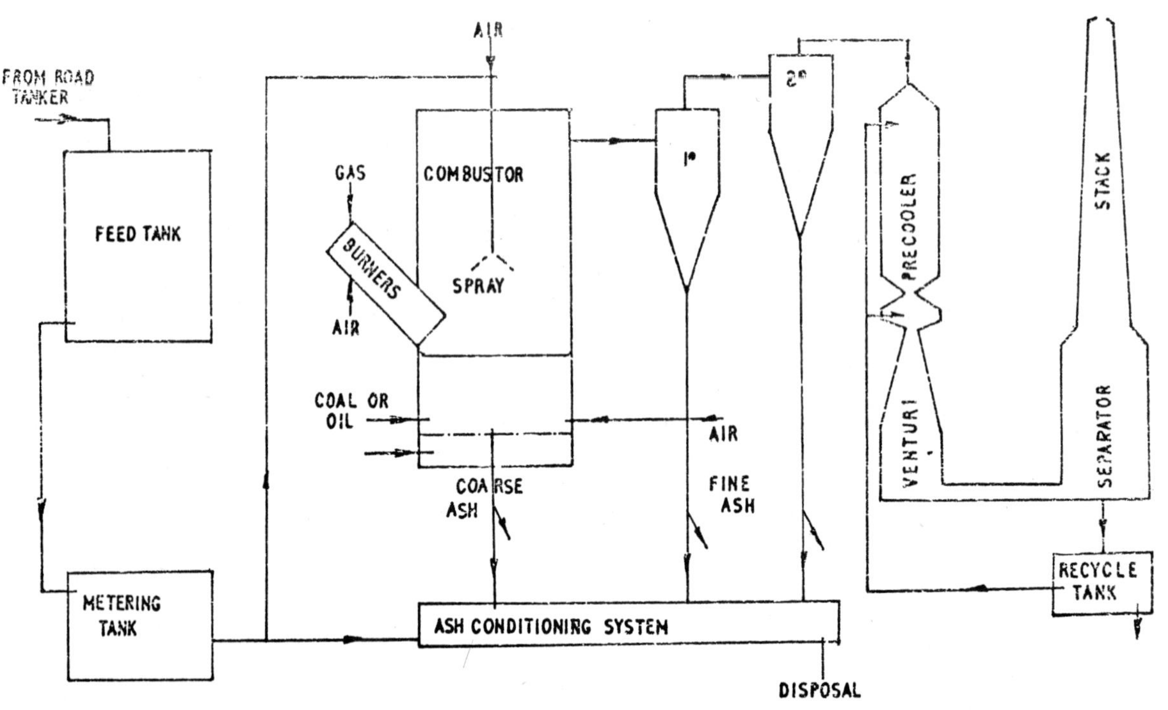

FIG.1. COMBUSTION OF COLLIERY TAILINGS – PROCESS SCHEMATIC.

Fig. 2 Fluidised bed tailings combustor.

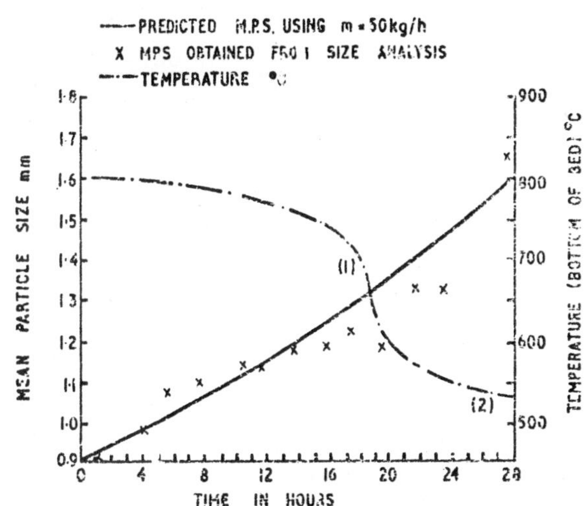

FIG. 3 INCREASE IN MEAN PARTICLE SIZE (M.P.S.)

FLUIDIZED BED COMBUSTION OF FLOTATION TAILINGS

By W. Poersch and G. Zabeschek

Babcock-BSH Aktiengesellschaft
Parkstrasse 29, D-4150 Krefeld
W. Germany

SYNOPSIS

The subject of this report is the incineration of flotation slurries containing varying amounts of carbon in a pilot fluidized bed combustor. Thermally pre-dried flotation slurries yielded higher combustion efficiencies than mechanically dewatered slurries, the latter ones dispersed satisfactorily only at solids contents of not more than 55%. To keep the emission of sulphur dioxide in the flue gas at the lowest possible level, additives of varying mean particle size were admixed to the flotation slurries. By the addition of limestone fines in mole ratios of about 2, desulphurization efficiencies of 90 % could be safely obtained. In carrying out these tests, it was essential to control the composition of the ash obtained so as to allow its further processing in the production of building materials.

NOTATION

C_1, C_2	carbon content of feed material resp. burnt-out material
η_C	combustion efficiency
$(SO_2)_{ES}$,	highest possible concentration of sulphur dioxide in flue gas
$(SO_2)_A$	measured concentration of sulphur dioxide in flue gas
η_S	sulphur retention
ϑ_{Bed}	average fluidized bed temperature
λ	excess air coefficient
p	partial pressure

INTRODUCTION

Increased mechanization in coal mining has led to increased production of fines during the past years. The usual way of separating the very fine coal from the debris is by flotation. The rate of flotation slurries from such separation processes is continuously growing. They contain about 96 % of water, 3 % fine shale and about 1 % of coal. The particle size of the solids is generally below 1 mm with an average of up to 40 % by weight of very fine particles below 10 microns. In most cases, the slurries are either thickened in settling tanks and then dumped, or they are mixed with the washery refuse after mechanical removal of the water and then likewise dumped. In West German coal mining, the annual rate of flotation tailings that has to be dumped is now more than 2,7 million tons (Wilczynski et al, 1977).

The environmental controls are becoming so stringent that it will hardly by possible in future to fulfill the conditions for dumping the tailings. For this reason, investigations into new solutions are now being made.

One of the solutions is the incineration of the flotation tailings in a fluidized bed by utilizing the residual coal energy. However, this solution seems acceptable only if it is possible to make use of the ash residues from the incineration process which are still about 70 % by weight of the dry flotation tailings before incineration. The basic mineral/clay substance of the ash residues seems to make them suitable for the production of raw and building materials for the building industry.

Much has been written already about the incineration in a fluidized bed. The influences that must particulary be taken care of when trying to efficiently burn out high-ash coal,have been discussed in detail (Waters 1975). The incineration of thickened flotation tailings in a fluidized bed was investigated in detail by Cooke and Hodgkinson (1975). Wilczynsky et al. (1977) report on an incineration test made with liquid flotation tailings in an industrial fluidized bed combustor of 10 m^2 grate surface.

In all the above quoted cases, however, the residual carbon content was generally above 5 % whereas if these residual products are to be used as building materials their carbon content must be below 3 % and they must have a CaO-content of more than 7 %. The material of the flotation tailings on the other hand has a CaO-content of only 2 - 4 % after incineration.

It has therefore been the aim of this investigation

a) to incinerate flotation tailings to a residual carbon content below 3 %;
b) to desulphurize the combustion gas and at the same time increase the CaO-content of the ash residues to above 7 %

EXPERIMENTAL

The fluidized bed reactor of 5 m height used for burning the flotation slurries is shown schematically in figure 1.

Above the fluidizing chamber, which has an inside diameter of 0,6 m, there is an after-combustion chamber of 4 m height flared to a diameter of 1 m. To achieve an even distribution of the combustion air and of the natural gas required for the starting-up operation, a nozzled bottom plate was chosen.

Pumpable flotation slurry can be introduced into the unit by means of a slurry lance, whereas predried tailings have to be fed in by a water-cooled screw. For controlling the fluidized bed temperature, water can be sprayed into the fluidized bed by means of an injector nozzle (using compressed air). The coarse ash particles are discharged via an overflow pipe, whereas the fine portion is carried out in the hot flue gas at the head of the reactor, and is then collected in cyclones.

The temperatures in the test unit are controlled by thermocouples attached at different levels over the circumference of the unit. A small stream of gas for flue gas analysis is branched off by a probe at the reactor head. This stream of gas is filtered and cooled and passed through a series of three gas analyzers (Uras) for measuring the flue gas components SO_2, CO and CO_2 . A thermo-magnetic oxygen analyzer (Magnos) for ascertaining the concentration O_2 is the last item of the measuring chain. The H_2S and NO_x contents in the flue gas can be ascertained by tube indicators.

All solid, liquid and gas flows into the reactor are continuously measured. The ash obtained is collected in containers and weighed in regular intervals. Samples of the flotation sludge feed and of the ash obtained are analyzed in the laboratory for particle size distribution, ignition loss, carbon and sulphur content.

RESULTS: PRE-CONCENTRATED FLOTATION SLURRIES

The solids content and composition of the flotation slurries as received highly fluctuated. For this reason, they were homogenized in a Nauta mixer. The solids content of these homogenized slurries was always below 40 %. More water was then removed by passing the slurry through a solid-jacket centrifuge which increased the solids content to 72 %.

This 72 % flotation tailings concentrate could, however, no longer be pumped or fed into the fluidized bed by means of special-type lances for thickened products. Modifications to the compressed-air driven injector nozzle configuration and the enlargement of the central sludge pipe did not result in any improvement. In all cases the sludge (even after screening off the larger particles above 1,5 mm) could only be distributedthrough an injector lance and dispersed in the fluidized bed without agglomerating at a maximum solids concentration of 55 - 57 %. The sludge injector lance was water-cooled to avoid incrustations of the pre-thickened sludge at the mouth of the lance dipping into the 900 °C hot bed. The centrally introduced sludge was finly dispersed into the fluidized bed at the mouth of the lance by adjustable rates of compressed air flowing through a compressed air pipe which co-axially surrounded the sludge pipe.

The essential problem obviously is the distribution of the sludge into very small droplets followed by fast and homogeneous blending into the inert fluidized bed. Flotation sludge containing 72 % solids will no longer disperse into small droplets. Larger droplets fell down to the perforated bottom plate and affected proper distribution of the air. Other sludge lumps did not break up sufficiently, agglomerated with the inert fluidized bed material and sintered during the burning-out process. Only with solids contents of 55 - 57 % proper incineration was possible over prolonged test periods (more than 26 hours). Sludge was first of all sprayed into an inert sand bed by a mono-pump through the above mentioned compressed-air operated injector nozzle. This caused the sludge to intimately intermingle with the bed of sand and being dried, heated, gasified and burnt by the high heat capacity of this inert sand bed which had been heated to approx. 900 °C. The particles to be burnt were for the most part, more than 65 %, removed from the bed by the carrier gas, and in many cases were burnt in the free space above the bed, which was evidenced by the waste gas temperature above the fluidized bed being up to 100 °C higher.

The silica content in the burnt dust residue was too high for the dust being processed for building materials, because of the fine dust and abrasion products it contained from the inert quarz sand bed. It increased from about 55 % SiO_2 in the burnt-out feedstock to approx. 70 % SiO_2 in the burnt-out dust.

So as not to affect the basic substance of the burnt-out material by too high a silica content, the quarz sand of the inert fluidized bed was replaced by an inert bed of similar basic substance. To do this, the washery tailings were ground to a particle size below 2 mm and incinerated in the fluidized bed. This incinerated substance then served as the inert fluidized bed for all further tests.

THERMALLY DRIED FLOTATION TAILINGS

All tests showed that with even fluidizing conditions the burning-out efficiency was the better, the less humid the fine flotation tailings particles could be introduced into the fluidized bed. This is plausible, for with the same reaction period available (same fluidizing velocity of approx.1,2m/sec.) in the oven, the pre-dried particles need no longer be dried, and heating, gasification, and combustion of the combustible residual substance can start without delay.

For economical reasons, we hence dewatered the flotation slurry to aprox. 72 % solids and then pre-dried it in a flash dryer to below 2 % residual moisture. The carbon portion burnt in the flotation tailings dust fed dry into the approx. 900 °C hot fluidized bed was higher than that of the tailings fed in as a slurry.

If the heat is to be utilized for the production of energy, fluidized bed combustion processes using mechanically dewatered and thermally dried flotation tailings will always give more favourable results than a fluidized bed into which the flotation tailings are sprayed-in in their liquid form:

1) the higher mechanical dewatering process that is possible will save evaporation energies;
2) the residual carbon in the ash is lower, and hence burning-out losses are saved;
3) the waste gas temperatures are lower because the heat can be recovered for the drying process, and hence waste gas losses are reduced;
4) the heat exchangers for the generation of energy can be smaller with the same amount of energy being generated because of the heat being mainly released in the fluidized bed with the heat transfer coefficients being three times as high as in the combustion gas;
5) due to points 1 and 2, the flotation tailings with lower residual carbon content ($C_1 \sim 14$ %) will burn autothermally, which is not possible with direct infeed ($C_1 \sim 24$ %).

For the above reasons, the greater number of the investigations was made with pre-dried material. (Table 1)

If C_1 denotes the carbon content of the flotation slurry prior to combustion, and C_2 that of the ash obtained in the test, the combustion efficiencies η_C for all tests can be calculated by using the equation

$$\eta_C = \frac{C_1 - C_2}{C_1} .$$

Fig. 2 shows these results plotted above the pertinent mean combustor temperature. The filled-in symbols denote the results obtained with injected slurry, the open symbols those using pre-dried flotation slurry.

It can be seen that with increasing combustion temperatures higher efficiencies are in both cases obtained. When the combustion temperatures were kept constant, the combustion efficiencies with dried flotation slurry were higher than those with wet flotation slurry. The combustion efficiency is obviously subject to the initial carbon content C_1, which varied between 19 and 37 %. The lowest residual carbon content of C_2 = 0,5 - 2 % was obtained with dried flotation slurries having an initial carbon content C_1 of approx. 19 %.

For the results plotted in fig. 2, the fluidizing velocities were varied between 1,1 and 1,5 m/s; the excess air levels were between 10 and 70 %; the mean particle size of the feed material was found to be 120 microns to 260 microns. Since the combustion efficiency η_C is also affected by the above magnitudes, the relatively wide range of measured results presented in fig. 2 becomes understandable. A much higher number of results would be required for a detailed presentation of all interrelations.

FLUE GAS DESULPHURIZATION

The sulphur dioxide (SO_2) which is obtained when burning sulphur-containing coal is chemically bound by some oxides of the alkaline earth metals within certain temperature ranges. This phenomenon has been known for a long time and has been utilized in a technically simple way particularly for desulphurizing the exhaust gas from fluidized bed combustors. In the tests discussed here, limestone or dolomite were added to the dry flotation sludge in pre-determined rates multiple of the mole rations Ca/S and Mg • Ca/S. In the hot fluidized bed these additives are first of all decarbonized. The calcium and magnesium oxides that are released will then react under excess air with the SO_2 of the flue gas and form sulphate. Important indications on the binding of sulphur that is possible in the fluidized bed reactor in dependence of the reactor temperature, are obtained when studying the thermo-dynamic balances of these reactions. According to Ulich&Jost (1960) and Barin&Knacke (1973) for instance the decomposition pressures of the CO_2 in the carbonates and the SO_2 in the sulphates can be quantitatively calculated as a function of the reaction temperature from the temperature-dependent equilibrium constant of the reactions.

Fig. 3 shows the equilibrium curves for the reaction equations given. The test conditions prevailing here allowed estimation of the resulting maximum carbon dioxide and sulphur dioxide partial pressures in the flue gas to be $P_{CO_2} < 100$ Torr and $P_{SO_2} < 3$ Torr under atmospheric pressure.

Accordingly, the sulphate formation takes place if the partial SO_2 pressure in the flue gas is higher than the decomposition pressure of the sulphates and if the partial CO_2 pressure in the flue gas is below the decomposition pressure of the carbonates. If $CaCO_3$ is used as an additive, the flue gas could consequently be desulphurized in the temperature range from about 730 °C to 1200 °C. The $MgCO_3$, an essential component of the dolomite, decomposes at much lower temperatures. The formation of $MgSO_4$ however, will only start at reaction temperatures below 850 °C under the prevailing conditions.

Preliminary tests made with the flotation slurries supplied revealed that at reactor temperatures of more than c. 970 °C, slagging and sintering of the ash can be expected and will, as a consequence, affect the operation of the fluidized bed reactor. On the other hand, satisfactory combustion of the flotation slurries is no longer possible at fluidized bed temperatures below c. 770 °C. These two temperature limits are marked in fig. 3. In connection with the equilibrium curves and the P_{CO_2} and P_{SO_2} lines obtained in the tests, it will hence be possible to ascertain the ranges of effectiveness of the various additives by approximation. It is revealed that, particularly when using dolomite as an additive, the test conditions can theoretically be varied only within a narrow operating zone.

The maximum possible sulphur dioxide content $(SO_2)_{ES}$ in the dry flue gas is obtained under the assumption that the complete quantity of sulphur contained in the dried flotation slurry is converted to sulphur dioxide in the reactor. It can be easily calculated from the set air and product flows and the sulphur content of the flotation slurries which is between 1,07 and 1,24 % by weight.

If $(SO_2)_A$ denotes the measured SO_2 concentration in the dry flue gas obtained by the addition of additives, a theoretic desulphurization efficiency η_S can be defined as follows:

$$\eta_S = \frac{(SO_2)_{ES} - (SO_2)_A}{(SO_2)_{ES}}$$

This definition usual in the technical literature, however, neglects the known phenomenon that coal ashes may bind considerable quantities of sulphur even without any additives being used. The rate of desulphurization that can actually be attributed to the use of additives is obtained if in the above equation the SO_2 concentration found in the flue gas without using additives is taken as reference magnitude.

In fig. 4, the desulphurization efficiency η_S has been plotted versus the mole ratios Ca/S = 1 and Ca • Mg/S = 1, i.e. for simple stoichiometric relations in accordance with the reaction equations given in fig. 3, 3,12 kgs of calcium carbonate or 2,93 kgs of dolomite are required for removing 1 kg of sulphur from the flotation slurry. The additives admixed to the slurry were fine and coarse limestone with mean particle sizes of 15 microns and 350 microns respectively, the $CaCO_3$ content of the additives being 96,6 %. The mean particles size of the dolomite used was 300 microns, and it essentially consisted of 59 % of $CaCO_3$ and 40 % of $MgCO_3$. The relatively wide range of measuring obtained and the relatively wide range of measuring results obtained and represented in fig. 4 is again explained by the changes in the remaining parameters required for the tests, as already mentioned in the previous section.

The desulphurization efficiencies obtained by using dolomite as an additive were lower than those obtained with limestone. It could assumed that the desulphurization process is mainly performed by the calcium portion of the dolomite. This assumption is supported by the better thermodynamic equilibrium found for limestone in the temperature range under investigation. Fine limestone as an additive gave higher desulphurization efficiencies than coarse limestone. This fact may on the one hand be due to the smaller specific surface of the coarse limestone. On the other hand, it is know from the literature (G. Moss, 1975) that in the temperature range under investigation the reaction velocity is controlled by pore diffusion and decreases with increasing particle sizes. From fig. 4, and from the measured SO_2 concentrations in the flue gas without additive, it is revealed that 30 - 40 % of the sulphur contained in the flotation slurries under investigation is already bound by the ash.

From fig. 4 it can finally be seen that with a limestone additive in the mole ratio of Ca/S of about 2 at least 90 % of the sulphur can be bound. Any improvements in this efficiency which may be obtainable by adding higher limestone rates are probably no longer justifiable from the economic point of view.

MASONRY CEMENT FROM THE ASHES

The ashes obtained from the fluidized bed tests were investigated for their suitability in the production of masonry cement. The evaluation of the material was to be made on the basis of the German Standards (DIN 4211 and DIN 1164). These DIN standards specify the main components of the binding material (cement and stone dust), the values limiting particle size and sulphate content, the approximate values of the setting time and the soundness of the test pieces,and also include the test specifications.

Within the scope of these investigations it was found that ashes containing less than 3 % carbon and approx. 7 % CaO are best suited for masonry cement. With the addition of not more than about 10 % by weight of fine cement, setting of the ground ashes was achieved. The bending tension strength and the compressive strength specified were obtained after a storage time of 7 days and 28 days in the atmosphere or in water resp. It was interesting to note that the results for ashes from pre-dried flotation tailings were about 30 % better than when using a feedstock of mechanically predewatered flotation slurries.

Investigations into the possibility of using the ashes as basic material for the production of expanded clay of small particle size are being made.

ACKNOWLEDGEMENTS

The authors wish to express their thanks to the Bergbau-Forschung GmbH, Essen, particularly to Messrs. D. Leininger and Th. Schieder, for placing the test material at their disposal free of charge and for their financial contributions to the test programme.

REFERENCES

Barin, I. & Knacke, O. (1973) 'Thermochemical properties of inorganic substances' Springer-Verlag Berlin, Heidelberg, New York; Verlag Stahl-Eisen Düsseldorf.

Cooke, M. J. & Hodgkinson, N. (1975) Proceedings of the Fluidised Combustion Conference, London, Vol. 1.

Moss, G. (1975) Proceedings of the Fluidised Combustion Conference, London, Vol. 1.

Ulich, H. & Jost, W. (1960) 'Kurzes Lehrbuch der Physikalischen Chemie' Steinkopff Verlag, Darmstadt.

Waters, P. L. (1975) Proceedings of the Fluidised Combustion Conference, London, Vol. 1.

Wilczynski, P.,Kamman, W., Kozianka, H. & Köhling, R. (1977) Aufbereitungstechnik 18, 71.

Table 1-Runs with pre-dried flotation tailings. Experimental conditions and results (mean values and/or measuring ranges).

Name	Dimension	Additiv: limestone <0.1mm Run 1	Additiv: limestone <0.1mm Run 2	Additiv: limestone <0.1mm Run 3	Additiv: limestone <1mm Run 4	dolomite <0.5mm Run 5
Air	kg/h	360	292	303.5	292	331.5
Water	kg/h	25	24-30	62	6-13	59
Flotation tailings	kg/h	95	100	52	100	50
Additiv	kg/h	1.3-12 3)	6-12	0.76-3.6	6-12	0.45-2.7
Temperatures:						
Fluidized bed	°C	935	922	898	918	905
Flue gas (outlet)	°C	818	855	829	830	857
Inlet temperatures	°C	25	25	25	25	25
Fluidizing velocity	m/s	1.42	1.22	1.40	1.12	1.48
Excess air coeff. λ	-	1.27	1.40	1.17	1.70	1.32
Bed height	m	0.9	0.8	0.9	0.8	0.9
Analysis of feed material:						
Loss on ignition	W.-%	27	30	51	27.5	51
Carbon content $\overline{C}_1$	W.-%	17.9	20.9	37.7	18	37.7
Sulphur content	W.-%	1.5	1.22	1.08	1.31	1.08
Mean particle diam.	mm	0.26	0.24	0.12	0.24	0.12
Analysis of cyclone fines: 1)						
Loss on ignition	W.-%	1.5-2.5	2.83	7.6	1.4	7.6
Carbon content $\overline{C}_2$	W.-%	0.6-1.9	1.8	6.7	1.05	6.7
Sulphur content	W.-%	-	0.02	0.03	0.01	0.03
Mean particle diam.	mm	0.13	0.11	0.11	0.11	0.11
Analysis of flue gas:						
Oxigen	V.-%	3.3-6.0	5.9-6.4	2.9-3.8	8.3-9.2	4.4-5.7
Carbon dioxide	V.-%	14.6-16.7	14.2-15.0	15.1-16.2	11.6-12.7	13.5-14.9
Carbon monoxide	V.-%	0.03-0.06	0.03-0.06	0.04-0.06	0.02	0.02
Sulphur dioxide $(SO_2)_A$	g/m^3_n	0.2-3 3)	0.21-0.63	0.15-2.2	0.24-1.95	1.0-2.2
Hydrogen sulphide	ppm	- 2)	-	≈ 10	-	≈10
Nitrogen oxides	ppm	- 2)	-	≈ 80	-	≤80

Remarks: 1) mean values; carbon content of bed ashes not more than 0.5 %.

2) not measured.

3) measuring ranges ; see Fig. 4.

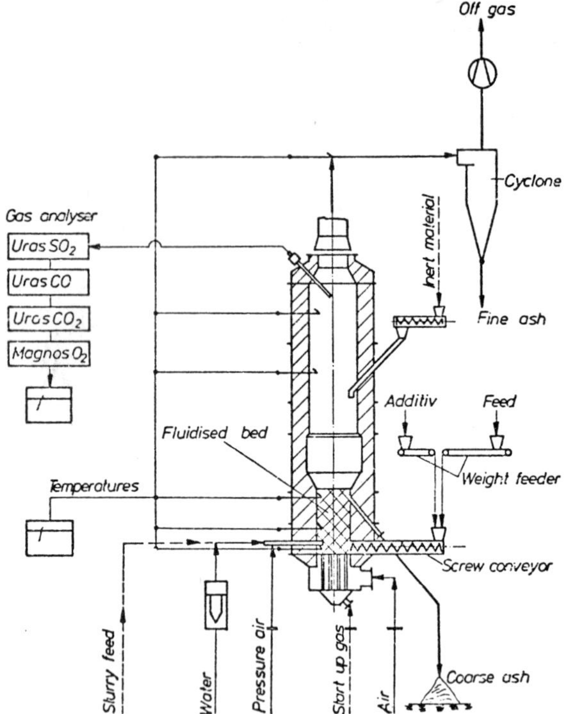

Fig. 1. Experimental apparatus

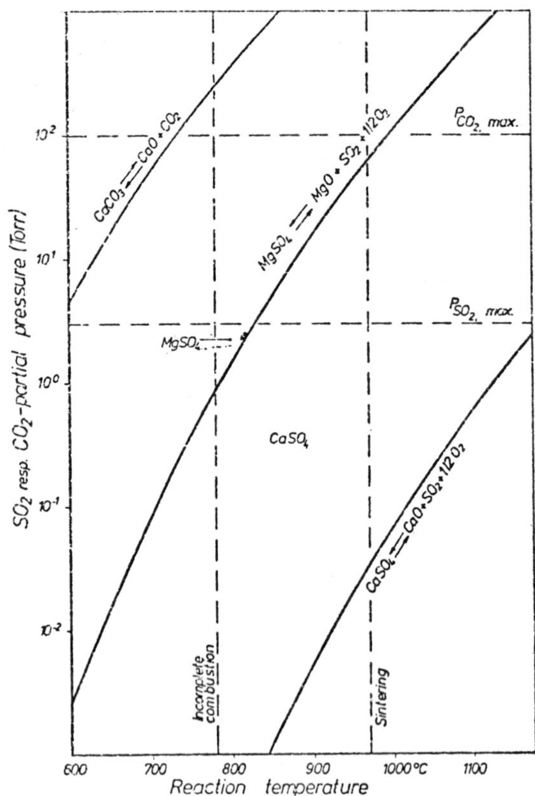

Fig. 3. Thermodynamic equilibrium curves and operating zones for flue gas desulphurization

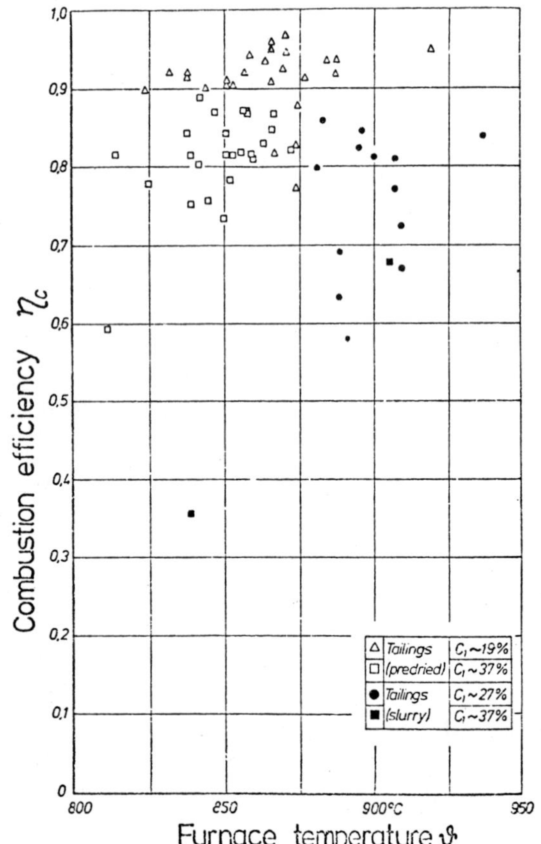

Fig. 2. Effect of furnace temperature, feeding procedure and carbon content of the flotation tailings on combustion efficiency

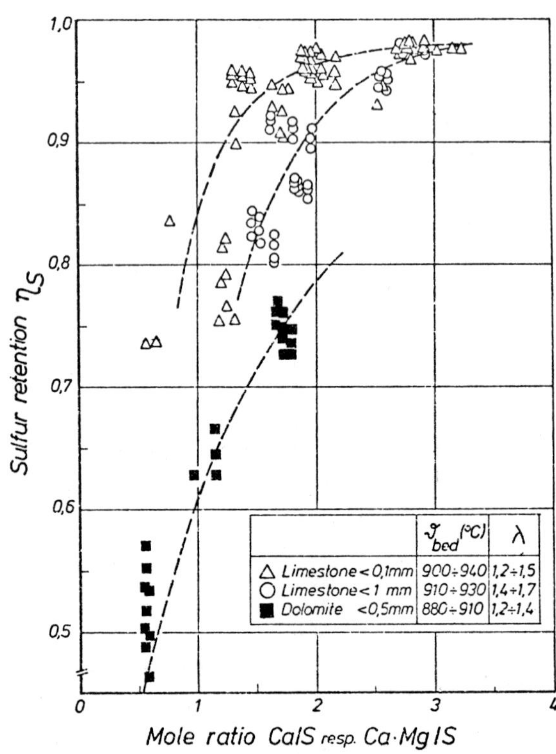

Fig. 4. Effect of mole ratios Ca/S and Ca Mg/S on sulfur retention for different additives

AN EVALUATION OF SOME FLUIDIZED BED REACTOR MODELS FOR SO_2 SORPTION ON COPPER OXIDE PARTICLES

By D. BARRETEAU, C. LAGUERIE and H. ANGELINO
L.A. 192 C.N.R.S.
Institut du Génie Chimique
Chemin de la Loge
31078 TOULOUSE CEDEX (France)

SYNOPSIS

Experimental results obtained on the non-catalytic gas-solid reaction between dilute sulphur dioxide and copper oxide supported on alumina particles continuously fed into a fluidized bed reactor are reported. Influence of various parameters i.e. gas flow rate, bed height, solid flow rate and temperature have been investigated. Experimental results have been compared with the theoretical results obtained from six different known models which have been slightly modified in order to take into account the continuous flow of solid. The Orcutt and Davidson's model provides the most accurate estimation for overall conversion.

NOTATION

A_1, A_2	pre, exponential factor for (1,2)
C_{AI}	inlet concentration of SO_2
C_{AB}	outlet concentration of SO_2
C_R	concentration of CuO-CuO in the reactor
C_{RI}	concentration of CuO-CuO at the inlet of the reactor
C_S	concentration of CuO-$CuSO_4$ in the reactor
C_T	concentration of $CuSO_4$-$CuSO_4$ in the reactor
D_e	bubble diameter
D_G	gas flow rate
D_S	solid flow rate
E_1, E_2	activation energies
g	acceleration owing to gravity
H_{mf}	height of solid at minimum fluidization conditions
h	height from the distributor
h_o	a constant characterising the distributor (h_o = o)
k_1, k_2	rate constante (1,2)
m	mean of errors
R	gas constant
s	standard deviation
T	temperature
U	superficial velocity
V_B	bubble velocity
U_{mf}	superficial velocity at minimum fluidization conditions

INTRODUCTION

Air pollution by sulphur components (SO_2, H_2S, COS etc...) is to be considered as a main problem in industry. Several processes, allow recovery of sulphur from effluents but the chemical reaction on copper oxide seems to be one of the most interesting of them (Dautzenberg, 1971, Mc Crea, 1970). In this reaction copper oxide deposited on alumina particles is transformed into cupric sulphate which has to be reduced by contact with light alkane such as methane in order to be regenerated into copper. During this step, a SO_2 concentrated gaseous flow may be obtained which is easily converted into sulphur, for example. The two reactions are at the same temperature, so that circulation of sorbent between two fluidized bed reactors would be very suitable.

Fluidized bed reactors have been studied mainly in the case of gas-solid catalytic reactions. A large number of models have been proposed and they differ in may aspects. In this paper, the reaction between cupric oxide and sulphur dioxide is studied in a fluidized bed reactor continuously fed with solid.

The experimental results have been compared with some of the models proposed in the literature (simple homogeneous models, Orcutt and Davidson, Kunii and Levenspiel, Partridge and Rowe models).

REACTION

Reaction between sulphur dioxide and cupric oxide with excess of oxygen has been studied (Best, 1974). The following mechanism has been selected as the best one to fit the experimental test runs.

$$CuO\text{-}CuO + \tfrac{1}{2}O_2 + SO_2 \longrightarrow CuO\ CuSO_4$$

$$CuO\text{-}CuSO_4 + \tfrac{1}{2}O_2 + SO_2 \longrightarrow CuSO_4\ CuSO_4$$

Each reaction is first order with respect to sulphur dioxide and to the active component of the solid. Rate constants may be expressed according to Arrhenius'law

$$k_1 = A_1\, e^{-\frac{E_1}{RT}} \quad (1) \qquad k_2 = A_2\, e^{-\frac{E_2}{RT}} \quad (2)$$

$$A_1 = 2.147\ m^3/mole \qquad A_2 = 0.143\ m^3/mole$$

$$E_1 = 17.62\ kJ/mole \qquad E_2 = 14.54\ kJ/mole$$

Decomposition of cupric sulphate is conveniently performed with methane. The reaction is stochiometrically

$$CuSO_4 + \tfrac{1}{2}CH_4 \longrightarrow Cu + \tfrac{1}{2}CO_2 + H_2O + SO_2$$

EXPERIMENTAL

The experiments were performed in the apparatus schematically shown in fig. 1.

The reactor consists of a 0.094 m inside diameter quartz tube. A low pressure drop gas distributor is used. It is made of several layers of iron spheres supported by a perforated plate (Hengl, 1975). The gas stream is preheated by flowing across an auxillary fluidized bed of alumina contained inside a metallic column. The withdrawal of solid is achieved by overflow. Sulphur dioxide is analysed by gas phase chromatography.

The experiments have been carried out with porous alumina particles 645 μm in diameter impregnated with 4.4 % in weight of copper. The concentration of sulphur dioxide at the inlet of the reactor has been 0.3 %.

Influence of bed temperature, solid flow rate, gas flow-rate and bed height have been studied.

MASS BALANCES

All the models tested have been developed mainly in the case of gas solid catalytic reaction for non circulating solid systems. Under these operating conditions, hydrodynamic behaviour of the solid does not need to be clearly indicated. On the contrary, for the system studied here, it is necessary to make some assumptions concerning the degree of solid mixing. The solid will be considered as perfectly mixed which is the case for high gas velocities.

Mass balances on solid and gaseous phases have been derived elsewhere (BARRETEAU, 1977). These balances express mass conservation between the two phases in the reactor : copper conservation

$$C_{RI} = C_R + C_S + C_T \qquad \dots \quad (3)$$

sulphur conservation

$$D_G(C_{AI} - C_{AB}) = D_S\,(2\,C_T + C_S) \quad \dots \quad (4)$$

Moreover, kinetic rate equations lead to another relation between solid reactant concentrations :

$$\frac{C_S}{C_R} = \frac{k_1}{k_1 - k_2}\left[\left(\frac{C_R}{C_{RI}}\right)^{k_2/k_1} - \frac{C_R}{C_{RI}}\right] \quad .. \ (5)$$

These three above equations are quite independant on the type of model chosen. The unknown quantities are the concentration at the oulet : C_{AB}, C_R, C_S, C_T. The reactor model will give an additional relation between these quantities. This system has to be solved simultaneously.

COMPARISON OF EXPERIMENTAL DATA WITH REACTOR MODELS

Theoretical results derived from six fluidized bed reactor models are compared with experimental data (see table 1) :

- Homogeneous reactor models with either perfect mixing or plug flow of gas

- Orcutt, Davidson models with either perfect mixing or plug flow of gas in the dense phase (Orcutt et al, 1962)

- Partridge and Rowe model (Partridge et Rowe, 1966)

- Kunii and Levenspiel "Bubbling Bed Model", (Kunii et Levenspiel, 1968a, 1968b, 1969).

Choice of these models results from a progression in mass transfer resistance between bubbles and dense phase.

Two resistances may be considered : the first is located between bubble and cloud, the second between cloud and dense phase.

Simple homogeneous reaction models ignore these two resistances. Orcutt and Davidson assume the first resistance as the main one. Partridge and Rowe consider only the second one in their model, and Kunii and Levenspiel take both into account.

1 Homogeneous reactor models (HRM)

In these models there is distinction between the bubbles and the dense phase.

a/ Perfect mixing

Conversions derived from the model have been calculated for each experiment. Results are presented in figure 2. Theoretical conversion is plotted versus experimental conversion. A fairly good agreement for the highest velocity of (U=0.6 m/s) can be noted.

b/ Plug flow of gas

Results are presented in figure 3. It can be seen that theoretical predictions for a 5cm high solid bed are not too different from the line representing experimental data. However, this model overestimates conversion for higher heights of solid.

2. Orcutt and Davidson reactor models-ODRM

The assumptions underlying the Orcutt et al. Models are the following (i) bubbles are of uniform size and are well distributed in all the bed, (ii), all the gas in excess of that required for incipient fluidization passes through the bed as bubble (iii) no solid exists in the bubble, so that reaction occurs only in the dense phase ; (iv) interphase mass transfer results from both through flow and molecular diffusion ; (v) gas in the dense phase is assumed to be either perfectly mixed or in plug flow.

Note that the mean bubble diameter has been calculated from Rowe's relation (Rowe, 1976) at the mid-point between the distributor and bed surface.

$$D_e = (U-U_{mf})^{\frac{1}{2}} (h+h_o)^{3/4}/g^{1/4} \quad .... \quad (6)$$

and the bubble velocity has always been determined from Davidson's relation :

$$U_B = U - U_{mf} + 0.711 \sqrt{g D_e} \quad ..... \quad (7)$$

others relations were tested but the influence of bubble diameter appeared to be very slight. For instance, a 10 per cent variation in bubble diameter results in a variation of 0.5 per cent for the conversion using Orcutt Davidson models.

a/ Dense phase perfectly mixed

Results are presented in figure 4. They appear to be nearly identical to that obtained from the homogeneous perfectly mixed reactor model

b/ Dense phase gas in plug flow

Overall conversions calculated using the plug flow assumption are in reasonable agreement with experimental results for the height of 5 cm (see figure 5). But they are overestimated for heights of 8 and 12 cm.

3. Partridge and Rowe model-PRM

In this model, bubbles and associated gas cloud are included in the same phase. The other assumptions are : (i) the visible bubble flow is given by the two phase theory of Toomey et Johnstone, (ii) gas in the dense phase is in plug flow, (iii) gas in bubble cloud phase is perfectly mixed ; (iv) cloud size is calculated from a semi-empirical modification of Murray's theory ; (v) interphase mass transfer between cloud and dense phase is identical as for rigid spheres at the same Reynolds number.

The results presented in fig. 6 show reasonable agreement over the whole range of operating conditions.

4. Kunii and Levenspiel "Bubbling bed model

The bubbling bed model of Kunii and Levenspiel presents two important features : the downflow of solid in the dense phase reduce the absolute percolation of gas and even may entrain the gas downward. Two distinct resistances to mass transfer between bubble and dense phase are taken into account. The assumptions of the model may be summarised as follows : (i) bubbles are of uniform size and well distributed in all the bed ; (ii) the associated cloud size is determined using Davidson's model. (iii) gas flow in the dense phase is neglected (iv) no reaction occurs in the bubbles, (v) mass transfer between bubble and cloud-wake is the same as for Orcutt et al model and transfer between the cloud-wake and the dense phase occurs by molecular diffusion.

Results are plotted in figure 7. It can be noted that this model leads to an underestimation of conversions at high gas flow rate and an overestimation at more important height of solid. It seems at any rate not to be convenient.

Discussion

Analysis of the above results requires a distinction into two class of experimental conditions according to the height of the bed. As it has been seen, for the bed height of 5cm and for any gas and solid flow rates the models based on the piston-flow of gas through the particles appear to be the most accurate. On the contrary, for bed heights of 8 and 12 cm, the models involving perfect mixing of gas in the dense phase leads to a better agreement with experimental data.

In order to select the most pertinent models in each case, means of errors and standard deviations between predicted and experimental conversion have been calculated. These values are given in table 2. It can be noted that for H = 5 cm, Orcutt-Davidson reactor model with plug flow of gas in the dense phase leads to the best prediction ($m = 0.05$, $s = 2.83$) while the simple homogenous perfect mixed reactor model gives a slight over estimation ($m = 0.99$).

For the heights of 8 and 12 cm, Orcutt-Davidson reactor model with perfect mixing of gas in the dense phase presents the best estimation but prediction appears to be sightly overestimated.

It might be concluded that the gas in the dense phase is in plug flow for low heights of bed and fairly well mixed for higher beds; but this conclusion cannot be asserted on the basis of such limited results since every model involves many other assumptions aside from those concerning gas mixing. The prediction of the Partridge and Rowe model for beds of 8 and 12 cm high are in fairly close agreement with those of Orcutt-Davidson reactor model with perfect mixing. This may appear very surprising since plug flow of gas in the dense phase is assumed. However it can be noted that the experiments have been carried out for low excess gas velocities $U-U_{mf}$. Under these conditions, cloud size is large and the dense phase fraction is reduced so that the reaction occurs mainly in the cloud. As gas in bubble-cloud phase is considered as perfectly mixed, this model is not too far from the homogeneous perfertly mixed reactor model.

As for the Kunii-Levenspiel model, it does not allow to predict conversion with a good accuracy but it should be remembered that this model is consistent for velocity ratio U/U_{mf} greater than 6 ; that was not the case in this work.

Besides, the effect of temperature, gas velocity and solid flow rate does not appear as determinent to choose a model in our operating conditions. The effect of each variable could be seen elsewhere (Barreteau, 1977a 1976b).

CONCLUSION

Six bed reactor models have been tested by comparing experimental results obtained with theoretical predictions. Orcutt and Davidson reactor model provides the most useful approximation of overall conversion. In the case of small bed height the assumption of plug flow of gas in dense phase has to be used and the assumption of perfect mixing of gas has to be employed for higher beds.

REFERENCES

Barreteau, D, (1977), Thèse de Docteur-Ingénieur, Toulouse, France.

Barreteau, D., and ANGELINO, H. (1976), Particle Technology, Nuremberg, to be published.

Best, R.J., (1974) Ph.D, Thesis University College, London, England.

Dautzenberg, F.M., Naber, J.E., Vanginneken, A.J.J. (1970), Chem. Eng. Prog., 67, 86.

Hengl, G.,(1975), Thèse d'Université, Toulouse, France.

Kunii, D., Levenspiel O. (1968a), Ind. Eng. Chem. Fundam., 7, 446.

Kunii, D., Levenspiel O. (1968b), Ind. Eng. Chem. Proc. Des. Dev., 7, 481.

Kunii, D., Levenspiel O. (1969) Fluidization Engineering, Wiley New York.

Mc Crea, D.H., Forney, A.J., Myers, J.G., J. Air. Pollution Control. Assoc. (1970) 20, 819.

Orcutt, J.C., Davidson, J.F., Pigford, R.L. (1962), Chem. Eng. Progr. Symp. Ser., 58, n° 38, 1.

Partridge, B.A., Rowe, P.N. (1966), Trans. Inst. Chem. Eng., 44, 335.

Rowe, P.N. (1976), Chem. Eng. Sci., 31, 285.

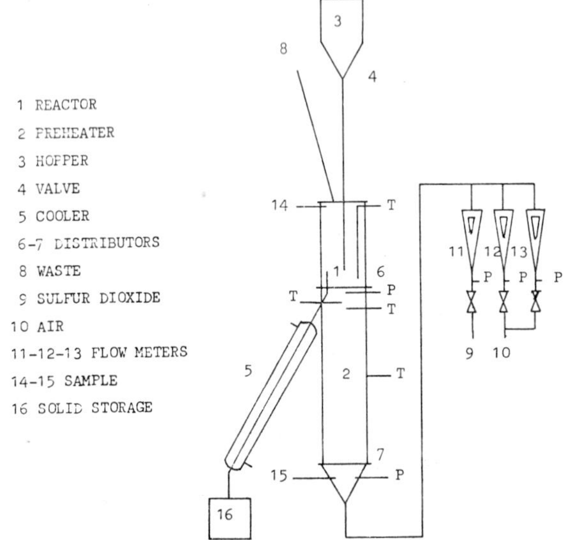

FIGURE 1 : APPARATUS

T °C	D_G m^3/h	D_S Kg/h	H_{mf} cm	$U-U_{mf}$ m/s	$D_e \times 10^2$ m	U_B m/s	X_{exp} %	HRM Perfect mixing	HRM Plug Flow	ODRM Perfect mixing	ODRM Plug Flow	KLM	RPM	Symbol	$U_B/U_{mf}/$ mf
340	7	1	5	0.45	4.0	0.89	25	24	26	24	25	23	23	+	2.2
320	7	1	5	0.43	3.9	0.86	27	24	25	23	24	22	23		2.1
285	7	1	5	0.39	3.6	0.81	23	23	23	22	23	21	21		2.0
255	7	1	5	0.36	3.4	0.77	22	22	22	20	21	20	20		1.9
320	7	2	5	0.43	3.8	0.86	32	30	33	30	32	28	28	v	2.1
300	7	2	5	0.41	3.7	0.83	28	29	32	28	30	27	27		2.0
275	7	2	5	0.38	3.5	0.79	27	27	30	27	28	25	26		1.9
340	7	3	5	0.45	4.0	0.89	42	35	40	35	37	32	32	∧	2.2
340	7	3	5	0.45	4.0	0.89	34	35	40	35	37	32	32		2.2
320	7	3	5	0.43	3.9	0.86	33	33	38	34	35	31	31		2.1
280	7	3	5	0.39	3.5	0.80	30	30	33	30	32	28	28		2.0
328	5	1	5	0.26	2.7	0.62	38	32	35	32	34	33	34	x	1.5
295	5	1	5	0.23	2.5	0.59	36	30	32	30	32	32	32		1.4
265	5	1	5	0.21	2.3	0.55	31	28	30	28	30	30	30		1.3
340	5	2	5	0.27	2.8	0.64	46	40	47	40	45	43	44	>	1.6
320	5	2	5	0.25	2.6	0.61	44	39	44	39	43	42	42		1.5
310	5	2	5	0.24	2.5	0.60	45	38	43	38	42	41	41		1.4
290	5	2	5	0.23	2.5	0.58	34	36	41	36	40	40	40		1.4
270	5	2	5	0.22	2.4	0.56	40	34	39	34	38	38	38		1.3
310	5	3	5	0.24	2.5	0.60	44	41	42	41	46	45	45	<	1.4
275	5	3	5	0.22	2.4	0.56	38	37	41	37	42	42	42		1.4
260	5	3	5	0.21	2.3	0.56	36	36	40	36	40	41	41		1.3
340	5	2	8	0.27	3.7	0.69	43	49	58	49	55	54	50	o	1.7
310	5	2	8	0.24	3.5	0.66	46	47	54	47	52	52	48		1.6
290	5	2	8	0.23	3.4	0.64	42	45	52	45	50	50	47		1.5
310	5	2	12	0.24	4.6	0.72	58	54	64	54	61	61	53	ϕ	1.7
290	5	2	12	0.23	4.4	0.70	51	52	62	52	59	59	52		1.7
260	5	2	12	0.21	4.1	0.66	53	50	58	50	56	57	51		1.6

TABLE 1 : Results

$U_{mf} = 0.182$ m/s

$\varepsilon_{mf} = 0.44$

Model		HRM perfect mixing	HRM plug flow	ODRM perfect mixing	ODRM plug flow	PRM	KLM
H = 5cm	m	- 2.42	0.99	-2.51	0.05	-1.53	-1.67
	s	3.76	3.08	3.76	2.83	3.70	3.73
H = 8 and 12cm	m	2.35	11.00	2.27	8.72	3.21	8.33
	s	4.29	11.50	4.25	9.38	5.30	9.10

TABLE 2 : Mean of error and standard deviation

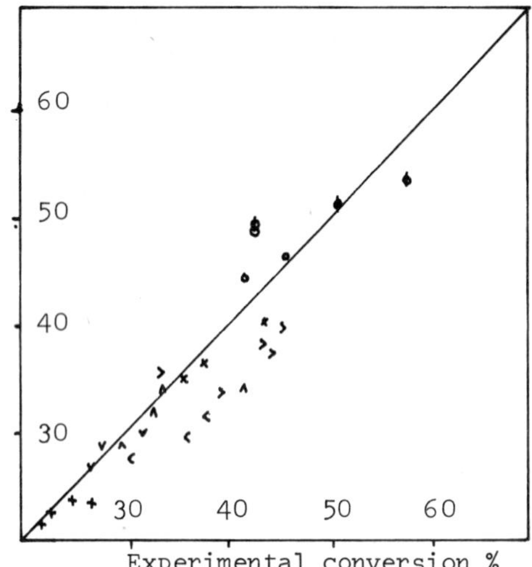

FIG. 2 - HRM PERFECT MIXING

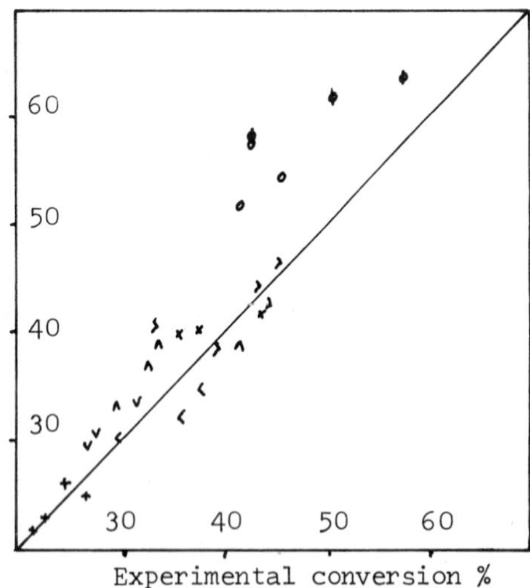

FIG. 3 - HRM PLUG FLOW

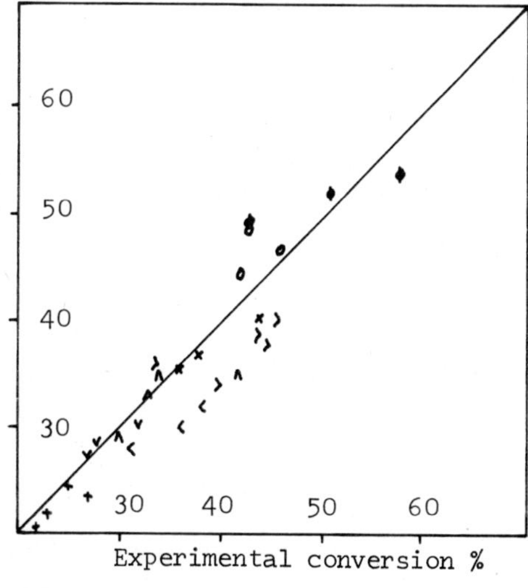

FIG. 4 - ODRM PERFECT MIXING

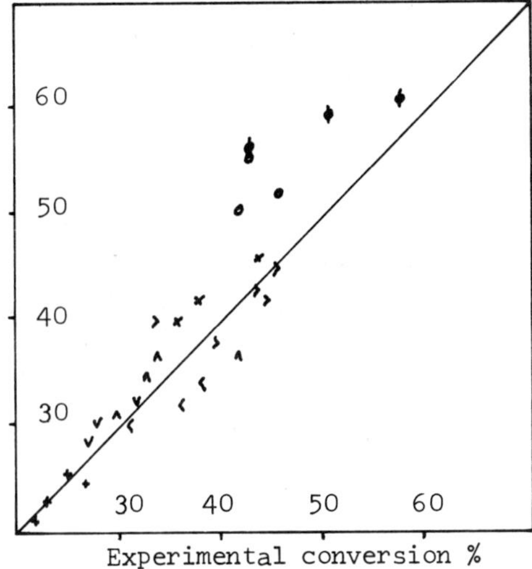

FIG. 5 - ODRM PLUG FLOW

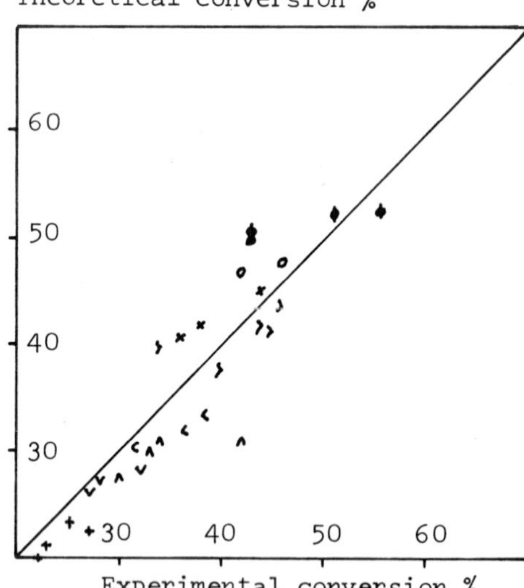

FIG. 6 - RPM

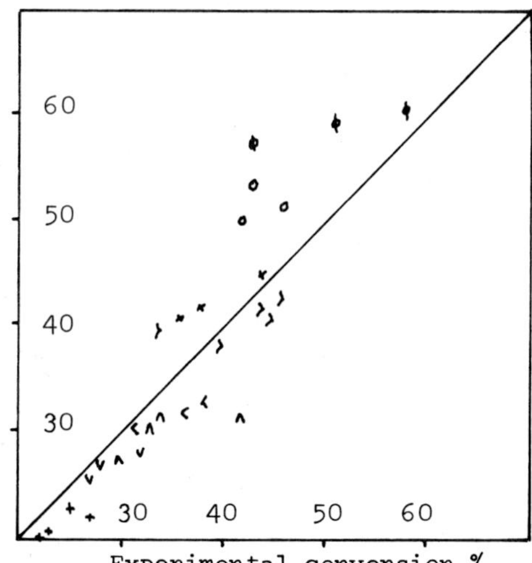

FIG. 7 - KLM

FLUIDIZED BED AKSO-PROCESS FOR REMOVING SO_2 FROM GASEOUS MIXTURES

By L. Neužil and F. Procháska

Department of Chemical Engineering
Prague Institute of Chemical Technology
Suchbátarova Street 3 , 166 28 Prague 6
Czechoslovakia

SYNOPSIS

We examined the preparation of active Na_2CO_3 from $NaHCO_3$ by thermal decomposition in a fluidized bed followed by the reaction of active soda with SO_2 contained in a low concentration air mixture. This process is termed the AKSO-process and it is suitable for separating sulphur dioxide from gaseous effluents. The work proper is aimed mainly at the actual desulphurization process, and in it we examined the influence of temperature, particle size of the solid phase and steam content. The results obtained have been used for verifying three selected models of a non-uniform fluidized bed. The suitability of a fluidized bed reactor for the AKSO-process has been proved.

NOTATION

k	effective kinetic constant for (4)
$n_{S,o}$	moles of Na_2CO_3 at the beginning of process B
$\dot{n}_{Z,in}$	molar flowrate of SO_2 at reactor inlet
p	total pressure in the system
R	reaction rate
t	time since the beginning of the process B
x_S	molar fraction of Na_2CO_3
$x_{S,o}$	molar fraction of Na_2CO_3 at the beginning of process B
y_Z	molar fraction of SO_2 in the gas
Y	relative outlet concentration of SO_2 for (7)
ε	bed porosity
θ	dimensionless time defined by (8)
ξ	degree of conversion of the solid phase, defined by (6)
ρ_f^+	molar density of fluid
ρ_s^+	molar density of solid
Ψ	degree of gas conversion, defined by (7)

INTRODUCTION

In a fluidized bed reactor we followed a set of reactions termed as the AKSO-process, suggested for SO_2 separation from gaseous mixtures (Erdös et al., 1968). This process is based on using a very reactive form of sodium carbonate (so-called active soda), prepared by the thermal decomposition of sodium hydrogen carbonate (process A):

$$2\ NaHCO_3 \rightarrow Na_2CO_3 + CO_2 + H_2O \qquad [130\ kJ\ mol^{-1}] \quad ....(1)$$

Resulting active soda is then used for desulphurization of the gas, assuming the following reactions to take place (process B):

$$Na_2CO_3 + SO_2 \rightarrow Na_2SO_3 + CO_2 \quad [-56.5 \text{ kJ mol}^{-1}] \quad ....(2)$$

$$2\,Na_2SO_3 + O_2 \rightarrow 2\,Na_2SO_4 \quad [-294 \text{ kJ mol}^{-1}] \quad ....(3)$$

Reaction given in (1) has been studied several times in the past (Ciborowski & Czarnota, 1963; Calistru & Ifrim, 1974), its kinetics were suggested and the course of the reaction has been described under different conditions in the reactor, both for a fixed, and for a fluidized bed. In the work of Hehl et al.(1977) a mathematical model of the fluidized bed reactor where reaction (1) takes place has been given.

We have less knowledge of the reactions (2) and (3). The kinetics of the process B were described (Erdös & Bareš, 1965) and have been verified in an integral laboratory reactor with a fixed bed (Bareš et al., 1970). A mathematical model has been presented for this reactor type. The fluidized bed AKSO-process had not been however studied to such detail as yet. The present contribution is based on our experimental results obtained in a batch fluidized bed reactor, partially published earlier (Procháska & Neužil, 1977).

MATHEMATICAL MODELLING FLUIDIZED BED REACTOR

Three models described in the literature have been adapted for modelling the batch fluidized bed reactor and process B. All models used set out from the assumption of the two-phase theory of the non-uniform fluidized bed, viz. its division into a bubble "phase" and an emulsion "phase":

1. Simple two-phase model (Gomezplata & Shuster, 1960).
2. The physical model (Orcutt et al., 1962).
3. Bubble assemblage model (Kato & Wen, 1969).

For evaluation of the models it is necessary to know the values of following parameters: mass transfer coefficient of the gas reacting component between bubbles and emulsion "phase", minimum fluidization velocity, bubble rising velocity, bubble diameter, number of bubbles in a unit volume or bubble frequency. They were computed using relations published by the authors of the original models and from other sources (Borodulya & Gupalo, 1976; Davidson & Harrison, 1963; Kunii & Levenspiel, 1969), and some values were obtained from our own experiments (cf. the following text).

Gas phase reaction rate R was expressed by means of the published kinetic equation of the process B (Erdös & Bareš, 1965). By adjustment of original equation we obtained

$$R \equiv dy_Z/dt = k\, C\, y_Z\, \xi^{1/2}\, (1 - \xi)^{3/2} \quad ....(4)$$

$$C \equiv -\, p\, \varrho_s^+ (1 - \varepsilon)/\, \varrho_f^+ \varepsilon \quad ....(5)$$

$$\xi \equiv (x_{S,o} - x_S)/x_{S,o} \quad ....(6)$$

Further, we introduced the degree of conversion at the reactor outlet for the gaseous phase

$$\psi \equiv 1 - Y \equiv (y_{Z,in} - y_{Z,out})/y_{Z,in} \quad ....(7)$$

and the dimensionless time

$$\Theta \equiv \dot{n}_{Z,in}\, t/n_{S,o} \quad ....(8)$$

Values of the degree of the solid phase conversion have been determined for the evaluation of (4) by numerical integration

$$\xi = \int_0^\Theta \psi \, d\Theta \quad ....(9)$$

EXPERIMENTAL

The laboratory batch fluidized bed reactor consisted of an 0.16 m ID glass column, 0.5 m high provided with a multiorifice distributing grid. Experimental plant was described in detail by Procháska & Neužil (1977). Sodium hydrogen carbonate, analytical grade, was used as the initial material for the preparation of active soda. Four narrow fractions have been prepared for the experiments, having the equivalent grain diameters 0.26, 0.36, 0.45 and 0.65 mm, resp. During the process B, temperature was kept constant at two levels: 120 and 140 °C. Shallow bed was used in all experiments (fixed bed height about 50 mm). An identical relative gas velocity was used throughout, equal to the four-fold of minimum fluidization velocity. Air was used in the process B for fluidization, containing a low percentage of SO_2 (0.2% v/v). This amount is usual for industrial effluents. The experimental gaseous mixture contained different amounts of steam.

The process A and B have been in the experiments arranged in three different ways:

1. Process B followed immediately the process A.
2. Process A took place separately - active soda prepared was stored in tightly closed bottles and then used for the process B.
3. The processes A and B took place simultaneously.

During the experiments we analyzed the gas composition at inlet and outlet and the solid phase, we measured pressure and temperature in the reactor and the relative humidity of the gas mixture. Apart from that, we determined the bubble parameters (diameter, rising velocity, frequency) at different heights in the bed using a needle capacitance sensor (Pešan, 1977). The microstructure of the solid phase was observed during the AKSO-process using a scanning electron microscope.

This activity was aimed mainly at:

- ▷ verifying the possibility of carrying out the AKSO-process in a fluidized bed,
- ▷ disclosing the influence of temperature, particle size and steam content in the gas on the desulphurization process,
- ▷ verifying 3 selected models using experimental conversion data from a fluidized bed reactor,
- ▷ verifying the possibilities of storing the active soda,
- ▷ finding out the possibility of carrying out the processes A and B simultaneously,
- ▷ comparing the measured parameters of bubbles with values resulting by calculation from published relationships.

DISCUSSION

The temperature and particle size influence has been evaluated from the dependencies of $\xi(\Theta)$ at different conditions using a factorial experiment design. The evaluation of this design indicated that there is no significant influence of temperature in the interval $\langle 120, 140 \rangle$ °C and of the particle size in the interval $\langle 0.26, 0.65 \rangle$ mm on the degree of conversion of the solid phase. This fact follows also from Fig.1, where we plotted the experimental points for different solid fractions and both temperatures. The influence of the steam content on solid phase conversion was tested at two levels corresponding to partial pressure of H_2O (g) 4 and 15 Torr, respectively and this is given in Fig.2. From the graph and from the results of the factorial experiment design it follows that the increase of steam content does have a pronounced positive influence on the solid phase conversion.

Experimental results have shown that the AKSO-process can be carried out successfully in a fluidized bed. We can see from the plot of $Y(\Theta)$ in Fig.1 that in the time interval since the start of the process until a degree of conversion in the solid phase equal approx. to 0.75 the degree of gas purification was higher than 95%. The break-through occured approximately at a dimensionless time equal to 0.9. After having achieved this degree of conversion, the purification worthened during a short time due to the influence of the break-through of SO_2 through the bed. We found out that storing of active soda in a dry medium (closed bottles) does not decrease substantially its reactivity. Experiments have substantiated the possibility to carry out simultaneously the processes A and B in a single fluidized bed. Until the break-through point, even in such an arrangement the degree of purification was higher than 95%. Incomplete purification is obviously also due to the short-circuiting of the low bed by bubbles. Using a horizontal baffle (a leakage grid), the degree of purification could have been increased up to 99%.

Comparing the experimental diameter and rising velocity of bubbles with values obtained from empirical correlations recommended in the literature, we arrived at the conclusion that the relationships published in the literature are suitable for the given conditions, however, experimental values were slightly higher than the predicted ones. We could also substantiate the assumption of bubble growth at increased distance from the grid and at increasing density of the solid particles. However, we could not confirm the validity of the assumption of the growth of bubbles with increasing particle diameter. The relationship for calculating the bubble frequency (Kunii & Levenspiel,

1969) was not in our case suitable as it gave values substantially higher than the experiment.

Values resulting on testing the selected models are given in Fig.3 and 4 for the solid and the gaseous phase, resp. Comparing the models with data on conversion resulting from the experiments it follows that both bubble models (No. 2 and 3) yielded better results, explaining the course of the process nearly in the entire region investigated. More significant deviations have been observed only in the conversion range in the solid phase higher than 0.9, here experimental values were lower. Introducing the assumption of bubble growth (model No. 3) did not bring any better fit of the model; this may be due to the low bed height used. We can state that these adapted models are suitable for the prediction of the behaviour and conversion in a laboratory fluidized bed reactor.

The microphotographs made at different process stages show that the high reactivity of active soda is caused largely by a porous structure having a large surface area. As an important conclusion we can therefore present the fact that due to the absence of the disintegration of initial crystal, there is only little elutriation.

CONCLUSIONS

The AKSO-process can be carried out successfully in a fluidized bed reactor and a high degree of desulphurization can be achieved (over 95%) at high apparatus efficiency. In the range of experimental conditions investigated we observed no influence of the temperature and of particle size on the process. There is a positive influence of the increase of steam contents in the gas. Both processes, A and B, can be carried out simultaneously. Modified bubble models of non-uniform fluidized beds can be used with good results for mathematical modelling of the B-process. The AKSO-process seems to be of advantage for the separation of sulphur dioxide from industrial exhalations mainly in the glass and paper technology.

REFERENCES

Bareš, J., Mareček, J., Mocek, K. & Erdős, E. (1970) Coll. Czech. Chem. Commun. 35, 1628.

Borodulya, V.A. & Gupalo, J.P. (1976) Mathematic Models of Chemical Reactors with Fluidized Bed, Nauka i Tekhnika, Minsk, USSR (in Russian).

Calistru, C. & Ifrim, L. (1974) Bull. Inst. Polit. Iasi. 20, 45.

Ciborowski, J. & Czarnota, T. (1963) Przemysl Chemiczny 42, 238.

Davidson, J.F. & Harrison, D. (1963) Fluidised Particles, Cambridge University Press, Cambridge.

Erdős, E., Bareš, J., Mareček, J. & Mocek, K. (1968) Czechoslovak Patent No. 129889.

Erdős, E. & Bareš, J. (1965) 5th Intern. Symp. on the Reactivity of Solids Vol. 5, p. 676.

Gomezplata, A. & Shuster, W.W. (1960) AIChE J. 6, 454.

Hehl, M., Helmrich, H., Kröger, H., Schügerl, K., Wippern, D. & Wittmann, K. (1977) Proceedings of 3rd Conference on Applied Chemistry, Unit Operations and Processes, pp. 365-369, Veszprém, Hungary.

Kato, K. & Wen, C.Y. (1969) Chem. Eng. Sci. 24, 1359.

Kunii, D. & Levenspiel, O. (1969) Fluidization Engineering, John Wiley, New York.

Orcutt, J.C., Davidson, J.F. & Pigford, R.L. (1962) Chem. Eng. Progr. Symp. Ser. 58, 1.

Pešan, B. (1977) Ph. D. Dissertation, Prague Inst. of Chemical Technology.

Procháska, F. & Neužil, L. (1977) Proceedings of 3rd Conference on Applied Chemistry, Unit Operations and Processes, pp. 461-466, Veszprém, Hungary.

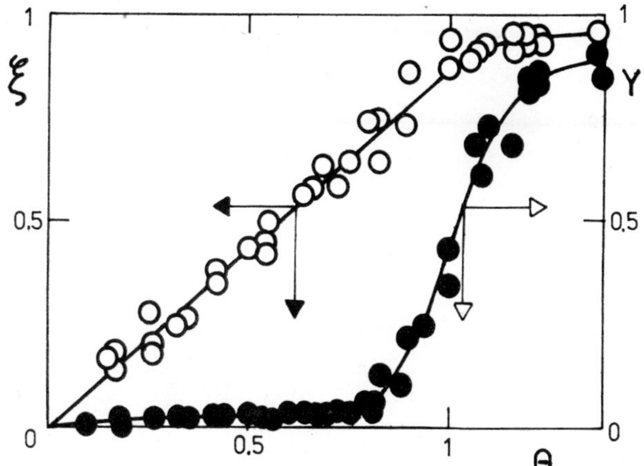

Fig. 1. Experimental data on solid phase coversion ξ-○ and relative outlet SO_2 concentration Y-● as a function of dimensionless time Θ.

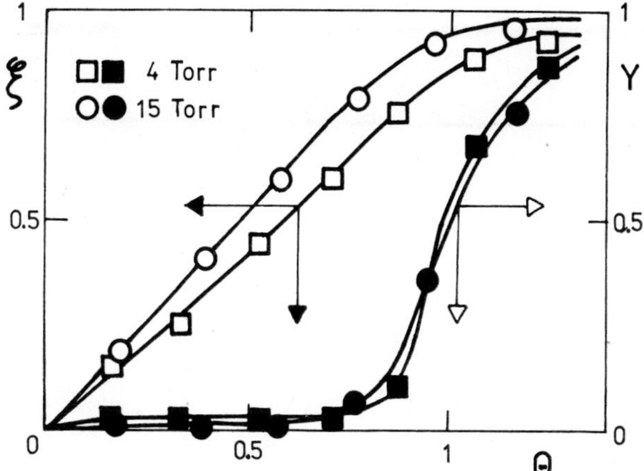

Fig. 2. Influence of steam content on the solid phase conversion ξ, and relative outlet SO_2 concentration Y at the desulphurization process.

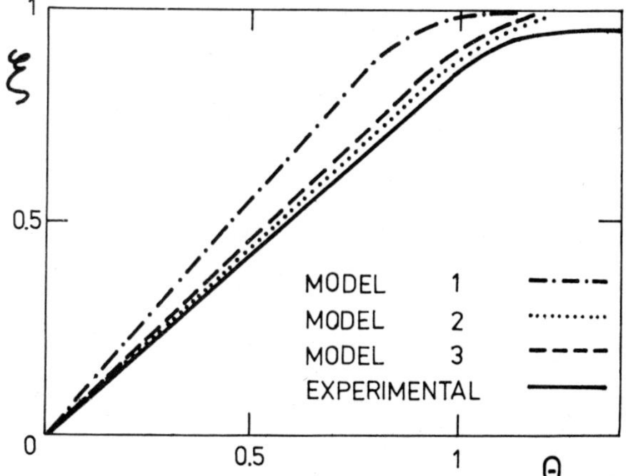

Fig. 3. Comparison of experimental values of conversion in the solid phase ξ as a function of dimensionless time Θ with values calculated from the simple 2-phase model 1, physical model 2 and bubble model 3.

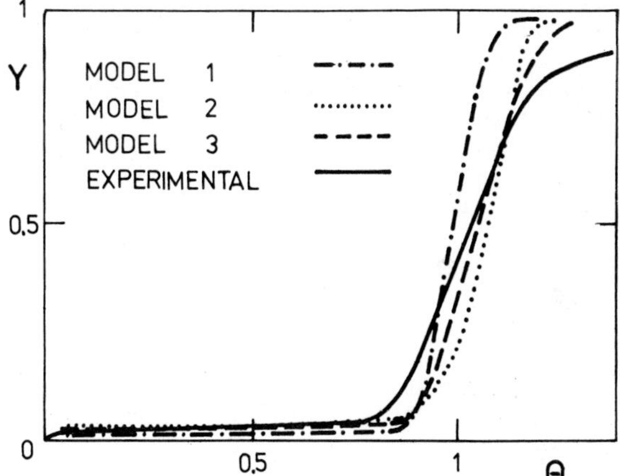

Fig. 4. Comparison of experimental relative outlet SO_2 concentration Y as a function of dimensionless time Θ with values calculated from theoretical considerations given by the models 1, 2 and 3.

REGENERATION OF THE SULFUR ACCEPTOR IN FLUIDIZED BED COMBUSTION

L. A. Ruth

Exxon Research and Engineering Company

Linden, NJ 07036

SYNOPSIS

A process is being developed for continuously regenerating sulfated limestone sorbent ($CaSO_4$) to CaO and SO_2 in a fluidized bed by reaction with a reducing gas at about 1100°C and 7-10 atm pressure. In preliminary experiments, batches of sulfated sorbent were regenerated and up to 3.7 mole percent SO_2 was produced, reduction of $CaSO_4$ to CaO was nearly complete, and agglomeration of the fluidized bed was avoided by careful control of temperature. The technical viability of continuous regeneration was demonstrated in a five day run during which hot sorbent circulated between the regenerator and a fluidized bed coal combustor, both at 7-1/2 atm pressure. Compared to a system without regeneration, make-up sorbent requirements were reduced by at least a factor of four.

INTRODUCTION

In a fluidized bed coal combustor, limestone or dolomite sorbent reacts with SO_2 and oxygen in the flue gas to form $CaSO_4$. However, large quantities of sorbent are required to achieve low SO_2 emissions when burning high sulfur coal. In order to reduce sorbent feed rates and minimize the amount of sulfated material which must be discarded, we have been developing a one-step process in which $CaSO_4$ is regenerated to CaO and SO_2 in a fluidized bed by reaction with a reducing gas at about 1100°C (2000°F) and 700-1000 kPa (7-10 atm abs.) pressure. Our objective is to determine if regeneration is technically feasible by studying a continuous combustion-regeneration system. Preliminary work involved batchwise regeneration of sulfated limestone in a fluidized bed vessel of eight cm diameter (Ruth, 1975). In this paper, we report on batchwise regeneration in a 22 cm diameter fluidized bed and continuous regeneration in a system consisting of this vessel coupled to a 32 cm diameter pressurized coal combustor. The combustor-regenerator combination has been referred to as the "miniplant." Operability of the miniplant has already been demonstrated in a run lasting over 100 hours during which sorbent was continuously recirculated between the combustor and regenerator.

REACTIONS AND FACTORS INFLUENCING SO_2 LEVELS

A reducing agent is used to regenerate $CaSO_4$ to CaO and SO_2. The principal reactions involved, for CO as the reductant, are:

$$CaSO_4 + CO = CaO + CO_2 + SO_2 \quad (1)$$

$$CaSO_4 + 4CO = CaS + 4CO_2 \quad (2)$$

Reaction (1) is endothermic and is favored by high temperature. Reaction (2) is a side reaction which is undesirable because it consumes large quantities of reductant without rejecting sulfur from the solids. $CaSO_4$ and CaS can react according to:

$$3CaSO_4 + CaS = 4CaO + 4SO_2 \quad (3)$$

but reaction (3) is not independent since it is a combination of reactions (1) and (2).

We use a dual zone fluidized bed regenerator to maintain the amount of CaS at low levels. Fuel is introduced at the bottom of the bed, creating a reducing zone. Ordinarily, fairly high levels of CaS would be present in this zone, in addition to CaO and $CaSO_4$. In order to minimize the amount of CaS present, a stream of secondary air is added about halfway up the bed, producing an oxidizing zone in the upper portion of the bed. In the oxidizing zone, the following reactions can occur which remove CaS:

$$CaS + 2O_2 = CaSO_4 \quad (4)$$

$$CaS + 3/2O_2 = CaO + SO_2 \quad (5)$$

Because solids mix vigorously in the fluidized bed, particles containing CaS are alternately exposed to oxidizing and reducing environments, thereby providing a mechanism for eliminating CaS.

Examining the thermodynamics of the above reactions shows that the maximum partial pressure of SO_2 is produced when all three solids ($CaSO_4$, CaO, CaS) are present, and that this partial pressure is a function only of temperature. The type of reductant, e.g. CO, H_2, C, CH_4, etc., does not affect the maximum partial pressure of SO_2 which can be produced. However, the SO_2 content of the gaseous phase, expressed as a mole fraction, is inversely proportional to total pressure. Figure 1 gives SO_2 pressure vs. temperature calculated from free energy data available in the literature. For example, at 1100°C (2012°F), the SO_2 pressure reaches about 50 kPa (0.5 atm). This partial pressure of SO_2 corresponds to mole fractions of 0.5 and 0.05 at total pressures of 100 kPa (one atm) and 1000 kPa (ten atm), respectively. Partial pressures of SO_2 that were determined experimentally (Curran et al. 1967) are given in Figure 1 for comparison. The experimental values are considerably lower than the calculated values. This could be due to the formation of solid solutions or to errors in the thermodynamic data.

The significance of thermodynamics is that it provides an upper theoretical limit to the concentration of SO_2 that can be produced in the off-gas from a regenerator. However, other factors such as reaction rates and material and energy balances may prevent the SO_2 level from reaching its thermodynamic limit. If reaction rates are slow, equilibrium may never be reached; however, the rate of regeneration is quite high, usually much faster than sulfation (Ruth, 1977). Material and energy balances may be a more crucial factor. It is obvious that in a regenerator which is continuously fed sulfated sorbent, no more sulfur can leave, as SO_2, than enters as $CaSO_4$. Fuel, of course, also enters the regenerator to provide the energy needed for the endothermic reduction reaction, to heat the reactants, and to balance heat losses. The SO_2 produced is "diluted" by the combustion products of the fuel. Thus, the concentration of SO_2 in the off-gas can be determined by the feed rate and sulfur content of the sorbent, and by the feed rates of fuel and air, which are set by the energy balance for the system.

We have already noted that when the regenerator is operated under conditions for which the partial pressure of SO_2 is determined by thermodynamic equilibrium, the mole fraction of SO_2 is inversely proportional to total pressure. Nevertheless, we have chosen to study regeneration at elevated pressure because, in the system being developed, sorbent is continuously circulated between the combustor and regenerator. Operating the regenerator at the same pressure as the combustor greatly facilitates transfer of sorbent between the two vessels. It should be appreciated that pressure has negligible effect when the SO_2 concentration is limited by material and energy balances. If reaction rates are limiting, then increasing pressure could increase the SO_2 concentration by allowing the reaction to proceed further to equilibrium.

EXPERIMENTAL EQUIPMENT AND PROCEDURES

The regenerator vessel is constructed from 46 cm (18 in) Schedule 40 steel pipe, refractory lined to an inside diameter of 22 cm (8.5 in). The fluidizing grid is a drilled stainless steel plate which is water cooled. The height from fluidizing grid to gas exit is 5.8 m (19 ft). Air and fuel (natural gas) are supplied in nearly stoichiometric proportions to a burner located beneath the fluidizing grid. Supplementary fuel and air are added just above the grid and about halfway up the bed, respectively. In the reducing zone below the supplementary air inlet, the air/fuel ratio is generally 60-85 percent of stoichiometric. Above the supplementary air inlet, the air/fuel ratio is about stoichiometric.

Gas leaving the regenerator is passed through a cyclone to remove particulates, a single pass double pipe heat exchanger to cool the gas, a control valve to reduce pressure, and a scrubber for cleanup before venting to the atmosphere. A sample stream for gas analysis is taken just after the pressure reducing valve.

When the regenerator is operated alone without the combustor, a batch of sulfated sorbent is charged and heated, under oxidizing conditions, to the temperature desired. Reducing conditions are then established by increasing the flow rate of supplementary fuel. Air and fuel flow rates are adjusted during the run to maintain nearly constant fluidized bed temperature. The concentration of SO_2 in the regenerator off-gas is recorded; the shape of the SO_2 vs. time curve is similar to what is shown in Figure 2. After the run, the solids are removed from the regenerator and analyzed for Ca, SO_4^{-2}, S^{-2} and total sulfur.

Combined operation of the combustor and regenerator required development of a transfer system to circulate sorbent between the two vessels. The system, shown schematically in Figure 3, was designed to accomplish this by utilizing high bulk density (stick-slip) flow of sorbent in transfer lines. Pressure in the regenerator is maintained slightly higher than that in the combustor. Solids in the regenerator-to-combustor transfer line move into the combustor when a pulse of nitrogen is applied to the lower end of the transfer line. The solids' flow rate is controlled by adjusting the frequency, duration, and intensity of the pulse. Two slide valves are used in the combustor-to-regenerator transfer line in order to prevent backflow of regenerator gas up the line. These automatic valves trap solids in the piping between them. Solids are discharged into the regenerator when the bottom valve is opened. The two solids' take-off plugs shown in Figure 3 are inserted into the take-off ports during startup to prevent solids from entering the transfer lines. Plugging could occur if the solids became wet due to water condensation during startup. The manual slide valve in the regenerator-to-combustor line is also closed during startup and in case of upsets.

RESULTS AND DISCUSSION

Batch Operation of Regenerator

Eleven runs were made in which batch charges of either sulfated limestone or dolomite, prepared in the fluidized bed coal combustor, were regenerated. Pressure was about 910 kPa (9 atm) and average bed temperature normally ranged from 1027-1120°C (1880-2050°F). Fluidized bed height was 1-1.5 m (3.3-4.9 ft) and gas contact time 1-2 seconds. The objective of these runs was to determine the extent of bed agglomeration, if any, the concentration of SO_2 produced in the off-gas from the regenerator, and the degree of reduction of $CaSO_4$ to CaO.

In experiments made at Exxon several years ago in a much smaller regenerator (8.3 cm diameter) agglomeration of the bed was a very serious problem (Ruth, 1975). Hence, it was of interest to see if agglomeration could be avoided in the 22 cm diameter miniplant, which was expected to provide an improved quality of fluidization because of its larger size. The results were encouraging. Agglomeration did occur occasionally but was invariably associated with excessively high bed temperatures (above 1120°C (2050°F)); when temperature was well controlled agglomeration did not occur. Controlling temperature was more difficult during batch than continuous operation because the rate of regeneration, and hence the heat requirements for the endothermic reduction (1300 kJ/kg (560 BTU/lb) of $CaSO_4$ converted), varied during the batch run. Hence, the fuel and air flow rates had to be continuously adjusted to compensate for the changing heat load as the charge of $CaSO_4$ was regenerated. Nevertheless, agglomeration occurred only slightly or not at all in runs made at average bed temperatures of up to 1100°C (2000°F).

SO_2 levels measured from the miniplant were also higher than those from the smaller unit. The highest mole fraction of SO_2 from the miniplant regenerator was 3.0 percent (3.7 percent on a dry basis), compared to 2.0 percent from the small regenerator. These are maximum levels corresponding to the peak shown in the SO_2 vs. time curve of Figure 2. The better performance of the miniplant is probably due to improved fluidization. Indeed, bed temperatures were much more uniform, indicating that the quality of fluidization was better than in the smaller unit.

Peak values of measured SO_2 concentration averaged about half of the calculated equilibrium concentrations corresponding to the temperature and pressure for each run. However, as shown in Figure 1, equilibrium SO_2 partial pressures obtained experimentally by Curran were considerably lower. Using Curran's results, SO_2 concentrations that we measured averaged about 80 percent of equilibrium. In fact, equilibrium was reached in several runs, according to Curran's data.

After each run the fractional conversion of $CaSO_4$ to CaO and CaS was calculated from the Ca, total S, SO_4^{-2}, and S^{-2} analyses. An average of 94 percent of the sulfate was reduced to oxide. The amount of CaS present was negligible.

Demonstration of a Fluidized Bed Combustion System with Continuous Sorbent Regeneration

After batch operation of the regenerator was improved to the point of becoming routine, we moved on to our main objective, which was to demonstrate and characterize combined operation of the combustor and regenerator with recirculation of sorbent between the two vessels. After linking the two vessels, a run was made to demonstrate continuous operation. Operating conditions are given in Table 1. Conditions (especially regenerator temperature) were conservative so as to provide the best chance of reaching our goal of 100 hours continuous operation. All variables were kept constant except that makeup limestone sorbent was added to the combustor at the rate necessary to make up for losses caused by attrition and entrainment from the fluidized beds.

SO_2 emissions from the combustor are given in Figure 4. To establish baseline operation, the regenerator was operated under oxidizing conditions for the first 24 hours with sorbent recirculating between combustor and regenerator. Emissions gradually increased to about 550 ppm (1.1 lbs $SO_2/10^6$ BTU). Since the SO_2 emissions would have been about 1330 ppm at zero retention, 550 ppm corresponds to about 60 percent retention. This is a much lower level of SO_2 emissions than would have been expected had the combustor been operated at the same conditions, but without sorbent recirculating to the regenerator. The low combustor emissions can be explained because, even though no regeneration was occurring during this period, the regenerator was acting as a calciner and supplying freshly calcined sorbent to the combustor.

Within about four hours after reducing conditions were established in the regenerator by increasing the flow of supplementary fuel, SO_2 emissions from the combustor fell from 550 ppm to below 200 ppm. After about 50 hours into the run (25 hours under reducing conditions) SO_2 levels began to increase and reached about 550 ppm when the run was terminated after 125 hours. The increase in emissions is, for the present, assumed to have been caused by a gradual decline in the activity of the regenerated sorbent. About fifteen cycles of sulfur sorption and regeneration should have occurred during the 100 hour period during which the regenerator was in reducing conditions.

The average feed rate of makeup sorbent was equivalent to a Ca/S ratio of 0.55; however, the makeup rate was varied over the Ca/S range of 0-1.3 in order to maintain constant bed levels. It should be emphasized that this variation in makeup sorbent rate was very small compared to the rate at which recirculated sorbent entered the combustor, which was equivalent to a Ca/S ratio of about 17. There was no apparent effect of changing the makeup rate on the emissions of SO_2 from the combustor.

SO_2 levels measured in the off-gas from the regenerator were nearly steady throughout the run and averaged 0.53 mole percent (dry basis). This is very close to the value predicted by a sulfur mass balance based on the feed rate and sulfur content of the coal entering the combustor. Higher SO_2 levels would probably have resulted had more coal of a higher sulfur content been used. The calculated equilibrium level for conditions prevailing in the regenerator was 2.9 mole percent.

Samples of bed were taken from the combustor and regenerator after the run. The combustor bed contained 35.8 percent CaO, 18.5 percent $CaCO_3$, and 45.7 percent $CaSO_4$ (molar). The regenerator bed was 80.5 percent CaO, 2.3 percent $CaCO_3$, and 17.3 percent $CaSO_4$. Because air was blown through the hot regenerator bed during shutdown, the composition of these solids may have changed. Any CaS, if present, would have been converted to $CaSO_4$, and possibly CaO.

A sulfur mass balance for the 125 hour run was calculated as part of the data-workup. Table 2 gives the results. Recovery of sulfur was 103.5 percent. The sulfur balance is very sensitive to the sulfur content of the coal, which would have needed to be only 0.07 percent higher to obtain a sulfur recovery of exactly 100 percent.

CONCLUSIONS

The batch runs made in the regenerator were extremely useful in establishing procedures and providing operating experience necessary to improve the performance and reliability of equipment. Batches of sulfated limestone and dolomite were regenerated under well controlled conditions to produce a gaseous effluent containing up to 3.7 mole percent SO_2 (dry basis). Calcium sulfate was converted almost completely to calcium oxide with negligible formation of sulfide, using the technique of adjacent reducing and oxidizing zones. Agglomeration of the bed could be avoided by carefully controlling temperature so that no portion of the fluidized bed was hotter than about 1120°C (2050°F).

Continuous operation of the miniplant provides a realistic way of measuring the potential benefits of a regenerative system, compared to a once-through system. An important question is what reduction in sorbent requirements can be realized in a regenerative system. Since only one continuous run has been made thus far, and at non-optimum conditions, this question cannot be answered precisely. However, the average SO_2 emissions from the combustor were 310 ppm (0.63 lbs $SO_2/10^6$ BTU) at an average Ca/S ratio of 0.55. Figure 5, which gives SO_2 emissions for once-through operation of the combustor at both calcining and carbonating conditions as a function of Ca/S feed ratio, shows that <u>at least four times</u> this makeup rate would have been required in a once-through system, had the combustor been under calcining conditions. Actually, carbonating conditions prevailed in the combustor and Figure 5 shows that the observed emission level would not have been reached in a once-through system at any Ca/S makeup rate.

Additional combined combustor-regenerator runs are planned to study the effects on performance of the important variables, including sorbent recirculation rate, regenerator temperature, and type of sorbent. We hope to determine which combustor and regenerator operating conditions provide, simultaneously, low emissions of SO_x and other pollutants from the combustor, high concentrations of SO_2 in the off-gas from the regenerator, low makeup rates of fresh sorbent, and moderate recirculation rates between combustor and regenerator.

Problems that remain to be solved include recovering sulfur economically from the fairly low concentration of SO_2 in the regenerator off-gas and using coal (or low quality liquids) as fuel for the regenerator. Although there are a number of processes which can reduce SO_2 to sulfur, these processes operate at atmospheric pressure and require higher levels of SO_2 in the feed gas than are present in the regenerator. Work to develop a process to recover sulfur from the regenerator off-gas at 700-1000 kPa pressure would be worthwhile.

ACKNOWLEDGEMENT

This work was supported by the U.S. Environmental Protection Agency under contract 68-02-1312.

REFERENCES

Curran, G. P., Fink, C. E., and Gorin, E. (1967) Fuel Gasification, Adv. in Chem. Series 6, 141-165.

Ruth, L. A., (1975) Proc. Fourth Intl. Conf. on Fluidized Bed Combustion, 425-38, McLean, Virginia.

Ruth, L. A., (1977) Fluidized Bed Combustion Technology Exchange Workshop, Reston, Virginia, 12-15 April.

Table 1. Operating Conditions During Demonstration Run

	Combustor		Regenerator
Pressure, kPa	760		770
Bed Temperature, Average, °C	900		1010
Bed Height, Expanded, Avg., m	3.4		2.3
Superficial Gas Velocity, m/sec	1.5		0.61
Solids Recirculation Rate, kg/hr		45	
Residence Time of Solids, Avg., hr.	5		1-1/2
Makeup Acceptor Addition Rate,			
Equiv. Ca/S, Average		0.55	
Range		0-1.3	
Combustor Coal Feed Rate, kg/hr	79		
Coal Type	Champion (Pittsburgh Seam), 2.0% S		
Stone Type	Grove Limestone, BCR No. 1359		

Table 2. Sulfur Balance: Combustion-Regeneration Demonstration Run

		kg	% of Sulfur Entering
Sulfur Entering System			
Coal		194.8	100
	Total	194.8	100
Sulfur Leaving System			
Regenerator off gas		91.8	47.1
Combustor flue gas		39.5	20.3
Combustor bed reject		17.5	9.0
Combustor overhead solids (flyash)		22.6	11.6
Regenerator overhead solids		2.0	1.0
	Total	173.4	89.0
Sulfur Accumulated (Δ Inventory)			
Regenerator bed		3.1	1.6
Combustor bed		25.1	12.9
	Total	28.2	14.5

$$\% \text{ S Recovery} = \frac{\text{Sulfur Leaving + Accumulated}}{\text{Sulfur Entering}} \times 100\% = \frac{173.4 + 28.2}{194.8} \times 100\% = 103.5\%$$

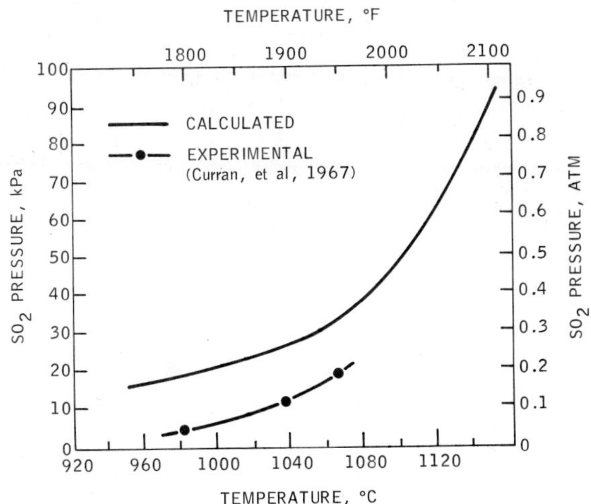

Figure 1. Equilibrium SO_2 Pressure Vs. Temperature

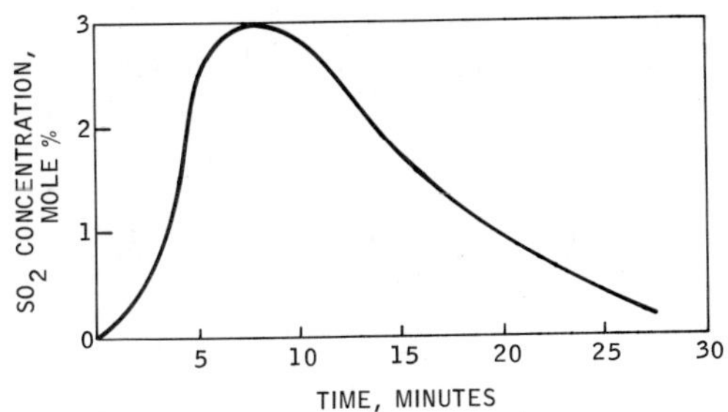

Figure 2. Typical SO_2 Curve for Batch Regeneration Runs

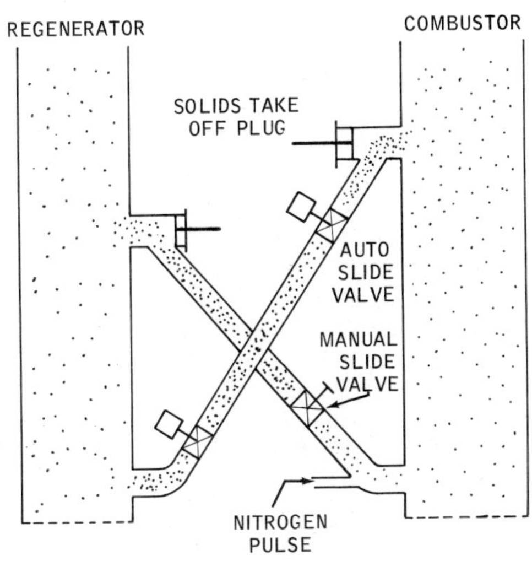

Figure 3. Miniplant Solids Transfer System

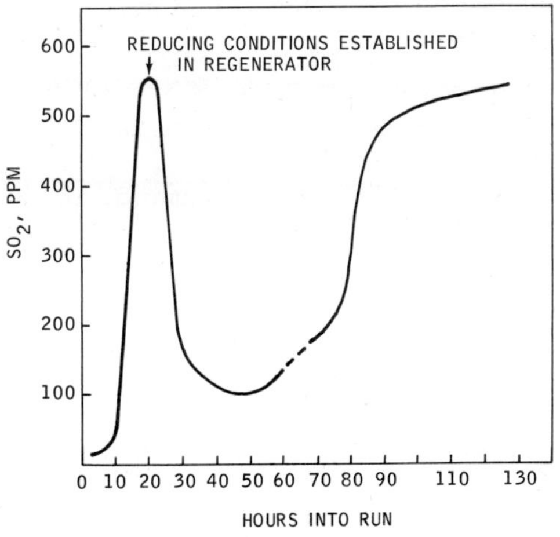

Figure 4. Combustor SO_2 Emissions for Operation of Combustor with Regenerator

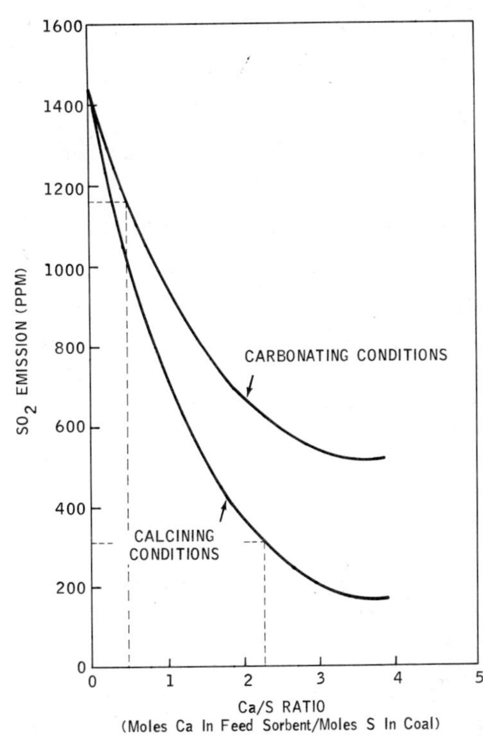

Figure 5. SO_2 Emissions Vs. Ca/S Ratio for Once-Through Operation of Combustor Using Grove No. 1359 Limestone and Champion Coal (2% S)

NITRIC OXIDE REDUCTION IN AN EXPERIMENTAL FLUIDIZED-BED COAL COMBUSTOR

By T. Furusawa, T. Honda, J. Takano & D. Kunii

Department of Chemical Engineering
Unversity of Tokyo
Bunkyo-ku, Tokyo
Japan 113

Synopsis

Continuous combustions of coal and char particles were carried out in an experimental fluidized-bed combustor. The significant reductions of "NO" and ammonia emissions were observed in the combustion of char under substoichiometric condition. The mechanism of the radical reduction of "NO" emission was discussed. The results obtained by the preliminary experiment of staged combustion indicated that the staged combustion of carbonized char is an effective combustion modification for the abatement of NO emission.

NOTATION

E : Emission index of "NO" = "NO" produced per unit mass of consumed coal or char, dry basis [moles/gram]

Z : Injection height of secondary air from the bed surface [cm]

λ : Stoichiometric ratio based on consumption
$$= \frac{\text{amont of air supplied}}{\text{Stoichiometric air for consumed fuel}}$$

λ_1: Stoichiometric ratio based on consumption for primary stage

INTRODUCTION

The fluidized-bed combustion technique is being developed as an alternative to the conventional combustion systems of coal, and is found to offer the potential of low sulfur dioxide emission even when using high sulfur coal (Hammons et al, 1972; Jonke et al, 1972). This technique can be considered to be an efficient combustion modification which can lower the combustion temperature. However, the major source of "NO" formed during the combustion of coal was determined to be the nitrogenous content of coal, and the reduction of "NO" emission from fluidized-bed combustor has proved difficult task (Pereira et al, 1974).

The purpose of the work reported in this paper is part of a larger program utilizing a small laboratory scale fluidized-bed combustor to examine the possibilities of controlling NOx emission from the combustion of carbonaceous materials through combustion modification. The specific goal of the present investigation is to propose the condition of the reduction of the "NO" emission and to explain its mechanism.

EXPERIMENTS

Fig. 1 is a schematic diagram of the experimental fluidized-bed combustion system. The combustor is a 5 cm diam. stainless steel vessel 58 cm long. The lower part of the vessel (15 cm) was packed with refractory materials and used for preheating this part. The fluidizing air or simulated air which consists of oxygen and argon enters the combustor through a multiorifice plate distributor into the bed of microspherical particles whose chemical and physical properties are given in Table 1.

The quality of fluidization is strongly influenced by the type of gas distributor used. Therefore, the multiorifice plate was designed so that a sufficient pressure drop to achieve homogeneous fluidization can be obtained. The pressure drop chosen in the present experimental condition ranges from 40 to 60 cm in water. In a series of experiments carried out to investigate the influence of air staging, primary stage of the bed is maintained at substoichiometric conditions, while the balance of air is introduced through the nozzles into the freeboard.

The static bed height was specified to be 10 cm. The fluidized bed combustor was externally heated by the use of electric furnace. The temperature of the bed was controlled by the conventional PID electronic controller.

Feeding of the employed carbonaceous materials was done by means of a solid feeder which had been developed in our laboratory. Fig.2 represents a schematic diagram of the solid feeder. The solid feeder consists of a moving bed hopper and small fluidized-bed. The desired rate could be obtained by adjusting the gas flow rate through both hopper and fluidized-bed and the intensity of the vibrator, which was stabilized by the regulated power supplier. Thus the continuous feed of the small flow rate of solids (such as 0.2 g/min) could be realized. The solids were sent into the fluidized-bed combustor at a point 3 cm above the distributor.

Ash was removed by elutriation and the elutriated solids were removed from the off-gas by a small cyclone separator. In a certain series of experiments, collected solids were used for chemical analysis to obtain the combustion efficiency.

Upstream from cyclone separator, (5 cm below the top cover of the combustor) the off-gas is continuously diverted to a gas-analysis system. Approximately one hundredth of the off-gas is supplied through a glass wool trap to a continuous chemiluminescent NOx analyzer. Gas chromatography provides intermittent analyses for H_2, N_2, CO, CO_2, O_2, CH_4 and C_2H_4.

Known gas mixtures were used to calibrate the gas chromatograph. Kitagawa ammonia low range detector tubes were used to analyze ammonia.

The experimental conditions employed were shown in Table 1. The carbonaceous materials employed for the present series of experiments are shown in Table 2.

RELATIVE CONTRIBUTION OF THERMAL AND FUEL NITRIC OXIDE TO THE TOTAL EMISSIONS

The relative contribution of thermal and fuel "NO" to the total "NO" emission was investigated in the initial series of the experiment by substituting an inert component, argon, for the nitrogen of the fluidizing air. This provides a mean of separating the fuel "NO" from the total "NO" emission. Fig. 3 indicates the results of the experiment with char. As the data indicate, only approximately 10% of the total "NO" emission was reduced when replacing the combustion air by a mixture of argon and oxygen. The same conclusion can be applied to the case of coal combustion.

Whether the observed value of the thermal "NO" was reasonable, was investigaed. Detailed discussions concerning the formation of thermal "NO" in fluidized-bed combustors will be given elsewhere (Furusawa et al, 1977).

REDUCTION OF "NO" EMISSION UNDER SUBSTOICHIOMETRIC COMBUSTION

When carbonaceous materials were burnt in excess air, the concentration of "NO" in the combustion products is known to be dependent not only upon the combustion temperature but also upon the stoichiometric ratio.

At a given stoichiometric ratio, the lower level of "NO" emission was attained by decreasing the combustion temperature. However, the temperature range of practical importance is determined by both the combustion efficiency and the efficiency of sulfur retention. Further reduction of "NO" might be possible only by reducing the stoichiometric ratio. This approach has not always been successful in the case of coal or gas combustion. However, the present investigation indicated that the reduction of "NO" emission with the decreased stoichiometric ratio was more pronounced in the case of char combusion. The results are shown in Fig. 4. This result is contrasted to the results obtained previously in the coal combustion.

MECHANISM OF "NO" REDUCTION IN SUBSTOICHIOMETRIC FLUIDIZED-BED COMBUSTION

The possible reaction which suppresses the "NO" emission might be considered as the reduction of "NO" by carbonaceous materials. Pereird and Beér (1975) also reported that the significant reduction of "NO" emission from fluidized-bed combustion of coal could be attributed to the reaction between char particles or volatiles and "NO". In order to estimate the effects of this reaction, the steady state concentration of carbon particles during the present experiment was measured. After terminating the feed of fuel solids, the amount of produced carbon dioxide and monoxide originating from the remaining carbon particles was measured by means of gas bag method. The obtained results are shown in Fig. 5. The steady state carbon concentration within the bed was found to be considerably small. If "NO" is reduced by the reaction with char, the reaction rate should be extremely high.

To verify the above explanation, a certain amount of char particles was injected into the fluidized-bed of inert particles fluidized by a gas mixture containing 860 ppm "NO" in nitrogen. The exit concentration was followed continuously by the NOx analyzer described previously. The concentration of "NO" was extremely low, therefore, the holdup of the carbon particles could be assumed constant during the experiments. The results obtained are shown in Fig. 6. It was found that char was very effective in reducing "NO" emission. Even when the bed is operated under a deficiency of oxygen, the local stoichiometric ratio in the vicinity of the coal feed point is much larger than unity, therefore, local "NO" emission is comparable to that of $\lambda>1$. The "NO" produced in the lower part of the bed can be supposed to be reduced to zero by chemical destruction higher in the bed.

The other feature of the data shown in Fig. 5 is that the steady state concentration of char within the bed at given feed rates of air and solids was lower in the case of coal combustion. This appears to correspond to the considerable "NO" emission observed in coal combustion at $\lambda<1$. This is because the reduction of "NO" by contacting carbonaceous solids within the bed is enhanced in the case of char combustion.

The lower steady state concentration of char obtained in the combustion of coal may be explained as follows. The combustion of coal can be considered to consist of two stages, volatilization and the following combustion of produced char and volatiles. The combustion of gas phase volatiles, especially gas of small molecules, is retarded within the bed. The apparent stoichiometric ratio of air used for the combustion of char produced from coal is much larger. Therefore, the concentration of solids within the bed is lower in the case of combustion of coal.

EMISSION OF NITROGENOUS COMPOUNDS WITH A DEFICIENCY OF AIR

In the previous section, it was shown that the "NO" emission was significantly reduced when the char was burnt under a deficiency of air. In the staged combustion, the balance of the combustion air is injected into the freeboard. Therefore, nitrogen oxides will be again formed from nitrogenous compounds other than "NO" which may be formed in the reducing atomosphere of the bed. The typical nitrogenous compounds may be considered as ammonia, $(CN)_2$ and HCN. Therefore, the concentration of ammonia was followed intermittently by the detector tube method. The results are shown in Fig.7. The temperature rating of the detector, was 40°C. Because of the relatively high dew point of flue gas, the ammonia emission from coal combustion appeared to be underestimated. The measurement of ammonia emission from char was much alleviated because the dew point of flue gas is extremely low.

The reduction of ammonia emission from both char and coal appeared to be achieved by raising the combustion temperature. The significant reduction of ammonia emission observed in the combustion of char may be related to lower hydrogen contents of char. The hydrogen contents of char can be reduced by raising the pyrolysis temperature of coal.

PRELIMINARY INVESTIGATION OF EFFECTS OF STAGED AIR FIRING ON "NO" EMISSION

Previous results indicated that the "NO" destruction was promoted in the absence of oxygen, and the emission of ammonia was significantly reduced in the case of substoichiometric combustion of char at relatively higher temperature. These findings suggest that the staged combustion of char is considered as an effective combustion modification for the reduction of "NO" emission. A series of experiments were carried out to check the possibilities of "NO" reduction in staged air firing.

In this operation, the combustion of air was separated into a primary air stream which was supplied for fluidization through the distributor, and the second air stream, which was injected above the bed surface into the freeboard. All carbonaceous materials were injected into the primary stage at the point 3 cm above the distributor. The bed could be operated under a deficiency of air to promote "NO" destruction by carbonaceous materials, while the overall stoichiometric ratio was maintained at excess air condition.

The secondary air was supplied by a 6 mm stainless steel tube and injected into radial direction through 12 holes (2 mm) drilled on the tip of the tube.

Before starting the staged air firing, only the primary air was supplied and the steady state concentration of "NO" at substoichiometric condition and the primary combustion products were measured. Then the secondary air was injected at several levels above the bed surface and the concentration of "NO" and the secondary combustion products were measured.

The data obtained in the combustion of char II and coal were shown in Table 3.

The extent of "NO" reduction shown in equation:

$$X = 1 - \frac{E\,(\text{"NO" emission index of staged operation})}{E_0\,(\text{"NO" emission index of conventional operation})}$$

E_0 employed for the above calculation is obtained by experiment.

Experimental results demonstrated that approximately more than 90% reduction of "NO" emission was achieved in staged combustion of char. Other data not listed in the table also showed that more than 85% reduction was attained in the case of char combustion at 850°C. However, the maximum extent of "NO" reduction attained in the staged combustion of coal was only 50%. The main difference concerning"NO" emission between combustion of coal and that of char may be explained by the difference of emission level of nitrogenous compounds from the primary stage. As shown in Fig. 7, the emission of ammonia was not detected in the flue gas from the primary stage of combustion of char at 1000°C. Therefore the considerable low "NO" emission observed was supposed to be formed by the combustion of carbonaceous fines in the freeboard. The formation of the thermal "NO" in the freeboard might be considered insignificant because the temperature rise caused by the combustion of gas products from the primary stage was found to be less than 50°C. The formation of fuel "NO" through the combustion of char within the bed by the oxygen introduced by the emulsion downflow was also negligible in the present operation condition.

The above preliminary experiments suggest that the staged combustion of char is an effective combustion modification for the abatement of the "NO" emission from the fluidized combustion of coal.

COMPARISON WITH THE CONVENTIONAL FLUIDIZED-BED COMBUSTION

The temperature chosen for this study was higher than the normal temperature of the conventional fluidized-bed combustion from the point of view of SOX emission control. However, the present research aimed at a exploratory investigation to develop a new modification of fluidized-bed combustion for the future rather than an analysis of the existing systems. The approach may be supported by the following facts that the present sulfur retention system by the use of dolomite appears to be facing with problems of industrial waste in the future. Therefore, the development of the new process which combines the coal carbonization and the combustion of produced char is supposed to solve the difficult task of the complete abatement of NOX and sulfur dioxide emissions without being involved in the problem of indestrial wastes since the significant fraction of organic sulfur and pyrite sulfur can be eliminated by the carbonization. Furthermore, the sulfur compound remaining in coal char was known to be stable calcium sulfate.

CONCLUDING REMARKS

Continuous combustions of coal and char particles were carried ont in an experimental fluidized-bed of inert silica-alumina particles at temperature ranging from 700 to 1000°C. The major source of "NO" formed during fluidized-bed combustion was found to be nitrogenous compounds of coal and char. The strong dependence of "NO" emission on the stoichiometric ratio was observed in the case of combustion of char with a deficiency of air. The reduction of ammonia emission observed in combustion of char suggested that "NO" emission might be significantly reduced by employing the two stage combustion of char. The mechanism of "NO" reduction was discussed.

Priliminary experiments concerning the effects of air staging demonstrated that the staged combustion of char was an effective combustion modification for abatement of "NO" emission.

ACKNOWLEDGEMENT

The authors wish to express their thanks to Science Rrsearch Grant of Ministry of Education, Japan (Grant No. 265281), and the Kawakami Foundation for their financial supports. We wish to acknowledge also Dr. Sadakata (Univ. of Tokyo) for his helpful discussions, which were of enormous benefit in conducting this research.

REFERENCES

Furusawa, T., Honda, T. & Kunii, D. (1977) "NOx produced in Fluidized-bed Combustion of Carbonaceous Materials," Proceedings of 2nd. Pacific chemical Engineering conference, Denver, Colorado, 1236.

Gibbs, B.M. et al (1976) "The Influence of Air Staging of the No Emission from a Fluidized-bed Coal Combustors," Paper Presented to the 16th. Symposium (International) on Combustion.

Hammons, G.A. and Skopp, A. (1972) "A Regenerative Limestone Process for Fluidized-bed Coal Combustion and Desulfurization," Chem. Eng. Prog. Symp. Series 68, No. 126, 252.

Jonke, A.A. et al (1972) "Pollution Control Capabilities of Fluidized- Bed Combustion," Chem. Eng. Prog. Symp. Series 68 No. 126, 241.

Pereira, F.J. and Beér, J. (1974) "NOx Emission from Fluidized-bed Combustors," 15th International Symp. on Combustion, Tokyo 1149.

Table 1. Scope of experiment

Inert particles:
Microspherical particles
S_iO_2: 8.93%·Al_2O_3: 90.61%·Fe_2O_3: 0.46%

Surface mean particle diameter: 580 microns

Bulk density: 0.57 g/cm^3 ;Total amount: 113g

Temperature of fluidized-bed: 700°C - 1000°C

Static height of the bed: 10 cm

Diameter of coal and char particles: 500-710μ

Flow rate of fluidizing air and simulated air: 4.2 ~ 8.1 l/min.

Feed rate of fuel particles: 0.5 ~ 1.7 g/min.

Table 2. Proximate and ultimate analysis of coal and char particles

	Proximate analysis (wt%)				Ultimate analysis (dry%)					
	Volatile matter	fixed carbon	ash	moisture	C	H	N	S	O	ash
Char II	2.74	66.00	24.47	6.69	71.66	1.03	0.61	0.01	0.34	26.35
Coal	43.3	39.10	12.7	4.9	66.9	5.4	1.4	0.1	13.2	13.0

Coal : Taiheiyo coal (Hokkaido, Japan), char II : produced from Taiheiyo coal, pyrolysis temperature 800°C.

Table 3. Effect of air staging on nitric oxide emission

Z	λ_1	λ	NO (ppm)	E x 10^4 (mol/g)	Extent of "NO" reduction	COMBUSTION PRODUCTS CO_2 %	CO %	H_2 %	O_2 %	CH_4 %	C_2H_4 %
				Char II		1000°C					
-	0.71	-	0	0.0	-	10.23	18.05	1.19	0.	0.	
1	0.71	1.12	63	0.195	0.878	18.03	0.03	tr.	2.00	0.	
10	0.71	1.20	31	0.103	0.936	16.85	0.01	tr.	3.41	0.	
-	0.80	-	0.5	0.001	-	14.02	11.10	0.69	0.	0.	
5	0.80	0.96	16	0.042	0.974	18.91	2.10	0.08	0.01	0.	
5	0.80	1.05	27	0.078	0.951	19.02	0.11	tr.	0.96	0.	
20	0.80	1.12	17	0.053	0.967	17.67	0.27	tr.	2.48	0.	
				Coal		1000°C					
-	0.83	-	20	0.049	-	13.00	7.96	2.39	0.0	0.47	0.13
5	0.83	1.22	245	0.880	0.267	14.76	0.	tr.	3.46	0.	0.
-	0.63	-	7	0.013	-	12.77	14.22	6.69	0.0	1.15	0.26
10	0.63	1.13	200	0.665	0.335	15.91	0.	tr.	3.88	0.	0.

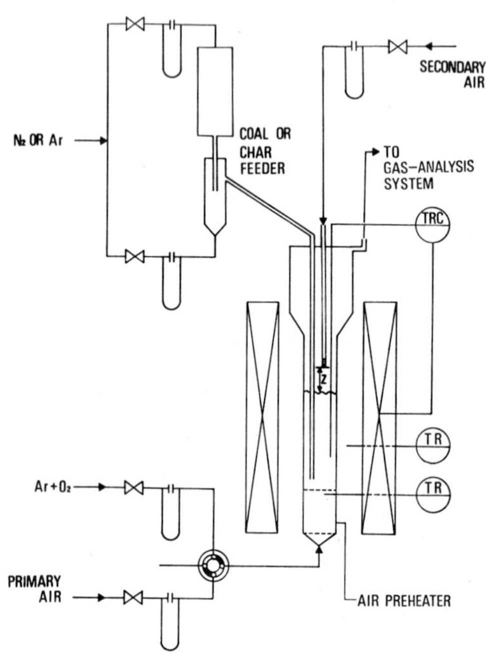

Fig. 1. Experimental fluidized-bed combustor system

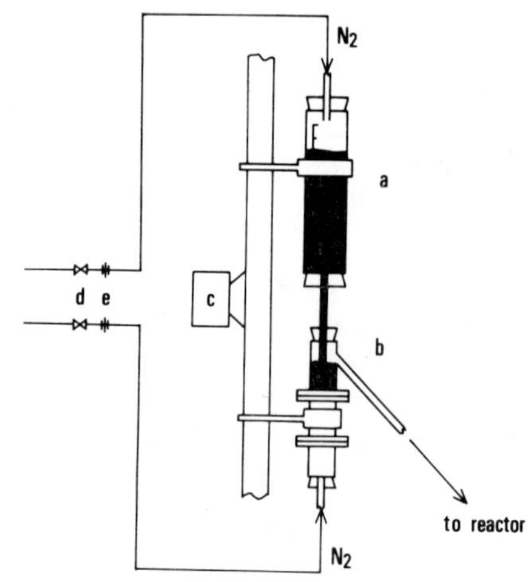

Fig. 2. Continuous coal or char feeder
a: hopper b: fluidized-bed c: vibrator
d: mass flow valve e: needle valve

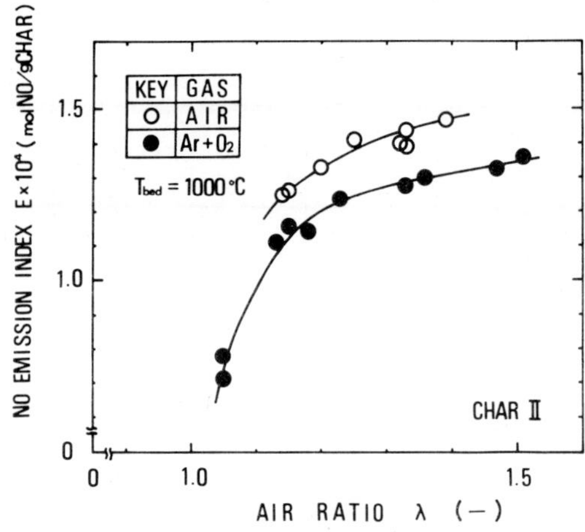

Fig. 3. Relative contributions of thermal and fuel "NO" emissions from char combustion

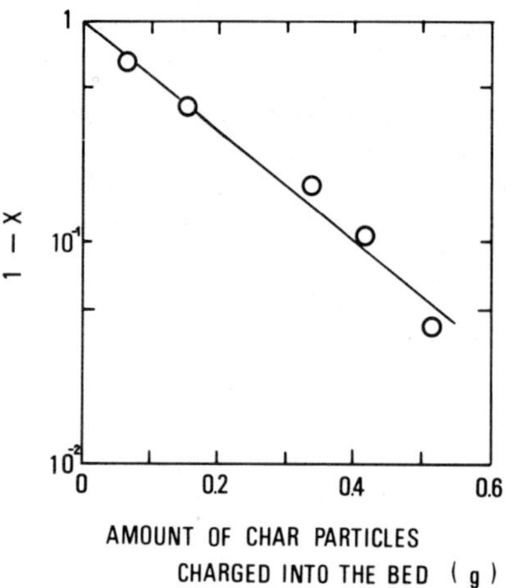

Fig. 6. Extent of "NO" reduction in diluted fluidized-bed of char; Superficial velocity of gas 19.9 cm/sec. Temperature: 1000°C

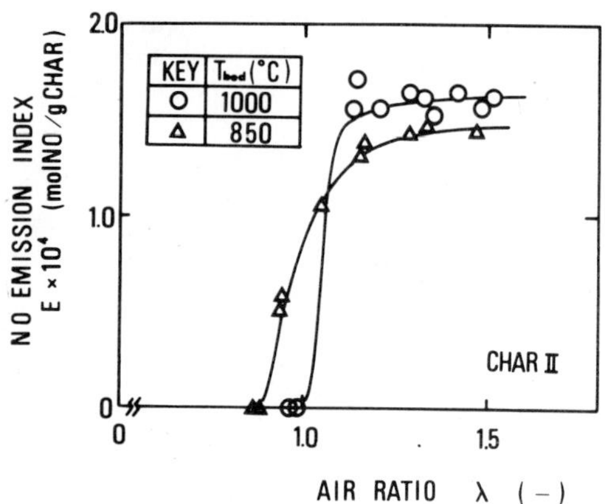

Fig. 4. Dependence of "NO" emission level on the stoichiometric ratio for char combustion

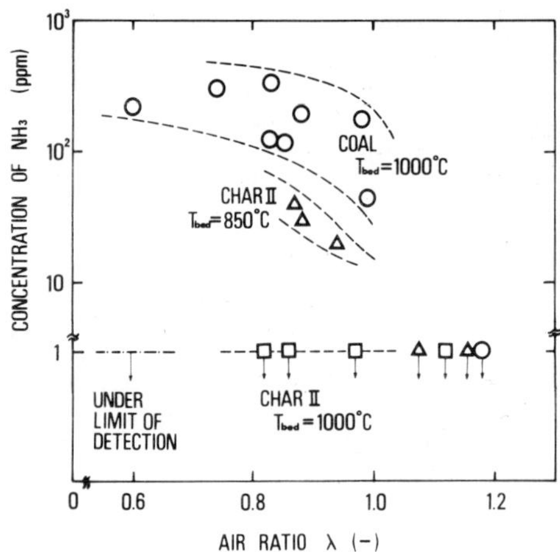

Fig. 7. Ammonia emissions from combustion of char and coal

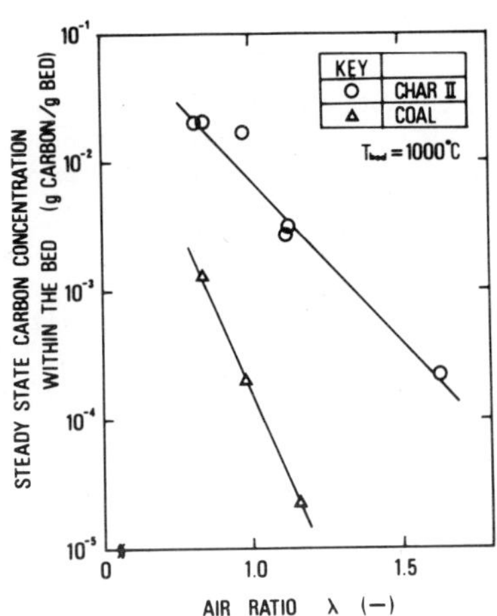

Fig. 5. Steady state char concentration within the fluidized-bed

FLUIDIZED BED HEAT TRANSPORT BETWEEN PARALLEL, HORIZONTAL TUBE-BUNDLES

By R. A. Newby and D. L. Keairns

Westinghouse R&D Center

Beulah Road

Pittsburgh, Pennsylvania 15235

SYNOPSIS

A fluidized bed heat transfer facility is described and results related to advanced fluidized bed heat exchanger concepts are discussed. Bed-to-tube heat transfer rates within compact tube-bundles and heat transport rates by bed mixing between parallel, horizontal tube-bundles are considered based on preliminary test results. The compact tube-bundles improve bubble distribution in the air-fluidized bed of 80 μm magnesia powder at velocities up to 30 cm/sec. Tube-to-bed heat transfer coefficients for centrally located tubes are generally increased by the presence of compact tube-bundles. A coefficient of bed transport is estimated for the tube-bundle configurations.

NOTATION

A	cross-sectional area of vessel
C_p	specific heat of bed material
d	outer diameter of tube
h	bed-to-tube surface heat transfer coefficient
h_o	bed-to-tube surface heat transfer coefficient with no internals in the bed
k	thermal conductivity of air
m	mass rate of transport of bed material
Q	rate of heat transport between the tube bundles by bed mixing
S_h	horizontal tube pitch
S_v	vertical tube pitch
U	superficial fluidization velocity
U_{mf}	minimum fluidization velocity based on the bed pressure drop equal to the bed weight per unit area
U_{me}	minimum expansion velocity
U_t	total fluidization velocity
ΔP	bed pressure drop
ΔP_t	theoretical bed pressure drop, bed weight per unit area
$\Delta P/\Delta P_t$ (max)	maximum value of the pressure drop ratio
ΔT_B	temperature difference (mean) between the two tube-bundle regions
α	bed transport coefficient, defined in Equation 2
ρ	static bed bulk density

INTRODUCTION

Westinghouse is evaluating advanced applications of fluidized bed heat transfer for electric-power generation. One concept being assessed utilizes a fluidized bed as an inert heat transfer media to transport heat by bed mixing between parallel, horizontal tube-bundles. The Liquid Metal Fast Breeder Reactor (LMFBR) steam generation system represents one application of this basic concept, where the fluidized bed heat exchanger results in improved economics and safety compared to conventional steam generators (Keairns, et al., 1975; Keairns, et al., 1976). A conceptual design of an LMFBR fluidized bed steam generator is shown in Figure 1 (for magnesia or nickel powders fluidized by helium).

Engineering studies of this concept have identified the critical design and operating parameters in terms of economic sensitivity, manufacturing feasibility and safety. In general, system optimization requires that the flow rate of fluidizing gas should be limited ($u < 30$ cm/sec) and the tube-bundles should be compact ($1.5 < S_h/d \lesssim 6$), while the rates of tube-to-bed heat transfer are maximized ($h \gtrsim 600$ J/sec $m^2 {}^\circ C$) and the rate of heat transport between the tube-bundles by bed mixing is maintained sufficiently high that vertical bed temperature gradients are small (the steam superheat temperature must be achieved and uniform and stable steam generation must be realized).

The existing state-of-knowledge of fluidized bed heat transfer and mixing within compact tube-bundles is inadequate to evaluate the concept without performing both experimental simulation and scale-up modeling. While several programs described in the literature have addressed heat transfer within tube-bundles in fluidized bed combustion systems, the operating and design conditions are not applicable to the present concept - for example, large particle diameters (> 1000 μm), vs. fine particles (< 100 μm), high velocities (> 100 cm/sec) vs. low velocities (< 30 cm/sec), large tube pitches vs. compact tube spacing, single tube-bundle vs. multiple tube-bundles. (Fitzgerald, et al., 1976; Canada & Staub, 1977; Hodges, et al., 1976; Horio & Wen, 1976; Bazan, 1975; Howe & Aulisio, 1976; McLaren & Williams, 1969; Wright, et al., 1970). Other general studies of heat transfer within horizontal tubes bundles do not extend to the proposed conditions of this concept, nor do they address heat transport between tube-bundles (Gelperin, et al., 1969; Gelperin & Einstein, 1971; Baskakov & Pakhaluev, 1973; Lese & Kermode, 1972; Bartel, 1971).

A fluidized bed heat transfer facility has been constructed to develop the information required to evaluate the concept and to provide concept simulation. Initial runs have been completed with two tube-bundle configurations, using reference design tube diameters and bed material. The runs were carried out at low temperatures (< 200°C) and with air fluidization. The initial runs are discussed and preliminary conclusions are presented.

HEAT TRANSFER FACILITY

The heat transfer facility, shown in Figure 2, consists of a rectangular vessel, 37 cm in width, 15 cm in length in the direction of the tube axes, and of about 2.5 m in total height. The maximum operable bed depth is about 1.4 m (static). The vessel is made up of interchangeable tube-bundle sections and spacers, a gas inlet plenum and a bed freeboard region. The vessel is constructed of 304 SS and designed to operate at atmospheric pressure and temperatures up to 600°C in air, helium or other noncorrosive gases. Electric power controls (up to 150 kw), cooling water controls, fluidizing gas controls and a data acquisition system to record bed pressure profiles, bed temperature profiles, tube surface temperatures and fluidizing gas and water inlet and outlet temperatures are also shown in Figure 2.

The vessel was selected to be of a size sufficient to yield results valid for simplified scale-up to commercial dimensions. Based on previous cold-modeling of fluidized beds with immersed compact tube-bundles it was concluded that the control of bubble size and distribution promoted by the tube-bundles would permit simulation of commercial fluidized beds in a vessel of relatively modest size.

Electrically heated tubes were selected to represent heat sources (sodium tubes in the LMFBR steam generator) and water-cooled tubes were selected to represent heat absorbers (steam tubes in the LMFBR steam generator). A comparison between the heat transfer facility characteristics and a commercial LMFBR steam generator design is shown in Table 1. Emphasis was placed on the capability of simulating the distributor jet inlet velocity and the specific power input to the bed.

FLUIDIZED BED DYNAMICS

A series of cold-tests were performed with 80 μm diameter (> 8 μm, < 322 μm) magnesia fluidized by air to develop an understanding of the influence of compact tube-bundles upon bed dynamics. Fluidization curves, bed pressure drop and bed expansion, were generated in the bed with no internals and compared with results for four tube-bundle configurations. The water tube-bundle was placed above the heater tube-bundle in all cases with a 30 cm separation between the bundles. Motion pictures of the bed surface and bed elutriation rates were compared. Bed depths of 1.3 m (static) were used at velocities up to 30 cm/sec. Heater tubes were 3.18 cm in diameter, water tubes were 1.59 cm in diameter.

The general nature of the fluidization curves is illustrated in Figure 3, defining several fluidization characteristics. The hysteresis behavior of the curves require a set of characteristics for both descending and ascending velocities. Table 2 lists values for the fluidization characteristics for each of the configurations. The hysteresis effect is magnified and the minimum fluidization velocity increases as the tube-bundles become more compact. The maximum pressure drop ratio becomes very large, showing great resistance to initial bed expansion, for the cases where $S_h/d=2$, but is unity for the other cases. The fully fluidized value of the pressure drop ratio, $\Delta P/\Delta P_t$, was very close to unity for all the configurations.

Bed surface motion pictures showed a dramatic influence of tube-bundles upon bubble size and bubble distribution. Bed slugging which was evident in the bed with no internals (bubbles up to 0.3 m diameter) was eliminated when bundles having $S_h/d=4$ and 2 were in the bed and small bubbles (6 to 10 cm diameter) were uniformly distributed across the surface. A reduction in the amplitude of bed pressure fluctuations was also noted when tube-bundles were present in the bed. Bed elutriation rates with a 1 m freeboard height were about 50% less when tube-bundles ($S_h/d=2$), were present in the bed and the static bed depth was about 10 cm above the top row of tubes. The elutriation rate was reduced by an order-of-magnitude when the expanded bed depth was even with the top row of tubes.

The information gathered in the cold tests are of interest to practical heat exchanger design and the selection of operating conditions. The observations of bubble distribution at the bed surface adds evidence that the facility will generate data applicable to large-scale operation.

TUBE-TO-BED HEAT TRANSFER RATES

Heat transfer coefficients for the 3.18 cm diameter heater tubes and the 1.59 cm water tubes were measured for the air-fluidized bed of 80 μm magnesia powder with no internals present and with two tube-bundle configurations: the water tube-bundle placed above the heater bundle with a 30 cm separation between bundles and tube pitches of S_v=1.5d and S_h=4d (staggered) in one configuration and S_v=1.5d and S_h=2d (inline) in the other. The subject tubes (one of each diameter) were placed at the center of each bundle in both configurations and were placed in the identical positions in the bed without internals. In order to smooth the data, taken over a wide range of bed temperatures, and to make a more direct comparison between the tests, the heat transfer coefficients were corrected to a standard temperature of 21°C by correcting for the estimated effect of the gas thermal conductivity on the heat transfer coefficient. A factor of $(k)^{0.6}$ was applied to account for this influence of temperature, resulting in a 5 to 20% reduction in the measured heat transfer coefficients (Gelperin, 1971).

Table 3 presents the results for the tests. The maximum values of the heat transfer coefficient at the top, side, and bottom tube surface locations and of the average of the local coefficients are tabulated for the bed without internals. The ratio of the corresponding coefficient with the tube-bundle in place to the maximum value without internals (both at the same fluidization velocity) shows the influence of the tube-bundle presence.

The tubes characterized in Table 3 do not necessarily represent the performance of all the tubes. Tubes located on either side of the center show only a small reduction in heat transfer coefficient (on the order of 10%), while the elevation of the tube in the bundle has an even smaller influence.

Gelperin's correlation for heat transfer rates within tube-bundles predicts a reduction in the maximum average heat transfer coefficients of 10% and 16% for the configurations $S_v/d=1.5$ with $S_h/d=4$ and $S_v/d=1.5$ with $S_h/d=2$, respectively, independent of the tube diameters (Gelperin, 1971). Gelperin's correlation is not applicable to the conditions of this investigation, but does appear to correctly predict the behavior of the water tube in the tube-bundle with $S_h/d=4$. Further investigation is required to understand the large increase in the local heat transfer coefficient at the top and side of the 3.18 cm diameter heaters.

HEAT TRANSPORT BY BED MIXING

Bed temperature profiles were measured with the two tube-bundle configurations described in the previous section. Temperature gradients in the heater bundle and water tube-bundle are shown for the configurations in Figure 4 as a function of the fluidization velocity. The heater bundle temperature gradient becomes negligible for velocities greater than 14 cm/sec., while significant gradients remain in the water tube-bundle over the range of velocities tested. The temperature gradients in the more compact configuration ($S_h/d=2$) are greater than those in the less compact configuration at low velocities, but the gradients converge to the same curve for velocities greater than 10 cm/sec. These results were found to be reproducible in duplicate runs.

An approximate expression for the rate of heat transport between the two tube-bundles is

$$Q = mC_p\Delta T_B \qquad \text{.... (1)}$$

The fluidizing air removes only 2 to 10% of the input power, depending upon the fluidization velocity, so Q is nearly equal to the electrical power input (heat losses being negligible at these low bed temperatures).

Bed mixing has frequently been related to a factor α representing the fraction of the bubble volume consisting of the entrained wake region (Kunii & Levenspiel, 1969):

$$m = \alpha A(U-U_{mf})\rho \qquad \text{.... (2)}$$

The bed temperature profile data has been applied to Equations (1) and (2) to estimate values of α for the tube-bundle configurations. The results are shown in Figure 5, based on the U_{mf} values in Table 2 (for descending velocities), a bulk density of 1.67 gm/cm^3 and a specific heat of 1.26 kJ/kg°K. The values of α for the two tube-bundle configurations fall on the same curve and level off to about 0.1 at velocities greater than 10 cm/sec. These values are in the same range, but somewhat lower than estimates of α reported in the literature for beds without internals, 0.6-0.2 (Rowe & Partridge, 1965; Baeyens & Geldart, 1974).

The results indicate little difference between the two tube-bundle configurations in terms of their resistance to the transport of energy by bed mixing. It appears that the mixing is more restricted than in a bed without internals. Higher power inputs and bed operating temperatures are required to simulate the commercial operating conditions.

CONCLUSIONS

The following conclusions of practical design importance have been developed based on the initial data collected over the range of conditions investigated:

1. Improved bubble distribution, and reduced bed slugging and elutriation is characteristic of fluidization with compact tube-bundles. The totally-fluidized bed pressure drop and bed expansion are not significantly influenced by the presence of compact tube-bundles, though the bed start-up characteristics are greatly modified.
2. For the single, central subject-tube locations considered, the compact tube-bundles had a limited effect on the bed-to-tube heat transfer coefficients for the 1.59 cm diameter water tubes, but caused a significant increase in the coefficient for the 3.18 cm diameter heaters (10 to 20%). Further investigation is required to understand the increase in the coefficient on the top and side surfaces of the heaters in the presence of tube-bundles. The overall performance of the tube-bundles must also be determined.
3. The simple model applied provides evidence that the heat transport by bed mixing is nearly identical for the two tube-bundle configurations considered, but is less efficient than in a bed without internals. A value for the transport coefficient, α, of 0.1 is projected for the compact tube-bundles.
4. The heat transfer facility appears to generate data which will provide insight into large-scale performance. Further modeling is required to permit interpretation of the heat transport data and to develop commercial-scale design relationships.

ACKNOWLEDGEMENT

The outstanding performance and ingenuity of Edward W. Hawk, Jr. in the construction and operation of the heat transfer facility is gratefully acknowledged.

REFERENCES

Baeyens, J. & Geldart, D. (1974) Proceedings of the International Symposium on Fluidization and Its Applications, pp. 182-195, Toulouse.

Bartel, W. J., (1971) Ph.D. Thesis, Montana State University.

Baskakov, A.P. & Pakhaluev, V.M. (1973) Int. Chem. Eng. 13 (1), 48.

Bazan, J.A. (1975) Proceedings of the Fourth International Conference on Fluidized-Bed Combustion, pp. 271-277, McLean, Virginia.

Fitzgerald, T., Junge, D.C. & Levenspiel, O. (1976) EPRI Project No. RP 315-1, First Annual Report.

Gelperin, N.I., Ainshtein, V.G. & Korotyanskaya, L.A. (1969) Int. Chem. Eng. 9 (1), 137.

Gelperin, N.I. & Einstein, V.G. (1971) "Fluidization" (J.F. Davidson & D. Harrison Eds.), pp. 471-517, Academic Press, London.

Hodges, J.L., Hoke, R.C. & Bertrand, R. (1976) ASME-AIChE Heat Transfer Conference, St. Louis, MO.

Horio, M. & Wen, C.Y. (1976) 69th Annual AIChE Meeting, Chicago, IL.

Howe, W.C. & Aulisio, C. (1976) 69th Annual AIChE Meeting, Chicago, IL.

Keairns, D.L., Newby, R.A. & Archer, D.H. (1976) "Fluidization Technology, Vol. II" (D.L. Keairns, Ed.), pp. 569-576, Hemisphere Pub. Corp., Washington, D.C.

Keairns, D.L., Newby, R.A., Cooper, M.H., Adkins, C.R. & Bieberback, G. (1975) Record of the Tenth Intersociety Energy Conversion Engineering Conf., pp. 338-341, Newark, Delaware.

Kunii, D. & Levenspiel, O. (1969), "Fluidization Engineering", pp. 150-156, John Wiley & Sons, Inc., New York.

Lese, H.K. & Kermode, R.I. (1972) Canadian J. Chem. Eng. 50, 44.

McLaren, J. & Williams, D.F. (1969) J. Inst. Fuel, 42, 303.

Rowe, P.N. & Partridge, B.A. (1965) Trans. Inst. Chem. Engrs. 43, T157.

Wright, S.J., Hickman, R. & Ketley, H.C. (1970) British Chem. Eng. 15 (12), 1551.

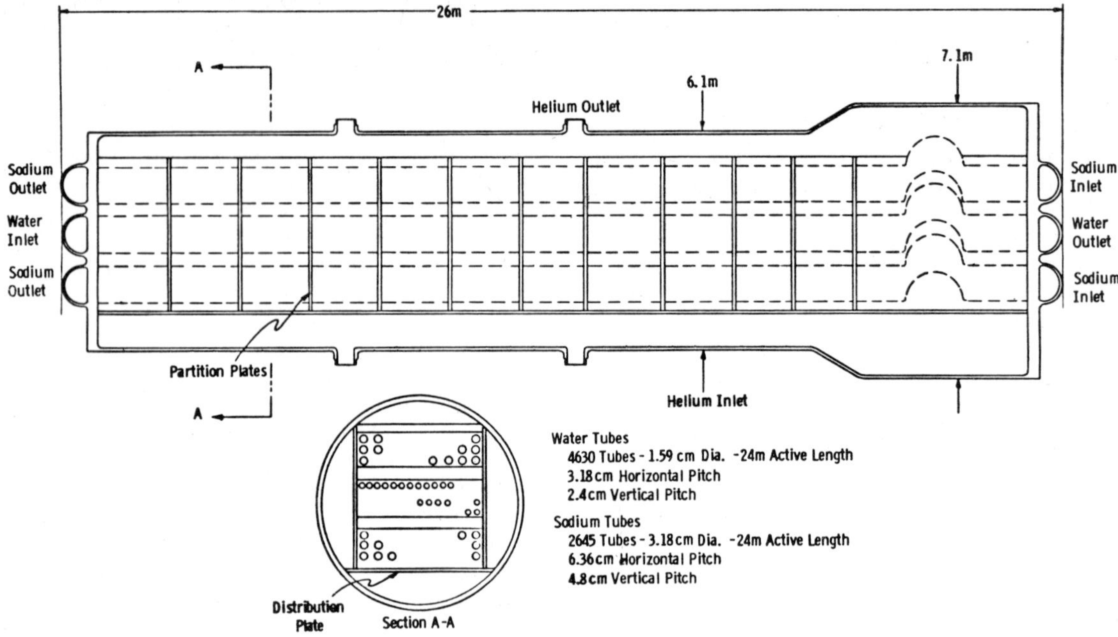

Fig. 1 - Conceptual design of stacked bundles fluidized bed steam generator-evaporator

Fig. 2 - Fluidized Bed Heat Transfer Facility

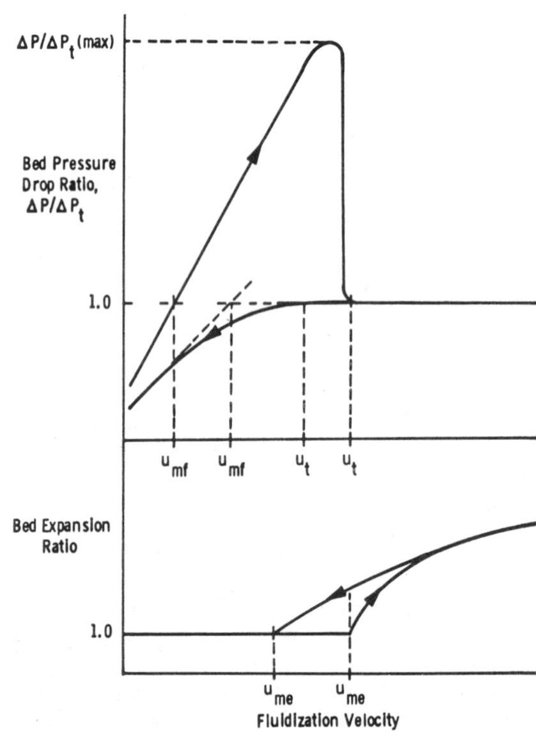

Fig. 3 — Illustration of fluidization characteristics

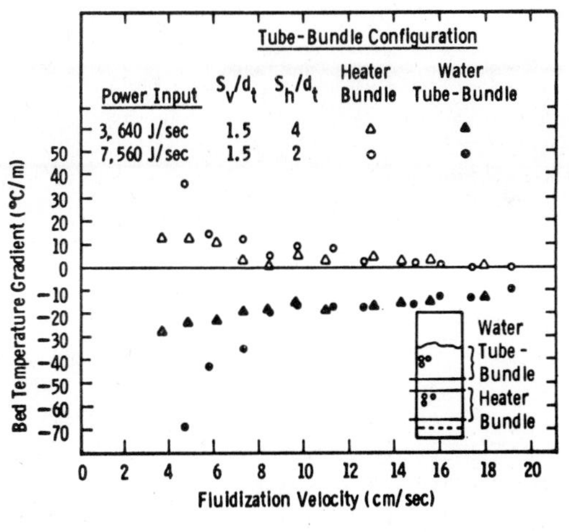

Fig. 4 – Bed temperature gradients

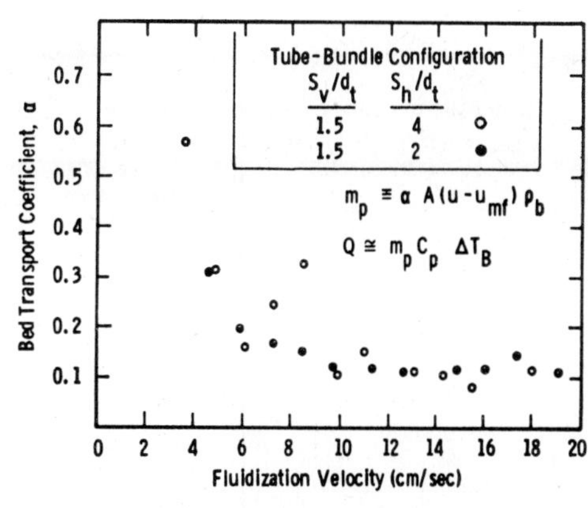

Fig. 5 – Bed transport coefficient

Table 2 - Influence of Tube-Bundle Configuration on Fluidization Characteristics

BUNDLE CONFIGURATION HEATERS/(WATER TUBES)				VELOCITY INCREASED FROM INITIAL STATIC BED				VELOCITY DECREASED FROM INITIAL FLUIDIZED BED		
Number	S_v/d	S_h/d	Arrangement	U_{mf} $\frac{cm}{sec}$	U_t	U_{me}	$\Delta P/\Delta P_t$(max)	U_{mf} $\frac{cm}{sec}$	U_t	U_{me}
0	- -	- -	- -	1.1	1.6	1.6	1.0	1.2	1.6	1.1
(0	- -	- -	- -)							
22	1.5	4	Staggered	1.7	2.0	2.0	1.0	1.6	2.1	2.0
(0	- -	- -	- -)							
22	1.5	4	Staggered	1.7	2.0	2.0	1.0	1.7	2.3	2.0
(66	1.5	4	Staggered)							
45	1.5	2	Inline	1.5	3.5	3.5	2.1	2.0	2.5	2.1
(66	1.5	4	Staggered)							
45	1.5	2	Inline	1.4	4.4	4.4	2.6	2.8	3.3	2.8
(132	1.5	2	Inline)							

Table 3 - Tube-To-Bed Heat Transfer Coefficient Results

Bed Without Internals	Heater Tube (3.18 cm) Maximum Heat Transfer Coefficient (J/sec m²°C) Position on surface –			Average	Water Tube (1.59 cm) Maximum Heat Transfer Coefficient (J/sec m²°C) Position on surface –			Average
	top	side	bottom		top	side	bottom	
	400	270	290	310	570	485	485	510

Heat transfer coefficient for subject tube in tube-bundle/Maximum heat transfer coefficient without internals

Tube-Bundle Configuration		Heater Tube Position on surface –			Average	Water Tube Position on surface –			Average
S_v/d	S_h/d	top	side	bottom		top	side	bottom	
1.5	4	1.3	1.3	1.0	1.2	0.94	0.92	0.91	0.91
1.5	2	1.0	1.3	0.8	1.1	1.04	1.04	1.08	1.04

Table 1 - Comparison of Commercial Unit and Heat Transfer Facility Characteristics

Characteristic	Commercial Fluid Bed Unit[a] Evaporator	Superheater	Test Facility
Bed width, m	4.6	3.7	0.37
Bed depth, m	4.9	4.0	1.1-1.4
Bed length[b], m	0.92	0.92	0.15
Number of tube-bundle layers[c]	3 (1 steam bundle surrounded by 2 sodium bundles)	3 (1 steam bundle surrounded by 2 sodium bundles)	2
Separation between tube-bundle layers, m	0.15-0.3	0.15-0.3	0.15-0.3
Number of tubes:			
Sodium	2645	2181	45
Steam/water	4630	3740	30
Blanks	---	---	102
Tube diameter, cm			
Sodium	3.18	3.18	3.18
Steam/water	1.59	1.59	1.59
Blanks	------	------	1.59
Tube pitch, (in line) cm			
Sodium (S_h, S_v)	6.35-4.77	6.35-4.77	6.35-4.77
Steam/water (S_h, S_v)	3.18-2.39	3.18-2.39	3.18-2.39
Tube material			
Sodium	2 1/4 Cr-1 mo.	2 1/4 Cr-1 mo.	SS 304
Steam/water	2 1/4 Cr-1 mo.	2 1/4 Cr-1 mo.	SS 304
Heat transfer fluids[d]	Sodium, steam/water	Sodium, steam	Electric heat, water
Type of gas distributor	Orifice	Orifice	Orifice
Gas distributor pressure drop, KPa[e]	4.8 - Nickel 1.4 - MgO	3.5 - Nickel 1.0 - MgO	2.8-6.9 - Nickel 0.69-1.38 - MgO
Separation between gas distributor and first row of tubes, m (ft)	0.15-0.3	0.15-0.3	0.21-0.22
Powder[f]	Magnesium oxide, nickel	Magnesium oxide, nickel	Magnesium oxide, nickel
Fluidizing gas[g]	Helium	Helium	Air, helium
Fluidizing gas pressure, KPa (psia)	207-345	207-345	138-207
Fluidizing gas superficial velocity, m/sec	0.092-0.31	0.092-0.31	Up to 0.46
Fluidizing gas inlet temperature, °C	~315	~315	27
Bed temperature, °C	344-427	400-510	Up to 593
Heat transfer fluid temperature, °C			
Sodium (electrical)	385-496	496-538	Up to 700
Steam/water	306-328	328-485	27-49
Heat transfer fluid temperature change per length[h], °C/m			
Sodium	4.6	2.7	0
Steam/water	0.86	10.2	73
Tube outer surface temperature, °C			
Sodium	371-482	482-538	Up to 700
Steam/water	315-371	371-538	150-260
Power input[i], MJ/hr	63,000	36,400	227
Power input per unit volume, MJ/hr³	3,100	2,750	3,310
Power input per unit tube length[v], MJ/hr m	26.0	18.3	33.0
Power input per unit bed flow cross-sectional area[k], MJ/hr m²	7,550	5,460	4,030

a) Reference design for 2 evaporator modules and 2 superheater modules per 4560 $\frac{MJ}{hr}$ heat transfer loop
b) Distance between vertical baffles in commercial unit.
c) In model sodium bundle may be either the lower or upper tube-bundle
d) Electrically heated tubes represent sodium (all sodium tubes heated), low-pressure water cooled tubes and blank tubes represent steam/water tubes in model.
e) Based on distributor pressure drop of 10% of the bed pressure drop in commercial unit.
f) Magnesium oxide is the reference bed powder
g) The model will use house air normally and bottled helium for short term tests.
h) Average value for units.
i) Maximum value for 45 tube reference design. Alternate bundle arrangements permit up to 108 kW_t.
j) Also per tube; average value for commercial unit.
k) Commercial value divided by 2 because there are three stacked tube bundles

WALL-TO-BED HEAT TRANSFER IN A FLUIDIZED BED

By J.H.B.J. Hoebink and K. Rietema

Eindhoven University of Technology
Postbus 513, Eindhoven, Netherlands

SYNOPSIS

The length of the heater on the heat transfer coefficient between a fluidized bed and the vessel wall was studied. The bed had a diameter of 0.70 m and bed heights up to 4 m. A heating panel consisting of thirty copper blocks was placed flush with the wall. For each block separately the heat transfer coefficient could be determined. It was found that this coefficient decreases from the top of the heating panel downwards and approaches a constant value which indicates a downwards solids flow along the wall.

NOTATION

λ_{eff}	effective heat conductivity
$(\rho C_p)_{eff}$	volumetric heat capacity
δ	layer thickness
h	height coordinate
α	heat transfer coefficient
v_o	superficial gas velocity
v_d	velocity of the dispersed phase
K	$\lambda_{eff}/(v_d\ \delta^2\ (\rho C_p)_{eff})$

INTRODUCTION

Theoretical models on heat transfer in fluidized systems are mostly based on the concepts of:

a) conduction of heat through the dense phase of solid particles which is in contact with the heat transferring object.

b) convection of dense phase packets from the heat transferring object towards the bulk of the fluidized bed, thus transporting the heat which they have accumulated near the heat transferring object.

Theoretical models differ mainly in the way in which the two concepts a) and b) are combined while a minor distinction holds whether or not a contact resistance is accounted for.

In the model of Mickley and Fairbanks (1955) there is a distinct sequence in the mechanisms: first conduction then after a certain residence time convection.

In the model of Van Heerden (1953) further elaborated by Yoshida e.a. (1969) it is assumed that the heat transfer resistance arises because of a layer of distinct thickness which moves along the heat transferring object. In this model conduction and convection happen simultaneously in such a way that at longer residence times the contribution of conduction finally dominates.

There is also a difference in conception in which the convection takes place:

b1) through chaotic movements of the solids induced by bubbles which are colliding with the heat transferring object or at least come during their rise path in very close neighbourhood of this object.

b2) through coordinated movement of the solids as a consequence of existing flow patterns of solids. In this connection it should be memorized that several authors working in the field of bubble flow patterns and solids mixing (Werther and Molerus, 1973;, Marshal and Gomezplata, 1965; Ohki and Shirai, 1975) have concluded that along the wall of a fluidized bed such flow patterns exist: sometimes upward but mostly downwards. This latter case seems in accordance with different other observations:

1)Solids are moving upwards in the wake of bubbles; this upward flow, probably increased by the well-known upward drift of particles in the neighbourhood of the bubbles should be compensated with a down flow elsewhere (Kunii and Levenspiel, 1969 Bayens and Geldart, 1973). Since bubbles can not pass the wall it should be expected that near the wall the downward flow would dominate.

2)There is a tendency for bubbles to move towards the centre of the bed (Werther and Molerus, 1973). As a result the average bed density in the central part of the bed is lower than near the wall, which induces an overall circulation of solids (Rietema, 1967). Equivalent flow patterns have been observed in bubble columns (Rietema and Ottengraf, 1970) and in liquid-liquid spray columns (Wijffels and Rietema, 1972).

As yet there is no or only little agreement between the different authors working in the field of heat transfer which mechanisms of solids movement finally are controlling the really observed heat transfer. This holds especially for heat transfer to the wall of the bed.

The reason of this lack of agreement must be sought in several circumstances:

1. most research on heat transfer in fluidized beds is performed in too small equipment and with too small heat transferring objects.
2. In case of heat transfer to the wall only average heat transfer coefficients have been determined instead of local heat transfer coefficients as function of a geometrical coordinate.
3. Not always clear distinction is made between heat transfer at the wall and that on immersed objects.

In order to get more clarity in this situation experiments have been set up at Eindhoven in a large scale fluidized bed of spent cracking catalyst. So far these experiments were confined to the problem of heat transfer at the wall of the bed.

EXPERIMENTAL EQUIPMENT

Experiments were carried out in a fluidized bed with a diameter of 0.7 m and with a packed bed height up to 4 m. The fluidized material was spent cracking catalyst with an average particle size of 66 μm. As gas distributor fine filter cloth was used in several layers upon eachother, the whole being kept in place between two coarse grids. Superficial gas velocities were varied up to 0.13 m/s.

A special heat transfer panel was designed which enabled us to determine the local heat transfer coefficient over the total height of the panel. It consists of 30 red copper blocks placed in an isolation plate of pertinax material. The blocks are situated in a two-dimensional array of ten rows above and three columns next to each other. The heat transferring surface area of each block was 2 x 3 cm (see figure 1). The panel is placed flush with the wall of the bed. The rear side of the blocks is isolated with pertinax and glass wool while the spacing in between adjacent blocks is of pertinax as well. All blocks are heated independently by means of soldering-iron elements which have been put in a small hole in the blocks from the rear side. Moreover each block contains a thermocouple for measurement of its temperature.

Three positions were available to mount the heating panel in the wall of the bed: at 0.5, 1.5 and 2.3 m above the distributor measured from the bottom side of the panel. In tangential direction the distance between the three positions was 120 °.

During the experiments all blocks were kept at the same temperature by adjusting the heating power of each block via the electrical circuit; this way of operation facilitated the interpretation of the heat transfer data while furthermore heat leakages from the inner vertical row of blocks were made as small as possible. For the outer blocks of the panel corrections for heat losses could be made; for this purpose the heat conductivity of the pertinax isolation was determined experimentally. Nevertheless, for interpretation of the measurements only the data from the inner column were used.

The thermocouples of the blocks could be switched alternately in contact with the reference thermocouple which was placed inside the fluidized bed. The heating panel could be replaced by a transparent perspex window, which also was flush with the bed wall. In this way it was observed that the solid particles moved down along the wall. The largest bubbles bursting through the bed level were about 15 cm Ø.

PROCESSING OF DATA

From the measurements of power input and temperature driving force the heat transfer coefficient of each block was calculated.

These experimental results were interpreted in terms of the theoretical models mentioned before, in view of the observation that the solids moved continuously downwards along the wall.

For such conditions Mickley and Fairbanks (1955) derived that the local heat transfer coefficients α_h should decrease with the distance h (measured in the direction of the solids flow) according to:

$$\alpha_h = \sqrt{\frac{\lambda_{eff}\,(\rho C_p)_{eff}\,v_d}{\pi\, h}} \tag{1}$$

Yoshida et al. (1969), assuming a limited thickness of the downward moving layer, derived:

$$\frac{\alpha_h}{\alpha_\infty} = 1 + 2\sum_{n=1}^{\infty} \exp\left[-\frac{n^2\pi^2\,\lambda_{eff}\,h}{(\rho C_p)_{eff}\,v_d\,\delta^2}\right] \tag{2}$$

in which α_∞ is the constant value of the heat transfer coefficient, which is reached at infinite distance and which corresponds with the steady state conduction through a layer of thickness δ:

$$\alpha_\infty = \lambda_{eff}/\delta \tag{3}$$

From the experimental data α_∞ could be determined with a reasonable accuracy by extrapolation; this value was used to calculate δ from equation 3. The velocity of the solids moving downwards along the wall was determined with the aid of figure 2. Here α_h/α_∞ has been plotted versus $K = \lambda_{eff}/((\rho C_p)_{eff}\ v_d\ \delta^2)$ for discrete values of h that correspond with the height coordinates of the center of the copper blocks numbers 2, 4 and 6 (counted from the top of the panel). From graph 2 the value of K was determined, which was used to calculated v_d since the other factors are known. λ_{eff} was determined as 0.087 W/m °C, using the method of Mittenbühler (1965) for a packed bed with a correction for the expanded dense phase in the fluidized bed (Schumann and Voss, 1934). $(\rho C_p)_{eff}$ was taken as $0.7\ 10^6$ J/m^3 °C.

EXPERIMENTAL RESULTS

Figure 3 shows the observed heat transfer coefficients of all thirty blocks of the panel at specific conditions. The transfer coefficients of individual blocks decrease from the top block downwards. Such an effect was found in all experiments, and corresponds with the downward solids flow along the wall, observed through the perspex window.

It is seen (figure 3) that the coefficients of blocks in the outer columns are larger than those of blocks in the inner column, which should be ascribed to extra heat transfer possible in lateral directions. As mentioned before the coefficients obtained with the inner column only were used in our calculations. Because of similar effects the heat transfer coefficient of the lowest block of the inner column does not fit the general trend that α decreases with increasing h. In figure 4 the influence of the superficial gas velocity u_0 on the heat transfer coefficients of the blocks in the middle column is presented. All coefficients increase at increasing gas velocity, as could be expected.

Table 1 summarises the data obtained under various experimental conditions, as well as some derived quantities like α_∞, δ and v_d.

DICUSSION

All experimental results reveal that α approaches to a constant non-zero value at the lowest blocks (large values of h), which is not in agreement with the Mickley-Fairbanks theory (see formula 1). The latter is illustrated once more in figure 5, in which α/α_9 has been plotted versus $\sqrt{h_9/h}$, α_9 and h_9 referring to the transfer coefficient and coordinate of block number 9. The drawn line in figure 5 corresponds to equation 1, and demonstrates that the concept of heat diffusion into a semi-infinitely deep layer does not fit the experimental data. Therefore the data have been interpreted further with the model of Yoshida e.a. (1969); table 1 presents the velocity v_d and thickness δ of the downward moving layer calculated according to that model from the measurements.

From table 1 the layer thickness δ appears to have an order of magnitude of 1 mm, which is in agreement with the well-known fact that the heat transfer resistance is concentrated in a narrow film along the wall. However from a viewpoint of solids recirculation in the bed one should not expect that the particles are moving downwards in such a thin layer only. For this reason it may be suggested that downward movement of solids occurs in a laminar sublayer, which controls the heat transfer resistance, and in an upper layer which is disturbed from time to time by some rising bubbles.

As far as the dependance of the velocity and thickness of the laminar sublayer on the process conditions is concerned no conclusions will be drawn at this moment in view of the limited accuracy of these results.

CONCLUSIONS

1. The heat transfer coefficient between the bed wall and the fluidized bed depends on the heater length, because of a downward movement of particles along the wall.
2. The dependence between heat transfer coefficient and heater length is not described by heat diffusion into a semi-infinite layer, as suggested by Mickley and Fairbanks (1955). The theory of Yoshida e.a. (1969) who assumed a limited thickness of the transport resisting layer seems a better approach.
3. The downward moving layer of solids should be considered to consist of a laminar sublayer, controlling the transfer resistance, and an upper layer which is stirred from time to time.

REFERENCES

Bayens,J., Geldart, D. (1973) Proceedings Int. Symp. "Fluidization and its applications", Toulouse, p. 182 - 195.

Van Heerden,C., Nobel, A.P.P., Van Krevelen, D.W. (1953) Ind.Eng.Chem, 45, p.1237

Kunii,D., Levenspiel, O. (1968) Ind.Eng.Chem. Fund. 7, 446

Marsheck, R.M., Gomezplata, A. (1965) A.I.Ch.E.J., 11, 167

Mickley, H.S., Fairbanks D.F. (1955) ibid 1, 374

Mittenbühler, A. (1965) Ber.Dtsch.Keram.Ges. 41, 15

Ohki, K., Shirai, T. (1975) Int.Fluidization Conference, Asilomar, California, Session I, p. 95 - 110.

Rietema, K. (1967) Proc.Int.Symp. "Fluidization", Eindhoven, p. 209

Rietema, K., Ottengraf, S.P.P. (1970) Trans. Inst.Chem.Engrs. 48, T54

Schumann, T.B.W., Voss, V. (1934) Fuel, 13, 249

Werther, J., Molerus, O. (1973) Int.J.Multiphase Flow 1, 103
Wijffels, J.B., Rietema, K. (1973) Trans. Inst.Chem.Engrs., 50, 224, 233
Yoshida, K., Kunii, D., Levenspiel, O. (1969) Int.J.Heat Mass Transfer 12, 529

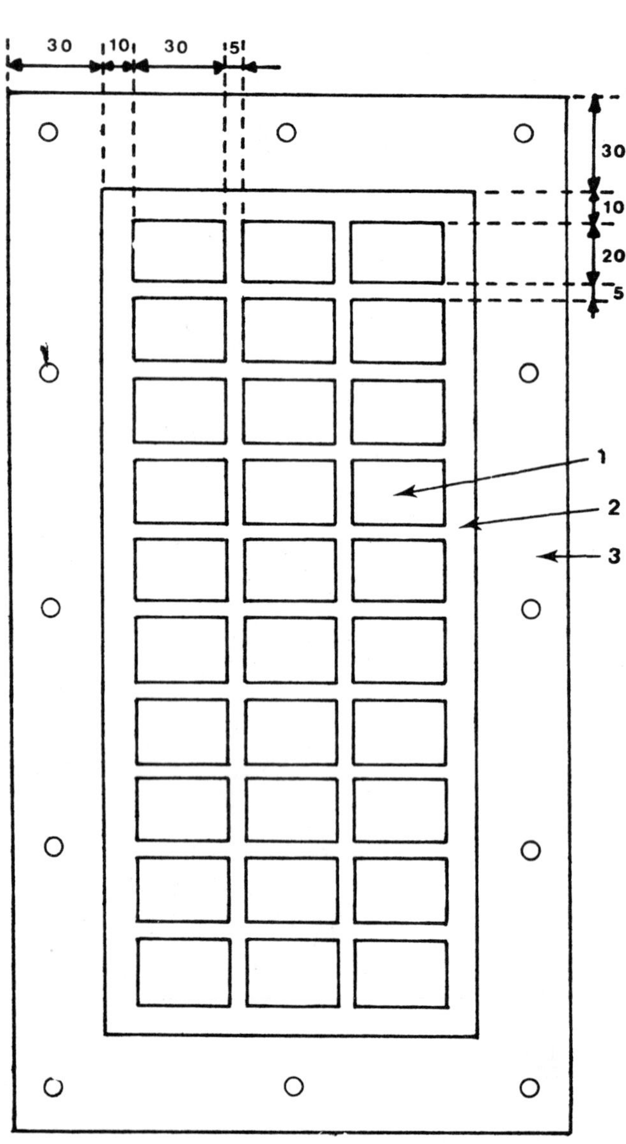

1 copper block
2 pertinax isolation
3 plate for connection to bed-wall
(measures in mm)

Fig.1. Bed side view of the heating panel.

Table 1

Run	h_p cm	h_b cm	v_o cm/s	ΔT °C	α_2	α_4	α_6	α_∞	$\overline{K}$ m^{-1}	δ cm	v_d cm/s
					kcal/m^2 hr °C						
1	50	300	2.80	47	147	109	82	68	1.52	0.11	12.3
2			7.1	46	169	122	95	73	1.33	0.10	17.0
3			12.9	46	188	142	106	84	1.39	0.09	20.0
4	150	300	1.35	46	126	98	80	66	1.73	0.11	10.8
5			2.80	50	136	103	83	70		0.12	
6			10.1	47	139	101	81	70	1.78	0.11	10.5
7			12.9	41	176	136	108	92	1.78	0.08	19.8
8	230	300	1.35	38	129	96	78	65	1.70	0.12	9.2
9			2.80	45	110	78	63	55	1.78	0.14	6.5
10			6.75	42	111	80	63	57	1.91	0.13	7.0
11	230	400	1.35	46	110	76	63	56	1.90	0.13	7.0
12			2.80	46	111	76	63	56	1.89	0.13	7.1

h_p = height of panel above distributor
h_b = packed bed height
ΔT = temperature driving force
α_i = heat transfer coefficient of block i (inner column)
$\overline{K}$ = mean value of K for blocks 2, 4 and 6

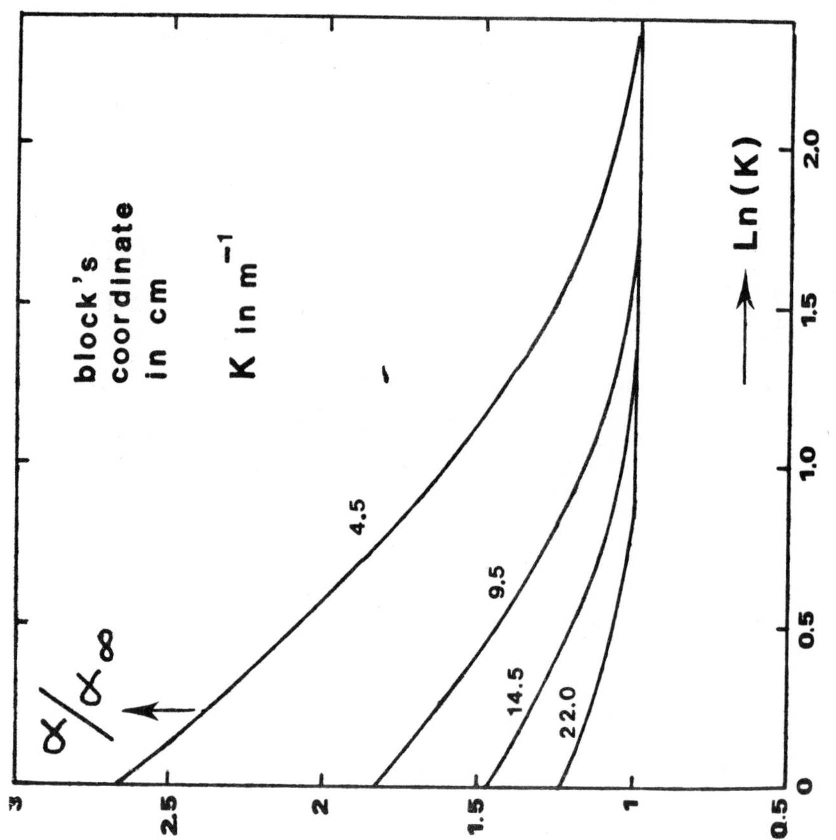

Fig.2. Plot of eq. 2 for four values of h (in cm's).

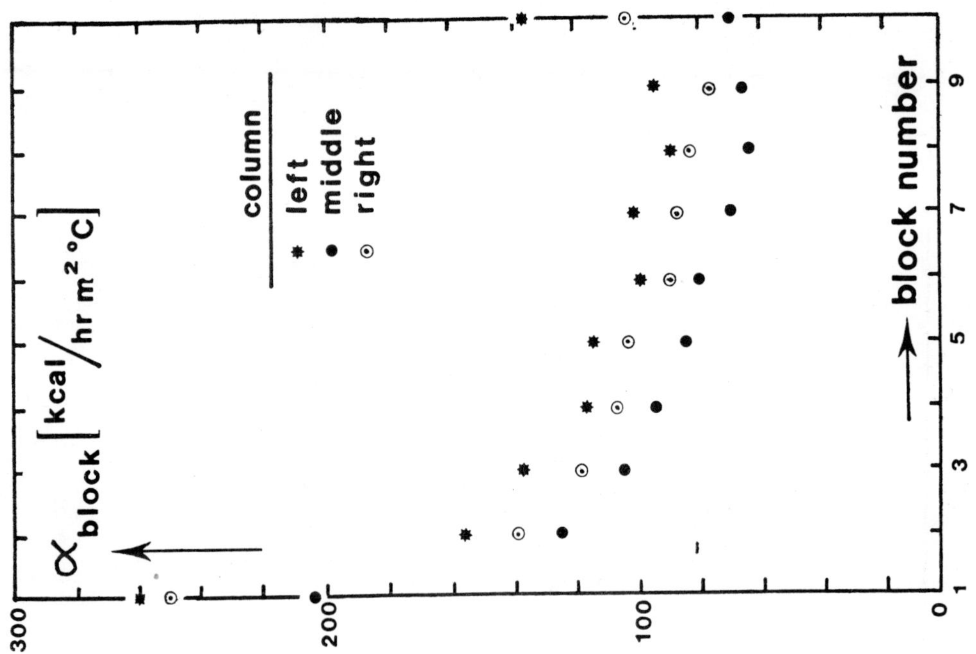

Fig.3. Data for run nr. 4 (all blocks).

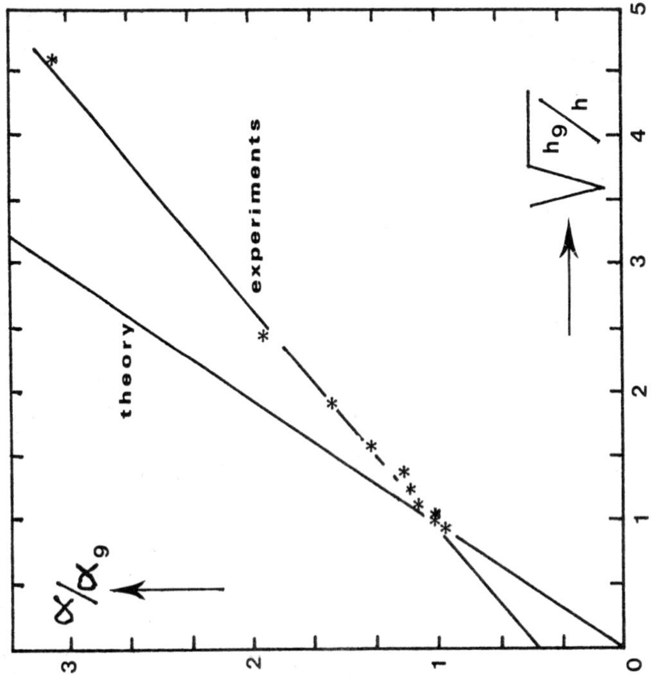

Fig.5. Comparison or run nr. 4 with eq. 1.

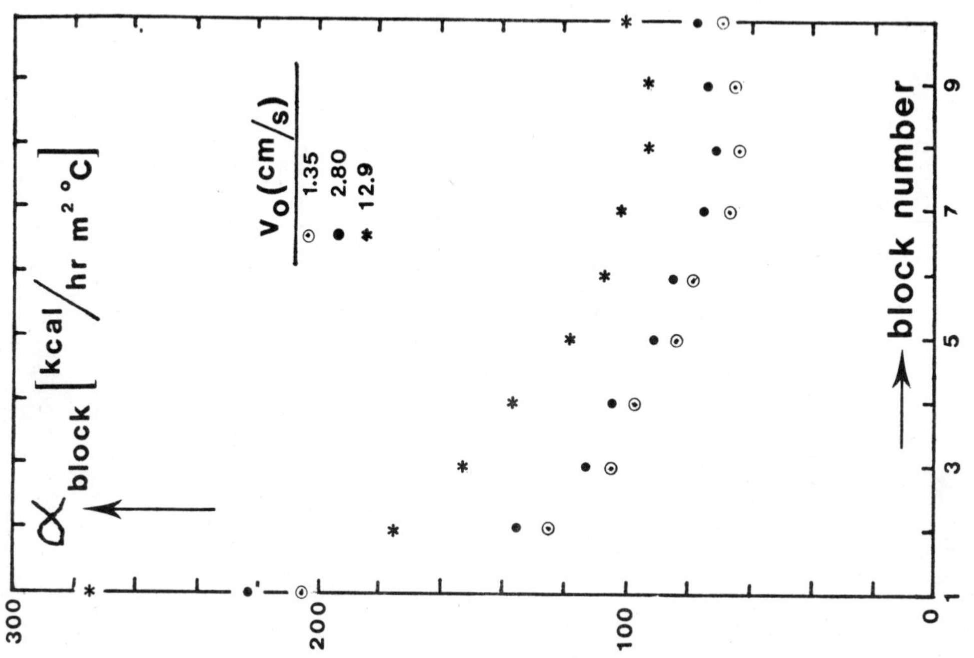

Fig.4. Data for runs nr. 4, 5, 7
(blocks of the inner column only).

HEAT TRANSFER TO SURFACES IMMERSED IN FLUIDISED BEDS, PARTICULARLY TUBE ARRAYS

By A.M.Xavier and J.F.Davidson

Department of Chemical Engineering

Pembroke Street, Cambridge

England CB2 3RA

SYNOPSIS

A model is proposed for the convective heat transfer between a slug flow fluidised bed and an immersed surface. The overall convective heat transfer coefficient is approximated as the sum of two components: (i) the particle convective and (ii) the gas convective heat transfer coefficients.

The particle convective heat transfer is predicted from the packet model of Mickley & Fairbanks, modified to account for gas film resistance adjacent to the surface.

The gas convective heat transfer coefficient is obtained from theory and experiment for heat transfer in packed beds.

Dimensionless groups derived from an analytical expression are used as the basis for tentative correlation of heat transfer data from bubbling fluidised beds.

The theory is applied to a bed with tube inserts for which new data on heat transfer and bed expansion are presented.

NOTATION

C_g	gas heat capacity
C_{mf}	particulate phase heat capacity
C_s	solid heat capacity
D	column diameter
d_p	particle diameter
d_T	diameter of immersed tube
H	bed height
H_{max}	H at maximum expansion
H_{mf}	H at incipient fluidisation
h_c	overall convective heat transfer coefficient
h_{cmax}	maximum value of h_c
h_{film}	gas film heat transfer coefficient
h_{gc}	gas convective heat transfer coefficient
h_{pc}	particle convective heat transfer coefficient
h_{pcmax}	maximum h_{pc} with no gas film resistance
k_e^o	effective packed bed thermal conductivity; no fluid motion
k_g	gas thermal conductivity
k_{mf}	particulate phase thermal conductivity
k_s	solid thermal conductivity
L	immersed surface length
U	superficial gas velocity
U_{mf}	U at minimum fluidisation
u_B	rise velocity of single bubble or slug
V_T	fraction of bed volume occupied by tube inserts
V_{Tmax}	V_T when $H = H_{max}$
V_{Tmf}	V_T when $H = H_{mf}$
x	$(U-U_{mf})/u_B$
y	$h_{film}/(h_{pcmax}\sqrt{\pi x})$
δ	average gas film thickness
ε_{mf}	voidage fraction at minimum fluidisation
ρ_g	density of gas
ρ_{mf}	density of particulate phase
ρ_s	density of solid
τ	contact time for particle packet

INTRODUCTION

There is a need for reliable design equations to predict surface to bed heat transfer in a variety of conditions which may not be readily attained in the laboratory, e.g. high pressure and temperature. With this in mind the present paper gives methods based on first principles which lead to a tentative correlation for predicting heat transfer.

To a first approximation, the heat transfer to or from surfaces immersed in gas-fluidised beds may be considered to be composed of three additive components (Botterill, 1977):

(i) the heat transfer to aggregates of particles that circulate within the bed under the influence of bubbles (particle convective component);

(ii) the heat transfer to the gas percolating through the bed of particles at minimum fluidising conditions (gas convective component);

(iii) the heat transfer due to radiation which becomes significant at high operating temperatures (radiant component).

This paper is concerned with components (i) and (ii). A model is proposed to predict the overall convective heat transfer coefficient between a slug flow fluidised bed and an immersed surface. The possibility of correlating published surface to bed heat transfer data for bubbling fluidised beds is investigated in the light of this model. The application of the model to the case of a fluidised bed filled with horizontal tubes is examined. An important question is the bubble velocity through the tube array: this is partly answered by measurement of the relation between bed height and fluidising velocity.

PARTICLE CONVECTIVE HEAT TRANSFER

It is generally accepted that particle convective heat transfer is due to 'surface renewal': stirring due to bubbles causes packets of particles near the surface to be replaced by other packets. This mechanism was first formulated by Mickley & Fairbanks (1955) who developed the 'packet model' of heat transfer. Baskakov (1964) modified the model by including the thermal contact resistance arising from the higher voidage fraction in the region adjacent to the surface.

The overall particle convective heat transfer coefficient h_{pc} is hard to predict from first principles, because it depends on τ, the contact time of a particle packet. However, for slug flow, τ can be predicted, because of the well-defined flow characteristics of this regime. Davidson (1975) pointed out that, for slug flow past a vertical surface of length L,

$$\tau = L/(U-U_{mf}) \tag{1}$$

where U is the fluidising velocity, and $U = U_{mf}$ at incipient fluidisation. From the two-phase theory, $(U-U_{mf})$ is the upward velocity of the particulate phase between slugs. The application of the two-phase theory of fluidisation to the packet model of Mickley & Fairbanks gives (neglecting the heat transfer from the surface when it is in a slug)

$$h_{pc} = 2\left[\frac{k_{mf}\rho_{mf}C_{mf}(U-U_{mf})}{\pi L}\right]^{\frac{1}{2}} \cdot \frac{u_B}{U-U_{mf}+u_B} \tag{2}$$

where k_{mf}, ρ_{mf} and C_{mf} are respectively the thermal conductivity, density and heat capacity of the particulate phase. The rise velocity of a single slug, u_B, is $0.35(gD)^{\frac{1}{2}}$, g being the acceleration due to gravity and D the column diameter.

Eq.(2) holds good for fine particles ($d_p < 0.1$mm) and long contact time between surface and particulate phase (L > 100mm). For short contact times, the influence of the gas film thickness (even for small particles) is appreciable.

Eq.(2) reaches a maximum value when $U-U_{mf} = u_B$, giving

$$h_{pcmax} = \left[k_{mf}\rho_{mf}C_{mf}u_B/(\pi L)\right]^{\frac{1}{2}} \tag{3}$$

which is the maximum value, in the absence of any contact resistance.

The heat transfer between a fluidised bed and a surface is observed to decrease as the particle diameter increases, although for very large particles ($d_p > 2$mm) the opposite occurs; (e.g. Gelperin & Einstein, 1971, p.494). The dependence of h_{pc} on particle diameter may be obtained by considering a contact resistance at the surface, due to a gas film of conductivity k_g and average thickness δ; δ is proportional to particle diameter and the resulting heat transfer coefficient is

$$h_{film} = k_g/\delta \tag{4}$$

Applying the two phase theory to the model of Baskakov and taking the contact resistance as $1/h_{film}$, the following expression is obtained, for slug flow:

$$\frac{h_{pc}}{h_{pcmax}} = \frac{\sqrt{\pi x}}{y(1+x)}(\text{erfc}\, y \cdot \exp y^2 - 1) + \frac{2\sqrt{x}}{1+x} \tag{5}$$

where $y = h_{film}/(h_{pcmax}\sqrt{\pi x})$, $x = (U-U_{mf})/u_B$.

For the particulate phase: (i) the density may be taken as $\rho_s(1-\varepsilon_{mf})$; (ii) the heat capacity may be taken as that of the solid particles; (iii) the thermal conductivity can be estimated from relationships developed for thermal conductivity in packed beds (Yagi & Kunii, 1957; Yagi *et al.* 1961) and may be written as

$$k_{mf} = k_e^o + 0.1\ \rho_g C_g d_p U_{mf} \qquad (6)$$

where the stagnant bed conductivity, k_e^o, can be calculated following the method of Kunii & Smith (1960) or Baskakov (Gelperin & Einstein 1971, p.487).

The average gas film thickness lies in the range $d_p/6$ (Zabrodsky, 1966, p.226) $< \delta <$ $d_p/10$ (Botterill & Williams, 1963; Botterill & Butt, 1968; Botterill & Desai, 1972). The minimum fluidising velocity may be calculated using the Ergun correlation (Richardson, 1971, p.35).

GAS CONVECTIVE HEAT TRANSFER

Assuming that the gas convective heat transfer coefficient, h_{gc}, is that of a bed at incipient fluidisation (Botterill & Denloye, 1975), the model of Gabor (1970) provides a simple theoretical prediction for h_{gc}. Thus, for a flat surface

$$h_{gc} = \left[4k_{mf}\rho_g C_g U_{mf}/(\pi L)\right]^{\frac{1}{2}} \qquad (7)$$

OVERALL CONVECTIVE HEAT TRANSFER COEFFICIENT

We assume that (5) and (7) can be added, so that

$$h_c = h_{pc} + h_{gc} \qquad (8)$$

gives the overall convective heat transfer coefficient. This simple addition is obviously an over-simplification: we might expect the actual value of h_c to be somewhat less than the sum of the two components.

EXPERIMENT AND THEORY COMPARED: SLUG FLOW

Heat transfer experiments were performed in two columns with internal diameters of 229mm and 102mm. The distributors were sintered metal plates. Heaters of length 40, 60 and 200mm were placed vertically, in the centre of each column. The materials used were sand, gravel, glass beads and iron shot, with mean particle diameter in the range 0.1-8.0mm. The superficial air velocity ranged from fixed bed to slugging conditions. The initial bed height was sufficient to ensure slug flow (H/D > 2).

Experimental and theoretical results are compared in detail elsewhere (Xavier, 1977). Fig.1 shows a typical relation between h_c and U for an iron shot-air system. To calculate the theoretical curves, material properties were from standard tables (e.g. International Critical Tables, 1933). The stagnant bed conductivity was calculated by the method of Kunii & Smith (1960), for a bed voidage of 0.4.

Heat transfer data obtained in beds of various particle sizes suggest that $d_p/10$ is a suitable average gas film thickness

CORRELATION OF DATA: BUBBLING BEDS

The solution obtained for the convective surface to bed heat transfer in slugging beds suggests that two sorts of dimensionless group can be derived,

(i) h_c/h_{pcmax} and h_{pcmax}/h_{film}

suitable for small particles, and

(ii) h_c/h_{gc} and h_{gc}/h_{film}

suitable for large particles.

Fig.2 compares theoretical curves with experimental results on maximum surface to bed heat transfer in slug flow fluidised beds, using the dimensionless groups (i) above. Note that the primary variable is particle size: small particles are represented by points towards the left hand side of the diagram and in this region the heat transfer is mainly due to surface renewal of particle packets.

Fig.3 is a tentative correlation of published data on maximum surface to bed heat transfer. A plot of k_e^o/k_g against k_s/k_g for a bed voidage of 0.4 (Kunii & Smith, 1960) is also included in this figure. Uncertainties in the values of some variables involved in the calculation of $h_{pcmax} = (k_{mf}\rho_{mf}C_s u_B/\pi L)^{\frac{1}{2}}$ may have contributed to the scatter in Fig.3. The curve through the points in Fig.3 is empirical: note that at the left hand side of the diagram the ordinate tends to 2.0 as compared with 1.0 in Fig.2 for slugging conditions.

Fig.4 shows an alternative way of correlating heat transfer data, suitable for beds of large particles (d_p > 0.5mm) using groups (ii) above. This can be justified from the fact that h_{gc} becomes of increasing importance as d_p increases.

BEDS WITH TUBE ARRAYS

The prediction of the surface to bed heat transfer in fluidised beds containing tubes depends on the bubble velocity, u_B. According to the two-phase theory (Davidson & Harrison, 1963)

$$(H_{max} - H_{mf})/H_{mf} = (U-U_{mf})/u_B \qquad (9)$$

and the bubble velocity may be determined from observations of maximum bed height H_{max}.

The maximum bed height is achieved when a bubble reaches the break-through position, because prior to the break through the top surface is moving upwards with velocity $U-U_{mf}$.

In the case of fluidised beds with tube inserts, the volume of tubes will influence the bed expansion and (9) becomes (Xavier, 1977)

$$\frac{H_{max}(1-V_{Tmax}) - H_{mf}(1-V_{Tmf})}{H_{mf}(1-V_{Tmf}) + H_{max}V_{Tmax}} = \frac{U-U_{mf}}{u_B} \qquad (10)$$

where V_{Tmax} and V_{Tmf} are the fraction of bed volume occupied by tubes when $H=H_{max}$ and $H=H_{mf}$, respectively.

Experiments on bed expansion and heat transfer were performed in a 305mm square bed with 28mm diameter rod inserts on a 76mm triangular pitch. The bed distributor consisted of a perforated plate containing 900 holes 2.5mm diameter and a fine mesh glued to the top to prevent the particles from falling through the holes. Sands of mean particle size 0.158mm, 0.385mm and 0.885mm were fluidised by air. A horizontal cylindrical heating unit with an external diameter of 28mm replaced a rod insert positioned at 400mm above the distributor, in the centre of the bed.

Fig.5 shows typical bed expansion results obtained with the 0.385mm sand/air system.

The mean bubble velocity in the 305mm square bed with rod inserts was found to be of the order of 1m/s, ± 0.2m/s. The bubble velocity was found to be remarkably independent of particle size.

Fig.6 shows that the experimental heat transfer results may be represented theoretically when the estimated value of u_B and a gas film thickness $\delta = d_p/4$ are used in the proposed model: the curves in Fig.6 were calculated by using these values with (3), (4), (5), (6), (7) and (8) to get h_c. Observations of continuous formation of gas pockets beneath the horizontal tubes may justify the use of a larger value of the effective gas film thickness, to correlate heat transfer data in fluidised beds with horizontal tube arrays: thus $\delta = d_p/4$ was appropriate for the horizontal tubes (Fig.6), whereas $\delta = d_p/10$ for the vertical surfaces (Figs 1, 2, 3).

CONCLUSIONS

(1) An analytical expression was obtained to calculate the convective surface to bed heat transfer coefficient in slug flow fluidised beds. The main variables involved are (i) particle properties, (ii) gas properties, (iii) surface dimensions, (iv) bubble velocity and (v) average thickness of the gas film, which is proportional to particle diameter.

(2) Dimensionless groups derived from the proposed expression were shown to be of value to correlate surface to bed heat transfer data in bubbling fluidised beds.

(3) From observations of maximum bed height the bubble velocity in a 305mm square fluidised bed containing tubes was estimated to be about 1 m/s.

(4) The effective thickness of the gas film adjacent to the immersed surface was taken as $d_p/10$ for single vertical surfaces, and about $d_p/4$ for horizontal surfaces in tube arrays.

ACKNOWLEDGEMENT

One of us (AMX) would like to acknowledge the support given by the British Council and the National Coal Board.

REFERENCES

Baskakov, A.P. (1964) Int.Chem.Eng, 4, 320

Botterill, J.S.M. (1977) Paper to European Congress "Transfer Processes in Particle Systems", Nuremberg, 28-30 March

Botterill, J.S.M. & Butt, M.H.D. (1968) Brit.Chem.Eng, 13, 1000

Botterill, J.S.M. & Denloye, A.O.O. (1975) Chisa '75, Lecture D, Prague

Botterill, J.S.M. & Desai, M. (1972) Powder Technol., 5, 231

Botterill, J.S.M. & Williams, J.R. (1963) Trans.Instn Chem.Engrs, 41, 217

Davidson, J.F. (1975) Chisa '75, Lecture D, Prague

Davidson, J.F. & Harrison, D. (1963), Fluidised Particles, Cambridge University Press

Gabor, J.D. (1970) Chem.Eng.Sci.,25, 979

Gelperin, N.I. & Einstein, V.G. (1971) in Fluidization, edited by Davidson & Harrison, Academic Press, London and New York

Henriksen, H.K. (1973) Report, Department of Chemical Engineering, University of Cambridge

International Critical Tables (1933) McGraw-Hill

Kunii, D. & Smith, J.M. (1960) A.I.Ch.E.J., 6, 71

Mickley, H.S. & Fairbanks, D.F. (1955) A.I.Ch.E.J., 1, 374

Morgan, C. (1967) Ph.D. dissertation, University of Cambridge

Richardson, J.F. (1971) in Fluidization, edited by Davidson & Harrison, Academic Press, London and New York

Xavier, A.M. (1977) Ph.D. dissertation, University of Cambridge

Yagi, S. & Kunii, D. (1957) A.I.Ch.E.J., 3, 373

Yagi, S., Kunii, D. & Wakao, N. (1961) Int. Develop.Heat Transfer, Part 4, p.742, A.S.M.E., N.Y.

Zabrodsky, S.S. (1966) Hydrodynamics and Heat Transfer in Fluidised Beds, MIT Press, Cambridge, Massachusetts and London.

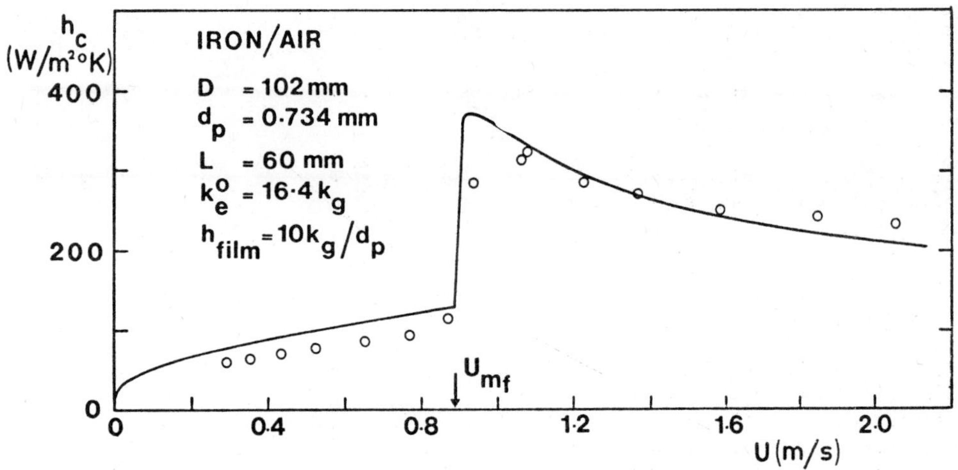

Fig.1 : Typical comparison between experimental points and theoretical curve - 0.734 mm iron shot-air system

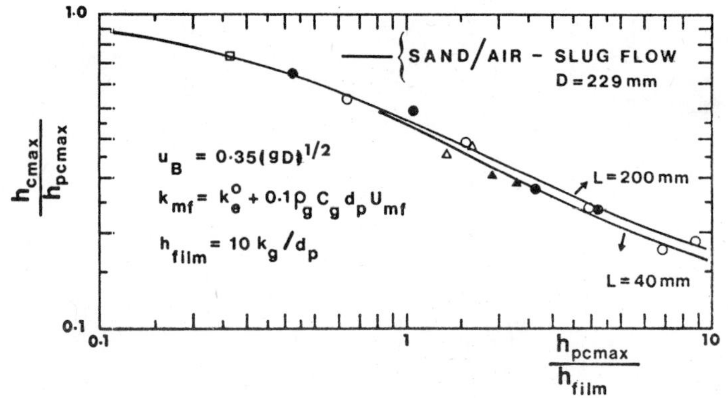

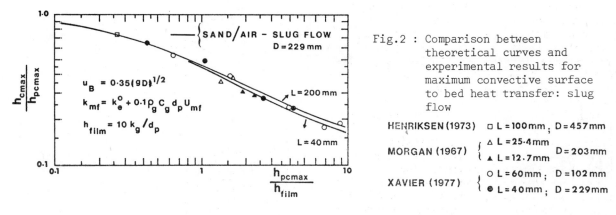

Fig.2 : Comparison between theoretical curves and experimental results for maximum convective surface to bed heat transfer: slug flow

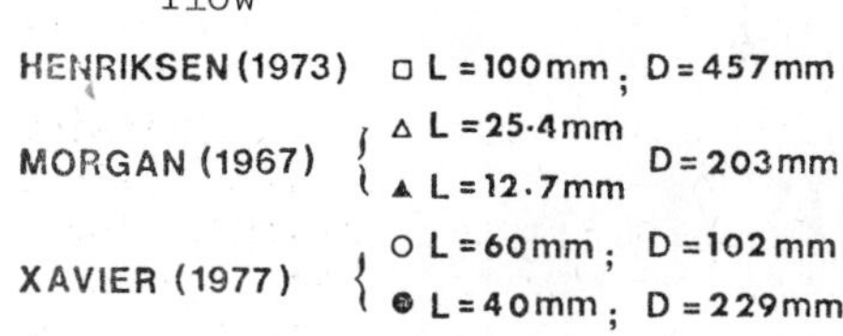

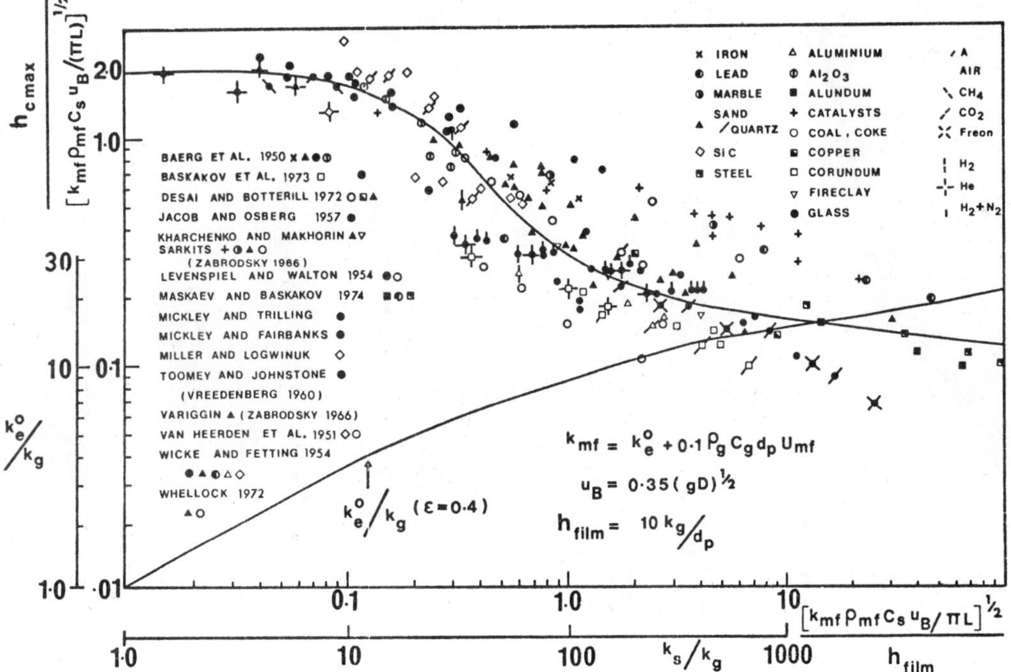

Fig.3 : Tentative correlation of data for bubbling beds on maximum surface to bed heat transfer

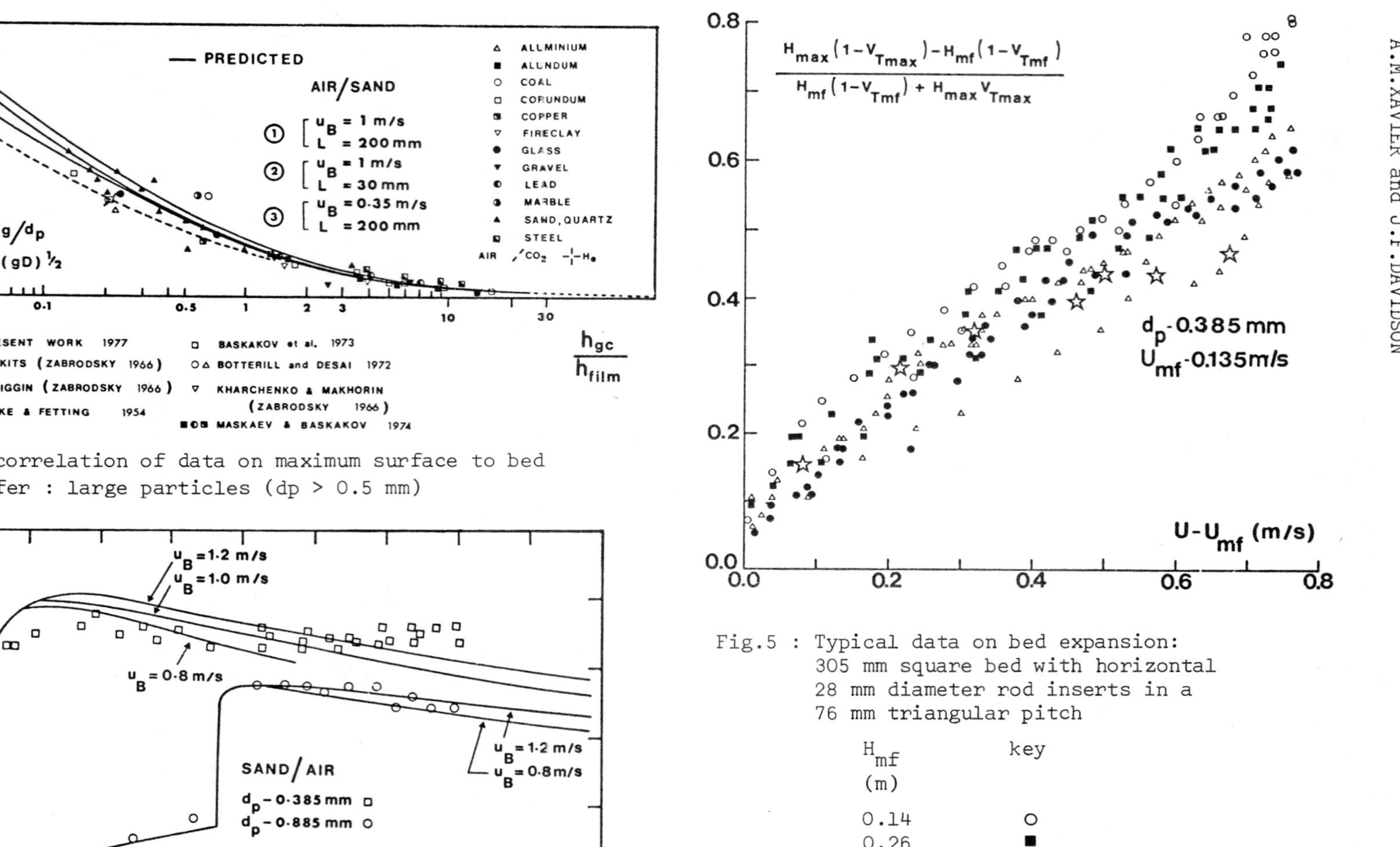

Fig.4 : Tentative correlation of data on maximum surface to bed heat transfer : large particles (dp > 0.5 mm)

Fig.6 : Heat transfer from one of the tubes of an horizontal tube array: 305 mm square bed with 28 mm diameter tube inserts in a 76 mm triangular pitch

Fig.5 : Typical data on bed expansion: 305 mm square bed with horizontal 28 mm diameter rod inserts in a 76 mm triangular pitch

H_{mf} (m)	key
0.14	○
0.26	■
0.39	△
0.48	●
0.48	☆ (film analysis)

N.B. The references on Figures 3 and 4 are cited in detail by Xavier (1977).

EFFECT OF TUBE BANK AND GAS DENSITY ON FLOW BEHAVIOR AND HEAT TRANSFER IN FLUIDIZED BEDS

By F.W. Staub and G.S. Canada

General Electric Corporate Research and Development

Schenectady, New York 12301

SYNOPSIS

Large particle (600-2600 μm) fluidization in a square 1/3 m and square 2/3 m facility has been carried out with and without bare and finned horizontal tube banks between pressures of one and ten atmospheres. A generalized correlation technique for average bed expansion is given. Flow regime maps and dynamic bed behavior are shown with and without tube banks. Heat transfer measurements using electrically heated bare and finned tubes were employed throughout the superficial velocity range investigated to determine the variation of heat transfer with particle size and gas density. The effectiveness of the finned tube is discussed.

NOMENCLATURE

C_o	distribution parameter or slope of velocity plot
H_{st}	static bed height (cm)
LG,SG	large and small particles respectively
U_{sg}	superficial gas velocity (m/s)
U_T	terminal velocity of particle (m/s)
U_{mf}	minimum fluidization velocity of particle (m/s)
$\bar{\alpha}$, $\bar{\alpha}_{mf}$	average bed void fraction and average void fraction at minimum fluidization respectively
ΔP ratio	see Fig. 1 title

INTRODUCTION

The gas fluidized bed flow behavior described here has been observed during part of a larger scale effort aimed at improving the understanding and predictability of such beds as they would be applied to generate steam or to carry out combustion for power generation. The purpose of this effort is to generate improved modeling and scaling correlations of the flow behavior and heat transfer to bare and finned tube banks immersed in fluidized beds. To permit the direct observation and measurements needed to generate sufficient supporting data, a temperature range below 150°C has been employed. The work reported here is limited to the flow behavior and tube bank heat transfer information with 650 μm and 2600 μm spherical particles (with specific gravities of 2.48 and 2.92 respectively), using a bubble cap type distributor in a 0.305 m x 0.305 m and a 0.61m x 0.61m test bed with and without tube banks. Current work includes heat transfer with mixed particle sizes and the determination of criteria for the onset of tube erosion. A particle flow and heat transfer model has been formulated based on the newly observed flow regime (Canada & Staub, 1977). Data are now being taken to validate this model using a newly developed probe for measuring particle number density and momenta.

It was felt necessary to first examine bed flow behavior without tube banks to assist our subsequent understanding of the effects of tube banks and in consideration of the fact that significant portions of the fluidized beds of interest do not contain tube banks. It was also felt necessary to determine the flow regimes and their transition points to aid in correlating the average bed expansion, to relate their effect on bed pressure and flow oscillations, and to allow the formulation of improved flow and heat transfer models. Special emphasis has been placed on higher superficial gas velocities due to the economic penalties of very large cross-sectional area beds and since little available published data exist on the flow and heat transfer at large fractions of the terminal velocity of the particles or for large particles in general. For the same reason emphasis has been placed on the performance of finned tube banks in consideration of the high heat removal per unit volume needed at high superficial velocities.

EXPERIMENTAL EQUIPMENT

The equipment and instrumentation employed is described by Canada *et al.* (Chicago, 1976) with further details by Brzozowski *et al.* (March, 1976), Canada *et al.* (July, 1976), and Canada *et al.* (October, 1976). Dynamic bed pressure drop, local capacitance probes (local void behavior) and local particle momentum probes (local particle flow rate and direction) are employed in addition to still and motion picture photography. The tube bank data reported here were taken with 5 row or 10 row staggered tube banks at 5.7 cm axial pitch and 10.2 cm transverse pitch. The finned tubes are 2.5 cm O.D. with 1.3 cm high, 0.27 cm thick continuous helical fins at 0.98 fins per cm spacing. Limited data with a finned tube having 0.15 cm thick fins at 2.8 fins per cm are given (see also Fig. 8). The bare tube banks have 3.2 cm O.D. tubes to permit the same minimum flow area between tubes in the finned or bare tube banks.

RESULTS AND DISCUSSION

Flow Regimes

While direct observation and photography are the most obvious methods for describing the flow regimes encountered, the ratio of oscillating-to-average steady state bed pressure drop (see Fig. 1), the bed dynamic pressure drop signal character and the local void fraction signal character (Brzozowski *et al.* March, 1976; Canada *et al.* October, 1976) all assisted and supported the flow regime identification. In Fig. 1, for example, the bed pressure drop ratio data without tube banks are shown at 1 atm pressure as the hatched area for the 2600 μm particles employed. The bed height oscillation and rate of bed material loss were proportional to the ordinate in Fig. 1. Above about 3.66 m/s superficial velocity a flow regime change from bubbly-slug to turbulent flow took place that causes a decreased bed oscillation and material loss at higher velocities.

The ratio of the oscillating-to-average bed pressure drop for 650 μm particles in a 40 cm deep bed without tube banks for 1 and 10 atm is shown in Fig. 2. The peak pressure drop ratio at atmospheric pressure occurs at a velocity ratio (here defined by U_{sg}/U_T), of approximately 0.5. At 10 atm the peak pressure drop ratio is reduced and occurs over a relatively wide range of velocity ratio (from around 0.5 to 0.75). The peak pressure drop ratio for large particles is at lower velocity ratio than that for the smaller particles. The flow regime transitions implied by these data are summarized below.

The oscillating pressure drop behavior with tube banks is similar to that without tube banks at 1 atm pressure although lower in the amplitude of the oscillations (see Fig. 1). If one allows an increasing expanded bed height above the tube bank the oscillation amplitude increases again.

The flow regimes, in beds without tube banks, as detected by the various probes and as directly observed for each of the particle sizes tested, are shown in the suggested regime maps in Figs. 3 and 4. The relative independence of flow regime transition with bed height persisted (Canada *et al.* Chicago,1976; Canada *et al.* July 1976). The maximum velocity tested was limited by particle loss due to our fairly short freeboard or, for large particles, by excessive distributor pressure drop.

A bubbly flow regime and a turbulent flow regime were observed for both particle sizes over the pressure range tested. The transition from bubbly to turbulent flow has been studied by Davidson *et al.* (1971) and Yerushalmi *et al.* (1976), who also observed and discusses the turbulent flow regime and who ascribes its first identification to Lanneau (1960). While Yerushalmi *et al.* (1976) and Lanneau (1960) deal with very small particles, where the turbulent regime is accompanied by significant bed loss, hence recycle, our observed large particle transition to turbulent flow is at a gas superficial velocity well below the particle terminal velocity. The transition to turbulent flow occurred at a velocity ratio of 0.55-0.65 for the small glass particles and 0.30-0.35 for the large particles. These values are reasonably independent of pressure and in general agreement with the decreasing pressure drop ratio in Figs. 1 and 2. For both particle sizes the bubbly flow regime extends to somewhat higher values of velocity ratio as the pressure increases although the mean bubble size decreases.

The most significant change demonstrated in the high pressure data relates to the intermediate velocity flow regime or regimes. At one atmosphere, a distinct slug flow regime with ∿1 Hz pressure oscillations was evident. At both 5 and 10 atm, with a tendency toward smaller void sizes, (a) the transitions (both low and high velocity) are much more gradual, and (b) while some signs of intermittent slugging do appear, a pure slug flow region is not observed. This central velocity region is really a mixed flow regime. For both particle diameters, the progression is from bubbly to a mixture of bubbly slug, to a mixture of turbulent slug, and finally to turbulent flow.

The presence of tube banks shifts the turbulent regime to lower superficial velocities within the tube bank. Recent data show that the particle flow in a tube bank is predominantly of a quasi-steady state counter-current flow nature with a larger number of small local circulating eddies at lower velocities and is not of a bubble nature.

A flow and heat transfer model based on this flow behavior is suggested by Canada & Staub (March 1977).

Bed Expansion

A generalized model for the correlation of the average bed expansion, taken from its successful application in gas-liquid systems, is described by Canada *et al.* (Chicago, 1976) with further supporting data at high pressure and with tube banks by Canada *et al.* (Oct. 1976) and Canada & Staub (March 1977). This model is based on (1) where C_o was found to vary ± 12% in a continuous manner with particle size and gas density.

$$\frac{U_{sg}}{\bar{\alpha}} = C_o U_{sg} + \bar{V}_{gu} \qquad \ldots(1)$$

where

$$\bar{V}_{gu} = \frac{1-\bar{\alpha}_{mf}}{\bar{\alpha}_{mf}} U_{mf}$$

Fig. 5 shows the correlation lines for 650 μm particles of different static bed heights between 1 atm and 10 atm in the absence of tube banks. In each case the value of the ordinate intercept is that calculated by (1). The decreasing slope (15% between 1 atm and 10 atm) is similar to that observed in gas-liquid systems at increasing gas densities. The break in the slope at U_{sg}/U_T=.55 and 1 atm was correlated with the transition to turbulent flow (Canada *et al.* Chicago 1976). A similar break at increased pressure is seen at about the same reduced velocity. Similar data with 2600 μm particles are also well correlated by (1) (Canada *et al.* Chicago 1976; Canada *et al.* July 1976) and the effect of the tube banks tested is shown to be negligible (Canada *et al.* Oct. 1976) in view of the small ratio of tube-to-bed volume.

Figure 6 shows that, when the surface mean particle diameter is employed (Canada & Staub, Dec. 1976) in a given size mix, the average value of C_o=1.05, between C_o=1.15 for 650μm particles and C_o= .95 for 2600 μm particles, correlates the mixed size data. Characteristic particle data are summarized in Table 1.

In summary, it can be said that, for our data to date, i.e., for particle sizes between 650 μm and 2600 μm and pressures between 1 atm and 10 atm, (1) correlates the average bed expansion within ± 12% using C_o=1.05. Improved accuracy is obtained if the best value of C_o is employed as given here and in Canada *et al.* (July 1976), Canada *et al.* (Oct. 1976) and Canada & Staub (Dec. 1976), and if, for the small particles, the extrapolation line to $U_{sg}=U_T$ is employed above U_{sg}/U_T=0.55.

Table 1
Characteristic Particle Data

Pressure (atm)	Particle Dia. (μm)	U_{mf} (m/s)	U_T	$\bar{\alpha}_{mf}$
1	650	0.31	4.60	0.4
5	650	0.21	2.78	±.02
10	650	0.16	1.97	↓
1	2600	1.34	13.47	
5	2600	0.63	6.02	
10	2600	0.45	4.25	

Heat Transfer

At higher superficial velocities the required bed height can be dictated by the amount of heat transfer surface needed to hold proper combustion bed temperatures. Finned tubes, if sufficiently effective, can then reduce bed height to more tolerable values. Alternatively, if turndown suggests the use of significantly different heat removal capability with tube bank height or if too close a tube pitch is harmful to other bed operating parameters, selective use of finned tubes may again be required for satisfactory performance. The first aim of our current work with finned tube banks is to determine their thermal performance relative to bare banks.

The majority of the tube bank heat transfer data taken to date has been measured in the .305 m x .305 m test section. Preliminary data taken in the 0.61 m x 0.61 m apparatus do not show any change within experimental accuracy, due to apparatus size. The circumferentially averaged heat transfer coefficient for the 3.27 cm OD bare tube bank at 1 atm pressure with 650 μm particles, is shown in Fig. 7 as a function of the reduced velocity. With 2600 μm particles, Canada & Staub (March 1977) showed this coefficient to be about 25% lower. The range of coefficients obtained at different tube locations in the bed is shown by the cross-hatched areas in Fig. 7. The ten row .98 fin-per-cm finned tube bank coefficient, based on the total heat transfer area and after correction for the fin efficiency, is shown to be about 46% lower in Fig. 7 than the bare tube bank coefficient at the same minimum cross-sectional flow area between tubes. On the other hand, the 2.8 fins-per-cm tubes yielded coefficients on the same basis equal to those with the bare tubes (see Fig. 7). It was noted that not only fin pitch but also the relative location of adjacent tubes had an influence on particle circulation between fins. This suggests that one should not extrapolate finned tube data beyond the specific geometries tested in fluidized beds.

The effect of air density on finned tube performance for the .98 fin-per-cm tube bank is shown in Fig. 8 for both the large and small particles. Unlike the bare tube bank, these finned tubes at 1 atm pressure yielded the same heat transfer coefficients for both particle sizes. The heat transfer coefficients in Fig. 8, again corrected for fin effectiveness, increase at a power of 0.3 of the gas density with the 2600 μm particles and a power of about 0.2 of the gas density with the 650 μm particles. Our recent data with the bare tube bank also show a gas density effect to a power of 0.2. These data therefore point out that the variation of heat transfer performance with gas density is affected by both particle size and the specific tube geometry if the tube is finned.

Using the 1 atm data in Fig. 7, a comparison of the bare and finned tube thermal performance per unit length of tube is shown in Table 2.

Table 2
Thermal Performance per Unit Tube Length (Fig. 7 Conditions)

	Watts/°C per cm
Bare Tube	.19
.98 Fin-per-cm Tube	.45
2.8 Fin-per-cm Tube	1.32

Since the above results include the fin effectiveness it is clear that, under the conditions tested above, the use of conventional finned tubes can result in a several fold increase in thermal performance. Rough calculations indicate that even a 2.1 fin per cm tube, with a fin structure that will not exceed a maximum fin tip temperature of 535°C in a 900°C combustion bed, can yield a three fold increase in thermal duty per unit length compared to that of a bare tube bank whose outer diameter is equal to the fin outer diameter of the finned bank. Such finned tubes would, of course, require hot evaluation including the question of tube fin erosion or corrosion.

CONCLUSIONS

1. In the particle size range of increasing interest to FB combustion the turbulent flow regime appears to dominate with a predominantly counter-current streaming particle flow within tube banks.
2. The turbulent regime is characterized by lower bed pressure drop oscillations, lower particle loss, improved particle mixing and less gas by-pass. These properties suggest its increased application in heat and mass transfer operations.
3. A simple generalized correlation of the average bed expansion has been found valid over a large range of particle sizes, gas densities and static bed heights.
4. The heat transfer coefficient to tube banks is proportional to the 0.2 to 0.3 power of gas density depending on the particle size and tube geometry over the conditions tested.
5. The large performance increase per unit tube length of finned vs. bare tube banks lends promise to the use of finned tubes at high energy densities and to obtain improved turn-down performance.

ACKNOWLEDGEMENT

The authors express their appreciation to the General Electric Corporate Research and Development Center and to the Electric Power Research Institute for their support of this effort and to Michael H. McLaughlin for his many valuable contributions.

REFERENCES

Canada, G.S., McLaughlin, M.H. & Staub, F.W. (Chicago, 1976) 'Flow Regimes and Void Fraction Distribution in Gas Fluidization of Large Particles in Beds Without Tube Banks', paper 103d, 69th Annual AIChE Mtg., to be published in AIChE Symp. Series.

Brzozowski, S.J., McLaughlin, M.H., & Staub, F.W. (March, 1976) Two Phase Flow and Heat Transfer in Fluidized Beds, 3rd Quarterly Report, SRD-76-041, EPRI Contract RP 525-1.

Canada, G.S., McLaughlin, M.H. & Staub, F.W. (July, 1976) Two Phase Flow and Heat Transfer in Fluidized Beds, 4th Quarterly Report SRD-76-073, EPRI Contract RP 525-1.

Canada, G.S., McLaughlin, M.H. & Staub, F.W. (October 1976) Two Phase Flow and Heat Transfer in Fluidized Beds, 5th Quarterly Report, SRD-76-114, EPRI Contract RP525-1.

Canada, G.S. & Staub, F.W. (December, 1976) Two Phase Flow and Heat Transfer in Fluidized Beds, 6th Quarterly Report, SRD-77-011. EPRI Contract RP 525-1.

Canada, G.S. & Staub, F.W. (March, 1977) Two Phase Flow and Heat Transfer in Fluidized Beds, 7th Quarterly Report, SRD-77-068, EPRI Contract RP 525-1.

Davidson, J.F., Kehoe, P.W.K., (1971) Inst. Chem. Eng. Symp. Series, 33, 97 (London)

Yerushalmi, J. et al, (Nov. 29, 1976), paper No. 119a, 69th Annual Meeting AIChE, Chicago, Ill.

Lanneau, K.P., (1960) Trans. Inst. Chem. Eng. 38, 125.

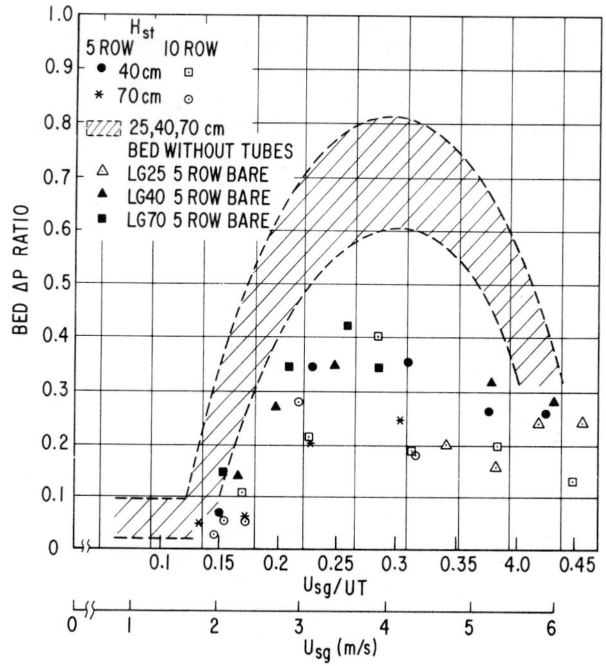

Fig. 1. Ratio of oscillating component (peak-to-peak) of bed pressure drop to the average value of bed pressure drop for beds of large glass particles with and without tube banks.

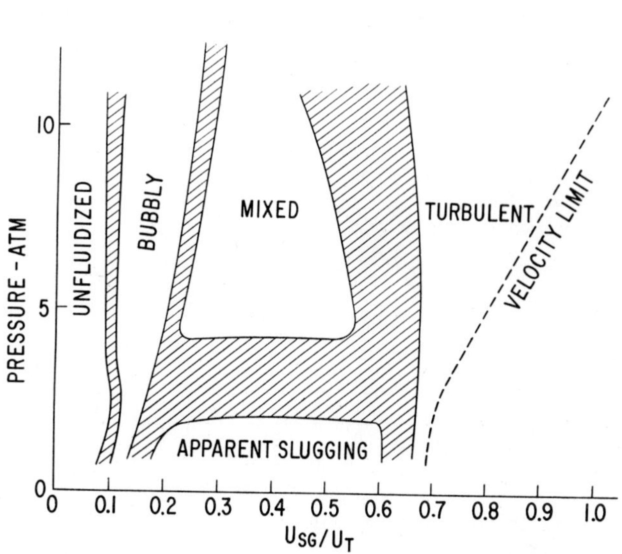

Fig. 3. Flow regime map for 650 μm particles, H_{st} from 25 cm to 70 cm (vel. limited by apparatus).

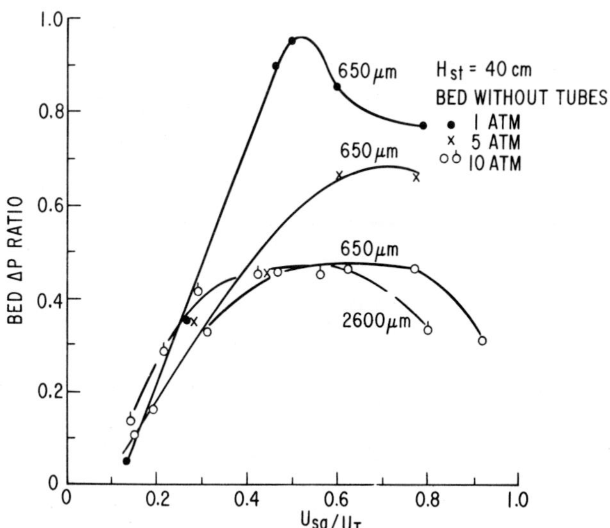

Fig. 2. Effect of pressure and particle size on bed pressure drop ratio (see Fig. 1).

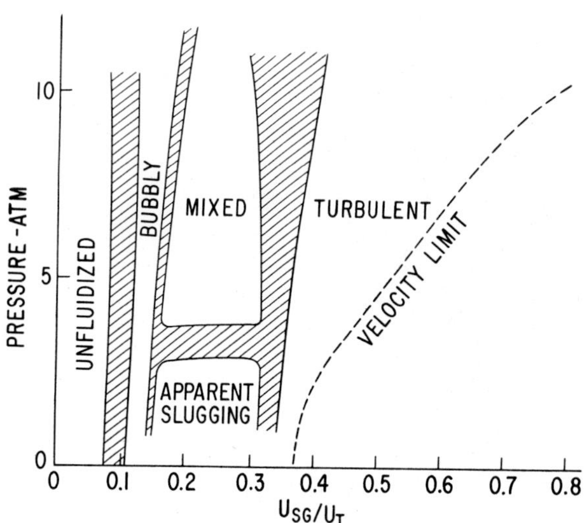

Fig. 4. Flow regime map for 2600 μm particles, H_{st} from 25 cm to 70 cm (vel. limited by apparatus).

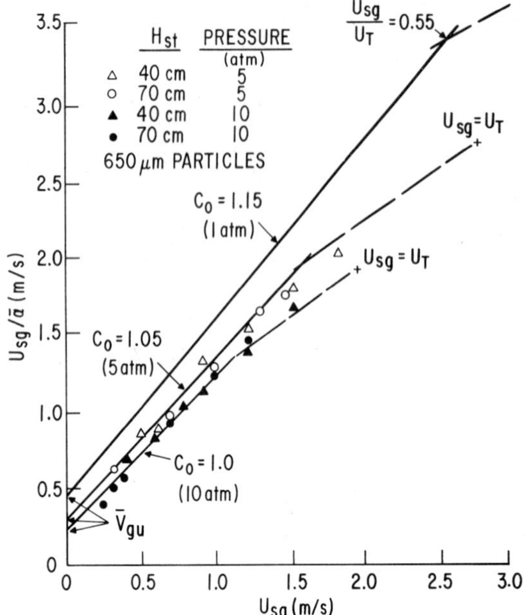

Fig. 5. Correlation of average bed void fraction for 650 μm particles at elevated pressure. Ordinate intercept calculated (Equ. [1]). See Ref. 1 for 1 atm data.

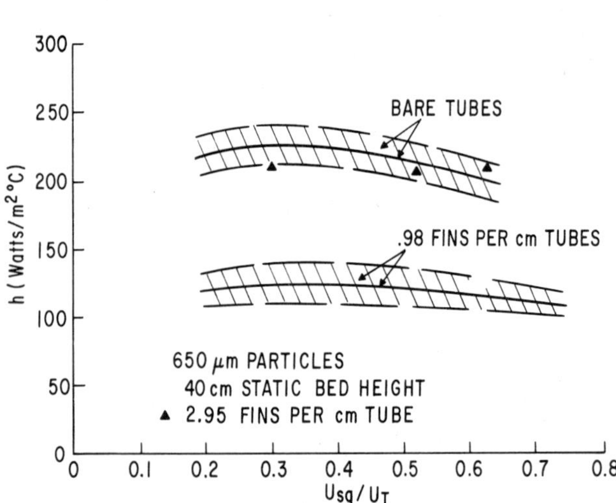

Fig. 7. Bed-to-tube heat transfer coefficient at 1 atm for 10 row bank of bare and finned tubes. Hashed area shows variation within tube bank (see exp. equip. and Fig. 8).

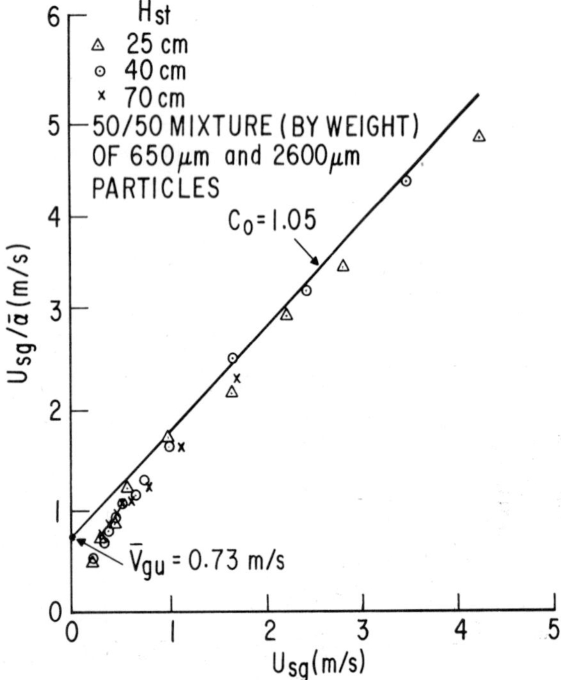

Fig. 6. Correlation of bed void fraction for mixed particle sizes at 1 atm using Equation [1].

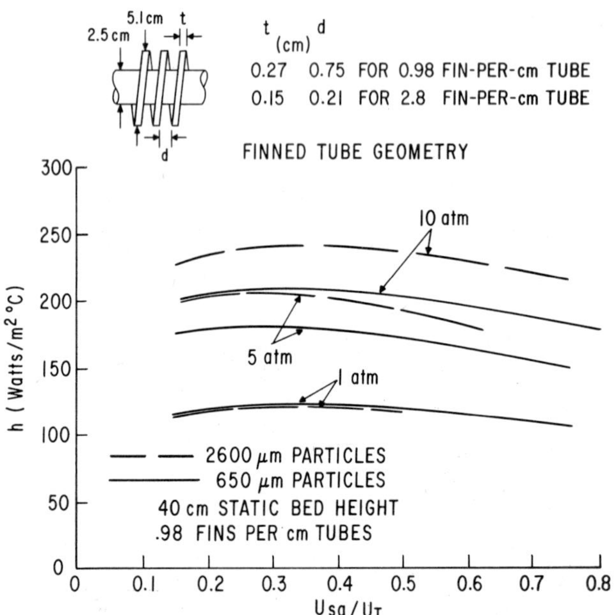

Fig. 8. Effect of pressure and particle size on bed-to-tube heat transfer coefficient for 10 row finned tube bank.

HEAT TRANSFER FROM FLATTENED HORIZONTAL TUBES

By B.R. Andeen, L.R. Glicksman and R. Bowman

Department of Mechanical Engineering
Massachusetts Institute of Technology
Cambridge, Massachusetts 02139

SYNOPSIS

Horizontal tubes in fluidized beds have been observed to have stagnant particles residing on the top surface and a region of low voidage directly below the tube. Heat transfer from the tube to the bed will be inhibited in these two regions. The overall heat transfer to the tubes should be increased if the effects of the two regions are minimized. To evaluate this postulate, film coefficients were experimentally measured in a shallow bed using flattened tubes. Flattened tubes have a smaller percent of their surface area at the top and bottom, and should, therefore, experience higher tube average coefficients. Heat transfer tests were conducted with flattened tubes using sands which had mean diameters ranging from 710 μm to 360 μm. When compared with results for round tubes of the same circumference, the flattened tubes were found to have average heat transfer coefficients one-third larger. The data was correlated by an expression similar in form to that proposed for round tubes.

NOTATION

c	specific heat of particle
d	diameter
F	force
G	mass velocity
g	gravitational constant
h	heat transfer coefficient
k	thermal conductivity
Pr	Prandl number
q	heat transfer rate
T	temperature
v	velocity
V_o	volume
ε	void fraction
μ	absolute viscosity
ρ	density
τ	thermal time constant of particle

SUBSCRIPTS

g	gas
p	particle
t	tube

INTRODUCTION

Many applications of fluidized beds require heat transfer surfaces within the bed. The use of horizontal tubes for the heat transfer surface has several advantages: they are easier to install, support and replace. Further, several investigations indicate that horizontal tubes tend to enhance bubble break-up and reduce the tendency for large bubbles or slugs to be formed in the bed.

However, the average film heat transfer coefficient for horizontal tubes are less than those for verticle tubes. This is due, in part, to the circumferential variation of heat transfer coefficients around the tubes. Measured coefficients are highest on the sides, lower on the bottom and lowest on the tops of the tubes (1,2). On the bottom of the tube the heat transfer is reduced because a region of low particle concentration forms under the tube. On top of the tube slow moving particles settle out of the wake and accumulate to form a barrier to heat transfer. Bowman (3) measured the size of this stagnant region and found that roughly 90° of the top surface of round tubes was covered by this stagnant cap region. For single tubes, Gel'perin (4) has shown that the relative magnitude of these circumferential coefficients change with fluidizing velocity. Along the sides of the tube, where particle action is most vigorous the heat transfer coefficient remains about the same, while local coefficients at the top and bottom of the tube increase with velocity. Film coefficients on the sides of the tubes can deteriorate with close tube spacing. Lese and Keremode (5) measured such deterioration and observed a region of high void between closely aligned horizontal tubes. This void is probably due to high local gas velocities. This effect disappeared with increased tube spacing.

Hence, the decreased thermal performance of a horizontal tube bank is caused by two effects: (1) loss of particle activity on top of the tube and less particles below the tube, and (2) increased voidages between tubes caused by blockage of the flow area by tubes.

Both these effects can be reduced by flattening the tube and alligning the major axis vertically. Flattening decreases the fraction of the tube area in the stagnant or high voidage region. For a given ratio of the tube surface area to bed frontal area, flattening also increases the spacing between tubes, thereby decreasing the gas velocities between tubes.

This paper presents experimental heat transfer results for flattened tubes, and a comparison is made to round tubes for the same conditions.

TEST APPARATUS

Heat transfer measurements were made in a 0.61m square bed. Upstream of the test section was a flow straightening section, consisting of cloth restrictions, packed soda straws, and screens. Air velocities across the flow straightening section were uniform to + 6 to 8% when measured at velocities between 1.5 to 4m/sec. The distributor plate, located directly above the flow straightening section consisted of a 60 mesh brass wire cloth backed with one to two layers of cotton cloth. The cloth ensured an adequate pressure drop across the distributor to maintain bed stability. The pressure drop of the distributor ranged between approximately thirty to seventy percent of the total pressure drop across the bed for the tests performed.

An instrumented, heated tube was used to measure heat transfer coefficients. Identical unheated tubes were used to form the remainder of the tube bank. The flattened tubes were 3.18 cm by 1.27 cm in cross section, with a circumference equal to a 2.54 cm round tube, see Figure 1. The tubes were arranged in a single row on 5.08 cm centers. The bottom of the tubes were 1.1 cm above the distributor plate.

The instrumented tube was made of solid aluminum to maintain a constant surface temperature. The center of the tube was drilled to receive a nichrome heating element. Two 7.6 cm long wooden spacers of the same cross-section as the tube were used to insulate the heated tube from the walls of the test section. Thus, measurement and interpretation of the average heat transfer from the tube was not obscured by flow variations over the ends of the tubes. Twelve thermocouples were placed in groves on the tube surface to detect any circumferential or axial temperature differences. The thermocouple grooves were filled with epoxy to preserve a smooth outside surface.

The air temperature was measured with a single thermocouple placed at the bottom of the adjacent tube in order to isolate it from the heater. Air velocities were measured with a hot wire placed upstream of the test section and calibrated to measure the superficial air velocity entering the test section.

The fluidized particles consisted of Ottawa Standard sand, ASTM designation C-190 (20-30 mesh); and Graded Ottawa sand, ASTM designation C-190, which was graded to cuts of 30-40 mesh and 40-50 mesh. The different gradings sort particles into three groups, 840-590 μm, 590-420 μm, and 420-300 μm. The mean diameter for each group was taken to be 710 μm, 510 μm, 360 μm respectively.

RESULTS

Figures 2, 3 and 4 show the average heat transfer coefficient measured as a function of superficial velocity. Data taken in the same bed with the same sand and 2.5 cm round tubes (6) is also shown in Figures 2 and 4. The total amount of particles in the bed was varied; data points for different particle loads are identified by different symbols. The terminal velocity for the mean particle size is also shown; it is based on the minimum cross-sectional area for air flow.

For the particle sizes tested, the flattened tubes have a thirty-three percent improvement in average heat transfer coefficient over round tubes with the same circumference. When the total particle load in the bed is small, heat transfer from the tube is poor until the velocity is high enough to cause the expanded bed to cover the tubes. The small particle results shown on Figure 2 illustrate the significant difference between the heat transfer characteristics for round and flattened tubes. The round tube exhibits a very narrow peak in the heat transfer coefficient while the results for the flattened tube falls from a peak to a relatively constant value. The modest decrease in the film coefficient for the flattened tube suggests that the conditions along the sides of the tubes change very little. Rather, a high voidage region is created along the leading edge of the tube. The round tube exhibits a more marked decrease in coefficient since a larger fraction of the surface area is covered by the high voidage region. Note that the peak heat transfer for the round tube occurs at about the same superficial velocity as the flattened tube.

The region of high voidage cannot be solely explained by the trapping of bubbles rising from below it. If it was, the results for larger particles, Figures 3 and 4, would also show the characteristic maximum. The high voidage region can also be explained by the following: since the leading edge is a stagnation point, the zone adjacent to it will have low velocity gas which cannot sustain particle fluidization. As the gas flows around the tube it is accelerated. If the local gas velocity around the side of the tube approaches or exceeds the terminal velocity of the particles, it will inhibit particle motion downward which would tend to fill in the void beneath the tube. With large particles, the effect of the high voidage region may be less severe since the void will be refilled by large particles of high momentum which tend to enter the region from below the tubes rather than follow the gas streamlines. In addition, for large particles the excess velocity, $u-u_{mf}$, becomes much larger before the terminal velocity is approached, implying the existence of a large number of bubbles. These bubbles can also displace particle into the high voidage region as they pass. Confirmation of these hypothesis will require more detailed local measurements.

The overall heat transfer results for the 710 μm particles cannot be attributed to the balancing of packet replacement frequency with average bed voidage along the sides of the tubes. The thermal time constant of the particles along the tube wall can be estimated as,

$$\tau = \frac{(\rho c V_o)\ \text{particle}}{(q/\Delta T)\ \text{Tube to Particle}} \qquad (1)$$

If the particle is close to the wall, the major resistance to heat transfer is conduction through the intervening gas layer. For particles close to the wall q can be estimated as,

$$q/\Delta T \simeq \frac{K_g}{d_{p/6}} \frac{\pi d_p^2}{4} \qquad (2)$$

Combining equations 1 and 2,

$$\tau \simeq \frac{\rho c\ d_p^2}{9\ K_g} \qquad (3)$$

and for the 710 μm sand, τ is approximately three and one-half seconds. For the 510μ and 360μ particles the time constants are 1.8 sec and 0.9 sec, respectively. In bubbling flow the observed point bubble frequencies are about one Hertz or above (7). This agrees with observed fluctuations around the walls (8). Thus, during the time the large particles are at the tube wall they are essentially isothermal and variations in the replacement frequency around one Hertz will not effect the average particle-to-wall heat transfer rate. Whereas, for smaller particles the replacement frequency does have a significant influence on the heat transfer.

Since heat transfer with large particles should be independent of replacement frequency, the major variable will be the voidage around the tube circumference. As the superficial velocity increases, the average voidage around the sides of the tubes should increase, decreasing the heat transfer. This may be counteracted by a lowering of the voidage in the leading edge region and more rapid replacement of the stagnant particles on top of the tube. For the intermediate particle size, increases of the replacement frequency with superficial velocity appears to balance the effect of the voidage increase so that the heat transfer coefficient remains constant. With the smallest particle size, voidage effects predominate and the coefficient falls with superficial velocity.

Figure 5 shows the results plotted using the modified form of the Vreedenberg relation (9) for large particles as presented by Andeen and Glicksman (6). For the flattened tube, d_t is defined as the tube circumference divided by π, or the diameter of a round tube with the same surface area. The seven data points below the correlation line correspond to low particle loadings where the tube was not entirely covered by the bed. A least square fit of the data, excluding the low points yields

$$\frac{hd_t}{K_g} = 1240\ (1-\varepsilon) \left(\frac{Gd_t\rho_p}{\rho_g\mu} \frac{\mu^2}{dp^3\rho_p^2 g} \right)^{.374} \qquad (4)$$

If the group on the left-hand side were written as

$$\frac{h\ d_p}{K_g(1-\varepsilon)}$$

the correlating expression would be almost independent of particle diameter. In the correlation, the void was calculated from Leva's relationship (10). Rearranging the dimensionless groups in equation 4 results in the following form of the correlation

$$\frac{hd_p}{K_g(1-\varepsilon)} = 1240 \left(\frac{Vd_p\mu}{\rho_p g\ d_p^3} \right)^{0.37} \left(\frac{dp}{dt} \right)^{.63} Pr^{0.3} \qquad (5)$$

The first dimensionless group on the right-hand side is the ratio of drag to gravity forces when the particle is within the Stokes flow regime. This form of the correlation is physically more satisfying when it is remembered that the variation of heat transfer is primarily due to the particle behavior at the bottom and at the top of the tube.

CONCLUSIONS

Flattened tubes increase the average heat transfer coefficient over that for a round tube; in the case at hand the increase was thirty-three percent. The increase is due to two mechanisms: a decrease of the upstream and downstream area of the tubes which have low heat transfer rates, and a decreased flow blockage yielding lower voidage between the tubes.

Heat transfer results with small particles exhibits a pronounced maximum followed by a rapid decrease, the latter may coincide with the creation of a low voidage region beneath the tube. When the bed contains large particles, 700 μm or above, variations of the particle replacement at the tube wall will not have a major influence on the heat transfer to the tube wall as long as the frequency is at least one Hertz.

ACKNOWLEDGEMENTS

This work was sponsored, in part, by the Empire State Electric Energy Research Corporation.

REFERENCES.

1. Gel'perin, N.I., Ainshtein, V.G., & Aronvich, F.D., "The Effect of Screening on Heat Transfer in a Fluidized Bed", International Chemical Engineering, Vol. 3, No. 2, pp. 185-190, April 1963.

2. Novack, R., "Lokaler Warmeubergang an horizontalen Rohren in Wirbelshichten", Chemie-Ingenieru Technik, Vol. 42, pp. 371, 1970.

3. Bowman, R., Masters Thesis, Massachusetts Institute of Technology, Department of Mechanical Engineering, 1974.

4. Gel'perin, N.I., Einstein, V.G., Korotjanskaja, L.A. & Perevozchkova, J.P., Teor, Osnovy, Khim, Tekhnol. Vol. 2, p. 430.

5. Lese, H.K. & Kermode, R.I., "Heat Transfer from a Horizontal Tube to a Fluidized Bed in the Presence of Unheated Tubes", The Canadian Journal of Chemical Engineering, Vol. 50, pp. 44-48, 1972.

6. Andeen, B.R. & Glicksman, L.R., "Heat Transfer to Horizontal Tubes in Shallow Fluidized Beds", ASME Paper 76-HT-67.

7. Kunii, D. & Levenspiel, O. Fluidization Engineering, John Wiley, New York, 1969.

8. Chen, T., Heat Transfer to Tubes in Fluidized Beds, ASME Paper No. 76-HT-75, Presented at the 16th National Heat Transfer Conference, 1976.

9. Vreedenberg. H.A., "Heat Transfer Between a Fluidized Bed and a Horizontal Tube", Chemical Engineering Science, Vol. 9, pp. 52-60, 1958.

10. Leva, M., "Correlations of the Dense Phase Fluidized State and Their Applications", The Canadian Journal of Chemical Engineering, Vol. 35, pp. 71-76, August 1957.

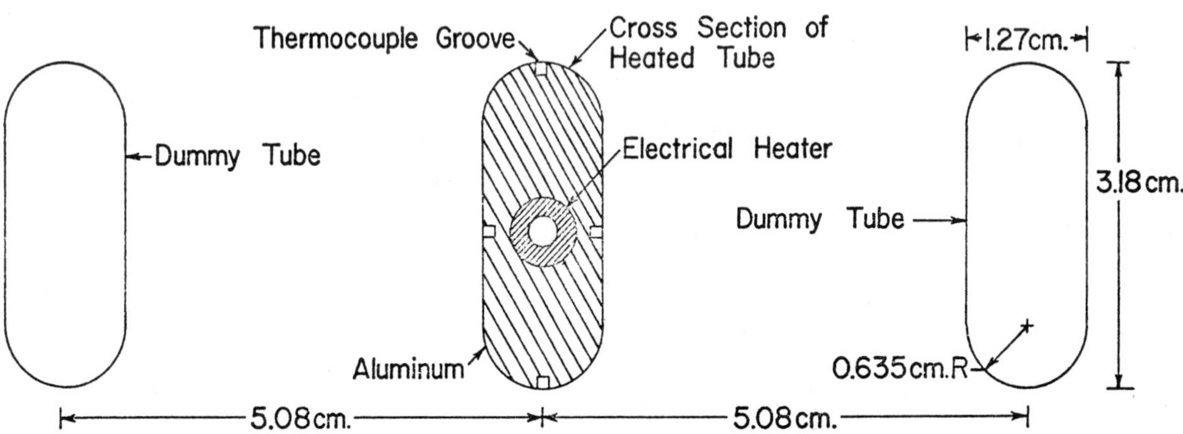

Figure 1 A Portion of the Flattened Tube Array

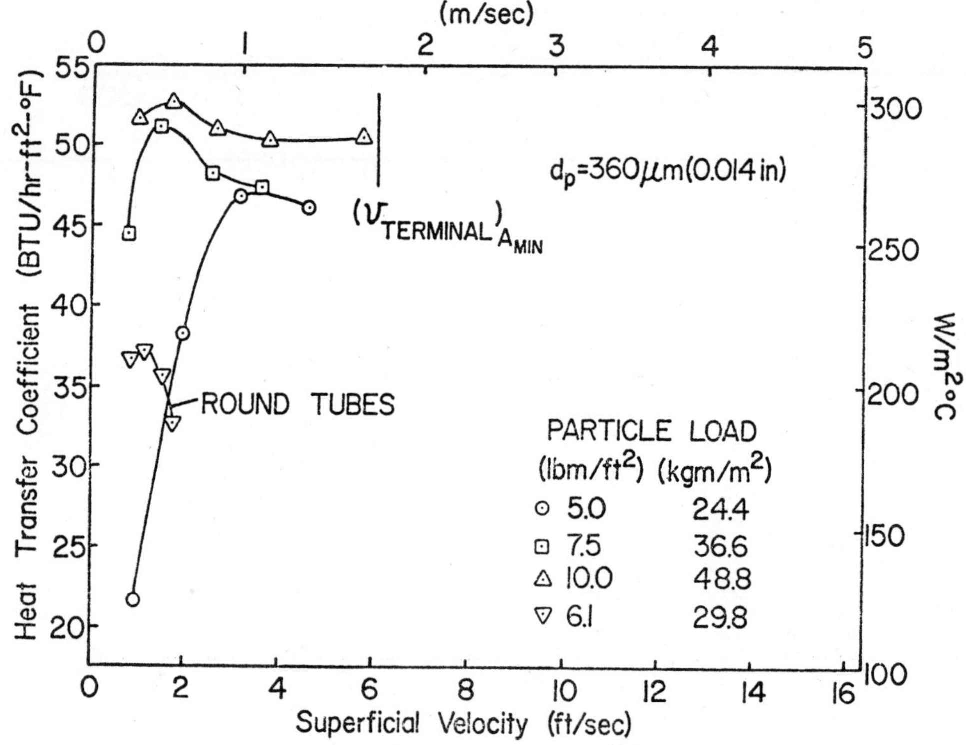

HEAT TRANSFER COEFFICIENT VS. SUPERFICIAL VELOCITY

Figure 2

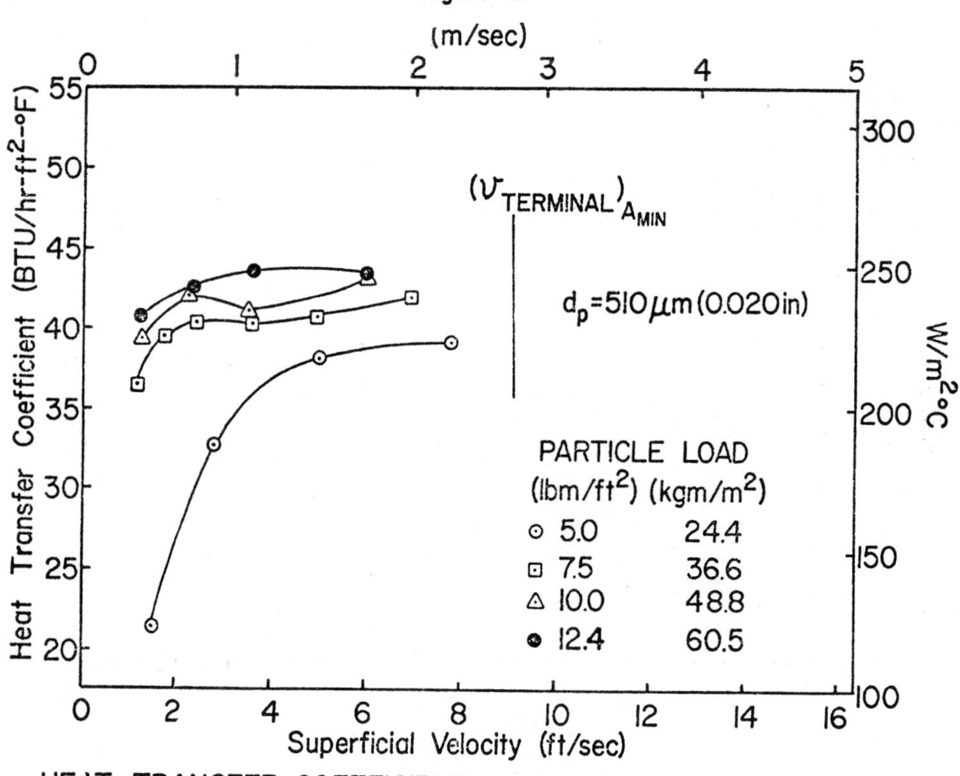

HEAT TRANSFER COEFFICIENT VS. SUPERFICIAL VELOCITY

Figure 3

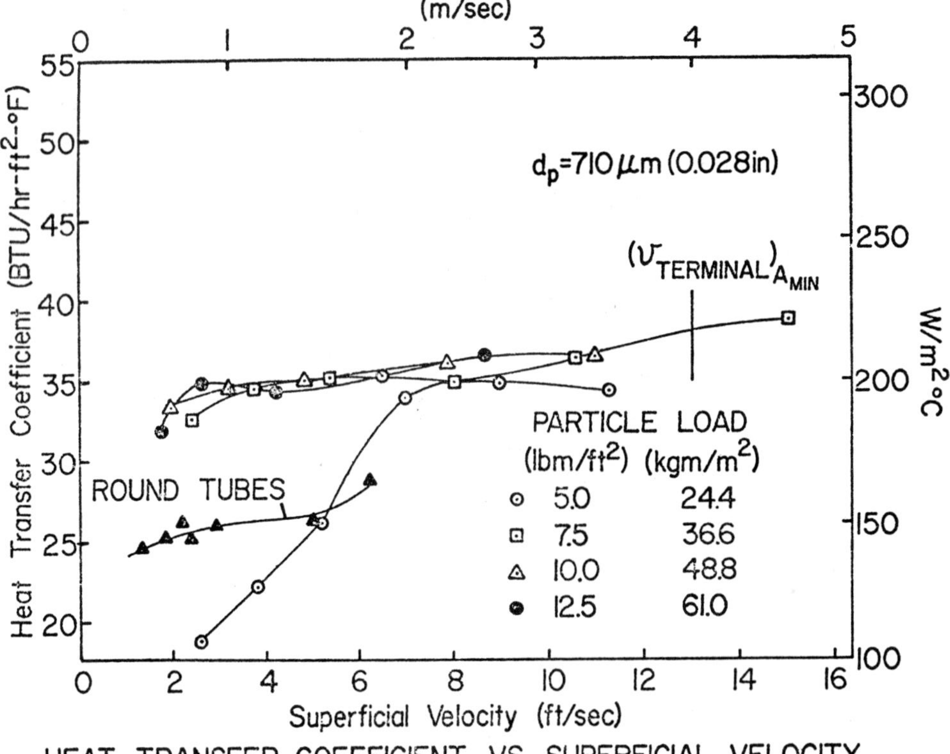

HEAT TRANSFER COEFFICIENT VS. SUPERFICIAL VELOCITY

Figure 4

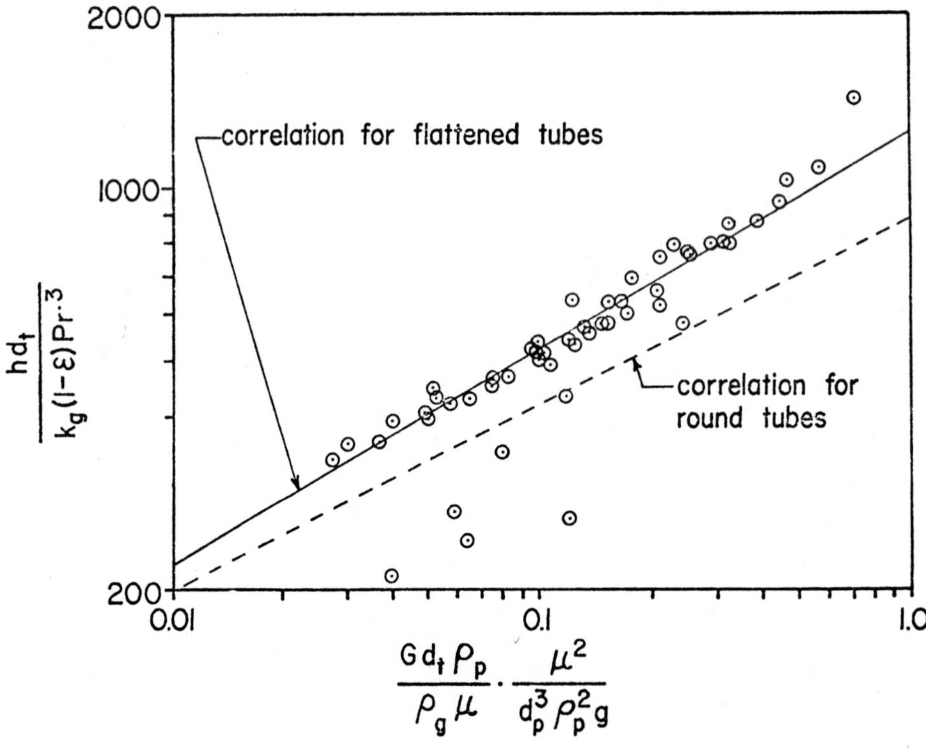

Figure 5 Correlation of Flattened Tube Data

HEAT TRANSFER BETWEEN A FLUIDISED BED AND A SMALL IMMERSED SURFACE

by A. R. Khan, J. F. Richardson and K. J. Shakiri

Department of Chemical Engineering
University College of Swansea
Singleton Park
Swansea SA2 8PP

SYNOPSIS

Measurements have been made of the heat transfer between a small electrically heated surface and a fluidised bed. Both liquid-solid and gas-solid systems have been studied using particles of several sizes and densities. Liquid viscosity has been varied 600 fold, and for the gas-solid systems molecular weight has been varied 60 fold and pressures have ranged from 0.03 to 1.10 MN m^{-2} .

INTRODUCTION

Although there have been many studies of the rate of heat transfer between a boundary or immersed surface and a fluidised bed,there are no reliable methods of predicting heat transfer coefficients, particularly for gas-solid systems, even though individual workers have been reasonably successful in correlating their own results. The basic mechanism which give rise to good heat transfer in fluidised beds are first the convective heat transfer caused by the particles and secondly the reduction in the effective thickness of the thermal boundary layer arising from the hydrodynamic effects of particle movement near the surface. The first mechanism is of particular importance in gas-solid systems where the presence of the particles in responsible for a very large increase in the heat capacity per unit volume of the system. The second mechanism is equally important in gas-solid and in liquid-solid systems.

Except for particles of very high density, liquid-solid systems give reasonably uniform fluidisation provided that a good distributer is incorporated into the base of the bed, though circulation patterns tend to become established. In gas-solid systems, however, the flow pattern is very much more complex, especially because most systems are capable of only a limited expansion before bubbling occurs. Mickley and Fairbanks (1955) have demonstrated just how important is the behaviour of the bubbles and their results have shown that there are high frequency variations in heat transfer coefficient, depending on whether the relevant part of the heat transfer surface is covered by gas bubbles or by fluidised solids and how long the solids have remained in that location.

The present work is a development of that already reported (Latif and Richardson,1972, Romani and Richardson, 1974, Richardson et al, 1976), in which a small electrically heated element is immersed in the bed. The previous work has been concerned with a single liquid and a variety of solid particles. Now several liquids have been used in order to give a range of values of Prandtl number. In addition, measurements have been made with gas-solid systems in order to obtain not only time-averaged but also instantaneous values of the transfer coefficient.

EQUIPMENT

Two separate experimental equipments have been used - one for liquid-solid and the other for gas-solid systems.

For liquid-solid fluidised beds, the equipment already described has been used with suitable modifications to take account of the flammability of the liquids. It consists of a fluidisation column 102 mm in diameter and 2 m tall incorporated into a recirculation loop, with facilities for flow and pressure measurement and for adjustment of the temperature of the liquid.

The equipment for gas-solid fluidisation has also been described in detail previously. It consists of a perspex fluidisation column 102 mm diameter and 1 m tall fitted with a porous sintered stainless steel distributor plate. It is equipped with flow meters and control valves and can be used either for a single passage of gas with the effluent gas discharged to atmosphere or as a closed circulation loop. In the latter case the gas is circulated using a small rotary compressor and the absolute pressure in the system can be adjusted by connection to an external pressure reservoir (Godard 1968).

In both the liquid-solid and the gas-solid system the heat transfer element consisted of a 25 mm square tufnol sheet wound with 48 turns of tungsten wire of 0.129 mm diameter the total length of wire being 2.9 m. The element was connected by means of thick copper leads to a Wheatstone bridge network in which the element forms one arm. The temperature of the wire T_E is determined from its resistance R_E. The rate of heat generation in the wire Q is a function of the voltage V applied across the bridge and is given by

$$Q = \frac{V^2}{(\Sigma R)^2} R_E \qquad (1)$$

where ΣR is the total resistance of the element and the arm with which it is in series. $R_E/(\Sigma R)$ remained constant to $\pm$ 1%. At equilibrium, the heat generated must be equal to the heat given up to the fluidised bed, expressed as the product of the heat transfer coefficient h, the external area of the element A_E and the difference in temperature between the wire and the fluid $T_E - T_B$.

Thus at equilibrium

$$Q = h A_E (T_E - T_B) = \frac{V^2}{(\Sigma R)^2} R_E \qquad (2)$$

$$\text{or } T_E = T_B + \frac{1}{hA_E} \frac{R_E}{(\Sigma R)^2} V^2 \qquad (3)$$

Thus, by plotting T_E against V^2 the heat transfer coefficient h is obtained from the slope and the bed temperature T_B from the intercept on the temperature axis.

MATERIALS

For the gas-solid fluidisation, air, hydrogen, carbon dioxide and difluoro-dichloromethane were used at pressures ranging from 3×10^4 to 1.1×10^6 Nm^{-2} . The properties of the fluids at the conditions employed are given in Table 1, and of the solids in Table 2.

For the liquid system a series of mixtures of kerosene and lubricating oil was made up to give a range of viscosities and hence of Prandtl numbers - density and specific heat capacity remaining substantially constant. The relevant properties of the liquids are given in Table 3, and of the solid particles in Table 4.

EXPERIMENTAL RESULTS

GAS-SOLID SYSTEMS

All experimental results are shown as heat transfer coefficient plotted against u/u_{mf}, the ratio of the superficial fluidising velocity to the minimum fluidising velocity. The curves are characterised by four factors:

(a) An approximately constant coefficient for the fluidised bed ($u/u_{mf} < 1$).

(b) A slow followed by a rapid rise to a sharp maximum for u/u_{mf} between about 1 and 4.

(c) A subsequent fall to a minimum.

(d) A final gradual increase of heat transfer coefficient with u/u_{mf}.

The results obtained for glass 20 fluidised by various gases at atmospheric pressure are given in Fig.1. Fig. 2 shows the effect of pressure for the fluidisation of glass 20 with air. The effect of solids properties is illustrated in Fig. 3 which refers to the fluidisation of a range of solid particles with air at atmospheric pressure.

The substantially constant values for heat transfer coefficient in the fixed bed region, with gas velocity exerting little influence, suggests that at these low velocities natural convection is playing a predominant role. As fluidisation begins the heat transfer improves as a result of particle movement which increases as small bubbles form within the bed. With increasing gas flow, the average size of the bubbles increases, the mixing becomes less effective, and the heat transfer surface becomes covered by bubble gas for a significant period. As a result, the heat transfer coefficient having passed through a maximum falls off quite rapidly. At high gas rates, where there is violent movement in the bed, the coefficient generally tends to rise again as velocity is increased.

By reference to Fig. 1, it is seen that the heat transfer coefficient at atmospheric pressure for both the fixed and fluidised bed regions for glass 20 is notably greater for hydrogen than for the other gases, and this appears to be related to its very much higher thermal conductivity. The curves do show a general tendency for heat transfer coefficient to be strongly dependent on thermal conductivity.

The effect of gas pressure is seen for the same solids fluidised by air in Fig.2. As pressure is increased, higher gas velocities can be used before large bubbles start to form and it is observed that the maximum coefficient has a higher value and occurs at a higher value of u/u_{mf} for elevated pressures, compared with atmospheric pressure. Furthermore, the minimum on the curves at higher values of u/u_{mf} are not as clearly defined as at lower pressures. The experimental results at sub-atmospheric pressure are markedly lower than those at higher pressures and a substantially lower value of maximum coefficient is obtained. Beyond the maximum, however, the results are very close to those for atmospheric pressure. There is very little variation in thermal conductivity with pressure, and differences in these experiments are therefore attributable to the general improvement in uniformity of fluidisation as pressure is increased. The similarity of the heat transfer characteristics at the three higher pressures is in line with observed similarities in fluidisation behaviour.

Glass particles of three different sizes were fluidised with air at atmospheric pressure and, from Fig. 3, are seen to show similar behaviour. The peaks and troughs become less sharply defined as particle size decreases (glass 15 → 18 - 20). Results for diakon which has a much lower particle density show a much smaller variation of coefficient with u/u_{mf}. Up to the maximum, occurring at u/u_{mf} 1.7, the bed is evenly expanding, but then bubbles start to form, the bed contracts, and the coefficient falls.

The nature of the variation of heat transfer coefficient with operating variables gives a clear indication of the reason why it has been so difficult to compare results obtained under different conditions and to give a reliable method of calculation of coefficient from a given system. However, the maximum values of coefficient, h_{max}, expressed as particle Nusselt Numbers, Nu_{max}, show a regular variation with Galileo number

$$\left\{ Ga = \frac{g\, d_p^3 \rho(\rho_s - \rho)}{\mu^2} \right\}$$

as seen in Fig.4, from which:

$$Nu_{max} = 0.157\ Ga^{0.475} \qquad (4)$$

The results given above are essentially time-average values, but because of the low heat capacity of the element it has been possible to measure fluctuations of element temperature, and hence of the heat transfer coefficient, by connecting a sensitive recorder in place of the null detector of the bridge. A typical series of traces for glass 15 fluidised with air at atmospheric pressure is given in Fig.5, in which the peaks are associated with packets of solids and the troughs with bubbles in contact with the element. These results are found to be useful in relating the variation of average heat transfer coefficient with bed behaviour since the behaviour tends to be characterised by a parameter P where

$$P = \text{(average frequency of peaks and troughs in } s^{-1}) \times \text{(mean amplitude in mm)} \qquad (5)$$

Then if h is the average heat transfer coefficient at a given value of u/u_{mf} and h_{mf} is the value at u_{mf}:

$$h/h_{mf} = 0.58\ P^{\frac{1}{2}} \qquad (6)$$

LIQUID-SOLID SYSTEMS

Experimental results for both bed expansion characteristics and for heat transfer coefficients were obtained for each liquid (viscosity range 15-940 $mNsm^{-2}$) and solid particle at 303K for a voidage range of from 0.40 to 0.91. In Fig. 6 are given results for heat transfer coefficient as a function of voidage (e) for each solid in a particular liquid (viscosity 3.19 $mNsm^{-2}$) and in Fig. 7 the results for 9 mm glass spheres fluidised by each of ten liquids. The curves are all characterised by a maximum, which is not in most cases sharply defined, occurring at a voidage between 0.5 and 0.75. Because of the wide range of physical properties of the liquids, the results span the region in which natural convection dominates, the laminar flow region and the turbulent flow region. In this paper attention is focussed, except where stated otherwise, on the region of forced convection and turbulent flow.

For liquid flowing alone, the heat transfer coefficient expressed in the form of Nusselt number (Nu_x, in which distance x is the length of the surface in the direction of flow) is represented in terms of the corresponding Reynolds number (Re_x) and Prandtl number (Pr) by:

$$Nu_x = 0.38\ Re_x^{0.6}\ Pr^{0.2} \qquad (7)$$

For the liquid-solid fluidised systems, all results for the turbulent region are represented to within ±10% by the following relation between Stanton number (St) Prandtl number (Pr) Galileo number (Ga) and bed voidage e:

$$St.Pr^{0.62} Ga^{0.22} = 11.2 \times 10^{-1.85e} \qquad (8)$$

This relation is similar in form to that developed previously (Richardson et al 1976) for fluidisation with dimethyl phthalate, but results are not directly comparable because the flow regime is different.

The curves of heat transfer coefficient against bed voidage show maxima because two opposing effects come into play as the voidage (or velocity is increased). First the number of particles per unit volume decreases and hence the potential for a beneficial contribution from the particles becomes less; and secondly the high liquid velocities themselves give rise to thinner boundary layers. For this reason, it is convenient to use an expression which takes both of these effects into account, using the particle Reynolds number (Re_p) which increases with liquid velocity and the term (1-e) which decreases with liquid velocity to express the particle Nusselt number (Nu_p). Thus:

$$Nu_p = 11.8\, Re_p^{0.48} (1-e)^{0.56} Pr^{0.14} (d_p/d_c)^{0.38} \tag{9}$$

The factor d_p/d_c is the ratio of particle diameter to column diameter.

By using the relation between fluidising velocity and bed voidage obtained during the experiments on bed expansion, it is possible to substitute for fluidising velocity in terms of voidage in equation (9) and to obtain the maximum value of Nusselt number by differentiating with respect to voidage and putting $d\,Nu_p/de$ equal to zero. The maximum value of Nu_p so obtained, Nu_{max}, is found to be related to the Galileo number (Ga) in the following way

$$Nu_{max} = 0.085\, Pr^{0.4} Ga^{0.4} \tag{10}$$

A comparison of the experimental results with equation (10) can be seen from Fig. 8.

Although the argument has been developed on the basis of results obtained for the turbulent flow regime, eqn. (10) is also found to be applicable to results for the streamline and transition region. The results for these latter regions will form the basis of a later communication.

DISCUSSION

Measurements of heat transfer coefficient to two similar immersed surfaces were made in gas-solid and liquid-solid fluidised systems. Because of the very different behaviour of the two types of fluidised bed, a direct comparison of coefficients is difficult. However, maxima occur in the values of the heat transfer coefficient in both cases as velocity is increased, though for rather different reasons.

For liquid-solid systems, equation (10) gives the maximum value of Nusselt number. If the Prandtl number is included in the same manner in the equation (4) for gas-solid systems. This becomes:

$$Nu_{max} = 0.18\, Pr^{0.4} Ga^{0.475} \tag{11}$$

It will thus be seen that Nu_{max} varies in the two cases in a similar manner.

Further work is in progress on both liquid-solid and gas-solid systems.

REFERENCES

Godard K. and Richardson J.F. (1968) I.Ch.E. Symposium Series No. 30 Symposium on Fluidisation 126.

Latif B.A.J. and Richardson J.F. (1972) Chem.Eng.Sci. 27,1933

Mickley H.S. and Fairbanks D.F. (1955) A.I.Ch.E.J. 1, 374

Richardson J.F., Romani M.N., Shakiri K.J. (1976). Chem. Eng. Sci. 31, 619

Romani M.N. and Richardson J.F. (1974 Letters in Heat and Mass Transfer 1 55.

TABLE 1. PROPERTIES OF GASES

Gas	Pressure (MNm^{-2})	Density ($kg\ m^{-3}$)	Viscosity (Nsm^{-2})	Thermal Conductivity ($Wm^{-1} K^{-1}$)	Prandtl No.
H_2	0.1	0.0833	880	0.182	0.73
Air	0.03	0.36	1830	0.0252	0.73
	0.1	1.20	1830	0.0253	0.73
	0.44	5.28	1840	0.0254	0.73
	0.77	9.24	1885	0.0255	0.73
	1.10	13.2	1976	0.0256	0.73
CO_2	0.1	1.83	1500	0.0157	0.77
CF_2Cl_2	0.1	5.12	1250	0.0090	0.87

TABLE 2. PROPERTIES OF SOLIDS

Solid	Size Range (μ_m)	Mean Size (d) (μ_m)	Density ($kg\ m^{-3}$)	Minimum Fluidised Velocity in air at atms.pressure ($mm\ s^{-1}$)
Glass 20	25-54	40	2600	1.8
Glass 18	39-73	58	2630	3.4
Glass 15	53-101	77	2740	8.4
Diakon (Perspex)	-	96	1180	6.2

TABLE 3. PROPERTIES OF LIQUIDS (KEROSENE-LUBRICATING OIL MIXTURES)

	Density $kg\ m^{-3}$	Viscosity Nsm^{-2}	Thermal Conductivity $Wm^{-1}K^{-1}$	Specific Heat Capacity J/kgK	Prandtl No.
1.a	815	0.00155	0.140	1968	21.9
2.b	818	0.00319	0.139	1965	45.1
3.c	825	0.00648	0.138	1956	92.0
4.d	848	0.0157	0.134	1930	225
5.e	865	0.0287	0.132	1911	416
6.f	872	0.0572	0.131	1904	833
7.g	876	0.122	0.130	1900	1780
8.h	882	0.219	0.129	1893	3210
9.i	890	0.522	0.128	1884	7690
10.j	902	0.940	0.126	1872	13930

TABLE 4. PROPERTIES OF SOLIDS

	Mean Diameter mm	Density kg m^{-3}	Thermal Conductivity W m^{-1} K^{-1}	Specific Heat Capacity J/kg K
1.Polyacrylic	6.0	1186	0.162	1450
2.Glass	3.0	2788	0.936	1130
3.Glass	6.3	2568	0.936	1130
4.Glass	8.9	2499	0.936	1130

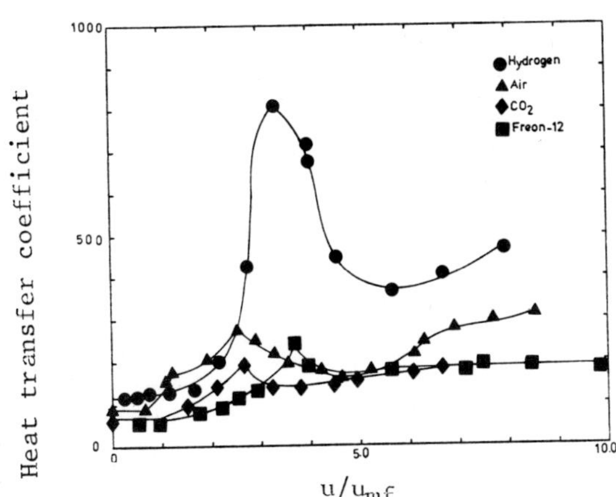

Fig. 1 Heat transfer coefficients for various gases (glass 20)

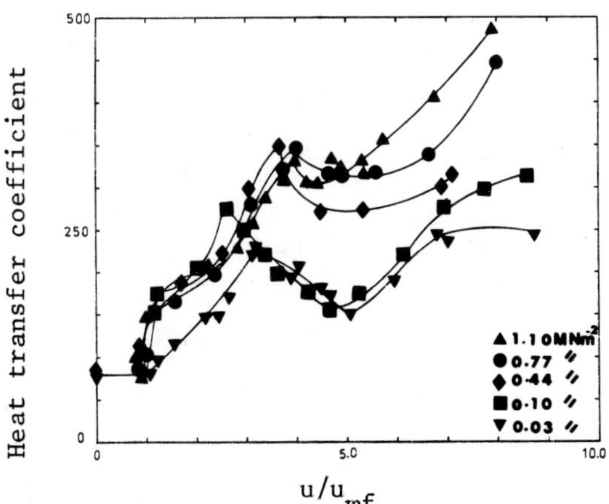

Fig. 2 Effect of pressure on heat transfer coefficient (air-glass 20)
(1)1.10 (2)0.77 (3)0.44 (4)0.10 (5)0.03 MNm^{-2}

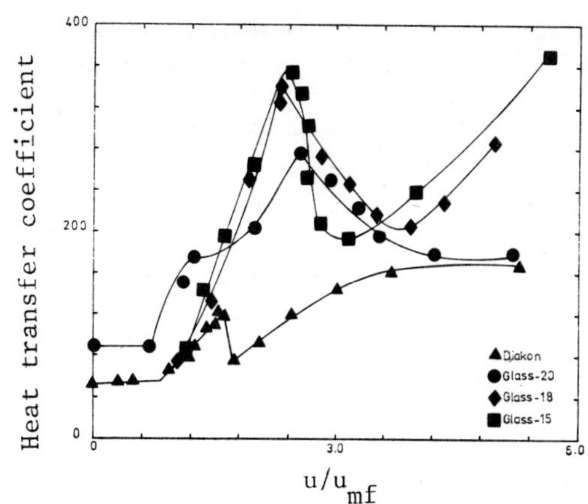

Fig.3 Effect of solid properties on heat transfer coefficient
(air at atmospheric pressure)

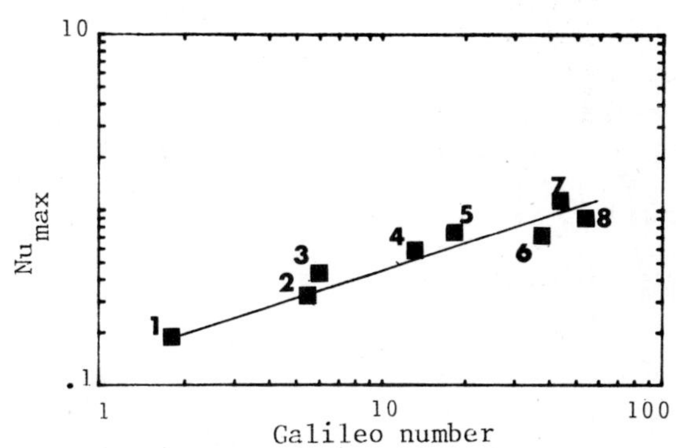

Fig.4 Maximum value of Nusselt number as a function of Galileo number

1)Glass 20-H_2 2)Synclate-AIR 3)Glass 20-AIR 4)Glass 20-CO_2 5)Glass 18-AIR 6)Diakon-AIR 7)Glass 15-AIR 8)Glass 20-F12
(at atmospheric pressure)

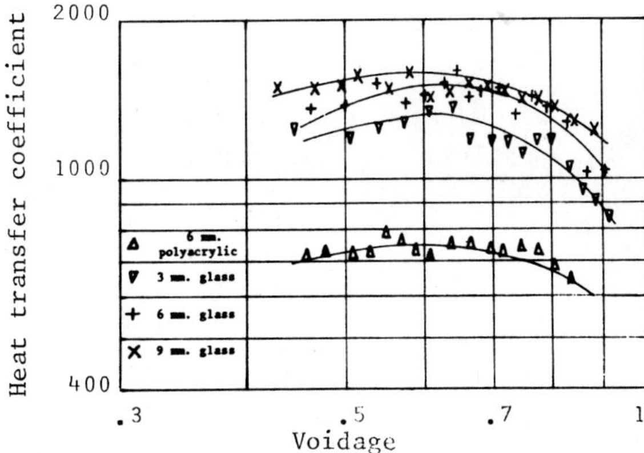

Fig. 6 Heat transfer coefficient as function of voidage for various solids
(liquid viscosity 3.19 mNsm^{-2})

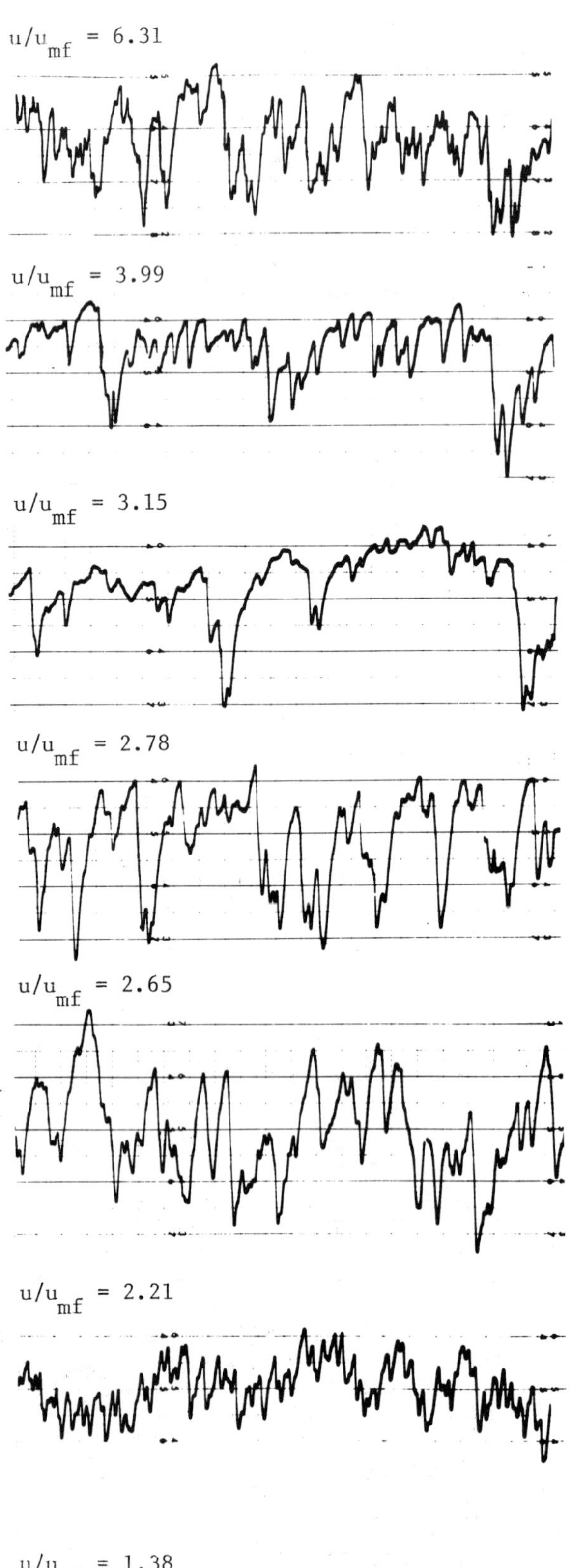

Fig. 5 Instantaneous values of heat transfer coefficient (glass 15 - air at atmospheric pressure)

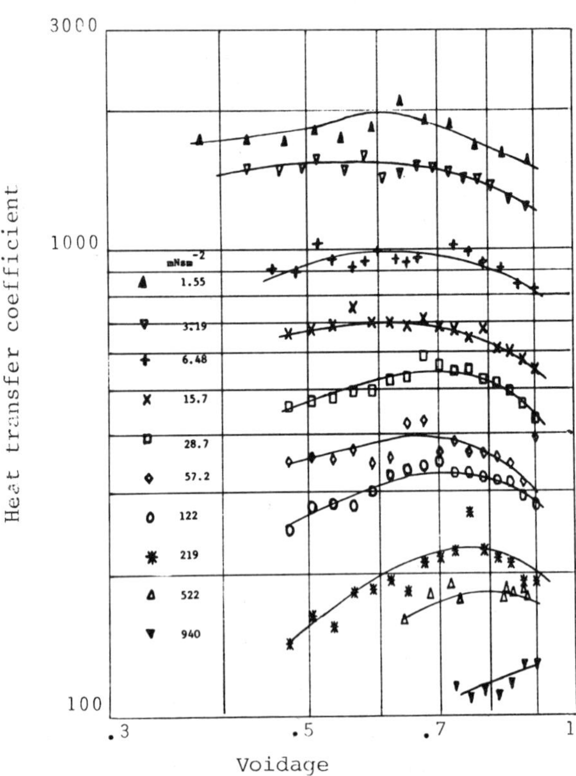

Fig. 7 Heat transfer coefficient as function of voidage showing effect of viscosity (9mm glass)

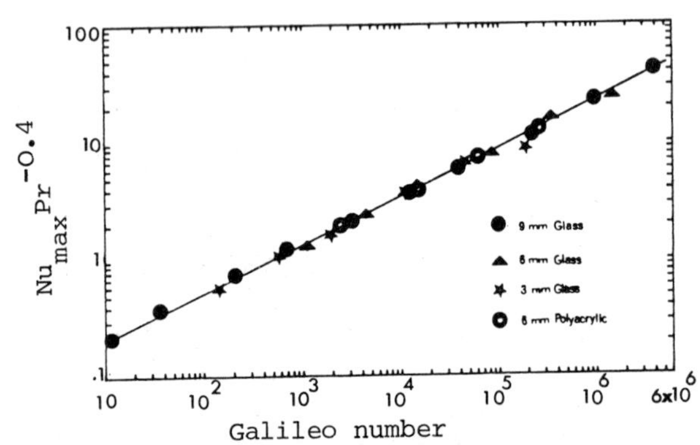

Fig. 8 Maximum value of Nusselt number (Nu_{max}) as function of Galileo number (Ga) and Prandtl number (Pr) for various solids

MECHANICAL STIRRING OF FLUIDIZED BEDS : POTENTIAL IMPROVEMENTS IN THE CONTROL OF HEAT TRANSFER

By G. Rios and H. Gibert

Laboratoire de Génie Alimentaire

Université des Sciences et Techniques du Languedoc

Place Eugène Bataillon

34060 - MONTPELLIER CEDEX

SYNOPSIS

After a brief recall about the discovery and the first applications of mechanical stirring of fluidized beds, the authors give the main conclusions of their own studies on the subject. The possible application of such a device to achieve very accurate control of fluidized bed temperatures is suggested. As an example, the authors discuss the principle of a fluidized bed heat exchanger regulation system using the rotation speed of the agitator as a single command. Ideas are proposed to enlarge this practice to temperature control of gas-solid reactors.

NOTATION

Symbols	Description
A	Heat transfer area
C	Specific heat
f	Frequency
G	Gas flow rate
h	Heat transfer coefficient
K	Static gain
M	Mass of bed solid
N	Rotating speed
P	Laplace variable
t	Time
T	Time constant
U	Overall heat transfer coefficient
W	Solid flow rate
Θ	Temperature
Λ	Dynamical gain
Φ/A	Heat flux
Ψ	Total heat of reaction
ω	Radian frequency
ΔH_R	Heat of reaction per unit mass
Δt_{min}	Minimum time lag

Subscript	Refers to :
c	Bed
e,g	Inlet gas, gas flow rate
p	Column wall
s	Solid

Superscript	Refers to :
$*$	Jacket fluid
$-$	Mean value
$\hat{}$	Objective value

INTRODUCTION

About 1960, Botterill et al. proposed the idea of using mechanical stirring of gas solid fluidized beds in order to control solid particles renewal close to the wall (Botterill 1963). Using such a device, the authors expected to get new experimental data that would allow them to check a wall to bed heat transfer model set up from the "Packet Theory" (Mickley 1955). In fact, after experimentation, it was found that the technique was unsuitable for guaranteeing an accurate control of particle residence time at wall ; however, it appeared that it was possible to improve greatly the heat transfer coefficients between the wall and the bed core :

$$h_c = \Phi/A \, . \, (\Theta_c - \Theta_p) \qquad \ldots(1)$$

by only increasing the agitator rotating speed.

Enlarging the range of the experimental variables, Angelino et al. (Angelino 1966, Angelino 1967) stressed the very significant effect of gas flow-rate on the intensity of wall-to-bed heat transfer ; more particularly they outlined the sharp increase of h vs.N curves when the mean apparent fluid velocity, based on the empty column, approaches the minimum fluidization velocity of solid particles. For example, with a rotating speed of 80 r.p.m., the "thermal gain" - which is defined as the ratio of the heat transfer coefficient between the fluidized bed and the wall with and without stirring - may be five or six in these flow conditions, whereas it does not exceed two when the gas velocity is two or three times greater.

The studies, that we have already reported on this subject, using the same device, confirm such a result (Rios 1977 a, Rios 1977 b). However, our interpretation, based only on mechanical grounds, is quite different from that given by Angelino ; it starts from the consideration that, if a stirred bed remains wholly fluidized when the fluid velocity is strongly in excess of the minimum fluidization velocity, it does not stay so close to this limiting value : a closely-packed zone appears near the wall while the initially flat bed surface becomes deformed by a vortex ; as a consequence, the bed temperature, which was uniform when the system was fully fluidized, becomes heterogenous and a temperature gradient develops ; because the true temperature of the bed near the wall is not taken into account, heat transfer coefficients defined by equation (1) are no longer representative of the single micromixing mechanism put forward by Botterill ; it is then necessary to consider both the micromixing induced by the blades movement near the wall and the macromixing promoted in the whole bed by the agitator.

As a consequence, it is quite obvious that the mechanical stirring of a fluidized bed in the minimum fluidization velocity range is less promising than previous authors thought (Angelino 1966, Angelino 1967) ; it seems better to operate with larger gas velocities in order to get a uniform whole bed temperature. If, in that last range of operating conditions, mechanical stirring does not allow a very sharp increase of wall-to-bed heat transfer, it nevertheless gives the possibility of changing h_c in such a way that the technique can be used to design efficient systems to control fluidized bed temperature.

In many industrial applications of fluidization, such as gas quenching or temperature control of gas-solid reactors, it is well known that temperature control of systems is an important problem which is not always solved easily. Usually, a solution is found by changing either the flow-rate or the inlet temperature of a fluid circulating in a heat exchanger inside the bed or in a jacket. In this way, it is possible to change at every time the thermal flux and control the temperature of the bed core. In the paper, we propose a new way of achieving the control by means of a mechanical stirrer that allows the agitation of the bed close to the containing wall without changing the wholly fluidized mechanical behaviour of the system ; in this solution, flow rate and temperature of jacket liquid are kept at fixed levels during all the operations. As a first analysis, one can expect two advantages from this new device as compared with the traditional one :

- a reduced thermal inertia of the governed systems ; this comes at first from the very fast modification of the mechanical behaviour of the bed at the wall when N is changed, so that the heat transfer coefficient is transformed immediatly ; on the other hand the fluid temperature is kept at a constant value so that steady state thermal behaviour prevails at any time in the jacket ;
- the possibility of proposing simple systems models which allow easy calculations and design of industrial scale plants.

PROBLEM ANALYSIS

In order to represent the thermal behaviour of batch, mechanically stirred fluidized beds in the high flow rate range, a first order equation may be used between Θ_c the bed temperature, Θ^* the fluid temperature in the jacket and Θ_e the inlet gas temperature (Rios 1976) :

$$T.\frac{d\Theta_c}{dt} + \Theta_c = K_e.\Theta_e + K^*.\Theta^* \quad ...(2)$$

where :

$$T = \frac{MC_s}{UA+GC_g} \; ; \; K_e = \frac{GC_g}{UA+GC_g} \; ; \; K^* = \frac{UA}{UA+GC_g}$$

and :

$$\frac{1}{U} = \frac{1}{h} + \frac{1}{h^*}$$

In this equation, the heat transfer coefficient h_c is, to a first approximation, a linear function of N and does not depend on gas flow-rate G (Rios 1977 b) ; so, it appears possible to know the dynamical behaviour characteristics of gas-solid systems, excited by input signals N, by simply integrating the previous model. Especially, if the input signals are step signals, it is easy to reach the analytical solution with the two main dynamical characteristics of the first order linear processes considered : the final steady state temperature $\Theta_c(\infty)$ and the time constant T. Obviously, these quantities will be dependent on the values $\bar{M}$, $\bar{\Theta}_e$, $\bar{\Theta}^*$ and $\bar{G}$ of the main design parameters of selected stirred fluidized systems.

Now if it is assumed that N is put at a given value - $\bar{N}$ -, and that Θ_e, Θ^* and/or G fluctuate during the operating time, round chosen mean values : $\bar{\Theta}_e$, $\bar{\Theta}^*$, G with weak instantaneous amplitudes : $\delta\Theta_e$, $\delta\Theta^*$, δG, one can calculate the resulting variation on the temperature of the bed, Θ_c, using equation (2) ; one gets :

$$\bar{T} \cdot \frac{d(\delta\Theta_c)}{dt} + (\delta\Theta_c) = \bar{K}_g \cdot (\delta G) + \bar{K}_e \cdot (\delta\Theta_e) + K^* \cdot (\delta\Theta^*) \quad \ldots(3)$$

where :

$$\bar{K}_g = \frac{C_g \cdot (\bar{\Theta}_e - \bar{\Theta}_c)}{(\bar{U}A + \bar{G}C_g)}$$

In this expression, $\delta\Theta$, is nothing but "instantaneous distance" which separates steady-state temperature $\bar{\Theta}_c$ of the ideal undisturbed system from the actual disturbed system temperature. Equation (3) is also appropriate when δG, $\delta\Theta_e$ or $\delta\Theta^*$ are not weak, provided that $\delta\Theta_e$ and δG are not simultaneous and : $\bar{U}A$ is far greater than : $\bar{G}C$.

By using the Laplace transform, equation (3) becomes :

$$\delta\Theta_c = \frac{\bar{K}_g}{1 + Tp} \cdot \delta G + \frac{\bar{K}^*}{1 + Tp} \cdot \delta\Theta^* + \frac{\bar{K}_e}{1 + Tp} \cdot \delta\Theta_e \quad ..(4)$$

on condition that at the initial time t = 0 :

$$\delta\Theta_c(t = 0) = 0$$

This last expression (4) gives the form of the filtering characteristics of a batch stirred fluidized system, that is, the frequency response functions that relate changes in system temperature Θ_c to changes in gas-flow rate G, jacket liquid temperature Θ^* and gas inlet temperature Θ_e. More particularly, it appears that, in the face of disturbances in G, Θ^* and /or Θ_e, stirred systems are behaving as first order low-pass filters which block (at 20 db/octave) all the signals δG, $\delta\Theta^*$ and $\delta\Theta_e$ that exhibit frequencies higher than f = $1/\bar{T}$. Proportionnal - or static - gains of these filters are respectively : $\bar{K}_g$, $\bar{K}^*$ and $\bar{K}_e$.

SYNTHESIS AND PERFORMANCES OF "ON-OFF CONTROL SYSTEM"

Synthesis

Let us now examine an entire temperature control loop in order to see what sort of dynamical behaviour is exhibited.

The control objective is to regulate the bed temperature of a batch heat exchanger pilot-plant (fig. 1), used in previous experimental work (Rios 1977 b), round : $\hat{\Theta}_c$ = 60°C and within a control band of ± 1°C. The mean operating conditions - or static characteristics - of the control system are related to the geometrical and physical characteristics of this heat exchanger : A = 0.121 m^2 ; $\bar{\Theta}^*$ = 18°C ; $\bar{M}$ = 18 kg ; C_g = 0.240 kcal/kg°C ; C_s = 0.175 kcal/kg°C ; h^* = 1000 kcal/h.m^2°C ; as well as the particular choice of the mean inlet gas temperature : $\bar{\Theta}_e$ = 300°C, and the mean gas flow-rate : $\bar{G}$ = 20 kg/h ; such a flow-rate value is consistent with :

- uniformity of the bed structure and temperature for all speeds of rotation up to 80 r.p.m.
- $\hat{\Theta}_c$ lies in the temperature range limited by extreme values $\bar{\Theta}_c(\infty)$ [0 r.p.m.] and $\Theta(\infty)$ [80 r.p.m.].

On figure 2, are plotted $\overline{\Theta(\infty)}$ vs.N and T vs.N curves calculated from the given values of the system characteristics parameters and from the knowledge of h vs.N : h(0)=250kcal/hm^2°C $h_c(80)$ = 450 kcal/h.m^2°C. It can be seen that Θ_c = $\bar{\Theta}_c(\infty)$ [17 r.p.m.] and that it is possible to vary the final steady-state temperature of the undisturbed system $\bar{\Theta}_c(\infty)$ within ± 4.7°C by simply changing N from 17 r.p.m. to 0 or 80 r.p.m. respectively.

Considering both the main features of the system's dynamical behaviour and the practical interest of using a step signal N for governing the process, one can think that "on-off regulation" - or "bang-bang system" - will be a very well adapted technique for solving this type of control problem.

Performances

Let us check this fact studying a particular control loop built on this model by making use of a relay with hysteresis as an agitation-circuit breaker (fig 3-a). If we neglect all transient mechanical behaviour of the individual elements (transducer, relay, motor moving agitator) as against the transient thermal behaviour of the process (a very good approximation), it is possible to represent the general characteristics of a "relay system" giving U vs.Θ = $\hat{\Theta}_c$ - $\Theta_c(t)$, and using the two rotating speeds of agitator : 0 and 41 r.p.m., as indicated in fig 3-b.

As regards the results and conclusions of "problem analysis", it will be possible with the proposed regulation system to counterbalance the effects of low frequency noise operating with Θ_e between 272°C and 336°C, and G between 17.6 and 22.9 kg/h. In the face of such signals, the dynamical gain of the heat exchanger (Gould 1969), $\bar{\Lambda} = 1/\sqrt{1 + \omega^2 . \bar{T}^2}$, will approach unity, so that :

$$\delta\Theta_c/\delta\Theta_e = \bar{K}_e \cdot \bar{\Lambda} \cong \bar{K}_e$$
$$\delta\Theta_c/\delta G = \bar{K}_g \cdot \bar{\Lambda} \cong \bar{K}_g \quad \ldots(5)$$

By computing values of $\bar{K}$ that correspond to rotational speeds of 0 and 80 r.p.m., one can check the previous numerical results. The regulation range may appear to be very large ($\Delta\Theta_e/\bar{\Theta}_e$ = 10 %, $\Delta G/G$ = 15 %), but it should be remembered that it holds only for separate fluctuations $\delta\Theta_e$ and δG.

In the same way, the device appears to be able to counterbalance low frequency fluctuations in Θ^*, provided that $|\bar{\Theta} - \Theta^*(t)| \leqslant 5.5$°C. Because with higher frequency noise ($f > 1/\bar{T}$) Λ becomes lower than unity, it will be possible to compensate fast fluctuations of much larger amplitude by this mean.

When a low frequency disturbance of maximum amplitude alters Θ_e or G at the same time that N changes from 0 to 41 r.p.m., one can calculate the minimum value of the time lag, Δt_{min}, between two relay-system switchings. Taking a step signal ΔN as the fluctuation, it follows from the computing that : Δt_{min} = 67 s. Such a value which is much larger than the time constants of all the individual elements, guarantees good operating conditions for the regulation loop, i.e. any oscillations of the entire loop. It proves also the accuracy of choice of an "on-off control technique" in order to solve problems of thermal fluidized beds regulation.

GENERAL CONSIDERATIONS ON CONTROL SYSTEM DESIGN

One can extend the solution to other thermal gas-solid fluidized bed control problems. To do so, it may be necessary to modify equation (2) in order to get a good model of the new process.

As an example, if one deals with a more complex heat exchanger involving a continuous feed of solid into the fluidized bed reactor it will be necessary to insert a new group that will account for the heat carried by solid circulation : $W_s \cdot C_s \cdot (\Theta_s - \Theta_c)$, in the second member of equation (2). In the case of a batch gas-solid catalytic reactor, equation (2) will be transformed by introducing a group in the second member :

$$\Psi = G \cdot \Delta H_R$$

taking into account the reaction heat generation.

Whatever the problem to be solved, the mathematical analysis will be performed in the same way. The only differences will be the number of disturbed parameters considered in that case.

CONCLUSION

As a conclusion of the present work, one can say that the new way proposed for temperature regulation of a fluidized bed system using mechanical stirring, exhibits, as expected, two main advantages as compared with the traditional one :

- On-off control systems built on this pattern appear to be very powerful as regards both to the dynamical behaviour and the high degree of accuracy which may be reached. It is to be outlined that the control is performed at two different levels : a static level, relevent to general process design, and a dynamical level, concerned with building of control loop round selected parameter mean values.
- The proposed solution permits simple modeling of phenomena, potentially valuable for extrapolating pilot plants results and for the design of new industrial scale plants.

REFERENCES

Botterill J.S.M. and Williams J.R. (1963) Trans. Inst. Chem. Engrs., 41, 217.

Couderc J.P., Angelino H. and Enjalbert M. (1966), Chem. Engr. Sci., 21, 233.

Couderc J.P., Angelino H. and Enjalbert M. (1967), Chem. Engr. Sci., 22, 109.

Gould L.A. (1969) "Chemical Process Control : Theory and Applications" - Chapter 3 and 5 Addision Wesley Publ. Comp.

Mickley H.S. and Fairbanks D.F. (1955), A.I.Ch.E. Journal, 1, 374.

Rios G., Gibert H., Couderc J.P. (1977 a) Chem. Engr. J., 13, 2, 101-109

Rios G., Couderc J.P. and Gibert H. (1977 b) Chem. Engr. J., 13, 2, 111 - 118

Rios G. (1976) Thèse de Docteur-Ingénieur U.S.T.L. MONTPELLIER

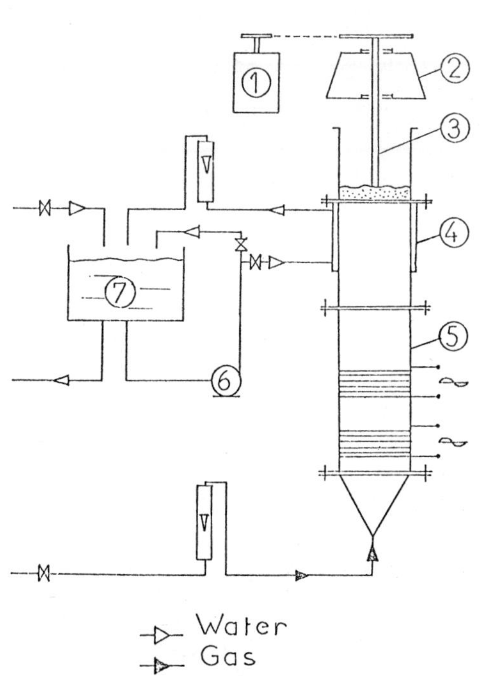

Fig. 1 : Diagrammatic view of the pilot plant : 1 - motor ; 2 - agitator support ; 3 - agitator; 4 - fluidized heat exchanger ; 5 - air preheater ; 6 - pump ; 7 - buffer-tank.

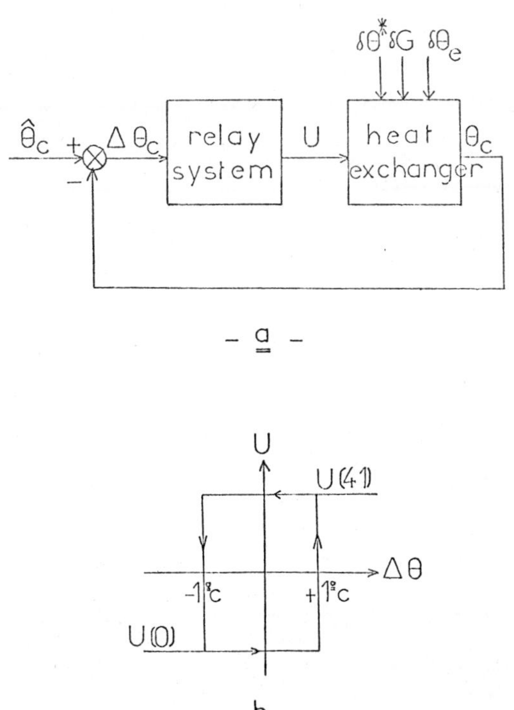

Fig. 3 : a - schematic drawing of an entire "on-off control" loop ; b - characteristics of a circuit breaker built from a relay with hysterisis.

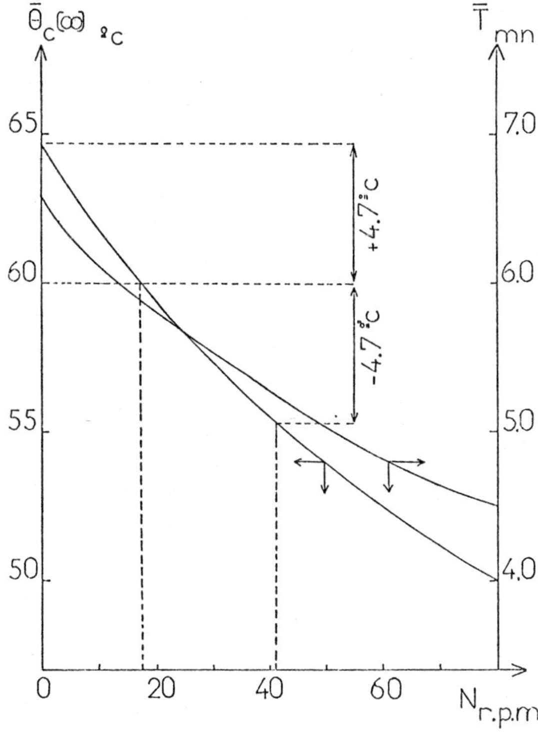

Fig. 2 : Variations of the equilibrium temperature and time constant of the system against the agitator rotating speed.

ON THE DEPENDENCE OF FLUIDIZED BED-WALL HEAT TRANSFER COEFFICIENTS ON THE THERMAL CONDUCTIVITY AND VOLUMETRIC HEAT CAPACITY OF THE PARTICLES

S.S.Zabrodsky, Yu.G.Epanov and D.M.Galershtein
The Luikov Heat and Mass Transfer Institute,
Minsk, U.S.S.R.

SYNOPSIS

An attempt is made in this paper to compare theoretical models of fluidized bed-to-surface heat transfer with experimental evidence and to explain the character of the dependence of heat transfer coefficient on thermal conductivity and volumetric heat capacity of solid particles. Based on the features specific for heat transfer at fluidized combustion, it is shown that there is a tendency to a decrease in the effect of volumetric heat capacity on heat transfer, as compared to some other conditions.

NOTATION

b	gap between a solid particle and heat transfer surface (wall) during particle exposure at the wall
c_f	specific heat of fluid
c_s	specific heat of a solid particle
d_p	particle diameter
H	height of fluidized bed
H_o	static bed height
h_{cc}	the component of h_w due to fluid heat conduction and convection
h_{cd}	the component of h_w due to heat conduction
h_{cvf}	component of h_w due to fluid convection
h_{max}	maximum value of h_w
h_{max}^{com}	maximum value of h_w for the whole industrial scale unit
h_r	the radiative component of h_w
h_w	the fluidized bed-to-surface (wall) heat transfer coefficient
h_{wf}	the interparticle fluidizing medium-to-surface (wall) heat transfer coefficient
k_f	thermal conductivity of fluid
K_s	thermal conductivity of a particle
t_{ex}	time of particle exposure at the heat transfer surface (wall)
U_{mf}	minimum fluidizing velocity
v_s	velocity of solids
δ_f	effective thickness of the fluid layer between the wall and first (adjacent) row of particles
ε	average porosity of the fluidized bed
	porosity of the static bed
ρ_f	fluid density
ρ_s	density of a particle

S.S.Zabrodsky, Yu.G.Epanov and D.M.Galershtein

INTRODUCTION

Theoretical fluidized bed-to-surface heat transfer equations derived on the basis of different models include $c_s \rho_s$, and the calculations based on Botterill's model (cf. Botterill, 1975) manifest strong dependence of the heat transfer coefficient on the thermal conductivity, k_s, of particles at small times of their exposure at the surface. (In order to shorten the list of references, monographs and surveys are referred to rather than the original references). This has raised hopes of possible sharp enhancement of fluidized bed heat transfer by the choice of materials with adequate thermophysical properties. If it is difficult to change $c_s \rho_s$ markedly (due to often observed adverse change of c_s and ρ_s with the change of material, for example from glass to steel), then other materials can be easily found (aluminum, copper, for example) with k_s ten times as much as that for sand and hundreds of times as much as that for glass.

However, the convincing experiments of Wicke and Fetting, (cf. Zabrodsky, 1966) with fluidized beds of sand and aluminum particles show that change to aluminum does not yield any enhancement in the maximum value of the heat transfer coefficient. Also Neukirchen and Blenke's (1973) experiments on heat transfer between fluidized beds and immersed horizontal tube bundles exhibited no marked enhancement of h_m from the change of quartz particles by aluminum. As regards the $c_s \rho_s$ effect on the maximum heat transfer coefficient, it was reduced quite successfully to the effect of particle density only in the empirical formula (cf. Zabrodsky, 1966, p.234).

Below, we endeavour to look into these and other apparent contradictions and explain the reasons hampering sharp enhancement of heat transfer due to choosing materials with promising thermophysical properties. The understanding of these reasons will allow more deliberate and purposful search for real ways of h_w enhancement in commercial scale equipment. In conclusion, some properties characteristic of heat transfer at low-temperature fluidized bed combustion are considered from this standpoint.

CONTRIBUTION OF MATERIAL THERMAL CONDUCTIVITY

When maintaining direct proportionality of h_w to k_s of particles contacting the heat transfer surface at small times of exposure, Botterill has made a considerable reservation for this to hold only at negligible changes of k_s. Then at every instant only few of the adjacent particles really contact the heat transfer surface, most of the particles being spaced from the surface. This diminishes k_s influence on h_w. Additionally, of importance is that real times of particles (or their "packets") exposure at the heat-transfer surface are, in the main, greater than those when there is a considerable effect of k_s, according to Botterill's predictions. If in the first approximation, the frequency of packets or individual particles replacement near the heat transfer surface is assumed equal to the frequency of bubbles, then the time of exposure will be 0.5–0.17sec, as, according to Kunii and Levenspiel (1969, p. 123), the dominating bubble frequencies are 2 to 6 Hz in the rather shallow beds typically used in fuel combustion.

If the replacement of particles near the heat transfer surface is related to the rate of descending emulsion circulation rather than to the velocity of large bubbles, then even for lower heat transfer surfaces (of the type of horizontal tubes) the time of exposure will not be very short, neither the particle flow will be nonseparated.

In Fig.1, adapted from Botterill (1975), it is evident that at t_{ex}/d_p^2 10^{-2} even with most asymmetric heating (particles are exposed touching the wall), $h_w d$ for copper and glass particles is almost equal, i.e. for very fine particles the effect of k_s does not show. It will not be observed for a fluidized bed of larger particles either because, when a packet of particles is exposed at the wall (especially at the first instant), there is close contact only for some particles, while most of them, just as in Zabrodsky's model (cf. Zabrodsky, 1966), are separated from the wall by additional distance "b" (Fig.2) and are exposed to more symmetric heating. Relying upon their own experimental data, Botterill and Butt (cf. Botterill, 1975) have estimated the mean value of such a gaseous gap for the particles of the first row as 0.1 d. Herein, according to Fig.1, at $t_{ex}/d^2 = 10^{-4}$, being by 2 orders less than at close contacting operation, the effect of k_s disappeared. With even less pronounced particle separation from the surface due to centrifugal or inertia forces (than is indicated by small heat transfer coefficients Botterill and Butt have obtained), one can still expect low sensitivity of h_w to k_s at t_{ex}/d^2_p of the order of 10^{-3}. For the 200 μm diameter particles this corresponds to t_{ex} = 0.04 sec, i.e. to the frequency 25 Hz.

Note that the assumption of the gap between the particles of the first row and heat transfer surface (wall) leads to the conclusion of a nonmonotonic dependence of the " packet" heat transfer coefficient on the time of exposure, and h decreases to the heat transfer coefficient for "pure" gas at very small times of exposure /$t_{ex} \rightarrow 0$/, when the thermal field would be concentrated in a thin gas layer near the wall being "insensible" to the solid phase effect, (cf. Todes et al., 1970). In Fig. 3 plotted according to the work of Zabrodsky et al. (1973) demonstrating such nonmonotony, the curve at small t_{ex}/d^2_p tends to 0 rather than to $h_{wg}d_p$. This is because the relevant calculations have operated with the heat received by the layer (but not released by the wall), beginning from the first solid interlayer in the model similar to the known Gabor model. But this nonmonotony is consistent with high rates of particle replacement that cannot be attained in an ordinary fluidized bed, and is of theoretical interest only.

Additionally, the effect of k_s in real fluidized beds can be weaker than that in Botterill's predictions (cf. Botterill, 1975) due to the nonspherical shape of particles destroying their "point heating".

Let us revert, however, to the possibility for observing experimentally the effect of k_s on "packet"-wall heat transfer. When working with very large particles such an opportunity, perhaps, may be expected with heat exchangers used by Ernst (1960) which, unfortunately, are but slightly promising in practice. Ernst operated with a dense bed, slipping down under gravity, such a bed being swept into contact with a low (5 mm) heating belt being flush with the wall. There was neither centrifugal nor inertial separation of solid bed particles from the wall. On the contrary, the wall levelled the packing of the first row of particles before the bed element has reached the belt. Hence, we may suppose that most of the particles of this row have been exposed touching the wall. The evidence that the contact is good is afforded by high heat transfer coefficients obtained by Ernst. These are about 1340 W/m²deg for a bed of quartz sand particles 100-200 μm in diameter and above 700 W/m²deg for the 300-500 μm fractions. The material velocity achieved 540 m/hr, i.e. the minimum time of the bed exposure at 5 mm belt was about 0.03 sec.

THE PART OF THE VOLUMETRIC HEAT CAPACITY OF PARTICLES

It has been established in the previous section that, in an ordinary fluidized bed for practically encountered times and tightness of the particles (packets) contact with the heat transfer surface, k_s has exerted no essential effect on heat transfer, i.e. it is not necessary to allow for the temperature history inside the particles. Hence, we may suppose, that particle heating is gradientless, as it has been assumed for the model suggested earlier, cf. Zabrodsky (1966), p.227, and make use of the expression for the conductive wall-fluidized bed heat transfer component in order to analyse the effect of different factors. This expression is

$$h_{cd} = 0.807 d_p c_s \rho_s \frac{(1-\varepsilon)^{2/3}}{t_{ex}}\left[1-e^{-n}\right]. \quad . \quad . \quad (1)$$

where
$$n = \frac{1.5 k_f t_{ex}}{\delta_f c_s \rho_s d_p}$$

In a general case, from (1) the dependence of h_{cd} of the fluidized bed on the volumetric heat capacity of particles $c_s \rho_s$ is evident. It has been shown, however, cf. Zabrodsky (1966), that under the conditions of maximum h_{cd} onset at n = 0.03, i.e. if $c_s \rho_s$ is high enough, while $k_s t_{ex}$ is sufficiently small, any dependence of h_{max} on $c_s \rho_s$ disappears. Herein, for one particle heating cycle its temperature increases only slightly. Therefore, change to a bed of particles of greater volumetric heat capacity and their more rapid replacement will hardly result in considerable increase of the effective temperature head applied to the principal thermal resistance, i.e. the resistance of the gas interlayer between a particle and the wall. Let us evaluate tentatively through the aid of formula (1) the effect of $c_s \rho_s$ without the above limitation, i.e. in a wider range of exposure times up to 1 sec and extensive choice of $c_s \rho_s$ (for example, from 850 to 4200 kJ/m^3deg). Since heat removal by conduction from the particles of the first row to the following ones for the exposure time has been neglected in the formula derivation, i.e. for this time the first row has been considered as if thermally insulated of the rest of the bed, then heat is removed by shift of the particles. For this reason the effect of $c_s \rho_s$, which formula (1) gives, will be greatly overestimated at long t_{ex}. So the estimation of the effect of $c_s \rho_s$ will be extremely "conservative" for rapidly heated (cooled) fine particles at the wall. We shall, therefore, apply this formula for the particles larger than 0.5 mm. In any case, they will be preferable in fluidized combustion of fuels as regards the possibility of operating with high gas velocities.

In Fig.4 the results of calculations by formula (1) are presented as h_{cd} against $c_s \rho_s$ for a very thin gas interlayer between o.5 mm particles and the wall ($\delta_f = d_p/6$); the mean bed porosity corresponding to its expansion $H/H_o=1.1$. In this case, the replacement of particles being slow (t_{ex} = 1 sec), the transition from $c_s \rho_s$ = 850 to $c_s \rho_s$ = 4200 would allow great increase of h_{cd} (by 1.47 times), while with quick replacement (t_{ex} = 0.1 sec) the same change of the material would give the gain of 1.04 times only. In Fig.4 the calculation is also presented for larger particles (d_p = 1 mm). Comparison between the curves demonstrates a sharp reduction of the $c_s \rho_s$ effect on h_{cd} with increasing particle diameter. Predictions for $d_p \approx 2$ mm particles showed the independence of h_{cd} on $c_s \rho_s$ for the exposure times up to 1 sec. For this instant, all the curves degenerated into horizontal lines.

In Fig.5 the calculation results are plotted for poorer particle-wall contacting operation, namely, for the great thickness of the gas interlayer $\delta_f = 0.5\ d_p$; comparison with Fig.4 shows deterioration of heat transfer at the same t_{ex}, as well as further decrease of the $c_s \rho_s$ effect. Here, for 500 μm particles the gain due to change of the material would not exceed 14% even at t_{ex} = 1 sec, and for particles 1 mm in diameter, would be quite negligible.

How will the role of $c_s \rho_s$ be affected by the situation that due to convective, and sometimes radiative components, the overall heat transfer coefficients of large particles will be much higher than conductive components calculated by formula (1), their values achieving 230-290W/m^2deg? Though radiative and convective heat transfer is in no direct dependence on $c_s \rho_s$, they may increase the role of $c_s \rho_s$ indirectly because at the same t_{ex} the first row of particles will be heated (cooled) more intensely. Using formula (1) one can allow for such h_{cvf} and h_r effects, by increasing conventionally k_f up to some effective k_{ef} yielding h_{cd} = h_w=230-290W/m^2deg at small exposure times. In Fig.6 the calculation results are plotted for such conventionally increased k_f to show that in these conditions, when working with the beds of 1 mm particles, the part played by $c_s \rho_s$, especially with developed fluidization (small t_{ex}), would hardly justify a special choice of material with high $c_s \rho_s$. Note that in such a way we largely overestimate the influence of $c_s \rho_s$ at large t_{ex}.

SOME SPECIFIC FEATURES OF HEAT TRANSFER IN LOW-TEMPERATURE FLUIDIZED COMBUSTORS AND THE PART PLAYED BY THE BED MATERIAL

The design of low-temperature fluidized-bed combustors is known to have the tendency to (1) construction of equipment with large unit capacity consisting of rather big blocks each with the gas-distributor area about 40 m^2; (2) high gas velocities about several meters per second, i.e. the use of large particles and (3) extensive "packing" of the fluidized bed with steam-generating surfaces, i.e. to employment of rather closely packed tube bundles immersed in the bed.

It is known, cf. Whitehead and Dent (1967), that nonuniformity of initial gas distribution increases with increasing grid area. This is especially pronounced in the case of low fluidization numbers when regions of poor fluidization or even complete defluidization of the material appear at some points. The defluidized (stagnant) regions disappear at higher fluidization numbers, but the local gas superficial velocities are not levelled completely. This should lead, in principle, to reduction of total

h_{max} in large scale heat exchangers (rather than that of individual tubes or probes) in contrast to h_{max} obtained in laboratory scale devices. Different parts of tube bundles in large scale heat exchangers are operated at different hydrodynamic conditions. Thus, some of these are located in the regions of dense downward circulating flows of material, other portions in the regions of more rarefied upward flows, others in the regions of interacting jets that leave the orifices of gas distributors. Therefore, if for the given rate of a gas flow, the conditions of maximum heat transfer are achieved for one portion of a bundle (bundles), this may not be the case for others and

$$h_{max}^{com} < \sum x_i h_{max\ i} \quad ....(2)$$

where h_{max} are maximum heat transfer coefficients for surface portions, and x_i is their fraction in a total surface. The greatest deviation of h_{max} from "partial" maxima would be observed when the latter are distinct as sharp peaks in the region of relatively small fluidization numbers. Such are the maxima for narrow fractions reported by Varygin and Martyushin (cf. Zabrodsky (1966), p.272). It is the comparison of the laboratory h_{max} with smaller h_{max}^{com}, achieved in large scale devices operating with the same average superficial velocities and possible existence of the defluidized (quiescent) regions, that seems to excite scepticism as regards the possibility of extending the laboratory experimental data on h_{max} to commercial scale conditions in general. The scale-up conditions, however, are improved if laboratory experimental data are obtained for real polydispersed materials whose maxima are more gentle. Further, levelling of the maxima is observed in coarse particle beds. This trend is seen in Fig.7 too. Besides, fluidized-bed combustion requires operation at forced conditions without formation of defluidized (quiescent) zones. Herein, due to rapid mixing of the bed, i.e. small t_{ex}, the effect of $c_s \rho_s$ becomes insignificant.

Additionally, some other attractive heat transfer features may be observed in the fluidized beds of coarse particles. With increasing particle diameter and the whole complex $U_{mf}d\ c_f \rho_f$ (cf. Zabrodsky, p.234) and Baskakov (1968), the convective heat transfer component increases. Since the conductive heat transfer component diminishes with increasing diameter, this component looks like being replaced by that of gas convection, with the total conduction-convection heat transfer coefficient being remained the same in a wide range of particle sizes at atmospheric pressure, as Traber et al have already shown experimentally (cf. Zabrodsky, p.278, 1966). Here, for comparatively light BAV catalyst ($\rho_s \approx 900$ kg/m^3) h_{max} is completely unchangeable in the range of d_p from 0.8 to 3.3 mm. For heavier ($\rho_s \approx 2200$ kg/m^3) alumonickel catalyst with higher fluidization velocity, the increase in the convective component has proved more pronounced than the fall of the conductive component. Nevertheless, within ± 7.5% the overall h_{max} may be assumed independent of d_p in the range shown in Fig.8.

The data of Fig.8 first seem to be sharp contradiction with the authors' claim that the role of the volumetric heat capacity of particles, $c_s \rho_s$, decreases with the increase in their diameter. Indeed, here h_{max} for the material with $c_s \rho_s$ = 2030 kJ/m^{3}deg is much higher than for the material with $c_s \rho_s$ = 870 kJ/m^{3}deg. Besides, the curves tend to diverge with further increase in the particle diameter. However, this is only seemingly so since ρ_s alone, and not the volumetric heat capacity $c_s \rho_s$, contributes to the increase in the gas convective, rather than conductive, heat transfer component. It is the gas convective component which becomes dominating.

Note, that the approximate constancy and even the increase of h_{max} with increasing d_p for heavy materials in this region makes larger particles more attractive, their use yielding the gain in the gas rate, without decrease in the heat transfer coefficient. Additionally, materials of higher density promise an increase in the heat transfer coefficient of large particle beds, but their employment involves also undesirable increase in the bed pressure drop unless a smaller height of the bed can be used.

The desire to design a fluidized bed packed with tube bundles with the bundles reasonably closely spaced needs that further investigations be conducted. In closely packed bundles there occur rarefied zones where h_{cd} is even more diminished, while the behaviour of h_{cvf} may be different.

In these zones the part of $c_s \rho_s$ will decrease with decreasing h_{cd}. When working at "hot" conditions, the importance of radiative heat transfer may increase considerably, that is extremely hot remote particles will be "visible" to heat transfer surfaces. Note, that the radiative heat transfer of particles which are not in contact with the tubes will not enhance the part of $c_s \rho_s$, contrary to the radiative heat transfer of the first-row particles. The heat transfer in closely-spaced bundles must therefore be studied under both cold and hot conditions, though there are no reasons to overestimate the part of the radiative heat transfer at low temperature combustion.

CONCLUSIONS

1. The effect of k_s on h_{cd} is strong only at very low t_{ex}, while, conversely, that of $c_s \rho_s$ is large at very high t_{ex} and sufficiently fine particles. As a result, under forced fluidization conditions encountered in the beds of large particles ($d_p > 1$ mm), the effects of k_s and $c_s \rho_s$ become insignificant.
2. The effect of $c_s \rho_s$ may be important in the case of poor fluidization of the bed (for example , with bad gas distributor and far too low fluidization numbers) when there are defluidized (stagnant) zones or those with low material mobility (large t_{ex}).
3. In the case of larger fluidized particles the contribution of h_{cd} to the overall heat transfer coefficient is reduced, while the contribution of the convective component, directly dependent on d_p U_{mf} $c_f \rho_f$ and, hence, increasing with the growth of particle size and their density, increases.

REFERENCES

Todes, O.M., Antonishin, N.V., Simchenko, L.E. & Lushchikov, V.V. (1970) J.Eng.Phys. 18 , 815.

Baskakov, A.P. (1968) 'Skorostnoi Bezokislitelnyi Nagrev i Termicheskaya Obrabotka v Kipyashchem Sloe' Metallurgiya, Moskva.

Zabrodsky, S.S., Borodulya, V.A., Ganzha, V.L. & Kovensky, V.I. (1973) Izv. AN BSSR , ser. Fiz.-Energet. Nauk, 3 , 103.

Botterill, J.S.M (1975) 'Fluidized Bed Heat Transfer', Academic Press.

Ernst, R. (1960) Chemie Ing. Techn. 12 , 17.

Kunii, D., Levenspiel, O. (1969) 'Fluidization Engineering', J.Wiley.

Neukirchen, B., Blenke, H. (1973) Chemie Ing.Techn. 45 , 307.

Gelperin, N.I., Einshtein, V.G. (1971) 'Fluidization', Chapter 10 (J.F.Davidson et al. Eds) Academic Press.

Werther. J. (1976) Powder Technology 15 , 155.

Whitehead, A.B., Dent , D.C. (1967) Proceedings of the International Symposium on Fluidization, p.802, Session 9, 6-9 June, Eindhoven, Netherlands.

Zabrodsky, S.S. (1966) 'Hydrodynamics and Heat Transfer in Fluidized Beds', MIT Press.

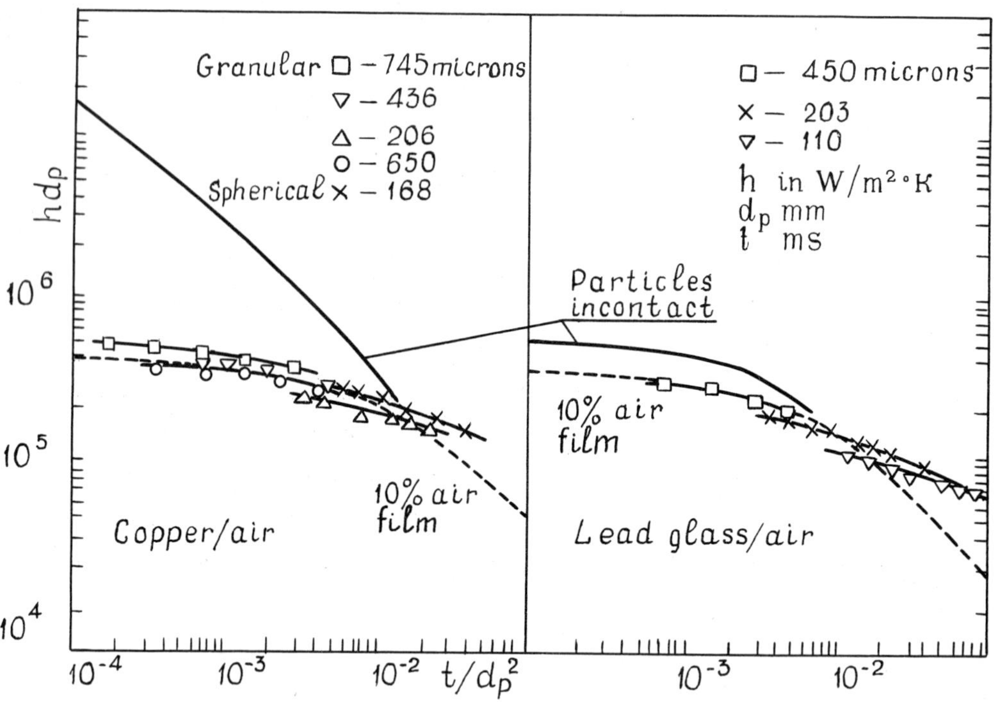

Fig. 1. The comparison of theoretical and experimental data on heat transfer between a wall and beds of moving particles, Botterill at al (cf.Botterill, 1975).

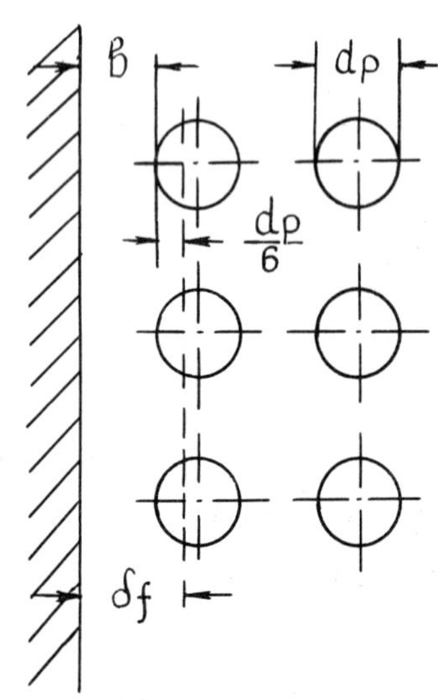

Fig. 2. The scheme to Zabrodsky's model of fluidized bed-to-wall heat transfer (cf.Zabrodsky,1966).

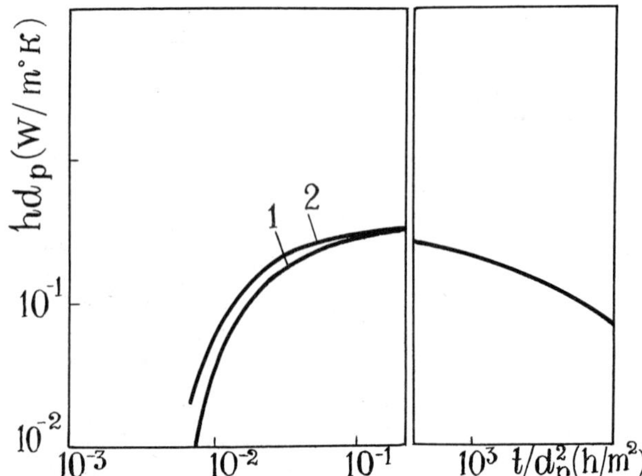

Fig. 3. The $hd_p = f(t_{ex}/d^2_p)$ plot for beds of graphite particles, conduction and radiation being taken into account (cf. Zabrodsky, 1973). 1 – d_p = 4 mm , 2 – d_p = 2 mm. T_w = 1000°C, T_b = 900°C, $h = h_{cd} + h_r$

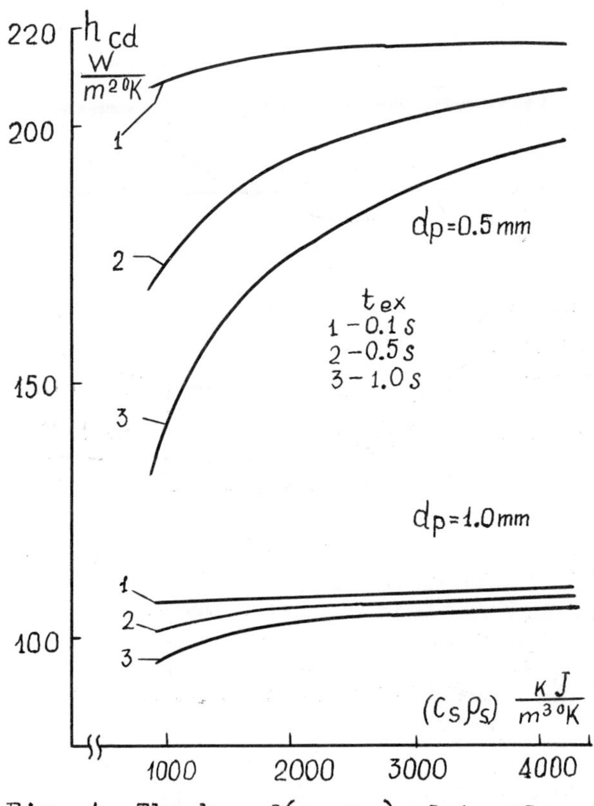

Fig. 4. The $h_{cd}=f(c_s\,\rho_s)$ plot calculated by Equation (1) at $\varepsilon = 0.476$, $H/H_o=1.1$, $\delta = d_p/6$, $k_f=2.47\cdot10^{-2}$.

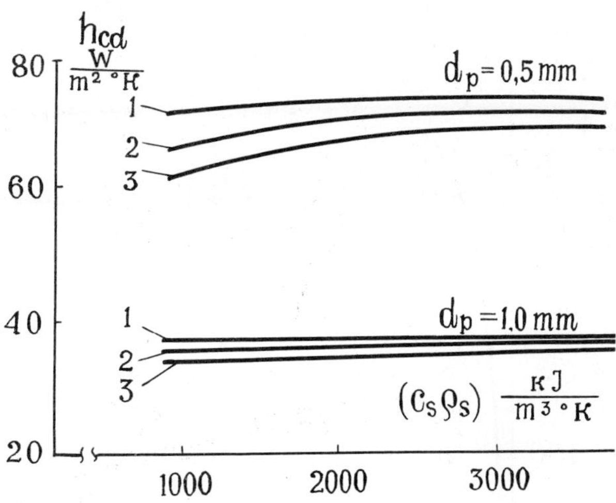

Fig. 5. The $h_{cd} = f(c_s\,\rho_s)$ plot by Equation (1) at $\delta = d_p/2$, other parameters being the same as in Fig.4 caption.

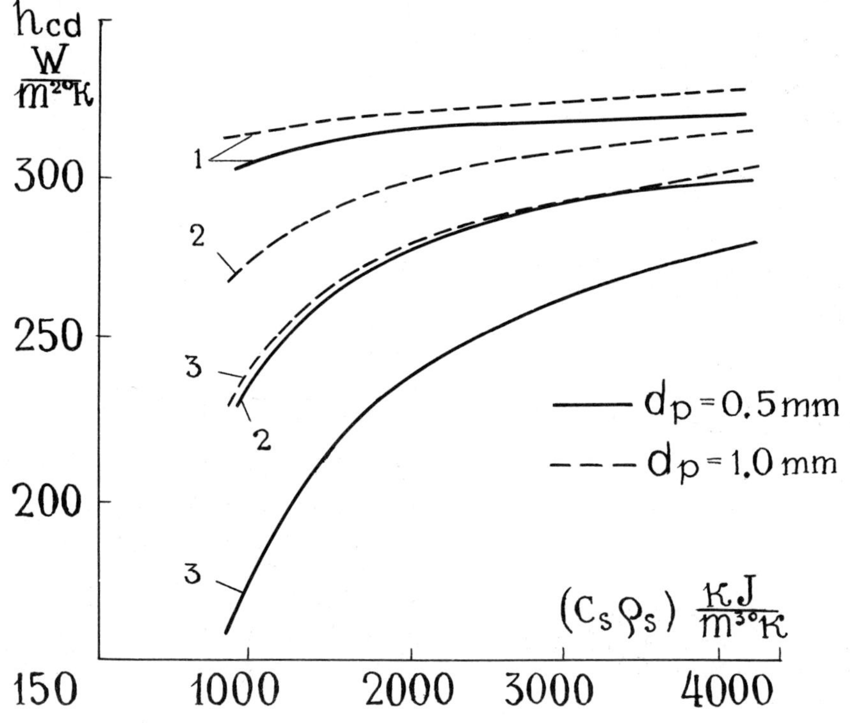

Fig. 6. The conventional $h_{cd}=f(c_s\,\rho_s)$ plot by Equation (1). $k_f=3.7\cdot10^{-2}$ for $d_p=0.5$ mm, $k_f=7.4\cdot10^{-2}$ for $d_p=1$ mm, other parameters being the same as for Fig.4.

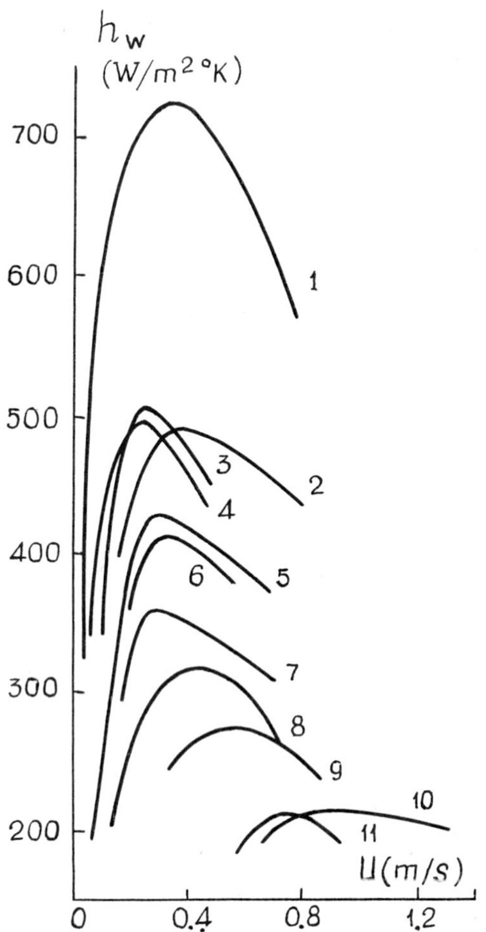

Fig. 7. Wall heat transfer coefficients as functions of the superficial fluid velocity, data of Varysin and Martyushin (cf. Zabrodsky, 1966).
Particle sizes in microns: 1) ferrosilicon, d_p=82.5, 2) hematite, d_p=173, 3) carborundum, d_p=137, 4)-10) quartz sand, d_p=140, 198, 216, 428, 515, 650, 1100 respectively; 11) glass spheres, d_p=1160.

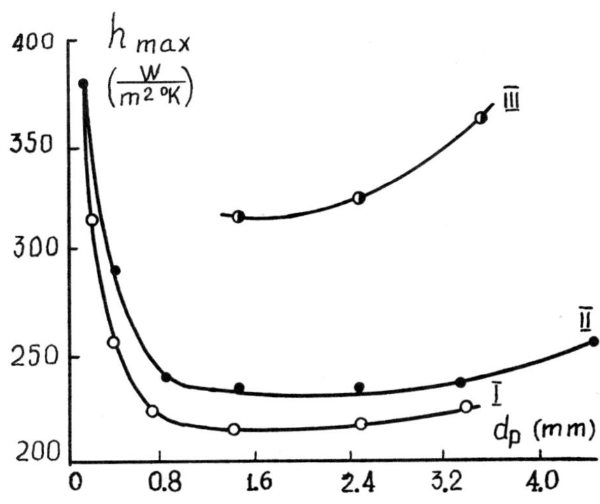

Fig. 8. h_m=f(d_p) by Traber et al. (cf. Zabrodsky, 1966)
I. BAV catalyst, bed diameter D_T=49 mm
II. BAV, D_T=73 mm
III. alumina-nickel catalyst, D_T=49 mm.

ON PARTICULATE PHASE MOTION AROUND A VOID IN A FLUIDIZED BED

by R.Y.Qassim (COPPE/UFRJ) and M.Saddy (Centro de Tecnologia/Promon)

C.P.1191 - 20.000 Rio de Janeiro

Brazil

SYNOPSIS

The Davidson-Harrison theory for particulate phase motion in gas-fluidized beds is used to solve problems of bubble motion near several types of bed boundary encountered in practice. In particular, the effects of bed side wall, bed distributor, and bed upper surface, on bubble motion are studied.

NOTATION

ϕ_s	solid velocity potential
p_f	fluid pressure
x,y,z	Cartesian coordinates
η,ξ,z	bi-cylindrical coordinates
η_o	coordinate curve representing bubble boundary
U_o	incipient fluidization velocity
U_b	bubble velocity
a	distance in bi-cylindrical coordinate system
k	Darcy equation parameter
p_o	bubble pressure
m	infinite series index

INTRODUCTION

In their pioneering work on the fluid-mechanical aspects of fluidization, Davidson & Harrison (1963) showed that several aspects of the motion of bubbles in gas-fluidized beds can be described by a theory based on the premise that the solid particles and fluid outside the void behave as a mixture of two coexisting interacting continuous media. Jackson (1963a, 1963b) relaxed the constant voidage assumption in the Davidson-Harrison theory and showed that the assumptions of constant voidage and irrotational solid flow are valid everywhere except in a narrow region near the upper surface of the bubble. Collins (1965) using the Davidson-Harrison equations studied the effect of bubble shape on the gas and solid flow and concluded that only regions very close to the bubble boundary are affected. Further, Collins (1965) showed that a bubble moving on the axis of a finite two-dimensional bed displayed the same type of behaviour as a bubble in an infinite bed. Most of the features of bubble motion in a fluidized bed are described satisfactorily by the Davidson-Harrison theory in contradistinction to bubble shape. Grace (1970) using an empirical relationship between bubble included angle and particulate phase viscosity concluded that the latter lies in the range 4 to 13 poise, indicating a highly viscous medium outside the bubble. Davidson *et al.* (1977) provided some theoretical basis for Grace's calculations but noted correctly that such an anomaly can only be rigorously resolved by a solution of the free boundary problem, a not inconsiderable task for a single phase, let alone for a two-phase mixture such as the particulate phase in a fluidized bed.

In this paper, the Davidson-Harrison theory is used to study three problems of bubble boundary interaction, the boundaries representing the bed side wall, the bed distributor, and the bed upper surface.

BUBBLE BOUNDARY INTERACTION

As noted above, in so far as the prediction of solid and fluid motions in the particulate phase are concerned, the Davidson-Harrison equations

$$\nabla^2\phi_s = 0 \quad , \tag{1}$$

$$\nabla^2 p_f = 0 \quad , \tag{2}$$

along with the assumption of a circular (spherical) bubble shape are adequate. The remaining variables of interest such as the solid and fluid velocities are obtained in a straightforward manner from a knowledge of the solid velocity potential and the fluid pressure. Three problems of bubble boundary interaction are considered.

(a) Bubble-impermeable rigid boundary
(b) Bubble-porous rigid boundary
and (c) Bubble-porous deformable boundary.

These three cases represent the effect on single bubble motion of bed side wall, bed distributor, and bed upper surface respectively. The bubble is assumed to be at a uniform pressure p_o, and rising vertically at a speed U_b. Far away from bubble and boundary alike, the bed is assumed to be at incipient fluidization with zero solid velocity and an upward velocity U_b.

Using bi-cylindrical coordinates (ξ,η,z) given by

$$x = \frac{a \sinh \eta}{\cosh\eta + \cos\xi}, \quad y = \frac{a \sin \xi}{\cosh\eta + \cos\xi}, \quad z = z, \tag{3}$$

the boundary conditions of these three problems may be set as follows (*cf.* Figures 1-3):

(a) bubble-impermeable rigid boundary.

$$\eta = 0, x,y \text{ bounded}; \; (\nabla\phi_s)_\eta = 0, \; (\nabla p_f)_\eta = 0,$$

$$\eta = 0, x\to\infty \text{ or } y\to\pm\infty; \; (\nabla\phi_s)_\eta \;\; 0, \; (\nabla p_f)_\eta \to -\frac{U_o}{k}$$

$$\eta = \eta_o; \; (\nabla\phi_s) = \frac{U_b \sin\xi \sinh\eta}{(\cosh\eta + \cos\xi)}, \; p_f = -p_o. \tag{4}$$

(b) bubble-porous rigid boundary.

$$\eta = 0, \; (\nabla\phi_s) = 0,$$

$$\eta = \eta_o, \; (\nabla\phi_s)_x = U_b, \; (\nabla\phi_s)_y = 0 \tag{5}$$

(c) bubble-porous deformable boundary.

$$\eta = 0; \; (\nabla\phi_s) = \frac{U_o}{\cosh\eta + \cos\xi} + k(\nabla p_f),$$

$$\eta = \eta_o; \; (\nabla\phi_s)_x = -U_b, \; (\nabla\phi_s)_y = 0, \; p_f = -p_o. \tag{6}$$

De Moraes *et al.* (1974) using the method of separation of variables, found the exact solution for the fluid pressure field problem (b). Utilizing the same method, we found the exact solutions for the solid velocity potential field of problems (a), (b) and (c), and for the fluid pressure field of problems (a) and (c). We summarise our solutions along with that of De Moraes below:

(a) bubble-impermeable rigid boundary:

$$\phi_s = 2aU_b \sum_{m=1}^{\infty} (-1)^m e^{-m\eta_o} \frac{\cosh m\eta}{\sinh m\eta_o} \sin m\xi$$

$$p_f = p_o - \frac{aU_o}{k} \frac{\sin\xi}{\cosh\eta+\cos\xi} - 2a\frac{U_o}{k} \sum_{m=1}^{\infty} (-1)^m e^{-m\eta_o} \times \frac{\cosh m\eta}{\cosh m\eta_o} \sin m\xi \tag{7}$$

(b) bubble-porous rigid boundary:

$$\phi_s = -2aU_b \sum_{m=1}^{\infty} (-1)^m e^{-m\eta_o} \frac{\cosh m\eta}{\sinh m\eta_o} \cos m\xi$$

$$p_f = -\frac{aU_o}{k}\left(1+2\sum_{m=1}^{\infty} (-1)^m \frac{\sinh m(\eta_o-\eta)}{\sinh m\eta_o} \cos m\xi\right) \tag{8}$$

(c) bubble-porous deformable boundary:

$$\phi_s = \sum_{m=1}^{\infty} 2aU_o(-1)^m e^{-m\eta_o} \frac{\sinh m\eta}{\sinh m\eta_o} \cos m\xi + \sum_{m=1}^{\infty} 2aU_b(-1)^m e^{-m\eta_o}\left(1 - \frac{U_o}{U_b} \coth m\eta_o\right) \times \frac{\cosh m\eta}{\sinh m\eta_o} \times \cos m\xi$$

$$p_f = \frac{aU_o}{k}\left| \frac{\sinh\eta}{\cosh\eta+\cos\xi} + \left(\frac{p_o k}{aU_o} - 1\right)\frac{\eta}{\eta_o} - 2\sum_{m=1}^{\infty} (-1)^m e^{-m\eta_o} \frac{\sinh m\eta}{\sinh m\eta_o} \cos m\xi \right| \tag{9}$$

As required by the continuity property of the solutions, each of these solutions tend to the Davidson-Harrison solution for a bubble in an infinite fluidized bed as the bubble-boundary distance tends to infinity.

CONCLUSION

These solutions provide the basis for the determination of several effects of finite beds such as the side wall, distributor, and top surface. In each case, it is possible to determine the existence of gas clouds around the bubble. For the bed surface, the solution provided may serve as a basis for determining the initial deformation of the top surface as a bubble approaches it. As noted by Davidson *et al.*, the problem of bubble shape prediction awaits an improvement in the constitutive equations proposed for the particulate phase.

REFERENCES

Collins, R. (1965) Chem.Eng.Sci. 20, 747.
Davidson, J.F., Harrison, D. & Guedes de Carvalho, J.R.F. (1977) Ann.Rev.Fluid Mech. 9, 55.
Grace, J.R. (1970) Can.J.Chem.Eng. 48, 30.
Jackson, R. (1963a) Trans.Inst.Chem.Engrs 41, 13.

Jackson, R. (1963b) Trans.Inst.Chem.Engrs 41, 22.
De Moraes, F.F., Qassim, R.Y. & Saddy, M. (1974) - II National Meeting on Flow in Porous Media - Rio Claro - Brazil.

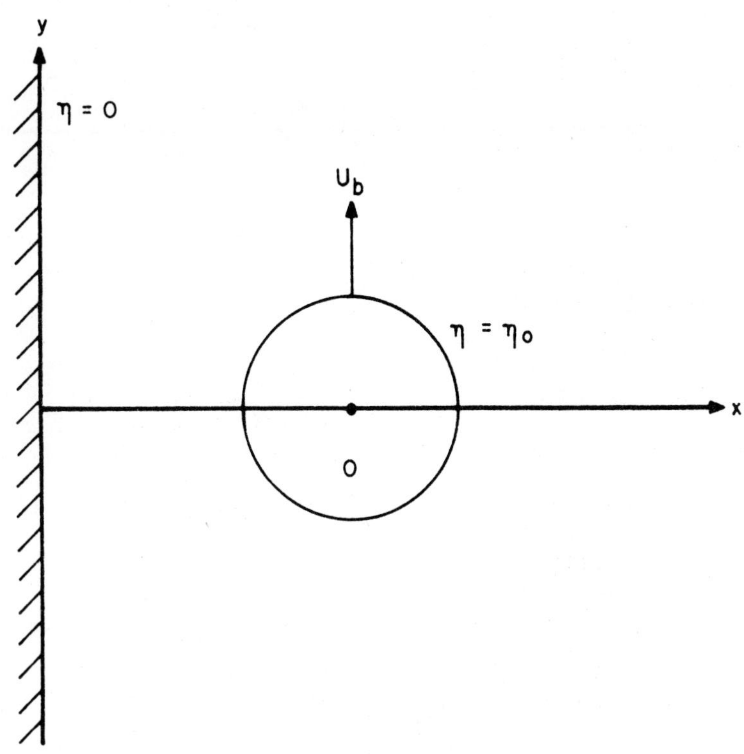

FIG. I - Void - Impermeable boundary problem.

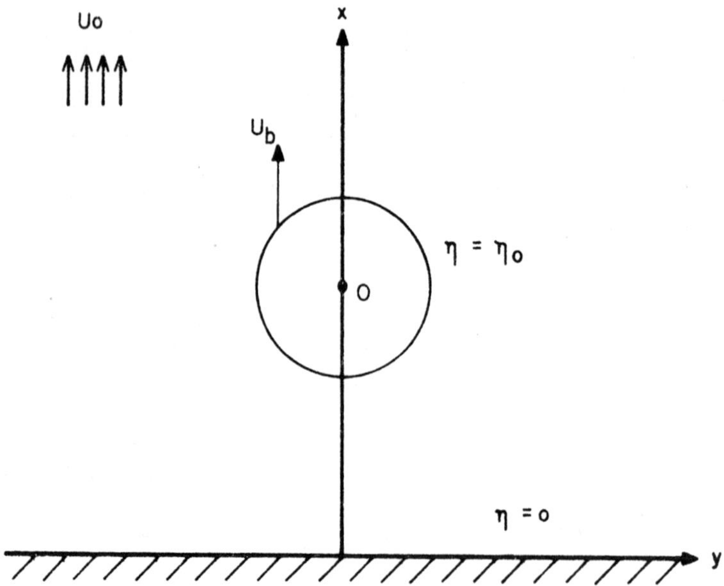

FIG. 2 - Void - Porous boundary problem.

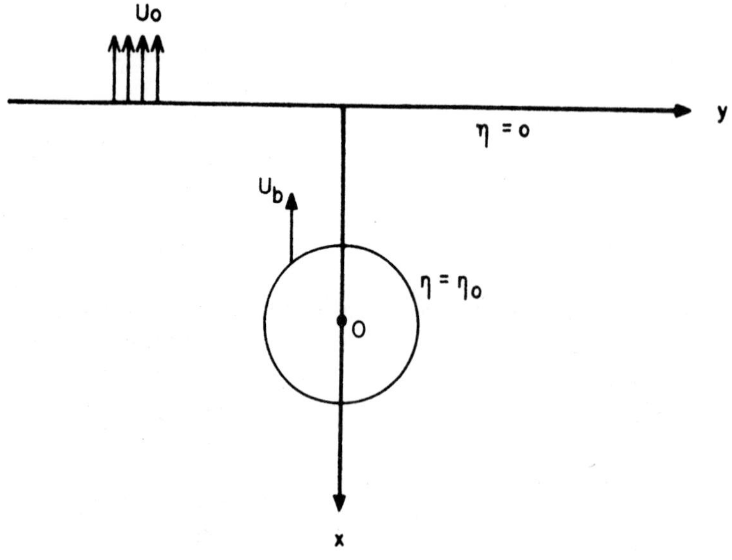

FIG. 3 - Void - Deformable Boundary Problem.

MECHANISMS OF MOMENTUM AND HEAT TRANSFER BETWEEN GAS JETS AND FLUIDIZED BEDS

By S. Donadono and L. Massimilla

Istituto di Chimica Industriale e Impianti Chimici, Università

Laboratorio di Ricerche sulla Combustione, C.N.R.

P.le V. Tecchio - 80125 Napoli (Italy)

SYNOPSIS

Gas injection in fluidized beds of coarse, as well as of fine particles, is characterized by high rates of momentum and heat transfer between the jet and the particulate phase. As shown by experiments with a jet in a two-dimensional bed, this can be related to the entrainment of solids into the expanding jet and to the effect of particles acting as exchange centers throughout the gas stream. With coarser materials gas entrainment from the bed might become comparatively relevant.

NOTATION

b_o	nozzle half-width
D_p	particle diameter
T_b	bed temperature
T_{gm}	axial gas temperature
T_{go}	inlet gas temperature
U_f	gas fluidization velocity
U_{mf}	minimum gas fluidization velocity
U_m	jet axial gas velocity
V_m	jet axial particle velocity
x	distance from the nozzle
x_p	jet potential core penetration
y	radial distance from jet axis
W_{go}	inlet jet gas mass flow-rate
W_s	mass flow-rate of entrained solids
γ	volume solids concentration in the jet
ε_b	bed voidage
θ	jet cone half-angle
ρ	particle density

INTRODUCTION

The dispersion of gas jets in fluidized beds is characterized by high rates of momentum and heat transfer. This is shown by the rapid decay of axial gas velocity downstream the nozzle (Schakhova, 1968; Behie et al., 1971; Schakhova & Minayev, 1972; Minayev,1975) as well as of temperature, when gas is injected in non-isothermal beds (Schakhova & Lastovceva, 1973; Behie et al., 1975; Donsì et al., 1975).

De Michele et al. (1976, 1977) have recently shown that momentum and heat transfer occurring in the dispersion of gas jets in fluidized beds of fine powders might be consistently related to the entrainment of particles into the jet. The model they propose is based on the turbulent jet theory (Abramovich, 1963). It accounts for the rates at which the expanding jet is charged with solids, and heat is exchanged between gas and solids, with the assumptions that entrained particles approximate local gas velocities and that the influence of gas entrainment from the bed is negligible. The model implies that a quasi-stationary, isobaric jet develops downstream from the nozzle and that the jet main region expands rectilinearly. It also assumes that "similar" radial profiles of jet gas velocities and temperatures, and solids velocities and concentrations develop beyond a certain distance x from the nozzle. Thus, once the rate of jet expansion is known, integration of these profiles over the jet cross-section gives mass, momentum and enthalpy of jet admixture as a function of x. Model equations are derived from:

- the conservation of mass flow rate of gas
- the conservation of momentum rate of jet admixture
- the conservation of enthalpy flow-rate of jet admixture
- the change of gas enthalpy flow-rate related to heat transfer to entrained solids.

Jet core penetrations and axial temperature decays evaluated by means of this model agree with results of experiments of heat transfer between gas jets and beds of fluid cracking catalysts ($D_p \simeq 60 \mu m$), carried out by Behie et al. (1975) and Donsì et al.(1975) in three-dimensional apparatus with various inlet gas chemical compositions, temperatures and velocities, and nozzle diameters.

This paper summarizes work carried out in a two-dimensional fluidization unit with the objective of:

- studying the solid motion induced by gas injection in fluidized beds
- testing assumptions of the model proposed by De Michele et al. (1976, 1977), when applied to beds made of coarse materials.

Details of experimental results and theoretical calculations shall be reported elsewhere (Donadono et al., 1978).

EXPERIMENTAL

Investigation was carried out in a 300x15 mm rectangular cross-section column with transparent walls. The bed was fully fluidized by air, using a porous plate as distributor. Air jet discharged vertically upwards, at nozzle velocity of 90 m/sec. The nozzle, 15x2 mm rectangular cross-section, was placed, flush to the distributor, in the column mid-plane.

Properties of fluidized solids are given in Table I, with symbols used in Figures 2-5 for each material. Bed heights were between 30 and 50 mm. With such shallow beds, bubble formation at the end of the jet was prevented, and its steadiness much improved.

Pitot tube and high speed photography techniques were used for measuring gas and particle velocities, respectively. The tip of the Pitot tube, 0.7 mm OD, was positioned inside the jet by means of a micrometrical x-y slide. Static pressures were also measured inside and in the neighborhood of the jet. Motion pictures were taken of particles moving in the jet and in the surrounding fluidized bed. The latter measurement, in particular, could be influenced to some extent by wall effects. An impact probe, 1 mm OD, was used for measuring the number of particles flowing per unit time through the jet cross-section. This and the measured average particle velocity enabled the evaluation of γ, the local solids volume concentration. The strain gage transducer was effective in detecting the particle shock on the plane surface of the probe in the case of bronze, limestone and plastic beads. With alumina the kinetic energy involved in each collision was smaller than required by the present state of probe development. An independent evaluation of γ was also made from the photographic study of the jet in the case of plastic beads.

In heat exchange experiments, the inlet gas was at 90°C, with the bed at about room temperature. Temperature measurements were made by means of thermocouples.

RESULTS

A typical pattern of solid flow induced by the jet in a fluidized bed of coarse material is shown in Figure 1. Trajectories of limestone particles are more or less horizontal, similar to those describing the entrainment of fluid particles in an homogeneous jet. Solid flow pattern when the unit is operated as a spouted bed (U_f=0) is presented for comparison. Deviations of fluidized particles upwards are assumed to spot, at various levels, the places where the solid material is entrained into the jet. These are fairly well aligned on a straight line, which, for this material, is inclined at an angle θ=22° to the vertical. Figure 1 confirms that jet expansion is almost rectilinear, and that a linear increase of the jet cross-section with x can be assumed in model calculations.

Similar conclusions apply to other materials, except alumina, for which the rate of expansion decreased somewhat towards the top of the jet. Also, with this material the jet showed some fluctuation, in spite of the shallowness of the bed.

Jet cone half-angles have been checked by independent measurements of gas velocities. The range of θ in Table I is comprehensive of results obtained in the exploration of jet border line by means of Pitot tubes. For alumina it also accounts for the non-linearity of jet expansion and its fluctuation. In agreement with previous results by Merry(1975), θ tends to increase as the particle size and density increase.

TABLE I

Material	D_p, μm	ρ, g/cm^3	U_{mf}, cm/sec	U_f, cm/sec	θ
Alumina ●	212-300	1.55	2.8	16	13°-19°
Bronze ■	355-420	8.5	41	51	19°-21°
Limestone ▼	710-950	2.6	48	78	21°-23°
Plastic beads ▲	2800-3000	1.2	89	124	27°-30°

Static pressures inside the jet followed rather closely pressure profiles in the bed. Some deviation was noticed with plastic beads, but not so large as to invalidate the assumption of the isobaricity of the jet.

The penetration of jet potential core x_p for materials tested is between 1 and 4 b_o. This is shorter than that in an homogeneous jet, but, in the average, not as much as that with gas jetting in fine fluid cracking catalyst, where x_p was about the nozzle radius. (De Michele et al., 1976). Also, rather surprisingly, the largest value of x_p was obtained with alumina, which was the finest of the materials used. More experiments are required to clarify this behaviour, considering that, apart from the difference in particle size, previous experiments with catalyst powder were carried out with fluctuating jets in a deep three-dimensional bed.

Analysis of radial distributions of particle and gas velocity in the jet also show that for coarse, as well as for fine particles, the hypothesis of similarity of radial profiles of jet properties is reasonable at $x > 1.5x_p$. As a first approximation, Schlichting type radial profiles for homogeneous, turbulent jets might also be assumed for gas jets in beds of coarse solids.

These results suggest that, as a first approximation, the turbulent jet theory can be considered in gas jetting in beds of coarse particles. Figures 2a and 5 give axial jet gas velocity and temperature profiles. As expected from previous experiments with fine particles (Behie et al., 1971, 1975; Donsì et al., 1975), decays are noticeably fast considering the two-dimensional geometry of the jet.

Figure 2 shows that there is a considerable difference throughout the jet between gas and particle velocities. The latter increases rapidly at first, then tends to level off at shorter distances from the nozzle, the greater the size and density of bed particles are. Changing from alumina to plastic beads, particle velocities at this stage of acceleration vary from about 8 to 2 m/sec.

On the contrary, velocities of solids moving horizontally in the dense bed phase towards the jet increases with the size and density of bed particles, being, on average, about 0.05 and 0.18 m/sec for alumina and plastic beads, respectively. This is shown in Figure 3, in terms of mass flow-rates of entrained solids. In passing from particle velocities to solids flow-rates, bed voidage ε_b=0.5 has been assumed.

Figure 4 presents data of volume solids concentration as obtained along the jet axis from experiments. It appears that values of γ increase from 0.03 to 0.2 from bronze to plastic beads. This brings axial gas velocity decays in Figure 2a close to each other for limestone and bronze, and for plastic beads, in spite of the considerably greater jet aperture with this material (Table I).

The consistency of experimental results presented above has been tested by means of theoretical calculations, whose results are given by curves in Figures 2 to 4. These calculations have been carried out by assuming Schlichting type radial profiles for gas and solids velocities, and rectangular radial profiles for solids concentrations. Theoretical evaluations have been restricted, for each material, to the main jet region.

Curves in Figure 2a have been obtained on the basis of observed jet cone half-angle and volume solids concentration from the equation of mass flow rate of gas. In the calculations for alumina, γ (not available from experiments) has been considered negligible. For bronze, limestone and plastic beads discrepancies between calculated and experimental gas velocities might be related to the gas entrai nment into the jet observed by Merry (1976). With alumina discrepancy should be related to the delay in the development of the jet main region observed for this material and to the non-linearity of the jet expansion.

For simplicity, calculated particle velocities on the jet center line shown in Figure 2b have been evaluated disregarding solid to solid interaction, and assuming vertical motion. At this stage of the study, drag coefficients were obtained from standard curve for single sphere, without considering the influence of turbulence and change in gas to particle relative velocity. From Figures 2a and b it appears that relative velocities $(U_m - V_m)$ approach terminal particle velocities at x/b_o above 20.

Curves of mass flow rates of entrained particles in Figure 3 have been obtained from the equation of conservation of momentum rate of jet admixture on the basis of observed axial velocities of gas and solids. In agreement with experimental observation, these calculations show that solids entrainment increases with the size and density of particles. However, the calculated values for plastic beads (not reported in Figure 3) are larger than the experimental by a factor of about 2. This might reflect the underextimation of gas velocity by the Pitot tube in the suspension of large particles at high solids volume concentration.

Curves of volume solids concentration given in Figure 4 have been obtained from the equation of conservation of solids flow rate on the basis of measured mass flow rate of entrained solids, axial particle velocity and jet cone angle.

Some results of heat transfer experiments at inlet gas velocity of 90 m/sec are presented in Figure 5. Changing particle size and density affects: gas and particle residence time in the jet, solids mass entrained per unit mass of injected gas, exposed particle surface per unit mass of entrained solids,gas to particle heat transfer coefficient. In addition, gas entrainment should be considered. Consequently, there is a fast initial decay of axial temperature profiles particularly for limestone and plastic beads. With the latter materials, heat transfer resistances inside the particle also play a role. This might explain the inversion in temperature profiles between bronze, limestone and plastic beads. The delay in the development of the jet main region for alumina is also shown by these heat transfer results.

Experimental data given in Figures 2 and 3 have been used in the theoretical evaluation of gas temperature decays shown for plastics and bronze by curves in Figure 5. Calculations have been based on the equations of the conservation of enthalpy flow rate of jet admixture and of the change of gas enthalpy flow rate related to heat transfer to entrained solids. For plastic beads the heat transfer internal resistance has been taken into account. The assumption of Schlichting type profiles for gas and solids velocities as well as temperatures has been made throughout the calculations.

CONCLUSIONS

The general trend of solids flow pattern indicates that the turbulent jet theory might be a basis for studying momentum and heat exchange mechanisms between the jet and the bed also in the case of coarse bed materials.

Increase in particle size and density produces increases in both jet cone angle and solids concentration in the jet. In addition, with larger particles the contribution of gas entrainment from the bed to transfer mechanisms seems to be relatively important in respect to that of solids entrainment.

Slip velocity between gas and solid is high throughout the jet for all the materials tested. Thus, in respect to the model by De Michele et al. (1976, 1977), a separate evaluation has to be made of momentum contribution of the two phases of the jet admixture and another equation considered in order to account for the finite rate of momentum exchange between the gas and solids. This equation expresses the change of gas momentum rate due to momentum transfer, by drag, to the entrained particles. Promising studies to modify the original model along this line are in progress.

ACKNOWLEDGMENT

One of us (S. Donadono) would like to thank the ENEL/CRTN (Milan) for support during the course of this research work.

REFERENCES

- Abramovich, G.N. (1963) The theory of turbu

lent jets, Mass. Inst. Technol. Press, Cambridge.
- Behie, L.A., Bergougnou, M.A., Baker, C.G.J. & Base, T.E. (1971) Can. J. Chem. Eng. 49, 557.
- Behie, L.A., Bergougnou, M.A. & Baker, C.G.J. (1975) Can. J. Chem. Eng. 53, 25.
- De Michele, G., Elia, A. & Massimilla, L. (1976) Ing. Chim. Ital. 12, 155.
- De Michele, G., Donsì, G. & Massimilla, L. (1977) European Congress Transfer Processes in Particle Systems, Nüremberg, 28-30 March vol. II.
- Donadono, S., Maresca, A. & Massimilla, L. (1978), paper to be submitted to Ing. Chim. Ital.
- Donsì, G., Massimilla, L. & Russo, G. (1975) Fluidized Combustion Conference, London, September, vol. I.
- Merry, J.M.D. (1975) AIChE J. 21, 507.
- Merry, J.M.D. (1976) AIChE J. 22, 315.
- Minayev, G.A. (1975) Khim. prom. 9, 694.
- Shakhova, N.A. (1968) Inzh. Fiz. Zh. 14, 61.
- Shakhova, N.A., Minayev, G.A. (1972) Heat Transfer - Soviet Research, 4, 133.
- Shakhova, N.A., Lastovceva, I.N. (1973) Inzh. Fiz. Zh. 25, 581.

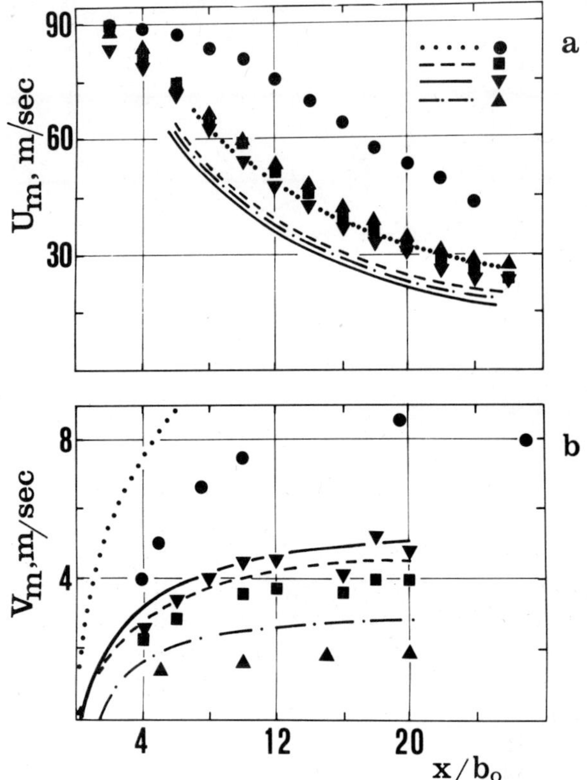

Fig. 2. Jet axial gas (a) and particle (b) velocity profiles.

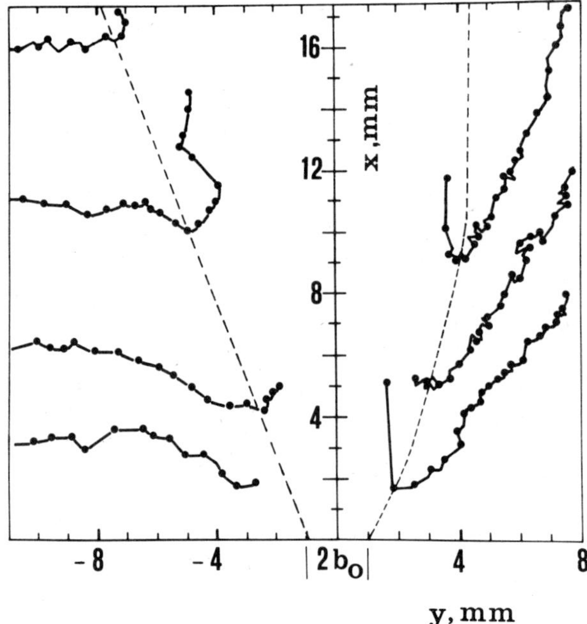

Fig. 1. Motion of solids induced by a jet discharging at 90 m/sec in a fluidized (left) and an unfluidized (right) bed of limestone particles. Time interval between two points about 5.10^{-3} sec.

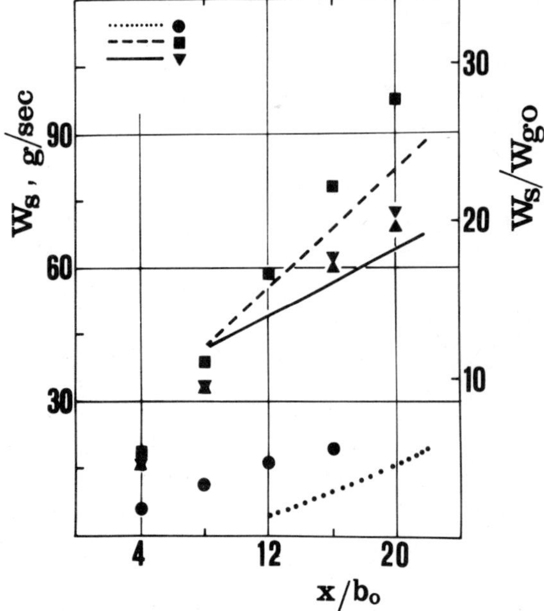

Fig. 3. Mass flow-rates of entrained particles (or mass flow-rates of entrained particles per unit mass flow-rate of injected gas) at different distances from the nozzle.

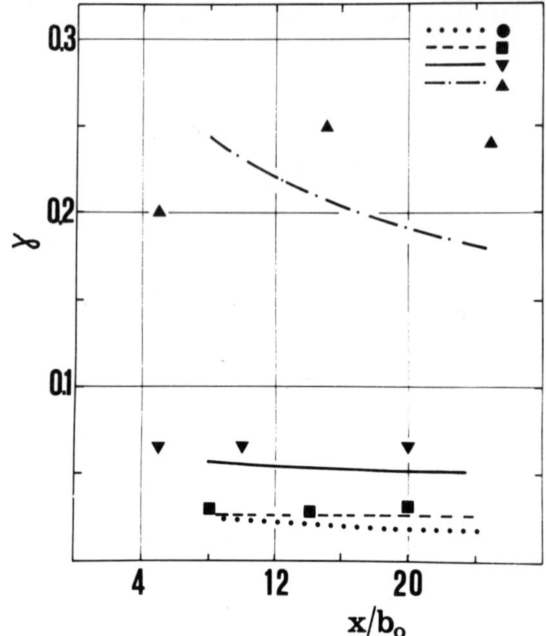

Fig. 4. Jet axial profiles of volume solids concentration.

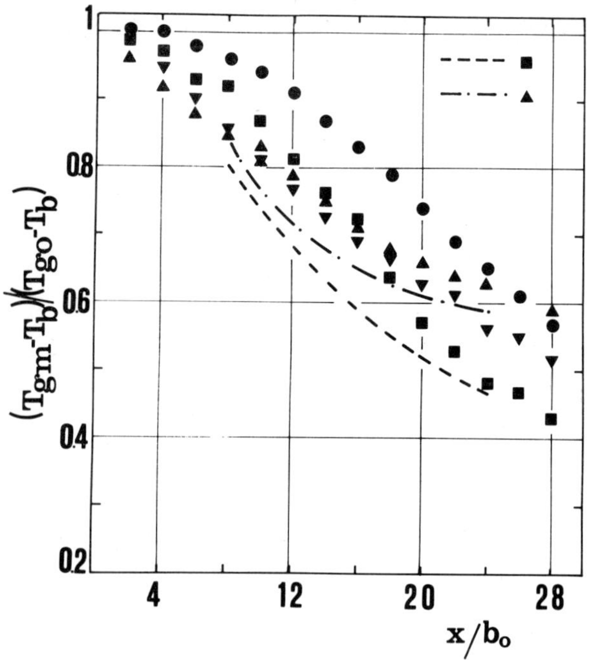

Fig. 5. Jet axial gas temperature profiles.

A METHOD FOR PREDICTING THE RELATIONSHIP BETWEEN THE SPOUT AND INLET TUBE RADII IN A SPOUTED BED AT ITS MAXIMUM SPOUTABLE HEIGHT

Howard Littman
Department of Chemical and Environmental Engineering
Rensselaer Polytechnic Institute
Troy, N. Y. 12181, USA

Morris H. Morgan III
General Electric Corporate R&D Center
Schenectady, N.Y. 12301, USA

Dragoljub V. Vuković and Fedor K. Zdanski
Faculty of Technology and Metallurgy
University of Belgrade
Belgrade, Yugoslavia

Željko B. Grbavčić
Institute for Chemistry, Technology
and Metallurgy
Belgrade, Yugoslavia

SYNOPSIS

A method is presented for predicting the ratio, r_S/r_i, in spouted beds at their maximum spoutable height which shows that the ratio is a function of two parameters r_c/r_i, and $A = [\rho_f/(\rho_p - \rho_f)][u_{mF}u_T/gr_i]$. Experimental values of r_S/r_i differ from predicted ones by only about 6% on average.

As a part of this development, it is shown that

$$\frac{H_m r_i}{r_c^2} = [0.435 + (0.020/A)]$$

which provides a means for calculating H_m if r_c, r_i and A are known.

NOTATION

A	= $[\rho_f/(\rho_p - \rho_f)][u_T u_{mF}/gr_i]$		K	= a parameter
Ar	= Archimedes number = $d_p^3(\rho_g - \rho_f)g/\nu^2\rho_f$		m	= $H_m r_i/r_c^2$
c_D	= drag coefficient		r_i	= spout inlet tube radius
d_i	= spout inlet tube diameter		r_c	= column radius
d_p	= particle diameter		r_S	= spout radius
d_S	= spout diameter		Re_{mF}	= Reynolds number = $d_p u_{mF}/\nu$
D_c	= column diameter		Re_T	= Reynolds number = $d_p u_T/\nu$
g	= gravitational acceleration		u_i	= fluid velocity in inlet tube
H_m	= maximum spoutable height			

u_{mF} = minimum fluidizing velocity
u_S = minimum spouting velocity
u_T = terminal velocity of a particle
ε_A = voidage in the annulus
ε_{mF} = voidage at the minimum fluidizing velocity
ρ_f = fluid density
ρ_p = particle density
ν = kinematic viscosity of the fluid

INTRODUCTION

In a recent paper, a theory was presented for predicting the maximum spoutable height in a spouted bed where fluidization of the annular solids limits the penetration of the fluid jet entering the bed (Littman, et al, 1977). That theory is based on an analysis of flow in the annulus of a spouted bed and the pressure and velocity distributions there were calculated using continuity, the vector form of the Ergun equation (Stanek and Szekely, 1974) and appropriate boundary conditions. The spout diameter was calculated from McNab's correlation (McNab, 1972).

The relationship between H_m and d_S for spherical particles determined from that analysis is

$$\frac{H_m d_S}{[D_c^2 - d_S^2]} = 0.345\left(\frac{d_S}{D_c}\right)^{-0.384} \quad ....(1)$$

When available experimental values for d_S (Lefroy and Davidson, 1969); Littman, et al, 1977); were inserted into equation (1), the calculated values of H_m differed from the experimental ones by only 8.5% on average. Thus equation (1) provides a self-consistent relationship between H_m and d_S when the spout terminates as described above.

A relationship is developed in this paper between H_m and the inlet tube radius, which is then combined with equation (1) to obtain the ratio of the spout and inlet tube radii.

There is not a great deal in the literature regarding the prediction of this ratio or the variables which determine it. Based on the assumption that the ratio of inlet jet momentum to the force associated with the spout pressure drop is 0.5, Lefroy and Davidson concluded that r_S/r_i is 2.13.

THEORY AND BASIC ASSUMPTIONS

The momentum integral written for a spouted bed of height, H_m, shows that a relationship must exist between H_m and the inlet jet momentum. If we assume that H_m is proportional to that momentum then

$$H_m = K\rho_f u_i^2 r_i^2 \quad ....(2)$$

where K is an empirical parameter.

When fluidization of the annular solids limits the penetration of the inlet fluid jet, $u_{mS} = u_{mF}$ and the flow to the bed is $u_{mF}\pi r_c^2$ (Grbavčić, et al, 1976). By continuity, $u_i = u_{mF}(r_c^2/r_i^2)$. Inserting this relationship into equation 2 and normalizing H_m, we obtain

$$\frac{H_m r_i}{r_c^2} = m \quad ....(3)$$

where $m = K\rho_f u_{mF}^2 r_c^2/r_i$

Literature data show that m is not a constant for different fluid-particle systems. It was therefore assumed to be a function of a single parameter whose form is suggested by the equation for m given above. Specifically, m is assumed to be a function of the ratio of the minimum inlet jet momentum per unit area required to maintain a spout to the pressure drop across the spout at minimum spouting in a bed of height, H_m. The minimum inlet jet momentum per unit area is assumed to be $\rho_f u_T^2$ since the inlet fluid velocity should at the very least equal the terminal fall velocity of a bed particle. For the spout pressure drop, the Lefroy and Davidson result was used. Thus

$$m = F\left[\frac{\rho_f u_T^2}{\frac{2}{\pi}(1 - \varepsilon_{mF})\,(\rho_p - \rho_f)\,g\,H_m}\right] \quad ....(4)$$

Multiplying the numerator and denominator of the function on the right hand side of equation 4 by r_i/r_c^2 and noting that if the inlet fluid velocity is u_T then by continuity,

$u_T = u_{mF}(r_c^2/r_i^2)$ and

$$m = F\left[\frac{\rho_f u_T u_{mF}}{\frac{\pi}{2}(1 - \varepsilon_{mF})(\rho_p - \rho_f)g\,r_i(H_m r_i/r_c^2)}\right] \quad (5)$$

Combining equation 3 and 5 and neglecting the variation of ε_{mF}

$$\frac{H_m r_i}{r_c^2} = f(A) \quad ....(6)$$

where $A = \rho_f/(\rho_p - \rho_f)(u_{mF}u_T/g\ r_i)$

$$= (Re_{mF}\ Re_T/Ar)(d_p/r_i) \qquad ....(7)$$

Values of A for systems described in the literature and tabulated in Tables 2 and 3, show that the range of A is 0.021 to 0.29. A plot of those data as m A vs A is linear (Figure 1) showing that the relationship between m and A is

$$m = 0.435 + (0.020/A)\ ;\ \text{std. dev.} = 0.0235 \qquad ....(8)$$

and therefore that

$$\frac{H_m r_i}{r_c^2} = 0.435 + (0.020/A) \qquad ....(9)$$

Equations 1 and 9 can be combined to show that the ratio, r_S/r_i, is a function of r_c/r_i and A. Thus

$$\frac{r_S}{r_i} = f_1\left[\frac{r_c}{r_i}, A\right] \qquad ....(10)$$

Calculations were made for r_S/r_i in terms of the parameters in equation 10 and the results are presented in Figure 2. In performing the calculations, r_c/r_i and A are specified and r_S/r_i is calculated using equations 1 and 9.

RESULTS AND DISCUSSION

Figure 2 shows that r_S/r_i increases with r_c/r_i and A. Since r_c/r_i generally takes values from 6 to 12, r_S/r_i can vary widely.

The minimum value of $(r_c/r_i)_{min} = (u_T/u_{mF})^{1/2} = 2.9$ assuming $c_D = 0.44$ and $\varepsilon_{mF} = 0.42$ (Richardson, 1971).

Experimental and calculated values of r_S/r_i are compared in Table 3. Overall the average difference is 6% which is probably better than one can measure the ratio. For the Lefroy and Davidson data, the average difference is 5.1%. For the air-glass and water-glass particle data, the average differences are 3.6 and 9.8% respectively.

The A parameter increases with particle size and for large particles Re_{mF} and Re_T are large enough for $Rc_T Rc_{mF}/Ar$ to become constant. Thus asymptotically $A = 0.358\ d_p/r_i$ assuming again $c_D = 0.44$ and $\varepsilon_{mF} = 0.42$. This result predicts that both r_S/r_i and H_m should asymptotically become independent of the fluid used for spouting. This is essentially true experimentally as can be seen from the data in Table 3 for 5.05mm glass particles spouted with air and water.

For large particles ($Re_{mF} > 1000$)

$$H_m = 0.435\ r_c^2/r_i + 0.0559\ r_c^2/d_p \qquad ....(11)$$

a result in reasonable agreement with Lefroy and Davidson's data for peas in Figure 9 of their paper.

The slope of an H_m vs r_i plot for data taken in a given column with particles of a given size is $0.435\ (r_c/r_i)^2$.

The results of this work show that the Lefroy and Davidson theory oversimplifies the prediction of r_S/r_i. Their value of 2.13 is not, in general, a reliable estimate of the actual ratio.

The McNab correlation also leads to erroneous conclusions with respect to the effect of r_c/r_i on r_S/r_i as can be seen in the following example. Consider two identical columns with the same superficial velocity but with inlet tube radii in the ratio of 2 to 1. Since McNab's correlation predicts that r_S would be the same in both cases, the r_S/r_i ratios would be in the ratio of 1 to 2. This is certainly different from the behavior depicted in Figure 2.

If McNab's correlation is substituted into equation (1), the resulting equation predicts that H_m always decreases with d_p but the magnitude of that effect depends dramatically on the particle Reynolds number. Assuming that $(d_S/D_c)^2 \ll 1$, H_m will change from being proportional to $d_p^{-1.35}$ ($Re_{mF} \leq 10$) to being proportional to $d_p^{-0.338}$ ($Re_{mF} \geq 1000$). Literature data (Mathur and Epstein, 1974), however show that a particular fluid-particle system H_m increases initially with particle diameter, passes through a maximum and then decreases. Littman, et al have shown from gas phase spouting experiments with glass beads that the inlet jet terminates due to the formation of slug in the spout where dH_m/dd_p is positive. When the same particles were spouted with water, H_m always decreased with d_p and the spout terminated due to fluidization at the top of the annulus. Throughout this work only those data for which the termination mechanism was probably fluidization of the annular solids were used to compare $(r_S/r_i)_{expt}$ with $(r_S/r_i)_{calc}$ in Table 3.

CONCLUSIONS

1. The ratio, r_S/r_i, is a function of two parameters, r_c/r_i and A. The ratio increases as each of these parameters in-increase.
2. The proposed method predicts available data on r_S/r_i to about 6% on average.
3. The maximum spoutable height can be calculated from the equation

$$\frac{H_m r_i}{r_c^2} = 0.435 + [0.020/A]$$

ACKNOWLEDGMENTS

This work was performed under NSF Grant ENG76-18573.
The calculations were performed at the General Electric Corporate R&D Center.
The experimental work reported here was performed in Yugoslavia.

REFERENCES

Grbavčić, Ž.B., et al (1976) Can. J. Chem. Eng., 54, 33.
Grbavčić, Ž.B., private communication April 20, 1976.
Kunii, D., Levenspiel, O. (1969) Fluidization Engineering, J. Wiley, New York
Lefroy, G. A. (1966) The Mechanics of Spouted Beds, Ph.D. thesis, Univ. of Cambridge, Cambridge, England.
Lefroy, G. A., Davidson, J. F. (1969), Trans. Instn. Chem. Engrs., 47, T120.
Lim, C.J. (1975) Gas Residence Time Distribution and Related Flow Patterns in Spouted Beds, Univ. of British Columbia, Vancouver, Canada.
Littman, H., et al. (1977) Can. J. Chem. Eng. (in press)
Malek, M. A. and Lu, B.C.Y. (1964) Can. J. Chem. Eng., 42, 14.
Mathur, K.B., Epstein, N. (1974), Spouted Beds Academic Press, New York.
McNab, G.S. (1972) Brit. Chem. Eng. and Proc. Technol., 17, 532.
Richardson, J.F. (1971) Fluidization Chapt. 2 (Davidson, J.F. & Harrison, D., Ed.) Academic Press, New York.
Stanek, V., Szekely, J. (1974) A.I.Ch.E. Journal, 20, 974.

Table 1. Flow and Particle Properties of Fluid Particle Systems

System	d_p mm	ρ_p kg/m^3	u_{mF} m/s	ε_{mF} or ε_A	u_T* m/s
air-sand	0.787	2656	0.515	0.460	6.44
air-timothy seeds	0.890	1134	0.323	0.450	4.12
air-brucite	0.991	2664	0.854	0.455	8.16
air-millet	1.98	1174	0.655	0.398	7.66
air-polythene (Malek & Lu)	3.38	921.6	0.788	0.358	8.86
air-glass (Lim)	2.93	2960	1.46	0.420	14.8
air-glass (Grbavčić et al., Grbavčić)	1.25	2700	0.786	0.414	9.23
	2.11	2480	1.19	0.420	11.5
	2.50	2500	1.45	0.422	12.6
	3.09	2510	1.54	0.425	14.0
	4.11	2550	1.91	0.430	16.3
water-glass (Grbavčić et al.)	1.25	2700	0.0120	0.405	0.212
	2.11	2480	0.0250	0.416	0.303
	2.50	2480	0.0320	0.423	0.338
	3.09	2510	0.0365	0.432	0.377
	4.11	2550	0.0500	0.444	0.440
CO_2-kale seeds (Lefroy)	1.74	1000	0.515*	0.420	5.17
air-kale seeds	1.74	1000	0.600	0.42	6.62
air-polythene	3.50	890	0.830	0.39	8.66
air-peas (Lefroy and Davidson)	9.00	1350	1.70	0.38	17.50
air-glass (this work)	1.55	2600	0.820	0.417	10.1
	1.84	2600	0.950	0.418	11.0
	2.11	2480	1.03	0.420	11.5
	3.09	2510	1.33	0.430	14.0
	4.11	2550	1.59	0.435	16.3
	5.05	2520	1.77	0.447	17.8
water-glass (this work)	1.55	2600	0.0166	0.409	0.274
	1.84	2600	0.0216	0.410	0.299
	2.11	2480	0.0250	0.416	0.333
	3.09	2510	0.0360	0.427	0.377
	4.11	2550	0.0487	0.440	0.440
	5.05	2520	0.0584	0.449	0.479

*Calculated using equations in Kunii and Levenspiel, 1969.

Table 2. Values of m and A Calculated from Literature Data

System	d_p mm	r_c mm	H_m mm	r_i mm	m expt.	A expt.
Air-sand	0.787	50	660	4.77	1.26	0.0322
Air-timothy seeds	0.890	50	711	6.25	1.78	0.0231
Air-brucite	0.991	76	508	9.53	1.19	0.0268
Air-millet	1.98	50	290	4.77	0.553	0.110
		50	221	12.7	1.12	0.0414
		76	965	4.77	0.796	0.110
		76	767	9.53	1.26	0.0551
		76	600	12.7	1.32	0.0414
		76	457	15.9	1.26	0.0331
		76	244	19.1	0.805	0.0276
		76	102	22.3	1.64	0.0276
		115	114	19.1	1.78	0.0231
Air-polythene	3.38	76	581	4.76	0.479	0.196
		76	404	9.53	0.666	0.0977
		76	366	12.7	0.805	0.0733
		76	192	19.1	0.633	0.0490
(Malek and Lu)		76	164	25.4	0.722	0.0367
Air-glass (Lim)	2.93	76	420	9.55	0.694	0.076
Air-glass (Grbavčić et al.)	1.25	25.2	106	4.65	0.776	0.0710
		40.0	264	5.15	0.850	0.0641
		55.0	276	7.30	0.666	0.0452
	2.11	25.2	90.0	4.65	0.659	0.144
		40.0	200	5.15	0.644	0.130
		55.0	254	7.30	0.613	0.0924
	2.50	55.0	242	7.30	0.585	0.122
	3.09	25.2	80.0	4.65	0.586	0.237
		40.0	173	5.15	0.557	0.205
		55.0	228	7.30	0.574	0.144
	4.11	40	160	5.15	0.515	0.291
Water-glass	1.25	55	273	7.30	0.659	0.021
(Grbavčić et al.)	2.11	40	197	5.15	0.634	0.103
		55	257	7.30	0.620	0.073
	2.50	55	242	7.30	0.584	0.101
	3.09	40	171	5.15	0.550	0.181
		55	224	7.30	0.541	0.127
	4.11	40	158	5.15	0.509	0.281
CO_2-kale seeds	1.74	38	265	4.75	0.872	0.111
(Lefroy)	1.74	38	208	6.35	0.914	0.083
Air-peas	9.00	152.5	250	42.50	0.457	0.064
(Lefroy and Davidson)		152.5	850	17.50	0.640	0.155
		152.5	900	12.75	0.493	0.213

Table 3. Comparison of Experimental and Calculated Values of r_s/r_i

System	d_p mm	r_c mm	H_m mm	r_i mm	m expt.	A expt.	r_s mm	r_s/r_i expt.	r_s/r_i calc.
Air-kale seeds	1.74	38	250	4.75	0.822	0.103	8.0	1.68	1.84
		38	180	6.35	0.793	0.0723	9.5	1.50	1.61
		152.5	1800	12.75	0.987	0.0383	19.0	1.49	1.57
		152.5	1225	17.5	0.922	0.0279	24.0	1.38	1.26
Air-polythene	3.5	152.5	1300	12.75	0.713	0.0797	25.0	1.96	1.96
		152.5	1100	17.51	0.828	0.0581	31.5	1.80	1.63
Air-peas (Lefroy and Davidson)	9.0	152.5	900	12.75	0.493	0.221	34.0	2.67	2.35
		152.5	750	25.5	0.822	0.111	42.5	1.67	1.68
Air-glass (this work)	1.55	55	315	6.75	0.703	0.0579	12.0	1.77	1.63
	1.84		295		0.658	0.0731	11.5	1.70	1.70
	2.11		283		0.632	0.118	11.5	1.77	1.85
	3.09		258		0.576	0.130	13.5	2.00	1.93
	4.11		220		0.491	0.181	13.0	1.93	2.04
	5.05		183		0.409	0.221	14.0	2.07	2.07
Water-glass (this work)	1.55	55	325	6.75	0.725	0.0430	12.5	1.85	1.44
	1.84		305		0.681	0.0610	11.5	1.70	1.63
	2.11		275		0.658	0.0728	11.5	1.70	1.70
	3.09		250		0.565	0.128	12.0	1.78	1.93
	4.11		210		0.469	0.203	12.0	1.78	2.07
	5.05		180		0.402	0.264	13.5	2.00	2.15

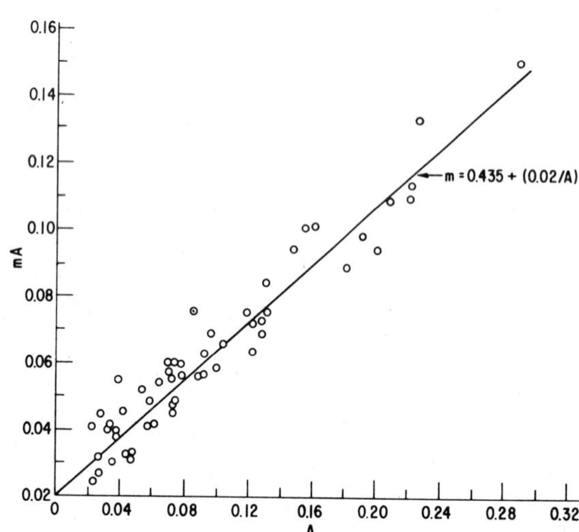

Figure 1. Evaluation of the A Parameter

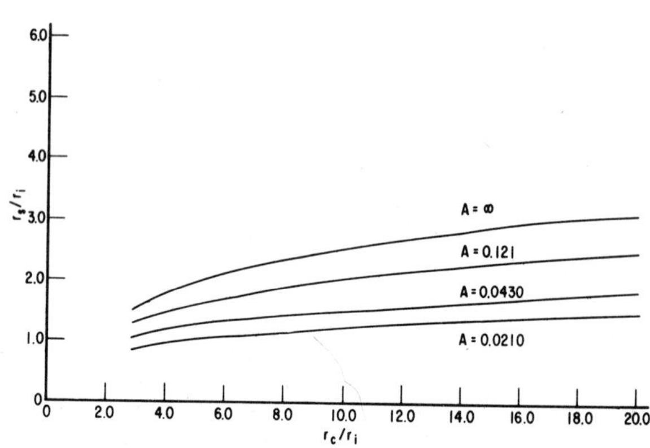

Figure 2. Plot of r_s/r_i as a function of r_c/r_i and A

GAS BACKMIXING IN HIGH VELOCITY FLUIDIZED BEDS

By N.T. Cankurt and J. Yerushalmi

Department of Chemical Engineering
City College, City University of New York
New York, N.Y. 10031 U.S.A.

SYNOPSIS

This paper presents the results of experiments on the extent of gas backmixing in a bed of catalyst fluidized at gas velocities spanning the bubbling, turbulent and the fast bed regimes. In a typical experiment, a tracer gas is continually injected at a given elevation and radial position, and axial and radial concentration profiles are measured upstream. Beyond the bubbling regime, where its extent is high, gas backmixing diminishes with gas velocity, and is practically negligible in the fast fluidized bed.

NOTATION

C_o reference tracer concentration

C_c tracer concentration in the center

C_w tracer concentration near the wall

x axial distance upstream

U_g superficial gas velocity

U_s mean solid velocity

ε gas voidage

INTRODUCTION

This paper reports the results of experiments on the extent of gas backmixing in a bed of cracking catalyst fluidized at gas velocities spanning the bubbling, turbulent, and the fast bed regimes. The underlying characteristics of the turbulent and the fast fluidized beds, which are species of high velocity fluidization, and instances of their commercial use to date have been described elsewhere (Yerushalmi et. al., 1976a,b; 1977). Further observations relevant to the experiments reported herein are furnished below.

EXPERIMENTAL

The experiments were conducted in a 6-inch I.D. column. A schematic of the entire apparatus is shown in Fig. 1. On the left is the 6-inch test column; it is 28 ft tall. On the right is a 13.5-inch I.D. vessel in which catalyst is maintained in low-velocity, bubbling fluidization. The so-called "slow" bed doubles as a standpipe and a storage for the solid. Solid can be circulated through the 6-inch column in dilute-phase flow, or in the form of a dense suspension typical to the high velocity fluid beds, or the solid can be maintained in the bubbling state with or without some net flow upwards. When the solid is circulating, it flows by gravity from the bottom of the slow bed into a U-tube, 6-inch I.D., which extends to the bottom of the 6-inch column. The solid rate is controlled by a slide valve installed in the U-tube just below the slow bed. The method we use to measure the solid rate has been described by Yerushalmi, et. al. (1976a).

To study the extent of gas backmixing in a given fluid bed regime, a tracer gas, methane, is from a given moment continually injected at a known rate through a 3/16-inch stainless steel tube entering the bed at any desired elevation. A stream of sample gas is at the same time withdrawn from locations below (upstream of) the injection point, and its methane concentration is continually monitored by a suitable analyzer, and recorded on a strip chart recorder. The tracer may be injected and gas may be sampled at any radial position in the bed.

The experiments were conducted at room conditions with air as the fluidizing gas. The fluid cracking catalyst (FCC) used in 99% minus 150 microns and has a volume-surface mean diameter of 55 microns. Its particle density is 67 lbs/cu.ft.

RESULTS AND DISCUSSION

1. Fluidization Regimes

With the slide valve fully opened, we observed the behavior of a bed of catalyst, established in the 6-inch column, as the velocity of the gas supplied to it was gradually raised. The observations were made with a relatively short bed (about 5 ft) to allow sufficient freeboard space. It should be appreciated that with the slide valve fully opened, solid carried over from the bed is matched by solid flow from the slow bed to the bottom of the 6-inch column.

For the test catalyst, bubbling is evident at a gas velocity around 0.1 fps, and slugging is in force at 1 fps. As the velocity of the gas is slowly raised beyond 1 fps, the slugs lengthen and the expansion and contraction of the bed increases in violence (Fig. 2). At a superficial gas velocity around 2.0 fps, the slugs break down, giving way to a turbulent state in which large voids or bubbles are absent and which the naked eye indeed perceives as rather uniform in structure; solid mixing is vigorous and extensive refluxing of large dense packets of particles is evident. The sharpness of the transition and the greater degree of homogeneity of the turbulent regime that lies beyond it are also reflected in the fluctuations of the manometer that serves to measure the pressure drop across the bed (Fig. 2).

As the velocity is raised beyond 2.0 fps, a series of ostensibly similar states of fluidization are traced. There is, however, evidence that between 2.0 fps and a gas velocity around 5.5 fps a turbulent fluidization regime exists that is distinct from the fast fluid regime that lies at higher gas velocities. In the turbulent regime, as indeed in the bubbling regime as well, part of the solid always defines (or settles in) a dense configuration (the bubbling zone itself for example), and the fluidized density in the dense zone depends primarily on the gas velocity and only slightly on the rate at which solid may be fed to or withdrawn from the bed. In the fast fluidized bed, and indeed in all vertical transport regimes, solid holdup, at any gas velocity, varies strongly with the solid rate. Also, in the turbulent fluid bed, though it is rather fuzzy, an upper bed level may be perceived nonetheless, and a sharp increase in carryover rate at a velocity around 5 - 6 fps appears to mark the transition to the fast bed regime.

In Fig. 3, we have assembled the data we have obtained in the bubbling, turbulent and the fast bed regimes in the form of a fluidization map depicting the slip velocity versus the solid volumetric concentration at the bottom of the bed. We use the slip velocity rather than the superficial gas velocity since it is the former that provides a relevant measure of the interaction between gas and solid. The figure attempts to capture some of the points made above: In the bubbling and the turbulent regimes, the solid concentration is only slightly affected by the solid rate, and the relation between the slip velocity and the solid concentration is fairly unique. Not so in the fast fluid bed, where corresponding to each solid rate, a different function connecting these quantities exists.

2. Gas Backmixing

Fig. 4 gives the radial concentration profile across the bed two inches below the injection point at different gas velocities. The tracer was injected at the center. These results anticipate the conclusion that the degree of gas backmixing is high in the bubbling regime (0.7 fps); it is considerably lower, though not insignificant, in the turbulent regime (2.7, 3.9 and 5.0 fps); and gas is essentially in plug flow in the fast fluidized bed (5.9 fps and beyond).

This conclusion is reinforced by Figs. 5 - 9 which provide complete concentration profiles (radial and axial) upstream of the injection point at gas velocities of 0.7, 2.7, 3.9, 5.0 and 5.9 fps. A striking aspect of these profiles is the increased concavity of the radial profiles with gas velocity up to 5.0 fps. This suggests a physical picture of a mixing pattern dominated by the downward flow of dense solid along the walls, and a fast-moving gas flowing upwards through a leaner core.

Fig.s 10 and 11 add further details to this picture. Figure 10 shows radial profiles obtained at three upstream locations when tracer was injected near the (left) wall of the bed fluidized at 5.0 fps. Figure 11 gives corresponding axial profiles measured along the left wall, the opposite (right) wall, and along the center of the bed. This mixing pattern is typical of beds of small diameter and reflects in a large measure the influence of the walls.

The influence of the walls was no doubt particularly pronounced in our experiments since in order to measure the extent of gas backmixing we have employed tall beds (around 10 - 15 ft). Carryover rates from these beds were rather high producing corresponding high rates of solid circulation in the 6-inch column. At a given slip velocity, a higher solid rate appeared to enhance the curvature of the radial concentration profiles.

The physical picture sketched above serves as the basis for a theoretical analysis of the mixing of gas and solid in a turbulent fluidized bed (contained in narrow vessels) we intend to offer for publication shortly. Further experiments are planned to measure the radial density profiles to provide additional input to the theoretical model.

We have noted the pronounced concavity of the radial profiles and its increase with gas velocity up to 5.0 fps. At some velocity beyond that, a sharp change occurs. At 5.9 fps, the curvature of the profile has become rather small, and at higher gas velocities, the profile is practically flat. This is illustrated very clearly in Fig. 12 in which the ratio of the concentration near the wall to that at the center is plotted versus the superficial gas velocity. The sharp change that is bracketed by 5.0 and 5.9 fps is most likely associated with a fundamental change in the fluidization regime. We regard it as the transition between turbulent and fast fluidization.

In the fast bed regime, the extent of gas backmixing is very small. Further, we have found that the concentration of tracer measured at any point -- 2 inches, 5 ft or even 15 ft, though not less than 1 inch -- upstream is substantially the same. Fig. 13 records this concentration (the solid lines) as a function of the solid rate at 7 and 13.5 fps. It might be argued that this concentration arises from a tracer gas introduced by the circulating solid. Indeed, at gas velocities above 3.9 fps, solid rate become appreciable, and the circulating solid carries with it, trapped in is interstices, a quantity of tracer. We have accordingly provided for measurement of the tracer concentration in the U-tube just below the slide valve, and the results are shown in Fig. 14. Raising the gas velocity in the slow bed provided some stripping action, and the tracer concentration was correspondingly reduced (see the starred data point).

To obtain the true concentration of tracer arising from gas backmixing, one also needs to know the rate of the total gas flowing through the U-tube into the 6-inch column. Assuming this rate to arise from the interstitial volume of the packed solid at a given circulation rate, we obtained the dashed lines shown in Fig. 13. In this case, the results would suggest that there is some backmixing of gas, owing probably to refluxing of closely packed strands and clusters of particles. At any rate, the very low concentrations recorded in this regime detract from the importance of resolving this question.

CONCLUSIONS

The extent of gas backmixing is high in the bubbling regime; it diminishes, though it may still be appreciable, over the turbulent regime, and gas is essentially in plug flow in the fast fluidized bed.

ACKNOWLEDGEMENTS

Work on high velocity fluidization is supported by ERDA under grant No. E(49-18)-2340. We are grateful to E.I. duPont de Nemours & Company for donating the gas analyzer. Dr. Samuel Dobner, now with Union Carbide Corporation, provided assistance in setting up the experiment.

REFERENCES

Yerushalmi, J., Turner, D.H. and Squires, A.M. (1976a) I&EC Proc. Des. & Dev. 15, 47.

Yerushalmi, J. Cankurt, N.T., Geldart, D. and Liss B. (1976b) "Flow Regimes in Vertical Gas-Solid Contact Systems", paper presented at AIChE Meeting, Chicago, Illinois; December 1 (to be published in an AIChE Symposium Series).

Yerushalmi, J. and Cankurt, N.T. (1977) "High Velocity Fluidized Beds", paper presented at International Powder & Bulk Solids Handling and Processing Conference, Chicago, Illinois; May 12 (to be published in CHEMTECH).

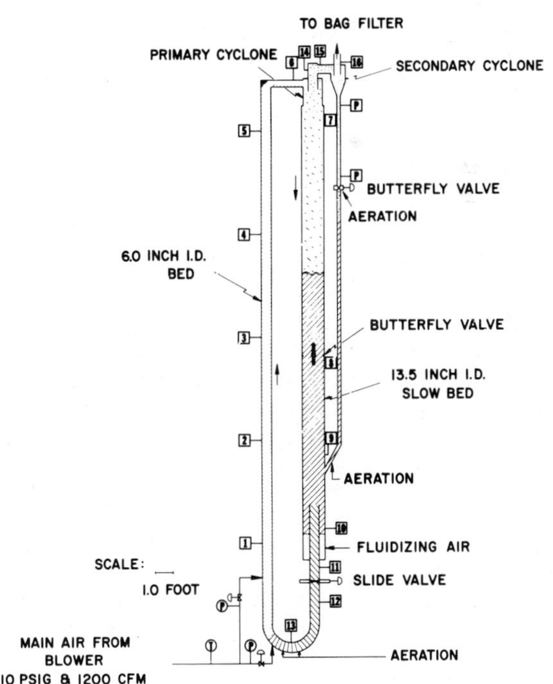

Fig. 1. Schematic of the 6-inch system.

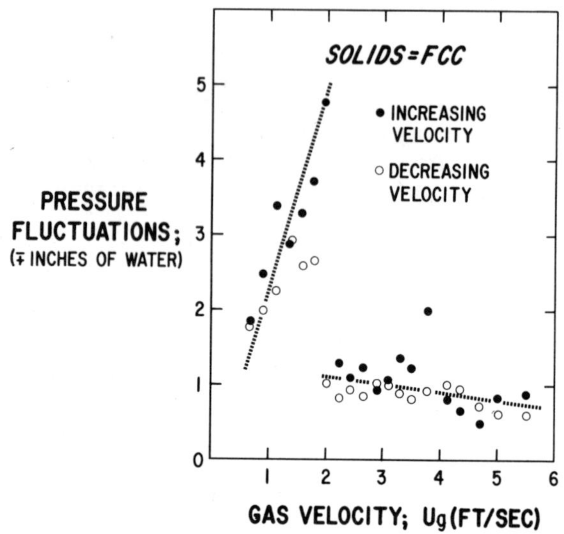

Fig. 2. Pressure fluctuations as recorded from a manometer measuring the pressure drop across the bottom of the 6-inch bed during transition from bubbling to turbulence and back.

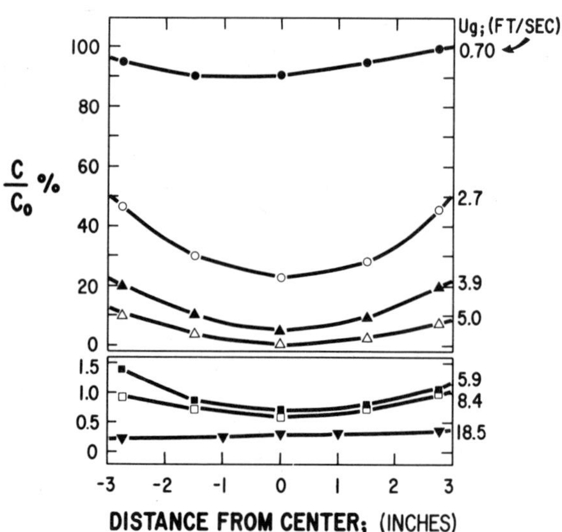

Fig. 4. Radial concentration profiles measured 2 inches upstream of the injection point at different gas velocities. Tracer was injected at center. Fluidized densities in the order of increasing velocities are 24.8, 16.6, 13.3, 11.8, 10.6 and 1.2 lbs/cu.ft.

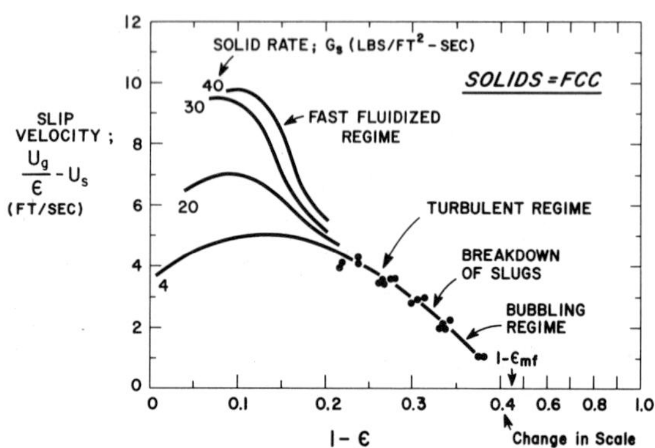

Fig. 3. Slip velocity vs. the volumetric concentration of solid at the bottom of a bed of FCC.

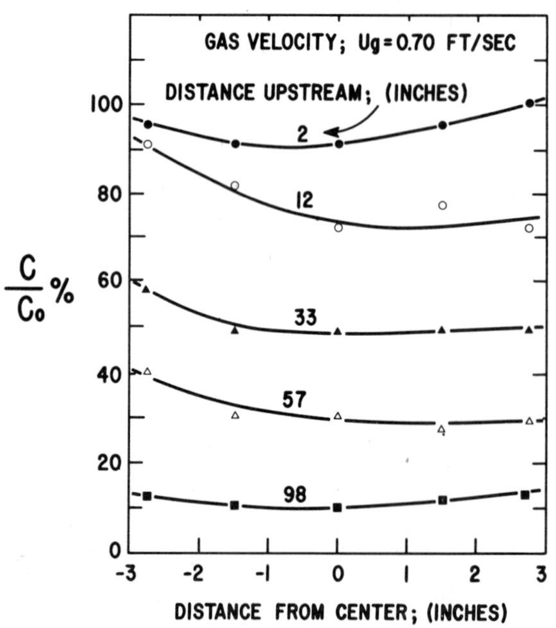

Fig. 5. Tracer concentration profiles at 0.7 fps. Tracer injection at center. Fluidized density = 24.8 lbs/cu.ft.

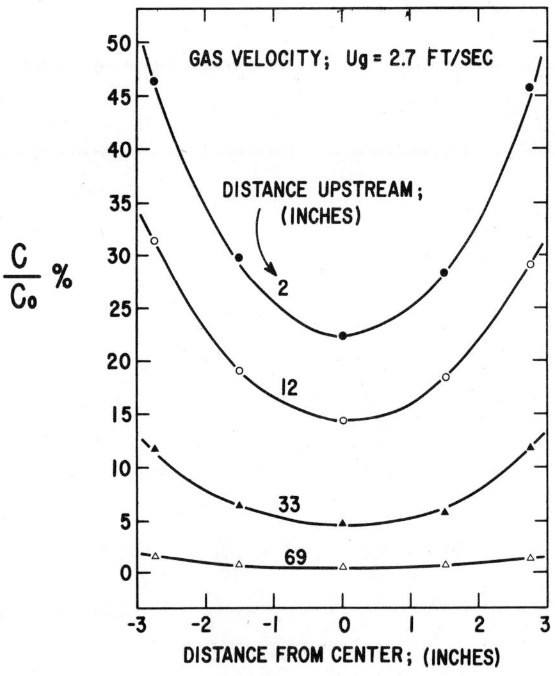

Fig. 6. Tracer concentration profiles at 2.7 fps. Tracer injection at center. Fluidized density = 16.6 lbs/cu.ft.

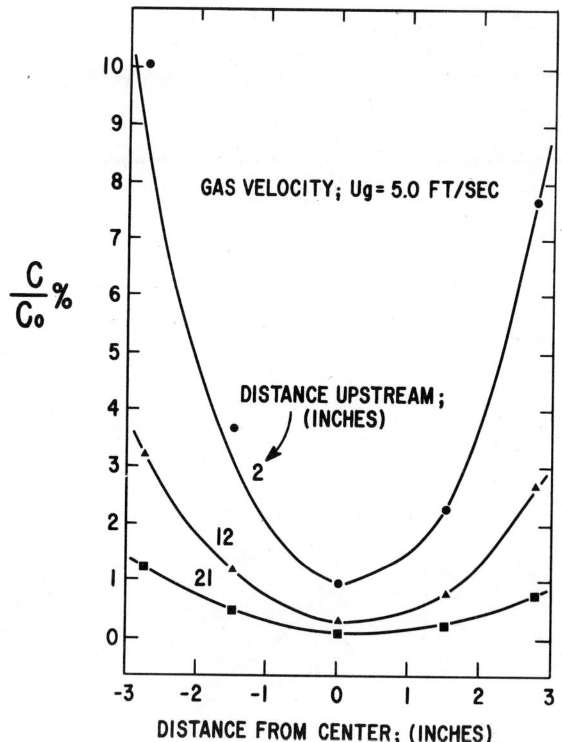

Fig. 8. Tracer concentration profiles at 5.0 fps. Tracer injection at center. Fluidized density = 11.8 lbs/cu.ft.

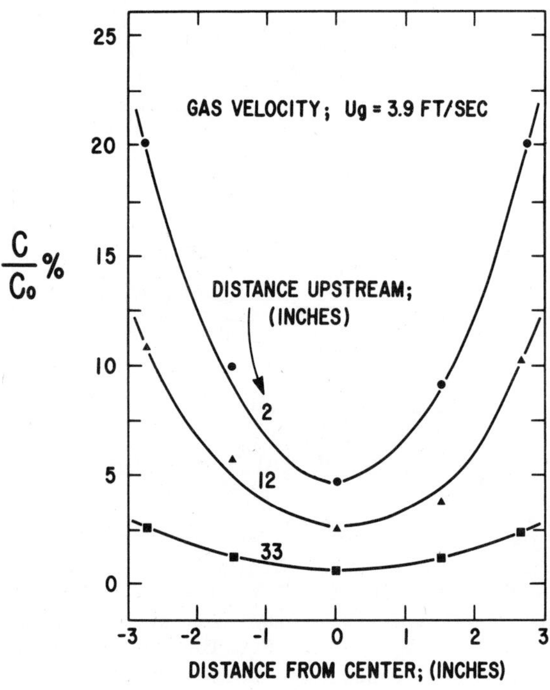

Fig. 7. Tracer concentration profiles at 3.9 fps. Tracer injection at center. Fluidized density = 13.3 lbs/cu.ft.

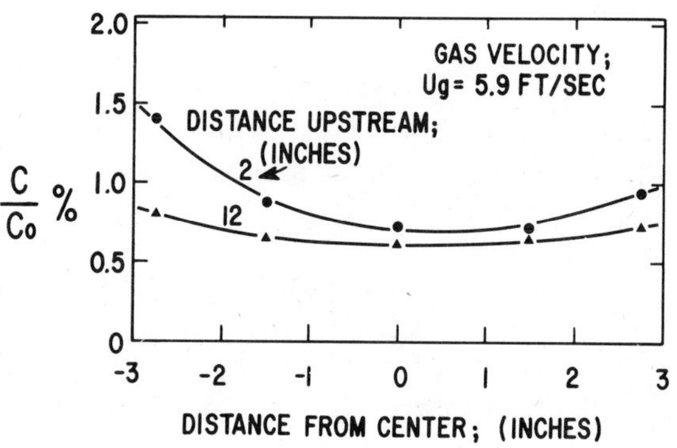

Fig. 9. Tracer concentration profiles at 5.9 fps. Tracer injection at center. Fluidized density at the bottom = 10.6 lbs/cu.ft.

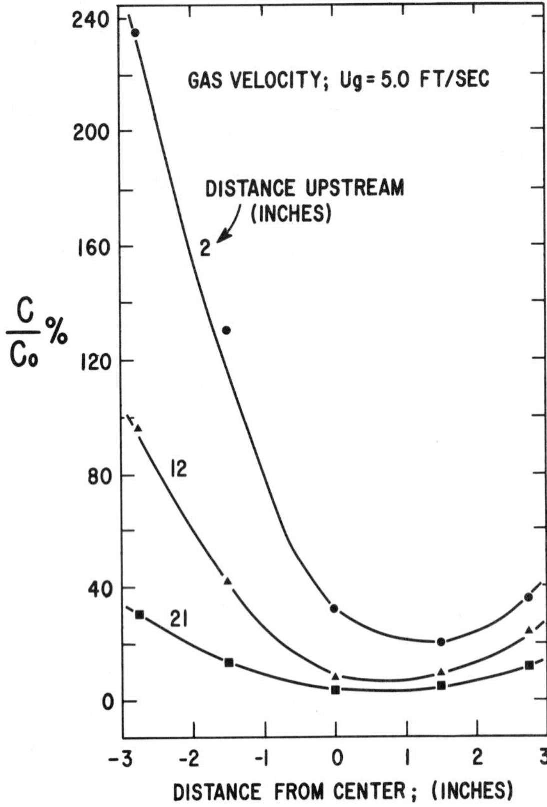

Fig. 10. Tracer concentration profiles at 5.0 fps. Tracer injected near the left wall.

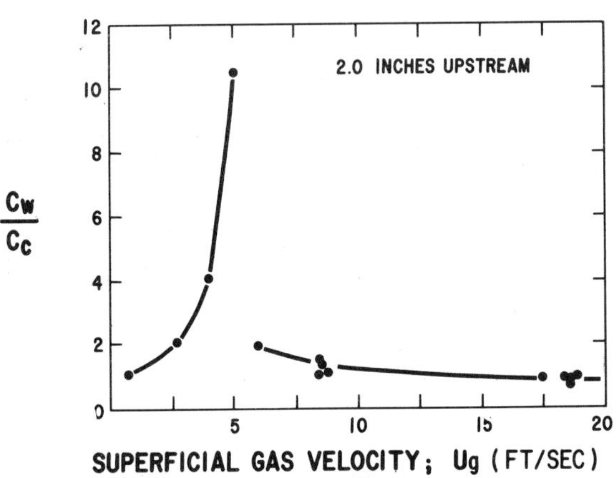

Fig. 12. Ratio of tracer concentration near the wall to that at the center, measured 2 inches upstream vs. superficial gas velocity. Tracer injection at center.

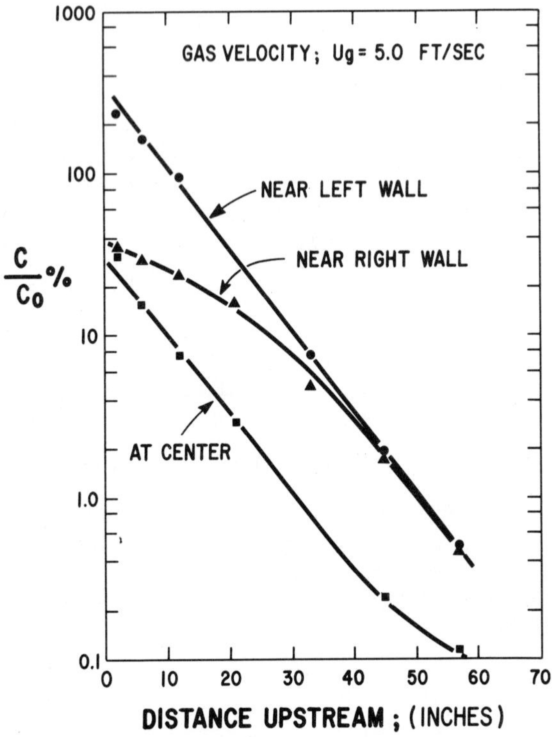

Fig. 11. Axial concentration profiles at 5.0 fps. Tracer injected near the left wall.

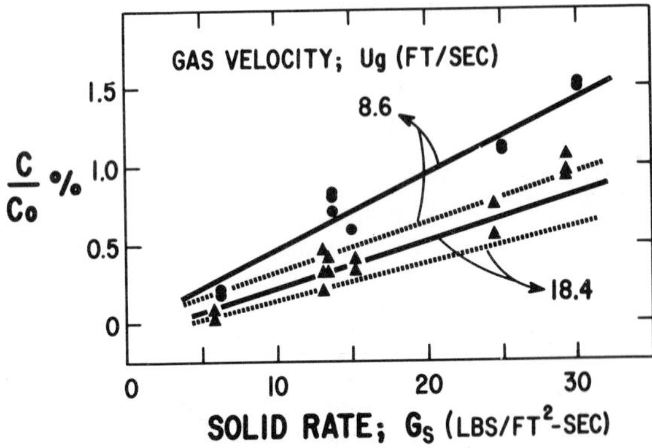

Fig. 13. Upstream tracer concentration measured in a fast fluidized bed at 8.6 and 18.4 fps vs. the solid rate. At 8.6 fps, the fluidized density varies from about 4 lbs/cu/ft. at G_S = 15 to 8.6 lbs/cu/ft. at G_S = 30 lbs/sq.ft.-sec. The corresponding densities at 18.4 fps are 1.4 and 3 lbs/cu.ft.

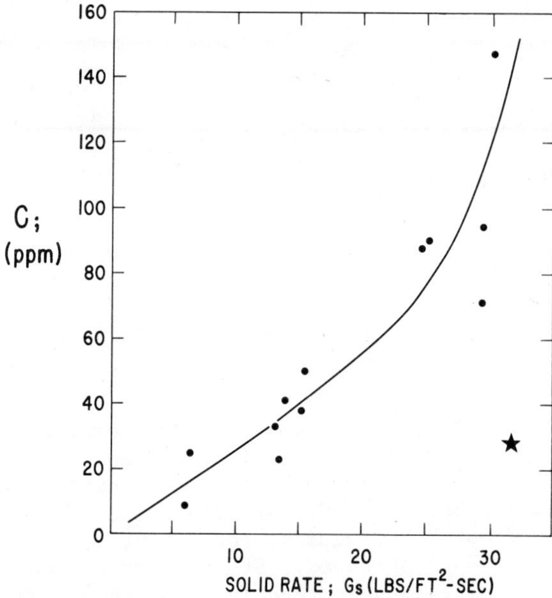

Fig. 14. Tracer concentration below the slide valve vs. the solid rate. C_o varied from 1040 to 1350 ppm.

EXPERIENCES IN REGENERATING SULFATED LIMESTONE FROM FLUIDIZED-BED COMBUSTORS

By G. J. Vogel, J. Montagna, W. Swift, A. A. Jonke

Argonne National Laboratory
Argonne, Illinois 60439

SYNOPSIS

The experiences at Argonne National Laboratory in developing a process for regenerating the partially sulfated limestone product from a fluidized-bed combustion process are described. Studies of candidate processes and the further development of the reductive-decomposition process wherein $CaSO_4$ is converted to CaO in a fluidized bed at 1100°C are discussed. Reductive decomposition reaction rates are rapid—less than 10 minutes is required for effective regeneration. A gas containing SO_2 at concentrations near 10% is produced. Stones were cycled batchwise between the combustor and regenerator to demonstrate that reactivity of the stone can be sufficiently maintained and that attrition rate of particles in the fluid bed is low. Regeneration can significantly reduce the quantity of fresh stone required in the combustion step.

NOTATION

cm	centimeter
°C	degree Celsius
dia	diameter
F	mean feed sorbent particle size, μm
hr	hour
ID	inside diameter
kPa	kilopascal
m	meter
min	minute
%	percent
R	total reducing gas concentration, %
s	second
T	operating temperature, °C
V_d	defluidization velocity, m/s
wt	weight

INTRODUCTION

The quantity of pollutants--SO_2, NO_x, and particulate solids--permitted to be released in the flue gas from boilers in the United States is now regulated. Control of SO_2 is the most difficult since NO_x release can be kept within specifications by modification of boiler equipment and particulates can be controlled by proper choice of equipment. Currently, low SO_2 concentrations are achieved by burning low-sulfur-content fossil fuels (either naturally occurring fuels or fuels from which sulfur has been removed prior to combustion) or by combustion of high-sulfur fuels accompanied by scrubbing of the flue gas. New techniques for pollution control are being developed. In one of these, fluidized-bed combustion, coal is burned in a bed of granular particles, limestone ($CaCO_3$) or a dolomite ($CaCO_3 \cdot MgCO_3$), that are kept in suspension by an upflowing stream of combustion air. The calcium in the bed particles combines with the sulfur released during coal combustion. Development of fluidized-bed combustion systems is proceeding rapidly--a 20-MWe, atmospheric pressure combustor awaits commissioning (Mesko, 1976); a 20-MWe, pressurized combustor is being designed (Energy Daily, 1977).

Although the release of sulfur compounds to the atmosphere is effectively controlled by this method, the impact of quarrying large quantities of the limestone and of handling the $CaSO_4$-containing solids generated in the process must be examined more thoroughly. In the case of a typical bituminous coal containing 4% sulfur, the quantity of rejected sulfated stone can be greater than the quantity of ash that must be discarded (Johnson, 1977; Vogel *et al.*, 1974; Pope, Evans and Robbins, Inc., 1975). The rejected sulfated stone and ash can probably be used as landfill or disposed of in open pits in an environmentally safe manner (Nat. Res. & Dev. Conf., (1973); Westinghouse, 1974). Methods for commercially utilizing the material may also be developed. Studies are underway to determine if the available lime and sulfate in the reject stone can be used for agricultural purposes (Minnick, 1977) or as a component in the manufacture of cement blocks (Westinghouse, 1974). The quantity of reject that can be used in these applications is uncertain and, since a large quantity may have to be disposed of, options for reducing the quantity of stone used in the process should be examined. Five options are available, as follows:

1. Since stones vary in reactivity (Vogel *et al.*, 1977), use only the reactive stones.

However, reactive stones may be unavailable within an economic distance of the combustion plant.

2. Modify nonreactive stones by changing the porosity, thereby increasing the quantity of sulfur that can be captured. Two methods have been demonstrated--(1) a slow calcination of the stone (O'Neill, 1976) and (2) salt addition (Pope, Evans and Robbins, 1975). In the slow calcination method, the stone would probably be precalcined in a separate vessel specially constructed for this purpose. In the second method, additives NaCl, Na_2CO_3, Na_2SO_4, KCl, and $CaCl_2$, have effectively increased the utilization of the calcium by 25 to 200% (Johnson, 1977). The chlorides of sodium and potassium appear to be the most effective. If the porosity of a stone before treatment is satisfactory for collecting sulfur [pores of a size larger than 0.3 μm are the effective collectors of the sulfur oxide (Potter, 1969)], the treatment may change the pore structure unacceptably. Studies have started at ANL to determine the possible corrosive effects of these additives on metals present in the boiler and ancilliary components.

3. Tailor-make a synthetic stone that has the porosity characteristics for good sulfur oxide sorption and which can be reused, *i.e.*, can be regenerated by a process to be described later. The best synthetic sorbent developed, containing 20% CaO impreganted on alumina, was about 60% as reactive as Greer limestone, one of the more reactive naturally occurring limestones (Snyder, 1977). Preparation of synthetics is expensive, ∿$200/tonne, and unless a cheaper preparation method can be developed, this option may not be feasible.

4. Adjust boiler operating conditions using higher bed temperatures, lower gas velocities, and longer particle residence times, which will help utilize more of the calcium in the particle. However, these conditions often conflict with other requirements for the boiler or the process and generally cannot be applied.

5. A fifth method is to regenerate the sulfated stone for reuse in the combustor, which is the topic of this paper.

SELECTION OF THE REGENERATION PROCESS

Thermodynamic and kinetic information were obtained from the literature on processes for converting calcium sulfate to calcium oxide (Swift, 1976). Of the processes examined, thermal decomposition of calcium sulfate was not considered to be a viable process because of the high temperature, >1200°C, needed to obtain a high concentration of SO_2 in the off-gas. At this temperature, ash and sulfated stone would fuse to form unusable clinkers.

Another process, studied in laboratory-scale equipment, consisted of two steps. CaS produced in the first step by reacting $CaSO_4$ with a reducing gas at ∿870°C is reacted with steam/CO_2 at ∿560°C to produce $CaCO_3$ and H_2S. Cyclic processing by alternate sulfation and regeneration showed that the extent of regeneration of the $CaSO_4$ decreased significantly in succeeding cycles (Vogel, 1972). Attempts to understand the mechanism and to develop the process futher were abandoned when it became necessary to select a process for larger-scale development. Selected for further development was the reductive decomposition process in which $CaSO_4$ is heated in a fluidized bed to ∿1100°C in the presence of reductant gases. The products are CaO, which is reused in the combustion step, and a SO_2 containing off-gas, which can be processed for its sulfur content. Calcium sulfide can also form in the bed, but its concentration is maintained below 0.1% by circulating particles into an oxidizing zone where CaS is converted to $CaSO_4$.

EQUIPMENT

Figures 1 and 2 illustrate the Process Development Unit (PDU) combustion and regeneration systems used in this investigation. The combustor is 15-cm ID and approximately 3.4 m high. The regenerator consists of a nominal 20-cm-dia pipe, refractory-lined to an ID of 7.5 cm in the early experiments and to 10.8 cm in later experiments. Bubble-type gas distributor plates are flanged to the bottom of each reactor and accommodate fluidizing-air inlets, thermocouples for monitoring of bed temperatures, solids feed and removal lines. In each system, the coal and sorbent are metered separately to a single pneumatic transport line which discharges these solids into the fluidized bed above the gas distributor plate. Expanded-bed heights of ∿0.9 m in the combustor and ∿46 cm in the regenerator are achieved with overflow pipes.

MATERIALS

Different coals and sorbent stones were used. The two sorbents tested were Tymochtee dolomite and Greer limestone. The dolomite contained ∿50 wt % $CaCO_3$, 39 wt % $MgCO_3$, and 2.1 wt % Si as received. The limestone contained 41.2 wt % CaO, 32 wt % CO_2, and 4.3 wt % Si. The nominal size distribution of each limestone was -14 +30 U.S. mesh in most, but not all, experiments.

Arkwright mine, Pittsburgh seam coal contained ∿2.8 wt % S, ∿7.7 wt % ash, and ∿2.9 wt % moisture and had a heating value of 7,600 kcal/kg and an average particle size of 320 μm.

In cyclic studies with Tymochtee dolomite, Triangle coal, a high-volatile, bituminous coal with a high ash-fusion temperature, 1390°C, was combusted in the regenerator. It contained 1.0% S and 9.4 wt % ash.

Sewickley coal, as received, contained ∿4.3 wt % S, 12.7 wt % ash, and 1.1 wt % moisture and had a heating value of 7,200 kcal/kg. Particle size range was -6 +14 mesh.

DEVELOPMENT OF THE REDUCTIVE DECOMPOSITION PROCESS

Methane gas was used in early experiments and coal in later experiments as a source of the reducing gases for the reaction. Coal is considered attractive since it is cheaper than methane. Also, it would be available at the combustion plant. The initial experimental work using methane reductant was carried out in a 7.5-cm-dia regenerator and generally with a 0.46-m-high bed. The results showed that the use of a shallower bed (0.46 m instead of 0.76 m), a lower fluidizing gas velocity(0.67 m/s instead of 0.92 m/s), and higher reaction temperature (1100°C instead of 1010°C) generally increased both the extent of regeneration and the SO_2 concentration in the effluent gas (Montagna & Lenc, 1977). Attrition losses ranged from 5 to 15% of the solids processed. Partial agglomeration of the bed particles occurred when the bed temperature was higher than 1100°C, the reducing gas concentration was greater than 5%, and when also the fluidization velocity was inadequate.

In the major experimental effort, using coal as the source of the reducing agent, the effects of solids residence time (7-35 min), regeneration temperature (1000-1100°C), and pressure (115-153 kPa) on the extent of calcium oxide regeneration and the SO_2 concentration in the off-gas were investigated in a 10.8-cm-dia regenerator (Montagna & Swift, 1977). The extent of regeneration improved at higher bed temperature and with longer particle residence time. Calculated results using an equation developed for regenerating sulfated Tymochtee dolomite using Triangle coal have been plotted in Fig. 3 (the data have been extrapolated for bed temperatures above the maximum temperature used in these experiments, 1100°C) Studies with sulfated Greer limestone/Arkwright coal have shown that extent of regeneration of the limestone is about the same as for the dolomite at similar operating conditions (Vogel *et al.*, 1977).

The SO_2 concentration in the off-gas increased at higher temperature, at shorter solids residence time, and at lower pressure. Typically, the SO_2 concentration in the off-gas ranged from 8-10%. Calculated concentrations obtained from an equation developed using the experimental results from the Tymochtee dolomite and Triangle coal experiments are plotted in Fig. 4. The reaction rate is rapid at ~1100°C and, with a few percent reducing gas in the bed, production rates greater than a half-tonne per day were achieved.

Bed defluidization velocity, *i.e.*, minimum velocity required to prevent particle agglomeration, was studied in a statistical experiment performed with Greer limestone and Sewickley coal (Montagna & Nunes, 1977). The effects on defluidization velocity of bed temperature (1050 and 1100°C), total reducing gas concentation (2.5 and 5.0% and feed sorbent particle size range (-10 +30 mesh and -14 +30 mesh) were determined. Defluidization velocities in this study ranged from 0.9 to 1.6 m/s. Defluidization velocity as a function of temperature and reducing gas concentration is shown in Fig. 5 along with minimum fluidization velocities. The effect of particle size is minimal. A least-squares fit of the results is the basis for the following equation:

$$V_d = 4.05 - 3.61 \times 10^{-3}T - 2.6R + 2.54 \times 10^{-3}TR + 5.26 \times 10^{-4}F \quad (1)$$

CYCLIC STUDIES

The feasibility of the regeneration process depends on the ability to recycle the sorbent a sufficient number of times without loss of sorbent reactivity for either sulfation or regeneration and without severe decrepitation of the particles. If either requirement is not met, the sorbent makeup rate will be too high to justify regeneration.

Two series of experiments were performed to evaluate the effects of repeated utilization on the reactivity and attrition resistance of two stones--Tymochtee dolomite and Greer limestone. Each series consisted of ten combustion/regeneration cycles. Arkwright coal was used in the experiments made with Tymochtee dolomite and Sewickley coal in the Greer limestone experiments. The dolomite series of experiments simulated the concept of combustion at elevated pressure (8-atm) coupled with regeneration at atmospheric pressure; the limestone series, combustor at 3-atm pres., simulated combustion at atmospheric pressure (the $CaCO_3$ calcines at the combustion conditions as it does in an atmospheric pressure combustor) and regeneration at atmospheric pressure.

Cyclic Sorbent Life Study with Tymochtee Dolomite

Conditions in the combustion step were a 900°C bed temperature, a 810 kPa system pressure, a constant CaO/S mole ratio of 1.5 (ratio of unsulfated calcium in sorbent to sulfur in coal), ~17% excess combustion air, a 0.9 m/s fluidizing gas velocity, and a 0.9 m bed height.

The regeneration step of each cycle was performed at a system pressure of 158 kPa, a bed temperature of 1100°C, a fluidizing gas velocity of 1.2 m/s, a fluidized-bed height of ~46 cm, and a solids residence time of ~7 1/2 min.

Changes in reactivity of the sorbent from cycle to cycle were reflected in changes in the SO_2 levels in the flue gas of the combustor (Vogel *et al.*, 1976, Sept. 1976, Dec. 1976).

Sulfur Acceptance and Release. Sulfur dioxide levels increased from ∿330 ppm in cycle 1 to ∿950 ppm in cycle 10 (see Fig. 6). This represents a decrease in sulfur retention from ∿88% in cycle 1 to ∿55% in cycle 10. It appears that the reactivity of the sorbent for sulfur retention decreased linearly with combustion cycle over the 10-cycle experiment.

There was no apparent loss in regenerability over the ten utilization cycles. The extent of CaO regeneration based on solids analysis varied from 67 to 80%. The SO_2 concentration in the dry off-gas varied from 6 to 9%.

Sorbent Makeup Requirements to Meet EPA Sulfur Emission. Based on the data from the cyclic combustion/regeneration batch experiment, an analysis was made to estimate the sorbent makeup rate which would be required in a continuous recycle operation to meet the EPA sulfur emission limit.

Adaptation of a procedure developed by Nagier (Nagier, 1964) resulted in an analytical expression for the age distribution of the combustor charge as a function of makeup rate assuming a continuous recycling between the regenerator and combustor.

The result of combining the data in Fig. 6 with the expression is presented in Fig. 7. As an example of using Fig. 7, an α of ∿0.2 (makeup CaO/total CaO) indicates that a makeup CaO/S ratio of ∿0.27 and a total CaO/S ratio of ∿1.5 are required for a sulfur retention of 75%. In comparison to the once-through CaO/S ratio of ∿0.93 for 75% sulfur retention, the makeup of 0.2 for a cyclic process corresponds to an estimated savings of 78% of the fresh limestone requirements.

Particle Porosity Changes. It has been reported that most sulfation takes place in larger pores and that pores smaller than 0.3 μm are relatively easy to plug. During sulfation of CaO, the pores shrink as a result of molecular volume changes.

The porosity of sulfated dolomite was relatively unaffected by number of cycles, although the sulfur content decreased from ∿10 wt % in the first cycle to ∿7 wt % in the tenth cycle. However, the porosity of the regenerated dolomite consistently decreased with each cycle, as did the sulfur content of the regenerated stones. As a result of this decrease in porosity with use, effectiveness of the sorbent as an SO_2 acceptor decreased.

Coal-Ash Buildup. Based on silicon enrichment of samples, coal-ash buildups were calculated. After the tenth cycle, it was found that for every 100 g of starting dolomite, ∿13 g of coal ash had accumulated in the sorbent. Photomicrographs clearly reveal that even the once-sulfated stones were beginning to be coated with coal ash.

A petrographic examination revealed the buildup of a vitreous crust surrounding most of the particles. X-ray diffraction analyses of the crust revealed the presence and accumulation of $Ca(Al_{0.7}Fe_{0.3})_2O_5$ with increasing number of cycles.

Electron microprobe analyses also confirmed the existence of the coal-ash shell. The ash which encapsulated the particle was strongly enriched with iron but contained very little sulfur. Sulfur profiles in scans of regenerated particles indicate that nearly complete regeneration occurred. Therefore the presence of a coal-ash shell did not prevent sulfur from escaping during regeneration.

Attrition and Elutriation Losses. Sorbent losses for each cycle are given in Table 1. Although the first cycle combustion loss was quite large, losses during the remaining combustion-regeneration cycles were reasonably small, averaging about 8% per cycle. The lower losses during regeneration can probably be attributed to the very short solids residence time (∿7.5 min) in this reactor, as compared with the much longer solids residence time (∿5 hr) for sulfation in the combustor.

Table 1. Attrition and Elutriation Losses for Tymochtee Dolomite during Combustion and Regeneration in the Cyclic Utilization Study.

Cycle No.	Loss During Combustion (wt %)	Loss During Regeneration (wt %)
1	16	1.9
2	4	1.7
3	5	3.0
4	3	1.2
5	3	3.5
6	6	3.7
7	4	2.0
8	7	2.6
9	6	0.9
10	4	1.3

Cyclic Sorbent Life Study with Greer Limestone

The combustion conditions were a 308-kPa system pressure, 855°C bed temperature, ∿17% excess combustion air, a 1.0 m/s fluidizing gas velocity, a 0.9 m bed height, and a constant sulfur retention of ∿84% by the sorbent. In this study, in which the concentration of SO_2 in the flue gas was held constant by adding limestone as required (in the dolomite cyclic study, the CaO/S ratio was held constant), changes in sorbent reactivity were reflected in changes in the CaO/S ratio required to achieve the constant concentration (in the dolomite cyclic study, retention was the dependent variable of reactivity).

The nominal operating conditions during the regeneration step in each cycle were a system pressure of 128 kPa, a bed temperature of 1100°C, a fluidizing-gas velocity of ∿1.2 m/s, a total reducing gas concentration of ∿3.0% in the dry off-gas, and a fluidized-bed height of ∿46 cm. The residence time of the sorbent was nominally 7 min (Vogel *et al.*, Mar. 1977).

Sulfur Acceptance During Combustion. The cyclic calcium utilization (*i.e.*, the percentage of CaO that was sulfated during each combustion step in the cycle) and the calcium present as $CaSO_4$ in the sulfated sorbent product at the end of each combustion half cycle are shown in Fig. 8. The calcium utilization decreased from ∿30% in the first cycle to ∿9% in the tenth cycle. Thus, in order to maintain a constant sulfur retention of 84%, it was necessary to increase the CaO/S ratio (which was ∿2.9 during the first combustion cycle) by a factor of ∿3 over the ten utilization cycles. This loss of reactivity agrees closely with the loss of reactivity observed for the experiments with Tymochtee dolomite.

Sulfur Release During Regeneration. Results for the regeneration step of the ten cycles were similar to those for the Tymochtee experiments. The SO_2 concentration in the dry off-gas varied from 6 to 9%. The regenerability of the limestone remained acceptable in all ten cycles ranging from 49 to 73%.

Porosity of Sorbent. Porosity effects were essentially the same as those observed in the Tymochtee series of experiments.

Coal-Ash Buildup. In the tenth cycle, for every 100 g starting limestone, ∿25 g of coal ash had accumulated, higher than in the Tymochtee dolomite experiment. Arkwright coal, however, which was used in the dolomite experiment, contains less ash, 7.7 wt %, than does Sewickley coal, 12.7 wt %.

Limestone particles from the first cycle were observed to contain some adhered coal ash. Particles from the tenth-cycle appeared to have more adhering ash than the first-cycle particles. However, not all particles were encapsulated with coal ash, as was the case with particles from the cyclic dolomite experiments. Many of the tenth-cycle limestone particles were visually identical to first cycle particles, which would indicate that the ash layer thickness was not increasing and that much of the coal ash was presented as individual particles in the bulk bed material. The maximum buildup expected when using Greer limestone and Sewickley coal is ∿20 wt % in the utilized stone.

Attrition and Elutriation of Particles. The limestone losses caused by attrition were ∿2.0% during each regeneration step. During sulfation, the loss was ∿20% in the first cycle and steadily decreased to ∿4% per cycle in the final cycles. The larger attrition loss in the first sulfation step can be attributed to calcination. In subsequent cycles, the resistance of the particles to attrition increased because of (1) sulfate hardening and (2) the partial sintering which occurs at the regeneration temperature.

The losses during sulfation were slightly higher in the dolomite experiment. However, combustion conditions differed in the two series of experiments. The Greer limestone was fully calcined at the system pressure of 307 kPa and the bed temperature of 855°C, whereas the Tymochtee dolomite was not fully calcined at 810 kPa and 900°C.

The combined losses average ∿10% per cycle. The makeup rate would have to be at least ∿10% to replenish losses and a makeup rate greater than 10% will be required to maintain the SO_2-sorption reactivity of the fluidized bed solids.

ACKNOWLEDGMENTS

We gratefully acknowledge the support of this program by the Energy Research and Dedevelopment Administration and by the Environmental Protection Agency.

REFERENCES

Johnson, I., Vogel, G. J., Montagna, J., Shearer, J., Snyder, R., Swift, W., & Jonke, A. A. (1977) Twelfth Intersociety Energy Conversion Engineering Conference, 28 August - 2 September.

Mesko, J. E. (1976) ERDA Report No. FE-1237-Q76-3.

Minnick, L. S. (1977) FE-2540-3.

Montagna, J., Lenc, J. F., Vogel, G. J., & Jonke, A. A. (1977) Ind. Eng. Chem. Process Des. Dev., 16, No. 2, 230-236.

Montagna, J., Swift, W., Smith, G., Vogel, G., & Jonke, A. A. (1976) American Institute of Chemical Engineers Meeting, Chicago, Illinois, 28 Nov. - 2 Dec.

Montagna, J., Nunes, F., Smith, G., Vogel, G., & Jonke, A. (1977) American Institute of Chemical Engineers Meeting, New York City, 13-17 November.

Nagier, M. F. (1964) International Series of Monographs on Chemical Engineering, Vol. 3, MacMillan, NY

National Research and Development Corp. (1973) OCR Contract 14-32-0001-1511, November 1973, 50-53.

O'Neill, E. P., Keairns, D. L., & Kittle, W. F. (1976) Thermochim Acta 14, 209.

Pope, Evans & Robbins, Inc. (1975) FE-1237-39.

Potter, A. (1969) Amer. Ceram. Soc. Bull., 48, 855.

Snyder, R., Wilson, I., Johnson, I., & Jonke, A. (1977) Argonne National Laboratory, ANL/CEN/FE-77-1.

Swift, W., Panek, A. F., Smith, G., Vogel, G., & Jonke, A. (1976) Argonne National Laboratory, ANL-76-122.

The Energy Daily, Feb. 9, 1977.

Vogel, G. *et al.* (1972) Argonne National Laboratory, ANL/ES-CEN-1005.

Vogel, G. *et al.* (March 1976) Argonne National Laboratory, ANL/ES/CEN-1015.

Vogel, G. *et al.* (1976) Argonne National Laboratory, ANL/ES/CEN-1016.

Vogel, G. *et al.* (Sept. 1976) Argonne National Laboratory, ANL/ES/CEN-1017.

Vogel, G. *et al.* (Dec. 1976) Argonne National Laboratory, ANL/ES/CEN-1018.

Vogel, G. *et al.* (1977) Argonne National Laboratory, ANL/CEN/FE-77-3.

Vogel, G. *et al.* (Mar. 1977) Argonne National Laboratory, ANL/ES/CEN-1019.

Westinghouse (1974) Monthly Report Oct. 1974, 6-11.

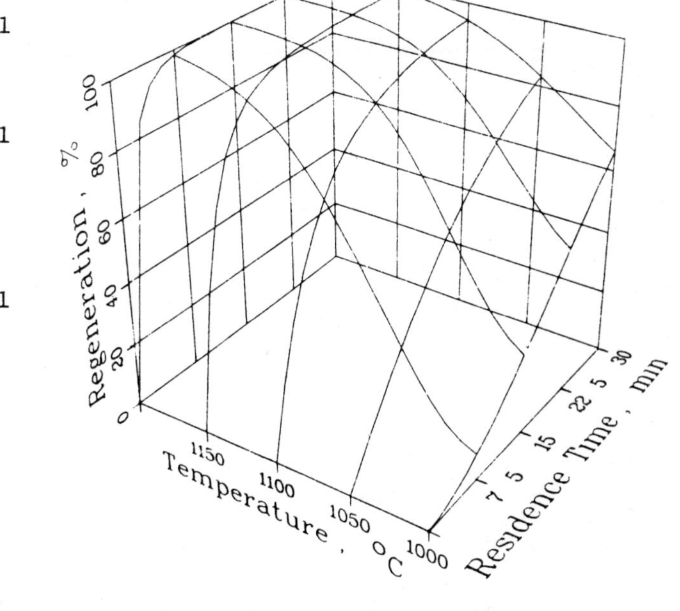

Fig. 3. The Extent of CaO Regeneration for Tymochtee Dolomite as a Function of Temperature and Residence Time

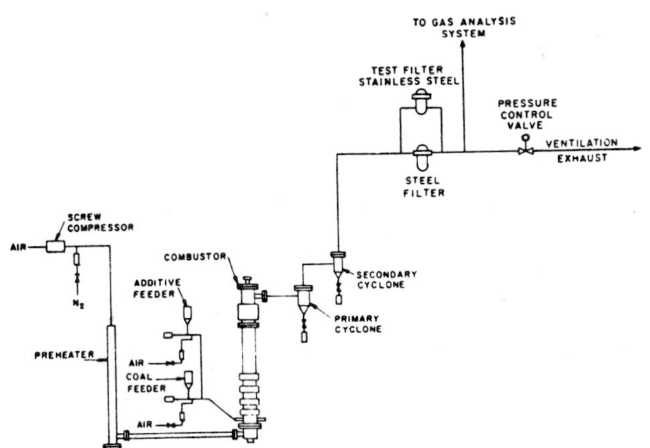

Fig. 1. Simplified Equipment Flowsheet of Fluidized-Bed Combustion Process Development Unit and Associated Equipment.

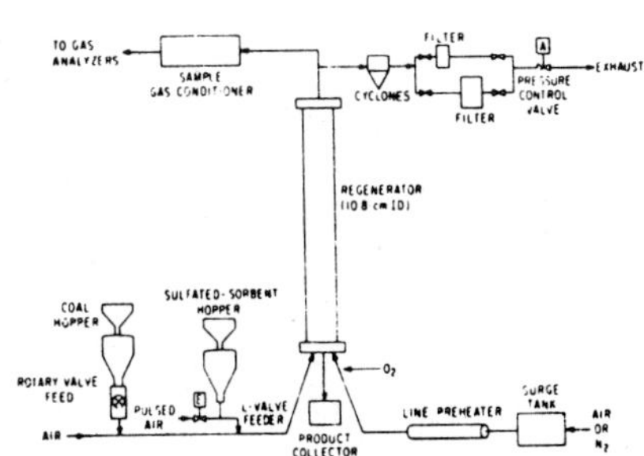

Fig. 2. Experimental Sorbent Regeneration System.

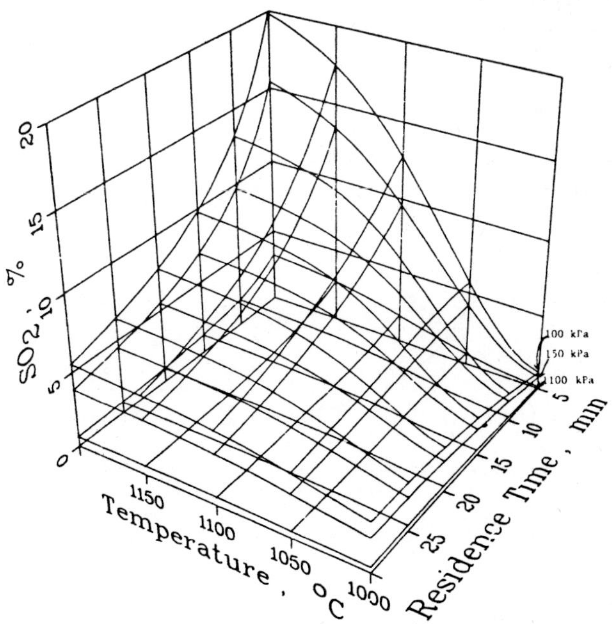

Fig. 4. SO_2 Concentration in Off-Gas from Regeneration of Sulfated Tymochtee Dolomite

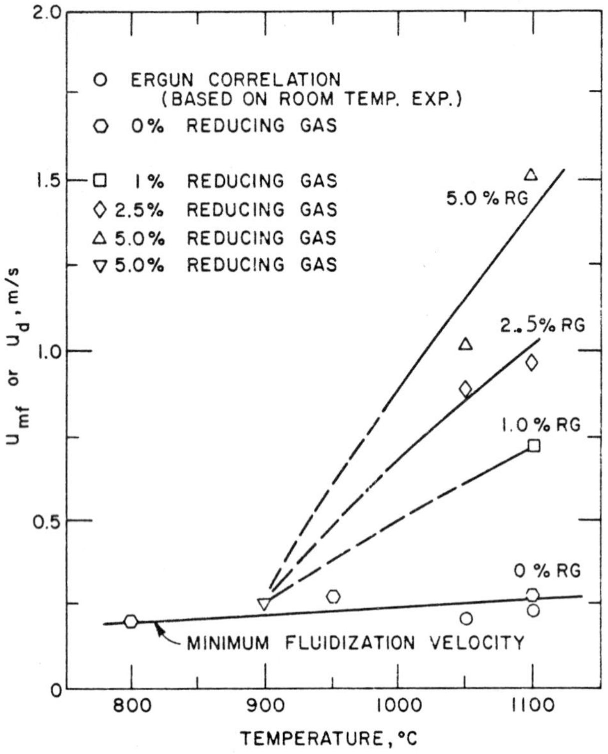

Fig. 5. Defluidization and Minimum Fluidization Velocity vs. Temperature

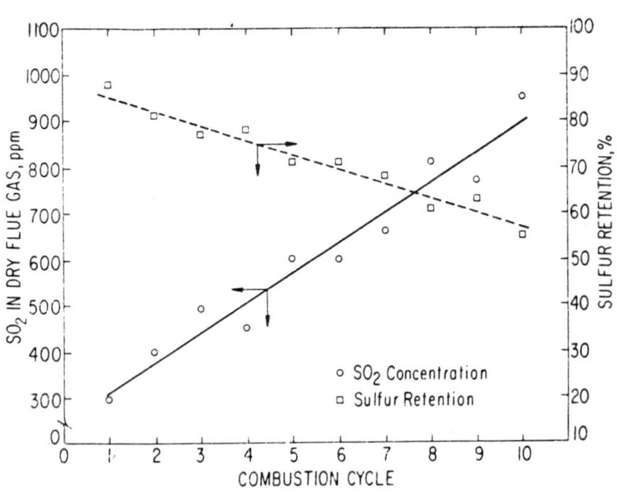

Fig. 6. Sulfur Retention in Bed and SO_2 Concentration in Flue Gas as a Function of Cycle Number

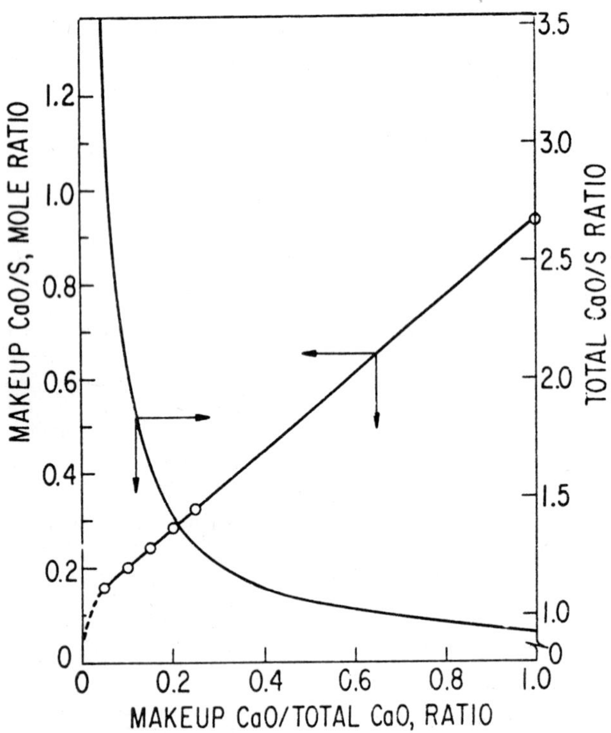

Fig. 7. Calculated Makeup and Total CaO/S Ratios Required to Achieve 75% Sulfur Retention as a Function of the Makeup CaO to Total CaO Ratio. Sulfation Conditions: Temp, 871°C; Pres, 810 kPa; Sorbent, Tymochtee dolomite.

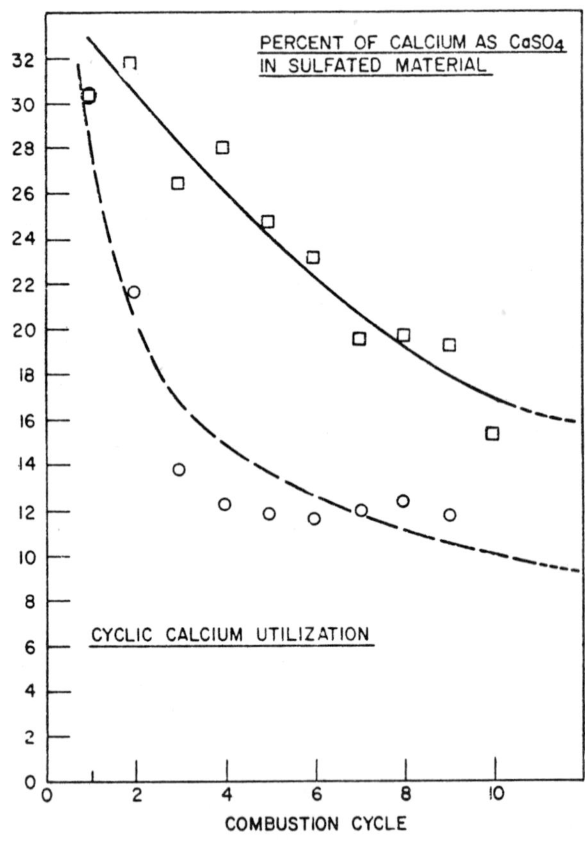

Fig. 8. Cyclic Calcium Utilization and Calcium Present as $CaSO_4$ for Greer Limestone at 84% Sulfur Retention

A MATHEMATICAL MODEL OF NO FORMATION AND DESTRUCTION IN FLUIDIZED COMBUSTION OF COAL

By F. J. Pereira* and J. M. Beér
Department of Chemical Engineering
Massachusetts Institute of Technology
Cambridge, Massachusetts USA

SYNOPSIS

Results of experimental studies are first summarized with the objective of the quantitative description of the significant processes involving fuel bound nitrogen in fluidized combustion of coal.

Mass balances written for fuel-nitrogen and NO for a volume element of the bed are used as the basis for deriving a system of simultaneous ordinary differential equations. The numerical solution of this system of differential equations yields the spatial distribution of NO in the fluidized bed as a function of bed operating variables, fuel nitrogen content and bed temperature. Comparison is made of computed data with those obtained by measurement and it is shown that this simple model can serve as the first approximate step in the prediction of NO emission from fluidized bed combustions. An extension of this theory by accounting for the NO reduction reaction by char in the fluidized bed and the freeboard is in progress at M.I.T.

NOMENCLATURE

A	cross section area of bed cm^2
C_1, C_2	concentration of species 1,2 mol cm^{-3}
H	bed height (fluidized)
N	number of bubbles per unit volume of bed, also excess air factor
Q	total rate of gas volume exchange between bubble and particulate phase, per bubble ($= g + k_G S$), $cm^3\ sec^{-1}$
U_1, U_o	superficial fluidizing velocity (bubbling bed, bed at minimum fluidizing vel) cm sec^{-1}
V	bubble volume cm^3
X	number of transfer units $\left(= \frac{QH}{VU}\right)$
y	vertical coordinate
δ	NV dimensionless
ε_o	voidage of particulate phase dimensionless
ε	dimensionless bed height (= y/H)
ϕ	molar rate of fuel-nitrogen release in the particulate phase mol $cm^{-3}\ sec^{-1}$

Subscripts: P - particulate
B - bubble phase

INTRODUCTION

Nitric oxide emissions from fluidzied bed combustion are expected to be low because of the low operating temperatures. Due to the high rates of conversion of fuel bound nitrogen to NO in the bed, however, NO emissions comparable with those from high temperature coal combustors were found (Robinson et al. (1972), Jonke (1969)). Recent experimental studies have shown that NO is both being formed and also destroyed in fluidized combustion of nitrogen bearing fuels (Gibbs et al. (1975)). NO is formed when coal volatiles containing some of the nitrogenous compounds burn, and more NO is formed during the combustion of the residual char. The NO formed may react with coal volatiles (typically NH_3) and with char in regions of low oxygen concentration with the result of reducing the NO concentration and producing N_2, CO, CO_2 and H_2O. The results of these studies can be summarized as follows: Under the operating conditions of atmospheric

*Fernando J. M. Antunes Pereira, Universidade de Louvenco Marques, C.P. 257 Mozambique (East Africa)

fluidized combustors, up to 90% of the NO_x is formed from the nitrogeneous compounds in the coal and 10% is due to the fixation of nitrogen from the combustion air. Nitric oxide is generated both during the volatile and the char combustion. The nitrogen left in the residual char decreases with increasing devolatilization (bed) temperature. NO is formed preferentially near the distributor where the O_2 concentration is high. There are two significant NO destruction reactions active in fluidized coal combustors: a homogeneous reaction between NO and gaseous nitrogen compounds (e.g. NH_3) from coal volatiles, and a heterogeneous reaction between NO and char. Both reactions lead to the formation of N_2. Approximate values of the rates of the above reactions are available from experimental studies by de Soete (1973) and Pereira (1975) respectively.

Figure 1 shows results of experiments in which gaseous species concentration distributions were determined in the bed and the freeboard of a one square foot fluidized combustor (Gibbs et al. (1975)). The significant reduction of the NO formed near the distribution plate, higher in the bed and in the freeboard can be explained in terms of the NO-NH_3 and NO-char reactions mentioned above.

For the description of these processes by a mathematical model it was necessary to have information on the hydrodynamic behavior, mixing pattern of the fluidized coal combustor, the rate of evolution of fuel-N compounds from the coal, the chemical kinetics of the reactions in which the fuel-N compounds participate, and the temperature, gas composition-history of the fuel-nitrogen in the fluidized bed. It was assumed that the fluid dynamic-mixing behavior of the fluidized combustor can be described with good approximation by the two phase model developed by Davidson and Harrison (1963). For the oxidation of fuel-N to NO and the destruction of the formed NO by fuel-N compounds the mechanisms proposed by de Soete (1973) were assumed. These involve the two competitive reactions

$$NH_3 + O_2 \longrightarrow NO + \ldots\ldots \tag{1}$$

$$NH_3 + NO \longrightarrow N_2 + \ldots\ldots \tag{2}$$

where ammonia simulates fuel-nitrogen compounds. The rate constants of the above reactions are given by de Soete as

$$k_1 = 2.544\ 10^3 T \exp(-15098/T) \mathrm{cm}^3\ \mathrm{mole, s}$$

and

$$k_2 = 6.564\ 10^{11} T^{0.5}\ \exp(-13583/T) \mathrm{cm}^3/\mathrm{mole,s}$$

Regarding the temperature, gas composition history of the fuel-N, it was assumed that two extreme cases may be considered for the fraction of nitrogen bearing compounds evolved with coal volatiles: they either mix quickly with the combustion air before the volatile flame ignites (premixed flame), or begin to burn before any significant mixing occurs (diffusion flame). This difference is important because it can be expected that the yield of NO from fuel-N conversion is lower in the case when no premixing occurs (Sarofim et al. (1972)). In this case intermediate fuel-N fragments will recombine into the stable N_2 molecule in regions of low oxygen concentration (the fuel side of the flame front in the diffusion flame).

It has been further assumed that the fuel-N in the residual char evolves within the particle boundary layer during the combustion of the char.

These assumptions about the physical-chemical processes of NO formation have led to the development of two models: a "homogeneous" model in which the reactions involving the fuel-nitrogen are assumed to take place in the gas phase outside the particle boundary layer, and a "droplet" model where the reactions are assumed to occur within the particle boundary layer.

In the following development of the first of these models, the "homogeneous" model is described; balance equations are written for the fuel-nitrogen species and NO in a volume element of the fluidized bed and a set of simultaneous ordinary differential equations is derived and solved numerically. Results of computations of NO concentration distributions in the fluidized combustor and of NO emissions as a function of bed temperature are presented and compared with experimental data.

THE "HOMOGENEOUS" MODEL OF NO EMISSION

Four cases were studied in which the state of mixing in the particulate phase was assumed to be either plug flow or stirred flow, and plug flow in the bubble phase. Chemical reaction involving NO was assumed to occur in the particulate phase, and only exceptionally in the bubble phase.

In the following one of these cases is considered in some detail. The assumptions are: the mixing state of the gas is plug flow both in the particulate and bubble phases, reactions involving NO occur both in the particulate and bubble phase, the devolatilization time of the coal particles ($d < 3$mm) is commensurable with their mixing time in the fluidized bed and therefore the release of the fuel-N during devolatilization takes place uniformly over the bed volume.

The molar fluxes of fuel-nitrogen compounds and NO associated with an element of the bed are illustrated in Figs. 2 and 3. Table 1 indicates the general procedure used to obtain the differential equations directly from the mass balances, and Table 2 lists the differential equations involving the concentrations of fuel-N and NO species.

Due to the mixed products of the dependent variables ($C_1 \cdot C_2$) in these equations, an analytical solution of the differential equation system is not possible. A numerical procedure was therefore chosen for the integration of these simultaneous ordinary differential equations; the numerical integration was carried out by a combined two and three-point predictor-corrector method (Lapidus and Seinfeld (1971)). This method has the advantages of low computing time, rapidity of convergence and stability.

The fuel-N (NH_3) and NO concentration distribution was computed for different levels of the input parameters and their combinations as shown in Table 3. For the first 16 cases listed, the fuel-N was assumed to be released uniformly throughout the bed, for cases 17 and 18 instantaneous release of fuel-N at the coal feed point was assumed.

The O_2 profile in the bubble phase was determined from experiments; the relationship expressing C_{oB} as a function of bed height, ε, has been calculated by regression analysis of the experimental data (Pereira (1975)) and is given in Table 3.

For the particulate phase O_2 concentration three arbitrary assumptions were made:

a) $C_{oP} = C_{oB}$ (cases 1-5)

b) $C_{oP} = \frac{.121}{T}\ 10^{-3}$ mole cm^{-3} (~1% v/v) (cases 6-11 and 17)

and

c) $C_{oP} = \frac{.608}{T}\ 10^{-3}$ mole cm^{-3} (~5% v/v) (cases 12-16 and 18)

In cases 1 to 5 in Table 3 the assumed O_2 concentration in the particulate phase is most likely to have been overestimated. The actual concentration of O_2 is thought to be nearly constant and very low, between 1 and 5% depending upon the carbon loading of the bed.

Throughout the numerical calculations a bubble size of 10 cm was assumed based partly on the observation that the NO was formed primarily in the lower part of the bed and also on empirical correlations (Geldart (1972)) which gave this bubble size for that region of the fluidized bed.

In Fig. 4 the normalized experimental NO concentration distribution is compared with that computed for the cases of 1% and 5% oxygen concentration in the particulate phase, and uniform or instantaneous fuel-N release in the bed. The best agreement was obtained by assuming 1% O_2 in the particulate phase and instantaneous fuel-N release near the coal inlet point (case 17 in Table 3).

The normalized experimental profile in Fig. 4 has been determined from statistical average of data obtained in fourteen experimental runs in which the bed temperature and the overall stoichiometry were varied within the ranges of 700-850°C and 0.82-1.20 respectively while the fluidizing velocity (90.8 cm/s), the particle size (< 1676 μm) and the bed height (60 cm) were maintained constant.

Fig. 5 represents the comparison of calculated NO emission data with those determined from experiments covering wide ranges of input air/fuel ratios, as a function of bed operating temperatures. The agreement is good in the temperature region below 800°C but the calculated curves do not follow the falling off of the experimental NO emission data at temperatures in excess of 800°C. This is because in the "homogeneous" model no account was taken of the NO reduction by char, a reaction which is known to become significant at temperatures above 800°C.

CONCLUSIONS

A mathematical model predicting NO emission from coal burning fluidized beds has been developed. The model is based on the "two phase" theory of fluidization and on experimental information on the formation and destruction of NO in fluidized combustion. Computed NO concentration distributions along the height of the bed were in good agreement with experimental data. Predicted NO emission as a function of bed temperature gave also good agreement with experiment in the temperature range of 700°-800°C but predicted NO emission went on rising above 800°C while experimental data levelled off. This is because in the model no account was taken of the NO reduction reaction with char particles, a reaction the rate of which was found to be strongly increasing above 800°C. The comparisons of computed and experimental data show that the relatively simple model presented may serve as a first approximate step toward the prediction of NO emission from fluidized combustors. Work to extend this model so as to incorporate the heterogeneous reaction between NO and char, and an improved coal devolatilization and char combustion model, is in progress at M.I.T.

ACKNOWLEDGEMENT

The research was carried out at the University of Sheffield England. The authors thank Dr. Nevin Selcuk and Dr. Bernard Gibbs for valuable discussions. Financial support to one of us (FJP) from the Gulbenkian Foundation and from the National Coal Board for the experimental studies is gratefully acknowledged.

REFERENCES

Davidson, J. F., and Harrison, D., (1963), "Fluidized Particles," Cambridge Univ. Press.

Geldart, D., (1972), Powd. Techn., 6, 201.

Gibbs, B. M., Pereira, F. J., and Beér, J. M., (1975), Institute of Fuel Symposium Series No. 1: Fluidized Combustion, D6, London.

Jonke, A. A., (1969), Argonne National Lab., Publication No. ANL/ES-CEN 1001.

Lapidus, L., and Seinfeld, J. H., (1971), "Numerical Solution of Ordinary Differential Equations," Academic Press, N. Y.

Pereira, F. J., (1975), Ph.D. Thesis, Univ. of Sheffield, England.

Robinson, E. B., Glenn, R. D., Ehrlich, S., Bishop, J. W., and Gordon, J. S., (Feb. 1972), EPA Contract CPA 70-10.

Sarofim, A. F., Williams, G. C., Modell, M., and Slater, S. M., (Nov. 1973), AIChE, 66th Ann. Meet., Phila. Pa.

de Soete, G., (1973), Combustion Inst. Europ. Symp. Sheffield, England, Academic Press.

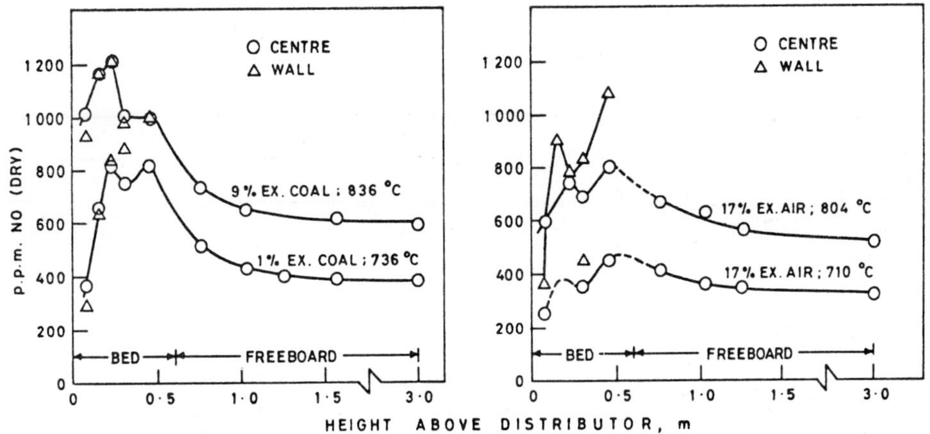

Figure 1. Axial NO concentration profiles at the wall and centre of the 30 cm x 30 cm combustor.

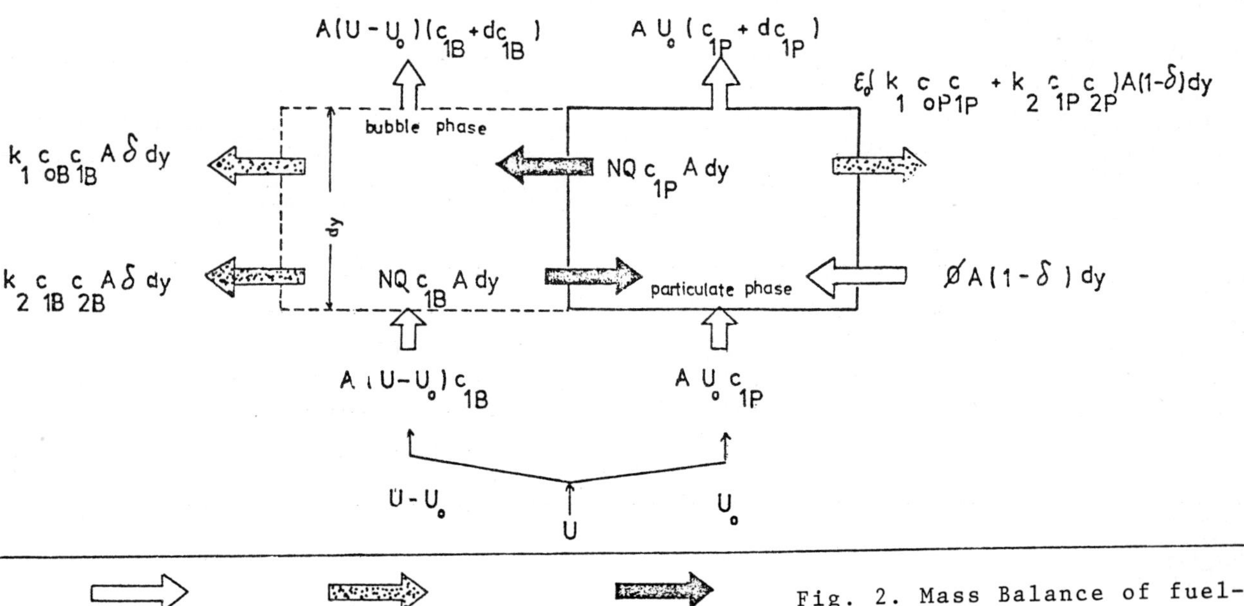

Fig. 2. Mass Balance of fuel-N

convective flow | chem. reaction | interphase mass transport

$A(U-U_o)(c_{2B}+dc_{2B})$ | $A U_o(c_{2P}+dc_{2P})$

$k_1 c_{oB} c_{1B} A \delta dy$ | $NQ c_{2P} A dy$ | $\epsilon_o k_2 c_{1P} c_{2P} A(1-\delta) dy$

dy

$k_2 c_{1B} c_{2B} A \delta dy$ | $NQ c_{2B} A dy$ | $\epsilon_o k_1 c_{oP} c_{1P} A(1-\delta) dy$

$U-U_o$ | U | U_o

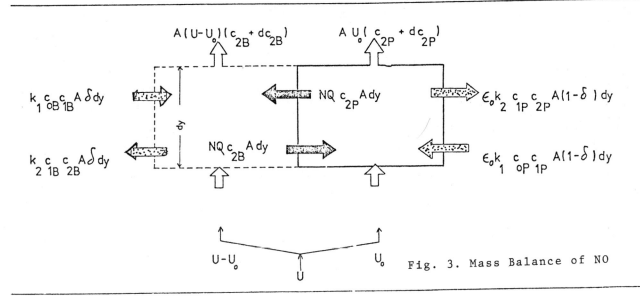

Fig. 3. Mass Balance of NO

Table 1 Mass balances of species 1 and 2.

Phase	Species	Mass balances
Bubble	1	IN : $A(U-U_o)c_{1B} + NQ\,c_{1P}\,A\,dy$. OUT : $A(U-U_o)(c_{1B}+dc_{1B}) + NQ\,c_{1B}\,A\,dy + (k_1c_{OB}c_{1B}+k_2\,c_{1B}c_{2B})\,A\,dy$. Equalizing, using $NQ = X(U-U_o)/H$, and simplifying: $dc_{1B}/dy = X(c_{1P}-c_{1B})/H - (k_1c_{OB}c_{1B}+k_2\,c_{1B}c_{2B})\,\delta/(U-U_o)$
	2	IN : $A(U-U_o)c_{2B} + NQ\,c_{2P}\,A\,dy + k_1\,c_{OB}\,c_{1B}\,A\,\delta\,dy$. OUT : $A(U-U_o)(c_{2B}+dc_{2B}) + NQ\,c_{2B}\,A\,dy + k_2\,c_{1B}\,c_{2B}\,A\,\delta\,dy$. Equalizing, using $NQ = X(U-U_o)/H$, and simplifying: $dc_{2B}/dy = X\,(c_{2P}-c_{2B})/H + (k_1c_{OB}c_{1B}-k_2c_{1B}c_{2B})\delta/(U-U_o)$
Particulate	1	IN : $A\,U_o\,c_{1P} + NQ\,c_{1B}\,A\,dy + \phi\,A\,(1-\delta)\,dy$ OUT : $A\,U_o\,(c_{1P}+dc_{1P}) + NQ\,c_{1P}\,A\,dy + (k_1c_{OP}c_{1P}+k_2c_{1P}c_{2P})\,A\,\varepsilon_o(1-\delta)\,dy$. Equalizing, using $NQ = X(U-U_o)/H$ and simplifying: $dc_{1P}/dy = X(U-U_o)\,(c_{1B}-c_{1P})/H\,U_o + (\phi-\varepsilon_o k_1c_{OP}c_{1P}-\varepsilon_o k_2c_{1P}c_{2P})(1-\delta)/U_o$.
	2	IN : $A\,U_o\,c_{2P} + NQ\,c_{2B}\,A\,dy + \varepsilon_o\,k_1\,c_{OP}\,c_{1P}\,A\,(1-\delta)\,dy$ OUT : $A\,U_o\,(c_{2P}+dc_{2P}) + NQ\,c_{2P}\,A\,dy + \varepsilon_o\,k_2\,c_{1P}\,c_{2P}\,A\,(1-\delta)\,dy$ Equalizing, using $NQ = X(U-U_o)/H$ and simplifying: $dc_{2P}/dy = X(U-U_o)(c_{2B}-c_{2P})/H\,U_o + \varepsilon_o(k_1c_{OP}c_{1P}-k_2c_{1P}c_{2P})\,(1-\delta)/U_o$

OBS: It is implicitly assumed that X is the same for both species 1 and 2, approximation that is justifiable by the similarity of the respective diffusion coefficients.

Table 2 Differential equations involving the concentrations of species 1 and 2.

$$\frac{dc_{1P}}{d\varepsilon} = X\left(\frac{U-U_o}{U_o}\right)(c_{1B}-c_{1P}) + (\phi-\varepsilon_o k_1 c_{OP}c_{1P}-\varepsilon_o k_2 c_{1P}c_{2P})\cdot\frac{H(1-\delta)}{U_o}$$

$$\frac{dc_{1B}}{d\varepsilon} = X\,(c_{1P}-c_{1B}) - (k_1c_{OB}c_{1B}+k_2c_{1B}c_{2B})\cdot\frac{H\delta}{U-U_o}$$

$$\frac{dc_{2P}}{d\varepsilon} = X\left(\frac{U-U_o}{U_o}\right)(c_{2B}-c_{2P}) + (k_1\,c_{OP}\,c_{1P}-k_2\,c_{1P}\,c_{2P})\,\frac{H(1-\delta)}{U_o}\,\varepsilon_o$$

$$\frac{dc_{2B}}{d\varepsilon} = X\,(c_{2P}-c_{2B}) + (k_1\,c_{OB}\,c_{1B}-k_2\,c_{1B}\,c_{2B})\,\frac{H\delta}{U-U_o}$$

with $\varepsilon = y/H$

Table 3 "Homogeneous" Model.

(D_e = 10 cm)

Case	temp. (°C)	Stoich. factor	C_{OP} (mole. cm^{-3})	C_{OB} (mole. cm^{-3})
1 2 3 4 5	600 700 800 900 1000	1.1	$C_{OP} = C_{OB}$	$C_{OB} = \frac{2.557}{T} 10^{-3} (.597 - .842\, \varepsilon + .495\, \varepsilon^2)$
6 7 8 9 10 11	600 700 800 900 1000 1100	-	$C_{OP} = \frac{.121}{T} 10^{-3}$ (~ 1 vol %)	
12 13 14 15 16	600 700 800 900 1000		$C_{OP} = \frac{.608}{T} 10^{-3}$ (~ 5 vol %)	
17	800		$C_{OP} = .121\ 10^{-3}/T$	
18	800		$C_{OP} = .608\ 10^{-3}/T$	

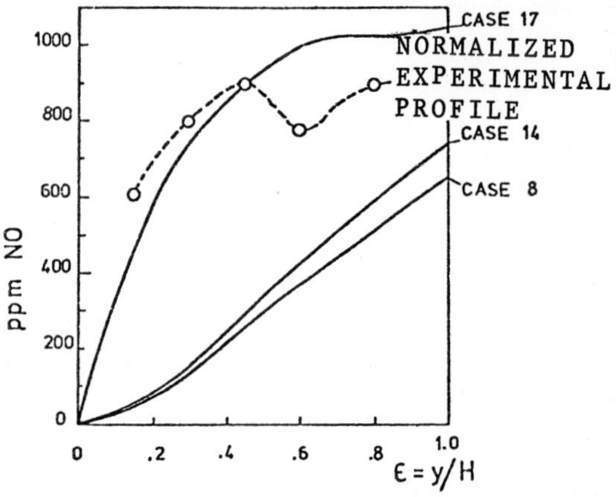

FIG. 4 "HOMOGENEOUS" MODEL. COMPARISON WITH EXPERIMENTAL DATA (BED CONCENTRATION PROFILES).

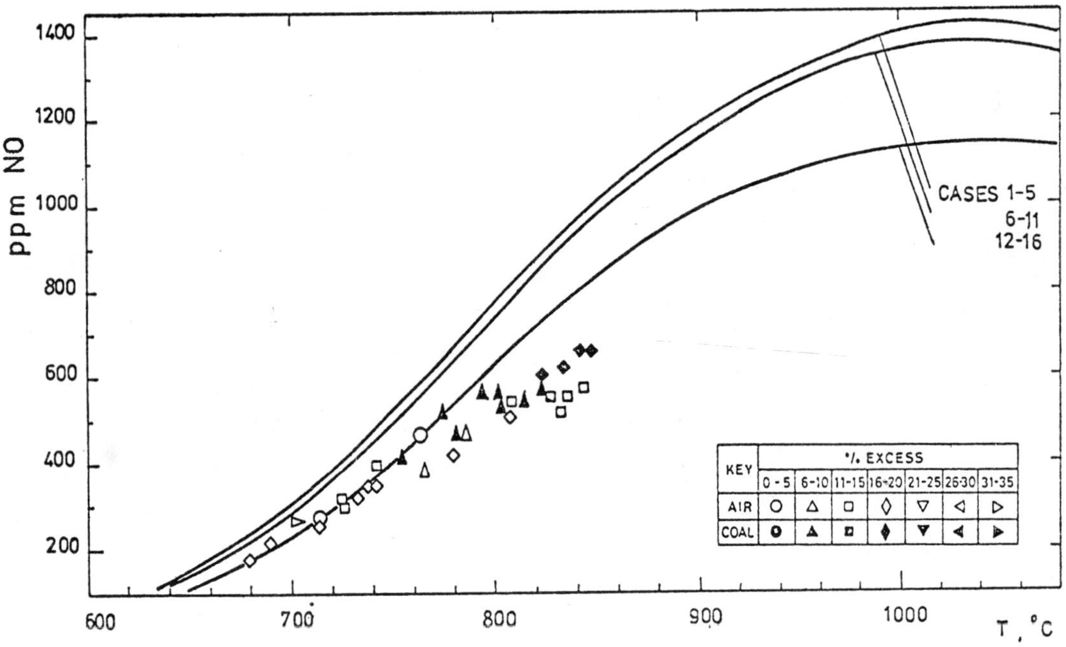

FIG. 5 "HOMOGENEOUS" MODEL. COMPARISON WITH EXPERIMENTAL DATA (FLUE EMISSION).

Fluidization, Cambridge University Press, 1978

Fluidization, Cambridge University Press, 1978